T0324164

Bioremediation and Bioeconomy

Bioremediation and Bioeconomy

Edited by
M.N.V. Prasad

Department of Plant Sciences
University of Hyderabad, Telangana, India

ELSEVIER

AMSTERDAM • BOSTON • HEIDELBERG • LONDON • NEW YORK • OXFORD
PARIS • SAN DIEGO • SAN FRANCISCO • SINGAPORE • SYDNEY • TOKYO

Elsevier
Radarweg 29, PO Box 211, 1000 AE Amsterdam, Netherlands
The Boulevard, Langford Lane, Kidlington, Oxford OX5 1GB, UK
225 Wyman Street, Waltham, MA 02451, USA

Copyright © 2016 Elsevier Inc. All rights reserved.

No part of this publication may be reproduced or transmitted in any form or by any means, electronic or mechanical, including photocopying, recording, or any information storage and retrieval system, without permission in writing from the publisher. Details on how to seek permission, further information about the Publisher's permissions policies and our arrangements with organizations such as the Copyright Clearance Center and the Copyright Licensing Agency, can be found at our website: www.elsevier.com/permissions.

This book and the individual contributions contained in it are protected under copyright by the Publisher (other than as may be noted herein).

Notices

Knowledge and best practice in this field are constantly changing. As new research and experience broaden our understanding, changes in research methods, professional practices, or medical treatment may become necessary.

Practitioners and researchers must always rely on their own experience and knowledge in evaluating and using any information, methods, compounds, or experiments described herein. In using such information or methods they should be mindful of their own safety and the safety of others, including parties for whom they have a professional responsibility.

To the fullest extent of the law, neither the Publisher nor the authors, contributors, or editors, assume any liability for any injury and/or damage to persons or property as a matter of products liability, negligence or otherwise, or from any use or operation of any methods, products, instructions, or ideas contained in the material herein.

Library of Congress Cataloging-in-Publication Data
A catalog record for this book is available from the Library of Congress

British Library Cataloguing in Publication Data
A catalogue record for this book is available from the British Library

For information on all Elsevier publications
visit our website at http://store.elsevier.com/

ISBN: 978-0-12-802830-8

Working together
to grow libraries in
developing countries

www.elsevier.com • www.bookaid.org

This book is dedicated to Challapalli Savithri.

Contents

SECTION 1 BIOPRODUCTS FROM CONTAMINATED SUBSTRATES (SOIL AND WATER)

SECTION 2   BIOMASS ENERGY, BIODIESEL, AND BIOFUEL FROM CONTAMINATED SUBSTRATES

SECTION 3 ORNAMENTALS AND CROPS FOR CONTAMINATED SUBSTRATES

SECTION 5 ALGAL BIOPRODUCTS, BIOFUELS, BIOREFINERY FOR BUSINESS OPPORTUNITIES

SECTION 6 BIOPROCESSES, BIOENGINEERING FOR BOOSTING BIO-BASED ECONOMY

SECTION 7 CASE STUDIES

CHAPTER 22 Bioremediation in Brazil: Scope and Challenges to Boost up the Bioeconomy 569
G. Labuto, E.N.V.M. Carrilho

CHAPTER 23 Phytoremediation of Soil and Groundwater: Economic Benefits Over Traditional Methodologies 589
Edward Gatliff, P. James Linton, Douglas J. Riddle, Paul R. Thomas

Contributors

S.Z. Ahammad
Indian Institute of Technology, Delhi, India

K. Amulya
Bioengineering & Environmental Sciences; CSIR-Indian Institute of Chemical Technology (CSIR-IICT), Hyderabad, India

D. Annapurna
University of Hyderabad, Hyderabad, Telangana, India

B. Barbosa
Universidade Nova de Lisboa, Caparica, Portugal

N.S. Bolan
University of Newcastle, Newcastle, NSW, Australia

A. Brandys
Institute of Natural Fibres and Medicinal Plants, Poznan, Poland

E.N.V.M. Carrilho
Universidade Federal de São Carlos, Araras, SP, Brazil

R. Chaturvedi
St. John's College, Agra, India

T.S. Chibrik
Ural Federal University named after First President of Russia B.N. Yeltsin, Ekaterinburg, Russia

J. Costa
Universidade Nova de Lisboa, Caparica, Portugal

Shikha Dahiya
Bioengineering & Environmental Sciences; CSIR-Indian Institute of Chemical Technology (CSIR-IICT), Hyderabad, India

K.C. Das
University of Georgia, Athens, GA, USA

V. Escande
FRE 3673 – Bioinspired Chemistry and Ecological Innovations, – CNRS, University of Montpellier 2, Stratoz – Cap Alpha, Avenue de l'Europe, 34830 Clapiers, France

P.J.C. Favas
University of Trás-os-Montes e Alto Douro, Vila Real, Portugal

A.L. Fernando
Universidade Nova de Lisboa, Caparica, Portugal

E.I. Filimonova
Ural Federal University named after First President of Russia B.N. Yeltsin, Ekaterinburg, Russia

Edward Gatliff
Applied Natural Sciences, Inc., Hamilton, Ohio, USA

M.A. Glazyrina
Ural Federal University named after First President of Russia B.N. Yeltsin, Ekaterinburg, Russia

S. Gopalakrishnan
International Crops Research Institute for the Semi-Arid Tropics (ICRISAT), Patancheru, Telangana, India

C. Grison
FRE 3673 – Bioinspired Chemistry and Ecological Innovations – CNRS, University of Montpellier 2, Stratoz – Cap Alpha, Avenue de l'Europe, 34830 Clapiers, France

U. Jena
Desert Research Institute, Reno, NV, USA

V. Kanaganahalli
International Crops Research Institute for the Semi-Arid Tropics (ICRISAT), Patancheru, Telangana, India

Srujana Kathi
Pondicherry University, Kalapet, Puducherry, India

J. Koelmel
University of Hyderabad, Hyderabad, Telangana, India

A. Kumar
Indian School of Mines, Dhanbad, Jharkhand, India

G. Labuto
Universidade Federal de São Paulo, São Paulo, SP, Brazil

G. Lemoine
Établissement Public Foncier Nord-Pas de Calais, Euralille, France

P. James Linton
Geosyntec, Clearwater, Florida, USA

N.V. Lukina
Ural Federal University named after First President of Russia B.N. Yeltsin, Ekaterinburg, Russia

S.K. Maiti
Indian School of Mines, Dhanbad, Jharkhand, India

M.G. Maleva
Ural Federal University named after First President of Russia B.N. Yeltsin, Ekaterinburg, Russia

O. Meesungnoen
Mahasarakham University, Maha Sarakham, Thailand

G. Mohanakrishna
VITO—Flemish Institute for Technological Research, Mol, Belgium

R. Naidu
University of Newcastle, Newcastle, NSW, Australia

W. Nakbanpote
Mahasarakham University, Maha Sarakham, Thailand

T.K. Olszewski
Organic Chemistry, Wroclaw University of Technology, Wroclaw

D. Pant
VITO—Flemish Institute for Technological Research, Mol, Belgium

E.G. Papazoglou
Agricultural University of Athens, Athens, Greece

M.S. Paul
St. John's College, Agra, India

C. Phadermrod
Padaeng Industry Public Co. Ltd. (Mae Sot Office), Tak, Thailand

M.N.V. Prasad
University of Hyderabad, Hyderabad, Telangana, India, and Ural Federal University named after First President of Russia B.N. Yeltsin, Ekaterinburg, Russia

J. Pratas
University of Coimbra, Coimbra, Portugal

M. Rajkumar
Central University of Tamil Nadu, Thiruvarur, Tamil Nadu, India

E.A. Rakov
Ural Federal University named after First President of Russia B.N. Yeltsin, Ekaterinburg, Russia

P. Srinivas Rao
International Crops Research Institute for the Semi-Arid Tropics (ICRISAT), Patancheru, Telangana, India

G.A. Ravishankar
Dayananda Sagar Institutions, Bengaluru, Karnataka, India

Douglas J. Riddle
RELLC, Mountain Center, California, USA

D. Rose
University of Hyderabad, Hyderabad, Telangana, India

H. Sarma
N N Saikia College, Titabar, Assam, India

A. Sathya
International Crops Research Institute for the Semi-Arid Tropics (ICRISAT), Patancheru, Telangana, India

K. Schmidt-Przewoźna
Institute of Natural Fibres and Medicinal Plants, Poznan, Poland

A. Sebastian
University of Hyderabad, Hyderabad, Telangana, India

B. Seshadri
University of Newcastle, Newcastle, NSW, Australia

V. Sivasubramanian
Phycospectrum Environmental Research Centre (PERC), Chennai, Tamil Nadu, India

T.R. Sreekrishnan
Indian Institute of Technology, Delhi, India

S. Srikanth
VITO—Flemish Institute for Technological Research, Mol, Belgium

S. Suthari
University of Hyderabad, Hyderabad, Telangana, India

O. Sytar
Kyiv National University of Taras Shevchenko, Kyiv, Ukraine

J.C. Tewari
Central Arid Zone Research Institute, Jodhpur, Rajasthan, India

R. Thangarajan
University of South Australia, Adelaide, SA, Australia

Paul R. Thomas
Thomas Consultants, Inc., Cincinnati, Ohio, USA

S. Venkata Mohan
Bioengineering & Environmental Sciences; CSIR-Indian Institute of Chemical Technology (CSIR-IICT), Hyderabad, India

S. Vidyashankar
M/s. DaZiran Health Products, Coimbatore, Tamil Nadu, India

H. Wang
Zhejiang A & F University, Hangzhou, Zhejiang, China

A.A. Warra
Kebbi State University of Science and Technology, Aliero, Nigeria

Foreword

Recent years have seen significant worldwide investment in the bioeconomy, in both political and economic terms. The bioeconomy encompasses the production of renewable biological resources and their conversion into food, feed, bio-based products and bioenergy via innovative and efficient technologies provided by industrial biotechnology. It is already a reality offering great opportunities and solutions to major societal, environmental and economic challenges, including climate change mitigation, energy and food security and resource efficiency.

Bioremediation, on the other hand, is a waste management technique that involves the use of organisms to remove or neutralize pollutants from a contaminated site. Bioremediation provides a technique for cleaning up pollution by enhancing the natural attenuation, a process that occurs in nature. Depending on the site and its contaminants, bioremediation may be safer and less expensive than alternative solutions such as incineration or landfilling of the contaminated materials.

This book on "Bioremediation and Bioeconomy" has the ambition to address the research gap at the interface between pollution mitigation and boosting the bio-based economy aiming at sustainable development. There is pressing need for pollution abatement and there is an even more pressing need to create viable value chains from contaminated substrates especially in developing economies.

This book will tackle a number of key questions: How can scientific innovations be more efficiently exploited to achieve bioremediation and bioeconomy? What are the needs of cutting edge environmental industries for scale-up and large-scale market penetration? What are important research priorities for industry and policymakers? and How can environmental sustainability be maintained and simultaneously economic sustainability be achieved.

The chapters of this book on bioproducts, biomass energy and biofuels, ornamentals and crops from contaminated substrates, brownfield development, algal bioproducts, bioprocesses and bioengineering, and on new biology specifically address the interface and synergies between bioremediation and the bioeconomy and will therefore contribute to turn into reality the promise to create viable value chains from contaminated substrates in the context of the emerging global bioeconomy.

Dr. Rainer Janssen
WIP Renewable Energies
Head of Unit Biomass Department
München
Germany

Preface

Environmental decontamination is an integral part of bioeconomy and sustainable development. Biodiversity is being used as raw material for environmental decontamination, and this field has grown phenomenally in recent years, having emerged less than 3 decades ago. On the other hand, the volume of contaminated substrates (water, soil, and air) is increasing due to anthropogenic and technogenic sources of organic and inorganic contaminants. Metals are the most prevalent inorganic pollutants/contaminants and are widely used for a wide variety of needs from building materials to information technology. Metal contamination is a global problem.

Today with a growing economy extensive industrialization and extraction of natural resources have resulted in environmental contamination and pollution. Large amounts of toxic waste have been dispersed in thousands of contaminated sites spread all over the globe. These pollutants belong to two main classes: inorganic and organic. The challenge is to develop innovative and cost-effective solutions to decontaminate polluted environments. In this direction, bioremediation is emerging as an invaluable tool for environmental cleanup.

Various strategies are being applied to reduce the levels of contamination. Cultivation of industrial and environmental crops in contaminated soils is one such option. If the contaminant concentration exceeds the permissible level in edible parts, it poses serious health concerns. Therefore, in such cases non-edible crop production and valorization to value chain and value additions is a feasible proposition.

The advancement in this field is toward production of diverse biofuels (solid, liquid, and gaseous). Essentially, bioremediation is centered on bioenergy and use of bioresources harvested from environmentally perturbed/stressed agro-ecosystems. Although bioremediation contractors must profit from the activity, the primary driver is regulatory compliance rather than manufacturing profit. It is an attractive technology in the context of bioeconomy. This book will address the bottlenecks and solutions to the existing limitations in field scale and the relevant techniques. Crucial aspects of biorefinery are also covered.

Globally, land and water resources are under immense pressure due to land degradation, pollution, population explosion, urbanization, and global economic development. Large amounts of toxic waste have been dispersed in thousands of contaminated sites, and bioremediation is emerging as an invaluable tool for environmental cleanup. *Bioremediation and Bioeconomy* addresses this challenge by presenting innovative and cost-effective solutions to decontaminate polluted environments, including usage of contaminated land and wastewater for bioproducts such as natural fibers, biocomposites, and fuels to boost the economy.

Bioremediation and Bioeconomy provides a common platform for scientists from various backgrounds to find sustainable solutions to these environmental issues. This book will also address all the topical issues crucial for understanding the ecosystem approaches for a sustainable development. It provides an overview of ecosystem approaches, conservation of natural resources, pollution abatement, and mitigation.

This book is a collective effort of 65 contributors from 14 countries: Australia, Belgium, Brazil, China, France, Greece, India, Nigeria, Poland, Portugal, Russia, Thailand, Ukraine and USA.

The book includes 26 chapters grouped in 8 sections:

1. Bioproducts from Contaminated Substrates (Soil and Water) (Chapters 1–4)
2. Biomass Energy, Biodiesel and Biofuel from Contaminated Substrates (Chapters 5–8)

3. Ornamentals and Crops for Contaminated Substrates (Chapters 9–12)
4. Brownfield Development for Smart Bioeconomy (Chapters 13–16)
5. Algal Bioproducts, Biofuels, Biorefinery for Business Opportunities (Chapters 17 and 18)
6. Bioprocesses, Bioengineering for Boosting Bio-Based Economy (Chapters 19–21)
7. Case Studies (Chapters 22 and 24)
8. New Biology (Chapters 25 and 26)

The book provides a comprehensive review of new information on bio-phyto-rhizoremediation. Moreover, I feel that this is the first attempt to link bio-phyto-rhizoremediation to bio-based economy citing several examples. The unique features of this books include (a) numerous color figures, flow diagrams, tables, and updated literature; (b) strategies to utilize contaminated susbtrates for producing bioresources and cogeneration of value chain and value addition products boosting bioeconomy; and (c) several success stories.

M.N.V. Prasad
Department of Plant Sciences
University of Hyderabad
Telangana, India
5 May 2015

Acknowledgments

I would like to thank all the authors of this volume for their cogent and comprehensive contributions.

I am thankful to Candice Janco, Senior Acquisitions Editor, Environmental Science; Laura Kelleher; Rowena Prasad; and Marisa LaFleur for excellent coordination of this fascinating project, suggestions; thanks are due to Mr Paul Prasad Chandramohan, and his team for excellent technical help in many ways that resulted in record time of publication of this edition.

I am grateful to Dr. Rainer Janssen, WIP Renewable Energies, Head of Unit Biomass Department, Sylvensteinstr. 2, D-81369 Muenchen, Germany, for preparing the Foreword at short notice.

My profuse thanks to Prof. Dr. Irina Kiseleva, Head, Department of Plant Physiology and Biochemistry, and all faculty members of the department; Ural Federal University, Ekaterinburg, Russia, for extending help in several ways.

Thanks are due to Dr James Philp (France), Organisation for Economic Co-operation and Development (OECD), Directorate for Science, Technology and Industry, Paris, and Anonymous reviewers from Germany and the United States for constructive comments that enhanced the outlook of this work.

Last but not least I wish to acknowledge the University of Hyderabad for granting sabbatical leave that enabled me to concentrate on this work, and the numerous colleagues from overseas, my students, and research associates for sharing knowledge, ideas, and assistance that helped in developing and shaping this book.

BIOPRODUCTS FROM CONTAMINATED SUBSTRATES (SOIL AND WATER)

PRODUCTION OF BIODIESEL FEEDSTOCK FROM THE TRACE ELEMENT CONTAMINATED LANDS IN UKRAINE

O. Sytar[1], M.N.V. Prasad[2]

Kyiv National University of Taras Shevchenko, Kyiv, Ukraine[1]
University of Hyderabad, Hyderabad, Telangana, India[2]

1 INTRODUCTION

The total area of Ukraine is about 60.37 million ha of which agricultural land is 41.76 million ha. Ukraine has several rivers and lakes. The prominent crops cultivated for grain production are winter rye, oats, corn, barley, and buckwheat (in recent years about 15% of the world's production has come from Ukraine). Flax, hops, sugar beets, and potatoes are other important crops in Ukraine. Soybean, rapeseed, and sunflower production has increased in recent years, boosting agro-based economy (Figure 1).

After the Chernobyl accident in 1986, the trace element pollution and contamination in Ukraine received considerable attention. In total, air pollution-related mortality represents about 6% of total mortality in Ukraine (Mnatsakanian, 1992; Strukova et al., 2006). In the Donets Basin (48.3°N, 38.0°E) of eastern Ukraine, there are about 1200 mine-waste dumpsites, mostly from coal mining. These have accumulated over more than 200 years of mining (Panov et al., 1999). Ore processing and the use of metal-enriched coal in coal-fired electric generation and home heating have led to widespread contamination of the environment (Panov et al., 1999; Kolker et al., 2009). Of the 26 elements considered to be pollutants in the Donets Basin, arsenic (As) and mercury (Hg) are of primary concern; most of the As-polluted soils in this region are near burning-coal waste heaps (Panov et al., 1999; Conko et al., 2013). For more than 50 years, the largest Hg production facility of the former Soviet Union operated in Horlivka (Gorlovka, Russian spelling), a city of approximately 300,000 residents in the Donets Basin. During this period, about 30,000 metric tons of Hg were produced from ore extracted from the adjacent Mykytivka (Nikitovka) mines.

The contaminated land that is unsuitable for food production and of little or no economic return could be used for the production of bioenergy. The problem of competition for fertile land can be alleviated by using contaminated land for the production of bioenergy plants.

Using biomass for fuel production is an attractive option for three main reasons. First, it is a renewable resource and thus its development could be sustained in the future. Second, because there is no net release of carbon dioxide and sulfur contents are low, it is environmentally attractive. Thirdly, given the rise in fossil fuel prices, it is also economically advantageous (Demirbas, 2009).

Bioremediation and Bioeconomy. http://dx.doi.org/10.1016/B978-0-12-802830-8.00001-0
Copyright © 2016 Elsevier Inc. All rights reserved.

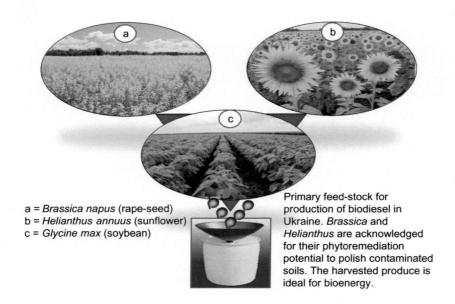

a = *Brassica napus* (rape-seed)
b = *Helianthus annuus* (sunflower)
c = *Glycine max* (soybean)

Primary feed-stock for production of biodiesel in Ukraine. *Brassica* and *Helianthus* are acknowledged for their phytoremediation potential to polish contaminated soils. The harvested produce is ideal for bioenergy.

FIGURE 1

Primary feed-stock for production of biodiesel in Ukraine.

This review chapter presents the nowadays situation of heavy metals pollution on Ukraine territory and the possibility of using oilseed crops (rapeseed, soybean, and sunflower) for 50% cleanup of the areas polluted by heavy metals.

2 MONITORING OF HEAVY METALS POLLUTION IN UKRAINE

The report of EEA-UNEP (2000) regarding the status of soil contamination in Europe (Figure 2) has shown that in Eastern Europe problems of diffuse soil contamination is greatest compared to other European countries. The high contamination of heavy metals is shown in Ukraine, especially in the Chernobyl area. Unfortunately, the international literature is still missing detailed analysis of heavy metals contamination in Ukraine.

The Ministry of Nature Protection of Ukraine reported that lead concentrations in the capital city of Kyiv were 4.6 times higher than the permissible limit. It was reported that the main source of lead in the air and soil is from automobile activities. Major sources of lead include gasoline combustion, also nonferrous smelting, and mining (Chiras, 2009).

About 90% of toxic metals accumulate from atmosphere to soil, where they migrate in groundwater, become absorbed by plants, and get into the trophic chains (Lozanovskaya et al., 1998; Anonymous, 1998; Trahtenberg, 1998; Alekseeva, 1987). According to coefficient of load factor, it has been established that in some regions of Ukraine the level of metal pollution of soil, including mobile forms, exceeds the permissible level by 2-14 times (Gaevsky and Pelypets, 1999; Nikolaychuk and Hrabovsky, 2000). In particular, the content of lead in soils of northern agricultural regions (Zhytomyr, Sumy, Rivne, Chernihiv, and Kyiv) exceeds the permissible level by 3 to 9 times (Shestapalov et al., 1996). A similar tendency is observed for soil in the region of Kyiv Polesye, where concentration of metal exceeds the natural level by 3 times (Brooks, 1982; Shestapalov et al., 1996) (Figure 3).

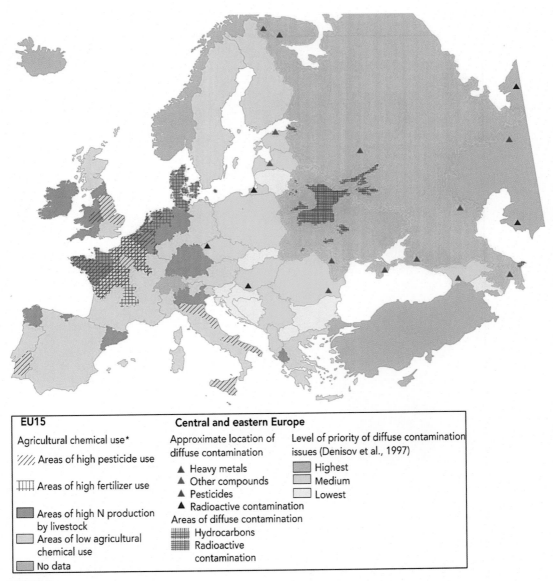

FIGURE 2

Probable problem areas of diffuse contamination in Europe **(EEA-UNEP, 2000)** * - agricultural areas which can use chemical during plant cultivation.

Second place among heavy metals highly contaminating Ukraine soil is nickel. It's known that the main anthropogenic source of nickel emission is combustion of fossil fuels and traffic (Krstić et al., 2007). Nickel is widely used in silver refineries, alloy, pigments electroplating, zinc-based casting, and storage batteries too, which also can have the effect of increasing nickel pollution.

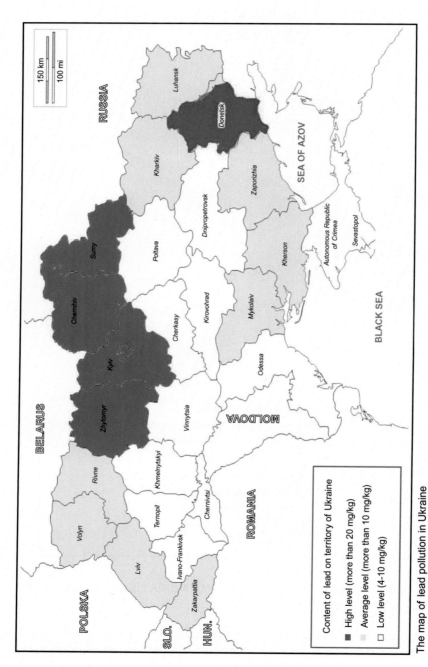

The map of lead pollution in Ukraine

FIGURE 3

The map of lead pollution on territory of Ukraine.

The greatest Ni content found in soils of the regions of Zhytomyr, Kyiv, Zaporizhia Donetsk, and Lugansk (in the soil 25-50 mg/kg when maximal permissible level 50 mg/kg) Figure 2. At these regions it is possible to get relatively safe crops with tolerance to Ni but in limited quantities. The agricultural land Volyn, Poltava, Vinnytsia, and partly other regions have pollution levels 5-10 and 10-15 mg/kg of soil areas for ecological agriculture nowadays. In these areas it is possible to grow vegetable raw materials suitable for the production of baby food. Soils with the average pollution level of Ni (15-25 mg/kg) (Figure 2) are mainly distributed in Chernihiv, Sumy, Zaporizhya, Ivano-Frankivsk, Rivne, and Transcarpathian region (Anonymous, 2004). High level of copper found in Kiev, Chernihiv, Zgytomyr, Volyn, Zakarpatiya regions. Chernivtsi, Kharkiv, Donetsk and Zaporizhya resgins has been characterized high level of chromium (Figure 3). In the Ivano-Frankivsk region, Pb concentrations exceed toxic levels in the areas close to industrial sectors (10% of the region territory). Local background levels of heavy metals are greatly exceeded in snow close to industrial regions. Cd and Mo accumulate in forest soils in the Ukrainian Carpathians region (Shparyk and Parpan, 2004) (Figure 3). A similar situation has been observed with pollution of soil with Zn and Co. The maximal permissible level of Zn and Co in air is 300 and 50 Clarke, respectively (Clarke is unit of each element and in soil it is 50 and 8 mg/kg, respectively). In general, for these three heavy metals (Ni, Zn, and Co) in many areas of Ukrainian territory their content in the soil is at less than maximal permissible level (Figures 4 and 5).

It's known that the maximum permissible level of Pb in soil is 30 mg/kg of dry soil (Samokhvalova et al., 2001). The limits of maximum permissible level depend on the complex of soil and ecological conditions, among which the most important are soil pH and content of hummus (Glazovskaya, 1994; Kabata-Pendias and Pendias, 1989; Obuhov et al., 1980). It has been proved that the main quantity of heavy metals is concentrated in the upper layer of arable soil (0-10 cm) (Glazovskaya, 1994; Trahtenberg, 1998). The period of heavy metal removal from soil is near 740 to 5900 years (for example, cadmium and lead) (Dobrovolsky, 1997). Most inorganic compounds of Pb and Cd are insoluble or poorly soluble in water. However, their dissolubility depends on soil pH. Dissolubility reaches a 4.0, but in the range of 6.5-8.0, dissolubility declines distinctly, but after further increasing of pH, slightly increases (Glazovskaya, 1994). The redox system, level of gross content, and the granulometric composition of soil have an influence on solubility of heavy metal compounds in the soil (Lysenko et al., 2001). Soils serve as a comprehensive geochemical barrier that keeps solid compounds and water-soluble compounds of heavy metals in the form of organo-mineral, mineral complexes, and compounds (Glazovskaya, 1994). It is known that soil has a buffer capacity as to toxic effect of pollutants, which provides for transformation of pollutants at not-soluble forms for plants. Granulometric structure of soil rich in organic matter and carbonate usually has the highest buffer capacity (Grakovsky et al., 1994).

3 ECONOMICAL BACKGROUND OF BIODIESEL PRODUCTION IN UKRAINE AND IN THE WORLD

It is known that humus-rich black soil of Ukraine is world famous. Since the dissolution of the Soviet Union with ongoing land reform and decreasing activity in the agriculture sector, Ukraine has not been making full use of its agricultural potential (Schaffartzik et al., 2014).

Ukraine is often singled out for its potential as a supplier of agricultural goods to the EU region. 42 of 60 Mha of Ukraine can be used for crop production (Elbersen et al., 2009). At 0.7 hectare per person (ha/cap), Ukraine has more land available than any other European country, except for Russia (0.9 ha/cap),

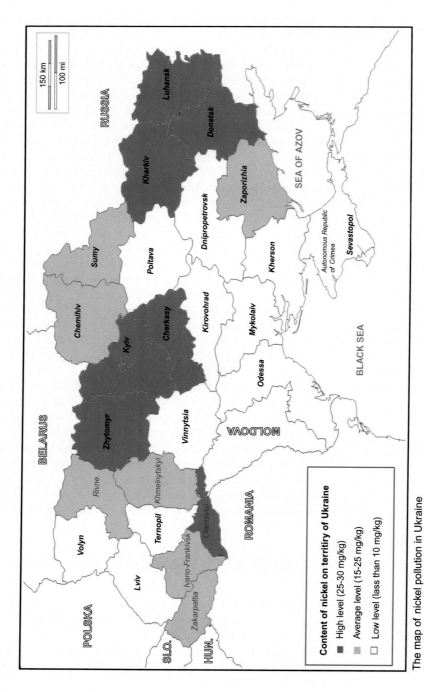

The map of nickel pollution in Ukraine

FIGURE 4

The map of nickel pollution on territory of Ukraine.

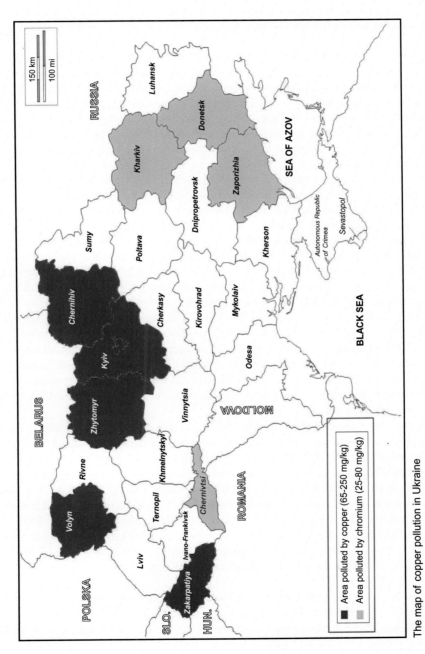

The map of copper pollution in Ukraine

FIGURE 5

The map of copper and chromium pollution on territory of Ukraine.

significantly surpassing France (0.3 ha/cap) and Germany (0.1 ha/cap) (FAOSTAT, 2012), for example. The country's geographical location as neighbor to the countries of the European Union and its access to the Black Sea is additionally advantageous to its role as a global supplier of biomass resources. Ukraine is a major exporter of agricultural primary products: In 2010 and in physical units, it ranked second among the international top exporters of barley, third among the exporters of rapeseed, seventh for exports of sunflower seed, and ninth among the exporters of wheat. Rapeseed production and exports drastically increased from 2004 onwards and grew by a factor of almost 20 until 2008 (FAOSTAT, 2012).

In Ukraine the sunflower, rapeseed, and soybean are the most widespread oilseed crops that can be used for phytoremediation process and production of biodiesel and biogas. In Eastern Europe, problems of diffuse soil contamination are greatest in Azerbaijan, Belarus, Moldova, Russia, and Ukraine (Denisov et al., 1997). In the total cultivation area of these crops in Ukraine, the area of heavy metals-contaminated soils is approximately 15-20%.

At the same time biofuels are currently not produced or consumed in relevant amounts within Ukraine. Yet in Ukraine potential is mainly as an alternative to liquid fossil fuel use and not to the use of natural gas and coal in industrial applications and electricity generation. The production of biofuel requires energy inputs. But the biggest advantage that biodiesel has over gasoline and petroleum diesel is its environmental friendliness. Biodiesel burns similar to petroleum diesel as it concerns regulated pollutants. On the other hand, biodiesel probably has better efficiency than gasoline. One such fuel for compression ignition engines that exhibits great potential is biodiesel. Diesel fuel can also be replaced by biodiesel made from vegetable oils. Biodiesel is now mainly being produced from soybean, rapeseed, and palm oils.

The higher heating values (HHVs) of biodiesels are relatively high. The HHVs of biodiesels (39-41 MJ/kg) are slightly lower than that of gasoline (46 MJ/kg), petrol diesel (43 MJ/kg), or petroleum (42 MJ/kg), but higher than coal (32-37 MJ/kg). Biodiesel has over double the price of petrol diesel (Demirbas, 2007). The major economic factor to consider for input costs of biodiesel production is the feedstock, which is about 80% of the total operating cost. The high price of biodiesel is in large part due to the high price of the feedstock. Economic benefits of a biodiesel industry would include value added to the feedstock, an increased number of rural manufacturing jobs, increased income taxes, and investments in plant and equipment.

The production and utilization of biodiesel is facilitated first through the agricultural policy of subsidizing the cultivation of nonfood crops (rapeseed, for example). Second, biodiesel is exempt from the oil tax. The European Union in 2005 got nearly 89% of all biodiesel production worldwide. Germany produced 1.9 billion liters or more than half the world total. In Germany biodiesel is also sold at a lower price than fossil diesel fuel. Biodiesel is treated like any other vehicle fuel in the UK. Other countries with significant biodiesel markets included France, the United States, Italy, and Brazil. All other countries combined accounted for only 11% of world biodiesel consumption. By 2004, the United States is expected to become the world's largest single biodiesel market, accounting for roughly 18% of world biodiesel consumption, followed by Germany (EBB, 2004). For Ukraine with its very low share of renewables in TPES, the promise of reduced environmental burdens is especially appealing (Schaffartzik et al., 2014).

Rapeseed, as an important feedstock for biofuel, contributed only about 2% to the agricultural Ukrainian harvest in 2008 but wheat, maize, sugar beets, and sunflower seeds (all also potential feedstock for biofuel production) figured much more prominently. Yet, while overall agricultural production decreased, rapeseed production grew by factor 65 between 1993 and 2008. Almost all of this production is designated for export. In 2008, 83% of Ukraine's direct material input of rapeseed was exported. In 2009, this share reached 99% (calculations by the authors based on data extracted from FAOSTAT (2012)).

The production of oilseeds in Ukraine has increased considerably during the last decade and with respect to sunflower seed in 2013 with total cultivation area 5051 thousand hectares and yield 21.7 centner per hectare (State Statistics Service of Ukraine, 2013). Ukraine is one of the biggest sunflower producers in the world besides Russia and the EU. From the mid-2000s onward, production of rapeseed also increased significantly (from about 0.14 million tons in 2000 to about 1.5 million tons in 2010) (van Leeuwen et al., 2012). The total oilseeds area harvested in Ukraine is projected to grow only slightly over the projection period. While the area for sunflower seeds is projected to actually decrease after 2015, more land will be allocated to rapeseed and soybeans (Figure 5).

Ukraine is considered an important (potential) source of rapeseed for the European market (Nekhay, 2012). Yields for this crop in Ukraine, however, are lower than in almost any other European country with the exception of Russia. While, in 2010, Germany was able to harvest 3.9 t/ha of rapeseed, harvest in Ukraine only amounted to 1.7 and in Russia to 1.1 t/ha. Özdemir et al. (2009) calculate biofuel yield in tons of oil equivalent per hectare and year as the gross energy yield in the period from 2000 to 2007 and found that Ukraine's rapeseed yield is consistently lower than that of the EU. In the dry southern part of Ukraine, the attainable biofuel energy yields are among the lowest in Europe at around 40 gigajoules per hectare (GJ/ha) (Fischer et al., 2010b). Although Ukraine managed to improve rapeseed yields after their initial decline following the collapse of the Soviet Union, the focus of expanding rapeseed production was on extensification rather than on the intensification required for improving yields (World Bank, 2008). The climatic risk of winter kill in many areas of the country adds to the decreasing attractiveness of this crop for farmers.

Figures 6–8 show the yield patterns of the oilseeds. Also based on former trends, oilseed yields are projected to grow faster than cereals yields.

Although there are currently rumors of Chinese investments into the distillation of bioethanol in Ukraine, and a cooperation agreement was signed in Spring 2013 between the Finnish and the Ukrainian governments to foster biofuel production (Schaffartzik et al., 2014), foreign

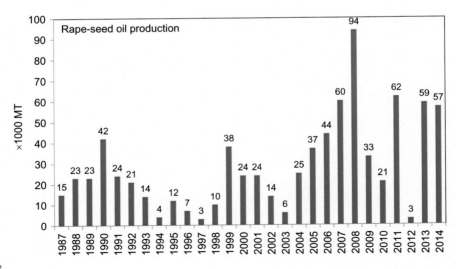

FIGURE 6

Rape-seed oil production in Ukraine (Myrna van Leeuwen et al., 2012).

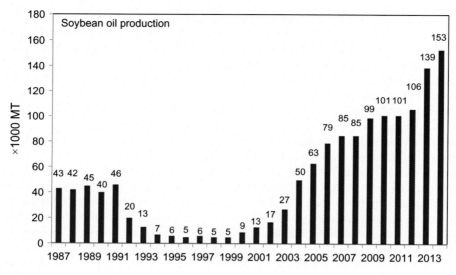

FIGURE 7

Soybean oil production in Ukraine (Myrna van Leeuwen et al., 2012).

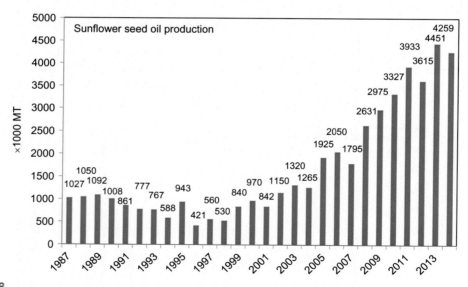

FIGURE 8

Sunflower oil production in Ukraine (Myrna van Leeuwen et al., 2012).

engagement and investment continue to be stronger for production feedstock than for fuels. For instance, in Ukraine there is no direct support for the production of fuels through the European Bank for Reconstruction and Development (EBRD); the bank is, however, an important credit grantor in the agricultural sector (EBRD, 2013). Moreover, Germany in particular has supported crop production in Ukraine in the past, for example through a GIZ project on sustainable biofuel production (IER, 2010) as well as through seminars and conferences on (sustainable) biofuel production embedded in the activities of the German-Ukrainian Agricultural Policy Dialogue (APD, 2013). This is due to the fact that according to the Renewable Energy Directive (European Parliament and Council, 2009) liquid biofuels have to follow certain sustainability criteria (e.g., sustainable crops must not be grown on areas with high biodiversity or with high existing carbon stock). The economic and social benefits that Ukraine can gain from its agricultural sector depend very strongly not just on the development of that sector itself but also on the benefits gained from the further use of this output. As a producer of rapeseed (and other crops) for the global market and thus as a producer of biofuel feedstock for other economies, Ukraine can follow the trajectory of other economies dependent on primary commodity exports (Schaffartzik et al., 2014).

4 BIODIESEL PRODUCED FROM OILSEED CROPS

Plant triacylglycerols are energy-rich compounds of reduced carbon available from nature. Most plant oils are derived from triacylglycerols stored in seeds (Srivastava and Prasad, 2000; Demirbas, 2006). During seed development, photosynthate from the mother plant is imported in the form of sugars, and the seed converts these into precursors of fatty acid biosynthesis. Given their chemical similarities, plant oils represent a logical substitute for conventional diesel, a nonrenewable energy source. However, as plant oils are too viscous for use in modern diesel engines, they are converted to fatty acid esters. Most plant labels have a viscosity range that is much higher than that of conventional diesel: 17.3-32.9 mm^2/s compared to 1.9-4.1 mm^2/s, respectively (Knothe and Steidley, 2005). The fatty acid methyl esters (FAMEs) found in biodiesel have a high energy density as reflected by their high heat of combustion, which is similar, if not greater, than that of conventional diesel (Knothe, 2005). Similarly, the cetane number (a measure of diesel ignition quality) of the FAMEs found in biodiesel exceeds that of conventional diesel (Knothe, 2005).

The higher oxygenated state compared to conventional diesel leads to lower carbon monoxide (CO) production and reduced emission of particulate matter (Graboski and McCormick, 1998). This latter air pollutant is especially problematic in European cities, motivating temporary curfews for diesel-powered vehicles. Biodiesel also contains little or no sulfur or aromatic compounds; in conventional diesel, the former contributes to the formation of sulfur oxide and sulfuric acid, while the aromatic compounds also increase particulate emissions and are considered carcinogens. In addition to the reduced CO and particulate emissions, the use of biodiesel confers additional advantages, including a higher flashpoint, faster biodegradation, and greater lubricity (EBB, 2004; Demirbas, 2007). Therefore, plant oil production needs to be greatly increased for biodiesel to replace a major proportion of the current and future fuel needs of the world.

The source of biodiesel usually depends on the crops amenable to the regional climate. Biodiesel is the monoalkyl esters of long-chain fatty acids derived from renewable feedstocks, such as vegetable oil or animal fats, for use in engine. Biodiesel is composed of FAMEs that can be prepared from triglycerides in vegetable oils by transesterification with methanol. The biodiesel is quite similar to conventional diesel fuel in its main characteristics (Meher et al., 2006). In the United States, soybean oil is the most commonly grown biodiesel feedstock, whereas the rapeseed (canola) oil and palm oil are the most common source for biodiesel in Europe and in tropical countries, respectively (Knothe, 2002). However, any vegetable oil—corn, cottonseed, peanut, sunflower, safflower, coconut, or palm—could be used to produce biodiesel (Demirbas, 2006). From a chemical point of view, oils from different sources have different fatty acid compositions. The fatty acids vary in their carbon chain length and in the number of unsaturated bonds they contain (Table 1). Fats and oils are primarily water-insoluble, hydrophobic substances in the plant and animal kingdom that are made up of one mole of glycerol and three moles of fatty acids and are commonly referred as triglycerides.

Twenty-one fatty acids are screened in all the samples. The fatty acids commonly found in vegetable oil and fat are stearic, palmitic, oleic, linoleic. Other fatty acids that are also present in many of the oils and fats are myristic (tetradecanoic), palmitoleic, acachidic, linolenic, and octadecatetraenoic. Many other fatty acids are also found in oils with the above-mentioned common fatty acids. Erucle fatty acid is found only in three oils: crambe, camellia oil, and *Brassica carinata*. In the oil of sunflower and soybean, the content of the fatty acid is different in the same plant species, which may be due to either the varietal or instrumental difference or in the different parts of plants.

Chemically the oil/fats consist of 90-98% triglycerides and small amount of mono- and diglycerides. Triglycerides are esters of three fatty acids and one glycerol. These contain a substantial amount of oxygen in their structures. When three fatty acids are identical, the product is simple triglycerides; when they are dissimilar, the product is mixed triglycerides, fatty acids which are fully saturated with hydrogen have no double bonds. Fatty acids with one missing hydrogen molecule have one double bond between carbon atoms and are called monosaturated. The fatty acids that have more than one missing hydrogen molecule or have more than one double bond are called polyunsaturated. Fully saturated triglycerides lead to excessive carbon deposits in engines. The fatty acids are different in relation to the chain length, degree of unsaturation, or presence of other chemical functions. Chemically, biodiesel is referred to as the mono alkyl esters of long-chain fatty acids derived from renewable lipid sources. Biodiesel is the name for a variety of ester-based oxygenated fuel from renewable biological sources. It can be used in compression ignition engines with little or no modification (Demirbas, 2002).

Biodiesel is made in a chemical process called transesterification, in which organically derived oils (vegetable oils, animal fats, and recycled restaurant greases) are combined with alcohol (usually methanol) and chemically altered to form fatty esters such as methyl ester. Biodiesel consists of alkyl (usually methyl) esters instead of the alkanes and aromatic hydrocarbons of petroleum-derived diesel. Diesel has no oxygen compound. It is a good quality of fuel (Singh and Singh, 2010).

Oil, ester, and diesel have different numbers of carbon and hydrogen compounds. On the other hand, in the case of vegetable oils oxidation resistance is markedly affected by the fatty acid composition. The large size of vegetable oil molecules (typically three or more times larger than hydrocarbon fuel molecules) and the presence of oxygen in the molecules suggest that some fuel properties of vegetable oil would differ markedly from those of hydrocarbon fuels (Goering et al., 1982).

Since biodiesel is produced in quite differently scaled plants from vegetable oils of varying origin and quality, it was necessary to create standards for fuel quality to guarantee engine performance without any difficulties. Austria was the first country in the world to define and approve the standards for rapeseed oil methyl esters as diesel fuel. As standardization is a prerequisite for successful market introduction and penetration of biodiesel, standards or guidelines for the quality of biodiesel has also been defined in other countries such as Germany, Italy, France, the Czech Republic, and the United States (Meher et al., 2006).

The parameters that define the quality of biodiesel can be divided into two groups. One group contains general parameters, which are also used for mineral oil-based fuel, and the other group describes the chemical composition and purity of fatty acid alkyl esters (Mittelbach,

Table 1 Fatty Acid Composition of Rapeseed, Sunflower, and Soybean Oil from Different Sources (%)

S. No.	Vegetable Oil	Fatty Acid Composition (wt.%)															6:0[a], 8:0[b], 10:0[c], and Others	Reference
		12:0	14:0	14:1	16:0	16:1	18:0	20:0	20:1	22:0	24:0	18:1	22:1	18:2	18:3	18:4		
1.	Rapeseed	–	–	–	3.5	–	0.9	–	–	–	–	54.1	–	22.3	8.2	–	–	Demirbas (2002, 2003) and Balat and Balat (2008)
2.	Rapeseed	–	0	–	3	–	1	0	–	0	0	64	0	22	8	–	–	Goering et al. (1982)
3.	Sunflower	–	–	–	6.4	0.1	2.9	–	–	–	–	17.7	–	72.9	–	–	–	Demirbas (2003) and Balat and Balat (2008)
4.	Sunflower	–	0	–	6	–	3	0	–	0	0	17	0	74	0	–	–	Goering et al. (1982)
5.	Soybean	–	–	–	13.9	0.3	2.1	–	–	–	–	23.2	–	56.2	4.3	0	–	Demirbas (2003)
6.	Soybean	–	–	–	14	–	4	–	–	–	–	24	–	52	–	6	–	Demirbas (2003)
7.	Soybean	–	0	–	12	–	3	0	–	0	0	23	0	55	6	–	–	Goering et al. (1982)

a,b,c data regarding fatty acids content were not available

1996). The second group contains the general and third group: the vegetable oil-specific parameters and the corresponding value of FAMEs according to standards of the countries mentioned (Tables 2 and 3).

Among the general parameters for biodiesel, the viscosity controls the characteristics of the injection from the diesel injector. The viscosity of FAMEs can reach very high levels, and hence it is important to control it within an acceptable level to avoid negative impacts on fuel injector system performance. Therefore, the viscosity specifications proposed are nearly the same as that of the diesel fuel. The flashpoint of a fuel is the temperature at which it will ignite when exposed to a flame or spark. The flashpoint of biodiesel is higher than that of petrodiesel, which is safe for transport purpose.

The possibilities of using different types of alternative fuels have been analyzed and investigated in the past 10 years at Khar'kov Polytechnic Institute National Technical University, Ukraine. Important attention has been focused on liquid and gaseous fuels from renewable resources: plant leaves, stems, or seeds. Most of these fuels differ significantly from traditional liquid hydrocarbon fuels in the physicochemical properties that affect both organization of operation and the technical and economic and environmental indexes of ICE.

The analysis showed that in the conditions of Ukraine, an alternative fuel produced by blending liquid hydrocarbon fuels with rape oil (RO) derivatives—methyl esters (ROME)—is promising for ICE (Table 4).

Biodiesel fuel (BDF) on the basis of ROs has passed tests and encouraging results have been received. The data of the specialized firms of Germany, Poland, the United States, and also the Ukrainian Agrarian Academy of Sciences, Research Transport Institute (Kiev), etc., are simultaneously generalized. Extensive researches on BDF with RO have also been carried out in Austria, France, Italy, Spain, and Great Britain (Topilin, 2002). The generalized data on the properties of RO and traditional diesel fuel are presented in Table 5.

As Table 5 shows, BDF on the basis of RO can be suitable for use in engines of all sorts without their constructive change. German BDF of the firm Tessol Stuttgart is made by dispersion from 20% of spirit, 25% gas, and 55% of coldly pressed filtered RO. Such fuel mix was widely used in diesel engines by the technical universities of Stuttgart, the Kaiser-Slautern and Hohenheim. Studies have established that emissions of the fulfilled gases under condition of transition to a rape methyl ester are reduced by 50% in case of the maximal loading of a diesel engine. Emissions of firm particles decrease by 20%. Ukraine had about 80 bio-ethanol factories with an average capacity of 670 millions liters per year per factory. The known domestic requirement is only 200 million liters per year from one factory. Therefore, it is rational to direct surplus of spirit to manufacture of fuel mixes for cars and other types of transport. Consumers have already been using liquid fuel from Odessa and Drogobych oil manufactures.

In Ukraine there are all opportunities for an organization of BDF manufacture on the basis of ROs. The appreciable effect can be received without radical reequipment of diesel engines of a batch production, having mastered technology of BDF production, consisting of a hydrodynamically activated mix of a petroleum origin with RO (Topilin, 2002). It is especially important for Ukraine, which has huge opportunities of rape growing (for example, on the polluted grounds of the Chernobyl zone) and which is characterized by considerable deficiency of mineral oil. Use of BDF will allow reducing needs for petroleum by 1-2 million tons (depending on the volume of BDF manufacture).

Table 2 General Parameters of the Quality of Biodiesel

Parameters	Austria (ON)	Czech Republic (CSN)	France (Journal Official)	Germany (DIN)	Italy (UNI)	USA (ASTM)	Brasil (ANP)
Density at 15 °C (g/cm³)	0.86–0.90	0.86–0.89	0.87–0.90	0.875–0.89	0.86–0.90	–	Like diesel
Viscosity at 20 °C (mm²/s)	6.5–9.0	6.5–9.0	–	–	–	–	–
Viscosity at 40 °C (mm²/s)	–	–	3.5–5.0	3.5–5.0	3.5–5.0	1.9–6.0	Like diesel
Flash point (°C)	≥55	≥56	≥100	≥110	≥120	≥130	≥100
CFPP (°C)	≤ −8	≤ −5/−15	–	0–10/−20	–	–	–
Pour point (°C)	–	≤ −8/−20	≤ −10	–	–	–	≥45
Cetane number	≥48	≥48	≥49	≥49	–	≥47	≥45
Methanol % mass	≤0.3	≤0.3	≤0.1	≤0.3	≤0.02	–	≤0.50
Acid value (mg KOH/g)	≤1	≤0.5	≤1	≤0.5	≤0.50	≤0.8	≤0.8
Conradson carbon residue (100%)	≤0.1	–	–	–	–	≤0.05	≤0.05
Water cont. (mg/kg)	–	≤1000	≤200	≤300	≤500	–	–
	ONORM (1991)	CSN (1994)	Journal official (2003)	DIN (1997)	UNI (2001)	D 6751. 02 (2002)	Ramos (2003)

Table 3 Vegetable Oil-Specific Parameters for the Quality of Biodiesel

Parameters	Austria (ON)	Czech Republic (CSN)	France (Journal Official)	Germany (DIN)	Italy (UNI)	USA (ASTM)	Brasil (ANP)
Ester content (% mass)	–	–	≥96.5	–	≥96.5	–	–
Monoglyceride (% mass)	–	–	≤0.8	≤0.8	≤0.80	–	≤1.0
Diglyceride (% mass)	–	–	≤0.2	≤0.4	≤0.20	–	≤0.25
Triglyceride (% mass)	–	–	≤0.2	≤0.4	≤0.20	–	≤0.25
Free glycerol (% mass)	≤0.03	≤0.02	≤0.02	≤0.02	≤0.02	≤0.02	≤0.02
Total glycerol (% mass)	≤0.24	≤0.24	≤0.25	≤0.25	≤0.25	≤0.24	≤0.38
Iodine number	–	–	≤115	≤115	≤120	–	Take note
	ONORM (1991)	CSN (1994)	Journal official (2003)	DIN (1997)	UNI (2001)	D 6751. 02 (2002)	Ramos (2003)

Table 4 Properties of diesel and rapeseed oils produced in Ukraine

Indexes	DF	ROME
Cetane number, min	45	48
Ignition point (°C), min	60	56
Solid point (°C), min	−10	−8
Carbone residue in 15 residue (%), max	0.5	0.3
Copper plate test	Passed	Passed
Acid number (mg KOH/g)	0.06	0.5
Content (%), max		
Sulfur	0.2	0.02
Ash	0.02	0.02
Water	None	None
Total glycerin content (%), max	–	0.3
Net beat value (MJ/kg)	42.5	37.5
	Abramchuk et al. (1992) and Marchenko and Semenov (2001)	Marchenko and Semenov (2001)

Table 5 Properties of rape Oil and Traditional Diesel Fuel

Fuel	Density at 20°C (kg/m³)	Kinematic Viscosity at 20°C (mm²/s)	Cetane Number	The Sinking Point	Point of Cold Filter Closing	Termoability MDZ/KG	Reference
Traditional diesel fuel	860	4-6	52	80	−11.0	46.4	Topilin et al. (2009)
Rapeseed oil ethyl ester	848	6.17-7.84	40	317	+15.0	39.5	Topilin et al. (2009) and Graboski and McCormick (1998)
Mix Tessol	898	3.2	39	–	−5.5	39.4	Topilin et al. (2008)
Rapeseed oil methyl ester	882	4.83	48	≥100	−8.0	40.0	Topilin et al. (2008) and Graboski and McCormick (1998)
Soybean oil ethyl ester	931	4.41-7.8	46	347	+18	43.7	Topilin et al. (2008) and Graboski and McCormick (1998)

5 CLEANUP OF HEAVY METALS FROM SOIL BY USING OILSEED CROPS

The remediation of metal-contaminated sites often involves expensive and environmentally invasive and civil engineering-based practices (Marques et al. 2008). A range of technologies such as fixation, leaching, soil excavation, and landfill of the top contaminated soil *ex situ* have been used for the removal of metals. Many of these methods have high maintenance costs and may cause secondary pollution (Haque et al., 2008) or have an adverse effect on biological activities, soil structure, and fertility (Pulford and Watson, 2003). Phytoremediation is an emerging technology that should be considered for remediation of contaminated sites because of its cost effectiveness, aesthetic advantages, and long-term applicability.

This technology is based on element accumulating *via* plants that grow on contaminated soil and later can be harvested in order to remove the specific elements from the soil (Salt et al., 1995; Alkorta et al., 2004; Witters et al., 2012). An alternative soil remediation technology has been proposed that uses rare, heavy metal tolerant-plant species that work as hyperaccumulators of heavy metals *via* the plants' roots and shoots.

The crops can be divided into four groups, which differ considerably in their ability to accumulate heavy metals: (1) low accumulators (maize and peas); (2) moderate accumulators (barley, lentils, gram, sunflower, sesame, fennel, coriander, dill, peppermint, basil, cotton, potatoes, datura) (Gholamabbas et al., 2009), (3) high accumulators (wheat, soybean, beans, rape, peanuts, anise, black mustard, flax, hemp, sugar beet, fodder beet), and (4) hyperaccumulators (*Salvia sclarea* L., *Talinum triangulare*, tobacco, etc.) (Kumar et al., 2012).

To date, a number of plants have been well characterized as hyperaccumulators (those that accumulate high levels of toxic contaminants versus nonaccumulator species) but have the distinct disadvantage of having slow growth rate and being low biomass producers. For example, these are *Thlaspi caerulescens*, *Arabidopsis halleri*, *Armeria maritima* (Baker and Brooks, 1989; Brown et al., 1994). Also, these plants are endemic to certain areas and do not perform well in other areas. The ideal characteristics of an effective plant-phytoremediator are fast growths, deep and extensive roots (phreatophytes), high biomass, easy to harvest.

The application of plants for phytoremediation has two advantages; the remediation of contaminated soil and the production of biofuel. The plant species is the most important factor that needs to be satisfied. The following information is required for effective remediation and the abundant production of biofuel: (1) heavy metal tolerance of species, (2) heavy metal accumulation capacity of the plants, and (3) biofuel production capacity per unit area. Not only sunflowers but also rape and soybean have been used as principal biofuel sources. Moreover, these plants have been studied for their potential in phytoremediation in order to clean up contaminated soils (Chen et al., 2004; Ru et al., 2004; Murakami et al., 2007). Sunflowers have good tolerance for various heavy metals such as Ni, Cu, As, Pb, and Cd (Lin et al., 2009; Turgut et al., 2004).

Use of sunflower (*Helianthus annuus* L.) for Cu phytoextraction and oilseed production on Cu-contaminated topsoils was investigated in a field trial at a former wood preservation site, and it has been confirmed that among six experimental cultivars, sunflower cultivar energic seed yield (3.9 Mg air-DW/ha) would be sufficient to produce oil. Phenotype traits and shoot Cu removal depended on sunflower types and Cu exposure (Kolbas et al., 2011).

The accumulation and distribution of Cd and Pb in plant parts and evaluation of the potential problem caused by heavy metals when biofuel is produced by plant seeds cultivated from Cd- and Pb-contaminated soil have been investigated. Based on the germination test results, sunflowers were selected as the target plant because of their high germination rate and IC50 compared to other plants. The trends of accumulated amounts and concentrations of Cd and Pb in sunflowers were not identical. The accumulation of Cd in the plant parts was in the order of: seeds > stem > leaves > petals > roots, with the accumulation of Pb: stem > roots > leaves > petals > seeds. Pb did not accumulate in the seeds. These findings show that there should not be any significant problem producing biofuel from sunflower seeds cultivated in Pb-contaminated soil (Lee et al., 2013). At the same time sunflower species can be used for revegetation of areas contaminated with As up to the safe limit of 150 mg As/dm^3 soil (Melo et al., 2012).

To identify oil crops that can be cultivated in Zn-contaminated soil for biodiesel production, the ability of Zn tolerance and accumulation of eight oil crops were evaluated under 200-800 mg Zn/kg sand substrates (DW) conditions. Results showed that all crops, except sunflower, could grow quite well under 400-800 mg/kg Zn stress. Among them, hemp, flax, and rapeseed showed small inhibitions in plant growth and photosynthetic activities, indicating these crops had a strong tolerance to high Zn concentrations and could be cultivated in Zn-contaminated soils. Peanut and soybean exhibited higher Zn concentrations in shoots, higher bioconcentration factor, and higher total Zn uptake, as well as higher biomass. These crops, therefore, are good candidates for the implementation of the new strategy of cultivating biodiesel crops for phytoremediation of Zn-contaminated soils (Gangrong Shi and Cai, 2010).

The soybean roots can accumulate the highest value of Pb—294.8 mg/kg, Zn—644.4 mg/kg, Cu—34.4 mg/kg, Cd—10.9 mg/kg (Angelova et al., 2003). In the roots of soybean were fixed and accumulated a great part of the heavy metals that had entered the soil, as soybean plants form a powerful root system with strong absorbing ability on depth as well as on width, including a great volume of soil. Soybean roots can accumulate nickel in high quantity (Malan and Farrant, 1998). Nickel uptake by plants is positively correlated with the metal's concentrations in soil solution.

The roots of sunflower are able to accumulate more quantity of Zn (2514.2 mg/kg), Cu (255.7 mg/kg), and Cd (53.2 mg/kg) (Sayyad et al. 2009) in comparison with soybean plants. We suggest that soybean plants can be a hyperaccumulator of lead and nickel. Therefore, these crops accumulate trace element beyond permissible level in their vegetative mass. Phytoextraction of metals, metalloids, and radionuclides is one of the promising options for contaminant cleanup. However, some schools have expressed reservations for application in field situations for environmental cleanup (Venselaar, 2004; Rascioa and Navari-Izzo, 2011).

At the same time, compost and vermicompost application led to effective immobilization of Pb, Zn, and Cd mobile forms in soil. The efficacy of the sunflower plant for phytoremediation of contaminated soils in the absence and presence of organic soil amendments (compost and vermicompost) has been studied (Angelova et al., 2012). A correlation was found between the quantity of the mobile forms and uptake of Pb, Zn, and Cd by the sunflower seeds. Tested organic amendments significantly influenced the uptake of Pb, Cu, Zn, and Cd by sunflower plants. Oil content and fatty acids composition were affected by compost and vermicompost amendment treatments. The compost and vermicompost treatments significantly reduced heavy metals concentration in sunflower seeds, meals, and oils, but the effect differed among them.

Trace element uptake by sunflower and its implications for biofuel and biochar production is actual topic of research activities (Prasad, 2008; Evangelou et al., 2010). The increasing demand for energy and the decreasing reserves of fossil fuels have turned the focus of governments to energy production from biomass (biofuels, biogas). As biomass is currently the only renewable source of organic carbon, it appears to be an ideal source for the production of transport fuels. However, because production of biofuel crops demands fertile agricultural land, it comes into conflict with an ever-increasing demand for food. This conflict is intensifying, as much arable land is lost through erosion and contamination. On the other hand, contaminated land that is unsuitable for food production and of little or no economic return could be used for the production of bioenergy. Phytoremediation, the use of vegetation for *in situ* restoration of contaminated soils, is generally considered a cost-effective and environmentally friendly approach (Arthur et al., 2005). However, it is increasingly recognized that the success of phytoremediation depends on its capacity to produce valuable biomass (Conesa et al., 2012).

Sunflower would be a suitable crop for the production of bioenergy on contaminated land. It can be used for the production of biodiesel (seed oil), as well as for biogas (straw). Since it has been used as a crop for centuries, cultivation techniques and infrastructure for harvesting and processing are already available.

However, cultivating metal-contaminated soils is complicated because, in most instances, several growth-limiting factors for the plants, such as a high phytotoxicity of the pollutants and poor fertility conditions, are acting simultaneously (Tordoff et al., 2000). Soil amendments are usually necessary to render the substrate suitable for the plant establishment (Brown et al., 2003; Houben et al., 2012). But the risks of contaminant transfer from the contaminated site into the environment, in particular water and food chains, need to be taken into account and limited to tolerable levels. Among others, trace elements uptake by the harvested plant parts must be kept at a minimum. This could be achieved by breeding for trace elements exclusion from uptake. Attention also has to be given to the residues of fuel burning, like ashes, and to by-products generated when the harvested biomass is converted into fuel or gas (Evangelou et al., 2012).

There are known beneficial effects of biochar application to contaminated soils on the bioavailability of Cd, Pb and Zn, and the biomass production of rapeseed (*Brassica napus* L.) (Ru et al., 2004; Houben et al., 2013). The *Brassicaceae* family includes numerous plant species cultivated in Ukraine and all around the world. Rapeseed (*Brassica napus*), brown mustard (*Brassica juncea*), dutch flax (*Camelina sativa*), and naven (*Brassica rapa*) are planted as oil crops (Hablaka and Chechenevab, 2009; State Checklist of Plant Cultivars Suitable for Distribution in Ukraine, 2011). *B. napus* L. accessions are suitable for phytoextraction of moderately heavy metal-contaminated soils (Grispen et al., 2006).

Treatment with 10% biochar proved equally efficient in reducing metal concentrations in shoots, but the biomass production tripled as a result of the soil fertility improvement. Thus, in addition to C sequestration, the incorporation of biochar into metal-contaminated soils could make it possible to cultivate bioenergy crops without encroaching on agricultural lands. Although additional investigations are needed, we suggest that the harvested biomass might in turn be used as feedstock for pyrolysis to produce both bioenergy and new biochar, which could contribute further to the reduction of CO_2 emission (Houben et al., 2013).

The study about phytoremediation of a copper mine soil with *Brassica juncea* L., compost, and biochar has shown that mustards extracted Ni efficiently from soils, suggesting that *B. juncea* L. is a good phytoextractor of Ni in mine soils. Amendments and planting mustards decreased the pseudototal concentration of this metal, reduced the extreme soil acidity, and increased the soil concentrations of C and TN. Both treatments also decreased the CaCl2-extractable Co, Cu, and Ni concentrations (Rodríguez-Vila et al., 2014).

It is now known that phytoextraction is not feasible in moderate to highly contaminated soils with naturally occurring metal hyperaccumulators for some of the disadvantages mentioned. Obviously agricultural crops that produce high biomass would be the right choice. Trace element contamination/pollution in agrocoenose, application of environmental crops (e.g., sunflower, rapeseed, and soybean for phytomanage are ments of trace element-contaminated soils, possible scope for development of phytoproducts, and establishment of sustainable agrocoenose (Figures 9 and 10).

FIGURE 9

(a) Sunflower and (b) rape seed are popular as environmental crops. These crops reported to be efficient in phytoremediation of inorganic and organic contaminants including radionuclides.

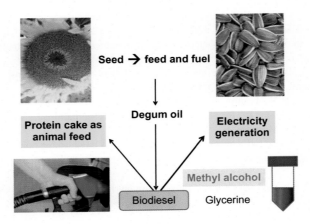

FIGURE 10

Production of biodiesel, bioproducts, glycerine, and electricity from the seeds of sunflower applied in phytoremediation.

6 CONCLUSIONS

The high concentration of heavy metal is harmful to human health through contaminated agro-production. The most heavy metals accumulate in soil after long-time addition. This chapter raised the issue of heavy metal pollution in the world and more specifically in Ukraine. The most dangerous aspect of this pollution is high levels of lead and nickel in soils of Ukraine territory. The proposed method for cleanup of polluted soils has been using heavy metal-tolerant plant species that are able to hyperaccumulate metals in plant roots and shoots. The widespread agricultures on Ukraine territory such as soybean, rapeseed, and sunflower can be accumulators of heavy metals. But the main part of harvest of these crops cannot be used for ecologically pure agricultural production according to international standards. For cleanup of contaminated soils, it will be better to use the part of this harvest on biodiesel production during one year. From the economic side this can be profitable also. At the same time it will be possible to clean up the most contaminated soils on the territory of Ukraine. But after a biogeochemical cycle, heavy metal will again appear in the environment. For full cleanup of soil by plants it is better to use the electrolysis method (placing the harvest in an oxide solution and using electrical power to gather metal under solid form).

ACKNOWLEDGMENTS

Thanks are due to the European Environment Agency and the UNEP Regional Office for Europe for authorization to use contaminated soil map. The authors would like to thank Paul Prasad Chandramohan, for his helpful comments and assistance.

REFERENCES

Abramchuk, F.I., Marchenko, A.P., Razleitsev, N.F., et al., 1992. Increased fuel economy and prolonged endurance. In: Shekhovtsov, A.F. (Ed.), Tekhnika, Kiev, Modern Diesels: [in Russian].

Alekseeva, U., 1987. Heavy metals in soil and plants. - L: Publishing house Science, 201 pp.

Alkorta, I., Hernarndez-Allica, J., Becerril, J.M., Amezaga, I., Albizu, I., Garbisu, C., 2004. Recent findings on the phytoremediation of soils contaminated with environmentally toxic heavy metals and metalloids such as zinc, cadmium, lead, and arsenic. Rev. Environ. Sci. Biotechnol. 3, 71–90.

Angelova, V., Ivanova, R., Ivanov, K., 2003. Accumulation of heavy metals in leguminous crops (bean, soybean, peas, lentils and gram). J. Environ. Prot. Ecol. 4 (4), 787–795.

Angelova, V.R., Ivanova, R.V., Ivanov, K.I., Perifanova-Nemska, M.N., Uzunova, G.I., 2012. Potential of sunflower (Helanthus annuus L.) for phytoremediation of soils contaminated with heavy metals. BALWOIS 2012-Ohrid Republic of Macedonia.

Anonymous, 1998. Trace Element in Environment. (P.A.Vlasjuka, Ed.). Publishing house Naucova dumka, Kyiv 57 p.

Anonymous, 2004. Scientific support of stable development of agriculture. Forest-steppe. Publishing house Alefa, Kyiv 2 volumes.

APD, 2013. German-Ukrainian Agricultural Policy Dialogue. http://www.ier.com.ua/en/agricultural_dialogue/ (accessed 11.07.13).

Arthur, E.L., Rice, P.J., Rice, P.J., Anderson, T.A., Baladi, S.M., Henderson, K.L.D., Coats, J.R., 2005. Phytoremediation—an overview. Crit. Rev. Plant Sci. 24 (2), 109–122.

ASTM, 2002. Standard specification for biodiesel fuel (b100) blend stock for distillate fuels. American Society for Testing and Materials D6751–02, .

Baker, A.J.M., Brooks, R.R., 1989. Terrestrial higher plants which hyperaccumulate metallic elements—a review of their distribution, ecology and phytochemistry. Biogeosciences 1, 81–126.

Balat, M., Balat, H., 2008. A critical review of bio-diesel as a vehicular fuel. Energy Convers. Manag. 49 (2008), 2727–2741.

Brooks, R.R., 1982. Pollution by trace elements, in chemistry of the environment. Misl' Publishing House, Moscow pp. 371–383, , in Russian.

Brown, S.L., Chaney, R.L., Angle, J.S., Baker, A.J.M., 1994. Phytoremediation potential of *Thlaspi caerulescens* and bladder campion for zinc- and cadmium-contamineted soil. J. Environ. Qual. 23, 1151–1157.

Brown, S.L., Henry, C.L., Chaney, R., Compton, H., DeVolder, P.S., 2003. Using municipal biosolids in combination with other residuals to restore metal-contaminated mining areas. Plant Soil 249 (1), 203–215.

Chen, Y., Li, X., Chen, Z., 2004. Leaching and uptake of heavy metals by ten different species of plants during an EDTA-assisted phytoextraction process. Chemosphere 57 (3), 187–196.

Chiras, D.D., 2009. Environmental Science. Jones and Bartlett Publishers, Sudbury. pp. 394.

Conesa, H.M., Evangelou, M.W.H., Robinson, B.H., Schulin, R., 2012. A critical view of current state of phytotechnologies to remediate soils: still a promising tool? Sci. World J. 2012, http://dx.doi.org/10.1100/2012/173829. pp. 10, Article ID 173829.

Conko, K.M., Landa, E.R., Kolker, A., Kozlov, K., Gibb, H.J., Centeno, J.A., Panov, B.S., Panov, Y.B., 2013. Arsenic and mercury in the soils of an industrial city in the Donets basin, Ukraine. Soil Sediment Contam. Int. J. 22 (5), 574–593. http://dx.doi.org/10.1080/15320383.2013.750270.

CSN 65 6507, 1994. Biopalivo pro vznetovÈ motory—methylestery řepkového oleje.

Demirbas, A., 2002. Diesel fuel from vegetable oil via transesterification and soap pyrolysis. Energy Sources 24, 835–841.

Demirbas, A., 2003. Biodiesel fuels from vegetable oils via catalytic and non-catalytic supercritical alcohol transesterifications and other methods: a survey. Energy Convers. Manag. 44, 2093–2109.

Demirbas, A., 2006. Biodiesel production via non-catalytic SCF method and biodiesel fuel characteristics. Energy Convers. Manag. 47, 2271–2282.

Demirbas, A., 2007. Importance of biodiesel as transportation fuel. Energy Policy 35 (9), 4661–4670.

Demirbas, A., 2009. Progress and recent trends in biodiesel fuels. Energy Convers. Manag. 50, 14–34.

Denisov, N.B., Mnatsakanian, R.A., Semichaevsky, A.V., 1997. Environmental Reporting in Central and Eastern Europe: a review of selected publications and frameworks. CEU/50-97.1. UNEP/DEIA/TR.97-6 Central European University/United Nations Environment Program.

DIN E 51606, 1997. Flussige Kraftstoffe - Dieselkraftstoff aus Fettsauremethylester (FAME). Mindestanforderungen. September 1997. http://rob-oil.com/biodiesel_DIN_51606.pdf

Dobrovolsky, V., 1997. Biospheric cycles of heavy metals and regulatory role of soil. Soil Sci. 4, 121–128.

EBB (European Biodiesel Board), 2004. EU: Biodiesel Industry Expanding Use of Oilseeds. European Biodiesel Board, Brussels.

EBRD, 2013. Projects. European Bank for Reconstruction and Development, Ukraine (Accessed 11-7-2013).

Elbersen, W., Wiersinga, R., Waarts, Y., 2009. Market scan bioenergy Ukraine. Report for the Dutch ministry of agriculture, nature and food quality (Wageningen).

European Parliament and Council, 2009. Directive 2009/28/EC of the European Parliament and of the Council of 23 April 2009 on the promotion of the use of energy.

Evangelou, M.W.H., Conesa, H.M., Fässler, E., Schulin, R., 2010. Trace Element (TE) uptake by sunflower and its implications for biofuel and biochar. In: Hughes, V.C. (Ed.), Sunflowers: Cultivation, Nutrition, and Biodiesel Uses. Nova publishers, New York, ISBN: 978-1-61761-309-8.

Evangelou, M.W.H., Conesa, H.M., Brett, H.R., Schulin, R., 2012. Biomass production on trace element–contaminated land: a review. Environ. Eng. Sci. 29 (9), 823–839. http://dx.doi.org/10.1089/ees.2011.0428.

FAOSTAT, 2012. FAO statistical database. Food and Agriculture Organization (FAO), Rome. http://faostat.fao.org/site/573/default.aspx#ancor (accessed 18.06.12).

Fischer, G., Prieler, S., van Velthuizen, H., Berndes, G., Faaij, A., Londo, M., de Wit, M., 2010. Biofuel production potentials in Europe: Sustainable use of cultivated land and pastures, Part II: Land use scenarios. Biomass and Bioenergy 34 (2), 173–187.

Gaevsky, V., Pelypets, M., 1999. Mobile forms of heavy metals in soils of Lviv region. Geol. Geochem. Combust. Miner. 3, 110–115.

Gangrong Shi, G., Cai, Q., 2010. Zinc tolerance and accumulation in eight oil crops. J. Plant Nutr. 33 (7), 982–997.

Gholamabbas, S., Majid, A., Sayed-Farhad, M., Karim, A.C., Mohammad, H.A., Brian, R.K., Schulin, R., 2009. Effects of cadmium, copper, lead, and zinc contamination on metal accumulation by sunflower and wheat. Soil Sediment Contam. 18 (2), 216–228.

Glazovskaya, M.A., 1994. Criterias of classification of soils by danger of pollution by lead. Soil Sci. 4, 110–120.

Goering, C.E., Schwab, A.W., Daugherty, M.J., Pryde, E.H., Heakin, A.J., 1982. Fuel properties of eleven oils. Trans. ASAE 25, 1472–1483.

Graboski, M.S., McCormick, R.L., 1998. Combustion of fat and vegetable oil derived fuels in diesel engines. Prog. Energy Combust. Sci. 24, 125–164.

Grakovsky, V., Sorokin, S., Fried, A., 1994. Sanitation of the polluted soils and recultivation their in Russia. J. Soil Sci. 4, 121–128.

Grispen, V.M.J., Nelissen, H.J.M., Verklei, J.A.C., 2006. Phytoextraction with *Brassica napus* L.: a tool for sustainable management of heavy metal contaminated soils. Environ. Pollut. 144 (1), 77–83.

Hablaka, S.G., Chechenevab, T.N., 2009. Prospectives for genetic improvement of the inflorescences structure in oil crops of the Brassicaceae family. Cytol. Genet. 43 (3), 161–163.

Haque, N., Peralta-Videa, J.R., Jones, G.L., Gill, T.E., Gardea-Torresdey, J.L., 2008. Screening the phytoremediation potential of desert broom (*Baccharis sarothroides* Gray) growing on mine tailings in Arizona, USA. Environ. Pollut. 153, 362–368.

Houben, D., Pircar, J., Sonnet, P., 2012. Heavy metal immobilization by cost-effective amendments in a contaminated soil: effects on metal leaching and phytoavailability. J. Geochem. Explor. 123, 87–94.

Houben, D., Evrard, L., Sonnet, P., 2013. Beneficial effects of biochar application to contaminated soils on the bioavailability of Cd, Pb and Zn and the biomass production of rapeseed (*Brassica napus* L.). Biomass Bioenergy 57, 196–204.

IER, 2010. Project "Supporting Sustainable Biomass Production and Use in Ukraine". Published by the Institute for Economic Research and Policy Consulting in Ukraine the international conference «Energy from a biomass». 242-243. http://www.ier.com.ua/en/biomass_project/ Accessed 11-7-2013.

Kabata-Pendias, A., Pendias, H., 1989. Microelements in soils and plants: translation from English - Moskow. Publishing house Mir 439 pp.

Knothe, G., 2002. Current perspectives on biodiesel. Information 13, 900–903.

Knothe, G., 2005. Dependence of biodiesel fuel properties on the structure of fatty acid alkyl esters. Fuel Process. Technol. 86, 1059–1070.

Knothe, G., Steidley, K.R., 2005. Kinematic viscosity of biodiesel fuel components and related compounds. Influence of compound structure and comparison to petrodiesel fuel components. Fuel 84, 1059–1065.

Kolbas, A., Mench, M., Herzig, R., Nehnevajova, E., Bes, C.M., 2011. Copper phytoextraction in tandem with oilseed production using commercial cultivars and mutant lines of sunflower. Int. J. Phytoremediation 13 (1), 55–76.

Kolker, A., Panov, B., Landa, E., Panov, Y., Korchemagin, V., Conko, K., Shendrik, T., 2009. Mercury and trace-element contents of Donbas coals and associated mine water in the vicinity of Donetsk, Ukraine. Int. J. Coal Geol. 79, 83–91.

Krstić, B., Stanković, D., Igić, R., Nikolić, N., 2007. The potential of different plant species for nickel accumulation. Biotechnol. Biotechnol. Equip. 21 (4), 431–436.

Kumar, A., Prasad, M.N.V., Sytar, O., 2012. Lead toxicity, defense parameters' and indicative biomarkers in *Talinum cuneifolium* grown hydroponically. Chemosphere 89, 1056–1065.

Lee, K.K., Cho, H.S., Moon, Y.C., Ban, S.J., Kim, J.Y., 2013. Cadmium and lead uptake capacity of energy crops and distribution of metals within the plant structures. KSCE J. Civ. Eng. 17 (1), 44–50.

Lin, C., Liu, J., Liu, L., Zhu, T., Sheng, L., Wang, D., 2009. Soil amendment application frequency contributes to phytoextraction of lead by sunflower at different nutrient levels. Environ. Exp. Bot. 65 (2–3), 410–416.

Lozanovskaya, I.N., Orlov, D.S., Sadovnikova, L.K., 1998. Ecology and biosphere protection at chemical pollution. M. Publishing House «Mir», 142 p.

Lysenko, L., Ponomarev, M., Kornilovich, B., 2001. Prospects of the decision of problem of soils pollution by heavy metals. Ecotechnol. Resour. Saving 4, 58–63.

Malan, H., Farrant, J., 1998. Effects of the metal pollutants cadmium and nickel on soybean seed development. Seed Sci. Res. 8, 445–453.

Marchenko, A.P., Semenov, V.G., 2001. Alternative biofuel from rape oil derivatives. Chem. Technol. Fuels Oils 37 (3), 183–185.

Marques, A.P.G.C., Oliveira, R.S., Rangel, A.O.S.S., Castro, P.M.L., 2008. Application of manure and compost to contaminated soils and its effect on zinc accumulation by Solanum nigrum inoculated with arbuscular mycorrhizal fungi. Environmental Pollution 151, 608–620.

Meher, L.C., Vidya Sagar, D., Naik, S.N., 2006. Technical aspects of biodiesel production by transesterification—a review. Renew. Sust. Energ. Rev. 10 (3), 248–268.

Melo, E.E.C., Guilherme, L.R.G., Nascimento, C.W.A., Penha, H.G.V., 2012. Availability and accumulation of arsenic in oilseeds grown in contaminated soils. Water Air Soil Pollut. 223, 233–240.

Mittelbach, M., 1996. Diesel fuel derived from vegetable oils, VI: specifications and quality control of biodiesel. Bioresour. Technol. 27 (5), 435–437.

Mnatsakanian, R.A., 1992. Environmental Legacy of the Former Soviet Republics. Centre for Human Ecology, University of Edinburgh, UK.

Murakami, M., Ae, N., Ishikawa, S., 2007. Phytoextraction of Cadmium by rice (*Oryza sativa* L.), soybean (*Glycine max* (L.) Merr.), and maize (*Zea mays* L.). Environ. Pollut. 145 (1), 96–103.

Nekhay, O., 2012. The agri-food sector in Ukraine: current situation and market outlook until 2025. European Commission Joint Research Centre scientific and policy report. pp. 74.

Nikolaychuk, V., Hrabovsky, O., 2000. Ecological problems of Carpathian mountains in connection with pollution of heavy metals biogeocenoses. In: Proceedings of International Conference "Problems of modern ecology", Zaporozhye, 20-22 September 2000. p. 30.

Özdemir, E.D., Härdtlein, M., Eltrop, L., 2009. Land substitution effects of biofuel side products and implications on the land area requirement for EU 2020 biofuel targets. Energy Policy 37 (8), 2986–2996.

Obuhov, A., Babyeva, I., Grin, A., 1980. The scientific basis of learning aid of maximum concentration limit of heavy metals in soils. In: Heavy Metals in Environment. Publishing House of Moskow University, pp. 20–27.

Official Journal L 094, 10/04/2003 P. 0001–0042. 2003/238/EC: Commission Decision of 15 May 2002 on the aid scheme implemented by France applying a differentiated rate of excise duty to biofuels (notified under document number C(2002) 1866).

ONORM, 1991. Vornorm C1190: Kraftstoffe—Dieselmotoren; Rapŝlmethylester; Anforderungen.

Panov, B., Dudik, A., Shevchenko, O., Matlak, E., 1999. On pollution of the biosphere in industrial areas: the example of the Donets coal basin. Int. J. Coal Geol. 40, 199–210.

Prasad, M.N.V., 2008. Trace Elements as Contaminants and Nutrients: Consequences in Ecosystems and Human Health. Publishing house: John Wiley & Sons, Inc., Hoboken, New Jersey. pp. 777.

Pulford, I.D., Watson, C., 2003. Phytoremediation of heavy metal contaminated land by trees: a review. Environ. Int. 29, 528–540.

Ramos, L.P., Kucek, K., Domingos, A.K., Wilhelm, H.M., 2003. Biodiesel: um projeto de sustentabilidade econùmica e sùcio-ambiental para o Brasil. Biotecnologia, Ciïncia e Desenvolvimento 31, 28–37.

Rascioa, N., Navari-Izzo, F., 2011. Heavy metal hyperaccumulating plants: how and why do they do it? And what makes them so interesting? Plant Sci. 180, 169–181.

Rodríguez-Vila, A., Covelo, E.F., Forján, R., Asensio, V., 2014. Phytoremediating a copper mine soil with *Brassica juncea* L., compost and biochar. Environ. Sci. Pollut. Res. 21 (19), 11293–11304. http://dx.doi.org/10.1007/s11356-014-2993-6.

Ru, S.H., Wang, J.Q., Su, D.C., 2004. Characteristics of Cd uptake and accumulation in two Cd accumulator oilseed rape species. J. Environ. Sci. 16 (4), 594–598.

Sayyad, G., Afyuni, M., Mousavi, S.-F., Abbaspour, K.C., Hajabbasi, M.A., Richards, B.K., Schulin, R., 2009. Effects of cadmium, copper, lead, and zinc contamination on metal accumulation by safflower and wheat. Soil and Sediment Contamination: An International Journal 18 (2), 216–228.

Salt, D.E., Blaylock, M., Kumar, N.P., Dushenkov, V., Ensley, B.D., Chet, I., Raskin, I., 1995. Phytoremediation: a novel strategy for the removal of toxic metals from the environment using plants. Biotechnology 13 (5), 468–474.

Samokhvalova, V., Miroshnichenko, M., Fadeev, A., 2001. Threshold level of toxicity of heavy metals for agricultural crops. J. Agric. Sci. 11, 61–65.

Schaffartzik, A., Plank, C., Brad, A., 2014. Ukraine and the great biofuel potential? A political material flow analysis. Ecol. Econ. 104 (2014), 12–21.

Shestapalov, V., Naboka, I., Bobyleva, O., 1996. A hygienic estimation of influence of heavy metals on an ecological situation of territories, damaged by Chernobyl accident. Rep. Natl. Acad. Sci. Ukr. 8, 156–163.

Shparyk, Y.S., Parpan, V.I., 2004. Heavy metal pollution and forest health in the Ukrainian Carpathians. Environ. Pollut. 130 (1), 55–630.

Singh, S.P., Singh, D., 2010. Biodiesel production through the use of different sources and characterization of oils and their esters as the substitute of diesel: a review. Renew. Sust. Energy Rev. 14 (1), 200–216.

Srivastava, A., Prasad, R., 2000. Triglycerides-based diesel fuels. Renew. Sust. Energ. Rev. 4 (2000), 111–133.

State register of cultivars which are suitable for distribution in Ukraine in 2011, (from 05.09.2011) URL: http://sops.gov.ua/uploads/files/documents/reyestr_sort/R2011_05.09.11.pdf.

State Statistics Service of Ukraine, 2013. The official website of the State Statistics Service of Ukraine: http://www.ukrstat.gov.ua

Strukova, E., Golub, A., Markandya, A., 2006. Air pollution costs in Ukraine. Working Papers from Fondazione Eni Enrico Mattei 120, http://ageconsearch.umn.edu/bitstream/12206/1/wp060120.pdf.

Topilin G., 2002. Small-sized installation for reception hydrodynamical fuel active.: the First in Ukraine the international conference «Energy from a biomass». 242-243.

Topilin, G., Uminski, S., Yakovenko, A., 2008. Biodiesel fuel for agricultural machinery. TEKA Kom. Mot. Energ. Roln.—OL PAN 8, 283–286.

Topilin, G., Yakovenko, A., Uminski, S., Nowak, J., 2009. Production of biodiesel fuel for self-propelled agricultural machinery. TEKA Kom. Mot. Energ. Roln.—OL PAN 2009 (9), 352–356.

Tordoff, G.M., Baker, A.J.M., Willis, A.J., 2000. Current approaches to the revegetation and reclamation of metalliferous mine wastes. Chemosphere 41 (1, 2), 219–228.

Trahtenberg, I., 1998. The priority aspects of problems of medical ecology in Ukraine. Mod. Probl. Toxicol. 1, 5–8.

Turgut, C., Pepe, M.K., Cutright, T.J., 2004. The effect of EDTA and citric acid on phytoremediation of Cd, Cr, and Ni from soil using *Helianthus annuus*. Environ. Pollut. 131 (1), 147–154.

UNI 10946, 2001. Automotive Fuels - Fatty Acid Methyl Esters (fame) for Diesel Engines - Requirements and test Methods. Free Catalogue Information Download. . http://www.freestd.us/soft4/4424834.htm.

Van Leeuwen, M., Salamon, P., Fellmann, T., Banse, M., von Ledebur, O., Salputra, G., Nekhay, O., 2012. The agri-food sector in Ukraine: current situation and market outlook until 2025. http://dx.doi.org/10.2791/89165 ISSN: 1831-9424.

Venselaar, J., 2004. Environmental protection: a shifting focus. Chem. Eng. Res. Des. 82 (12), 1549–1556.

Witters, N., Mendelsohn, R.O., Van Slycken, S., Weyens, N., Schreurs, E., Meers, E., Tack, F., Carleer, R., Vangronsveld, J., 2012. Phytoremediation, a sustainable remediation technology? Conclusions from a case study. I: energy production and carbon dioxide abatement. Biomass Bioenergy 39, 454–469.

World Bank, 2008. World Development Report 2008: Agriculture for Development. Oxford University Press for the World Bank, Washington.

ENERGY PLANTATIONS, MEDICINAL AND AROMATIC PLANTS ON CONTAMINATED SOIL

S.K. Maiti, A. Kumar

Indian School of Mines, Dhanbad, Jharkhand, India

1 INTRODUCTION

Pollution due to heavy metal contamination is the major concern for today's environment across the globe. Numerous physical, chemical, and biological remediation technologies are available to clean up the contaminated sites. However, physical and chemical methods are sometimes found expensive and of low efficiency in pollution cleanup. A cost-effective comparison of various methods available for remediation of contaminated/metalliferous soil is presented in Table 1.

Bioremediation is a new emerging biological technology defined as application of microbial consortium to reduce, degrade, and detoxify organic and inorganic contaminants. Phytoremediation, one of the most visible emerging techniques of bioremediation, is an eco-friendly, cost-effective, esthetically pleasing technology, gaining importance to remediate or reduce metal- and metalloid-contaminated sites. The different types of phytoremediation include phytostabilization, phytoextraction, phytovolatilization, phytodegradation, and rhizofiltration (Figure 1). However, phytoextration and phytostabilization are the two most important aspects of phytoremediation, which derives its scientific justification from the emerging concept of green chemistry and engineering.

Both the techniques involve metal accumulation, its uptake and translocation from soil to plant. Phytoextraction involves removal of metal from soil by accumulating and translocating the metal into the aerial/harvested part. These plant materials can be incinerated or burned to recover valuable material. The residue can be disposed of in suitable places under controlled conditions or in some cases recycled as fertilizer, whereas phytostabilization is the immobilization of metal in soil (see Figure 1).

Extensive studies over the last three decades have used varieties of native trees and grasses along with leguminous plants to enhance the economy of the local farmers as well as of the country. However, few plants such as lemon grass and vetiver grass are turned out to be the most widely acceptable and efficient plant for phytoremediation of metalliferous soils.

Bioremediation and Bioeconomy. http://dx.doi.org/10.1016/B978-0-12-802830-8.00002-2
Copyright © 2016 Elsevier Inc. All rights reserved.

Table 1 Cost-Effective Comparison of Various Methods Available for Remediation of Contaminated/Metalliferous Soil

Methods/Treatment	Costs[a]		
	Juwarkar et al. (2010)	Mulligan et al. (2001)	Glass (1999)
Physical	30-258	–	–
Solidification	26-260	75-425	90-870
Electrokinetics	–	70-170	20-200
Chemical	18-910	–	–
Thermal	45-1137	–	–
Biological/Bioremediation	8-258	15-200	–
Phytoremediation	–	–	5-40

[a]Approximate cost (US$ tons^{-1} of soil).

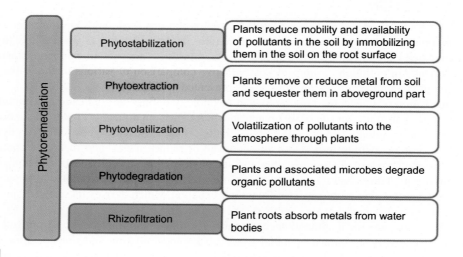

Phytoremediation	Phytostabilization	Plants reduce mobility and availability of pollutants in the soil by immobilizing them in the soil on the root surface
	Phytoextraction	Plants remove or reduce metal from soil and sequester them in aboveground part
	Phytovolatilization	Volatilization of pollutants into the atmosphere through plants
	Phytodegradation	Plants and associated microbes degrade organic pollutants
	Rhizofiltration	Plant roots absorb metals from water bodies

FIGURE 1

The five main types of phytoremediation.

2 AROMATIC, MEDICINAL, AND ENERGY PLANTS

Metal- or metalloid-contaminated soils are potentially toxic sites generally found with sparse vegetation of invasive or natural plants. These plants are often metal stress tolerant (they exhibit the property of an excluder, accumulator, or hyperaccumulator). However, because of low biomass these plants are not frequently used by the phytologists for remediation of contaminated sites. Aromatic plants and grasses are gaining importance in remediation because of their luxuriant biomass and metal tolerance capacity under stress conditions. Aromatic plants such as vetiver grass, lemongrass, tulsi, palmarosa, citronella, and geranium mint are ecologically feasible and viable and are widely used aromatic and medicinal plants. Of these, vetiver (*Chrysopogon zizanioides*) and lemon grass (*Cymbopogon citratus*)

Table 2 Plants and Grasses with Level of Metal Tolerance, Stress Tolerance, Medicinal Property, and Energy Production

Plant/Grass	Metal Tolerance	Stress Tolerance	Medicinal Property	Energy Production
Vetiver	✓✓✓	✓✓✓	✓	✓✓✓
Lemon	✓✓✓	✓✓	✓✓	✓✓
Tulsi	✓	✓	✓✓✓	✓
Stylo	✓	✓✓		✓✓

✓: Low; ✓✓: medium; ✓✓✓: high.

are the two most important grasses gaining keen attention throughout the world. Grasses and plants that have high metal tolerance, adaptability to environmental stress, and medicinal property along with capacity to produce energy after burning are presented in Table 2.

3 VETIVER AND LEMON GRASS

Vetiver and lemongrass are aromatic and perennial grass species belonging to the Poaceae family. Vetiver grass, commonly known as "wonder grass," is found growing throughout the world. It is considered as a miracle grass because it can tolerate and grow even in adverse conditions and on different soils, such as:

- Metal- and metalloid-contaminated soil
- Sandy soil
- Acid sulfate soil
- Sodic soil
- Mangrove soil
- Saline soil
- Lateritic soil.

The main features that promote the use of vetiver and lemon grass against other plants and grasses are listed in Table 3.

3.1 USES AND ECONOMIC IMPORTANCE OF VETIVER AND LEMON GRASS

Both live and dry masses of vetiver and lemon grass have great uses and economic importance (Table 4). Live plants are widely used in soil water conservation, disaster prevention, erosion control and stabilization, protection of terrestrial and aquatic environment, and oil production, whereas dry plants can be used in cosmetics, perfumes, medicines, and handicrafts.

3.2 ECONOMICS OF VETIVER AND LEMON GRASS

Vetiver Technology (VT) system has been established in many countries including India for bioremediation and resource generation. In many developing countries such as China, India, and Sri Lanka, where labor cost is very low, the VT system is very cost effective and popular. For example, in developing

Table 3 Silent Features and Properties of Vetiver and Lemon grass

Properties/Parameters	Vetiver Grass	Lemon grass
Luxuriant growth	✓✓✓	✓✓
Noninvasive	✓	✓
Root system and biomass	✓✓✓	✓✓
Metal accumulation in root	✓✓✓	✓✓✓
Metal accumulation in shoot	✓	✓
Drought resistant	✓✓✓	✓✓
Fire resistant	✓✓✓	✓
Flood resistant	✓✓✓	✓
Pathogen resistant	✓✓	✓✓✓
Soil as substrate to grow	✓✓	✓✓✓
Water as substrate to grow	✓✓	✗
Erosion control and steep slope stabilization	✓✓✓	✓✓
Explosive degradation	✓✓	✓
N and P removal	✓✓✓	✓
Dioxins and PAH removal	✓✓	✓✓
Pesticide tolerance and removal	✓✓✓	✓✓
Radionucleid removal	✓	✗
Carbon sequestration	✓✓	✓

✓: Low; ✓✓: medium; ✓✓✓: high.

Table 4 Commercial Importance of Vetiver and Lemon grass

Uses	Vetiver Grass	Lemon grass
Handicrafts (fans, cloth baskets, and hangers)	✓	✓
Home appliances (chairs, tables, stools, room partitions)	✓	✗
Soaps and detergents	✓	✓
Perfumes	✓	✓
Thatching of roofs	✓	✗
Medicines	✓	✓
Veti-concrete slabs	✓	✗
Insect repellent	✓	✓

countries overall propagation and production of vetiver grass is less than 80-90% of the total cost compared to conventional physical and chemical methods. In developed countries such as Australia, Thailand, and the United Kingdom, enhanced bioremediation of contaminated sites that was carried out by coupling of both soft engineering (VT system) with hard engineering (other physical and chemical methods) technologies, resulted in a 40-60% cost increase, compared to soft engineering technology alone. Even in developed countries such as Australia, where labor cost is high, soft engineering technology cost between 27% and 40% that of hard engineering solutions (Diti, 2003).

India is the largest producer (300-350 tons annum^{-1}) of lemon grass oil, 80% of which is exported to the developed countries; namely, China, France, Germany, Italy, Japan, Spain, the United Kingdom, and the United States (Lal et al., 2013). According to the study of Janhit Foundation (2014), 20-25 tons of fresh lemon grass per acre can produce 70-75 kg of oil in a year, and the profit level of farmers ranged between 80% and 120%.

4 PHYTOSTABILIZATION OF INTEGRATED SPONGE IRON PLANT WASTE DUMPS BY AROMATIC GRASS-LEGUME (LEMON-STYLO) MIXTURE AND ENERGY PLANTATION (*SESBANIA*)

4.1 GENERATION OF INTEGRATED SPONGE IRON PLANT WASTE

Direct-reduced iron (DRI), also called sponge iron, is produced from direct reduction of iron ore (hematite) in the solid state by a reducing gas produced from coal. In the DRI process, the Fe ore (generally having 65-70% of Fe) is heated at high temperature (about 1500 °C) to burn off its carbon and oxygen content, to produce briquettes contains 90-97% pure Fe. In 2013, approximately 87% of the total world DRI was produced by 14 countries (India, Canada, Mexico, Trinidad and Tobago, Argentina, Peru, Venezuela, Egypt, Libya, South Africa, Iran, Qatar, Saudi Arabia, and United Arab Emirates). In 2014, India ranked first, followed by Iran and Mexico. These three countries together contributes more than 30% of the world's DRI production (60.6 million metric tons) (World steel, 2014).

The principal raw materials required for sponge iron production are iron ore, non-coking coal, and dolomite. For an example, production of 100 tons of sponge iron requires 154 tons (65 wt% Fe) of iron ore and 120 tons (B grade) of coal and generates 45 tons of solid waste out of which 25 tons is char (dolochar). In the absence of good grade coal, sponge iron industries are using poor quality coal (F grade), which contains more than 40% by weight of ash, leading to more waste generation. Slag is also generated from the sponge iron plant or SMS (steel melting shop) plant in the process. Therefore, in an integrated sponge iron plant, large quantities of solid wastes generated in the form of dolochar, slag, and fly ash were dumped together due to scarcity of space to cover a large area exceeding 40-50 m. They caused severe air pollution problems during the summer, while in monsoon, the loose waste was easily carried away along with runoff to pollute the nearby water bodies.

4.2 CHARACTERISTICS OF WASTE DUMP MATERIALS

The waste was very homogenous, loose and devoid of nutrients; hence stabilization of this type of waste poses a severe challenge. Waste samples from the boilers and DRI plant of the sponge iron unit were collected for physicochemical analysis. They are highly alkaline in nature and consist of slag, fly ash, and dolochar. The detailed characteristics of individual waste and composite waste are given in Table 5. The waste had high electrical conductivity (EC), with high concentration of Na, K, and Ca and was very fine, loose, easily windborne from the surface of the dumps thus worsening the quality of air, while in the monsoon large amount of wastes were washed off into nearby water channels, polluting them. The representative samples of the waste were collected, air dried, thoroughly mixed, grinded if needed, and sieved down to micron size with standard sieves. Analyses of the physicochemical parameters were done according to standard procedures in seven replicates. The mineralogical and morphological characteristics of the

Table 5 Physico-chemical Characteristics of Individual Wastes and Composite Waste Generated from the Integrated Sponge Iron Unit ($n = 3$)

Parameters	Fly Ash from (AFBC Boiler)	Fly Ash from (WHRB Boiler)	Dolochar	Slag	Composite Waste
Color	Light gray	Dark gray	Black	Shiny black	Blackish gray
pH_{H2O} 1:1 (w/v)	8.69 ± 0.04	9.37 ± 0.02	11.77 ± 0.06	10.38 ± 0.06	9.66 ± 0.04
EC_{H2O} (1:1) (w/v) (dS m^{-1})	1.73 ± 0.04	2.25 ± 0.01	5.10 ± 0.27	0.14 ± 0.021	2.89 ± 0.05
WHC (%)	71.37 ± 0.82	63.81 ± 0.23	94.35 ± 0.12	33.94 ± 0.14	65.21 ± 2.62
LOI (%)	3.86 ± 0.03	5.91 ± 0.02	17.08 ± 2.53	–	6.08 ± 0.31
Ex. K (mg kg^{-1})	156 ± 11.7	113.1 ± 7.8	78 ± 3.9	59.86 ± 1.7	128.7 ± 1.95
Ex. Na (mg kg^{-1})	124.2 ± 1.38	117.3 ± 2.07	103.5 ± 0.69	31.24 ± 0.69	116.5 ± 0.69
Ex. Ca (%)	0.19 ± 0.01	0.28 ± 0.01	0.99 ± 0.01	0.01 ± 0.001	0.28 ± 0.01
EDS analysis					
CaCO$_3$ (%)	15.69 ± 1.23	25.5 ± 0.01	62.32 ± 3.11	6.09 ± 0.37	26.2 ± 0.09
SiO$_2$ (%)	56.82 ± 0.68	50.64 ± 2.36	27.55 ± 2.16	44.28 ± 1.23	48.61 ± 0.69
MgO (%)	0.69 ± 0.02	1.7 ± 0.27	0.56 ± 0.06	0.45 ± 0.02	0.79 ± 0.03
Al$_2$O$_3$ (%)	10.49 ± 0.17	7.67 ± 1.14	3.98 ± 0.43	7.63 ± 0.22	8.63 ± 0.33
FeS$_2$ (%)	0.34 ± 0.01	0.31 ± 0.01	0.54 ± 0.04	–	0.43 ± 0.01
Fe (%)	2.34 ± 0.04	6.41 ± 0.02	2.42 ± 0.02	25.09 ± 0.29	11.86 ± 0.12
Trace element analysis					
Cr (mg kg^{-1})	529.8 ± 1.23	341 ± 2.02	97.4 ± 0.12	353.4 ± 4.29	312 ± 2.58

Ex.: exchangeable; WHC: water-holding capacity; LOI: loss on ignition.

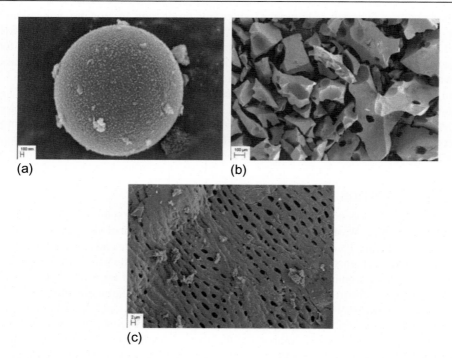

(a)

(b)

(c)

FIGURE 2

SEM-EDX analysis of (a) fly ash; (b) slag; and (c) dolochar.

individual wastes (fly ash, dolochar, and slag) as well as the composite waste were evaluated through (Energy Dispersive X-ray Spectroscopy) EDX analysis and SEM (Scanning Electron Microscopy) studies using FESEM (Field Emission Scanning Electron Microscope) (Model:Supra 55) (Figure 2). The finely grinded and dry samples were sprinkled onto double sided carbon tape mounted onto SEM stub to determine particle morphology, external surface structure, and elemental distribution.

Properties of fly ash mainly depend on the coal used in the thermal power plant. The samples in this study had alkaline pH of 8-9. Dolochar and slag samples had pH of 12 and 11, respectively, while that of the composite sample was approximately 9-10. The waste had high EC of $3\,dS\,m^{-1}$ due to presence of dolochar (nearly $5\,dS\,m^{-1}$) and fly ash (1.5-$2\,dS\,m^{-1}$). The composite waste consisted of fine-sized particles mainly due to the presence of fly ash and had a water-holding capacity of 65%, which was corroborated by the fact that fly ash and dolochar had high water-holding capacity of 63-70%. High water-holding capacity of the dolochar particles (94%) are substantiated by SEM studies, which showed that the particles had microporous structure and thus high surface area to withhold maximum amount of water. The unburned carbon in the samples was determined by loss on ignition (LOI) procedure and was found to be 3-6% in fly ash and 17% in dolochar samples while the composite sample had 6%. High LOI in dolochar was due to the presence of carbon particles, also showed by SEM and EDX studies. Ammonium acetate extract of the samples showed presence of high calcium mainly in fly ash (0.3%) and dolochar samples (1%) while the composite sample had 0.3% exchangeable calcium. Rapid pyrolysis of the raw materials in the DRI leads to its porous nature with complex mineralogical compositions as found through EDX analysis and elemental mapping. They showed mineral carbon inclusion in fine dolochar samples. The major constituents of inorganic inert material present in the char are $CaCO_3$ (62%) and SiO_2 (27%). The content of Al_2O_3, Fe, and MgO present in the sample were 4%, 2.5%, and 0.5%, respectively. EDX mapping showed elements in dolochar particles in the order $Ca > C > K > O > Si > Al > Mg > S$.

4.3 PHYTOSTABILIZATION OF WASTE DUMP

A challenging task was undertaken to ecologically restore a huge waste dump covering 7 ha ground area (top flat surface area of 5 ha), aerial height of 50-70 m, without any intermediate benching with steep slope (>70°). Practically no space was available to reduce the slope. The sequence of restoration operation was (1) regrading of the dump, (2) blanketing with topsoil, (3) covering the slope with coir-mat, (4) sowing of grass-legume mixture on the slope, (5) construction of drainage, (6) watering arrangement, and (7) aftercare and maintenance of the site for 3 years.

Use of good-quality topsoil is essential for the success of any ecorestoration program (Maiti, 2013). Sources of topsoil are either stockpiled or simultaneous excavation or reuse (i.e., concurrent use). In the present study, topsoil was used in a concurrent manner.

Geotextiles are used to protect slopes by preventing erosion and creating favorable soil conditions for revegetation, especially in the initial stage of slope restoration. Natural geotextile mat (jute mat or coir-mat) is preferred since it is more effective and more environment friendly (Shao et al., 2014). It can markedly increase vegetation cover, reduces evapotranspiration (Rickson, 1995; Mitchell et al., 2003), and helps to conserve moisture, which increases survival of young seedlings during hot summers (Sawtschuk et al., 2012). In the present study, as there was no scope to reduce the slope of the dump by terracing, the entire slope was covered with high-tensile-strength coir-mat (slope length 45-50 m).

Use of grass-legume mixtures to stabilize steep slopes has now become a widely used technique. Studies have shown that Dinanath grass (*Pennisetum pedicellatum*) along with forage legumes (*Stylosanthes humilis*) can successfully restore a degraded site (Maiti, 1997; Maiti and

Saxena, 1998; Caravaca et al., 2002; Lenka et al., 2012; Araujo and Costa, 2013). Field experiments showed that forage legume (*Stylosanthes hamata*) can be used as an initial colonizer during reclamation of coal mine-degraded sites, without application of topsoil (Maiti and Saxena, 1998). The combination of grass-legume mixture creates a nitrogen balance in the soil, and decomposition of dry plant parts creates nitrogen-rich litter and mulch for the soil. Grasses have extensive fibrous root systems that can reduce erosion by holding the loose soil particles, can tolerate adverse soil conditions, and form mulches after drying (Maiti, 2013). Generally, perennial forage-type legumes are used, but native species show greater improvement in soil fertility (Ledgard and Steele, 1992; Agbenin and Adeniyi, 2005). Thus, during the hydroseeding of grass-legume mixture, a higher percentage of legume seed are used, which will accelerate nutrient cycling, improve the quality of soils, biodiversity, and sustainability of the plant community (Maiti, 2013). The aims of the study are to investigate the: (1) changes in physical and chemical properties of waste dump surface due to the growth of grass-legume mixture, (2) effectiveness of use of coir-mat and grass-legume seeds for stabilization of steep slope, and (3) measurement of soil temperature amelioration due to mulch and litter accumulation.

4.4 STUDY AREA

The study was carried out on waste dumps of an integrated sponge iron unit located in Raigarh district, Chhattisgarh, India, which falls between latitudes 22°00′51″N and 22°02′10″N and longitudes 83°22′35″E and 83°23′27″E, covering an area approximately 7 ha. Wastes generated during the processes were dumped on the outskirts of the plant, which was surrounded by protecting forest. The height of the dump ranged between 40 and 50 m with steep slope (>70-80°), which was unstable, causing serious environmental problems. An aerial photograph of the original waste dump is shown in Figure 3.

The climate of the area is dry and tropical with three seasons a year such as cold winter (December-February), very hot summer (April-June), and rainy season (July-September). The minimum-maximum temperature range is 29.5-49 °C in summer, and 8-25 °C in winter. During summer fine particulate matters are easily windborne while in monsoon the loose waste materials easily wash off and drain to the nearby nallah/river, causing blackening of water. Moreover, the dump height exceeds the surrounding

FIGURE 3

Aerial view of waste dump before ecological restoration.

canopy height of 20 m, causing deterioration of esthetics. As there was no extra land available, the reduction of slope by terracing was practically impossible and the entire steep slope (slope length 50-70 m) was covered with coir-mat.

4.5 CHARACTERISTICS OF BLANKETING SOIL AND COIR-MAT

Topsoil blanketing of the waste dump was done to cover the waste and create a suitable substratum for vegetation growth and stabilization of surface. Topsoil, being of slightly acidic nature (4.7), will buffer the pH of loose alkaline solid wastes. Soil was sandy loam (73%), with moderate fertility, with cation exchange capacity (CEC) of 7.2 cmol(+) kg^{-1} and base saturation of 51.3%. About 50,000 m^3 of topsoil excavated from the forest area (which was to be used for the creation of fresh dump) was transported to the restored site, spread on the waste dump, and leveled by dozer (Figure 4a–d).

Coir-mat is a tough organic fiber having high tensile strength, rich in lignin (46%), about 1.5-2 cm thick, interweaved with coconut-coir, fixed by nylon net; about 1-m wide and 30-m/50-m length rolls were used for stabilization of the slope (manufacturer: Sri Venkateshwara Fibre Udyog Private Limited, Bangalore, India) (Figure 4e and f). Coir-matting was essential to stabilize the slope and prevent erosion of materials from the steep slope during monsoon. It was laid loosely from the berm of the slope and unrolled downwards without stretching, and stapled in each 1-m distance with a U-shaped iron nail (size: 30 cm arm length and 8 cm width between the arms).

4.6 COMPOSITION OF GRASS-LEGUME MIXTURE

Grass-legume seeds consisting of *P. pedicellatum*, *Stylosanthes hamata*, *Crotalaria juncea* and *Sesbania sesban*, and *Hibiscus sabdariffa* (high-biomass-yielding, non-leguminous undershrub) were used for the revegetation of slopes of the waste dump. As the slope was very steep, it was initially covered with coir-mat, then topsoil was spread over it to create a substratum for germination and growth of seeds.

Three rows of tillers of *Cymbopogon citratus* (after shorting) were planted along the berm (edge) of the dump (Figure 5a and b). Additionally, one row of *Azadirachta indica* seeds was sown at 3-m intervals on the berm. All these activities were carried out before the onset of monsoon to protect the soil and seed mixture from erosion. A seed germination test for grass-legume mixture was conducted in the laboratory/field conditions to test the viability of seeds and time taken for germination by standard test method (Maiti, 2013). Leguminous seeds have a low dormancy period and within 6 days, 60% of the seeds are germinated with the rest germinated in 10 days, whereas grass seeds took a longer time to germinate. Approximately 3350 kg of seed mixture was used and sown in three times at an interval of 1 week. The composition of grass-legume mixture is given in Table 6.

4.7 TECHNICAL RESTORATION OF THE SITE

The top surface of the dump was leveled by dozer and a 3° slope toward the central drainage system was provided (as shown in Figure 5b and C). The acidic nature of topsoil was helpful to ameliorate the soil pH, because waste material has an alkaline pH (11.9). The slope was blanketed with soil, seeded with grass-legume mixture, and covered with coir-mat. The mat (dimension: 1 m × 30 m and 50 m) encored

FIGURE 4

(a) Leveling of waste dump; (b) source of topsoil; (c) blanketing of waste dump with topsoil; (d) leveling of topsoil on waste dump using dozer; (e) fixing of coir-mat on the dump slope; and (f) waste dump fully covered with topsoil and coir-mat.

at the berm and spread from top to bottom. Next, seeds of grass-legume mixed with fine soil materials (seed:soil; 1:1) were broadcasted. All these activities were carried out before monsoon. About 2000 tillers of *Cymbopogon citratus* were planted along berm of the dump in 0.5 m × 0.5 m spacing, and toward the inner side one row of *A. indica* was developed by seeding. The vegetated slope was irrigated in the summer months. At the initial stage (first year), a piped water supply with sprinkler distribution system was used during winter and summer months (December-June). Plant growth monitoring and soil sampling were done after 7 months. Reinforced concrete drainage was provided to drain runoff from the top surface of the dump, while seepage from the slope was drained out through an earthen channel.

FIGURE 5

(a) Shorting of fresh lemon grass tillers; (b) planting of lemon grass on the berm of dump in a row; (c) growth of small grass-legume plants after a few month; (d) growth of *Sesbania sesban*; (e) growth of lemon grass; (f) growth of *Stylosanthes*; (g) constructed heli-pad on the top of the phytostabilized dump; and (h) fully restored and phytostabilized waste dump.

Table 6 Composition of Grass-Legume Species

Botanical Name (Family) (Common Name)	Plant Characteristics
Stylosanthes hamata (L.) Taub. (Fabaceae) (Caribbean stylo)	Creeping under shrub, perennial, nitrogen-fixing legumes
Crotalaria juncea L. (Fabaceae) (Shon)	Tall herb, annual, nitrogen fixing
Cymbopogon citratus (DC.) Stapf. (Poaceae) (Lemongrass)	Aromatic grass, perennial, propagated through tillers
Sesbania sesban (L.) Merr. (Fabaceae) (Egyptian pea)	Shrub, annual, nitrogen fixing, produces green manure
Hibiscus sabdariffa (L.) Roselle. (Malvaceae) (Rosemallow)	Under shrub, annual, high biomass yielding
Pennisetum pedicellatum Trin. (Poaceae) (Dinanath grass)	Grass, annual, soil stabilizer

4.8 GROWTH OF GRASS-LEGUME MIXTURE

The entire slope was found to be covered, stabilized, and erosion controlled during the monsoon with lush green growth of *Stylosanthes* legume (Figure 5f). A distant view of the restored dump is shown in Figure 5h. During the field survey in monsoon, it was observed that the creeping nature of *Stylosanthes* had caused it to form a 30-40 cm carpet above the coir-mat, and underneath decomposition of leaves (black in color) contributed soil organic matter (SOM) and initiated the formation of humus. Natural colonization of vegetation on the slope surface due to accretion of habitat by *Stylosanthes* legume was observed (Figure 6); for example, colonization of *Evolvulus alsiniodes*, which was not sown, rather colonized in the later stage, and coexisted with *Stylosanthes*. During monsoon season, at the foot of the restored dump, regermination and luxuriant growth of *Sesbania sesban* was observed. *Sesbania sesban* is a tall woody annual shrub (dies after one monsoon), excellent nitrogen fixer which stabilizes the bottom edges of the dump and controls erosion.

The dominance of *Stylosanthes* on the slope was observed, and with the passage of time (second year onward), *Stylosanthes* substantially reduced abundance of *P. pedicellatum* on the slope. Only a few patches of *P. pedicellatum* were noticed, which is an example of competition in which *Stylosanthes* eradicated

FIGURE 6

Distant view of restored waste dump with *Stylosanthes* and other legumes.

Stylosanthes
hamata

Sesbania sesban

Hibiscus sabdariffa

Pennisetum
pedicellatus

FIGURE 7

Luxuriant growth of *Stylsanthes hamata, Sesbania sesban, Hibiscus sabdariffa,* and *Pennisetum pedicellatus* on the phytostablized dump.

the *P. pedicellatum.* Highest total biomass production was observed under *Stylosanthes hamata* followed by *P. pedicellatum > Cymbopogon citratus > Sesbania sesban.* In terms of tiller production, one clump of *Cymbopogon citrates* contained 48-65 numbers, while in *P. pedicellatum,* ranges between 90 and 110 nos were found. Highest leaf litter production was observed under *Stylosanthes hamata* (841 g m^{-2}), while highest amount of roots was produced by grass species, which actually help the stabilization of the surface.

Along with *Stylosanthes,* germination of *Sesbania,* which was planted in the previous year (Figure 7), was also observed at the start of monsoon. Regermination and luxurious growth of *Hibiscus sabdariffa,* a prickly annual herb, at the foot of the restored dump during rainy season was also observed. Decomposition of biomass of *Sesbania sesban* and *H. sabdariffa* provides nutrients, and dead stems act as mulch materials, which ameliorate microclimate of the slope-surface. Dense vegetation cover on the berm of the restored dump is essential for direct draining of rainwater to the slope, which will also add protection and stabilization to the slope. At the berm of the restored dump, two rows of lemon grass planted in the previous year were found growing excellently. In the second year itself, a thick vegetative barrier has developed by healthy growth of lemon grass (two rows) and *S. sesban.* Invasion of open space available at the top surface of the dump by *Stylosanthes* was observed from the second year onward. The berm of the dump was stabilized with massive growth of *Cymbopogon citratus, Crotalaria juncea,* and *A. indica.*

4.9 SOIL CHARACTERISTICS IN RESTORED SITE

Sampling points were dispersed all over the slope of the dump and erosion of topsoil as well as mixing of wastes with the topsoil was observed in some places. Soil fraction ranged between 48% and 56%, which is less than nearby reference forest site (73%). Field moisture was found between 3.2% and 4.6% and the highest was observed under *S. hamata* (4.6%) due to the formation of a dense green cover over the surface which is still less than forest soil. Low moisture content was attributed to the sampling season (winter); however, during postmonsoon sampling, as high as 16% of moisture content was observed under *Stylosanthes hamata.* Paste soil pH (1:1; w/v) was found in the acidic range due to the acidic nature of the blanketing topsoil material while the soil under *Crotalaria juncea* and *P. pedicellatum* was found near neutral, which may be because topsoil erodes during rainfall and remaining soil is mixed with the waste.

EC was found higher than the forest soil due to the presence of soluble salts of alkaline metals (Ca^{2+}, Mg^{2+}, and K^+) present in fly ash, dolochar, and slag. Soil quality can be effectively judged by the presence of SOC. Highest SOC was observed in forest soils: $9.7 g kg^{-1}$ and that of the *C. citratus* and *S. hamata* rhizosphere ($4 g kg^{-1}$). The mineralizable nitrogen (N) content in the soil was also affected by the species used for vegetation and found highest in *S. hamata* rhizosphere ($72 mg kg^{-1}$), which is lower than the forest soils ($98.2 mg kg^{-1}$). There is a significant variation of available N in soil under grasses and legumes. Total N constitutes more than 90% of the soil nitrogen pool; while in forest soil, it constitutes more than 97% of the nitrogen pool. Higher percentage of mineralizable N in the restored dump is due to higher microbial activity and nitrogen fixation by legumes.

Nitrogen accumulation is controlled by the OC input and nitrogen fixation, and P content is determined by the organic matter, pH of the soil substrate, and the weathering process (Andrews et al., 1998). Under all the grass-legume species, available P was found lesser than in the forest soil ($3.26 mg kg^{-1}$) and was found to differ significantly at 5% level. Available K ranged between 114 and $153 mg kg^{-1}$, which indicates medium fertility. Available Ca and Mg values in the restored site were found higher than reference forest soil, which may be due to pulling up of Ca^{2+} and Mg^{2+} cations during evapotranspiration and mixing up of waste with topsoil. The CEC values were observed between 7.2 and $10.4 cmol(+) kg^{-1}$ of soil on the restored site, which is higher than reference forest soil. In some samples, higher CEC values were high may be due to mixing of some wastes having high concentration of Ca^{2+} and Mg^{2+}. The base saturation values varied widely between 25% and 92% due to similar reasons; however, in reference forest soil, CEC values were observed at 35% due to the acidic nature of the soil. Moreover, the reference forest site was dominated by *S. robusta*, which only grows in acidic soils with a sandy texture, porous in nature with high infiltration rate. This may be responsible for leaching of cations (Ca^{2+}, Mg^{2+}, K^+, and Na^+) (Maiti, 2013).

Increasing soil depth reduces the soil dehydrogenase activity (DHA) as highest microbial activity is observed in the surface layer of the soil profile (Levyk et al., 2007; Mukhopadhyay et al., 2013). It was found to be significantly higher in the top 0-5 cm layer (4-12 folds higher from the deepest soil profile studied) and decreased significantly at successive lower depths (5-10, 10-20 cm) in all sampling locations. DHA was found to decrease significantly (2-6 folds) at 5 cm depth from the top surface on the dump. Forest soil had highest DHA level ($8.1 \mu g TPF h^{-1}$ per gram of soil) while the rhizospheric soil of the dump berm under *C. citratus* and *S. hamata* vegetation showed the second highest level ($3.7 \mu g TPF h^{-1}$ per gram of soil) at 0-5 cm.

4.10 AMELIORATION OF SOIL TEMPERATURE DUE TO MULCH AND LITTER ACCUMULATION

Soil temperature is an important parameter that controls biochemical processes and thus nutrient cycling. Mulch properties like quantity and architecture also affect soil microenvironment. Mulching makes soil less prone to erosion, increases infiltration and biomass, and ameliorates soil surface temperature. Different types of mulches modify soil temperatures in different ways (van Donk and Tollner, 2000; Fehmi and Kong, 2012). Accumulation of dry mulch (stems, branches of *Crotalaria juncea*, *Sesbania sesban*, and *Stylosanthes hamata*) in a quantity 13-15 ton ha^{-1} was observed on the waste dump, which was higher than the reference forest site (8.16 tons ha^{-1}). Higher accumulation of dry mulch was due to the massive growth of the leguminous species (*Sesbania sesban* and *Crotalaria juncea*) and drying within 5 months. Their accumulation also had a significant influence in the reduction of rhizospheric temperature. A significant reduction in the rhizospheric temperature of 12-17% was observed under mulch cover, while it was hardly 4% decrease on bare surfaces (area covered with minimum vegetation). The nearby reference forest site also showed 12% reduction in soil temperature under litter (at 10 cm depth). This indicates dry mulches generated from *Stylosanthes hamata*, *Crotalaria juncea*, *Sesbania sesban*, and *H. sabdariffa* not only enhance SOC but also their dry parts ameliorate surface temperature during summer and helps in moisture conservation.

4.11 ESTIMATION OF THE COST OF ECOLOGICAL RESTORATION OF WASTE DUMP

Topsoil is regarded as a strategic resource for successes of any ecological restoration project. Before salvaging operation of topsoil for concurrent use, inventory (quality and depth) and time of scrapping is essential. In the present project, the major challenge was to stabilize the very steep slope (>70°) with 50-60 m straight height without any bench. The slope could have been easily stabilized if dump slope could have been reduced to <28-30° by creating two additional benches of 15 m height. As there was practically no space left between the edges of the dump and forest land, a terracing option to reduce the slope angle was ruled out.

In the present study, topsoil was excavated (about 1 m deep) from the newly constructed dumping site and transported to the waste dump, which constituted 31% of the total cost. The running cost of heavy earth-moving machinery (HEMM) accounted for 13% of the total cost, which again depended on efficiency of use of HEMM and supervision. The third major cost involved purchasing of coir-mat @ US $1.3 m^{-2}, which depends on the technical requirement and the type of coir-mat. Even though, for the present work, soil was available free of cost, the cost of topsoil was considered as US $ 0.50 m^{-3} for restoration work, which comes to 11% of total cost.

The cost of biological restoration comes close to 17%, of which aftercare and maintenance for 3 years constitutes half of the cost, and the rest was for the purchase of grass-legumes seed mixture and construction of watering facilities. The cost of grass-legume seed materials was Rs. 150 kg^{-1} (US $2.5 kg^{-1}), *C. citratus* tiller at Rs. 7.50 tiller^{-1} (US $0.125 tiller^{-1}), and *A. indica* seeds at Rs. 500 kg^{-1} (US $8.34 kg^{-1}). It is always advisable to test the real value or pure live seed count for seed lot (Maiti, 2013). The knowledge of seed dormancy and methods of seed treatment to overcome dormancy is essential to leguminous seeds. Seeds of *A. indica* do not have longer dormancy periods; therefore, for better results, the seeds should be sown immediately. The quantity of seeds to be ordered depends on the types of seeds, seed viability, methods of sowing, and time of broadcasting. It is advisable, rather than having a single broadcasting operation, to sow the seeds three or four times over an interval of 7-10 days.

The average cost of ecorestoration was US $39,730 ha^{-1} (i.e., US $4 m^{-2}), out of which the total biological reclamation components constituted 17%, which includes aftercare and maintenance for 3 years. A detailed breakdown of activities and costs incurred is given in Table 7.

4.12 CONCLUSION

Grass-legume mixture (*Stylosanthes-Pennisetum*) can be used as initial colonizers for stabilization of a very steep slope, after blanketing with topsoil and coir-mat. Addition of fast-growing, annual,

Table 7 Cost of Ecological Restoration of Sponge Iron Waste Dump by Blanketing with Soil, Coir-Mat, and Initial Vegetation Cover Development with Grass-Legume Mixture (area 5 ha)

Sl. No	Activity Quantity	Cost in US$
Technical or engineering restoration		
1.	Running cost of dozers, loader, pockland for grading of dump surface, salvaging of topsoil, transportation, topsoil application on slopes, compact soil on the dump surface and provision of drainage	26,667 (13.4[a])
2.	Excavation and transportation of soil for blanketing the flat surface of waste dump (80-100 cm thick) and in slope (20-50 cm thick)—45,000 m^3 (approx.) at US $1.067 m^{-3}	61,515 (31)
3.	Cost of soil materials up to a depth of 100 cm with moderate fertility from forest area[b]—45,000 m^3 at US $0.50 m^{-3}	22,500 (11.3)
4.	Blanketing of slope with coir-mat: 1.5-2 cm thick, interweaved with coconut-coir, fixed by nylon net; available in a roll of about 1 m width with 30 m/50 m length—30,000 m^2 (approx.)	39,000 (19.6)
5.	Hooks used to anchor coir-mat on slopes; U-shaped iron nails were used at an interval of 1 m on the slope of the dump to fix the coir-mat	3333 (1.7)
Biological restoration		
6.	Grass-legume seeds: *S. hamata, C. juncea, H. sabdariffa, S. sesban, P. pedicellatum,* etc.—3350 kg; Tillers of *C. citratus,* 5000 nos; seeds of *A. indica,* 5 kg	10,217 (5.1)
7.	Aftercare and maintenance of site for 3 years, which includes watering arrangement during lean seasons, provision of watchguard, day-to-day unskilled labor charges, running cost of pump for 2 h day^{-1}, etc.	19,167 (9.6)
8.	Construction of boring with 6-in. diameter (15.24 cm) GI pipe, depth of 150 ft (45.72 m); 5 hp pump (7.5 kWh, three stages)—submersible	3750 (1.9)
9.	Expert advice (professional cost of scientist/consultants)—lump sum	12,500 (6.3)
Total cost		198,649

Unit cost: US $4 m^{-2} (excluding the cost of soil and water).
[a]*As % of total cost; conversion 1 US $ = Indian Rs. 60.*
[b]*Approximate cost of soil materials (up to a depth of 100 cm with moderate fertility from forest area) considered as at US $0.50 m^{-3}, even though free soil was available for the present ecorestoration work.*

high-biomass-producing species is essential to increase the soil organic matter and moisture. Fast-growing species can form massive green cover in a very short time and play an important role in reducing erosion and conserving moisture. In the initial year, both *Stylosanthes-Pennisetum* colonized together, but from the second year onward, *Stylosanthes* covered the entire slope surface and eradicated the *Pennisetum*. Natural colonization of other herbaceous leguminous species (*Desmodium* spp., *Tephrosia*) was also observed. The short life cycle of *Sesbania sesban* and *Crotalaria juncea* can play a role of green manure for the soil as it adds organic carbon and nitrogen to the soil after drying. This would also be economic. *Stylosanthes hamata* is a very effective forage legume that uplifts the nitrogen economy of degraded soils in a short period of time. These legumes can effectively influence the nitrogen cycling in soils due to presence of root nodules, which fix atmospheric nitrogen. Aftercare and maintenance of the ecorestored site, particularly watering and protection of cattle, is essential.

5 PHYTOREMEDIATION OF CHROMITE-ASBESTOS MINE WASTE USING *CYMBOPOGON CITRATUS* AND *CHRYSOPOGON ZIZANIOIDES*: A POT SCALE STUDY

The abandoned chromite-asbestos mine waste of Roro Hills, West Singhbhum, Jharkhand, India, has neutral pH (7.56), low EC ($0.094\,dS\,m^{-1}$), and OC (0.13%) with higher concentration of toxic total metals (Cr: $2555\,mg\,kg^{-1}$ and Ni: $1160\,mg\,kg^{-1}$) as compared to control soil. However, only small proportions of these elements were found plant available (Cr: 0.07% and Ni: 0.038%). A pot study was conducted using vetiver grass (*Chrysopogon zizanioides*) and lemon grass (*Cymbopogon citratus*), and a combination of different organic amendments (farmyard manure, chicken manure, and garden soil) was used to find the most suitable treatment for higher production of both the grasses with maximum metal tolerance ability. Application of manures resulted in significant improvements of mine waste characteristics and plant growth, reduction in the availability of total extractable toxic metals (Cr and Ni), and increase in Mn, Zn, and Cu concentration in the substrate. The maximum growth and biomass production for *Chrysopogon zizanioides* and *Cymbopogon citratus* were found in T-IV combination comprising of mine waste (90%), chicken manure (2.5%), farmyard manure (2.5%), and garden soil (5%). Addition of T-IV combination also resulted in low Cr and Ni accumulation in roots and reduction in translocation to shoots. *Chrysopogon zizanioides* accumulated 62% of Cr in roots and 36.01% in shoots whereas *Cymbopogon citratus* had accumulated 82.01% and 44.63% in roots and shoots, respectively. Total dry biomass ($g\,pot^{-1}$) of *Cymbopogon citratus* and *Chrysopogon zizanioides* was found low in control (unamended MW), while production of aerial biomass for all the four treatments was found significantly higher. Total biomass of *Cymbopogon citratus* was found higher in all the treatments in comparison to *Chrysopogon zizanioides*. Study indicates that luxuriant growth of *Chrysopogon zizanioides* and *Cymbopogon citratus* can be used for phytostabilization of abandoned chromite-asbestos mine waste with amendments, and adding them enhances esthetic and commercial value (Kumar and Maiti, 2015).

6 PHYTOSTABILIZATION OF COPPER TAILINGS USING *CYMBOPOGON CITRATUS* AND *VETIVERIA ZIZANIOIDES*: A POT SCALE STUDY

The tailings of Rakha Cu mines, East Singhbhum, Jharkhand, India were permanently stored in tailings ponds that require vegetation to reduce their impact on the environment. A pot scale study was conducted to evaluate the suitability of *Vetiveria zizanioides* (L.) Nash (new name: *Chrysopogon zizanioides*) for the reclamation of Cu tailings and to evaluate the effects of chicken manure and soil-manure mixtures on the revegetation of such tailings. Application of manure and soil-manure mixtures resulted in significant increase in pH, EC, OC, CEC, and nutritional status of Cu tailings. The environmentally available and Diethylene Triamine Pentaacetic Acid (DTPA) extractable Cu and Ni concentration reduced in amended tailings, while Mn and Zn content increased significantly. Plants grown on amended tailings accumulated lesser amounts of Cu and Ni but higher amounts of Mn and Zn. Plant biomass increased proportionally to manure and soil-manure mixtures application rates. From the pot experiment, it can be suggested that application of chicken manure at the rate of 5% (w/w) results in luxuriant growth of vetiver and could be a viable option for reclamation (phytostabilization) of toxic tailings (Das and Maiti, 2009).

ACKNOWLEDGMENTS

The authors are grateful to Mr. P.S. Rana and Mr. M.L. Sahu, of Nalwa Steel and Power Limited (NSPL), Raigarh, India, for awarding the ecorestoration project to ISM, Dhanbad. The authors are also thankful to the Chhattisgarh Environmental Conservation Board (CECB), Raipur, for entrusting ISM Dhanbad with monitoring the site and submitting compliance monitoring of "Solid waste management plan and activity."

REFERENCES

Agbenin, J.O., Adeniyit, T., 2005. The microbial biomass properties of a savanna soil under improved grass and legume pastures in northern Nigeria. Agr. Ecosyst. Environ. 109, 245–254.

Andrews, J.A., Johnson, J.E., Torbert, J.L., Burger, J.A., Kelting, D.L., 1998. Minesoil and site properties associated with early height growth of eastern white pine. J. Environ. Qual. 27, 192–199.

Araujo, I.C.S., Costa, M.C.G., 2013. Biomass and nutrient accumulation pattern of leguminous tree seedlings grown on mine tailings amended with organic waste. Ecol. Eng. 60, 254–260.

Caravaca, F., Hernandez, M.T., Garcia, C., Roldan, A., 2002. Improvement of rhizosphere aggregates stability of afforested semi-arid, plant species subjected to mycorrhizal inoculation and compost addition. Geoderma 108, 133–144.

Das, M., Maiti, S.K., 2009. Growth of *Cymbopogon citratus* and *Vetiveria zizanioides* on Cu mine tailings amended with chicken manure and manure-soil mixtures: a pot scale study. Int. J. Phytorem. 11, 651–663.

Diti, A., 2003. VGT: A Bioengineering and Phytoremediation Option for the New Millennium. APT Consulting Group, Bangkok.

Fehmi, J.S., Kong, T.M., 2012. Effects of soil type, rainfall, straw mulch, and fertilizer on semi-arid vegetation establishment, growth and diversity. Ecol. Eng. 42, 70–77.

Glass, D.J., 1999. Economic potential of phytoremediation. In: Raskin, I., Ensley, B.D. (Eds.), Phytoremediation of Toxic Metals: Using Plants to Clean up the Environment. John Wiley & Sons, New York, pp. 15–31.

Janhit Foundation, 2014. Factsheet on lemon grass. www.janhitfoundation.in (accessed on 23.9.2014).

Juwarkar, A.A., Singh, S.K., Mudhoo, A., 2010. A comprehensive overview of elements in bioremediation. Rev. Environ. Sci. Biotechnol. 9, 215–288.

Kumar, A., Maiti, S.K., 2015. Effect of organic manures on the growth of *Cymbopogon citratus* and *Chrysopogon zizanioides* for the phytoremediation of chromite-asbestos mine waste: a pot scale experiment. Int. J. Phytorem. 17, 437–447.

Lal, K., Yadav, R.K., Kaur, R., Bundela, D.S., Khan, M.I., Chaudhary, M., Meena, R.L., Dar, S.R., Singh, G., 2013. Productivity, essential oil yield, and heavy metal accumulation in lemon grass (*Cymbopogon flexuosus*) under varied wastewater-groundwater irrigation regimes. Ind. Crop. Prod. 45, 270–278.

Ledgard, F.S., Steele, W.K., 1992. Biological nitrogen fixation in mixed legume/grass pastures. Plant Soil 141, 137–153.

Lenka, N.K., Choudhury, P.R., Sudhishri, S., Dass, A., Patnaik, U.S., 2012. Soil aggregation, carbon build up and root zone soil moisture in degraded sloping lands under selected agroforestry based rehabilitation systems in eastern India. Agric. Ecosyst. Environ. 150, 54–62.

Levyk, V., Maryskevych, O., Brzezinska, M., Wlodarczyk, T., 2007. Dehydrogenase activity of technogenic soils of former sulphur mines (Yavoriv and Nemyriv, Ukraine). Int. Agrophys. 21, 255–260.

Maiti, S.K., 1997. Nitrogen accumulation in coalmine spoils by legume (*Stylosanthus humilis*). Environ. Ecol. 15, 580–584.

Maiti, S.K., 2013. Ecorestoration of the Coalmine Degraded Lands. Springer, New York.

Maiti, S.K., Saxena, N.C., 1998. Biological reclamation of coal mine spoils without topsoil: an amendment study with domestic raw sewage and grass legumes mixture. Int. J. Surf. Min. Reclam. Environ. 12, 87–90.

Mitchell, D.J., Barton, A.P., Fullen, M.A., Hocking, T.J., Zhi, W.B., Yi, Z., 2003. Field studies of the effects of jute geotextiles on runoff and erosion in Shropshire, UK. Soil Use Manag. 19, 182–184.

Mukhopadhyay, S., Maiti, S.K., Masto, R.E., 2013. Use of Reclaimed Mine Soil Index (RMSI) for screening of tree species for reclamation of coal mine degraded land. Ecol. Eng. 57, 133–142.

Mulligan, C., Young, R., Gibbs, B., 2001. An evaluation of technologies for the heavy metal remediation of dredged sediments. J. Hazard. Mater. 85, 145–163.

Rickson, R.J., 1995. Simulated vegetation and geotextiles. In: Morgan, P.C., Jane, R. (Eds.), Slope Stabilization and Erosion Control: A Bioengineering Approach. Chapman & Hall, London, pp. 95–132.

Sawtschuk, J., Gallet, S., Bioret, F., 2012. Evaluation of the most common engineering methods for maritime cliff-top vegetation restoration. Ecol. Eng. 45, 45–54.

Shao, Q., Gu, W., Dai, Q., Makoto, S., Liu, Y., 2014. Effectiveness of geotextile mulches for slope restoration in semi-arid northern China. Catena 116, 1–9.

van Donk, S., Tollner, E.W., 2000. Measurement and modeling of heat transfer mechanisms in mulch materials. Trans. Am. Soc. Agric. Eng. 43, 919–925.

World Steel, 2014. Direct reduced iron (DRI) production. http://www.worldsteel.org/statistics/DRI-production.html (accessed on 23.9.2014).

PROSOPIS JULIFLORA (SW) DC: POTENTIAL FOR BIOREMEDIATION AND BIOECONOMY

3

M.N.V. Prasad[1], J.C. Tewari[2]

University of Hyderabad, Hyderabad, Telangana, India[1]
Central Arid Zone Research Institute, Jodhpur, Rajasthan, India[2]

1 INTRODUCTION

It is a general belief that nonnative or alien plants growing outside their natural territory can become invasive. These invasives are widely distributed in a variety of ecosystems throughout the world. Many invasive alien species support farming and forestry systems positively. However, some of alien species become invasive when they are introduced deliberately or unintentionally outside their natural habitats into new areas where they express the capability to establish, invade, and outcompete native species. According to the International Union for Conservation of Nature and Natural Resources (IUCN), the term *alien invasive species* denotes an exotic species that becomes established in natural or seminatural ecosystems or habitat, an agent of undesirable change that threatens the native biological diversity. Invasive species are therefore considered to be a serious hindrance to conservation and profitable use of biodiversity, with significant undesirable impacts on the services provided by ecosystems. Alien invasive species are supposed to demand huge resources that would ultimately bring several undesirable changes in the ecosystem functioning. However, the case of *Prosopis juliflora*, though an exotic, with proper management can be converted to an invaluable bioresource in dryland and desert ecosystems (Elfadl and Luukkanen, 2006; ElSiddig et al., 1998; Muthana and Arora, 1983; Muturi et al., 2009; Osmond, 2003; Pasiecznik et al., 1995, 2006a; Silva, 1990; Viégas et al., 2004).

2 ABOUT *PROSOPIS JULIFLORA*

Prosopis juliflora (Sw.) DC, (Velvet Mesquite) (Fabaceae, subfamily Mimosoideae) (henceforth referred to as *Prosopis*) contains 44 species, of which 40 are native to the Americas (Figure 1), three to Asia and one to Africa. *Prosopis* is conspicuously thorny with a wide, flat-topped crown with genetic variation (Harris et al., 2003; Pires et al., 1990). The root system of *Prosopis* consists of a deep taproot (phreatophyte), sometimes reaching to the unusual depth of 35 m, combined with extensive lateral roots. *Prosopis* is especially suitable for dry sites with annual rainfall between 150 and 700 mm (von Maydell, 1986). Taproots contribute to a stable anchoring of the tree and expand toward groundwater reserves. They are essential during periods of drought when only deepwater sources are available. The

Copyright © 2016 Elsevier Inc. All rights reserved.

Prosopis juliflora native range of distribution is shown in red color. Central America—Mexico, Caribbean Isles; Northern South America—Columbia and Venezuela

FIGURE 1

Distribution *of Prosopis juliflora* in its native region, the central and southern Americas.

depth of the roots depends on the quality and structure of the soil and the availability of soil water; it is also determined by the density of the stand (Wunder, 1966). Once the water source is reached, the roots extend horizontally in the direction of the water flow. *Prosopis* species are adapted to areas with low rainfall and long periods of drought once they are established and are able to tap groundwater or any other water source during the first years. The lateral roots play an important role during rainy seasons or periods of abundant water, for instance, in irrigated areas. The trees are also able to absorb moisture through their foliage during light rains or from dew or other atmospheric sources of moisture (Pasiecznik et al., 2001).

3 GLOBAL DISTRIBUTION

Prosopis is now found in Africa, Argentina, Australia, Brazil (northeast), Cameroon, the Caribbean, Central America, Egypt, Ethiopia, Hawaii, India (Figure 2), Kenya, Nigeria, Pakistan. Paraguay, Peru, Peruvian-Ecuador, Portugal, Senegal, Spain, Sri Lanka, Sudan, Uganda, the United States, and Yemen.

Prosopis has spread rapidly in the arid regions. People as well as nature have taken advantage of this alien plant. It has some biological characters that foster invasion; hence, appropriate management practices are needed to exploit its resource potential. In spite of its invasive nature, services provided by *Prosopis* are numerous. A large number of publications have appeared on various aspects of its bio-resource potential (Figure 3). The notable bioeconomic aspects are (1) phytochemistry, (2) allelopathy, (3) antioxidant, (4) food, (5) biopesticide, (6) phytoremediation, (7) bioethanol, (8) synthesis of nanoparticles, (9) timber and firewood, (10) improvement of livelihood of rural community, and (11) medicinal uses, etc. (Table 1 and Figure 4).

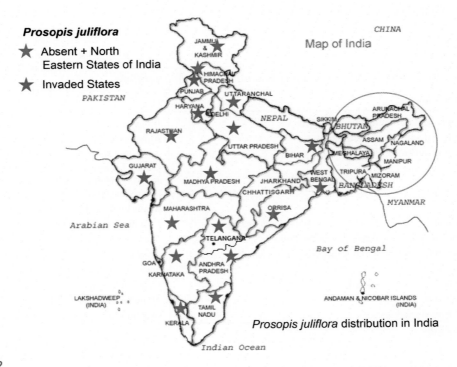

FIGURE 2

Prosopis juliflora's distribution in India. Green star, abundantly distributed in wastelands. Red star, absent (Jammu Kashmir, Himachal Pradesh, Uttaranchal, and northeastern states; viz., Sikkim, Assam, Meghalaya, Arunachal Pradesh, Nagaland, Manipur, Mizoram, and Tripura).

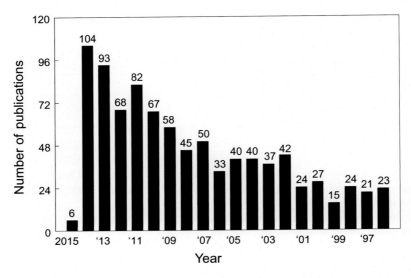

FIGURE 3

Number of articles published on various aspects of *P. juliflora*.

Data source: www.sciencedirect.com.

Table 1 Bioresource and Bioeconomy Potential of *Prosopis juliflora*

1. Phytochemistry	
Flavonoids; and flavonoid glycosides from the bark	Bragg et al. (1978), Shukla and Misra (1981)
Flavonols and anti-oxidant substances	Almaraz-Abarca et al. (2007)
Source of commercial galactomannans	Azero and Andrade (2002), Daas et al. (2000), López-Franco et al. (2014), Shukla and Misra (1981), Vieira et al. (2007), López-Franco et al. (2012)
Seed polysaccharide	Bhatia et al. (2014)
The alkaloid, "juliflorine" for Alzheimer's disease therapy	Choudhary et al. (2005)
Degradation of gum exudates from some *Prosopis* species	Churms et al. (1981)
Prosopis and its potential to improve livestock production	Riveros (1992)
Alkaloid cytotoxicity and stimulate NO	Silva et al. (2007)
Galactomannans from the seeds	Vieira et al. (2007)
Alkaloids from the leaves	Ahmad et al. (1989)
Phytochemical analyses	Ibrahim et al. (2013)
2. Allelopathy	
Ecophysiology of seed germination	El-Keblawy and Al-Rawai (2005, 2006)
Autotoxic potential of foliage	Warrag (1995)
Autotoxicity of mesquite (*Prosopis juliflora*) pericarps on seed germination and seedling growth	Warrag (1994)
Conditional allelopathic effects	Kaur et al. (2014)
Growth inhibitory alkaloids	Nakano et al. (2004)
Inhibit seed germination and seedling growth	Al-Humaid and Warrag (1998)
Allergy in Chennai	Bhuvaneswari et al. (2004)
Plant growth inhibitory alkaloids were isolated from the leaf extract	Nakano et al. (2004)
Allelopathic effects	Noor et al. (1995), Csurhes (1996)
Growth inhibitory alkaloids from mesquite	Hiroshi et al. (2004)
3. Antioxidant	
Antioxidant activity of polyphenolic extract of monofloral honeybee-collected pollen from mesquite	Almaraz-Abarca et al. (2007)
4. Food	
Physicochemicals of seed gum	Rincón et al. (2014)
In vitro digestibility seeds by mammalian digestive proteinases	Araüjo et al. (2002)
In situ ruminal and intestinal nutrient digestibilities of mesquite (*P. juliflora*) pods	Batista et al. (2002)
The gum had interesting emulsifying properties and good flavor encapsulation function and the microcapsules can be readily used as a food ingredient	Beristain et al. (2001)
Feed stock for biomass power plants and gasifier	Bhoi et al. (2006)
Prosopis pods as human food, with special reference to Kenya	Choge et al. (2007)
Chemical and nutritional studies on mesquite beans	Del Valle et al. (1983)
Fodder production	Ibrahim (1988)

Table 1   Bioresource and Bioeconomy Potential of *Prosopis juliflora*—cont'd

Physicochemical characterization and functional properties of galactomannans from mesquite seeds (*Prosopis* spp.)	Lee et al. (1992)
Pods as a feed for goats	Mahgoub et al. (2005c)
Pods are fed to goats. *P. juliflora* pod concentrate and date palm by-products are suggested to replace commercial concentrate for Omani sheep	Mahgoub et al. (2005a,b, 2007)
Partial substitution of barley grain with *P. juliflora* pods in lactating	Obeidat and Shdaifat (2013)
Effective utilization of *P. juliflora* pods by ensiling with dessert grass *Lasiurus sindicus*	Pancholy and Mali (1999)
Pods can be effectively used in silage (by ensiling with desert grass *Lasiurus sindicus*) as replacement for urea	Pancholy and Mali (1999)
Pods provide an alternate feed resource for feedlot lambs for meat production	Ravikala et al. (1995)
Growth efficiency in feedlot lamb	Ravikala et al. (1995)

5. Biopesticide

A proteinaceous inhibitor can be used as an effective defenses against insects and pathogens (bioinsecticide)	Araüjo et al. (2002), Oliveira et al. (2002)

6. Phytoremediation

Dune stabilization	Singh and Rathod (2002)
Revegetating fly ash landfills	Rai et al. (2004)
Appropriate species for establishing plantations with saline irrigation	Tomar et al. (2003)
Phytoremediation of chromium contaminated soil in the presence of *Glomus deserticola*	Arias et al. (2010)
Correlations between metals in tree-rings of *Prosopis julifora* as indicators of sources of heavy metal contamination	Beramendi-Orosco et al. (2013)
Sodic soil restoration in India	Bhojvaid and Timmer (1998)
Remediation of sodium toxicity and accelerated fertility restoration. Addition of amendments facilitated tree growth, root development, and biomass accumulation	Bhojvaid and Timmer (1998)
Biomonitoring studies on the lead levels in mesquite (*P. juliflora*) in the arid ecosystem	Bu-Olayan and Thomas (2002)
Appropriate tree for soil amelioration in semi-arid Senegal	Deans et al. (2003)
Seedlings bioassay at seawater salinity concentrations	Rhodes and Felker (1988), Velarde et al. (2003)
Mycorrhizal distributions in soil as inocula showed most promise for use in agroforestry systems	Ingleby et al. (1997)
Seeds have the potential for use as water treatment coagulants	Diaz et al. (1999)
Single-stemmed trees are appropriate for agroforestry systems	Elfadl and Luukkanen (2003)
Activated carbon remove dye	Gopal et al. (2014)
Removal of Pb(II) from aqueous solution by seed powder	Jayaram and Prasad (2009)
Activated carbon is used in oil, food, and pharmaceutical industries	Kailappan et al. (2000)
Production of activated carbon	Kailappan et al. (2000)
Sorption of chromium by *P. juliflora* bark carbon	Kumar and Tamilarasan (2013a, 2013b)

(Continued)

Table 1   Bioresource and Bioeconomy Potential of *Prosopis juliflora*—cont'd

Adsorption of aniline blue and removal of Victoria blue by using *P. juliflora* carbon	Kumar and Tamilarasan (2013a, 2014)
Sorption of chromium by *Prosopis juliflora* bark carbon	Kumar and Tamilarasan (2013b)
Reclamation of saline and sodic soils	Maliwal (1999), Singh et al. (1989)
Greening of pegmatite tailings	Nagaraju and Prasad (1998)
Appropriate plant for revegetating fly ash landfills with different amendments and Rhizobium inoculation	Rai et al. (2004)
Fly ash for soil amelioration	Ram and Masto (2014)
Rehabilitation of gypsum mine	Rao and Tarafdar (1998)
Phytoremediation—fluoride accumulator	Saini et al. (2012)
Appropriate for green technologies to decontaminate Cu and Cd contaminated soils	Senthilkumar et al. (2005)
Decontaminate heavy metal (Cu and Cd)	Senthilkumar et al. (2005)
Metal accumulation	Shukla et al. (2011)
Ecological restoration of degraded sodic land	Singh et al. (2012)
Reclaimation of high-pH soils	Singh (1996a, 1996b, 2009)
Lead/zinc mine tailings remediation	Solís-Domínguez et al. (2011)
Under arid field conditions, its soil host VAMF viz.., *Glomus mosseae, G. fasciculatum, Gigaspora margarita* which enhanced biomass and uptake of N, P, K, Cu, and Zn	Tarafdar and Kumar (1996)
Phytoextraction of Pb and Cd	Varun et al. (2011)
7. Bioethanol	
Production of ethanol from mesquite (*Prosopis juliflora* (SW) D.C.) pods mash by *Zymomonas mobilis* and *Saccharomyces cerevisiae*	da Silva et al. (2010)
Saccharification of *P. juliflora*	Kapoor et al. (2008)
Chemical pretreatment for lignocellulosic	Naseeruddin et al. (2014)
8. Synthesis of nanoparticles	
Toxicity and biotransformation of ZnO nanoparticles in the desert plants *Prosopis juliflora*-velutina, *Salsola tragus*, and *Parkinsonia florida*	De La Rosa et al. (2011)
Synthesis of silver nanoparticles	Raja et al. (2012)
Zinc absorption and distribution treated with ZnO nanoparticles	Hernandez-Viezcas et al. (2011)
9. Timber and fire wood	
Grading mesquite lumber	Felker and Anderson (1997)
Potential uses of leguminous trees for minimal energy input agriculture	Felker and Bandurski (1979)
Prosopis for firewood, poles, and lumber	Felker and Patch (1996)
Appropriate tree for fuelwood production on sodic soil	Goel and Behl (2001)
Most promising species for short rotation fuel wood forestry programs on alkaline soil sites because of faster growth rate, high biomass production potential and calorific value	Goel and Behl (1995, 1996), Singh et al. (1991)
Fuelwood production on sodic soil	Goel and Behl (2001)
Fuelwood production potential	Goel and Behl (1995)
Fuelwood quality of promising tree species for alkaline soil	Goel and Behl (1996)

Table 1 Bioresource and Bioeconomy Potential of *Prosopis juliflora*—cont'd	
Source of fuelwood, pulpwood, or timber products from plants grown on degraded land and saline sites	Mahmood et al. (2001)
10. Promote bioeconomy and improve livelihood	
An ellagic acid glycoside from the pods	Malhotra and Misra (1981a, 1981b, 1981c)
Is a source of low-cost mycorrhizal inocula that improved the growth of other agroforestry species for e.g., *Acacia tortilis*	Munro et al. (1999)
Provide income for facilitating rural livelihoods	Mwangi and Swallow (2008)
Improve livelihoods in Sri Lanka	Perera and Pasiecznik (2005)
Fuel, fodder, and food	Rawat et al. (1992)
Barks is source of novel natural cellulosic fiber	Saravanakumar et al. (2013)
Economic use	Sato (2013)
Glomus species promoted growth of *Prosopis*	Siddhu and Behl (1997)
Benefits, threats, and potential of *Prosopis*	Walter and Armstrong (2014)
Biological control	Zimmermann (1991), Warrag (1994)
11. Medicinal use	
Ethnomedicine—oral health care	Hebbar et al. (2004)
In vitro antiplasmodial activity of ethanolic extract	Ravikumar et al. (2012)
Diagnostic and therapeutic reagent for patients an allergenically related to other tree species	Dhyani et al. (2006)
Antimicrobial activity of alkaloids	Al Shakh Hamed and Al Jammas (1999), Aqeel et al. (1988)
Uncoupling of oxidative phosphorylation by juliprosopine on rat brain mitochondria	Maioli et al. (2012)
Pharmacological potentials of phenolic compounds	Prabha et al. (2014)
Nutrient evaluation and elemental analysis of plants of Khyber PakhtoonKhwa, Pakistan	Hussain et al. (2011)
Bactericidal and fungicidal effect	Kanthasamy et al. (1989)
Antibacterial potential of ethanolic leaf extract	Sathiya and Muthuchelian (2008)
Antibacterial properties of alkaloid rich fractions obtained from various parts	Singh et al. (2011)
Ethnobotanical importance in Loja and Zamora-Chinchipe, Ecuador	Tene et al. (2007)
Anti-pustule plant metabolites for pimple creams	Rajadurai Jesudoss et al. (2014)

4 RESTORATION OF CONTAMINATED/DEGRADED LAND, PHYTOREMEDIATION

Prosopis is an ideal species for stabilizing the pegmatitic tailings of mica mines in Nellore district of Andhra Pradesh, India (Nagaraju and Prasad, 1998). It is also helpful for reclamation of copper, tungsten, marble, and dolomite mine tailings and is a green solution to heavy metal-contaminated soils. An appropriate species for rehabilitation of gypsum mine spoil in arid zones and for restoration of sodic soils, *Prosopis* outperformed all other tree species in sand dune stabilization (Kailappan et al., 2000; Rai et al., 2004; Senthilkumar et al., 2005). *P. juliflora* seedlings cultivated in hydroponics are able to

FIGURE 4

Multiple uses of *P. juliflora*: (a) It is the most common tree in semiarid tropic regions. (b, c) Its biomass is used for carbonation technology to produce charcoal. (d) Charcoal is used as a carrier for plant growth-promoting bacteria; e.g., *Azospirillum*. (e) Its biomass is the major feedstock for biomass-based gasifier/power plant. (f) Drug deaddicting pellets are produced from its extract. (g, h) Animal feed and (i) furniture products from its timber.

bioaccumulate Ni, Cd, and Cr. Pods. Chromium (Cr) is an essential mineral for ruminants in tropical regions. Its pods contain high Cr concentration (up to 150 p.p.b.). Thus, supplements Cr in fodder for animals (goats) Cr requirement for animals is >0.1 p.p.m., while toxic level was 1000 p.p.m.

Prosopis was very helpful for reclamation of copper mine tailings; e.g., copper mines in Arizona, United States, and abandoned mine waste in Mexico (Figures 5 and 6). An appropriate species for rehabilitation of gypsum mine spoil in arid zone; restoration of sodic soils. *Prosopis* pure stands have been observed on soils industrially polluted with a wide variety (Gratão et al., 2005; Prasad, 2004). It outperformed all other tree species in sand dune stabilization. Arbuscular mycorrhizal inocula isolated from its rhizosphere, when inoculated to seedlings of other agroforestry and social forestry legumes, accelerated the growth of these seedlings on perturbed ecosystems. It is an ideal species for afforestation and helps in the reclamation of wastelands, the stabilization of technogenically contaminated soils, and to regulate soil erosion. It is an important tree for afforestation/restoration of perturbed ecosystems and grows satisfactorily without any amendments (Herrera-Arreola et al., 2007). Mycorrhizae are reported to greatly improve the growth of *Prosopis* on high pH soils (Velarde et al., 2005).

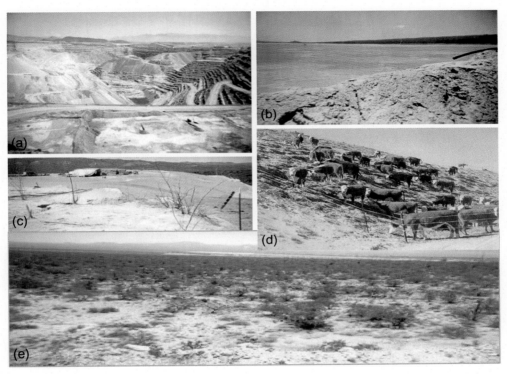

FIGURE 5

(a–d) In copper mines near Globe and Tucson, Arizona, USA (Arizona Ranch, Resource Management and Mine Reclamation; ASARCO Inc. Copper Operations), the ecosystem rehabilitation and mine reclamation program is primarily based on cattle. Herds of cows are impounded with electric fence on mine by providing fodder and water for varying durations. Cows not only stabilize the soil with their hooves but also enrich soil nutrient status via urination. They also augment microbes to the soil through the dung. This process is repeated at regular intervals in cycles. Thus, the presence of cattle accelerates rhizosphere development and improves plant root association via enriching soil microorganisms and nutrients. The results obtained with cattle for rhizosphere ecodevelopment are spectacular. (e) *P. juliflora* colonization on copper mine tailings Arizona, USA.

Prosopis and *Leptochloa fusca* association was successful for the restoration of salt lands. Therefore, grass-legume-tree association needs to be tested on different sites for remediation, if necessary with biotic and abiotic amendments (Figure 7). There are a few publications reporting its fly ash landfills' revegetating potential following different amendments and *Rhizobium* inoculation. However, *Prosopis* has some biological characters that foster invasion; hence, appropriate management practices need to be developed for recommending it for phytoremediation. In rainfed agriculture, leaves from multipurpose trees are traditionally one of the main sources of nutrients that maintain soil fertility. It has been suggested that the use of tree legumes like *Prosopis* has potential for a minimal input farming system because *Prosopis* has the ability to grow with little or no irrigation, produces high yields, is able to fix nitrogen, and produces large amounts of high-quality protein (Felker and Bandurski, 1979). However, there is competition between trees and crops for water, nutrients, and light. Another point to consider

FIGURE 6

(a, b) *P. juliflora* colonization on lead mine waste in Mexico (inset: *Prosopis* growing on smelting industry metalliferous waste, and *Prosopis* with bunch of mature pods).

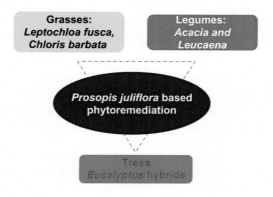

FIGURE 7

Tree-grass-legume association was found to be the best combination for restoration of mica, copper, tungsten, marble, dolomite, limestone, and mine spoils of Rajasthan Sate and elsewhere in India (Prasad, 2007).

is the number of years it takes before a constant improved soil status is reached for agricultural crop cultivation (Snapp et al., 1998).

A widespread traditional practice in the semiarid tropics is to grow trees scattered or dispersed on croplands known as "parklands." The degree of interaction between trees and crops depends on the density of trees. Trees in this system are rarely planted but occur and regenerate naturally. Farmers maintain these trees on their lands to provide fodder, fuelwood, timber, and nonwood products (Rao and Tarafdar, 1998). The improvement of soil properties comes as a side product, along with the provision of shade for people and livestock.

Prosopis is potentially useful for rehabilitating degraded saline soils in dry land ecosystems where degradation is enhanced by high temperatures and irregular precipitation. *Prosopis* planted on degraded sodic soils increases soil fertility through adding to and increasing the soil organic C, nitrogen (N), available phosphorus (P), exchangeable potassium (K), calcium (Ca), and magnesium (Mg) levels. In addition, decreases in the exchangeable sodium level (Na), pH, and EC can be observed. These processes have a positive effect on the rehabilitation of sodic soils through the improvement of nutrient cycling and detoxifying sodicity. As an effect of *Prosopis*, the crop productivity tested on wheat indicated higher germination, survival, plant growth, and grain yield (Bhojvaid and Timmer, 1998).

Prosopis is reported to reduce Electrical conductivity (EC) and Exchangeable sodium percentage (ESP) significantly in saline-sodic environment. *Prosopis* improved the saline-sodic soil where agriculture was possible (Maliwal, 1999). Likewise, alkali soils were determined suitable for forestry using *Prosopis* (Singh, 1996a, 1996b). The use of trees to remediate heavily metal-contaminated soils is profitable (Prasad, 2007; Pulford and Dickinson, 2006). *Prosopis* has high capacity to accumulate heavy toxic metals occurring in the soil (Senthilkumar et al., 2005). *Prosopis* is also able to accumulate lead (Pb) in various parts of the plant. The highest concentration of lead is found in the leaves, lower levels in the bark, and the lowest levels in the pods. The lead content in the plant can be used as an indicator to measure air pollution derived from industrial factories and vehicles (Bu-Olayan and Thomas, 2002). Because of the high amounts of heavy metals accumulated in *Prosopis* leaves and pods, it is recommended to keep livestock away from grazing on this species when it grows in heavy metal-polluted soils (Senthilkumar et al., 2005). *Prosopis* colonized the industrial effulent-laden oils produced by textile, paper products, tannery, chemical products, basic metal products, machinery parts, and transport equipments industry. *P. juliflora* and *Leptochloa fusca* association was successful for revegetating salt-laden lands (Singh, 1995). Therefore, grass-legume-tree association needs to be tested on different sites for remediation, if necessary with biotic and abiotic amendments. Restoration of fly ash landfills with *P. juliflora* following different amendments and Rhizobium inoculation yielded promising results (Rai et al., 2004). Mycorrhizae improved the growth of *P. juliflora* on high pH soils (Siddhu and Behl, 1997). *P. juliflora* seedlings growing in gypsum mine had high-frequency arbuscular mycorrhizal fungal infection. However, *P. juliflora* has some biological characters that foster invasion; hence, appropriate management practices need to be developed for recommending it for phytoremediation. In Colombia, *Prosopis* trees improved the on-soil microbial community and enzymatic activities in intensive silvopastoral systems (Vallejo et al., 2012).

Prosopis is also a suitable candidate for revegetating fly ash landfills. *Prosopis* has shown potential to accumulate in its tissues the metals contained in fly ash. Inoculation of *Prosopis* with fly ash-tolerant rhizobium (PJ-1) accelerates the translocation of metals from the soil to the above-ground growth. *Rhizobium* inoculation also increases the biomass of the plant in nitrogen-deficient fly ash, which confirms that *Prosopis* is beneficial for re-vegetation and decontamination of soil and landfills affected by fly ash (Rai et al., 2004; Sinha et al., 1997).

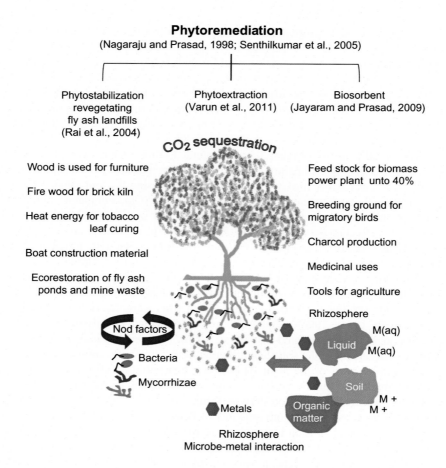

Phytoremediation
(Nagaraju and Prasad, 1998; Senthilkumar et al., 2005)

Phytostabilization
revegetating
fly ash landfills
(Rai et al., 2004)

Phytoextraction
(Varun et al., 2011)

Biosorbent
(Jayaram and Prasad, 2009)

CO₂ sequestration

Wood is used for furniture

Fire wood for brick kiln

Heat energy for tobacco
leaf curing

Boat construction material

Ecorestoration of fly ash
ponds and mine waste

Feed stock for biomass
power plant unto 40%

Breeding ground for
migratory birds

Charcol production

Medicinal uses

Tools for agriculture

Rhizosphere

Nod factors

Bacteria

Mycorrhizae

Metals

Liquid   M(aq)

M(aq)

Soil   M +

Organic
matter   M +

Rhizosphere
Microbe-metal interaction

FIGURE 8

Bioremediation potential and its economic uses.

Mycorrhizae are reported to greatly improve the growth of *P. juliflora* on high pH soils. *Prosopis* was able to grow satisfactorily without amendments up to pH 9. Arbuscular mycorrhizal inocula that have been isolated from its rhizosphere (low-cost agrotechnology) were found to accelerate the growth of other agroforestry and social forestry legumes in perturbed ecosystems (Siddhu and Behl, 1997; Singh, 1995). Mycorrhizae are reported to greatly improve the growth of *Prosopis* on high pH soils. *Prosopis* was able to grow satisfactorily without amendments up to pH 9 (Figure 8).

5 *PROSOPIS* AS LIVESTOCK FEED

The use of *Prosopis* pods as a livestock feed is of great advantage in the arid and semiarid areas of India (Figure 9). *Prosopis* grows naturally in drought-affected areas, requiring only small quantities of water for its growth. Malnutrition is one of the multiple hazards that livestock faces under

FIGURE 9

(a) Stocks of dried pods of *P. juliflora*. (b, c) Dried pods ready for sending to fruit grinding machine. Fruit powder for preparation of animal feed. (d) Fruit-grinding machine. (e) Output of b and c using pod grinding machine (d).

these harsh conditions (Tabosa et al., 2000). It has been observed that a mature 10-year-old tree grown under good conditions of soil and water availability is able to produce 90 kg of pods per year (Rawat et al., 1992). A pod yield of 2000 kg/ha of *Prosopis* was estimated for the unmanaged Arizona desert in North America and 4000-20,000 kg/ha in the arid Hawaiian savannas (Felker and Bandurski, 1979). In the northeast of Brazil, pod yields of 2-3 tons/ha are produced even on shallow stony soils with vegetation typical of semiarid regions and with no agricultural value. Pod production of *Prosopis* can be increased three to four times if irrigation is provided during the flowering period, which lasts 2 months. However, pod production may continue even after a drought period of 2-3 years (Riveros, 1992).

Prosopis pods contain 20-30% sucrose and about 15% crude protein (Del Valle et al., 1983). They can be used in various ways as animal feed without causing any adverse digestive effects when used properly. Although crushed and ground pods are suggested as an additional feed, the feasibility of these techniques is questioned due to limited availability of manpower for processing the pods, causing additional expenditure to the average livestock farmer. It is further suggested that the pods should be mixed with other feed and that continued feeding of pods to cattle as the only diet should be avoided (Rawat et al., 1992). Pods in well-preserved silage have produced excellent livestock feed. However, the cost and availability of those additives make their use difficult. The ensiling of *Prosopis* pods together with dry grass, like the desert sewan grass *Lasiurus sindicus* Henr., is able to improve the nutritional value of the feed without pretreatment (Pancholy and Mali, 1999).

The inclusion of *Prosopis* pods in the diet of Indian sheep, under a feedlot system raising lambs for meat production, indicated that the incorporation of 15-30% pods into the feed significantly lowered the feeding costs. A diet consisting of 15% pods added to other ingredients, like wheat straw (30%), alfalfa hay (20%), and other minor components like wheat bran, urea, mineral mixture, and salt (3.5%), showed superior feed efficiency (Ravikala et al., 1995). In whatever way the pods are used, crushed and ground or otherwise prepared as a mixture, whenever the seeds are destroyed in the process it has an influence on the control of the *Prosopis* invasion.

The use of *Prosopis* pods as livestock feed provides the farmer monetary savings. It is a low-cost solution compared with other commercial animal feeds. The pods have a high concentration of proteins and the tree easily adapts to semiarid conditions where hardly anything else is able to grow. Other products like honey and beeswax provide additional income to the families in these areas and thus contribute to diversifying their livelihood opportunities.

Very few animals graze on the foliage of *P. juliflora* because of its unpalatable leaves and long spines. The leaves of *P. cineraria*, on the other hand, are browsed on by various animals. *P. cineraria* (khejri), which is native to India, is a slow-growing tree that develops a height of 6 m only after 10-15 years, compared to a height of 12-15 m in 4-5 years in *P. juliflora*. Khejri is well adapted to low rainfall (150-500 mm annually) areas. The particular feature of the tree is that it produces its leaves, flowers, and pods during the hottest period of the year, between March and June. Thus the tree offers great opportunities as a forage resource in the extremely arid zones during the hottest season. Although khejri produces animal feed only after 10 years, under proper management regimes it continues production for up to 2 decades. Annually, the tree produces an average of 25-30 kg of dry leaf forage (Riveros, 1992).

6 *PROSOPIS* FOR FUELWOOD

The most common use of *Prosopis* is as fuelwood, especially among the landless and poor. Farmers who are economically better off use gas, but after the growing season is over, they often engage in cutting *Prosopis* for firewood. Hotels and restaurants use *Prosopis* as fuelwood processed as charcoal for preparing food (Figure 10).

People have been aware of the advantage that *Prosopis* is a free fuelwood for the poor, but the thorns cause injuries while collecting firewood. There are possibilities to genetically improve *Prosopis* through selective breeding and hybridization with other less thorny *Prosopis* species, in order to obtain a variety that would make the handling of the tree easier (Wojtusik et al., 1994; Tewari et al., 2000; Ewens and Felker, 2003).

Prosopis was introduced into the drylands of India one-and-a-half centuries ago mainly for the purpose of conservation but has meanwhile become the main source of fuelwood in the rural areas, fulfilling more than 70% of the firewood requirements of rural people (Reddy, 1978; Harsh and Tewari, 1998; Tewari et al., 2000). Today, the tree has established itself, spreading all over the arid and semiarid regions, and in some areas it has become an invasive weed.

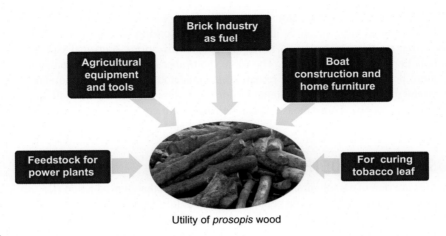

Utility of *prosopis* wood

FIGURE 10

Uses of *Prosopis* wood.

7 *PROSOPIS* FOR PRODUCTION OF TIMBER

Prosopis timber had high demand for manufacturing various products. *Prosopis* wood is being tested with partial substitution by gypsum for production of *Prosopis* wood composites for the purpose of construction and furniture (Abdelrhman et al., 2015). Large trees of more than 25 cm diameter are desirable and can be achieved through various stand management practices. There is no need to establish *Prosopis* plantations especially for this purpose. There are already large areas of invaded wastelands that can be managed to produce stem sizes that are needed in the wood manufacturing industry (Felker and Patch, 1996; Patch et al., 1998; Elfadl and Luukkanen, 2003).

The quality of *Prosopis* wood is similar to several other species of the genius *Prosopis* and can easily compete with Indian rosewood (*Dalbergia latifolia* Roxb.), Indian teak (*Tectona grandis* L.), and cocobolo (*Dalbergia retusa* Hemsl.), three commonly recognized fine hardwoods, in relation to physical and mechanical properties. Tangential and radial shrinkage values are similar in *Prosopis*, which indicate that the wood shrinks or swells equally in both directions. The hard surface of *Prosopis* wood makes it an ideal raw material for furniture. The wood of *Prosopis* is light brown when cut and after drying and aging it darkens, turning a dark reddish golden-brown. *Prosopis* can be classified as one of the world's most precious tropical hardwoods (Felker and Anderson, 1997).

In many South American countries, like Argentina and Peru where *Prosopis* species are indigenous, furniture and flooring made of *Prosopis* wood is highly appreciated and the technology to produce sawn timber has been developed during recent years to suit these species (Felker, 1998). In India, the potential for managing trees to be used as timber is still underexploited. It has been concluded that chainsaw milling that is portable and able to cut small-diameter, crooked, and small logs is the most suitable method for the conversion of *Prosopis* wood into timber (Pasiecznik et al., 2006b).

Small-scale enterprises using chainsaws to mill *Prosopis* would improve livelihoods by adding value to an already existing resource and creating new working places. At the same time, it has a controlling effect on *Prosopis* invasions (control through utilization). A critical aspect of profitable saw timber production is proper marketing. The more the timber fulfills the requirements of the potential user, the more profit can be made. As research has indicated, the most common sizes of timber that are used for the manufacture of furniture and flooring are less than 10 cm in width and 1.5 m in length (Pasiecznik et al., 2006b). Those measures can be easily extracted from *Prosopis* trees.

8 *PROSOPIS* FOR BIOETHANOL

Prosopis pods are rich in carbohydrates and considered for bioethanol production (Da Silva et al., 2010). Lignocellulosic ethanol and studies on pelletization and delignification of cellulosic biomass *P. juliflora* have been carried out via two stage pretreatments involving an alkali and an acid (National Bioresource Development Board, NBDB). Studies on *P. juliflora* resulted in 85% or higher delignification for all the three residues using 0.5% NaOH at room temperature for 24 h on the basis of holo-cellulose recovery and total reducing sugar yield per gram of initial substrate.

Two novel bifunctional cellulolytic enzymes with good activities were developed for application in lignocellulosic ethanol:

1. Endoglucanase/glucosidase chimera (EG5)
2. Endoglucanase/xylanase (Endo5A-GS-Xyl11D)

Also, an engineered *E. coli* strain (SSY10) is being used to ferment C5/C6 into ethanol via native pathway engineering. SSY10 produced ethanol from C5/C6 sugars at the rate of 0.7 g/l/h.

SSY10 also efficiently fermented lignocellulosic hydrolysates generated via acid treatment and via ammonia treatment to ethanol at neat theoretical maximum yield. *Paenibacillus* ICGEB2008, a strain that produces cellulolytic enzymes as well as ferments sugars into ethanol and 2,3-butanediol has been isolated. This strain could ferment glucose, xylose, cellobiose, and glycerol and produce ethanol.

9 *PROSOPIS* FOR HONEY PRODUCTION

Prosopis regularly produces an abundant amount of flowers that are used for forage by honeybees even during times of drought. After the introduction of *Prosopis* into Hawaii in the 1930s, the island became one of the largest producers of honey (Figure 11).

The honey produced from *Prosopis* flowers is claimed to be of excellent quality. A substantial amount of this honey (300 metric tons) was harvested and marketed during a 5-year period in the early 1990s in the state of Gujarat alone. The rare local honeybee species *Apis florea* (Fabricius) and *Apis cerana* (Fabricius) use the nectar and pollen of *Prosopis* to produce honey, which is also used in traditional medicine (Varshney, 1996; Pasiecznik et al., 2001).

A large number of beekeepers in South India earn their living from *Apis cerana* bees, which are known to be good pollinators. *Apis florea* (dwarf honeybee) has adjusted to the extreme climate of the arid and semiarid zones and occurs in large numbers in the arid zones of Gujarat; it is also

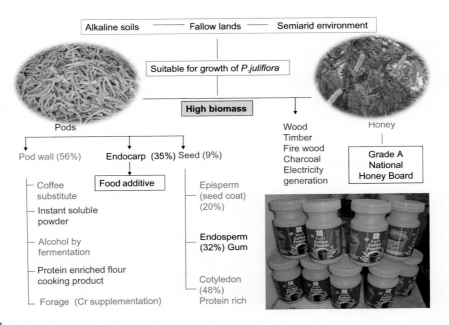

FIGURE 11

Multiple uses of *P. juliflora*.

present in smaller numbers in South India (Thomas et al., 2002). Beeswax is used as a pharmaceutical and industrial raw material, out of which candles, creams, and balms are manufactured (Varshney, 1996).

10 MEDICINAL USES

Several beneficial medicinal compounds have been reported from *P. juliflora*, some of the prominent being steroids, tannins (Makkar et al., 1990), leucoantho cyaniding, and ellagic acid glycosides. Extracts of *P. juliflora* seeds and leaves have several in vitro pharmacological effects such as antibacterial, antifungal, antiinflammatory properties (Al Shakh Hamed and Al Jammas, 1999; Kanthasamy et al., 1989). Flavonoids are water-soluble phytochemicals showing antioxidant, anticancer, and antiinflammatory activities. These prevent cells from oxidative damage and carcinogenesis. Flavonoids are also used to cure some heart-related diseases (Hussain et al., 2011). Flavonoids occur virtually in all parts of the plant, the root, heartwood, sapwood, bark, leaf, fruit, and flower. Alkaloids and their derivatives are used as basic medicinal agents for their analgesic, antispasmodic, and bactericidal activities. Alkaloid-rich fractions of *P. juliflora* are active antifungal and antibacterial agents (Aqeel et al., 1988) (Ahmad et al., 1989). In this study *P. juliflora* was taken as a plant source to screen its antipustule active compounds. Screened compounds were tested for their antipustule behavior by studying the inhibition levels of *Staphylococcus* sp. Combined effect of bioactive compound from *P. juliflora* along with commercially available acne creams were studied. Clinigex and Clingel were chosen as synthetic acne creams for the study.

10.1 CHARCOAL PRODUCTION

Charcoal, a low calorific value and low costs fuel is preferred in hotels and domestic sectors both in rural and the urban areas. Charcoal and coke are also used in industries for specialized purposes like activated charcoal. *Prosopis* wood in large quantities is most commonly used for production of charcoal (Figures 12 a-d).

Charcoal is produced by burning *Prosopis* wood in "bhatti" formations (the local name for the kiln used). In a, Bhattis the wood is arranged in a round circles with a top layer of coal dust that is sprinkled with water. The whole process, from "bhatti" formation to the end product takes 1-2 weeks depending upon the quantity of the wood to be converted into charcoal and control of burning process. About

FIGURE 12

Charcoal production from *Prosopis juliflora* a) harvested stem wood lots b) harvested wood arranged as a small "bhatti" for controlled burning c) charcoal from root wood and d) charcoal final product ready for domestic and industrial use.

4 tons of air dried wood would produce about 1 tons of charcoal. *Prosopis* roots are preferred for charcoal production due to high calorific value. In semi arid region of India, *Prosopis charcoal* production generate employment and revenue besides are tangible benefits

11 NATURAL PERSTICIDE (ALLELOPATHY)

Prosopis foliage has allelopathic effects on seed germination and seedling growth of bermuda grass (*Cynodon dactylon*): three cultivars of *Zea mays* L. (R 796, Gohar, EV 1081), four cultivars of *Triticum aestivum* L. (Inqalab, Chakwal, Pak 81, Rohtas), and *Albizia lebbeck* (L.). (Table 1, item 2)

12 *PROSOPIS* PODS AND SEEDS ARE A RICH SOURCE OF CARBOHYDRATE

The gum of *Prosopis* is a good encapsulating material. Cardamom (*Elettaria cardamomum*) essential oil microcapsules were produced using its gum.

Prosopis seeds serve as an alternate source of the galactomannans and possible to extract into water at room temperature. Galactomannans are water-soluble neutral polysaccharides, composed of a linear mannan backbone bearing side chains of a single galactose unit. These have wide industrial applications, mainly as thickening and stabilizing agents in a range of applications. Its leaves are ground with tobacco (*Nicotiana tabacum* L.) and lime and placed on painful tooth for relief (Hebbar et al., 2004). It is also well known for its antibiotic and antibacterial properties. Polyphenols and tannins of medicinal importance have been reported from the stem bark and pods.

13 EECOSYSTEM SERVICES: BIRDS BREEDING ON *PROSOPIS*

Prosopis provides shelter and breeding ground to migratory birds, thus contributing to environmental protection and conservation. A classic example is Uppalapadu wetland near Guntur, Andhra Pradesh, India. The Uppalapadu bird sanctuary is located in Guntur District, 7 km from Guntur town. The water tank on the edge of the village Uppalapadu is unique since they provide refuge to more than 30 migratory bird species throughout the year. Endangered species, such as the spot-billed pelican (*Pelecanus philippensis*), painted stork (*Mycteria leucocephala*), and white ibis (*Threskiornis melanocephalus*), have shifted from Nelapattu in the Nellore District to Uppalapadu to breed on *Prosopis* trees and bushes. The water tank is surrounded by dense *Prosopis* tree and bush formations, and the small island in the middle of the tank is totally covered with *Prosopis*. According to the guard of the sanctuary, it is estimated that the bird population of these water tanks amounts to 7000 and the number of birds stopping in the sanctuary during the peak season between September and February amounts to 20,000. Some migratory birds, like the rosy pastor (*Sturnus roseus*), travel hundreds of kilometers to come to the sanctuary. The villagers have agreed to keep this tank only for the birds while using other ponds nearby for their needs. The unique aspect of the place is that the birds can be spotted throughout the year in this small four-and-a-half-acre stretch of land (Figure 13).

FIGURE 13

P. juliflora providing shelter and breeding ground to migratory birds (Uppalapadu near Guntur, Andhra Pradesh, India), contributing to environmental protection and conservation.

ACKNOWLEDGMENTS

Thanks are due to Dr. Kurt Walter Viikki Tropical Resources Institute (VITRI), University of Helsinki, Finland for useful discussion and constructive suggestions. The authors are thankful to Director, Central Arid Zone Research Institute, Jodhpur, for providing valuable information on Value Chain on Value Added Products Derived from *Prosopis juliflora*.

REFERENCES

Abdelrhman, H.A., Paridah, M.T., Shahwahid, M., Samad, A.R.A., Abdalla, A.M.A., 2015. The effects of pretreatments, wood-cement ratios and partial cement substitution by gypsum on Prosopis chilensis wood composites. European Journal of Wood and Wood Products 73 (4), 557–559. http://dx.doi.org/10.1007/s00107-015-0909-x.

Ahmad, V.U., Azra, S., Qazi, S., 1989. Alkaloids from the leaves of *Prosopis juliflora*. J. Nat. Prod. 52 (3), 497–507.

Al Shakh Hamed, W.M.A., Al Jammas, M.A., 1999. The antimicrobial activity of alkaloidal fraction of *Prosopis juliflora*. Iraqi J. Vet. Sci. 12 (2), 281–287.

Al-Humaid, A.I., Warrag, M.O.A., 1998. Allelopathic effects of mesquite (*Prosopis juliflora*) foliage on seed germination and seedling growth of bermudagrass (*Cynodon dactylon*). J. Arid Environ. 38, 237–243.

Almaraz-Abarca, N., Campos, M.D.G., Ávila-Reyes, J.A., Naranjo-Jiménez, N., Corral, J.H., González-Valdez, L.S., 2007. Antioxidant activity of polyphenolic extract of monofloral honeybee-collected pollen from mesquite (*Prosopis juliflora*, Leguminosae). J. Food Compos. Anal. 20, 119–124.

Aqeel, A., Ahmad, K.A., Khan, S., Qazi, V.U., 1988. Antibacterial activity of an alkaloidal fraction of *Prosopis julflora*. Fitoterapia 59 (6), 481–484.

Araüjo, A.H., Cardoso, P.C.B., Pereira, R.A., Lima, L.M., Oliveira, A.S., Miranda, M.R.A., Xavier-Filho, J., Sales, M.P., 2002. *In vitro* digestibility of globulins from cowpea (*Vigna unguiculata*) and xerophitic algaroba (*Prosopis juliflora*) seeds by mammalian digestive proteinases: a comparative study. Food Chem. 78, 143–147.

Arias, J.A., Peralta-Videa, J.R., Ellzey, J.T., Viveros, M.N., Ren, M., Mokgalaka-Matlala, N.S., Castillo-Michel, H., Gardea-Torresdey, J.L., 2010. Plant growth and metal distribution in tissues of *Prosopis juliflora*-velutina grown on chromium contaminated soil in the presence of *Glomus deserticola*. Environ. Sci. Technol. 44 (19), 7272–7279. http://dx.doi.org/10.1021/es1008664.

Azero, E.G., Andrade, C.T., 2002. Testing procedures for galactomannan purification. Polymer Testing 21, 551–556.

Batista, A.M., Mustafa, A.F., McKinnon, J.J., Kermasha, S., 2002. *In situ* ruminal and intestinal nutrient digestibilities of mesquite (*Prosopis juliflora*) pods. Anim. Feed Sci. Technol. 100, 107–112.

Beramendi-Orosco, L.E., Rodriguez-Estrada, M.L., Morton-Bermea, O., Romero, F.M., Gonzalez-Hernandez, G., Hernandez-Alvarez, E., 2013. Correlations between metals in tree-rings of *Prosopis juliflora* as indicators of sources of heavy metal contamination. Appl. Geochem. 39, 78–84. http://dx.doi.org/10.1016/j.apgeochem.2013.10.003.

Beristain, C.I., García, H.S., Vernon-Carter, E.J., 2001. Spray-dried encapsulation of cardamom (*Elettaria cardamomum*) essential oil with mesquite (*Prosopis juliflora*) gum. Lebensm. Wiss. Technol. 34, 398–401.

Bhatia, H., Gupta, P.K., Soni, P.L., 2014. Structure of the oligosaccharides isolated *from Prosopis juliflora* (Sw.) DC. seed polysaccharide. Carbohydr. Polym. 101, 438–443. http://dx.doi.org/10.1016/j.carbpol.2013.09.039.

Bhoi, P.R., Singh, R.N., Sharma, A.M., Patel, S.R., 2006. Performance evaluation of open core gasifier on multi-fuels. Biomass Bioenergy 30, 575–579.

Bhojvaid, P.P., Timmer, V.R., 1998. Soil dynamics in an age sequence of *Prosopis juliflora* planted for sodic soil restoration in India. For. Ecol. Manag. 106, 181–193.

Bhuvaneswari, S., Vitta, B., Prakash, U.N., Janaki, T., Raju, V., 2004. Aerobiology and allergenicity of mesquite pollen in Chennai, India. J. Allergy Clin. Immunol. 113, S223.

Bragg, L.H., Bacon, J.D., McMillan, C., Mabry, T.J., 1978. Flavonoid patterns in the *Prosopis juliflora* complex. Biochem. Syst. Ecol. 6, 113–116.

Bu-Olayan, A.H., Thomas, B.V., 2002. Biomonitoring studies on the lead levels in mesquite (*Prosopis juliflora*) in the arid ecosystem of Kuwait. Kuwait J. Sci. Eng. 29 (1), 65–73.

Choge, S.K., Pasiecznik, N.M., Harvey, M., Wright, J., Awan, S.Z., Harris, P.J.C., 2007. *Prosopis* pods as human food, with special reference to Kenya. Water SA 33, (Special Edition).

Choudhary, M.I., Nawaz, S.A., Zaheer-ul-Haq, Azim, M.K., Ghayur, M.N., Lodhi, M.A., Jali, S., Khalid, A., Ahmed, A., Rode, B.M., Atta-ur-Rahman, Gilani, A.H., Ahmad, V.U., 2005. Juliflorine: a potent natural peripheral anionic-site-binding inhibitor of acetylcholinesterase with calcium-channel blocking potential, a leading candidate for Alzheimer's disease therapy. Biochem. Biophys. Res. Commun. 332, 1171–1179.

Churms, S.C., Merrifield, E.H., Stephen, A.M., 1981. Smith degradation of gum exudates from some *Prosopis* species. Carbohydr. Res. 90 (I), 261–267.

Csurhes, S. (Ed.), 1996. Mesquite (*Prosopis* spp.) in Queensland. Pest Status Review Series. Department of Natural Resources and Mines, Land Protection Branch, Queensland, Australia.

Da Silva, C.G., Stamford, T.L., de Andrade, S.A., de Souza, E., Araújo, J., 2010. Production of ethanol from mesquite (*Prosopis juliflora* (SW) D.C.) pods mash by *Zymomonas mobilis* and *Saccharomyces cerevisiae*. Electron. J. Biotechnol. 13. http://dx.doi.org/10.2225/vol13-issue5-fulltext-21. ISSN: 0717-3458. http://www.ejbiotechnology.info.

Daas, P.J.H., Schols, H.A., de Jongh, H.H., 2000. On the galactosyl distribution of commercial galactomannans. Carbohydr. Res. 329, 609–619.

De La Rosa, G., Lopez-Moreno, M.L., Hernandez-Viezcas, J.A., Montes, M.O., Peralta-Videa, J.R., Gardea-Torresdey, J.L., 2011. Toxicity and biotransformation of ZnO nanoparticles in the desert plants *Prosopis juliflora-velutina*, *Salsola tragus* and *Parkinsonia florida*. Int. J. Nanotechnol. 8, 492–506.

Deans, J.D., Diagne, O., Nizinski, J., Lindley, D.K., Seck, M., Ingleby, K., Munro, R.C., 2003. Comparative growth, biomass production, nutrient use and soil amelioration by nitrogen-fixing tree species in semi-arid Senegal. For. Ecol. Manag. 176, 253–264.

Del Valle, F.R., Escobedo, M., Munoz, M.J., Ortega, R., Bourges, H., 1983. Chemical and nutritional studies on mesquite beans (*Prosopis juliflora*). J. Food Sci. 48, 914–919.

Deng, H., Ye, Z.H., Wong, M.H., 2004. Accumulation of lead, zinc, copper and cadmium by 12 wetland plants species thriving in metal contaminated sites in China. Environ. Pollut. 132 (1), 29–40.

Dhillon, K.S., Varshney, K.R., 2013. Study of microbiological spectrum in acne vulgaris: an in vitro study. Sch. J. App. Med. Sci. 1 (6), 724–727.

Diaz, A., Rincon, N., Escoriuela, A., Fernandez, N., Chacin, E., Forster, C.F., 1999. A preliminary evaluation of turbidity removal by natural coagulants indigenous to Venezuela. Process Biochemistry 35, 391–395.

Dhyani, A., Arora, N., Gaur, S.N., Jain, V.K., Sridhara, S., Singh, B.P., 2006. Analysis of IgE binding proteins of mesquite (*Prosopis juliflora*) pollen and cross-reactivity with predominant tree pollens. Immunobiology 211, 733–740.

Elfadl, M.A., Luukkanen, O., 2003. Effect of pruning on *Prosopis juliflora*: considerations for tropical dryland agroforestry. J. Arid Environ. 53, 441–455.

Elfadl, M.A., Luukkanen, O., 2006. Field studies on the ecological strategies of *Prosopis juliflora* in a dryland ecosystem: 1. A leaf gas exchange approach. J. Arid Environ. 66, 1–15.

El-Keblawy, A., Al-Rawai, A., 2005. Effects of salinity, temperature and light on germination of invasive *Prosopis juliflora* (Sw.) D.C. J. Arid Environ. 61, 555–565.

El-Keblawy, A., Al-Rawai, A., 2006. Effects of seed maturation time and dry storage on light and temperature requirements during germination in invasive *Prosopis juliflora*. Flora – Morphol. Distribution Funct. Ecol. Plants 201, 135–143.

ElSiddig, E.A., Abdelsalam, A.A., Abdel Magid, T.D., 1998. Socio-Economic, Environmental and Management Aspects of Mesquite in Kassala State-Sudan. Sudanese Social Forestry Society (SSFS), Sudan 80 p.

Ewens, M., Felker, P., 2003. The potential of mini-grafting for large-scale production of Prosopis alba clones. J. Arid Environ. 55, 379–387.

Felker, P., 1998. The value of mesquite for the rural southwest. J. For. 96, 16–21.

Felker, P., Anderson, P., 1997. Grading Mesquite Lumber. Pamphlet Centre for Semi-Arid Forest Resources, Kingsville, TX 7 p.

Felker, P., Bandurski, R.S., 1979. Uses and potential uses of leguminous trees for minimal energy input agriculture. Econ. Bot. 33 (2), 172–184.

Felker, P., Patch, N., 1996. Managing coppice, sapling, and mature *Prosopis* for firewood, poles, and lumber. In: Felker, P., Moss, J. (Eds.), *Prosopis*: Semiarid Fuelwood and Forage Tree, Building Consensus for the Disenfranchised. Centre for Semi-Arid Forest Resources, Kingsville, TX.

Goel, V.L., Behl, H.M., 1995. Fuelwood production potential of six *Prosopis* species on an alkaline soil site. Biomass Bioenergy 8, 17–20.

Goel, V.L., Behl, H.M., 1996. Fuelwood quality of promising tree species for alkaline soil sites in relation to tree age. Biomass Bioenergy 10, 57–61.

Goel, V.L., Behl, H.M., 2001. Genetic selection and improvement of hard wood tree species for fuelwood production on sodic soil with particular reference to *Prosopis juliflora*. Biomass Bioenergy 20, 9–15.

Gopal, N., Asaithambi, M., Sivakumar, P., Sivakumar, V., 2014. Adsorption studies of a direct dye using polyaniline coated activated carbon prepared from *Prosopis juliflora*. J. Water Process Eng. 2, 87–95. http://dx.doi.org/10.1016/j.jwpe.2014.05.008.

Gratão, P.L., Prasad, M.N.V., Lea, P.J., Azevedo, R.A., 2005. Phytoremediation: green technology for the clean up of toxic metals in the environment. Braz. J. Plant Physiol. 17, 53–64.

Harris, P.J.C., Pasiecznik, N.M., Smith, S.J., Billington, J.M., Ramírez, L., 2003. Differentiation of *Prosopis juliflora* (Sw.) DC. and *P. pallida* (H. & B. ex. Willd.) H.B.K. using foliar characters and ploidy. For. Ecol. Manag. 180, 153–164.

Harsh, L.N., Tewari, J.C., 1998. *Prosopis* in the arid regions of India: some important aspects of research and development. In: Tewari, J.C., Pasiecznik, N.M., Harsh, L.N., Harris, P.J.C. (Eds.), *Prosopis* Species in the Arid and Semi-Arid Zones of India. Prosopis Society of India and the Henry Doubleday Research Association, Coventry, UK.

Hebbar, S.S., Harsha, V.H., Shripathi, V., Hegde, G.R., 2004. Ethnomedicine of Dharwad district in Karnataka, India – plants used in oral health care. J. Ethnopharmacol. 94, 261–266.

Hernandez-Viezcas, J.A., Castillo-Michel, H., Servin, A.D., Peralta-Videa, J.R., Gardea-Torresdey, J.L., 2011. Spectroscopic verification of zinc absorption and distribution in the desert plant *Prosopis juliflora-velutina* (velvet mesquite) treated with ZnO nanoparticles. Chem. Eng. J. 170 (1–3), 346–352. http://dx.doi.org/10.1016/j.cej.2010.12.021.

Herrera-Arreola, G., Herrera, Y., Reyes-Reyes, B.G., Dendooven, L., 2007. Mesquite (*Prosopis juliflora* (Sw.) DC.), huisache (*Acacia farnesiana* (L.) Willd.) and catclaw (Mimosa biuncifera Benth.) and their effect on dynamics of carbon and nitrogen in soils of the semi-arid highlands of Durango Mexico. J. Arid Environ. 69, 583–598.

Hiroshi, N., Eri, N., Syuntaro, H., Yoshiharu, F., Kosumi, Y., Hideyuki, S., Koji, H., 2004. Growth inhibitory alkaloids from mesquite (*Prosopis juliflora*, Swartz DC.) leaves. Phytochemistry 65 (5), 587–591.

Hussain, J., Khan, F., Ullah, R., Muhammad, Z., Najeeb-ur-Rehman, Shinwari, Z.K., Khan, I.U., Zohaib, M., Imad-ud-Din, Hussain, S.M., 2011. Nutrient evaluation and elemental analysis of four selected medicinal plants of Khyber Pakhtunkhwa, Pakistan. Pak. J. Bot. 43 (1), 427–434.

Ibrahim, K.M., 1988. Shrubs for fodder production. In: Habit, M., Saavedra, J.C. (Eds.), The Current State of Knowledge on *Prosopis juliflora*. II International Conference on Prosopis, Recife, Brazil, 25–29 August 1986. Food and Agriculture Organization of the United Nations (FAO), Rome, Italy.

Ibrahim, M., Nadir, M., Amanat, A., Ahmad, V.U., Rasheed, M., 2013. Phytochemical analyses of *Prosopis juliflora* swartz DC. Pak. J. Bot. 45 (6), 2101–2104.

Ingleby, K., Diagne, O., Deans, J.D., Lindley, D.K., Neyra, M., Ducousso, M., 1997. Distribution of roots, arbuscular mycorrhizal colonisation and spores around fast-growing tree species in Senegal. For. Ecol. Manag. 90, 19–27.

Jayaram, K., Prasad, M.N.V., 2009. Removal of Pb(II) from aqueous solution by seed powder of *Prosopis juliflora* (Sw.) DC. J. Hazard. Mater. 169, 991–997.

Kailappan, R., Gothandapani, L., Viswanathan, R., 2000. Production of activated carbon from *Prosopis* (*Prosopis juliflora*). Bioresour. Technol. 75, 241–243.

Kanthasamy, A., Subramanian, S., Govindasam, S., 1989. Bactericidal and fungicidal effects of *Prosopis juliflora* alkaloidal fraction. Indian Drugs 26, 390–394.

Kapoor, M., Nair, L.M., Kuhad, R.C., 2008. Cost-effective xylanase production from free and immobilized *Bacillus pumilus* strain MK001 and its application in saccharification of *Prosopis juliflora*. Biochem. Eng. J. 38, 88–97.

Kaur, R., Callaway, R.M., Inderjit, 2014. Soils and the conditional allelopathic effects of a tropical invader. Soil Biol. Biochem. 78, 316–325. http://dx.doi.org/10.1016/j.soilbio.2014.08.017.

Kumar, M., Tamilarasan, R., 2013a. Modeling studies: adsorption of aniline blue by using *Prosopis juliflora* carbon/Ca/alginate polymer composite beads. Carbohydr. Polym. 92, 2171–2180. http://dx.doi.org/10.1016/j.carbpol.2012.11.076.

Kumar, M., Tamilarasan, R., 2013b. Kinetics, equilibrium data and modeling studies for the sorption of chromium by *Prosopis juliflora* bark carbon. Arab. J. Chem.Available online 9 June 2013, http://dx.doi.org/10.1016/j.arabjc.2013.05.025.

Kumar, M., Tamilarasan, R., 2014. Removal of Victoria Blue using *Prosopis juliflora* bark carbon: kinetics and thermodynamic modeling studies. J. Mater. Environ. Sci. 5, 510–519.

Lee, S.G., Russell, E.J., Bingham, R.L., Felker, P., 1992. Discovery of thornless, non-browsed, erect tropical *Prosopis* in 3-year-old Haitian progeny trials. For. Ecol. Manag. 48, 1–13.

López-Franco, Y.L., Córdova-Moreno, R.R., Goycoolea, F.M., Valdez, M.A., Juárez-Onofre, J., Lizardi-Mendoza, J., 2012. Classification and physicochemical characterization of mesquite gum (*Prosopis* spp.). Food Hydrocoll. 26, 159–166 10.1016/j.foodhyd.2011.05.006.

López-Franco, Y.L., Cervantes-Montaño, C.I., Martínez-Robinson, K.G., Lizardi-Mendoza, J., Robles-Ozuna, L.E., 2014. Physicochemical characterization and functional properties of galactomannans from mesquite seeds (*Prosopis* spp.). Food Hydrocoll. 30, 656–660. http://dx.doi.org/10.1016/j.foodhyd.2012.08.012.

Mahgoub, O., Kadim, I.T., Forsberg, N.E., Al-Ajmi, D.S., Al-Saqry, N.M., Al-Abri, A.S., Annamalai, K., 2005a. Evaluation of Meskit (*Prosopis juliflora*) pods as a feed for goats. Anim. Feed Sci. Technol. 121, 319–327.

Mahgoub, O., Kadim, I.T., Johnson, E.H., Srikandakumar, A., Al-Saqri, N.M., Al-Abri AS Ritchie, A., 2005b. The use of a concentrate containing Meskit (*Prosopis juliflora*) pods and date palm by-products to replace commercial concentrate in diets of Omani sheep. Anim. Feed Sci. Technol. 120, 33–41.

Mahgoub, O., Kadim, I.T., Forsberg, N.E., Al-Ajmi, D.S., Al-Saqry, N.M., Al-Abri, A.S., Annamalai, K., 2005c. Evaluation of meskit (*Prosopis juliflora*) pods as feed goats. Anim. Feed Sci. Technol. 121 (3–4), 319–327.

Mahgoub, O., Kadim, I.T., Al-Busaidi, M.H., Annamalai, K., Al-Saqri, N.M., 2007. Effects of feeding ensiled date palm fronds and a by-product concentrate on performance and meat quality of Omani sheep. Anim. Feed Sci. Technol. 135, 210–221.

Mahmood, K., Morris, J., Collopy, J., Slavich, P., 2001. Groundwater uptake and sustainability of farm plantations on saline sites in Punjab province, Pakistan. Agric. Water Manag. 48, 1–20.

Maioli, M.A., Lemos, D.E.C.V., Guelfi, M., Medeiros, H.C.D., Riet-Correa, F., Medeiros, R.M.T., Barbosa-Filho, J.M., Mingatto, F.E., 2012. Mechanism for the uncoupling of oxidative phosphorylation by juliprosopine on rat brain mitochondria. Toxicon 60 (8), 1355–1362. http://dx.doi.org/10.1016/j.toxicon.2012.09.012.

Makkar, H.P.S., Singh, B., Negi, S.S., 1990. Tannin levels and their degree of polymerisation and specific activity in some agro-industrial by-products. Biol. Waste 31, 137–144.

Malhotra, S., Misra, K., 1981a. An ellagic acid glycoside from the pods of *Prosopis juliflora*. Phytochemistry 20, 860–861.

Malhotra, S., Misra, K., 1981b. 3,3′-di-O-Methylellagic acid 4-O-rhamnoside from the roots of *Prosopis juliflora*. Phytochemistry 20, 2043–2044.

Malhotra, S., Misra, K., 1981c. Ellagic acid 4-O-rutinoside from pods of *Prosopis juliflora*. Phytochemistry 20, 2439–2440.

Maliwal, G.L., 1999. Reclamation of saline and sodic soils through *Prosopis juliflora*. Indian J. For. 22, 132–135.

Munro, R.C., Wilson, J., Jefwa, J., Mbuthia, K.W., 1999. A low-cost method of mycorrhizal inoculation improves growth of *Acacia tortilis* seedlings in the nursery. For. Ecol. Manag. 113, 51–56.

Muthana, K.D., Arora, G.D., 1983. *Prosopis juliflora* (SW) DC, a fast-growing tree to blossom the dessert. In: Habit, M.A., Saavedra, J.C. (Eds.), The Current State Knowledge on *Prosopis juliflora*. FAO, Rome, Italy, pp. 133–144.

Muturi, G.M., Mohren, G.M.J., Kinani, J.N., 2009. Prediction of *Prosopis* species invasion in Kenya using geographical information system techniques. Afr. J. Ecol. 48, 628–636.

Mwangi, E., Swallow, B., 2008. *Prosopis juliflora* invasion and rural livelihoods in the Lake Baringo Area of Kenya. Conserv. Soc. 6 (2), 130–140.

Nagaraju, A., Prasad, K.S.S., 1998. Growth of *Prosopis juliflora* on pegmatite tailings from Nellore Mica Belt, Andhra Pradesh, India. Environ. Geol. 36, 320–324.

Nakano, H., Nakajima, E., Hiradate, S., Fujii, Y., Yamada, K., Shigemori, H., Hasegawa, K., 2004. Growth inhibitory alkaloids from mesquite (*Prosopis juliflora* (Sw.) DC.) leaves. Phytochemistry 65, 587–591.

Naseeruddin, S., Yadav, K.S., Sateesh, L., Manikyam, A., Desai, S.L., Rao, V., 2014. Selection of the best chemical pretreatment for lignocellulosic substrate *Prosopis juliflora*. Bioresour. Technol. 136, 542–549. http://dx.doi.org/10.1016/j.biortech.2013.03.053.

Noor, M., Salam, U., Khan, A.M., 1995. Allelopathic effects of *Prosopis juliflora* Swartz. J. Arid Environ. 31, 83–90.

Obeidat, B.S., Shdaifat, M.M., 2013. Partial substitution of barley grain with *Prosopis juliflora* pods in lactating Awassi ewes' diets: effect on intake, digestibility, and nursing performance. Small Rumin. Res. 111, 50–55. http://dx.doi.org/10.1016/j.smallrumres.2012.09.013.

Oliveira, A.S., Pereira, R.A., Lima, L.M., Morais, A.H.A., Melo, F.R., Franco, O.L., Bloch, C., Grossi-de-Sá, M.F., Sales, M.P., 2002. Activity toward bruchid pest of a Kunitz-type inhibitor from seeds of the algaroba tree (*Prosopis juliflora* D.C.). Pestic. Biochem. Physiol. 72, 122–132.

Osmond, R., 2003. Mesquite: Control and Management Options for Mesquite (*Prosopis* spp.) in Australia. The State of Queensland (Department of Natural Resources and Mines), Queensland, Australia 29 p.

Pancholy, R., Mali, P.C., 1999. Effective utilization of *Prosopis juliflora* pods by ensiling with dessert grass *Lasiurus sindicus*. Bioresour. Technol. 69 (3), 281–283.

Pasiecznik, N.M., Vera-Cruz, M.T., Harris, P.J.C., 1995. *Prosopis juliflora* withstands aridity and goat browsing in the Republic of Cap Verde. Nitrogen Fixing Tree Res. Reports 13, 89–91.

Pasiecznik, N.M., Felker, P., Harris, P.J.C., Harsh, L.N., Cruz, G., Tewari, J.C., Cadoret, K., Maldonado, L.J., 2001. The *Prosopips juliflora–Prosopis pallida* Complex: A Monograph. HDRA, Coventry, UK 162 p.

Pasiecznik, N.M., Harris, P.J.C., Trenchard, E.J., Smith, S.J., 2006a. The dilemma of *Prosopis*: is there a rose between the thorns? Biocontrol News Info. 27 (1), 1N–26N.

Pasiecznik, N.M., Choge, S.K., Muthike, G.M., Chesang, S., Fehr, C., Bakewell-Stone, P., Wright, J., Harris, P.J.C., 2006b. Putting Knowledge on *Prosopis* into Use in Kenya. Pioneering Advances in 2006. KEFRI/HDRA, Nairobi/Coventry 13 p.

Patch, N.L., Geesing, D., Felker, P., 1998. Suppressing of resprouting in pruned mesquite (*Prosopis glandulosa* var *glandulosa*) saplings with chemical or physical barrier treatments. For. Ecol. Manag. 112, 23–29.

Perera, A.N.F., Pasiecznik, N.M., 2005. Using Invasive *Prosopis* to Improve Livelihoods in Sri Lanka. HDRA, Coventry, UK.

Pires, I.E., Andrade, G.C., Aroújo, M.S., 1990. Genetic variation for growth characteristic in *P. juliflora* progenies. In: Habit, M.A., Saavedra, J.C. (Eds.), The Current State of Knowledge on *Prosopis juliflora*.

Prabha, D.S., Dahms, H.-U., Malliga, P., 2014. Pharmacological potentials of phenolic compounds from *Prosopis* spp.-a review. J. Coast. Life Med. 2 (11), 918–924.

Prasad, M.N.V., 2004. Phytoremediation of metals in the environment for sustainable development. Proc. Indian Natl. Sci. Acad. 70, 71–98.

Prasad, M.N.V., 2007. Phytoremediation in India. In: Willey, N. (Ed.), Methods in Biotechnology. Phytoremediation: Methods and Reviews, vol. 23. Humana Press, Totowa, USA, pp. 433–452.

Pulford, I.D., Dickinson, N.M., 2006. Phytoremediation technologies using trees. In: Prasad, M.N.V., Sajwan, K.S., Naidu, R. (Eds.), Trace Elements in the Environment. CRC Press, Boca Raton, London 726 p.

Rai, U.N., Pandey, K., Sinha, S., Singh, A., Saxena, R., Gupta, D.K., 2004. Revegetating fly ash landfills with *Prosopis juliflora* L.: impact of different amendments and Rhizobium inoculation. Environ. Int. 30, 293–300.

Raja, K., Saravanakumar, A., Vijayakumar, R., 2012. Efficient synthesis of silver nanoparticles from *Prosopis juliflora* leaf extract and its antimicrobial activity using sewage. Spectrochim. Acta A Mol. Biomol. Spectrosc. 97, 490–494. http://dx.doi.org/10.1016/j.saa.2012.06.038.

Rajadurai Jesudoss, R.P., Lakshmipraba, S., Gnanasaraswathi, M., Ganesh Kumar, S., Praveen Kumar, T.G., 2014. Screening of anti-pustule plant metabolites from *Prosopis juliflora* and their combined antipustule activity with synthetic pimple creams. In: National Conference on Plant Metabolomics (Phytodrugs – 2014). J. Chem. Pharm. Sci., 0974-2115.

Ram, L.C., Masto, R.E., 2014. Fly ash for soil amelioration: A review on the influence of ash blending with inorganic and organic amendments. Earth Science Review 128, 52–74.

Rao, A.V., Tarafdar, J.C., 1998. Selection of plant species for rehabilitation of gypsum mine spoil in arid zone. J. Arid Environ. 39, 559–567.

Ravikala, K., Patel, A.M., Murthy, K.S., Wadhwani, K.N., 1995. Growth efficiency in feedlot lambs on *Prosopis juliflora* based diets. Small Rumin. Res. 16, 227–231.

Ravikumar, S., Inbaneson, S.J., Suganthi, P., 2012. In vitro antiplasmodial activity of ethanolic extracts of South Indian medicinal plants against *Plasmodium falciparum*. Asian Pac. J. Trop. Dis. 2, 180–183. http://dx.doi.org/10.1016/S2222-1808(12)60043-7.

Rawat, M.S., Uniyal, D.P., Vakshasya, R.K., 1992. *Prosopis juliflora* (Schwartz) DC: fuel, fodder and food in arid and semi-arid areas: some observations and suggestions. Indian J. For. 15 (2), 164–168.

Reddy, C.V.K., 1978. *Prosopis juliflora*, the precocious child of the plant world. Indian Forester 104, 14–18.

Rhodes, D., Felker, P., 1988. Mass screening of *Prosopis*(mesquite) seedlings for growth at seawater salinity concentrations. For. Ecol. Manag. 24, 169–176.

Rincón, F., Muñoz, J., Ramírez, P., Galán, U., Alfaro, M.C., 2014. Physicochemical and rheological characterization of *Prosopis juliflora* seed gum aqueous dispersions. Food Hydrocoll. 35, 348–357. http://dx.doi.org/10.1016/j.foodhyd.2013.06.013.

Riveros, F., 1992. The genus *Prosopis* and its potential to improve livestock production in arid and semi-arid regions. In: Speedy, A., Pugliese, P. (Eds.), Legume Trees and Other Fodder Trees as Protein Sources for Livestock. FAO, Rome, pp. 257–276 Animal Production and Health, Paper 102.

Saini, P., Khan, S., Baunthiyal, M., Sharma, V., 2012. Organ-wise accumulation of fluoride in *Prosopis juliflora* and its potential for phytoremediation of fluoride contaminated soil. Chemosphere 89, 633–635. http://dx.doi.org/10.1016/j.chemosphere.2012.05.034.

Saravanakumar, S.S., Kumaravel, A., Nagarajan, T., Sudhakar, P., Baskaran, R., 2013. Characterization of a novel natural cellulosic fiber from *Prosopis juliflora* bark. Carbohydr. Polym. 92, 1928–1933. http://dx.doi.org/10.1016/j.carbpol.2012.11.064.

Sathiya, M., Muthuchelian, K., 2008. Investigation of phytochemical profile and antibacterial potential of ethanolic leaf extract of *Prosopis juliflora* DC. Ethnobot. Leaflets 12 (15), 1240–1245.

Sato, T., 2013. Beyond water-intensive agriculture: expansion of *Prosopis juliflora* and its growing economic use in Tamil Nadu, India. Land Use Policy 35, 283–292. http://dx.doi.org/10.1016/j.landusepol.2013.06.001.

Senthilkumar, P., Prince, W.S.P.M., Sivakumar, S., Subbhuraam, C.V., 2005. *Prosopis juliflora,* a green solution to decontaminate heavy metal (Cu and Cd) contaminated soils. Chemosphere 60, 1493–1496.

Shukla, R.V., Misra, K., 1981. Two flavonoid glycosides from the bark of *Prosopis juliflora*. Phytochemistry 20 (I), 339–340.

Shukla, O.P., Juwarkar, A.A., Singh, S.K., Shoeb Khan, S., Rai, U.N., 2011. Growth responses and metal accumulation capabilities of woody plants during the phytoremediation of tannery sludge. Waste Manag. 31, 115–123. http://dx.doi.org/10.1016/j.wasman.2010.08.022.

Siddhu, O.P., Behl, H.M., 1997. Response of three glomus species on growth of *Prosopis juliflora* Swartz at high pH levels. Symbiosis 23, 23–34.

Silva, S., 1990. *Prosopis juliflora* (Sw) DC in Brazil. In: Habit, M.A., Saavedra, J.C. (Eds.), The Current State of Knowledge on *Prosopis juliflora*. FAO, Rome, Italy.

Silva, A.M.M., Silva, A.R., Pinheiro, A.M., Freitas, S.R., Silva, V.D., Souza, C.S., Hughes, J.B., El-Bachá, R.S., Costa, M.F.D., Velozo, E.S., Tardy, M., Costa, S.L., 2007. Alkaloids from *Prosopis juliflora* leaves induce glial activation, cytotoxicity and stimulate NO production. Toxicon 49, 601–614.

Singh, G., 1995. An agroforestry practice for the development of salt lands using *Prosopis juliflora* and *Leptochloa fusca*. Agrofor. Syst. 29, 61–75.

Singh, G., 1996a. Effect of site preparation techniques on *Prosopis juliflora* in an alkali soil. For. Ecol. Manag. 80, 267–278.

Singh, G., 1996b. The role of *Prosopis* in reclaiming high-pH soils and in meeting firewood and forage needs of small farmers. In: Felker, P., Moss, J. (Eds.), *Prosopis*: Semiarid Fuelwood and Forage Tree; Building Consensus for the Disenfranchised. Centre for Semi-arid Forest Resources, Kingsville, TX.

Singh, D.K., 2009. Rationale for prescribing the requisite forest/tree covers in India. Academy of Forest and Environmental Sciences (AFES). Under contract of: Indian Council of Forest Research & Education (ICFRE). http://www.icfre.org/UserFiles/File/Education/Forest-&-Tree-Cover-Rationale-140809.pdf.

Singh, G., Rathod, T.R., 2002. Plant growth, biomass production and soil water dynamics in a shifting dune of Indian desert. For. Ecol. Manag. 171, 309–320.

Singh, G., Abrol, I.P., Cheema, S.S., 1989. Effects of gympsum application on mesquite (*Prosopis juliflora*) and soil properties in an abandoned sodic soil. For. Ecol. Manag. 29, 1–14.

Singh, G., Gill, H.S., Abrol, I.P., Cheema, S.S., 1991. Forage yield, mineral composition, nutrient cycling and ameliorating effects of Karnal grass (*Leptochloa fusca*) grown with mesquite (*Prosopis juliflora*) in a highly alkaline soil. Field Crop Res. 26, 45–55.

Singh, S., Swapnil, Verma, S.K., 2011. Antibacterial properties of alkaloid rich fractions obtained from various parts of *Prosopis juliflora*. Int. J. Pharm. Sci. Res. 2 (3), 114–120.

Singh, K., Pandey, V.C., Singh, B., Singh, R.R., 2012. Ecological restoration of degraded sodic lands through afforestation and cropping. Ecol. Eng. 43, 70–80. http://dx.doi.org/10.1016/j.ecoleng.2012.02.029.

Sinha, A.K., Pathre, U., Sane, P.V., 1997. Purification and characterization of sucrose-phosphate synthase from *Prosopis juliflora*. Phytochemistry 46, 441–447.

Snapp, S.S., Mafongoya, P.L., Waddington, S., 1998. Organic matter technologies for integrated nutrient management in smallholder cropping systems of southern Africa. Agric. Ecosyst. Environ. 71, 185–200.

Solís-Domínguez, F.A., Valentín-Vargas, A., Chorover, J., Maier, R.M., 2011. Effect of arbuscular mycorrhizal fungi on plant biomass and the rhizosphere microbial community structure of mesquite grown in acidic lead/zinc mine tailings. Sci. Total Environ. 409, 1009–1016. http://dx.doi.org/10.1016/j.scitotenv.2010.11.020.

Tabosa, I.M., Souza, J.C., Graça, D.L., Barbosa Filho, J.M., Almeida, R.N., Riet-Correa, F., 2000. Neuronal vacuolation of the trigeminal nuclei in goats caused by ingestion of *Prosopis juliflora* pods (mesquite beans). Vet. Hum. Toxicol. 42 (3), 155–158.

Tarafdar, J.C., Kumar, P., 1996. The role of vesicular arbuscular mycorrhizal fungi on crop, tree and grasses grown in an arid environment. J. Arid Environ. 34 (2), 197–203.

Tene, V., Malagón, O., Finzi, P.V., Vidari, G., Armijos, C., Zaragoza, T., 2007. An ethnobotanical survey of medicinal plants used in Loja and Zamora-Chinchipe, Ecuador. J. Ethnopharmacol. 111, 63–81.

Tewari, P., 1998. *Prosopis cineraria*: pods in the human diet. In: Tewari, J.C., Pasiecznik, N.M., Harsh, L.N., Harris, P.J.C. (Eds.), *Prosopis* Species in the Arid and Semi-Arid Zones of India. Prosopis Society of India and HDRA, Coventry, UK.

Tewari, J.C., Harris, P.J.C., Harsh, L.N., Cadoret, K., Pasiecznik, N.M., 2000. Managing *Prosopis juliflora* (Vilayati babul): A Technical Manual. CAZRI (Central Arid Zone Research Institute)/HDRA, Jodhpur, India/Coventry, UK.

Thomas, S.C., Coltman, D.W., Pemberton, M., 2002. The use of marker based relationship information to estimate the heritability of body weight in a natural population: a cautionary tale. J. Evol. Biol. 15, 92–99.

Tomar, O.S., Minhas, P.S., Sharma, V.K., Singh, Y.P., 2003. Performance of 31 tree species and soil conditions in a plantation established with saline irrigation. For. Ecol. Manag. 177, 333–346.

Vallejo, V.E., Arbeli, Z., Terán, W., Lorenz, N., Dick, R.P., Roldan, F., 2012. Effect of land management and *Prosopis juliflora* (Sw.) DC trees on soil microbial community and enzymatic activities in intensive silvopastoral systems of Colombia. Agric. Ecosyst. Environ. 150, 139–148.

Varshney, A., 1996. Overview of the use of *Prosopis juliflora* for livestock feed, gum, honey, and charcoal, as well as in combating drought and desertification: a regional case study from Gujarat, India. In: Felker, P., Moss, J. (Eds.), Prosopis: Semiarid Fuelwood and Forage Tree; Building Consensus for the Disenfranchised. Center for Semi-Arid Forest Resources, Kingsville, Texas, USA, pp. 6.35–6.4.

Varun, M., D'Souza, R., Pratas, J., Paul, M.S., 2011. Phytoextraction potential of *Prosopis juliflora* (Sw.) DC. with specific reference to lead and cadmium. Bull. Environ. Contam. Toxicol. 87 (1), 45–49. http://dx.doi.org/10.1007/s00128-011-0305-0.

Velarde, M., Felker, P., Degano, C., 2003. Evaluation of Argentine and Peruvian *Prosopis* germplasm for growth at seawater salinities. J. Arid Environ. 55 (3), 515–531.

Velarde, M., Felker, P., Gardiner, D., 2005. Influence of elemental sulfur, micronutrients, phosphorus, calcium, magnesium and potassium on growth of *Prosopis alba* on high pH soils in Argentina. J. Arid Environ. 62, 525–539.

Viégas, R.A., Fausto, M.J.M., Queiroz Rocha, J.E.I.M.A.R., Silveira, J.A.G., Viégas Pedro, P.R.A., 2004. Growth and total-N content of *Prosopis juliflora* (Sw.) D. C. are stimulated by low NaCl levels. Braz. J. Plant Physiol. 16, 65–68.

Vieira, I.G.P., Mendes, F.N.P., Gallão, M.I., de Brito, E.S., 2007. NMR study of galactomannans from the seeds of mesquite tree (*Prosopis juliflora* (Sw) DC). Food Chem. 101, 70–73.

von Maydell, J.H., 1986. Trees and Shrubs of the Sahel. Their Characteristics and Uses. Deutsche Gesellschaft für Technische Zusammenarbeit (GTZ), Eschborn, Germany 523 p.

Walter, K.J., Armstrong, K.V., 2014. Benefits, threats and potential of *Prosopis* in South India. Forests Trees Livelihoods 23, 232–247. http://dx.doi.org/10.1080/14728028.2014.919880.

Warrag, M.O.A., 1994. Autotoxicity of mesquite (*Prosopis juliflora*) pericarps on seed germination and seedling growth. J. Arid Environ. 27, 79–84.

Warrag, M.O.A., 1995. Autotoxic potential of foliage on seed germination and early growth of mesquite (*Prosopis juliflora*). J. Arid Environ. 31, 415–421.

Wojtusik, T., Boyd, M.T., Felker, P., 1994. Effect of different media on vegetative propagation of *Prosopis juliflora* cuttings under solar-powered mist. For. Ecol. Manag. 67, 267–271.

Wunder, W.G., 1966. *Prosopis juliflora* (*P. africana*): in the arid zone of Sudan. Pamphlet of the Forest Research Education Project.

Zimmermann, H.G., 1991. Biological control of mesquite, *Prosopis* spp. (Fabaceae), in South Africa. Agric. Ecosyst. Environ. 37, 175–186.

GIANT REED (*ARUNDO DONAX* L.): A MULTIPURPOSE CROP BRIDGING PHYTOREMEDIATION WITH SUSTAINABLE BIOECONOMY

A.L. Fernando[1], B. Barbosa[1], J. Costa[1], E.G. Papazoglou[2]

Universidade Nova de Lisboa, Caparica, Portugal[1]
Agricultural University of Athens, Athens, Greece[2]

1 INTRODUCTION

The worldwide expansion of anthropogenic activities such as the use of municipal waste, pesticides, fertilizers, emissions from waste incinerators, car emissions, the activity of metallurgical and petrochemical industries, mining, and construction is making large-scale changes in natural environments. This is also changing the rate of release of inorganic compounds such as heavy metals into the ecosphere, contributing significantly to worldwide degradation, contamination and pollution of air, water systems, and soils, where ultimately these compounds tend to accumulate (Alloway, 1995; Fergusson, 1991; Garbisu and Alkorta, 2003; He et al., 2005; Kabata-Pendias and Pendias, 2011; Zhu and Shaw, 2000). These factors and activities, alone or together, reduce soil and water quality, posing an imminent threat to humans, animals, and ecosystem services. Therefore it is necessary to find solutions that promote decontamination and remediation, if possible in a cost-effective way.

Several types of technologies have been used in recent years to remediate soil contaminated with heavy metals. The most common techniques used are: (a) excavation and disposal; (b) immobilization; (c) toxicity reduction; (d) vitrification; (e) encapsulation and cover with clean soil; and (f) washing (Fernando, 2005; Mulligan et al., 2001; Wuana and Okieimen, 2011). For water decontamination, common options to remove heavy metals include alkaline precipitation, ion-exchange, electrochemical removal, filtration, and membrane technologies. These soil and water remediation approaches are in most cases very expensive and, in some cases, may lead to adverse effects on ecosystems, often requiring other methods for waste disposal (Nsanganwimana et al., 2014).

Phytoremediation, the use of plants and their associated microbes for soil, water, and air decontamination, is a cost-effective, solar-driven, and alternative or complementary technology for

Bioremediation and Bioeconomy. http://dx.doi.org/10.1016/B978-0-12-802830-8.00004-6
Copyright © 2016 Elsevier Inc. All rights reserved.

physicochemical approaches (Baker, 1981; Bañuelos et al., 2000; Cunningham and Ow, 1996). Plants can be used for the extraction and stabilization of many heavy metals found in contaminated media, reducing their risks to humans and ecosystems. Much research has been conducted in order to determine which plant species have the double ability to produce high yields and tolerate and/or accumulate heavy metals (Cunningham and Ow, 1996; Huang et al., 1997; Meers et al., 2005). Energy crops have been recognized as tolerant to contaminated matrices, and the high-yielded biomass can be used for the production of energy, paper pulp, and biomaterials. Phytoremediation capabilities have been suggested for perennial grasses, such as *Miscanthus* and *Arundo*, short rotation coppices, such as poplar trees, and annual crops, such as hemp, flax, and kenaf among other energy crops (Fernando et al., 2014). The irrigation of energy crops with nonconventional water such as wastewater (WW), especially in water-scarce regions, or their cultivation on marginal lands should be implemented. Biomass from energy crops produced in marginal lands and wetlands, or with use of WW irrigation, is expected to support the growing market for renewable energy and byproducts as well as increase supply. This new rationale of considering full use of all available resources in a sustainable way is the foundation of a sustainable bioeconomy (Nsanganwimana et al., 2014), also being within the logic of "biorefinery."

The energy crop *Arundo donax* L. is a plant with genetic and physiological potential to remove or immobilize inorganic compounds from contaminated media, such as heavy metals contained in contaminated soils and WW. Phytoremediation using *A. donax* L. could offer owners and managers of contaminated sites a cost-effective option over physicochemical remediation techniques, while at the same time generating biomass for bioenergy, fiber, and other byproducts with economic value (Lasat, 2000; McIntyre, 2003; Pilon-Smits, 2005; Zema et al., 2012). The reuse of WW in the irrigation of *A. donax* L., as well as the use of marginal soils for its production, assumes increased importance in water-scarce regions such as the Mediterranean basin, where some economic and policy contexts, a surge in the human population, and competition for land and water between agriculture, industry, and urbanization are leading to desertification (Barbero-Sierra et al., 2013; Portnov and Safriel, 2004).

The main biomass and physiological characteristics of *A. donax* L., its ecological requirements, the main case studies of its cultivation under marginal soils (namely, contaminated soils with inorganic compounds such as heavy metals) or when irrigated with WW, as well as the main social, economic, and environmental risks and benefits related to this practice are reviewed in this chapter.

2 GIANT REED (*A. DONAX* L.)

Giant reed (*A. donax* L.) is an herbaceous, perennial, nonfood crop that produces high yields of dry biomass (El Bassam, 2010), showing high potential for phytoremediation purposes (Figure 1).

The plant belongs to the *Poaceae* family and to the *Arundineae* tribe, being the most common of the species of its genus, which also includes the plants *Arundo plinii*, *Arundo collina*, *Arundo mediterranea*, and *Arundo formosana* (Mariani et al., 2010). Some sources suggest an Asian origin (Angelini et al., 2005; Cosentino et al., 2006), but the species grows spontaneously throughout the Mediterranean basin (El Bassam, 1998) (Figure 2). Nowadays, the species is globally dispersed (Pilu et al., 2012). Despite being a species from subtropical and warm temperate climates, certain genotypes are well adapted to cooler climates, such as the ones of United Kingdom and Germany. It has been cultivated throughout Asia, southern Europe, and North Africa, and in the Middle East for thousands of years (El Bassam, 2010).

FIGURE 1

Illustration of *Arundo donax* L.

Source: Prof. Dr. Otto Wilhelm Thomé, Flora von Deutschland, Österreich und der Schweiz, 1885, Gera, Germany.

This robust plant with C3 cycle achieves photosynthetic rates comparable with those of C4 plants (Papazoglou et al., 2005), reaching growth rates of up to 7 cm day^{-1} between June and July in Mediterranean environments (El Bassam, 2010). In warm temperate regions, vegetative growth normally occurs between February and March, when new shoots emerge from the soil. The plant develops from February to October. Flowering occurs from August to November, when an inflorescence (panicle) appears in the apical leaf. Cessation of growth occurs between November and February, when stems start to lose moisture, leaves begin to enter senescence, and the panicle breaks. The seeds (kernels) are sterile; hence the species reproduces by vegetative propagation of the rhizome, a feature that gives it strength, important for phytoremediation purposes. Morphologically, *A. donax* L. possesses a dense, robust, branched, and woody rhizome, 5-50 cm long, usually growing near to the surface of the ground (5-10 cm), with a diameter of 1-10 cm, as well as having fibrous roots reaching 100 cm in length (Lewandowski et al., 2003; El Bassam, 2010) (Figure 3).

The stems are hollow, robust, and erect, growing in dense groups derived from the same rhizome, reaching 3.5 cm in diameter and 10 m high (El Bassam, 2010), with a thickness between 2 and 7 mm,

FIGURE 2

Giant reed growing in Greece.

FIGURE 3

Details of the *Arundo donax* L. rhizome and roots.

and internode distances that can reach 30 cm in length (Pilu et al., 2012) (Figure 4). The leaves are alternate, 5-8 cm wide and 30-70 cm in length (Lewandowski et al., 2003). The inflorescence is a branched apical panicle that can reach 60 cm in length (El Bassam, 2010).

This plant tolerates a high range of ecological conditions and easily adapts to virtually all types of soil, including clay and sandy soils, gravel, and saline soils (making it a halophyte), as well as infertile or heavy metal-contaminated soils. It grows commonly at the margins of agricultural land, near roads or watercourses (Papazoglou et al., 2005; Shatalov and Pereira, 2002). It prefers a well-drained soil with high humidity, where it achieves greater productivities, but it can survive for long periods of time under a dry environment (El Bassam, 2010). It occurs naturally in areas with 300-4000 mm of annual

FIGURE 4

Details of the *Arundo donax* L. stem.

precipitation, with annual temperatures of 9-28 °C and at up to 2300 m altitude (Lewandowski et al., 2003). Normally it does not tolerate low temperatures, except when dormant. Propagation is *via* rhizomes or by stem cuttings. Methods of *in vitro* propagation have also been reported and some research attempted to induce the development of roots on the stems in order to increase the efficiency of the stem-cutting technique (Cosentino, 2014). The crop does not have any special requirements in terms of soil preparation. Angelini et al. (2009) refer to sowing densities in the range of 20,000-40,000 plants ha^{-1}. The species responds positively to irrigation (El Bassam, 2010). Under dryland conditions, 44-108 mm of irrigation (at the time of fertilization) may be adequate (Lewandowski et al., 2003). Each year the plant may require between 375 and 560 mm (Nackley et al., 2014) provided by irrigation to offset the annual volume of water reached by precipitation. Borin et al. (2013) and Mantineo et al. (2009) apply *ca.* 450 mm by irrigation when rainfall is around 400-500 mm. In soils with low nitrogen content, annual applications of 100 kg ha^{-1} year^{-1} N are recommended during the initial stages of crop growth (El Bassam, 2010). When crop reaches maturity, fertilization with 40 kg ha^{-1} N is the most appropriate from an economic and environmental point of view (Christou et al., 2001). An annual application of *ca.* 230 kg ha^{-1} (P$_2$O$_5$) and 170 kg ha^{-1} (K$_2$O) (Barbosa, 2014) is usually necessary to maintain the fields, but the amount to be applied should also root on the soil's fertility and plant uptake.

A. *donax* L. is a species highly resistant to most pests; however, when attacked by *Sesamia* spp., it may die during the early stages of growth. It can also be attacked by *Tetramesa romana* and by *Rhizaspidiotus donacisand* species when the plant exhibits invasiveness. It does not require the application of herbicides, especially if the propagation technique is through rhizome cuttings (El Bassam, 2010; Lewandowski et al., 2003). Currently, A. *donax* L. is not cultivated on a large scale, but experimental field trials have been conducted (Ceotto and DiCandilo, 2010; Cosentino et al., 2006). It is considered one of the most profitable energy crops, since it is perennial and inputs required after the year of establishment are very low. Its natural populations may produce 40 Mg ha^{-1} (dry matter) (El Bassam, 2010). The average productivity of fresh and dry biomass can reach 59.8 and 32.6 Mg ha^{-1}, respectively, with high irrigation rates (700 mm year^{-1}) and 55.4 and 29.6 Mg ha^{-1} at low irrigation rates (300 mm year^{-1}) (El Bassam, 2010).

Depending on the biomass use, different types of harvest can be chosen. For materials used in protection shelters, for example, harvest should be carried out every 2 years in order to increase the durability of protection, but for energy production purposes, harvest should be performed annually. In the southern parts of Europe, harvest can be done either in autumn or in late winter; however, in a late harvest losses of *ca.* 30% (dry matter) can occur, especially if the winter is rigorous, and wind is a very prominent factor. The autumn harvest may result in a very wet biomass. Nonetheless, especially in semi-arid climates of the Mediterranean, weather conditions usually allow natural drying of its biomass in the field after cutting. However, harvest anticipation can also anticipate plant cycle, especially in warm regions in fertile soils where shoots emerge in early spring. At this stage, there is still frost, and many of these shoots may be destroyed. Its biomass can easily be stored, and normally no treatment is applied, but a 10-15% total biomass can be lost, mainly at leaf level but not with stems (El Bassam, 2010). Its cellulose and hemicellulose content is approximately 45% of dry biomass, while the content in lignin is about 25%. The calorific value of stems (dry matter) is around 17.3-18.8 and 14.8-18.2 MJ kg^{-1} (dry matter) in leaves. Ash content ranges between 4.8% and 7.4%, carbon between 17.7% and 19.4%, and moisture between 36% and 49% (autumn harvest) (Lewandowski et al., 2003; El Bassam, 2010). Venturi and Monti (2005) refer to contents of 0.74% N, 1.8 mg kg^{-1} S, and 2.63 mg kg^{-1} Cl in biomass of *A. donax* L. Correct time of the harvest period can reduce the ash content by about 20%, but Nasso et al. (2010) showed that although the winter harvest over the autumn harvest may reduce the silica and potassium content, the ash content of the biomass harvested in winter was higher.

Arundo biomass shows great application in the production of solid biofuels for direct combustion and co-combustion (Hoffmann et al., 2010), for gasification and pyrolysis (Ghetti et al., 1996), and in anaerobic digestion to produce biogas, or even when submitted to alcoholic fermentation for the production of bioethanol (Pilu et al., 2012). This crop shows an excellent energy balance, but biomass quality is rather poor when the aim is combustion. The chemical analysis of chopped biomass and pellets of *A. donax* L. was shown to have a high content of ash, chlorine, sulfur, total silica, and metals, which can contribute to reduce the combustion reactor life span (Nasso et al., 2010; Picco, 2010). In addition, the emissions derived from *Arundo* biomass combustion have high concentrations of dust and microparticles, producing compounds that can be harmful to human health and to the environment, such as oxides of nitrogen, hydrochloric acid, carbon monoxide, and sulfur dioxide (Nasso et al., 2010).

The robustness of its rhizomes allows its use in the support of sloping terrains and in erosion control, which is a beneficial trait of this crop, and with significance for its production in marginal soils. Considering the volume and ground cover that it can achieve, as well as its high water and nutrient use efficiencies (Angelini et al., 2009), its use can also be envisaged in combating desertification. Planting on marginal and degraded soils may increase the value of such soils, and additional revenue can be allocated to farmers through the use and processing of this biomass; for example, for fiber and other products. Stem material of *A. donax* L. is flexible and strong, and can be used in the manufacture of fishing rods, brass musical instruments, canes, construction supplies, paper pulp, as well as supports for climbing plants and vines (Perdue, 1958). *A. donax* L. embodies both the attributes required for ideal energy crops and those of typical invasive flora: rapid growth, low input of fertilizers, high water use efficiency (WUE) and nutrient use efficiency (NUE), and the absence of pests and diseases (Low et al., 2011). In many places of the world where it is nonnative, its dissemination assumes a dramatic character, and local institutions have invested many resources (not only economic) in its control, since its presence leads to many types of damage and perturbation in water supply for agriculture, biodiversity, access to rivers, risk of fire, and for the conservation of local ecosystems (California Invasive Plant Council, 2011). Its anticipated vegetative cycle and the large production of propagating material are the most important competitive advantages when compared with native plants, especially after extreme events such as floods and fires. Chemical control has been reported to be the most efficient method for its eradication, consisting basically of applying the herbicide glyphosate between August and November (Cosentino, 2014). Mechanical removal of the rhizomes and soil has also been reported as an effective technique for removing *A. donax* L. (Cosentino, 2014). Nevertheless, this is a technique that removes both plant and soil, and given the depth and extent of the *Arundo* root system, may leave small rhizome fragments in the soil that may persist. In order to prevent its spread in places where it might be potentially dangerous, and always under nonriparian ecosystems, precautionary measures during harvest, transport, and processing of giant reed biomass should be adopted.

What is actually the potential of *A. donax* L. for the phytoremediation of heavy metal-contaminated soils and for the phytodepuration of effluents? In the next sections, several case studies retrieved from literature will be presented and discussed.

3 PRODUCTION OF GIANT REED IRRIGATED WITH WASTEWATERS: CASE STUDIES

Water supply and water quality degradation are global concerns, and in water-scarce countries, marginal-quality water will become an increasingly important component of agricultural water supplies (Angelakis and Durham, 2008; Qadir et al., 2007; Trinh et al., 2013). However, although the reuse of WW should be encouraged, it should also be controlled since its application could lead to numerous nutrients, pathogens, and potentially toxic elements being added to soils (Marecos do Monte and Albuquerque, 2010). At the same time, the reuse of WW in the irrigation of energy crops, like *A. donax* L., should ensure environmental and public health. Considering the genetic and physiological characteristics of *A. donax* L., it is possible to irrigate with different types of WW fulfilling those needs.

Kausar et al. (2012) refer to the potential of *Arundo* for the removal of Cr from contaminated WW, and Bonano et al. (2013) suggest that *A. donax* L. can be used for the phytoextraction of Al, As, Cd, Cr, Cu, Hg, Mn, Ni, Pb, and Zn in sediments and water bodies. This crop accumulates metals preferentially in underground components and seems to follow metal exclusion mechanisms. Yet, in this latter study, *A. donax* L. exhibited toxic levels of Cr in all organs ($>0.5 \, \mathrm{mg \, kg^{-1}}$). Among all species tested by Bonano et al. (2013), giant reed was the one with the lowest bioaccumulation capacity.

Giant reed exhibits high biomass production and tolerance to different types of WW composition. In many of the studies cited here, it was found that this plant combines high yields with a high potential for the removal of many inorganic contaminants dissolved in many different types of WW, under both natural and artificial conditions. Heavy metal contamination in water environments may threaten not only aquatic ecosystems but also human health. Phytoremediation using plants like *A. donax* L. to remove, detoxify, or stabilize heavy metals can be considered an important tool for cleaning polluted water, with the concomitant production of a biomass that can be further valorized. In constructed wetlands, for example, heavy metals in WW can be removed by processes such as absorption, precipitation, and plant uptake. *Arundo* is able to preserve the main attributes shown under natural conditions or under noncontrolled conditions (situations where it shows invasiveness behavior) when produced with WW irrigation. Tuttolomondo et al. (2015) tested *A. donax* L. and *Cyperus alternifolius* in a constructed wetland system receiving domestic WW and observed that giant reed began vegetative growth and regrowth earlier than *C. alternifolius*, leading to greater plant growth during the crop development stage. During the 2-year experiment, average WUE for *A. donax* L. was greater than for *C. alternifolius* for all growth stages. Giant reed used water in a more efficient way thanks to greater above-ground biomass production and to higher photosynthetic activity, related to high leaf area index (LAI) and total chlorophyll levels. An important consideration concerning WUE for *A. donax* L. is the relationship between this parameter and the availability of water in the growing medium. *A. donax* L. tends to increase WUE when water availability is reduced, with values ranging from 6 to 10 g dm^{-3}, depending on the type of soil and climatic conditions (test conditions) (Christou et al., 2001; Tuttolomondo et al., 2015). Under constructed wetland systems, its performance in the removal of nutrients and heavy metals is highly dependent on the system's water balance. Sun et al. (2013) tested *A. donax* L. among other macrophyte species and found that the plant showed excellent Fe accumulation capacity, but did not accumulate Cr, Cu, and Zn.

In other cases, and depending on the conditions of the study, the plant showed high efficiency in the removal of many metals and other nutrients from WW. In a pot experiment performed in Portugal, Costa (2014) tested *A. donax* L. under irrigation with different types of swine effluents containing two different concentrations of Zn (10 and 20 g dm^{-3}) and Cu (1 and 2 g dm^{-3}) and under three different types of water supply regimes (950, 475, and 238 mm). The author found that reducing water supply by 75% reduced yields by 74%, a fact that shows the greater dependence of *A. donax* L. biomass production on water availability in the growing medium. However, it was verified that the tested concentrations of zinc or copper did not greatly affect biomass productivity or biomass quality. Yet a higher accumulation of Zn and Cu in the biomass irrigated with WW was observed (even if this increment was not meaningful). Zinc accumulation occurred mostly in rhizomes and leaves, and copper accumulation occurred mostly in leaves and roots. The soil-plant system retained over 90% of zinc/copper, but *Arundo* itself merely bioremoves 8% and 3%, respectively. Other important and interesting results of this study point to higher translocation of Zn and Cu to aerial components of *A. donax* L. when there was a lower WW supply.

Elhawat et al. (2014) tested the phytoextraction potentials of two biotechnologically propagated ecotypes (an American and a Hungarian) of *A. donax* L. in Cu-contaminated synthetic WW (0, 1, 2, 3, 5, 10 and 26.8 mg dm^{-3}) for 6 weeks. The increment of Cu concentration in the nutrient solution slightly reduced root, stem, and leaf biomass up to 26.8 mg dm^{-3}. Yet Cu removal from WW ranged between 96.6% and 98.8% for the American ecotype and 97-100% for the Hungarian ecotype. Accordingly, both populations of *A. donax* L. can be employed to treat water bodies contaminated with Cu up to 26.8 mg dm^{-3}.

Phytoextraction and phytovolatilization of As (0, 50, 100, 300, 600 and 1000 μg dm^{-3}) contained in synthetic WW was tested by Mirza et al. (2010, 2011). They observed some toxicity symptoms in giant reed biomass, which included the appearance of red spots on the roots, emergence of young leaves with red color, leaf yellowing, and necrotic leaves, as well as margin development, especially at the top concentration. The arsenic content recovered in plants was similar to its concentration in WW. Measurements of physiological and ultrastructural anatomical changes showed the presence of stomata on the stems of *A. donax* L. at 1000 μg dm^{-3} As, highlighting phytovolatilization as one of the mechanisms that the species uses to tolerate arsenic. The species can tolerate concentrations in the range of 600-1000 μg dm^{-3} As without suffering any toxicity symptoms. The appearance of certain symptoms at 1000 μg dm^{-3} As reveals that the plant cannot tolerate this concentration of arsenic, but still continues to accumulate and volatilize arsenic at concentrations above 600 μg dm^{-3} As.

A. donax L. is pointed out by Williams et al. (2008) as a suitable crop for the production of biofuels and for the pulp/paper industry in Australia, when irrigation with saline winery WW is performed and in saline soils. *A. donax* L. can produce biomass yields of 45.2 Mg ha^{-1} (dry weight) when receiving 21,000 m^3 of winery WW, a yield similar to the ones registered by Hidalgo and Fernandez (2001) (46 Mg ha^{-1} dry weight in Spain) or Picco (2010), Shatalov and Pereira (2002), and Venturi and Monti (2005) (30-40 Mg ha^{-1} dry weight), with low technical inputs under arable land and with irrigation using tap water. Due to its composition, the irrigation with winery WW also allows the fertilization of *A. donax* L. In the same study of Williams et al. (2008), *A. donax* L. removed large amounts of N, P, and K at rates of 528, 22, and 664 kg ha^{-1} year^{-1}, with salinity up to 9 dS m^{-1}, accumulating at the same time 20.6 Mg ha^{-1} of organic carbon, a result that highlights the possibility of the use of *A. donax* L. in carbon credit programs. High dry matter yields of *A. donax* L. biomass, as well as high nutrient and heavy metal removals from WW, in plants receiving swine effluents in a closed gravel hydroponic system were registered by Mavrogianopoulos et al. (2002) in the third year of the experiment, in which extra P was added to the WW solution. After 1 year of growth, annual stem production was 12-15 Mg ha^{-1} (dry weight), and during the third year, after extra P was added, the same parameter ranged from as high as 20-23 Mg ha^{-1} (dry weight). The same behavior was observed in the average infiltration rate for most elements (N, K, Ca, Mg, Fe, Mn, Zn, and Cu), which increased by 46%, and for P an increment of 169% was registered. This work shows that it is possible to combine biomass production with WW treatment and in particular mineral and metal removal. Idris et al. (2012) evaluated the performance of *A. donax* L. when treating WW from a dairy processing factory with a median electrical conductivity (EC) of 8.9 mS cm^{-1}, registering approximately 179 Mg ha^{-1} year^{-1} biomass (dry weight), during the 250 days of the growing season as well as a percentage of removal of 69%, 95% and 26% of biochemical oxygen demand (BOD), suspended solids and total N, respectively. They pointed out *A. donax* L. as a suitable crop for WW treatment, and the biomass produced as additional opportunity for secondary income streams through its utilization. Others tested many agronomic parameters under WW irrigation practices. Barbagallo et al. (2011) tested the irrigation of *A. donax* L. with domestic treated WW with 66% and 100% of crop evapotranspiration (ETc), registering 40 Mg ha^{-1} (dry weight) and 47 Mg ha^{-1} (dry weight), respectively, as well as an energy content of 8.0 MJ kg^{-1} in biomass. When crops were irrigated with 100% ETc, they obtained higher growth parameters such as higher culm density. The authors highlighted *A. donax* L. as a suitable crop for energy production in marginal lands of the Mediterranean basin, a water-scarce region. These results may also indicate a good performance for fiber purposes under this sort of irrigation, but further research is needed.

In Portugal, Calheiros et al. (2012) used *A. donax* L. and *Sarcocornia* in the treatment of tannery WW with high and variable concentrations of complex pollutants, combined with high salinity levels, in a wetland system composed of three beds. They discovered that constructed wetland systems planted with salt-tolerant plants such as *A. donax* L. are promising solutions for that type of WW. By the end of the trial, the average height of the plants was around 97 cm and the root system was deep and well developed. In the second bed, *A. donax* L. produced 108-365 g plant^{-1} (dry weight) and in the third bed 59-407 g plant^{-1} (dry weight). Giant reed had a higher capacity than *Sarcocornia* to take up nutrients. Mass removal rates were up to 615 kg chemical oxygen demand (COD) ha^{-1} d^{-1} and 363 kg BOD after 5 days ha^{-1} d^{-1}, and removal efficiencies were 40-93% for total P, 31-89% for NH$_4$$^+$, and 41-90% for total Kjeldahl nitrogen. The performance of giant reed across different constructed wetland types receiving different types of WW ranged between 50% and 90% for removing excess nutrients (N, P, K), total suspended solids, COD, and BOD (Nsanganwimana et al., 2014). Kouki et al. (2012) tested the potential of a polyculture of *A. donax* L. and *Typha latifolia* for growth and phytotreatment of rural WW. Giant reed had higher height elongation (288 cm; 9-month experiment), phytomass production (2.4 kg dry biomass m^{-2}), and nitrogen uptake (21.1 mg kg^{-1} dry weight).

Rhizofiltration is one of the phytoremediation methods that employ terrestrial plants and their root system or aquatic plants as a biofilter and which can be extraordinarily effective in the sequestration of metals and radionuclides from polluted water bodies. Dürešová et al. (2014) tested a rhizofiltration system composed of *A. donax* L. for the removal of ^{109}Cd and ^{65}Zn. The authors found complete Zn and Cd removal from a solution containing 0.28 µmol dm^{-3} CdCl$_2$ (78.9 kBq dm^{-3} ^{109}CdCl$_2$), and ZnCl$_2$ (66.6 kBq dm^{-3} ^{65}ZnCl$_2$) in deionized water by a rhizofiltration system assembled from 20 experimental units containing juvenile plants of *Arundo* and at a solution flow rate of 0.125 cm^3 min^{-1}. This work confirmed that *A. donax* L. has potential to be used in rhizofiltration systems for Cd and Zn removal from WW or contaminated liquids under continuous flow conditions.

Constructed wetlands can also be used as a preapplication treatment scheme in order to reduce nutrient concentration in the effluent, being implemented between secondary treatment in WW treatment plants and land treatment systems (LTS), a strategic way to promote WW resource management in water-scarce regions. *A. donax* L. may be a suitable plant to use in these systems, allowing water quality improvements as well as reducing irrigation risk for LTS. In this kind of system, the water treatment is achieved by utilizing vegetation that affects hydraulic loading and nutrient uptake. In Greece, Tzanakakis et al. (2009) tested *A. donax* L. among other species for that purpose, obtaining a total biomass of 72.81 Mg ha^{-1} at the end of 3 years of the experience. At the same time, the plants accumulated 967.2 kg N ha^{-1} and 56.66 kg P ha^{-1}. Effluents were applied at rates capable of satisfying ETc in a semiarid region of Greece, and biomass growing in LTS contributed to 35% of the nutrient recovery. These systems require additional management practices in order to approximate nutrient load with potential assimilation (Tzanakakis et al., 2009). Due to its capacity to control nitrate leaching, promoted by its deep and extensive root system and by maintaining a higher nitrification rate than trees, giant reed is a relevant candidate species for growing on poor soil irrigated by nutrient-enriched WW (Tzanakakis et al., 2011). These studies show the great adaptability of *A. donax* L. to different environments, as well as its potential for biomass use. The plant is well adapted to high disturbance dynamics in ecosystems, as well as to phytodepuration of waters containing high amounts of heavy metals such as Cd and Ni (Papazoglou, 2007). Nevertheless, Zema et al. (2012) reported a decrease in productivity of *A. donax* L. when irrigated with WW. Lower growth (reductions of 7.2% in height, 5.9% in stem diameter, and 3.1% in LAI) was recorded for plants irrigated with WW compared with plants irrigated with conventional water. Mantineo et al. (2009) also recorded lower productivities in 2 out of 5 years of experiment for *A. donax* L. treated with water with high dissolved-nutrient content. Another interesting point from the study of Zema et al. (2012) concerns the mean high heating value of 18.18 MJ kg^{-1}, a value in the range referred by Angelini et al. (2005) under conventional water application and in fertile soil (16 MJ kg^{-1} in the establishing year and 18 MJ kg^{-1} from the second to sixth year). *A. donax* L. irrigated with WW produced appreciable biomass and energy yields, but in some aspects its behavior and response to WW seemed to be independent and higher nutrient load and water did not in all cases lead to higher biomass yields for this crop. Another point that should be considered in this analysis refers to the fact that *A. donax* L. has a greater capacity to survive under WW irrigation—namely, after rhizome transplantation—when compared to other species like *Phragmites* and *Typha*. Zema et al. (2012) reported survival rates of up to 90% for *A. donax* L. and 75% and 65% for *Phragmites* and *Typha*, respectively. Christou et al. (2001) also referred to survival rates of around 100% for giant reed.

Even though in some cases this plant produces less biomass than others under similar conditions, the survival rate related to this crop may ensure higher overall biomass. Similarly, the mean low heating value was 50% lower than the maximum value measured for *Typha*, but since its biomass yield per hectare is much higher, the energy yield per unit of cultivated area (24.03 MJ m^{-2}) for *A. donax* L. was about 13 times that calculated for T*ypha*. *A. donax* L. is effective in controlling and removing pollutants such as nitrates and phosphates, as well as COD, from WW (Chang et al., 2012; Tam and Wong, 2014). In many of the studies cited here, *A. donax* L. had a higher capacity to control the percolation of salts and of organic and inorganic contaminants. Further research activities should be developed to accurately determine the agronomic aspects of its production under WW irrigation.

4 PRODUCTION OF GIANT REED IN CONTAMINATED SOILS: CASE STUDIES

The growing demand for biomass for bioenergy production and, in particular, the production of energy crops generates several conflicts concerning the use of land. Such conflicts can be solved by spatial segregation of the area for the production of energy crops—referred to as surplus land in Dauber et al. (2012)—which encompasses various types of marginal lands, including heavy metal-contaminated lands. The introduction of a perennial grass such as *A. donax* L. to poor-nutrient soils may bring many ecological advantages and provide a wide range of ecological services such as extensive vegetation cover and permanence in soils, and control of erosion dynamics, thereby contributing to carbon and water storage of soils, as well as to the reversal, control, and mitigation of desertification.

This section briefly introduces the studies that have used *A. donax* L. for the phytoremediation of heavy metal- and other inorganic-contaminated soils, as well as the main results concerning agronomic parameters, physiological processes, pollutant removal rates, and yields achieved in production on contaminated soils.

A. donax L. showed a great endurance when exposed to Cd and Ni soil concentrations of up to 973.8 and 2543.3 mg kg^{-1}, respectively. Plants showed no toxicity effects on biometric parameters and physiological processes (i.e., on growth, biomass production, net photosynthesis −Pn−, stomatal conductance, intercellular concentration of CO_2, stomatal resistance, chlorophyll content, and chlorophyll fluorescence) (Papazoglou et al., 2005, 2007). Plant productivity and WUE remained unaffected, indicating the extent of the tolerance to Cd and Ni of this plant species (Papazoglou, 2007). Higher Cd concentrations were determined in the lower leaves (13.7 mg kg^{-1}) and in the upper part of the stems (9.0 mg kg^{-1}), while in rhizomes Cd concentration was 9.4 mg kg^{-1}. Corresponding Ni values were 42.4 mg kg^{-1} in lower leaves, 16.0 mg kg^{-1} in upper stems, and 54.8 mg kg^{-1} in rhizomes (Papazoglou, 2009).

The studies of Guo and Miao (2010) and Miao et al. (2012) registered fast growth of giant reed as well as high yields in aerial biomass under As (254 mg kg^{-1}), Cd (76.1 mg kg^{-1}), and Pb (1552 mg kg^{-1}) contamination. The height of the plants and dry biomass were reduced slightly due to the presence of heavy metals, and metal accumulation was greater at root level. Soil amendments such as acetic acid, ethylenediaminetetraacetic acid (EDTA), and citric acid enhanced biomass production. The concentrations of As, Cd, and Pb in giant reed shoots were significantly increased when lower levels of acetic acid, citric acid, and sepiolite were applied and higher levels of EDTA. These amendments could be considered optimum for remediation systems using *A. donax* L. Han et al. (2005) refer to high yields of biomass, to a huge growth of the root system and to a high adaptability, tolerance, and accumulation of Cd and Hg when giant reed was tested for phytoextraction of Hg ($101\,mg\,kg^{-1}$) and Cd ($115\,mg\,kg^{-1}$) from soils. Han and Hu (2005) tested the effects of Cu^{2+}, Pb^{2+}, Cd^{2+}, Zn^{2+}, Ni^{2+}, Hg^{2+} ($100\,mg\,kg^{-1}$), and Cr^{6+} ($50\,mg\,kg^{-1}$), registering a decrease in chlorophyll content (20-56%) and effects on leaves. The plant did not tolerate this concentration of Cr^{6+} and the root system was damaged. Heavy metal concentration in soil decreased during the period of growth of the plant, probably due to translocation of metals within the *A. donax* L. biomass. Río et al. (2002) tested many plant species including *A. donax* L. for the phytoremediation of Pb, Cu, Zn, Cd, Tl, Sb, and As ($5000\,mg\,kg^{-1}$) in Doñana Natural Park, Spain. The plant accumulated $0\text{-}23\,mg\,kg^{-1}$ Pb, $133\text{-}147\,mg\,kg^{-1}$ Zn, $13\text{-}16\,mg\,kg^{-1}$ Cu, and $7\text{-}11\,mg\,kg^{-1}$ As. Under the conditions described above, *A. donax* L. did not assume a dominant status over all other heavy metal accumulator species, showing coexistence under wetland conditions, a type of environment where it normally displays invasiveness (California Invasive Plant Council, 2011).

Barbosa (2014) tested *A. donax* L. contaminated soils with Zn (450 and $900\,mg\,kg^{-1}$) and Cr (300 and $600\,mg\,kg^{-1}$) under three different types of water supply regimes (950, 475, and 238 mm). It was found that higher water supplies led to a higher absorption of elements such as Zn and Cr in the biomass (phytoextraction), as well as a higher LAI, leading to the increase of photosynthetic activity and biomass accumulation in the different plant components. The water availability is crucial to the success of phytoextraction of Zn and Cr. Nevertheless, the production of giant reed under contaminated soils conduces to higher ash content in the biomass, which may compromise its use for bioenergy production by combustion. Plants accumulated more Zn or Cr in their rhizomes, and low translocation factors were observed for both metals. Plants were able to remove a maximum of 0.4% of the total Zn fraction in soil and 2% of the bioavailable fraction (by the aerial components). Plants removed only 0.3% of the total Cr fraction in soil and a maximum of 1.2% of the bioavailable fraction by the aerial components. The low mobility of chromium in soils determined the overall reduced phytoextraction to above-ground biomass. Nonetheless, the production of giant reed can be performed in marginal soils such as those contaminated with Zn and Cr, releasing soil of better quality for food production and reducing conflicts associated with this issue.

Boularbah et al. (2006) refer to ecological restoration of soils situated in former mining sites in Morocco contaminated with Cd, Cu, Pb, and Zn, by using many plant species in its phytoremediation. *A. donax* L. was used among these species, with bioaccumulation factors of 0.01 for Cu, 0.004 for Pb, and 0.04 Zn. Total concentration of metals in the plant was $0.2\,mg\,kg^{-1}$ Cd, $7.1\,mg\,kg^{-1}$ Cu, $2.8\,mg\,kg^{-1}$ Pb, and $72\,mg\,kg^{-1}$ Zn. Sabeen et al. (2013) tested its use to treat Cd (0, 50, 100, 250, 500, 750, and $1000\,\mu g\,dm^{-3}$) for 21 days both in hydroponics and contaminated-soil environments. Roots were the plant organ with higher Cd accumulation, followed by stems and leaves. Better results were recorded in hydroponic culture than with Cd-contaminated soils: the maximum Cd content in root was $300\,\mu g\,g^{-1}$ during hydroponics experiments over $230\,\mu g\,g^{-1}$ in soil experiment. Under hydroponic conditions, translocation and bioaccumulation factors were always greater than 1, and above the reference value (1.0) for hyperaccumulation. In the soil, despite the low Cd uptake, the translocation factors were above the reference value (1.0); however, bioaccumulation values were below 1. At higher Cd exposure, plants showed some antioxidative stress as the concentration of antioxidants was increased with increasing Cd exposure. Alshaal et al. (2013) tested the phytoremediation of bauxite-derived red mud by giant reed. The presence of this plant species reduced the EC of red mud by 25% and that of mud-polluted soil by 6%. At the same time, giant reed promoted phytoextraction of Cd, Pb, Co, Ni, and Fe, with high translocation factors, especially for Ni. Its presence on the contaminated medium also improved other soil quality parameters such as pH, EC, organic carbon, microbial counts, and soil enzyme activities.

A. donax L. has a high potential to uptake more than one type of metal from polluted soil at significant rates, as well as to improve many properties of poor or contaminated soils.

5 BENEFITS AND CONSTRAINTS OF USING GIANT REED FOR PHYTOREMEDIATION PURPOSES

Giant reed can be used for phytoremediation purposes of different types of contaminated soil, sediments, and water bodies. *A. donax* L. has the advantages of many annual energy crops and being perennial allows the effects of these characteristics on the ecosystems in which it is implemented to endure. The implementation of this crop on marginal land may increase the yield and profitability of the soil over time, promoting the control of diseases and pests on site by increasing biological and landscape diversity, as well as providing a source of biomass for fiber, bioenergy, and other byproducts. Giant reed can also be introduced for the mitigation and reversal of desertification, because of low WUE and NUE, being able to generate commercial value for a given region. Its implementation on contaminated land may involve fewer problems for natural ecosystems and be integrated into a waste management strategy (Kassam et al., 2012; Laraus, 2004). Being a perennial crop, giant reed offers additional ecological advantages, providing a wide range of ecological services such as greater vegetation cover; greater permanence in the soil, which limits soil erosion; low susceptibility to diseases; reduced need for pesticides; and, due to its extensive and deep root system, the plant can be used for leaching control of many contaminants (Fernando, 2005; Fernando et al., 2010; Zhang et al., 2011). It is also a contributor crop for carbon sequestration (Monti and Zatta, 2009).

Due to its fast growth and high cellulose content, *A. donax* L. is considered one of the most promising energy crops for marginal lands, since its culms represent an important source of cellulose for the production of paper, second-generation ethanol, biodiesel, and biopolymers (Fiorentino et al., 2013). Its residual lignin content could be used for production of lignin-based resin coatings and composites. When produced in heavy metals- contaminated soils or with heavy metals-rich WW, there may be an increased accumulation of those elements in the biomass. If the levels of heavy metals contained in its biomass are high, the most environmentally safe solution is the production of energy from combustion or pyro-gasification (Lievens et al., 2008a,b), followed by metal recovery from fly ash using hydrometallurgical routes (Fiorentino et al., 2013). As metal toxicity may seriously limit the microbial-driven conversion of lignin cellulose to second-generation ethanol and biopolymers, these options are usually put aside. These studies validate the logic of giant reed use with WW and in contaminated soils by offering solutions for the recovery of contaminants such as heavy metals, as well as for the provision of biomass for bioenergy and other byproduct production. This is one of the main advantages of this approach and clearly meets the purposes of biorefinery and sustainable bioeconomy.

A. donax L. can also contribute to the restoration of soil ecosystems, along with biomass production, as indicated in the work of Alshaal et al. (2014), who tested the potential of *A. donax* under microbial communities in marginal soils. Giant reed showed considerable potential for recovering red mud-affected soils. Its cultivation increased the activities of most soil enzymes, especially of dehydrogenase, urease, and catalase. Regarding soil organic carbon, the authors did not find any considerable effect of its presence on soils but did find that giant reed has a special microbial community associated with its root system that helps to restore and improve many soil properties. Total fungi increased in soils after the presence of giant reed, but the total bacterial count decreased after its plantation by a range of 29-93%. Giant reed can also be used to restore and recover soil ecosystems after exposure to natural disasters such as bushfires. *A. donax* L. shares the advantages of many hyperaccumulator plant species used in phytoremediation approaches and can be used within *in situ* and *ex situ* applications, with additional gains. Phytoremediation technology requires several growing seasons to clean a contaminated site, and owners of contaminated lands may not wish to wait several growing seasons when hyperaccumulator plants are used to clean the sites due to lower yields and no profitable revenue from it. Energy/fiber crops such as giant reed offer the possibility of generating commercial value for the biomass produced in contaminated soils, overcoming this problem from the first year.

Giant reed, like other perennial crops, can tolerate soils contaminated with one or more contaminants, showing interesting yields as well as providing quality biomass for fiber or bioenergy purposes (Fernando, 2005). One of the disadvantages of its use in the phytoremediation of inorganic compounds is the limited amount of these compounds that can be extracted from soil, especially when compared with the amounts of the same compounds that can be extracted by physicochemical remediation methods, and the related time needed for remediation. Nevertheless, that disadvantage is clearly outweighed by the income obtained from the use of the biomass.

The use of a wild or nonindigenous plant could provoke its spread as well as leading to several risks for ecosystem functions if used near phytotreatment sites (McIntyre, 2003). In fact, the attributes required for optimal herbaceous energy/fiber crops correspond to those of typical weeds and invasive plants (Low et al, 2011); namely, rapid growth, low fertilizer inputs, high WUE and NUE, and an absence of pests and diseases. *A. donax* L. meets many of these requirements, being an herbaceous crop with high potential for invasiveness. Because of that, the implementation of this type of crop requires containment plans for their potential spread. Another disadvantage associated with its potential invasiveness is the limited information that is available in order to perform an appropriate risk assessment of the species (McIntyre, 2003), as well as appropriate methods for crop removal from the field after the remediation process. In fact, giant reed can reach huge population density and its fast growth can take control of the existing resources in the neighborhood, such as soil nutrients and access to light. Moreover, it can have more than one growth cycle in 1 year, and shows huge adaptability to different environments. On one hand, this could be useful in cases where there is need for a different genetic background to approach the remediation in a particular site (for example, inexistence of diseases and plagues or other trophic relations). On the other hand, this means environmental managers have to be aware of potentially harmful impacts in situations where the crops shows invasiveness and should promote its precautionary control. Because *A. donax* L. possesses a deep and extensive root system, its presence on slope terrains could also provide slope support for marginal lands, restraining erosion processes. In these sloping areas, as well as in space-restricted sites, the application of machinery and other traditional agricultural cropping techniques may not be possible (McIntyre, 2003). The application of energy and fiber crops in phytoremediation opens up the possibility of recycling all constituents contained in the phytoremediation biomass. Understanding the main pathways and main techniques to achieve that objective is fundamental to promoting phytoremediation in the remediation market, as well as for commercialization of its byproducts (McIntyre, 2003). In addition to the economic recovery when used for energy production or biomaterials, using *Arundo* in the phytoremediation of soils and WW can also provide additional benefits, such as the amount of carbon sequestered by the biomass (Monti and Zatta, 2009), or the water and mineral resources saved, if irrigation uses WW (Costa, 2014), or improvement of soil functions, when cropped in marginal land (Barbosa, 2014).

6 CONCLUSIONS AND RECOMMENDATIONS

Giant reed is a good candidate for marginal and wetland soils, for incorporation into LTS and in constructed wetlands, with an enormous potential for phytoremediation. The plant is able to improve the quality of water-polluted bodies, being able to remove COD, BOD, nitrates, ammonium and phosphate ions, and heavy metals. In contaminated soils, giant reed with its associated microorganisms and fungi improves many soil properties, whether at chemical or physical (slope terrains) levels, being able to remove heavy metals and some types of radionuclides, among other inorganic compounds. This crop can be used to control water, wind, and biological erosion on marginal lands, important for the mitigation and reversal of desertification. At the same time, the plant shows good physiological and biomass responses to stress in both polluted water and soil environments, returning biomass with high-quality parameters for production of fiber, energy, and other bioproducts. Under some specific conditions, precautionary control measures should be implemented because of its potential invasiveness.

REFERENCES

Alloway, B., 1995. Heavy Metals in Soils, 2nd edition. Blackie Academic and Professional, London.

Alshaal, T., Szabolcsy, É., Márton, L., Czakó, M., Kátai, J., Balogh, P., Elhawat, N., Ramady, H., Fári, M., 2013. Phytoremediation of bauxite-derived red mud by giant reed. Environ. Chem. Lett. 11, 295–302.

Alshaal, T., Szabolcsy, É., Márton, L., Czakó, M., Kátai, J., Balogh, P., Elhawat, N., Ramady, H., Geröes, A., Fári, M., 2014. Restoring soil ecosystems and biomass production of *Arundo donax* L. under microbial communities-depleted soil. Bioenerg. Res 7, 268–278.

Angelakis, A.N., Durham, B., 2008. Water recycling and reuse in EUREAU countries: trends and challenges. Desalination 218, 3–12.

Angelini, L., Ceccarini, L., Bonari, E., 2005. Biomass yield and energy balance of giant reed (*Arundo donax* L.) cropped in central Italy as related to different management practices. Eur. J. Agron. 22, 375–389.

Angelini, L., Ceccarini, L., Nasso, N., Bonari, E., 2009. Comparison of *Arundo donax* L. and *Miscanthus x giganteus* in a long-term field experiment in Central Italy: analysis of productive characteristics and energy balance. Biomass Bioenergy 33, 635–643.

Baker, A., 1981. Accumulators and excluders: strategies in the response of plants to heavy metals. J. Plant Nutr. 3, 643–654.

Bañuelos, G., Zambrzuski, S., Mackey, B., 2000. Phytoextraction of Se from soils irrigated with selenium-laden effluent. Plant Soil 224, 251–258.

Barbagallo, S., Cirelli, G.L., Consoli, S., Milani, M., Toscano, A., 2011. Utilizzo di acque reflue per l'irrigazione di biomasse erbacee a scope energetici. In: Convegno di Medio Termine dell'Associazione Italiana di Ingegneria Agraria, 22–24 settembre, 2011, Belgirate.

Barbero-Sierra, C., Marques, M.J., Ruíz-Perez, M., 2013. The case of urban sprawl in Spain as an active and irreversible driving force for desertification. J. Arid Environ. 90, 95–102.

Barbosa, B.M.G., 2014. Fitorremediação de solos contaminados com Zn e Cr utilizando *Arundo donax* L. (Ph.D. thesis). Universidade Nova de Lisboa, Lisbon.

Bonano, G., Cirelli, G.L., Toscano, A., Giudice, R., Pavone, P., 2013. Heavy metal content in ash of energy crops growing in sewage-contaminated natural wetlands: potential applications in agriculture and forestry? Sci. Total Environ. 452–453, 349–354.

Borin, M., Barbera, A.C., Milani, M., Molari, G., Zimbone, S.M., Toscano, A., 2013. Biomass production and N balance of giant reed (*Arundo donax* L.) under high water and N input in Mediterranean environments. Eur. J. Agron. 51, 117–119.

Boularbah, A., Schwartz, C., Bitton, G., Aboudrar, W., Ouhammou, A., Morel, J., 2006. Heavy metal contamination from mining sites in South Morocco: 2. Assessment of metal accumulation and toxicity in plants. Chemosphere 63, 811–817.

Calheiros, C.S.C., Quitério, P.V.B., Silva, G., Crispim, L.F.C., Brix, H., Moura, S.C., Castro, P.M.L., 2012. Use of constructed wetland systems with *Arundo* and *Sarcocornia* for polishing high salinity tannery wastewater. J. Environ. Manage. 95, 66–71.

California Invasive Plant Council, 2011. *Arundo donax* (giant reed): distribution and impact report March 2011. State Water Resources Control Board, Berkeley, CA, USA.

Ceotto, E., DiCandilo, M., 2010. Shoot cuttings propagation of giant reed (*Arundo donax* L.) in water and moist soil: the path forward? Biomass Bioenergy 34, 1614–1623.

Chang, J., Wu, S., Dai, Y., Liang, W., Wu, Z., 2012. Treatment performance of integrated vertical-flow constructed wetland plots for domestic wastewater. Ecol. Eng. 44, 152–159.

Christou, M., Mardikis, M., Alexopoulou, E., 2001. Research on the effect of irrigation and nitrogen upon growth and yields of *Arundo donax* L. Asp. Appl. Biol. 65, 47–55 (in Greece).

Cosentino, S., 2014. Management and yields of *Arundo donax* L. In: FIBRA Summer School, Lignocellulosic Crops as Feedstock for Future Biorefineries, 26-31 July 2014, Lisbon.

Cosentino, S., Copani, V., D'Agosta, G., Sanzone, E., Mantineo, M., 2006. First results on evaluation of *Arundo donax* L. clones collected in Southern Italy. Ind. Crop. Prod. 23, 212–222.

Costa, F.J.G., 2014. Fitoremediação de Águas Residuais Contaminadas com Zn ou Cu utilizando *Arundo donax* L. (Ph.D. thesis). Universidade Nova de Lisboa, Lisbon.

Cunningham, S., Ow, D., 1996. Promises and prospects of phytoremediation. Plant Physiol. 110, 715–719.

Dauber, J., Brown, C., Fernando, A., Finnan, J., Krasuska, E., Ponitka, J., Styles, D., Thrän, D., Groenigen, K., Weih, M., Zah, R., 2012. Bioenergy from "surplus" land: environmental and socio-economic implications. BioRisk 7, 5–50.

Dürešová, Z., Šuňovská, A., Horník, M., Pipíška, M., Gubišová, M., Gubiš, J., Hostin, S., 2014. Rhizofiltration potential of *Arundo donax* for cadmium and zinc removal from contaminated wastewater. Chem. Pap. 68, 1452–1462.

El Bassam, N., 1998. Energy Plant Species. James & James (Science Publishers), London.

El Bassam, N., 2010. Handbook of Bioenergy Crops—A Complete Reference to Species, Development and Applications. Earthscan, London.

Elhawat, N., Alshaal, T., Szabolesy, É., Ramady, H., Márton, L., Czakó, M., Kátai, J., Balogh, P., Sztrik, A., Molnár, M., Popp, J., Fári, M., 2014. Phytoaccumulation potentials of two biotechnologically propagated ecotypes of *Arundo donax* in copper-contaminated synthetic wastewater. Environ. Sci. Pollut. Res. Int. 21, 7773–7780.

Fergusson, J., 1991. The Heavy Elements: Chemistry, Environmental Impact and Health Effects. Pergamon Press, Oxford.

Fernando, A.L.A.C., 2005. Fitorremediação por Miscanthus x giganteus de solos contaminados com metais pesados (Ph.D. thesis). Universidade Nova de Lisboa, Lisbon.

Fernando, A.L., Duarte, M.P., Almeida, J., Boléo, S., Mendes, B., 2010. Environmental impact assessment of energy crops cultivation in Europe. Biofuels Bioprod. Biorefin. 4, 594–604.

Fernando, A.L., Boléo, S., Barbosa, B., Costa, J., Lino, J., Tavares, C., Sidella, S., Duarte, M.P., Mendes, B., 2014. How sustainable is the production of energy crops in heavy metal contaminated soils? In: Hoffmann, C., Baxter, D., Maniatis, K., Grassi, A., Helm, P. (Eds.), Proceedings of the 22nd European Biomass Conference and Exhibition, Setting the course for a Biobased Economy, 23-26 June 2014, Hamburg, Germany. ETA-Renewable Energies, pp. 1593–1596.

Fiorentino, N., Fegnano, M., Adamo, P., Impagliazzo, A., Mori, M., Pepe, O., Ventorino, V., Zoina, A., 2013. Assisted phytoextraction of heavy metals: compost and Trichoderma effects on giant reed (*Arundo donax* L.) uptake and soil N-cycle microflora. Ital. J. Agron. 8, e29.

Garbisu, C., Alkorta, I., 2003. Basic concepts on heavy metal soil bioremediation. Eur. J. Miner. Process. Environ. Prot. 3, 58–66.

Ghetti, P., Ricca, L., Angelini, L., 1996. Thermal analysis of biomass and corresponding pyrolysis products. Fuel 75, 565–573.

Guo, Z., Miao, X., 2010. Growth changes and tissues anatomical characteristics of giant reed (*Arundo donax* L.) in soil contaminated with arsenic, cadmium and lead. J. Cent. South Univ. Technol. 17, 770–777.

Han, Z., Hu, Z., 2005. Tolerance of *Arundo donax* to heavy metals. Chin. J. Appl. Ecol. 16, 161–165 (abstract).

Han, Z., Hu, X., Hu, Z., 2005. Phytoremediation of mercury and cadmium polluted wetland by *Arundo donax*. Chin. J. Appl. Ecol. 16, 945–950 (abstract).

He, Z., Yang, X., Stoffella, P., 2005. Trace elements in agroecosystems and impacts on the environment. J. Trace Elem. Med. Biol. 19, 125–140.

Hidalgo, M., Fernandez, J., 2001. Biomass production of ten populations of giant reed (*Arundo donax* L.) under the environmental conditions of Madrid (Spain). In: Kyritsis, S., Beenackers, A.A.C., Helm, P., Grassi, A., Chiaramonti, D. (Eds.), Biomass for Energy and Industry: Proceedings of the First World Conference, 5-9 June 2000, Seville, Spain, vol. 1. James & James (Science Publishers), London, pp. 1881–1884.

Hoffmann, G., Schingnitz, D., Schnapke, A., Bilitewski, B., 2010. Reduction of CO_2-emissions by using biomass in combustion and digestion plants. Waste Manage. 30, 893–901.

Huang, J., Chen, J., Berti, W., Cunningham, S., 1997. Phytoremediation of lead-contaminated soils: role of synthetic chelates in lead phytoextraction. Environ. Sci. Technol. 31, 800–805.

Idris, S.M., Jones, P.L., Salzman, S.A., Croatto, G., Allinson, G., 2012. Evaluation of the giant reed (*Arundo donax*) in horizontal subsurface flow wetlands for the treatment of dairy processing factory wastewater. Environ. Sci. Pollut. Res. 19, 3525–3537.

Kabata-Pendias, A., Pendias, H., 2011. Trace Elements in Soils and Plants, fourth ed. CRC Press, Boca Raton, FL.

Kassam, A., Friedrich, T., Derpsch, R., Lahmar, R., Mrabet, R., Basch, G., Gonzále-Sánchez, E., Serraj, R., 2012. Conservation agriculture in the dry Mediterranean climate. Field Crop Res. 132, 7–17.

Kausar, S., Mahmood, Q., Raja, I., Khan, A., Sultan, S., Gilani, M., Shujaat, S., 2012. Potential of *Arundo donax* to treat chromium contamination. Ecol. Eng. 42, 256–259.

Kouki, S., Saidi, N., Rajeb, A., M'hiri, F., 2012. Potential of a polyculture of *Arundo donax* and *Typha latifolia* for growth and phytotreatment of wastewater pollution. Afr. J. Biotechnol. 11, 15341–15352.

Laraus, J., 2004. The problems of sustainable water use in the Mediterranean and research requirements for agriculture. Ann. Appl. Biol. 144, 259–272.

Lasat, M.M., 2000. Phytoextraction of metals from contaminated soil: a review of plant/soil/metal interaction and assessment of pertinent agronomic issues. J. Hazard. Subst. Res. 2, 1–25.

Lewandowski, I., Scurlock, M.O.J., Lindvall, E., Christou, M., 2003. The development and current status of perennial rhizomatous grasses as energy crops in the US and Europe. Biomass Bioenergy 25, 335–361.

Lievens, C., Yperman, J., Vangronsveld, J., Carleer, R., 2008a. Study of the potential valorization of heavy metal contaminated biomass via phytoremediation by fast pyrolysis: part I. Influence of temperature, biomass species and solid heat carrier on the behavior of heavy metals. Fuel 87, 1894–1905.

Lievens, C., Yperman, J., Vangronsveld, J., Carleer, R., 2008b. Study of the potential valorization of heavy metal contaminated biomass via phytoremediation by fast pyrolysis: part II. Characterization of the liquid and gaseous fraction as a function of the temperature. Fuel 87, 1906–1916.

Low, T., Booth, C., Sheppard, A., 2011. Weedy biofuels: what can be done? Curr. Opin. Environ. Sustain. 3, 55–59.

Mantineo, M., Agosta, D., Copani, V., Patanè, C., Cosentino, S., 2009. Biomass yield and energy balance of three perennial crops for energy use in the semi-arid Mediterranean environment. Field Crop Res. 114, 204–213.

Marecos do Monte, H., Albuquerque, A., 2010. Reutilização de Águas Residuais. In: Série Guias Técnicos, vol. 14. Entidade Reguladora dos Serviços de Águas e Resíduos e Instituto Superior de Engenharia de Lisboa, Lisbon (in portuguese).

Mariani, C., Cabrini, R., Danin, A., Piffanelli, P., Fricano, A., Gomarasca, S., Dicandilo, M., Grassi, F., Soave, C., 2010. Origin, diffusion and reproduction of the giant reed (*Arundo donax* L.): a promising weedy energy crop. Ann. Appl. Biol. 157, 191–202.

Mavrogianopoulos, G., Vogli, V., Kyritsis, S., 2002. Use of wastewaters as a nutrient solution in a closed gravel hydroponic cultures of giant reed (*Arundo donax*). Bioresour. Technol. 82, 103–107.

McIntyre, T., 2003. Phytoremediation of heavy metals from soils. Adv. Biochem. Eng. Biotechnol. 78, 97–123.

Meers, E., Ruttens, A., Hopgood, M., Lesage, E., Tack, F., 2005. Potential of *Brassica rapa*, *Cannabis sativa*, *Helianthus annuus* and *Zea mays* for phytoextraction of heavy metals from calcareous dredged sediment derived soils. Chemosphere 61, 561–572.

Miao, Y., Xi-yuan, X., Xu-feng, M., Zhao-hui, G., Feng-yong, W., 2012. Effects of amendments on growth and metal uptake of giant reed (*Arundo donax* L.) grown on soil contaminated by arsenic, cadmium and lead. Trans. Nonferrous Met. Soc. China 22, 1462–1469.

Mirza, N., Mahmood, Q., Pervez, A., Ahmad, R., Farooq, R., Shah, M.M., Azim, M.R., 2010. Phytoremediation potential of *Arundo donax* in arsenic-contaminated synthetic wastewater. Bioresour. Technol. 101, 5815–5819.

Mirza, N., Pervez, A., Mahmood, Q., Shah, M., Shafqat, M., 2011. Ecological restoration of arsenic contaminated soil by *Arundo donax* L. Ecol. Eng. 37, 1949–1956.

Monti, A., Zatta, A., 2009. Root distribution and soil moisture retrieval in perennial and annual energy crops in Northern Italy. Agric. Ecosyst. Environ. 113, 252–259.

Mulligan, C.N., Yong, R.N., Gibbs, B.F., 2001. Remediation technologies for metal-contaminated soils and groundwater: an evaluation. Eng. Geol. 60, 193–207.

Nackley, L., Vogt, K.A., Kim, S., 2014. *Arundo donax* water use and photosynthetic responses to drought and elevated CO_2. Agric. Water Manage. 136, 13–22.

Nasso, N., Angelini, L., Bonari, E., 2010. Influence of fertilization and harvest time on fuel quality of giant reed (*Arundo donax* L.) in central Italy. Eur. J. Agron. 32, 219–227.

Nsanganwimana, F., Marchland, L., Douay, F., Mench, M., 2014. *Arundo donax* L., a candidate for phytomanaging water and soils contaminated by trace elements and producing plant-based feedstock. A review. Int. J. Phytoremediation 16, 982–1017.

Papazoglou, E.G., 2007. *Arundo donax* L. stress tolerance under irrigation with heavy metals aqueous solutions. Desalination 211, 304–313.

Papazoglou, E.G., 2009. Heave metal allocation in giant reed plants irrigated with metalliferous water. Fresenius' Environ. Bull. 18, 166–174.

Papazoglou, E.G., Karantounias, G., Vemmos, S., Bouranis, D., 2005. Photosynthesis and growth responses of giant reed (*Arundo donax* L.) to the heavy metals Cd and Ni. Environ. Int. 31, 243–249.

Papazoglou, E.G., Konstantinos, G., Bouranis, D., 2007. Impact of high cadmium and nickel soil concentration on selected physiological parameters of *Arundo donax* L. Soil Biol. 43, 207–215.

Perdue, R., 1958. *Arundo donax*—source of musical reeds and industrial cellulose. Econ. Bot. 12, 368–404.

Picco, D., 2010. Coltore energetiche per il disinquinamento della laguna di Venezia, Veneto Agricoltura Azienda Regionale per i Settori Agricolo. Forestale e Agroalimentare, Legnaro.

Pilon-Smits, E., 2005. Phytoremediation. Annu. Rev. Plant Biol. 56, 15–39.

Pilu, R., Bucci, A., Badone, F., Landoni, M., 2012. Giant reed (*Arundo donax* L.): a weed plant or a promising energy crop? Afr. J. Biotechnol. 11, 9163–9174.

Portnov, B.A., Safriel, U.N., 2004. Combating desertification in the Negev: dryland agriculture vs. dryland urbanization. J. Arid Environ. 56, 659–680.

Qadir, M., Sharma, B.R., Bruggeman, A., Choukr-Allah, R., Karajeh, F., 2007. Non-conventional water resources and opportunities for water augmentation to achieve food security in water scarce countries. Agric. Water Manage. 87, 2–22.

Río, M., Font, R., Almela, C., Vélez, D., Montoro, R., Bailón, A., 2002. Heavy metals and arsenic uptake by wild vegetation in the Guadiamar rivera area after the toxic spill of the Aznalcóllar mine. J. Biotechnol. 98, 125–137.

Sabeen, M., Mahmood, Q., Irshad, M., Fareed, I., Khan, A., Ullah, F., Hussain, J., Hayat, Y., Tabassum, S., 2013. Cadmium phytoremediation by *Arundo donx* L. from contaminated soil and water. BioMed Res. Int. http://dx.doi.org/10.1155/2013/324830. Article ID 324830, pp 9.

Shatalov, A., Pereira, H., 2002. Influence of stem morphology on pulp and paper properties of *Arundo donax* L. reed. Ind. Crop. Prod. 15, 77–83.

Sun, H., Wang, Z., Gao, P., Liu, P., 2013. Selection of aquatic plants for phytoremediation of heavy metal in electroplate wastewater. Acta Physiol. Plant. 35, 355–364.

Tam, N., Wong, Y., 2014. Constructed wetland with mixed mangrove and non-mangrove plants for municipal sewage treatment. In: 4th International Conference on Future Environment and Energy. In: IPCBEE, vol. 61. IACSIT Press, Singapore.

Trinh, L.T., Duong, C.C., Van Der Steen, P., Lens, P.N.L., 2013. Exploring the potential for wastewater reuse in agriculture as a climate change adaptation measure for Can Tho City, Vietnam. Agric. Water Manage. 128, 43–54.

Tuttolomondo, T., Licata, M., Leto, C., Leone, R., 2015. Effect of plant species on water balance in a pilot-scale horizontal subsurface flow constructed wetland planted with *Arundo donax* L. and *Cyperus alternifolius* L.—two-year tests in a Mediterranean environment in the West of Sicily (Italy). Ecol. Eng. 74, 79–92.

Tzanakakis, V.A., Paranychianakis, N.V., Angelakis, A.N., 2009. Nutrient removal and biomass production in land treatment systems receiving domestic effluent. Ecol. Eng. 35, 1485–1492.

Tzanakakis, V.A., Paranychianakis, N.V., Londra, P.A., Angelakis, A.N., 2011. Effluent application to the land: changes in soil properties and treatment potential. Ecol. Eng. 37, 1757–1764.

Venturi, G., Monti, A., 2005. Energia da colture dedicate: aspetti ambientali ed agronomici. In: Prima conferenza nazionale sulla politica energetica a cura di Ateneo di Bologna, AIGE Bologna. 18-19 aprile, 1-18.

Williams, C.M.J., Biswas, T.K., Schrale, G., Virtue, J.G., Heading, S., 2008. Use of saline land and wastewater for growing a potential biofuel crop (*Arundo donax* L.) irrigation. In: Irrigation Australia 2008, Conference May 20. South Australian Research and Development Institute, Melbourne.

Wuana, R.A., Okieimen, F.E., 2011. Heavy metals in contaminated soils: a review of sources, chemistry, risks and best available strategies for remediation. ISRN Ecol. http://dx.doi.org/10.5402/2011/402647. Article ID 402647.

Zema, D., Bombino, G., Andiloro, S., Zimbone, S.M., 2012. Irrigation of energy crops with urban wastewater: effects on biomass yields, soils and heating values. Agric. Water Manage. 115, 55–65.

Zhang, Y., Li, Y., Jiang, L., Tian, C., Li, J., Xiao, Z., 2011. Potential of perennial crop on environmental sustainability of agriculture. Procedia Environ. Sci. 10, 1141–1147.

Zhu, Y.G., Shaw, G., 2000. Soil contamination with radionuclides and potential remediation. Chemosphere 41, 121–128.

BIOMASS ENERGY, BIODIESEL, AND BIOFUEL FROM CONTAMINATED SUBSTRATES

SECTION

2

BIOMASS ENERGY,
BIODIESEL, AND
BIOFUEL FROM
CONTAMINATED
SUBSTRATES

BIOMASS ENERGY FROM REVEGETATION OF LANDFILL SITES*

B. Seshadri[1], N.S. Bolan[1], R. Thangarajan[2], U. Jena[3], K.C. Das[4], H. Wang[5], R. Naidu[1]

University of Newcastle, Newcastle, NSW, Australia[1]
University of South Australia, Adelaide, SA, Australia[2]
Desert Research Institute, Reno, NV, USA[3]
University of Georgia, Athens, GA, USA[4]
Zhejiang A & F University, Hangzhou, Zhejiang, China[5]

1 INTRODUCTION

Landfills provide the most economical and easiest means of disposing waste globally (Lamb et al., 2012). In many developed and developing countries, domestic and industrial wastes are managed mainly using landfills (Figure 1). Although there has been a significant increase in the reduction, reuse, and recycling of solid waste, disposal to landfill will inevitably remain the most widely used waste management method. While landfilling provides an economic means of waste disposal, it can lead to environmental degradation by releasing various contaminants if not managed properly. The major environmental challenges associated with the sustainable management of landfills are the surface and groundwater contamination, and greenhouse gas (GHG) and odor emissions (Albright et al., 2006).

Increasingly, phytocapping is practiced in traditionally managed landfills sites to mitigate the environmental impacts resulting from leachate generation and GHG emission (Lamb et al., 2012). Phytocapping also provides a major source of biomass for bioenergy production (Lamb et al., 2012; Venkatraman and Ashwath, 2009). Degraded land areas such as landfills and mine sites are increasingly being used to grow biomass crops as a renewable energy source (Brunner et al., 2009). This chapter provides an overview on the role of landfills as a potential biorefinery site for biomass energy production.

2 PHYTOCAPPING TECHNOLOGY TO MANAGE LANDFILL SITES

Landfill covers are used to reduce water percolation, thereby mitigating groundwater contamination risk of leachate generation and the environmental impacts of GHG emissions. Traditionally, landfill covers have been designed to minimize water entry through the use of low permeability layers (e.g., compacted clay caps and geosynthetic liners; Albright et al., 2006).

---

*This chapter is a condensed version of the following review paper: Bolan, N.S., Thangarajan, R., Seshadri, B., Jena, U., Das, K.C., Wang, H., Naidu, R., 2013. Landfills as a biorefinery to produce biomass and capture biogas. Biores. Technol. 135, 578–587.

Bioremediation and Bioeconomy. http://dx.doi.org/10.1016/B978-0-12-802830-8.00005-8
Copyright © 2016 Elsevier Inc. All rights reserved.

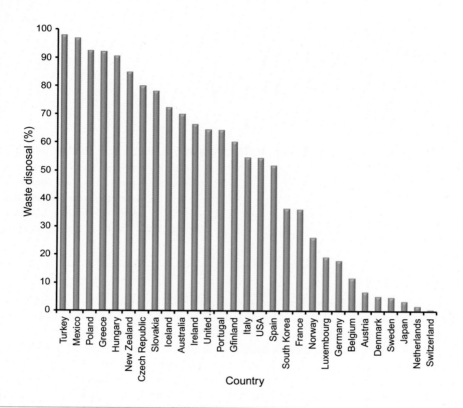

FIGURE 1

Percentage waste disposal in landfills in various countries (Chalmin and Gaillochet, 2009).

Landfill capping generally involves placement of a clay layer over the waste materials to minimize the percolation of rainwater or surface water into the waste (Scott et al., 2005). However, often this is not achieved due to formation of cracks as landfills age due to drying and wetting cycles associated with seasonal changes in rainfall and temperature (Albright et al., 2003). Alternative cover systems such as phytocaps are increasingly being considered for use at a range of waste disposal sites in many countries. Phytocapping of landfills generally takes two approaches (Venkatraman and Ashwath, 2009). One approach is to revegetate the area with endemic plant species for the creation of a wildlife corridor or habitat for local fauna (Venkatraman and Ashwath, 2009). The other is to establish plantations of high-biomass plant species for production of plant-based products.

Phytocapping involves placing a layer of soil material and growing dense vegetation on top of a landfill. In this system, soil and the associate root system act mainly as a "sponge" to hold and store water, and phytocapping minimizes water entry through the cover (Venkatraman and Ashwath, 2009). This alternative technology also enhances the aesthetic qualities of landfills that are mostly adjacent to urban communities and introduces economic benefits such as biomass generation for energy, timber, and fodder (Venkatraman and Ashwath, 2009).

3 LANDFILL BIOMASS FOR ENERGY PRODUCTION

The biomass produced from the landfill sites can be used in many ways including energy generation and biochar production. The abundance of biomass and scope for their continuous availability if managed effectively will make them one of the sustainable sources of renewable energy resources. Bridgewater (2012) defines biomass as any renewable source of fixed carbon emerging from wood, wood residues, agricultural crops, and their residues. If the landfill sites are managed effectively for the production of biomass, it has wider economic potential, apart from the environmental benefits.

Although phytocapping landfill sites were initially targeted to manage leachate generation, the practice also showed promise in producing biomass, which can be used as an energy source. For example, Venkatraman and Ashwath (2009) selected 19 different tree species and compared their shoot biomass yields, where the trees were established in thick soil covers (Figure 2). They observed the significance of soil cover in holding water, thereby reducing leachate generation and increasing biomass production. Table 1 presents the potential for biomass production from landfill sites from selected countries. The United States, China, and India are among the countries with highly prospective landfill sites available, in terms of area, for biomass production. However, there are issues concerning the growth of plants on landfill sites, which include insufficient nutrients, heavy metal abundance, unsuitable soil properties (e.g., pH), and selection of suitable plant species (Lamb et al., 2012).

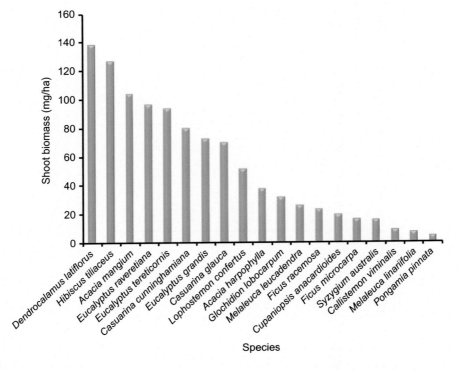

FIGURE 2

Dry matter production from a landfill site (Venkatraman and Ashwath, 2009).

Table 1 Potential Landfill Area and Volume of Biomass Produced from Landfill Sites in Various Countries

Region	Country	Landfill Area (ha)[a]	Potential Biomass Yield from Landfill Site (mg/year)[b]					
			Sunflower	Sugarcane	Giant Reed	Willow	Switch	Miscanthus
Oceania	Australia	296	17,265	16,456	20,992	3706	2550	6286
	New Zealand	38	2223	2118	2702	477	328	809
America	United States	5146	299,635	285,587	364,317	64,321	44,253	109,089
	Canada	289	16,822	16,033	20,453	3611	2484	6124
	Brazil	1308	76,158	72,588	92,598	16,349	11,248	27,727
Europe	United Kingdom	1260	73,358	69,919	89,194	15,748	10,834	26,708
Asia	Japan	1319	76,808	73,207	93,389	16,488	11,344	27,964
	India	788	45,866	43,716	55,767	9846	6774	16,699
	China	3398	197,843	188,567	24,0551	42,470	29,219	72,029
	Hong Kong	158	9186	8756	11,169	1972	1357	3345
	Taiwan	127	7408	7061	9008	1590	1094	2697
Africa	South Africa	170	9927	9462	12,070	2131	1466	3614
Biomass yield (mg/ha)			58.23	55.5	70.8	12.5	8.6	21.2

[a] Potential area of landfill (ha) = weight of waste collected in landfill (mg)/(bulk density of landfill waste (mg/m³) × average depth of landfill (6 m) × 10,000(m²/ha)). However, recommended depth of landfill differs for countries. For example, United States = 28 m, Australia = 20 m, and China = 22 m.
[b] Potential biomass production (mg/year) = Yield (mg/ha/year) × potential landfill site area (ha).

Typically landfills with clay caps have high bulk density and low nutrient supply. Application of inorganic and organic amendments can also improve the landfill soil physical and chemical properties. A number of studies examined the value of biosolids as an amendment to mitigate the issues associated with nutrients and heavy metals. For instance, Lamb et al. (2012) monitored the growth of *Arundo donax* L. (giant reed), *Brassica juncea* (L.) Czern. (Indian mustard), and *Helianthus annuus* L. (sunflower) on a landfill site with different biosolid amendment rates (0, 25, and 50 mg/ha). Cultivation on the landfill cap and amendment with biosolids significantly improved the characteristics of the soil. Growth of each plant species increased in response to biosolid addition (Figure 3). Giant reed produced the largest biomass at 50 mg/ha biosolid amendment rate (38 mg/ha dry weight).

Ni et al. (2006) divided biomass feedstocks for energy production into four general categories:

- Energy crops: herbaceous energy crops, woody energy crops, industrial crops, agricultural crops, and aquatic crops
- Agricultural residues and waste: crop waste and animal waste
- Forestry waste and residues: mill wood waste, logging residues, tree, and shrub residues
- Industrial and municipal wastes: municipal solid waste, sewage sludge, and industry waste

Some of these types of feedstocks may be produced in landfill sites and can be used in generation of heat/electrical energy or as transport fuels in the form of solid (e.g., wood), liquid (oilseeds), or gas (McKendry, 2002).

Feedstocks play a major role in energy production (Atadashi et al., 2011). The potential biomass production from landfill sites has been calculated for a number of crops (Table 1). Some of the prospective crops with high potential biomass production include oilseeds (e.g., sunflower), grains (e.g., maize and rapeseed), and grasses (e.g., sugarcane and giant reed).

Biomass is the most common form of renewable energy. Traditionally, biomass is used as a common domestic energy source in rural areas of the developing regions, but it plays a limited role in

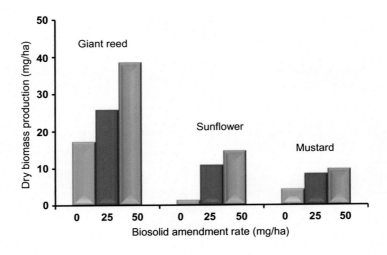

FIGURE 3

Effect of biosolid application on dry matter production from a landfill site (Lamb et al., 2012).

commercial energy production. Considerable efforts have been made to identify suitable biomass species with the aim to provide high-energy outputs and replace fossil fuel energy sources. The type of biomass required for industrial energy production is principally determined by the energy conversion process and the form in which the energy is required (McKendry, 2002). Numerous plant species have been tested for commercial energy farming internationally. Among those energy crops identified, perennial grasses/herbaceous plants may be suitable for biomass production on closed landfill sites where minimal water infiltration into buried waste is the primary objective of the capping soil. The C_4 plants, such as *Miscanthus* and switchgrass (Tao et al., 2011), have been proposed as main perennial grass species for energy production because of their high photosynthesis rates. In addition to grasses, woody plants could also be used for biomass production on closed landfill sites where a phytocap is built on purpose (Venkatraman and Ashwath, 2009). An ideal energy crop for a landfill site should have high yield, low input for maintenance and harvesting, and low nutrient demand.

4 TECHNOLOGIES FOR BIOMASS CONVERSION TO ENERGY

Biochemical and thermochemical technologies are the two main processes for conversion of biomass to energy (Figure 4). The preparation of biomass feedstocks before initiating the biomass conversion involves certain physical and chemical processes such as crushing, oil extraction, fermentation, and transesterification. In the case of biochemical processes, acid hydrolysis and enzyme activities are predominant reactions that facilitate conversion (Harmsen et al., 2010). After the initial preparation, most biomass feedstocks can be processed in a thermochemical conversion plant. The energy yield of a crop can be calculated based on the dry weight of biomass yield and the

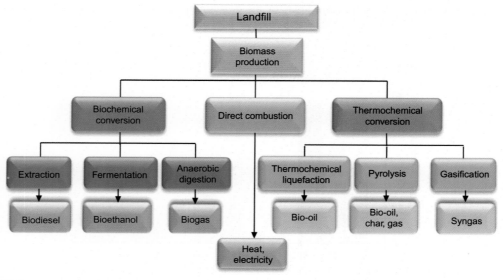

FIGURE 4

General routes of conversion of biomass and landfill waste into fuels and products.

heating value of the biomass. Most plant species have similar energy content of biomass (on a dry, ash-free basis), lying in the range 17-21 MJ/kg (McKendry, 2002). Therefore, biomass yield usually determines the energy yield.

4.1 BIOCHEMICAL PROCESSES

Some biomass species are better suited for biochemical conversion processes to produce gaseous or liquid fuels (e.g., bioethanol). Biochemical processes include fermentation, anaerobic digestion, and esterification. In a biochemical conversion process, the proportions of cellulose and lignin in biomass have significant influence on energy yield, because cellulose is much more biodegradable than lignin. The overall conversion of the plant material present as cellulose is greater than for plants with a higher proportion of lignin. Hence, ammonia pretreatment is used for the removal of lignin and decrystallization of cellulose (Kim and Lee, 2005). For ethanol production, a biomass feedstock with high cellulose/hemi-cellulose content is needed to provide a high yield (Tao et al., 2011). For example, up to 280 l/t of ethanol can be produced from switchgrass, because it has high content (up to 90%) of biodegradable cellulose/hemi-cellulose and low lignin content (5-20%). In contrast, 205 l/t ethanol can be produced from wood, largely due to the increased proportion of lignin (up to 30%) in wood (McKendry, 2002). Atabani et al. (2012) found that the cost of biomass feedstocks dominated the overall biodiesel production cost by around 75%, and therefore the selection of ideal feedstock is important for sustainable biodiesel production.

4.2 THERMOCHEMICAL PROCESSES

Thermochemical routes of biomass conversion are more attractive and have certain advantages such as higher productivity, complete utilization of feedstocks leading to multiple products, applicability to a wide range of feedstocks, independence of climatic conditions, and better control over the process relative to biological processes (Verma et al., 2012). Thermochemical processes include pyrolysis, liquification, and gasification.

Pyrolysis is the thermal decomposition of biomass to liquid (bio-oil), solid (char), and gaseous products in the absence of oxygen at 350-700 °C (Goyal et al., 2008; Mohan et al., 2006). Changes that take place during pyrolysis reactions are a combination of heat and mass transfer (Bridgewater, 2012; Mohan et al., 2006). Based on the operating temperature (low to high) and residence time (long to short), the process is known as slow pyrolysis, intermediate pyrolysis, or fast/flash pyrolysis (Table 2). Pyrolysis liquids are complex mixtures of oxygenated aliphatic and aromatic compounds (Meier and Faix, 1999) and consist of a tar fraction and pyroligneous acids. The tar components contain native resins, intermediate carbohydrates, phenols, aromatics, aldehydes, their condensation products, and other derivatives. Pyroligneous acids consist of 50% acetic acid (CH_3OH), acetone (C_3H_6O), phenols, and water. Biochar generally contains elemental C, H, and various inorganic species including N, P, K, Si, etc. (Goyal et al., 2008). Physical characteristics such as porosity, pH, affinity to metal sorption, and slow release of nutrients make biochar an attractive soil amendment. It helps improve soil quality and reduces cost of plant production by reducing fertilizer requirements. Since biochar has an estimated longevity of 100-1000 years in soil, it can be considered as a significant sequestration of CO_2 from the atmosphere to the soil. Pyrolysis gases mainly contain CO_2, CO, CH_4, H_2, C_2H_6, C_2H_4, and minor

Table 2 Typical Product Yields of Pyrolysis

Type	Conditions	Yields (%)		
		Liquid	Char	Gas
Slow (carbonization)	Low temperature (300-700 °C), long solids and vapor residence time (600-6000 s), Particle size: 5-50 mm	30	35	35
Intermediate	Moderate temperature (around 500 °C), moderate vapor residence time ~10-20 s	50	20	30
Fast	Moderate to high temperature (500-1000 °C), moderate vapor residence time ~1 s, particle size (<1 mm)	75	12	13

Compiled from Maschio et al. (1992), Mohan et al. (2006), and Zhang et al. (2010).

amounts of higher organics and water vapor (Goyal et al., 2008). These gases can be either combusted to generate process heat for the feedstock drying process or can be converted into liquid synfuels.

Another thermal approach for bioremediation of waste and biomass is their wet conversion via hydrothermal carbonization (HTC), a process that involves thermal processing of biomass organics in hot, compressed water (175-295 °C and 2-5 MPa pressure) (Libra et al., 2011; Hoekman et al., 2011). This process converts organics and wastes into a carbon-rich solid known as "hydrochar" that has higher energy and mass densities, homogeneity, friability, and hydrophobicity and is easier for handling, transportation, and storage (Hoekman et al., 2011). An advantage of HTC over many dry biomass conversion processes is that heterogeneous wet organic residues and waste streams can be processed without preliminary pretreatment steps such as separating and drying (Libra et al., 2011). The HTC process involves several reactions including hydrolysis, dehydration, decarboxylation, condensation, polymerization, and aromatization, resulting in a loss of some carbon, hydrogen, and oxygen and altering the thermal, chemical, and structural properties (Funke and Ziegler, 2010). The liquid effluents obtained in the HTC treatment can be sterilized, recycled, and reused for further HTC processes or can be used as low-value agricultural uses after inexpensive treatments. The HTC process is also an effective way of densifying the energy content of biomass, changing its physical, chemical, and thermal behavior, and sequestering CO_2 (Funke and Ziegler, 2010; Libra et al., 2011; Sevilla et al., 2011; Román et al., 2012; Reza et al., 2014). A range of feedstocks can be processed using HTC, including woody and herbaceous feedstocks (Yan et al;, 2010; Hoekman et al., 2011; Hoekman et al., 2012; Hoekman et al., 2013; Kalderis et al., 2014), algae biomass, (Broch et al., 2013), fecal biomass (Danso-Boateng et al., 2013), agricultural wastes (Román et al., 2012; Oliveira et al., 2013), municipality solid wastes, and food processing waste (Berge et al., 2011), and digested sludge from anaerobic digestion of agricultural residues and wastewater treatment sludge (Mumme et al., 2011). The HTC process results in 40-78% hydrochar that has 20-40% higher energy content than the raw biomass (Hoekman et al., 2011; Reza et al., 2014). The herbaceous feedstocks result in a lower yield of hydrochar than that of the woody feedstocks. However, an overall higher energy densification is reported in HTC in all feedstocks in the literature. The energy contents of HTC hydrochar from woody feedstocks and herbaceous feedstocks were reported as 28-30 MJ/kg and 23-25 MJ/kg, respectively. Another benefit of HTC is its ability to partition inorganic contents in the biomass into the aqueous phase. Large amounts of nitrogen and phosphorous and other organics have been found in the HTC of algae (Broch et al., 2013). A large amount of nonvolatile residues

(7-14%) have been reported in the HTC of woody and herbaceous feedstocks (Hoekman et al., 2011). The majority of these nonvolatile components could be due to disposition of inorganic elements (metals and nonmetals) from the biomass. The useful components could be recovered for plant growth and soil application. The HTC of digested sewage sludge has a product yield of 80-94%, an energy yield of $\geq$90%, energy densification of 1.05-1.36, and a fuel ratio (ratio of fixed carbon to volatile matter) of 0.13-0.27 (compared to 0.11 for raw sludge) (Kim et al., 2014). It is also characterized by higher porosity (larger micropores and mesopores) and improved surface area, suggesting it could be a better combustible material and have superior drying properties.

Thermochemical liquefaction (TCL) is the conversion of biomass in sub/supercritical water reactions. TCL resembles the pyrolysis process in some ways, where biomass is heated in an oxygen-free atmosphere in the presence or absence of a catalyst that enhances the hydrolysis, depolymerization, and repolymerization/condensation reactions to produce bio-oil (also called "bio-crude") as the major product along with gases, water solubles, and char as coproducts (Akhtar and Amin, 2011; Balat, 2008; Yan et al., 2010). However, unlike pyrolysis where the biomass conversion takes place at near or slightly above atmospheric pressure, in TCL, biomass is converted at significantly higher pressures (5-40 atm) and lower temperatures (280-370 °C) (Balat, 2008). A key difference is that TCL can process wet biomass without drying and is therefore energetically more efficient than pyrolysis (Jena and Das, 2011).

Gasification is the conversion of biomass into a combustible gaseous fuel by partial oxidation of biomass at high temperature in the range of 800-900 °C (Goyal et al., 2008). The process can take place in air, oxygen, or steam as the reaction medium (McKendry, 2002). The resulting gas, known as producer gas (or syngas), is a mixture of CO, CO_2, H_2, CH_4, and N_2 gases. The gas is more versatile than the original solid biomass and can be burned to produce heat and steam, or used in internal combustion engines or gas turbines to produce electricity. Syngas can be combusted in boilers or engines to generate electricity or can be converted to liquid synthetic fuels by Fischer-Tropsch reactions.

5 SUMMARY AND CONCLUSION

While landfilling provides a simple pathway for waste disposal, it causes environmental impacts including leachate generation and GHG emissions. Advanced phytocapping technique not only provides a major source of biomass for energy production, but also energy through various chemical conversion processes including pyrolysis, gasification, and cogeneration. Although the potential landfill area available for biomass production is relatively less compared to other marginal and degraded lands (e.g., mine sites), landfill sites are readily accessible, thereby providing an attractive option for biomass production.

By determining the biochemical characteristics of biomass crops, suitable microorganisms and enzymes can be selected for biochemical conversion of biomass to fuels (e.g., anaerobic digestion). Thermochemical conversion technologies are fast and efficient methods for treating solid wastes/high-moisture wastes and also provide scope for cogeneration of renewable fuels and chemicals. While pyrolysis can generate liquid energy (bio-oil), gasification can produce gas that can potentially be used for running engines/turbines to generate electricity.

REFERENCES

Akhtar, J., Amin, N.A.S., 2011. A review on process conditions for optimum bio-oil yield in hydrothermal liquefaction of biomass. Renew. Sustain. Energ. Rev. 15 (3), 1615–1624.

Albright, W.H., Benson, C.H., Gee, G.W., Abichou, T., Roesler, A.C., Rock, S.A., 2003. Examining the alternatives. Civil Eng. 73 (5), 70–75.

Albright, W.H., Benson, C.H., Gee, G.W., Abichou, T., McDonald, E.V., Tyler, S.W., Rock, S.A., 2006. Field performance of a compacted clay landfill final cover at a humid site. J. Geotech. Geoenviron. 132 (11), 1393–1403.

Atabani, A., Silitonga, A., Badruddin, I.A., Mahlia, T., Masjuki, H., Mekhilef, S., 2012. A comprehensive review on biodiesel as an alternative energy resource and its characteristics. Renew. Sustain. Energ. Rev. 16 (4), 2070–2093.

Atadashi, I., Aroua, M., Aziz, A.A., 2011. Biodiesel separation and purification: a review. Renew. Energy 36 (2), 437–443.

Balat, M., 2008. Mechanisms of thermochemical biomass conversion processes. Part 3: reactions of liquefaction. Energ. Source Part A 30 (7), 649–659.

Berge, N.D., Ro, K.S., Mao, J., Flora, J.R., Chappell, M.A., Bae, S., 2011. Hydrothermal carbonization of municipal waste streams. Environ. Sci. Technol. 45 (13), 5696–5703.

Bridgewater, A.V., 2012. Upgrading biomass fast pyrolysis liquids. Environ. Prog. Sustain. Energy 31 (2), 261–268.

Broch, A., Hoekman, S.K., Unnasch, S., 2013. A review of variability in indirect land use change assessment and modeling in biofuel policy. Environ. Sci. Policy 29, 147–157.

Brunner, A., Munsell, J., Gagnon, J., Burkhart, H., Zipper, C., Jackson, C., Fannon, A., Stanton, B., Shuren, R., 2009. Hybrid Poplar for Bioenergy and Biomaterials Feedstock Production on Appalachian Reclaimed Mine Land. Virginia Tech, Blacksburg, VA. Powell River Project Research and Education Program Progress Reports.

Chalmin, P., Gaillochet, C., 2009. From Waste to Resource. An Abstract of World Waste Survey, 2009. Cyclope and Veolia Environmental Services, Paris.

Danso-Boateng, E., Holdich, R., Shama, G., Wheatley, A., Sohail, M., Martin, S., 2013. Kinetics of faecal biomass hydrothermal carbonisation for hydrochar production. Appl. Energ. 111, 351–357.

Funke, A., Ziegler, F., 2010. Hydrothermal carbonization of biomass: a summary and discussion of chemical mechanisms for process engineering. Biofuel. Bioprod. Bior. 4 (2), 160–177.

Goyal, H., Seal, D., Saxena, R., 2008. Bio-fuels from thermochemical conversion of renewable resources: a review. Renew. Sustain. Energ. Rev. 12 (2), 504–517.

Harmsen, P., Huijgen, W., Bermudez, L., Bakker, R., 2010. Literature Review of Physical and Chemical Pretreatment Processes for Lignocellulosic Biomass. Wageningen UR Food & Biobased Research, Netherlands.

Hoekman, S.K., Broch, A., Robbins, C., 2011. Hydrothermal carbonization (HTC) of lignocellulosic biomass. Energy Fuels 25 (4), 1802–1810.

Hoekman, S.K., Broch, A., Robbins, C., Zielinska, B., Felix, L., 2012. Hydrothermal carbonization (HTC) of selected woody and herbaceous biomass feedstocks. Biomass Convers. Bioref. 3 (2), 113–126.

Hoekman, S.K., Broch, A., Robbins, C., Purcell, R., Zielinska, B., Felix, L., Irvin, J., 2013. Process development unit (PDU) for hydrothermal carbonization (HTC) of lignocellulosic biomass. Waste Biomass Valor. 5 (4), 669–678.

Jena, U., Das, K., 2011. Comparative evaluation of thermochemical liquefaction and pyrolysis for bio-oil production from microalgae. Energy Fuels 25 (11), 5472–5482.

Kalderis, D., Kotti, M., Méndez, A., Gascó, G., 2014. Characterization of hydrochars produced by hydrothermal carbonization of rice husk. Solid Earth Discuss. 6 (1), 657–677.

Kim, T.H., Lee, Y.Y., 2005. Pretreatment of corn stover by soaking in aqueous ammonia. In: Proc. of Twenty-Sixth Symposium on Biotechnology for Fuels and Chemicals.

Kim, D., Lee, K., Park, K.Y., 2014. Hydrothermal carbonization of anaerobically digested sludge for solid fuel production and energy recovery. Fuel 130, 120–125.

Lamb, D.T., Heading, S., Bolan, N., Naidu, R., 2012. Use of biosolids for phytocapping of landfill soil. Water Air Soil Poll. 223 (5), 2695–2705.

Libra, J.A., Ro, K.S., Kammann, C., Funke, A., Berge, N.D., Neubauer, Y., Titirici, M.-M., Fühner, C., Bens, O., Kern, J., 2011. Hydrothermal carbonization of biomass residuals: a comparative review of the chemistry, processes and applications of wet and dry pyrolysis. Biofuels 2 (1), 71–106.

Maschio, G., Koufopanos, C., Lucchesi, A., 1992. Pyrolysis, a promising route for biomass utilization. Biores. Technol. 42 (3), 219–231.

McKendry, P., 2002. Energy production from biomass (part 1): overview of biomass. Bioresour. Technol. 83 (1), 37–46.

Meier, D., Faix, O., 1999. State of the art of applied fast pyrolysis of lignocellulosic materials—a review. Bioresou. Technol. 68 (1), 71–77.

Mohan, D., Pittman, C.U., Steele, P.H., 2006. Pyrolysis of wood/biomass for bio-oil: a critical review. Energy Fuels 20 (3), 848–889.

Mumme, J., Eckervogt, L., Pielert, J., Diakité, M., Rupp, F., Kern, J., 2011. Hydrothermal carbonization of anaerobically digested maize silage. Bioresou. Technol. 102 (19), 9255–9260.

Ni, M., Leung, D.Y., Leung, M.K., Sumathy, K., 2006. An overview of hydrogen production from biomass. Fuel process. Technol. 87 (5), 461–472.

Oliveira, I., Blöhse, D., Ramke, H.-G., 2013. Hydrothermal carbonization of agricultural residues. Bioresour. Technol. 142, 138–146.

Reza, M.T., Andert, J., Wirth, B., Busch, D., Pielert, J., Lynam, J.G., Mumme, J., 2014. Hydrothermal carbonization of biomass for energy and crop production. Appl. Bioenerg. 1 (1), 2300–3553.

Román, S., Nabais, J., Laginhas, C., Ledesma, B., González, J., 2012. Hydrothermal carbonization as an effective way of densifying the energy content of biomass. Fuel Process. Technol. 103, 78–83.

Scott, J., Beydoun, D., Amal, R., Low, G., Cattle, J., 2005. Landfill management, leachate generation, and leach testing of solid wastes in Australia and overseas. Crit. Rev. Env. Sci. Tec. 35 (3), 239–332.

Sevilla, M., Maciá-Agulló, J.A., Fuertes, A.B., 2011. Hydrothermal carbonization of biomass as a route for the sequestration of CO_2: chemical and structural properties of the carbonized products. Biomass. Bioenerg. 35 (7), 3152–3159.

Tao, L., Aden, A., Elander, R.T., Pallapolu, V.R., Lee, Y., Garlock, R.J., Balan, V., Dale, B.E., Kim, Y., Mosier, N.S., 2011. Process and technoeconomic analysis of leading pretreatment technologies for lignocellulosic ethanol production using switchgrass. Bioresour. Technol. 102 (24), 11105–11114.

Venkatraman, K., Ashwath, N., 2009. Phytocapping: importance of tree selection and soil thickness. Water Air Soil Poll. Focus 9 (5–6), 421–430.

Verma, M., Godbout, S., Brar, S.K., Solomatnikova, O., Lemay, S.P., Larouche, J.P., 2012. Biofuels production from biomass by thermochemical conversion technologies. Int. J. Chem. Eng. 2012, 18.

Yan, W., Hastings, J.T., Acharjee, T.C., Coronella, C.J., Vásquez, V.R., 2010. Mass and energy balances of wet torrefaction of lignocellulosic biomass. Energy Fuels 24 (9), 4738–4742.

Zhang, M., Chen, H.-P., Gao, Y., He, R.-X., Yang, H.-P., Wang, X.-H., Zhang, S.-H., 2010. Experimental study on bio-oil pyrolysis/gasification. BioResources 5 (1), 135–146.

BIOENERGY FROM PHYTOREMEDIATED PHYTOMASS OF AQUATIC PLANTS VIA GASIFICATION

6

Srujana Kathi

Pondicherry University, Kalapet, Puducherry, India

1 INTRODUCTION

As the demand for energy sources continues to surge globally, the need for exploring new renewable sources of energy rises. While the transport sector continues to expand in the United States and Europe, growth in the emerging economies of India and China is predicted to be substantially greater, growing by at least 3% per year (IEA, 2006). Rapid depletion of currently used nonrenewable fossil fuels such as coal, petroleum, and natural gas is a much-debated issue. At the same time the liberation of trapped carbon from fossil fuels into the atmosphere in the form of carbon dioxide has led to increased discussion about global warming. The irregular distribution of fossil fuel resources geographically renders many countries severely dependent on oil imports (Kumar et al., 2009). There is clear scientific evidence that human activities coupled with fossil fuel combustion and land-use change causes emission of greenhouse gases (GHG) such as carbon dioxide, methane, and nitrous oxide, which are disturbing the Earth's climate (IPCC, 2007). Therefore, exploration for other renewable forms of energy appears inevitable. When compared with alternative technologies such as hydrogen and fuel cells, biofuels are currently the only fuels compatible with the existing engine technologies. Climate change, energy security, and rural development are three main drivers for the development of bioenergy and biofuels.

After failing to eradicate most terrestrial and aquatic weeds, people started exploring ways and means to make use of these weeds. Constantly harvesting aquatic macrophytes might keep them under check, thereby allowing a gainful use of them. Biogas generation has been found to be a sustainable option among several uses of these weeds (Deshmukh, 2012). Aquatic weeds possess the ability to fulfill the biomass requirements during biogas production (Mshandete, 2009). In recent years, phytoremediation has been increasingly used for improving the water quality in contaminated water bodies. Direct dumping, burning, high-temperature decomposition, chemical extraction, and phytomining have been suggested for the disposal of plants with high levels of heavy metals after phytoremediation. The disposal of plants used in remediation has been one of the main concerns.

Phytoremediation is a promising, cost-effective, plant-based technology designed to remove contaminants such as heavy metals, pesticides, and xenobiotics from wastewater, employing ecologically sustainable processes associated with natural wetland ecosystems (Dipu et al., 2011). The biomass produced in phytoremediation could be economically valorized in the form of bioenergy, contributing to an important environmental cobenefit. Biogas technology provides an approach to utilize certain categories

Bioremediation and Bioeconomy. http://dx.doi.org/10.1016/B978-0-12-802830-8.00006-X
Copyright © 2016 Elsevier Inc. All rights reserved.

of biomass for meeting partial energy needs. The Biomass Action Plan and the EU (European Union) Strategy for Biofuels set an ambitious vision for 2030, that up to a quarter of the EU's transport fuel needs could be met by clean and CO_2-efficient biofuels. In the United States, the Energy Independence and Security Act of 2007 established a renewable fuel standard, totalling 36 billion gallons (1 billion gallons biodiesel) by 2022 (Gomes, 2012). The development of renewable energy and reduction of GHG emissions are now established priorities within EU policy, the United States, China, and many other countries. The following targets are defined by the EU for 2020: to increase the proportion of renewable energy source in the EU's final energy consumption to at least 20%; to increase the proportion of biofuels in the road transportation sector to at least 10% in each member state (Ericsson et al., 2009).

Among thermochemical routes of digestion of lignocellulosic biomass, basing on high conversion efficiency, gasification can be preferred. Gasification is a high-temperature transformation of biomass utilizing any type of lignocellulosic material as feedstock. Hence this process can be preferred when phytoremediated phytomass from aquatic macrophytes is used as the source of biomass (Alauddin et al., 2010). This chapter presents a description of aquatic weeds acquainted with decontamination of water, analysis of biogas generation from phytoremediated phytomass of water weeds, mechanism of biogas generation including cleaning and upgrading, factors influencing biogas yield, bioeconomy of biogas generation, and constraints in the usage of phytoremediated aquatic macrophytic phytomass for biogas generation.

2 AQUATIC WEEDS FOR DECONTAMINATION OF WATER

Aquatic macrophytes play a key role in the structural and functional balance of aquatic ecosystems by altering water movement regimes, providing shelter to fish and aquatic invertebrates, serving as a food source and altering water quality by regulating oxygen balance, nutrient cycles, and accumulating heavy metals. The ability to hyperaccumulate heavy metals makes them interesting research areas, especially for the treatment of industrial effluents and sewage wastewater (Sood et al., 2012). Infestation of aquatic weeds in the water bodies leads to sedimentation and unsuitability for domestic use, interference with navigation, effects on fisheries, amplification in breeding habitat to disease-transmitting mosquitoes, blocking irrigation canals, and evapotranspiration. A lot of soil, nutrients, and heavy metals are carried away in the runoff in the lower reaches of the catchment areas. As a result, heavy aquatic weed invasion and silting occurs in different kinds of water bodies. Aquatic plants can be used for biological treatment and stabilization of contaminated water bodies (Khankhane and Varshney, 2009).

Both constructed and natural wetlands are currently being used for heavy metal removal and wastewater quality control. The diversified macrophytes growing in the wetlands enables the retention of contaminants such as metals from water passing through them. Aquatic macrophytes encompass many common weeds, enabling cost-effective treatment and remediation technologies for wastewaters contaminated with inorganic and organic compounds (Prasad, 2013). The use of aquatic plants to absorb pollutants from water bodies is termed as rhizofiltration or phytofiltration (Mahendran, 2014). Plants preserve the natural structure and texture of soil using solar energy; therefore, phytoremediation is an ecologically compatible tool practically suitable to clean up a wide array of environmental contaminants (Kathi and Khan, 2011).

Water hyacinth (*Eichhornia crassipes* (Martius)) is a member of family Pontederiaceae. It is a monocotyledonous freshwater weed, native to the Brazil and Ecuador region. It is used in traditional

FIGURE 1

Irrigation canals in India are choked by *Eichhornia crassipes* (water hyacinth). Phytoremediation function of water hyacinth has been elucidated in a number of research publications (Singhal and Rai, 2003; Bhattacharya and Kumar, 2010). Mechanical harvesting and use as feed stock for biogas production is a profitable solution.

medicine and even used to remove toxic elements from polluted water bodies (Figure 1a–c, Singhal and Rai, 2003; Bhattacharya and Kumar, 2010). They reproduce both asexually through stolons and sexually through seeds. The seeds remain viable for up to 20 years and therefore are difficult to control (Bhattacharya and Kumar, 2010). The rapid proliferation of water hyacinth can result in detrimental effects, including exhausting oxygen in water and interfering with navigation and recreation. Wang et al. (2011) while evaluating the potential of *E. crassipes* and *Lolium perenne* L. that serve as phytoremediating plants by floating bed on Guxin River in Hangzhou, concluded that the growing of *E. crassipes* can reduce turbidity of water by adsorbing a mass of silt from the river. Moreover, harvesting *E. crassipes* aids in carrying the contaminant away from the contaminated water body. The removal of water hyacinth from a polluted water body might contribute to water quality improvement. Studies on water hyacinth (*E. crassipes*), channel grass (*Vallisneria spiralis*), and water chestnut (*Trapa bispinnosa*) used as phytoremediating plants for remediating industrial effluents, demonstrated that the slurry of these plants produces significantly more biogas than the slurry of control plants grown in unpolluted water (Witters et al., 2012).

Pistia stratiotes L. is a hyperaccumulator, known to remove heavy metals, organic compounds, and radionuclides from water bodies. It purifies the polluted aquatic system contaminated with harmful metals. The lower size of the plant aids in the removal of heavy metals, which is an additional advantage as compared to water hyacinth. This plant can be successfully exploited for biofuel production thereby contributing to weed management, water pollution mitigation, relieving energy problems, and protecting the aquatic ecosystem (Khan et al., 2014). Several studies have proved reed canary grass (*Phalaris arundinacea* L.) to be a noteworthy substrate for bioenergy production (Lakaniemi et al., 2011; Kacprzak et al., 2012). There is need to evaluate the potential biogas yield of such aquatic macrophytes as *Sacciolepsis africana*, *Ipomea cornea*, *Vossia cuspidate*, and *Pista statiotes* (Adeleye et al., 2013).

3 BIOGAS GENERATION FROM PHYTOREMEDIATED BIOMASS OF WATER WEEDS

Biogas, a clean and renewable form of energy, could substitute for conventional sources that cause ecological and environmental issues (Yadvika et al., 2004). Biomass constitutes lignin, hemicellulose, cellulose, mineral matter, and ash. It possesses high moisture and volatile matter constituents, low bulk density, and high calorific value. The primary rationale for exploiting aquatic plant biomass is that their primary productivity rates are significantly higher than the terrestrial bioenergy feedstocks. The conversion of the biomass of harvested aquatic macrophytes into combustible biogas proved to be an inevitable option for the control and management of environmental pollution associated with aquatic macrophytes. Potential productivity of water hyacinth and water chestnut in nutrient-enriched wastewaters has led to its selection for phytoremediation of various industrial effluents and subsequently produced biomass as a feedstock for biogas production to achieve economic success in energy harvest (Verma et al., 2007). Unlike corn-based ethanol production, which can greatly affect the world food market, nonfood, crop-based, cellulosic biofuels have the advantage of not competing with food sources. Total plant biomass in order of concentration consists of cellulose, lignin, and hemicelluloses that are cellulosic materials, in addition to ash and extractives. Ash consists mostly of metallic salts, metal oxides, and trace mineral residues left over after burning plant biomass. When considering bioenergy and biofuel production from aquatic macrophytes, lignin and cellulose assumes significance as they are combustible parts of plant biomass and are consumed on burning (Witters et al., 2012).

Mitsubishi Heavy Industries, Ltd. (MHI) has been working to develop a variety of commercially viable technologies to support biomass energy. MHI successfully performed once-through operation of woody biomass gasification to methanol synthesis process at their demonstration plant with a throughput of $2\,\text{tons}\,\text{day}^{-1}$ (Hishida et al., 2011). The conversion of the biomass of harvested aquatic macrophyte water hyacinth from the Niger Delta into renewable energy demonstrated an inevitable option for the control and management of environmental pollution associated with aquatic macrophytes and their usability for poverty alleviation in the Niger Delta region of Nigeria (Adeleye et al., 2013). The saccharification of cell walls for the production of reduced sugars for conversion to value-added products or ethanol has been well described. Duckweed, the most investigated aquatic macrophyte, can produce a theoretical ethanol yield reaching $6.42 \times 10\,3\,1\,\text{ha}^{-1}$, 50% more ethanol when compared to maize, which is the main ethanol-producing feedstock in many countries (Miranda et al., 2014).

4 MECHANISM OF BIOGAS GENERATION

Biomass fuels and residues can be converted to more valuable energy forms via a number of processes including thermal, biological and mechanical, or physical processes. Thermal conversion contributes to multiple and complex products, in very short reaction times making use of inorganic catalysts to improve the product quality, unlike biological processing which is usually very selective and produces a small number of discrete products using biological catalysts. Three main thermal processes—namely, pyrolysis, gasification, and combustion—are available for converting biomass to a more useful energy form (Bridgwater, 2012). Among all thermochemical conversion processes, gasification is considered one of the promising ways to convert the energy content of biomass into a clean fuel, which can be used for synthesis of methanol and other liquid fuels (Kaewpanha et al., 2014). Steam gasification of biomass can produce a gaseous fuel with a relatively higher H_2 content having relatively low tar content (Shen et al., 2008; Figure 2).

The biomass gasification technology package consists of a fuel- and ash-handling system and gasification system: reactor, gas cooling, and cleaning system. Torrefaction is a biomass pretreatment method carried out at approximately 200-300 °C in the absence of oxygen. During this process, biomass is completely dried and partially decomposed, losing its fibrous structure. If the biomass is pretreated remotely combining with pelletization, very energy-dense fuel pellets are produced, which reduces transportation costs. Pretreatment of biomass prevents biological degradation, facilitating long-term storage. Process efficiency can be improved by dewatering and drying biomass prior to gasification. Plugging of feeders can be reduced by properly drying biomass, which improves air emissions. Many types of dryers are used in drying biomass, including direct and indirect fired-rotary dryers, conveyor dryers, cascade dryers, flash or pneumatic dryers, and superheated steam dryers. Selection of the appropriate dryer depends on several factors such as size and characteristics of the feedstock, requirements

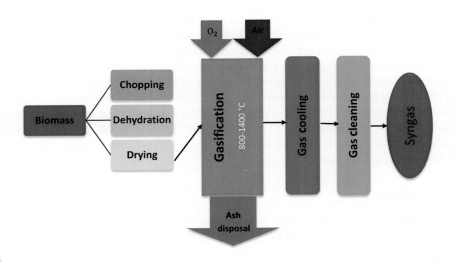

FIGURE 2

Schematic diagram for syngas generation from biomass.

of operation and maintenance, energy efficiency, environmental emissions, availability of waste heat sources, available space, potential fire hazard, and capital cost.

In a biomass gasifier, biomass is heated with controlled amounts of oxygen, air, and steam so that only a relatively small portion of the fuel burns. This "partial oxidation" process provides the heat. Rather than burning, most of the carbon-containing feedstock is chemically broken apart by the gasifier's heat and pressure, setting into motion chemical reactions that produce syngas. The producer gas/syngas consists of carbon monoxide (CO) and hydrogen (H_2) along with gaseous constituents such as methane (CH_4), carbon dioxide (CO_2), and nitrogen (N_2) (Das and Najmul Hoque, 2014). The resultant mixture of hydrogen and carbon monoxide is then cooled and converted at elevated temperature or via a Fischer-Tropsch reaction. Thus the key to gasifier design is to create conditions such that the biomass is reduced to charcoal and the charcoal is then converted at suitable temperature to produce CO and H_2.

4.1 GASIFICATION REACTIONS

The chemical reactions taking place in the process of gasification (Figure 3) are:

1. *Drying*: Biomass fuels usually contain up to 35% moisture. When the biomass is heated to around 100 °C, the moisture gets converted into steam.
2. *Pyrolysis*: Pyrolysis is the thermal decomposition of biomass fuels in the absence of oxygen process occurring at around 200-300 °C. Biomass decomposes into solid charcoal, liquid tars, and gases.
3. *Oxidation*: Air is introduced into a gasifier in the oxidation zone. The oxidation takes place at about 700-1400 °C, in which the solid carbonized fuel reacts with oxygen in the air, producing carbon dioxide and releasing 406 kJ/g mol of heat energy.
4. *Reduction*: At higher temperatures and under reducing conditions several reactions take place, which results in formation of CO, H_2, and CH_4 (Kishore, 2008). Reduction in a gasifier is accomplished by passing carbon dioxide (CO_2) or water vapor (H_2O) across a bed of red-hot charcoal. The carbon in the hot charcoal is highly reactive with oxygen. It has high oxygen affinity that strips the oxygen off water vapor and carbon dioxide and redistributes it to as many single bond sites as possible. The oxygen is more attracted to the bond site on the charcoal than to itself. Therefore, no free oxygen can survive in its usual diatomic O_2 form. All available oxygen will bond to available charcoal sites until all the oxygen is completed. When all the available oxygen is redistributed as single atoms, reduction stops. Through this process, CO_2 is reduced by carbon to produce two CO molecules, and H_2O is reduced by carbon to produce H_2 and CO. Both H_2 and CO are combustible fuel gases, and those fuel gases can then be piped off to do desired work elsewhere.

4.2 TYPES OF GASIFIERS

Biomass gasifiers are classified according to the way air or oxygen is introduced. The design of gasifier depends on the type of fuel used and whether the gasifier is portable or stationary. Gas producers are classified according to how the air blast is introduced in the fuel column. Gasifiers are classified as: updraft gasifier, downdraft gasifier, fluidized bed gasifier, crossdraft gas producer, etc.

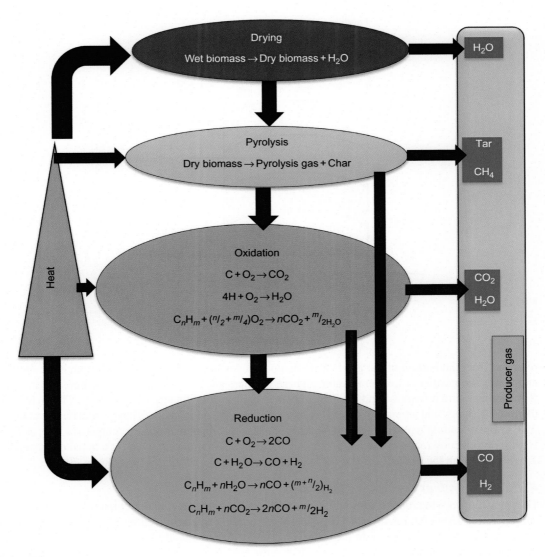

FIGURE 3

The reaction mechanism for thermal gasification of biomass.

4.2.1 Updraft or countercurrent gasifier

The oldest and simplest type of gasifier is the countercurrent or updraft gasifier. A diagram of a fluidized bed gasifier is shown in Figure 4. An updraft gasifier has clearly defined zones for partial combustion, reduction, and pyrolysis. In this gasifier, biomass is fed at the top of the gasifier; air enters from the bottom and acts as countercurrent to fuel flow. The gas thus produced is collected from the top. Ashes are

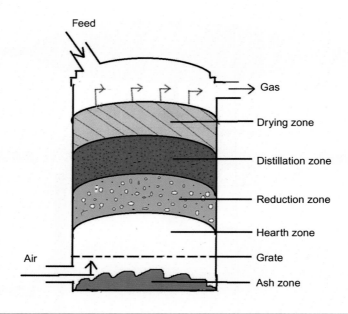

FIGURE 4

Updraft gasifier.

removed from the bottom of the gasifier. In the updraft gasifier the hot gas passes through the fuel bed and leaves the gasifier at low temperature. These gasifiers are best suited for applications where high-flame temperature is required and moderate amounts of dust in the fuel gas is acceptable. They have high thermal efficiency and are more tolerant for fuel switching. Updraft gasifiers have outlet temperatures of 250 °C and operating temperatures of 800-1200 °C. They can handle moisture contents as high as 55%. High tar production requires extensive cleaning of the syngas produced, which is a noted disadvantage. Tar removal from the product gas has been a major problem in updraft gasifiers (Roos, 2010).

4.2.2 Downdraft or cocurrent gasifiers

These gasifiers are the most commonly used and easy to control, having outlet temperatures of 800 °C and operating temperatures of 800-1200 °C. The resultant gas is removed at the bottom of the apparatus, so that fuel and gas move in the same direction, as schematically shown in Figure 5. One of the important drawbacks of downdraft gasifiers is that the feedstock must have a moisture content of about 20% or lower. Depending on the temperature of the hot zone and the residence time of the tarry vapors, a more or less complete breakdown of the tars is achieved. The tar-free gas thus produced is suitable for engine applications. Downdraft gasifiers produce syngas that typically has low tar and particulate content. They can produce as much as 20% char, but more typically char content is 2-10% (Roos, 2010).

4.2.3 Cross-draft gasifier

Cross-draught gasifiers, schematically illustrated in Figure 6, are an adaptation for the use of charcoal. Charcoal gasification results in very high temperatures (1500 °C and more) in the oxidation zone, which can lead to material problems. Air or air/steam mixtures are introduced into the side of the gasifier near

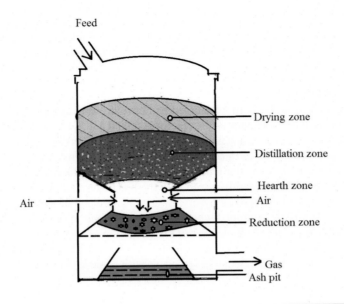

FIGURE 5

Downdraft gasifier.

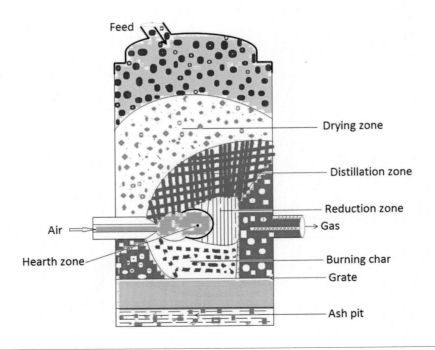

FIGURE 6

Cross-draft gasifier.

the bottom while the product gas is drawn off on the opposite side. Normally an inlet nozzle is used to bring the air into the middle of the combustion zone. They are simpler to construct and more suitable for running engines than the other types of fixed-bed gasifiers. However, they are sensitive to changes in biomass composition and moisture content.

4.2.4 Fluidized bed gasifiers

Unlike fixed-bed reactors, models with a fluidized bed have no distinct reaction zones such as drying, pyrolysis, and gasification in the reactor as the reactor is mixed and, so, close to isothermal. The design of fluidized bed gasifier is illustrated schematically in Figure 7. Here, air is blown through a bed of solid particles at a sufficient velocity in order to keep them in suspension state. The bed is externally heated. The feedstock is introduced after reaching a suitable temperature. The fuel particles introduced at the bottom of the reactor can be readily mixed with the bed material and can be instantly heated up to the bed temperature, resulting in very fast pyrolysis. Gasification and tar-conversion reactions occur in the gas phase. Most systems are equipped with an internal cyclone in order to minimize char blowout as much as possible. Ash particles are also carried over the top of the reactor and should be removed from the gas stream if the gas is used in engine applications. Problems with feeding, instability of the bed, and fly-ash sintering in the gas channels can occur with some biomass fuels. Feedstock flexibility

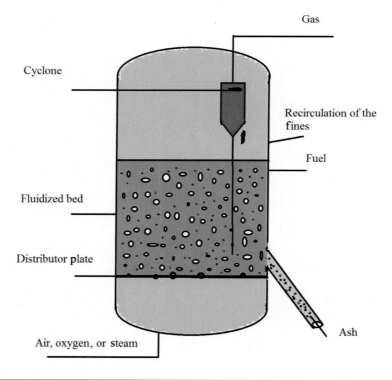

FIGURE 7

Fluidized bed gasifier.

is one of the important advantage of these gasifiers. It has the ability to deal with fluffy and fine-grained materials (sawdust, etc.) without the need of preprocessing.

Other biomass gasifier systems such as double fired, entrained bed, and molten bath are currently under development. Though there is a considerable overlap of the processes, each can be assumed to occupy a separate zone where fundamentally different chemical and thermal reactions take place.

5 CLEANING AND UPGRADING BIOGAS

The gas cleaning step is often considered an important phase in the commercialization of gasification process. In addition to the syngas, gasification produces several contaminants including tars, sulphur, and nitrogen compounds. These compounds must be abated to meet process requirements and emission standards as these gases might disrupt the subsequent catalytic conversion processes. Gas cleanup is technically complex where the output composition of the gaseous mixture is determined by the characteristics of the feedstock (Royal Society, 2008). Gasification is followed by passing the resultant gas through a cold trap and collecting in a gas bag. For the separation of carbon monoxide, methane, and carbon dioxide using helium as carrier gas, the gas compositions should be analyzed using a gas chromatograph (Kaewpanha et al., 2014). Since raw syngas produced resembles more conventional fuels like coal and oil residues, similar cleaning technologies can be adopted. There is a profitable transaction between gas cleaning and catalyst performance. Cleaning well below the specifications might be economically attractive for synthesis processes that use sensitive and expensive catalytic materials. The cleaning process includes a filter, rectisol unit, and downstream gas polishing to remove the traces. This process comprises zinc oxide and active carbon filtering. A water gas-shift reactor will also be part of gas conditioning because the H_2/CO ratio needs adjustment. The Rectisol unit combines the removal of the bulk of the impurities and the separation of CO_2 (van der Drift and Boerrigter, 2005).

Tar removal is a major challenge to make syngas valorization technologies technically and commercially feasible. Tar tolerance levels for gas engines equal $50\,mg\,Nm^{-3}$; gas turbines equal $5\,mg\,Nm^{-3}$, and fuel cells equal $1\,mg\,Nm^{-3}$ (Fjellerup et al., 2005). Tar removal methods can be divided into two categories: (1) primary methods/treatments inside the gasifier, and (2) secondary methods/hot gas cleaning. For advanced syngas applications such as usage in gas engines, primary methods are not sufficient. Primary methods can be used as a tool to optimize the gas composition for the secondary cleaning step. Secondary tar reduction methods can be divided into physical/mechanical and chemical methods. Several physical methods such as cyclones, filters (baffle, fabric/ceramic), rotating particle separators, electrostatic precipitators, and scrubbers (water/organic-liquid-based) are available. Chemical methods are further subdivided into thermal cracking, catalytic cracking, and plasma cracking methods (Bosmans et al., 2013). By adapting, demonstrating, and optimizing existing industrial processes as well as evolving innovative separation processes based on membranes such as pressure swing adsorption it is possible to upgrade synthesis gas. These upgrading processes can be integrated with carbon sequestration technologies to improve GHG balance. In the conversion of synthesis gas to biofuels it is necessary to develop, adapt, and demonstrate the Fischer-Tropsch processes as well as methanol/dimethyl ether and mixed alcohol processes, especially when biomass is utilized as the feedstock. Since thermochemical routes produce huge amounts of heat, the possibilities of using them in polygeneration systems should be explored (Public Consultation Draft, 2007). If biomass gasification is integrated with combined gas steam cycles, conversion efficiencies up to 50% might be achieved (Caputo et al., 2005).

6 FACTORS INFLUENCING BIOGAS YIELD

Gasification process is affected by a number of factors such as temperature, pressure and height of the reactor bed, fluidization velocity, gasifying medium, moisture content of feed material, particle size, equivalence ratio, air-to-steam ratio, and presence of catalysts. These parameters are quite interrelated and each of them affects the gasification rate, product gas heating value, process efficiency, and product distribution. Large variations in biogas yields were reported based on the composition of raw materials. The gas yield can also be affected by digestion temperature, retention time, load, digestion technology (codigestion; batch or continuous; one- or two-phase digestion), pretreatment of the raw materials, etc. (Berglund and Borjesson, 2006a).

The rate of formation of combustible gases can be increased by increasing the temperature and decreasing the yield of char, thus leading to complete conversion of the fuel. The amount of steam and reaction temperature have significant influence on the gasification performance of biomass. The yield of hydrogen increases with the rise of reaction temperature and amount of steam while excessive steam reduces the hydrogen yield (Kaewpanha et al., 2014). Hydrocarbon gases such as methane and ethylene emission increase with temperature, while the yields of higher hydrocarbons (C3-C8) decrease at temperatures above 650 °C. Energy content of the syngas steadily increases up to 700 °C and then decreases at still higher temperatures.

With increase in pressure the rate of char gasification and yields of methane increases, having significant impacts at high temperatures (900-950 °C). For a given reactor temperature, higher fuel beds the reaction time thereby increasing total syngas yields. Higher fluidization velocities increase the temperature of the fuel bed and lead to the production of syngas lower in energy content. The equivalence ratio means actual fuel-to-air ratio divided by the stoichiometric fuel-to-air ratio. The higher the equivalence ratio, the higher is the rate of syngas produced and vice versa. Increase in the steam-to-air ratio increases the energy content of the syngas (Sadaka, 2009). Thermal gasification is effective mostly for feedstock with low water content before the gasification process. Low nitrogen and alkali content as well as a feedstock particle size are other desirable properties. It is faster than anaerobic digestion but has a relatively low conversion efficiency rate of 20-40%. New technologies such as the integrated combined gasification cycle are expected to increase conversion efficiency up to 50%. Recently, cogasification with coal has become popular because it facilitates the gasification of smaller quantities of plant material with lower investment and higher conversion efficiency (Elbehri et al., 2013). Some studies have found that heavy metals associated with phytoremediated phytomass affect the activity of the microorganisms that participate in biogas digestion (Cao et al., 2015).

7 BIOECONOMY OF BIOGAS PRODUCTION

Bioeconomy is defined by the EU as the sustainable production and conversion of biomass for food, health, energy, and industrial products. Biofuels and the energy sector are one of the components of the newly emerging bioeconomy. It has increasingly been recognized globally that plant-based raw materials will eventually replace fossil reserves as feedstocks for industrial production, addressing both the energy and nonenergy sectors including chemicals and materials (EC, 2004). Bioeconomy brings together phenomena that have so far been unrelated, such as biomass and products for mainstream consumers; business and sustainability; ecosystem services and industrial applications. It is about sustainability, closed circles, cross-sectoral cooperation, and the use of biomass. Therefore, bioeconomy can be understood as a large, sociotechnical system that binds together technologies, markets, people, and policies defined by principles rather than the sectoral borderlines of existing industries (People in the Bioeconomy, 2014).

Considering international effects on energy balance, the biogas supply creates external economies substituting fossil fuels. At the same time it reduces import duties due to fossil fuel trading. Moreover, biogas replaces conventional fuels, contributing to sustainable and green environment. The price of supplied energy produced by biogas competes with distorted prices on the energy market. Therefore, under competitive conditions, a decentralized economically self-sufficient biogas unit provides its energy without market distortions. In this context it is appropriate to recognize that global markets for bioautomotive fuels are rapidly expanding with governmental support. Countries such as Brazil and the United States are supporting the usage of bioethanol, while Germany and Austria are encouraging biodiesel (derived from vegetable oils) for transportation purposes. From the viewpoint of economics, environment, land use, water use, chemical fertilizer use, etc., there is a strong inclination for the use of woody, grassy materials and agricultural residues, municipal wastes, and industrial wastes as a feedstock. Therefore, the production of the synthetic transportation fuels such as biomethanol, bioethanol, dimethyl ether, Fischer-Tropsch fuel, synthetic natural gas, and hydrogen via gasification and synthesis seems to be hopeful (Zhang, 2010). Furthermore, from the macroeconomic point of view, decentralized energy generation from biomass appears to be beneficial over centralized energy generation. This process reduces the costs of investment in additional networks such as transportation and the costs of losses on the transmission network (ISAT, 1999).

Comprehensive knowledge of the performance of various types of biofuels is essential to make choices between promising biofuels, whether to use as neat fuel or blended with existing fossil fuels. It is essential to guarantee the adaptability of engine configurations to a specific biofuel. This in turn has to be widely available at the fuel filling stations to make conversion worthwhile. The development of standards for clean liquid and gaseous biofuels as well as for fossil fuels blended with biofuels is a prerequisite (Public Consultation Draft, 2007). In addition, it may be necessary to develop advanced vehicle and engine technologies compatible with biofuel/biofuel blends (e.g., E85, BTL). Developing alternative biofuels can open up the potential to deliver specifically designed engine technologies. For example, Volkswagen has developed a combined combustion system coupled to fuel derived from the Fischer-Tropsch process (Royal Society, 2008).

Considering all biofeedstock opportunities, the bioeconomy is expected to save 1-2.5 billion tons of CO_2 emission (0.27-0.7 billion tons carbon) by 2030. To reach this target, if syngas can be used as a source of industrial carbon, aquatic plant biomass might prove to be a valuable feedstock (Kircher, 2012). Some of the possible barriers to biomass gasification in heat and power markets are lack of industrial-scale demonstration of biomass technology, availability of fewer number of suitable gas turbines and limited expertise in high-pressure biomass gasification, challenges in ash sintering and fouling in gasification of agro-biomass, high cost of the present gas-cleaning systems, limited knowledge in the recovery of valuable metals and recycling of nutrients from gasification ash, limited feedstock flexibility of small-scale gasifiers, low efficiency of small gas engines, and high investment costs in small-scale combined heat and power production (Girio et al., 2013).

In order to avoid heavy costs incurred in transporting the raw materials to the refineries, preprocessing of feedstocks can be carried out in the place of their availability. This provides an opportunity for local people to gain employment in these biorefineries. Providing sufficient renewable carbon is the biggest challenge in unfolding the bioeconomy (Bio-economy Innovation, 2010). In this process, governments should play a significant role in promoting biofuel technologies by providing incentives and tax benefits to all the stakeholders involved (such as feedstock providers, automobile industries, marketing sector, etc.). Around 70% of the global biofuel production occurred in the United States of America (45.4%) and Brazil (22.5%). The only other countries worth mentioning are Germany (4.8%), Argentina (3.8%), and France (3%), with the rest of the world lagging behind considerably with regard to overall biofuel production (Energy Academy and Centre for Economic Reform and Transformation, 2013).

Transition from a fossil-based economy to bioeconomy is justified by the need for an integrated response to several global trends such as high dependence on fossil-based resources. This can be addressed by strengthening energy security, which calls for a more diversified supply option range; increasing demand of biological resources for bio-based products; increasing sustainability concerns such as reduction of GHG emissions, moving toward a zero waste society, environmental sustainability of primary production systems, and increasing land-use competition, etc. In order to tackle all of these interconnected global drivers and constraints, an integrated management of renewable biological resources in agriculture, food, bio-based, and energy industries appears to be the most desirable (Nita et al., 2013).

8 CONSTRAINTS IN BIOGAS GENERATION FROM PHYTOREMEDIATING AQUATIC MACROPHYTES

Securing a stable supply of biomass is one of the obstacles to implementing this technology. Yields from different aquatic plants and sometimes the same aquatic plant grown under different environmental conditions can be quite variable. Water hyacinth produced greater biomass than hydrilla and *Azolla* spp. grown in the river Ganga (Wilkie and Evans, 2010). Differences in handling of the raw materials can increase the energy input in the biogas systems significantly. The environmental impact of biogas systems depends largely on the raw material digested, the energy efficiency in the biogas production chain, uncontrolled losses of methane, and the status of the end-use technology. The fuel-cycle emissions are normally significantly higher from biogas systems based on raw materials that require extensive handling—for example, dedicated ley crops—than from systems based on raw materials requiring limited handling, such as food industry waste at large industrial units. Even though an energy-rich raw material could improve the biogas yield, it might not be added if it makes the digestate unsuitable as fertilizer. When evaluating the impacts of biogas usage, it is therefore essential to include all emissions from the biogas fuel-chain, not only the end-use emissions but also to clearly define the production chain (Berglund and Borjesson, 2006a,b).

Minor quantities of certain metals such as Ni, Co, Mn, and Fe are found to enhance biogas potential stimulating the bacterial activities. On the contrary, metals such as Cu, As, and Pb may induce an inhibitory effect due to their toxic effect on the microbes. Certain heavy metal ions such as Cu, Pb, Cr (VI), and Zn can inactivate enzymes during biogasification (Selling et al., 2008). Heavy metals accumulated in the plant parts during phytoextraction might affect the activity of microorganisms that participate in biogas digestion (Cao et al., 2015). When bioslurry and residue produced during biogas generation from biomass are used as biofertilizers, they affect soil structure and soil microbial communities. The physiological and biochemical role of metal ions contained in residues on the soil microbiota need to be studied further. There are chances that the slurry because of high liquidity, if not managed well, might enter into water or groundwater through runoff or leaching, thereby causing water eutrophication (Feng et al., 2011). Biogas yield might be reduced by the presence of certain metals. When contaminant-enriched biomass crops are used for energy purposes, their impact on conversion efficiency as well as the energy should be considered. First, verifying that there is no impact of metals on the biomass production; second, investigating the impact of the elevated presence of trace elements on energy conversion efficiency; and third, the alternative use and safe disposal of the remaining product are the key issues to be addressed while utilizing phytoremediated plant biomass for energy production (Witters et al., 2012). Long-term effects of nickel, cobalt, and molybdenum on the biogas process should, however, be studied in continuous or semicontinuous reactors (Pobeheim et al., 2010).

Singhal and Rai (2003), who studied the use of water hyacinth and channel grass used for phytoremediation of industrial effluents in biogas production, suggested that there is a good possibility to use phytoremediation plants in biogas production. They also opined that arsenic content and speciation in sludge should be investigated to prevent its redistribution in the environment. More research is required to study the uptake of contaminants by various plants, combustion rests and emissions, content in biofuel under different conditions, and different technological processes in order to be confident that the end-products of phytoremediation can be used for bioenergy with minimal environmental impacts (Mahendran, 2014).

9 CONCLUSIONS

Biogas systems utilizing phytoremediated phytomass of aquatic macrophytes have the potential to be an effective strategy in combating some of the important environmental problems relating to climate change, phytoremediation, eutrophication, acidification, and water pollution. This study establishes that biogas production from phytoremediated biomass of aquatic macrophytes is a commercially potential and sustainable process. Aquatic macrophytic biomass is a globally available resource. Such a novel treatment system not only directly improves water quality but also benefits indirectly in the reduction of air pollution by replacing fossil fuels. The study of biogas systems is important in improving phytoremediation systems and can be referred to for sustainable disposal of plants used in the bioremediation. The availability of aquatic macrophytes for biogas production in water bodies needs to be explored. Proper combination of green technologies is most essential in an era of increased energy crisis, pollution, and global warming.

A critical analysis of literature reveals that thermal gasification is the most reliable technology to displace the usage of fossil fuels and to reduce CO_2 emission, especially when plant biomass is used as feedstock. These biofuels should mimic the properties of petrol and diesel such as in octane equivalent, energy density, cetane number, and hydrophobicity. It is essential to design and develop new engine technologies that can directly substitute fossil fuels with advanced biofuel delivering, in addition to significantly lowering GHG emissions. For vehicle manufacturers to make the investments needed, a long-term market for transport fuels containing a high blend of biofuels must be established. Economic and regulatory incentives are needed to accelerate the technology developments needed to deliver biofuel supply chains that can provide more substantial reduction of emissions.

The use of bioenergy resulting from phytoremediation is constrained by the lack of knowledge regarding the emissions that may be generated and the issues associated with pollution transfer, especially heavy metals. The key goal for bioeconomy of biofuels is to fulfill social, economic, and environmental criteria in a sustainable manner with the integration of desperate fields.

ACKNOWLEDGMENTS

The author would like to thank Pondicherry University for providing internet access to collect the information essential for writing this book chapter.

REFERENCES

Adeleye, B.A., Adetunji, A., Bamgboye, I., 2013. Towards deriving renewable energy from aquatic macrophytes polluting water bodies in Niger Delta region of Nigeria. Res. J. Appl. Sci. Eng. Technol. 5 (2), 387–391.

Alauddin, Z.A.B., Lahijani, P., Mohammadi, M., Mohamed, A.R., 2010. Gasification of lignocellulosic biomass in fluidized beds for renewable energy development: a review. Renew. Sust. Energy Rev. 14 (9), 2852–2862.

Berglund, M., Borjesson, P., 2006a. Assessment of energy performance in the life-cycle of biogas production. Biomass Bioenergy 30, 254–266.

Berglund, M., Borjesson, P., 2006b. Environmental systems analysis of biogas systems—Part I: fuel-cycle emissions. Biomass Bioenergy 30, 469–485.

Bhattacharya, A., Kumar, P., 2010. Water hyacinth as a potential biofuel crop. Electron. J. Environ. Agric. Food Chem. 9 (1), 112–122.

Bio-economy Innovation, 2010. Bioeconomy Council Report: Research and Technology Development for Food Security, Sustainable Resource Use and Competitiveness. Research and Technology Council, Berlin.

Bosmans, A., Wasan, S., Helsen, L., 2013. 2nd International Enhanced Landfill Mining Symposium, Houthalen-Helchteren.

Bridgwater, A.V., 2012. Review of fast pyrolysis of biomass and product upgrading. Biomass Bioenergy 38, 68–94.

Cao, Z., Wang, S., Wang, T., Chang, Z., Shen, Z., Chen, Y., 2015. Using contaminated plants involved in phytoremediation for anaerobic digestion. Int. J. Phytoremediation 17 (3), 201–207. http://dx.doi.org/10.1080/15226514.2013.876967.

Caputo, A.C., Palumbo, M., Pelagagge, P.M., Scacchia, F., 2005. Economics of biomass energy utilization in combustion and gasification plants:effects of logistic variables. Biomass Bioenergy 28, 35–51.

Das, B.K., Najmul Hoque, S.M., 2014. Assessment of the potential of biomass gasification for electricity generation in Bangladesh. J. Renewable Energy 2014. http://dx.doi.org/10.1155/2014/429518. Article ID 429518, 10 pp.

Deshmukh, H.V., 2012. Economic feasibility and pollution abetment study of biogas prodution process utilizing admixture of *Ipomoea carnea* and distillery waste. J. Environ. Res. Dev. 7 (2), 633–642.

Dipu, S., Kumar, A.A., Salom Gnana Thanga, V., 2011. Potential application of macrophytes used in phytoremediation. World Appl. Sci. J. 13 (3), 482–486.

EC, 2004. Towards a European Knowledge-Based Bioeconomy. Workshop Conclusions on the Use of Plant Biotechnology for the Production of Industrial Biobased Products. EUR 21459. European Commission, Directorate-General for Research, Brussels, Belgium. Available online at http://ec.europa.eu/research/agriculture/library_en.htm.

Elbehri, A., Segerstedt, A., Liu, P., 2013. Biofuels and Sustainability Challenge: A Global Assessment of Sustainability Issues, Trends and Policies for Biofuels and Related Feedstocks. Food and Agriculture Organization of the United Nations, Rome. pp. 42.

Energy Academy and Centre for Economic Reform and Transformation, 2013. BP Statistical Review of World Energy. Heriot-Watt University. www.bp.com/.../bp/.../statistical-review/statistical_review_of_world_energy.

Ericsson, K., Rosenqvis, H., Nilsson, L.J., 2009. Energy crop production costs in the EU. Biomass Bioenergy 33 (11), 1577–1586.

Feng, H., Qu, G-f., Ning, P., Xiong, X.-f., Jia, L-j., Shi, Y-k., Zhang, J., 2011. The resource utilization of anaerobic fermentation residue. Procedia Environ. Sci. 11, 1092–1099.

Fjellerup, J., Ahrenfeldt, J., Henriksen, U., Gobel, B., 2005. Formation, Decomposition and Cracking of Biomass Tars in Gasification. Technical University of Denmark (DTU) Biomass Gasification Group, Denmark.

Girio, F., Kurkela, E., Kiel, J., Lankhorst, R.K., 2013. Longer term R&D needs and priorities on bioenergy – bioenergy beyond 2020. European Energy Research Alliance, European Industrial Bioenergy Initiative Workshop Report.

Gomes, H.I., 2012. Phytoremediation for bioenergy: challenges and opportunities. Environ. Technol. Rev. 1 (1), 59–66. http://dx.doi.org/10.1080/09593330.2012.696715.

Hishida, M., Shinoda, K., Akiba, T., Amari, T., Yamamoto, T., Matsumoto, K., 2011. Biomass syngas production technology by gasification for liquid fuel and other chemicals. Mitsubishi Heavy Ind. Tech. Rev. 48 (3), 37–41.

IEA, 2006. World Energy Outlook 2006. International Energy Agency, Paris. Chapters 2, 3, 14.

IPCC, 2007. In: Solomon, S., Qin, D., Manning, M., Chen, Z., Marquis, M., Averyt, K.B., Tignor, M., Miller, H.L. (Eds.), Climate Change 2007: The Physical Science Basis. Contribution of Working Group 1 to the Fourth Assessment Report of the Intergovernmental Panel on Climate Change. Cambridge University Press, Cambridge and New York.

ISAT, 1999. Biogas Digest, Vol. III: Biogas – Costs and Benefits and Biogas. Information and Advisory Service on Appropriate Technology, Frankfurt. pp. 2–61.

Kacprzak, A., Matyka, M., Krzystek, L., Ledakowicz, S., 2012. Evaluation of biogas collection from reed canary grass, depending on nitrogen fertilisation levels. Chem. Process Eng. 33 (4), 697–701.

Kaewpanha, M., Guan, G., Hao, X., Wang, Z., Kasai, Y., Kusakabe, K., Abudula, A., 2014. Steam co-gasification of brown seaweed and land-based biomass. Fuel Process. Technol. 120, 106–112.

Kathi, S., Khan, A.B., 2011. Phytoremediation approaches to PAH contaminated soil. Indian J. Sci. Technol. 4 (1), 56–63.

Khan, M.A., Marwat, K.B., Gul, B., Wahid, F., Khan, H., Hashim, S., 2014. *Pistia stratiotes* L. (Araceae): phytochemistry, use in medicines, phytoremediation, biogas and management options. Pakistan J. Bot. 46 (3), 851–860.

Khankhane, P.J., Varshney, J.G., 2009. Possible use of giant reed, *Arundo donax* for phytoremediation of runoff water in a catchment area. In: National Consultation on Weed Utilization, Paper presented in National Consultation on Weed Utilization, 20-21 October, 2009, DWSR Jabalpur, p. 28.

Kircher, M., 2012. The transition to a bio-economy: emerging from the oil age. Biofuel. Bioprod. Bior. 6, 369–375. http://dx.doi.org/10.1002/bbb.

Kishore, V.V.N. (Ed.), 2008. Renewable Energy Engineering and Technology a Knowledge Compendium. The Energy and Resources Institute, New Delhi.

Kumar, A., Jones, D.D., Hanna, M.A., 2009. Thermochemical biomass gasification: a review of the current status of the technology. Energies 2, 556–581. http://dx.doi.org/10.3390/en20300556.

Lakaniemi, A.M., Koskinen, P.E.P., Nevatalo, L.M., Kaksonen, A.H., Puhakka, J.A., 2011. Biogenic hydrogen and methane production from reed canary grass. Biomass Bioenergy 35 (2), 773–780.

Mahendran, R.P., 2014. Phytoremediation – insights into plants as remedies. Malaya J. Biosci. 1 (1), 41–45.

Manninen, J., Nieminen-Sundell, R., Belloni, K. (Eds.), 2014. People in the Bioeconomy 2044. VTT Technical Research Centre of Finland, pp. 8–9. www.vtt.fi/publications/index.jsp.

Miranda, A.F., Muradov, N., Gujar, A., Stevenson, T., Nugegoda, D., Ball, A.S., Mouradov, A., 2014. Application of aquatic plants for the treatment of selenium-rich mining wastewater and production of renewable fuels and petrochemicals. J. Sust. Bioenergy Syst. 4, 97–112. http://dx.doi.org/10.4236/jsbs.2014.41010.

Mshandete, A.M., 2009. The anaerobic digestion of cattailweeds to produce methane using American cockroach gut microorganisms. ARPN J. Agric. Biol. Sci. 4 (1), 45–57.

Nita, V., Benini, L., Ciupagea, C., Kavalov, B., Pelletier, N., 2013. Bio-economy and sustainability: a potential contribution to the Bio-economy Observatory. European Commission, Joint Research Centre, Institute for Environment and Sustainability, European Commission.

Pobeheim, H., Munk, B., Johansson, J., Guebitz, G.M., 2010. Influence of trace elements on methane formation from a synthetic model substrate for maize silage. Bioresour. Technol. 101, 836–839.

Prasad, M.N.V., 2013. Heavy Metal Stress in Plants: From Biomolecules to Ecosystems. Springer Science & Business Media. 2nd edition, p. 365.

Public Consultation Draft, 2007. Biofuels Technology Platform Report, European Biofuels Technology Platform. www.biofuelstp.eu/.../070926_PublicConsultationDraftBFTPReport.pdf.

Roos, C.J., 2010. Clean Heat and Power Using Biomass Gasification for Industrial and Agricultural Projects. US Department of Energy, Clean Energy Application Centre, Northwest.

Royal Society, 2008. Sustainable Biofuels: Prospects and Challenges. The Royal Society, London. https://royalsociety.org/~/media/Royal_Society_Content/.../7980.pdf. Policy document (2008).

Sadaka, S., 2009. Gasification, Producer Gas and Syngas. University of Arkansas, Little Rock. www.uaex.edu/publications/PDF/FSA-1051.pdf.

Selling, R., Hakansson, T., Bjornsson, L., 2008. Two-stage anaerobic digestion enables heavy metal removal. Water Sci. Technol. 57 (4), 553–558.

Shen, L., Gao, Y., Xiao, J., 2008. Simulation of hydrogen production from biomass gasification in interconnected fluidized beds. Biomass Bioenergy 32, 120–127.

Singhal, V., Rai, J.P.N., 2003. Biogas production from water hyacinth and channel grass used for phytoremediation of industrial effluents. Bioresour. Technol. 86, 221–225.

Sood, A., Uniyal, P.L., Prasanna, R., Ahluwalia, A.S., 2012. Phytoremediation potential of aquatic macrophyte, Azolla. Ambio 41, 122–137. http://dx.doi.org/10.1007/s13280-011-0159-z.

van der Drift, A., Boerrigter, H., 2005. Synthesis gas from biomass for fuels and chemicals. In: SYNBIOS Conference, Stockholm.

Verma, V.K., Singh, Y.P., Rai, J.P.N., 2007. Biogas production from plant biomass used for phytoremediation of industrial wastes. Bioresour. Technol. 98, 1664–1669.

Wang, H., Zhang, H., Cai, G., 2011. An application of phytoremediation to river pollution remediation. Procedia Environ. Sci. 10, 1904–1907.

Wilkie, A.C., Evans, J.M., 2010. Aquatic plants: an opportunity feedstock in the age of bioenergy. Biofuels 1 (2), 311–321.

Witters, N., Mendelsohn, R.O., Van Slycken, S., Weyens, N., Schreurs, E., Meers, E., Tack, F., Carleera, R., Vangronsveld, J., 2012. Phytoremediation, a sustainable remediation technology? Conclusions from a case study. I: Energy production and carbon dioxide abatement. Biomass Bioenergy 39, 454–469.

Yadvika, S., Sreekrishnan, T.R., Kohli, S., Rana, V., 2004. Enhancement of biogas production from solid substrates using different techniques—a review. Bioresour. Technol. 95, 1–10.

Zhang, W., 2010. Automotive fuels from biomass via gasification. Fuel Process. Technol. 91, 866–876.

JATROPHA CURCAS L. CULTIVATION ON CONSTRAINED LAND: EXPLORING THE POTENTIAL FOR ECONOMIC GROWTH AND ENVIRONMENTAL PROTECTION

A.A. Warra[1], M.N.V. Prasad[2]

Kebbi State University of Science and Technology, Aliero, Nigeria[1]
University of Hyderabad, Hyderabad, Telangana, India[2]

1 INTRODUCTION

Jatropha curcas (Euphorbiaceae), a multipurpose, drought-resistant perennial, has been gaining considerable importance recently beyond its use as a biofuel (Dyer et al., 2012; Edrisi et al., 2015; Ehrensperger et al., 2014; Gubitz et al., 1999; Heller, 1996; Jongschapp et al., 2007; Maes et al., 2009). It is a tropical plant that can be grown in low to high rainfall areas either on farms as a commercial crop or as hedge boundaries to protect fields from grazing animals and to prevent erosion (Rodríguez-Acosta et al., 2010). Most commonly known as "physic nut," other names used to describe the plant vary by region or country: In Mali it is known as *pourghere*, in the Ivory Coast as *bagani*, in Senegal as *tabanani*, in Tanzania as *makaen/mmbono* (Anonymous, 2006). In Nigeria it is known as *binidazugu/cinidazugu* and *lapa lapa* in the Hausa and Yoruba languages, respectively, (Blench, 2007). The seed contains 40-50% viscous oil known as curcas oil (Verma and Gaur, 2009). *J. curcas* L. is a subtropical shrub unique for its fruits, which have a high content of nonedible oil suitable for the preparation of cosmetics. Cosmetic potential, Soxhlet extraction, physicochemical, analysis, and cold process saponification of Nigerian *J. curcas* L. seed oil (Tomar et al., 2015; Warra, 2012; Warra et al., 2012) and the moisture-dependent physical properties of *Jatropha* seed has been reported (Garnayak et al., 2008).

Some recent researches predominantly focused on *Jatropha* for cultivation and establishment of nurseries instead of investigating in the entire *Jatropha* value chain, which turned out to be a challenge. Growing a productive *Jatropha* crop was much more complex than initially anticipated (Slingerland

Bioremediation and Bioeconomy. http://dx.doi.org/10.1016/B978-0-12-802830-8.00007-1
Copyright © 2016 Elsevier Inc. All rights reserved.

and Schut, 2014). There is also a lack of information on basic agronomic properties of *J. curcas* L. cultivation on marginal lands in semiarid regions. Evaluation of agronomic performance of identified elite strains of *J. curcas* in marginal lands would be of paramount importance for addressing gap areas in their agronomic properties and subsequently for harnessing their optimum economic potentials (Ahamad et al., 2013) (Figures 1 and 2).

FIGURE 1

J. curcas plantation on constrained land.

FIGURE 2

High density *J. curcas* plantation on constrained land. In set harvested seeds for biodiesel and value addition and value chain products.

1.1 SITE REQUIREMENTS

J. curcas prefers well-drained sandy or gravelly soils with good aeration. It does not withstand heavy clay soils and all soils with risk of even ephemeral waterlogging. Soil depth should be at least 45 cm. Investigations are in progress to assess the yield potential of the crop in the different agroecological zones of Senegal. According to the preliminary results obtained on the field and the data mentioned in the literature, a minimum annual rainfall of 500 mm seems to be necessary to obtain a profitable seed yield in rainfed agriculture conditions (Saverys et al., 2008). Influence of land size on adoption of *J. curcas* in Yatta District, Kenya, it was reported that only 15.4% of farmers had adopted *J. curcas* cultivation. It was also found that total land size owned was not a major factor influencing land size allocated to *Jatropha* cultivation since total land size owned accounted for only 28% variance of land size allocated to *Jatropha* cultivation, and 78% of respondents were not willing to convert their pasture lands to *Jatropha* farms. Yatta District, which covers 246,700 ha, lies in the arid and semiarid regions of the country, making it suitable for *J. curcas* cultivation. *Jatropha* requires extensive land for high yields to be realized, and therefore marginal lands and wastelands are deemed appropriate to avoid competition with food. The study revealed that there was a strong negative correlation between the size of land a farmer owned and the size of land allocated for *J. curcas* cultivation, implying that the more land a farmer has, the less likely they are to engage in *Jatropha* cultivation. Farmers with lesser land size were more likely to take up *J. curcas* cultivation because they mostly seek a more economically rewarding enterprise to boost their low incomes as opposed to their counterparts with bigger land sizes. Size of land owned was not a key factor contributing to increase in adoption of *J. curcas* (Munyao et al., 2013). The marginal land with potential for planting energy plants including *Jatropha* was identified across Asia. The results indicated that the areas with marginal land suitable for *Cassava*, *Pistacia chinensis*, and *J. curcas* L. were established to be 1.12, 2.41, and 0.237 million km^2, respectively. Shrub land, sparse forest, and grassland are the major classifications of exploitable land (Fu et al., 2014). In China, the total area of marginal land exploitable for development of energy plants including *J. curcas* L. was on a large scale, about 43.75 million ha. As China has fairly limited cultivated land resources, it is widely acknowledged that the development of energy plants should not affect food security and the environment; therefore, bioenergy development in China may mainly rely on marginal land (Zhuang et al., 2011). To avoid the food versus fuel debate, the use of "marginal" land for biofuel feedstock production (*Jatropha*) has emerged as a dominant narrative. But both the availability and suitability of marginal land for commercial-level *Jatropha* production is not well understood nor has it been well examined, especially in Africa. Using a case study of large-scale *Jatropha* plantation in Ethiopia, report examines the process of land identification for *Jatropha* investments, and the agronomic performance of large-scale *Jatropha* plantation on marginal land (Wendimu, 2013). When marginal lands are exploited, the impact on the soil seems to be positive, depending on use practices and type of soil, but its contribution to soil restoration might be obtained at the expense of biodiversity loss (Negussie et al., 2015; Mergeai, 2008). However, a report examined whether it would be possible to sustainably produce *J. curcas* L. seeds on the marginal land situated close to the Senegal River. A 6-ha pilot plantation was cultivated under drip irrigation between September 2007 and November 2011, close to the village of Bokhol (lat. 16°31′N, long. 15°23′W). A series of tests were conducted on this plot in order to identify the best cultivation methods for the area (date, density and method of planting, appropriate type of pruning, fertilizers to be applied, irrigation method, etc.). The average yields obtained at this site after 4 years of cultivation (less than 500 kg ha^{-1} of dry seed), using the best known production techniques, are significantly lower than anticipated, compared to the available figures for the irrigated cultivation of *Jatropha* in other parts of the world. The main causes of this failure are the plant's limited useful vegetation period of 6 months per year, instead of 12, and the scale of attacks by a soil-borne vascular disease, which destroyed over 60% of the plantation within 4 years (Terren et al., 2012). It was also reported that *Jatropha* could be easily integrated into the microcatchments for water harvesting already used by most farmers (Maingi, 2010).

1.2 PROPAGATION AND PLANTATION ESTABLISHMENT

Direct seeding can give good results when it is carried out with good-quality seeds at the beginning of the rainy season in a place where the rains last longer than 4 months. The sowing of three seeds per hole is recommended with an early refilling of the missing holes. In the case of direct seeding, regular weeding operations are compulsory to avoid the complete disappearance of *J. curcas* plantlets due to the heavy concurrence of weeds during the growing season. Intercropping of *J. curcas* with annual crops (peanut, pearl millet, and okra) helps achieve this goal. An efficient solution to succeed in the establishment of a *J. curcas* plantation at a low cost is to precultivate seedlings for 2 months (till the plantlets reach the fifth true leaf stage) in seed beds under nursery conditions and to extract and transplant bare root seedlings in the field. The best results for nursery bed preparation were obtained by mixing in equal volumes sand and local soil found under trees (Saverys et al., 2008; Silva et al., 2015) (Figures 3 and 4).

FIGURE 3

(a) High yielding stem cuttings; (b) Sapling with fresh flush of leaves; (c) Plantation in constrained land; (d-f) Plants bearing clusters of fruits.

FIGURE 4

J. curcas plantation on constrained land (a and b) yielding high quality fruits and seeds (c and d).

1.3 CULTIVATION ON UNSUITABLE LAND

Generally, establishing *Jatropha* on natural areas has caused a significant loss of carbon stored in the biomass and an initial loss of soil carbon (observed in 3- to 4-year-old plantations; soil carbon stock might, however, increase again in the longer term) (Caldeira et al., 2004). If relatively low carbon stock areas are transformed into *Jatropha* plantations, the carbon stock generally increases. Nevertheless, the potential for carbon sequestration (the uptake and storage of carbon-containing substances, in particular carbon dioxide) is low due to limited biomass buildup caused by continuous pruning and short rotation length (Zah et al., 2013). An attempt to show how greenhouse gas (GHG) emissions from land use change associated with the introduction of large-scale *J. curcas* cultivation on Miombo woodland in East Africa led to an estimated carbon debt (the amount of biomass and soil carbon released due to land conversion) of more than 30 years, using data from extant forestry and ecology studies (Romijn, 2011). A study indicates that moisture stress is the key factor in the failure of many large-scale *Jatropha* plantations in Ethiopia (Wendimu, 2013). *Jatropha* performance in semiarid zones has been satisfactory (Ahamad et al., 2013; Behera et al., 2010).

2 CULTIVATION AND ECONOMIC GROWTH - VALUE ADDITIONS AND VALUE CHAIN PRODUCTS

Jatropha biodiesel has attracted global hype to local opportunity (Achten et al., 2008, 2010a–c). *Jatropha* value chains can potentially enable more than 40% savings in GHG emissions relative to a fossil reference. Climate change mitigation can only be achieved, however, if *Jatropha* is cultivated on land with initially low carbon stocks and if no trees are cut. The carbon sequestration potential of *Jatropha* is limited, due to its low growth rate and the necessity of continuous pruning. It is only possible to achieve a relatively low carbon stock, which is comparable to fallow land. Consequently, cultivating *Jatropha* solely for carbon payments is inadvisable. The cultivation of *Jatropha* as a living fence allows maximum GHG savings due to low resource inputs and minimal land use transformation (Zah et al., 2013).

   Jatropha has attracted huge interest as a wonder crop that could generate biodiesel oil on marginal lands in semiarid areas. Its promise appeared especially great in East Africa, where many projects were launched to grow *Jatropha* in plantations or within contract-growing schemes. Today, however, *Jatropha*'s value in East Africa appears to lie primarily in its multipurpose use by small-scale farmers (Anonymous CDE Policy Brief, 2014). Assessment of the opportunities and challenges of scaling up *J. curcas* for environmental management and enhanced community livelihoods in Mtito Andei in Kenya was reported (Maingi, 2010). Rural development obtained through the setup of decentralized *Jatropha* production and marketing chains should be socially, economically, and environmentally more sustainable than the benefits generated by large-scale estates even if job creation in those centralized systems was to comply with national and international labor standards (Mergeai, 2008). Indonesia focuses *Jatropha* research on four fields including plant breeding, development of cultivation techniques, processing technology, and byproduct utilization. Researches done by Indonesian scientists in plant breeding and cultivation have resulted in the improvement of *Jatropha* productivity by 60% (from 5 to 8 tons ha^{-1} year^{-1}) through the improved population method.

2.1 POTENTIAL FOR REMEDIATION

Environmentally, *J. curcas* has great potential for soil enrichment. Its leaves and branches can replace synthetic fertilizers for coconut trees. It contributes to carbon sequestration thereby aiding in the mitigation of climate change. It also has the potential of retaining marginal and degraded soil by anchoring the soil with substantial roots, reducing possibility of soil erosion. A study also established that *J. curcas* has the potential of remediating heavy metal and hydrocarbon-contaminated soils (Agbogidi et al., 2013).

   In Ethiopia, approximately 90% of the population lives in areas marked by land degradation and reduced agricultural productivity. A case study in Ethiopia showed that planting *Jatropha* can be an effective prevention and mitigation measure against soil erosion (Bach, 2012). The buildup of biomass and collection of seeds to secure local energy needs also allows for GHG savings. Thus, *Jatropha* has the potential to alleviate soil degradation, increase carbon stocks, and contribute to energy security (Zah et al., 2013).

   The same features that make *Jatropha* useful as hedges make it a good choice for rehabilitating gullies and for planting along erosion-prone slopes (Anonymous CDE Policy Brief, 2014) (Figures 5 and 6). To stop a gully from expanding, it is necessary to slow the flow of water down the gully during

FIGURE 5

J. curcas plantation on mine waste.

rainstorms. The usual way to do this is to build stone walls across the floor of the gully at intervals. These check dams trap sediment behind them, which gradually builds up to form a series of steps. The pockets of soil behind the dams are level and relatively fertile, and because they are on a natural water-course, they remain reasonably moist even during the dry season. Many farmers find them a good place to grow crops (Ehrensperger et al., 2014).

2.2 ROLE OF PRODUCTS FROM *J. CURCAS* L. IN RURAL POVERTY ALLEVIATION

Recent investigations carried out all over the world have demonstrated that *J. curcas* could contribute drastically to the improvement of the living conditions of rural populations in the least developed countries of our planet. *J. curcas* is a multiple-function hardy shrub that can be used for medicinal purposes, to prevent and/or control erosion, to reclaim land, to produce pesticides, to contain or exclude farm animals when grown as a living fence, and to be planted as a commercial crop. The seed contains a high rate of nonedible oil that can be used for soap making in the cosmetic industry (Saverys et al., 2008).

2.3 OIL EXTRACTION AND COSMETIC POTENTIAL

Extraction of oil from *J. curcas* has aroused interest worldwide; the extracted oil is regarded as a potential source for local production in developing countries for use in cosmetic preparations (Naik et al., 2015). Rodríguez-Acosta et al. (2010) reported the composition of seed oil of three endemic Mexican

Jatropha curcas for ecodevelopment

Remediation — Biomaterials

(a) Restoration of fly ash (Jamil 2009; Jamil et al., 2009) (b) Phytormediation of lindane metals (Abhilash et al., 2013) and heavy metals (Yadav et al., 2009; Mangkoedihardjo and Surahmaida, 2008) (c) Rhizoremediation of soils contaminated with lubricating oils (Agamuthu et al., 2010)	(a) Biodiesel (Foidl et al., 1996; Openshaw 2000) (b) Source of feed, fertilizer (King et al., 2009; Gunaseelan 2009) (c) Medicinal value (Heller 1996; Samy et al., 1998; Osoniyi and Onajobi 2003; Mujumdar and Misar 2004) (d) Industrial poducts and applications (Bar et al., 2009)

CO₂ sequestration

FIGURE 6

J. curcas for carbon dioxide sequestration and eco-development (Achten et al., 2008; Ogunwole et al., 2008; Openshaw, 2000; Srivastava, 2010).

species of *Jatropha*. Oils were analyzed through the formation of their corresponding methyl esters, and their yield and composition were found to be close to that of the oil of *J. curcas* L. Akbar et al. (2009) and Adebowale and Adedire (2006) reported a high oil content (64.4%), dominant triacylglycerol lipid species (88.2%), and an appreciable percentage (47.3%) of linoleic acid after extraction from samples of *J. curcas* seeds obtained from markets in five different towns in Nigeria. Table 1 shows some physicochemical properties of the Malaysian *J. curcas* seed oil compared to the Nigerian *J. curcas* seed oil (Salimon and Abdullah, 2008; Akintayo, 2004). The parameters measured and the values obtained are in favor of utilization of the *J. curcas* seed oil for cosmetic production (Figure 7).

2.4 PROSPECTS FOR COSMETIC PRODUCTS

Although the current boom for *Jatropha* production is based mainly on the incentive of producing biofuel, the possible range of products that can be derived from *Jatropha* is much broader (Grass, 2009). Expert opinions from 39 countries describe a range of cosmetics and soap products based on use of *Jatropha*'s seed oil (GEXSI, 2008). Kakute Ltd, one of the Tanzanian organizations promoting

Table 1 Physicochemical Characteristic of Nigerian *J. curcas* L Seed Oil in Comparison to Malaysian Product

Parameter	Increase ↑/Decrease↓ (%)	Actual Value for Nigerian Product
Iodine value (mg/g)	−12.72↓	105.20 ± 0.70
Acid value (mg KOH/g)	+40↑	3.50 ± 0.10
Saponification value (mg/g)	−2.37↓	198.85 ± 1.40

Data source: Akintayo (2004).

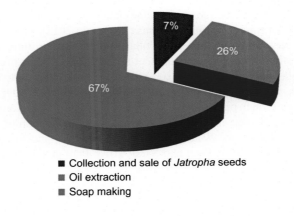

■ Collection and sale of *Jatropha* seeds
■ Oil extraction
■ Soap making

FIGURE 7

Jatropha seed oil is profitable in cosmetic and soap industry than for biodiesel.

Jatropha for oil production, erosion control, and soap making, conducted an evaluation in 2003 of the profitability of *Jatropha*-related activities. It found soap making to be more profitable than oil extraction, which, in turn, was more profitable than seed collection or production. Soap produced from *Jatropha* is sold as a medicinal soap, effective in treating skin ailments. It is noted that *Jatropha* soap is sold in dispensaries at a higher price than other soaps on the market (Henning, 2004). The case of the German Technical Cooperation Agency [Geselleschaft fur International Zusammenarbeit (GIZ)] project in Mali will be used to demonstrate the economic benefits of physic nut cultivation. The *Jatropha* system in Mali was based on existing hedges that were used to fence in fields and to control erosion. The project promoted this system by creating a market for physic nut products. The oil was used as a raw material (Teo et al., 2014, 2015; Tiwari et al., 2007) for soap production that generated income for the producers, who were local women (Heller, 1996). In Senegal, in a project carried out by Appropriate Technology International, Senegal ATI (now Enterprise Works) an American NGO, in the region of Ties, *Jatropha* hedges were planted and *Jatropha* oil was extracted with a ram press. The oil was used to run diesel engines (for flour mills) and to make soap. *Jatropha* soap is perceived in Malawi as a medicated soap. The soap from unrefined oil is considered, like neem-based soap, to be an effective, gentle, anti-scabies wash (Pratt et al., 2002). In former times Portugal imported *Jatropha* seeds from Cape Verde Islands to produce soap (Anonymous, 2006). In India, Nepal, and Zimbabwe, the price of tallow or the price of *Jatropha* and other plant oils is at least 2.5 times the selling price of diesel. Obviously, selling *Jatropha* oil for soap making is far more profitable in these countries than using it as a diesel or kerosene substitute (Openshaw, 2000).

2.5 RECOMMENDATIONS FOR FURTHER RESEARCH

Participatory data acquisition is needed to build up a consolidated and reliable database that includes documented metadata along the entire land use system. The focus should be primarily on yield data, carbon stock data, and non-CO_2 GHG emissions, which typically show large variations and are highly context specific. It would also be useful to consolidate knowledge related to land use dynamics across both croplands and their surrounding areas. In addition, most studies about the potential for climate change mitigation focus on established *Jatropha* value chains, and therefore updated models are required to understand the impact of future *Jatropha* systems (Zah et al., 2013). In order to develop *Jatropha* as a commercially profitable commodity, research collaboration with institutions in countries concerned with upstream to downstream *Jatropha* development is necessary. Genetic diversity assessment for optimizing the yield (Basha and Sujatha, 2007, 2009; Bhering et al., 2015; Carvalho et al., 2008; Maravi et al., 2015; Sunil et al., 2008, 2011; Tao et al., 2015; Tatikonda et al., 2009; Wang et al., 2015), seed source variation for better growth performance (Ginwal et al., 2004, 2005; Kumar et al., 2003; Srivastava et al., 2011, 2015) and postharvest management (Gour, 2006), application of plant growth and health-promoting rhizobacteria (Jha et al., 2012a,b), progeny evaluation, and influence of soil ameliorants for growth promotion (Padma, 2007; Patel and Saraf, 2013; Patil et al., 2015) need to be investigated (Figure 8 and Tables 2 and 3).

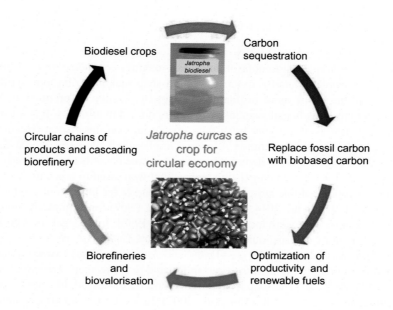

FIGURE 8

J. curcas as industrial crop for circular economy.

Table 2 *Jatropha curcas*—**Environmental Remediation**

References	Research Topics
Anonymous (2014)	CDE Policy Brief Beyond biofuels: *Jatropha*'s multiple uses for farmers in East Africa, pp. 1–4
Abhilash et al. (2013)	Remediation of lindane by *J. curcas* L.: utilization of multipurpose species for rhizoremediation of lindane
Agbogidi et al. (2013)	Health and environmental benefits of *J. curcas* Linn.
Wendimu (2013)	Jatropha potential on Marginal land in Ethiopia: reality or myth
Zah et al. (2013)	Can *J. curcas* contribute to climate change mitigation?
Zah et al. (2013)	*J. curcas* contribution to climate change mitigation
Bach (2012)	*J. curcas* as a multipurpose crop for s soil and water conservation
Firdaus and Husni (2012)	Planting *J. curcas* on constrained land
Abhilash et al. (2011)	An oil plant of multiple benefits
Srivastava (2010)	Evaluation of soil carbon sequestration potential of *J. curcas* L. plantation growing in varying soils conditions
Maingi (2010)	The potential role of *J. curcas* L. for environmental management and sustainable livelihoods in Kibwezi, Kenya
Jamil et al. (2009)	A potential crop for phytoremediation of coal fly ash
Sreedevi and Wani (2009)	Jatropha and Pongamia rainfed plantations on wastelands in India for improved livelihoods and protecting environment
Kumar et al. (2008)	Growth of *J. curcas* on heavy metal contaminated soil amended with industrial wastes and *Azotobacter*
Mangkoedihardjo and Surahmaida (2008)	Phytoremediation of lead and cadmium polluted soils
Ogunwole et al. (2008)	Contribution of *J. curcas* to soil quality improvement in a degraded Indian entisol
Chaudharry et al. (2007)	Changes in soil characteristics and foliage nutrient content in relation to stand density in Indian wasteland
Ogunwole et al. (2007)	Improvement of quality of a degraded entisol with *J. curcas* L. under Indian semiarid conditions
George et al. (2005)	A concept for simultaneous wasteland reclamation, fuel production, and socio-economic development in degraded areas in India
Francis et al. (2005)	Potential and perspectives of Jatropha plantations for simultaneous wasteland reclamation, fuel production, and socio-economic development in degraded areas in India
Foidl and Kashyap (1999)	Potential of *J. curcas* in rural development and environmental protection
Jones and Miller (1992)	*J. curcas*: a multipurpose species for problematic sites

Table 3 *Jatropha curcas* **for Bioeconomy**

References	Research Topics
Liu et al. (2015)	Proteomic analysis of oil bodies in mature *J. curcas* seeds with different lipid
Lopes et al. (2015)	Bioconversion of *J. curcas* seed cake to hydrogen by a strain of *Enterobacter aerogenes*
Sánchez et al. (2015)	Production of esters from Jatropha oil using different short-chain alcohols

(Continued)

Table 3 *Jatropha curcas* for Bioeconomy—Cont'd

References	Research Topics
Sánchez-Arreola et al. (2015)	Biodiesel production and de-oiled seed cake nutritional values of a Mexican edible *J. curcas*
Altei et al. (2014)	Jatrophidin I, a cyclic peptide from Brazilian *J. curcas* L.: isolation, characterization, conformational studies, and biological activity
Baka and Bailis (2014); Henning (2004)	Wasteland energy-scapes
Cordova-Albores et al. (2014)	Chemical compounds of a native *J. curcas* seed oil from Mexico and their antifungal effect on *Fusarium oxysporum* f. sp.
García-Dávila et al. (2014)	*J. curcas* L. oil hydroconversion over hydrodesulfurization catalysts for biofuel production
Liu (2014)	Extraction of oil from *J. curcas* seeds by subcritical fluid extraction
Luu et al. (2014)	Production of biodiesel from Vietnamese *J. curcas* oil by a co-solvent
Méndez et al. (2014)	*J. curcas* leaves as a mineral source for low sodium diets
Ahamad et al. (2013)	Performance of *J. curcas* L. in semiarid zone
Contran et al. (2013); Roy et al. (2014)	*J. curcas* productive chain: from sowing to biodiesel and by products
Bach (2012)	*J. curcas* as a multipurpose crop for sustainable energy
Achten et al. (2010a)	Life cycle assessment of *Jatropha* biodiesel as transportation fuel in rural India
Achten et al. (2010b)	*Jatropha*: from global hype to local opportunity
Arvidsson et al. (2010)	Life cycle assessment of hydrotreated vegetable oil from rape, oil palm, and *Jatropha*
Biswas et al. (2010)	Biodiesel from *Jatropha*
Jain and Sharma (2010)	Biodiesel from *Jatropha* in India
Li et al. (2010)	System approach for evaluating the potential yield and plantation of *J. curcas* L. on a global scale
Akbar et al. (2009)	Characteristic and composition of *J. curcas* oil seed from Malaysia and its potential as biodiesel feedstock
Bar et al. (2009)	Green synthesis of silver nanoparticles using latex of *J. curcas*
Gunaseelan (2009)	Biomass estimates and biochemical methane potential
King et al. (2009)	Potential of *J. curcas* as source of renewable oil and animal feed
Makkar and Becker (2009)	*J. curcas*, a promising crop for the generation of biodiesel and value added coproducts
Achten et al. (2008)	*Jatropha* biodiesel production and use
Chaudharry et al. (2008)	Soil characteristics and mineral nutrient in wild *Jatropha* population of India
Kumar and Sharma (2008); Gofferjé et al. (2014)	Multipurpose oil seed crop for industrial feed stock
Saverys et al. (2008)	Contributions of *J. curcas* L. to rural poverty alleviation in Senegal

Table 3 *Jatropha curcas* for Bioeconomy—Cont'd	
References	**Research Topics**
Chaudharry et al. (2007)	Changes in soil characteristics and foliage nutrient content in *J. curcas* plantation in relation to stand density in Indian wasteland
Adebowale and Adedire (2006)	Chemical composition and insecticidal properties of the underutilized *J. curcas* seed oil
Mangaraj and Singh (2006)	Engineering properties of *Jatropha* for use as biodiesel
Akintayo (2004)	Characteristic and composition of *J. curcas* oils and cakes.
Pramanik (2003)	Use of *J. curcas* oil and diesel fuel blends in compression ignition engine
Augustus et al. (2002)	Evaluation and bioinduction of energy components of *J. curcas*
Cano-Asseleih et al. (1998)	Purification and partial characterization of the hemagglutination from seeds of *J. curcas*
Aker (1997)	Growth and reproduction of *J. curcas*
Foidl and Eder (1997)	Biofuels and industrial products from *J. curcas*
Lopez et al. (1997)	Production of biogas from *J. curcas* fruitshells
Sharma et al. (1997)	Future source of hydrocarbon and other industrial products
Staubmann et al. (1997)	Production of biogas from *J. curcas* seed press cake
Foidl et al. (1996)	*J. curcas* L. as a source for the production of biofuel in Nicaragua

REFERENCES

Abhilash, P., Srivastava, P., Jamil, S., Singh, N., 2011. Revisited *Jatropha curcas* an oil plant of multiple benefits: critical research needs and prospects for the future. Environ. Sci. Pollut. Res. Int. 18, 127–131.

Abhilash, P., Singh, B., Srivastava, P., Schaeffer, A., Singh, N., 2013. Remediation of lindane by *Jatropha curcas* L.: utilization of multipurpose species for rhizoremediation of lindane. Environ. Exp. Bot. 74, 127–130.

Achten, W.M.J., Verchot, L., Franken, Y.J., Mathijs, E., Singh, V.P., Aerts, R., Muys, B., 2008. *Jatropha* bio-diesel production and use. Biomass Bioenergy 32, 1063–1084.

Achten, W.M.J., Almedia, J., Fobelets, V., 2010a. Life cycle assessment of *Jatropha* biodiesel as transportation fuel in rural India. Appl. Energy 87, 3652–3660.

Achten, W.M.J., Maes, W.H., Aerts, R., Verchot, L., Trabucco, A., Mathijs, E., Singh, V.P., Muys, B., 2010b. *Jatropha*: from global hype to local opportunity. J. Arid Environ. 74, 164–165.

Achten, W.M.J., Maes, W.H., Reubens, B., Mathijs, E., Singh, V.P., Verchot, L., Muys, B., 2010c. Biomass production and allocation in *Jatropha curcas* L. seedlings under different levels of drought stress. Biomass Bioenergy 34, 667–676.

Adebowale, K.O., Adedire, C.O., 2006. Chemical composition and insecticidal properties of the underutilized *Jatropha curcas* seed oil. Afr. J. Biotechnol. 5 (10), 901–906.

Agamuthu, P., Abioye, O.P., Aziz, A.A., 2010. Phytoremediation of soil contaminated with used lubricating oil using *Jatropha curcas*. J. Hazard. Mater. 179, 891–894. http://dx.doi.org/10.1016/j.jhazmat.2010.03.088.

Agbogidi, O.M., Akparobi, S.O., Eruotor, P.G., 2013. Health and environmental benefits of *Jatropha curcas* Linn. Unique Res. J. Agric. Sci. 1 (5), 76–79.

Ahamad, S., Joshi, S.K., Arif, M., Ahmed, Z., 2013. Performance of *Jatropha curcas* L. in semi-arid zone: seed germination, seedling growth and early field growth. Not. Sci. Biol. 5, 169–174.

Akbar, E., Yaakob, Z., Kamarudin, S.K., Ismail, M., Salimon, J., 2009. Characteristic and Composition of *Jatropha curcas* oil seed from Malaysia and its potential as biodiesel feedstock. Eur. J. Sci. Res. 29 (3), 396–403.

Aker, C.L., 1997. Growth and reproduction of *J. curcas*. In: Gubitz, G.M., Mittelbach, M., Trabi, M. (Eds.), Biofuels and Industrial Products from *Jatropha curcas*. dbv-Verlag fur die Technische Universitat Graz, Graz, pp. 2–18.

Akintayo, E.T., 2004. Characteristic and composition of *Parkia biglobbossa* and *Jatropha curcas* oils and cakes. Biosour. Technol. 92, 307–310.

Altei, W.F., Picchi, D.G., Abissi, B.M., Giesel, G.M., Flausino Jr., O., Reboud-Ravaux, M., Verli, H., Crusca Jr., E., Silveira, E.R., Cilli, E.M., Bolzani, V.S., 2014. Jatrophidin I, a cyclic peptide from Brazilian *Jatropha curcas* L.: isolation, characterization, conformational studies and biological activity. Phytochemistry 107, 91–96.

Anonymous (2006) Handbook om Jatropha curcas. Fact Foundation. http://www.globalbioenergy.org/uploads/media/0603_FACT_Foundation_-_Jatropha_Handbook.pdf

Anonymous, 2014. Centre for Development and Environment (CDE). Policy Brief. Beyond biofuels: Jatropha's multiple uses for farmers in East Africa, pp. 1–4.

Arvidsson, R., Persson, S., Froling, M., Svanstrorm, M., 2010. Life cycle assessment of hydrotreated vegetable oil from rape, oil palm and *Jatropha*. J. Clean. Prod. 19, 129–137.

Augustus, G.D.P.S., Jayabalan, M., Seiler, G.J., 2002. Evaluation and bioinduction of energy components of *Jatropha curcas*. Biomass Bioenergy 23, 161–164.

Bach, S., 2012. Potentials and Limitations of *Jatropha curcas* as a Multipurpose Crop for Sustainable Energy Supply and Soil and Water Conservation. A Case Study in Bati, Ethiopia, Using the WOCAT Approach. (Master's thesis) University of Bern, Bern.

Baka, J., Bailis, R., 2014. Wasteland energy-scapes: a comparative energy flow analysis of India's biofuel and biomass economies. Ecol. Econ. 108, 8–17.

Bar, H., Bhui, D.K., Sahoo, G.P., Sarkar, P., De, S.P., Misra, A., 2009. Green synthesis of silver nanoparticles using latex of *Jatropha curcas*. Colloids Surf. A Physicochem. Eng. Asp. 339, 134–139.

Basha, S.D., Sujatha, M., 2007. Inter and intra-population variability of *Jatropha cucrcas* L. characterized by RAPD and ISSR markers and development of population-specific SCAR markers. Euphytica 156, 775–786.

Basha, S.D., Sujatha, M., 2009. Genetic analysis of *Jatropha* species and interspecific hybrids of *Jatropha curcas* using nuclear and organelle specific markers. Euphytica 168, 197–214.

Behera, S.K., Srivastava, P., Tripathi, R., Singh, J.P., Singh, N., 2010. Evaluation of plant performance of *Jatropha curcas* L. under different agro-practices for optimizing biomass—a case study. Biomass Bioenergy 34, 30–41.

Bhering, L.L., Filho, J.E.A., Peixoto, L.A., Laviola, B.G., Gomes, B.E.L., 2015. Plateau regression reveals that eight plants per accession are representative for *Jatropha* germplasm bank. Ind. Crop. Prod. 65, 210–215.

Biswas, P.K., Pohit, S., Kumar, R., 2010. Biodiesel from *Jatropha*: can India meet the 20% blending target? Energy Policy 38, 1477–1484.

Blench, R., 2007. Hausa names for plants and trees. Accessed at. http://www.rogerblench.info/Ethnoscience/Plants/General/Hausa%20plant%20names.pdf.

Caldeira, K., Granger Morgan, M., Baldocchi, D., Brewer, P.G., Arthur Chen, C.T., Nabuurs, G., Nakicenovic, N., Robertson, G.P., 2004. A portfolio of carbon management options. In: Field, C.B., Raupach, M.R. (Eds.), The Global Carbon Cycle: Integrating Humans, Climate and the Natural World. Island Press, Washington, pp. 103–130.

Cano-Asseleih, L.M., Plumbly, R.A., Hylands, P.J., 1989. Purification and partial characterization of the hemagglutination from seeds of *Jatropha curcas*. J. Food Biochem. 13, 1–20.

Carvalho, C.R., Clarindo, W.R., Praca, M.M., Araujo, F.S., Carels, N., 2008. Genome size, base composition and karyotype of *Jatropha curcas* L., an important biofuel plant. Plant Sci. 174, 613–617.

Chaudharry, D.R., Patolia, J.S., Ghosh, A., Chikara, J., Boricha, G.N., Zala, A., 2007. Changes in soil characteristics and foliage nutrient content in *Jatropha curcas* plantation in relation to stand density in Indian wasteland. In: Proceedings of the Expert Seminar on *Jatropha curcas* L. Agronomy and Genetics, Wageningen, The Netherlands, 26-28 March 2007. FACT Foundation.

Chaudharry, D.R., Ghosh, A., Chikara, J., Patolia, J.S., 2008. Soil characteristics and mineral nutrient in wild Jatropha population of India. Commun. Soil Sci. Plan. 39, 1476–1485.

Contran, N., Chessa, L., Lubino, M., Bellavite, D., Roggero, P.P., Enne, G., 2013. State of the art of the *Jatropha curcas* productive chain: from sowing to biodiesel and by products. Ind. Crop. Prod. 42, 202–215.

Cordova-Albores, L.C., Rios, M.Y., Barrera-Necha, L.L., Bautista-Baños, S., 2014. Chemical compounds of a native *Jatropha curcas* seed oil from Mexico and their antifungal effect on *Fusarium oxysporum* f. sp. Gladioli. Ind. Crop. Prod. 62, 166–172.

Dyer, J.C., Stringer, L.C., Dougill, A.J., 2012. *Jatropha curcas:* sowing local seeds of success in Malawi? In response to Achten et al. (2010). J. Arid Environ. 79, 107–110.

Edrisi, S.A., Dubey, R.K., Tripathi, V., Bakshi, M., Srivastava, P., Jamil, S., Singh, H.B., Singh, N., Abhilash, P.C., 2015. *Jatropha curcas* L.: a crucified plant waiting for resurgence. Renew. Sustain. Energy Rev. 41, 855–862.

Ehrensperger, A., Bach, S., Lyimo, R., Portner, B., 2014. Beyond Biofuels: Jatropha's Multiple Uses for Farmers in East Africa. CDE Policy Brief, No. 1 CDE, Bern.

Firdaus, M.S., Husni, M.H.A., 2012. Planting *Jatropha curcas* on constrained land: emissions and effects from land use change. Sci. World J. 2012, 7.

Foidl, N., Eder, P., 1997. Agro-industrial exploitation of *Jatropha curcas*. In: Gubitz, G.M., Mittelbach, M., Trabi, M. (Eds.), Bio-Fuels and Industrial Products from *Jatropha curcas*. dbv-Verlag, Graz, pp. 88–91.

Foidl, N., Kashyap, A., 1999. Exploring the potential of *Jatropha curcas* in rural development and environmental protection. Rockefeller Foundation, New York.

Foidl, N., Foid, G., Sanchez, M., Mittelbach, M., Hackel, S., 1996. *Jatropha curcas* L. as a source for the production of biofuel in Nicaragua. Bioresour. Technol. 58, 77–82.

Francis, G., Edinger, R., Becker, K., 2005. A concept for simultaneous wasteland reclamation, fuel production, and socio-economic development in degraded areas in India: need, potential and perspectives of Jatropha plantations. Nat. Resour. Forum 29, 12–24.

Fu, J., Jiang, D., Huang, Y., Zhuang, D., Wei Ji, W., 2014. Evaluating the marginal land resources suitable for developing bioenergy in Asia. Adv. Meteorol. 2014, 1–10.

García-Dávila, J., Ocaranza-Sánchez, E., Rojas-López, M., Muñoz-Arroyo, J.A., Ramírez, J., Martínez-Ayala, A.L., 2014. *Jatropha curcas* L. oil hydroconversion over hydrodesulfurization catalysts for biofuel production. Fuel 135, 380–386.

Garnayak, D.K., Pradhan, R.C., Naik, S.N., Bhatnagar, N., 2008. Moisture dependent physical properities of Jatropha seed (*Jatropha curcas* L.). Ind. Crop. Prod. 27, 123–129.

George, F., Raphael, E., Klaus, B., 2005. A concept for simultaneous wasteland reclamation, fuel production and socio-economic development in degraded areas in India: need, potential and perspectives of *Jatropha* plantations. Nat. Resour. Forum 29, 12–24.

Global Exchange for Social Investment (GEXSI), 2008. Global market study on Jatropha. Final report; prepared for the World Wide Fund for Nature (WWF), London/Berlin. http://www.jatropha-alliance.org/fileadmin/documents/GEXSI_Global-Jatropha-Study_ABSTRACT.pdf.

Ginwal, H.S., Rawat, P.S., Srivastava, R.L., 2004. Seed source variation in growth performance and oil yield of *Jatropha curcas* L. in Central India. Silvae Genet. 53, 186–192.

Ginwal, H.S., Phartyal, S.S., Rawat, P.S., Srivastava, R.L., 2005. Seed source variation in morphology, germination and seedling growth of *Jatropha curcas* L. in Central India. Silvae Genet. 54, 76–80.

Gofferjé, G., Motulewicz, J., Stäbler, A., Herfellner, T., Schweiggert-Weisz, U., Flöter, E., 2014. Enzymatic degumming of crude Jatropha oil: evaluation of impact factors on the removal of phospholipids. J. Am. Oil Chem. Soc. 91, 2135–2141.

Gour, V.K., 2006. Production practices including post-harvest management of *Jatropha curcas*. In: Singh, B., Swaminathan, R., Ponraj, V. (Eds.), Proceedings of the Biodiesel Conference Toward Energy Independence—Focus on *Jatropha*, Hyderabad, India, 9-10 June. Rashtrapati Bhawan, New Delhi, pp. 223–251.

Grass, M., 2009. *Jatropha curcas* L: visions and realities. J. Agric. Rural Dev. Trop. 110 (1), 29–38.

Gubitz, G.M., Mittelbach, M., Trabi, M., 1999. Exploitation of the tropical oil seed plant *Jatropha curcas* L.. Bioresour. Technol. 67, 73–82.

Gunaseelan, V.N., 2009. Biomass estimates, characteristics, biochemical methane potential, kinetics and energy flow from *Jatropha curcas* in dry lands. Biomass Bioenergy 33, 589–596.

Heller, J., 1996. Physic Nut. *Jatropha curcas* L.. Promoting the conservation and use of underutilized and neglected crops. 1 Institute of Plant Genetics and Crop Plant Research, Gatersleben/International Plant Genetic Resources Institute, Rome, ISBN: 92-9043-278-0.

Henning, R.K., 2004. The Jatropha system, integrated Rural Development by Utilization of *Jatropha curcas* L. (JCL) as raw material and as renewable energy. http://www.jatropha-alliance.org/fileadmin/documents/knowledgepool/Henning_The_Jatropha_System.pdf.

Jain, S., Sharma, M.P., 2010. Prospects of biodiesel from Jatropha in India: a review. Renew. Sustain. Energy Rev. 14, 763–771.

Jamil, S., Abhilash, P.C., Singh, N., Sharma, P.N., 2009. *Jatropha curcas*: a potential crop for phytoremediation of coal fly ash. J. Hazard. Mater. 172, 269–275.

Jha, C.K., Annapurna, K., Saraf, M., 2012a. Isolation of Rhizobacteria from *Jatropha curcas* and characterization of produced ACC deaminase. J. Basic Microbiol. 52, 285–295.

Jha, C.K., Patel, B., Saraf, M., 2012b. Stimulation of growth of the *Jatropha curcas* by the plant growth promoting bacterium *Enterobacter cancerogenus* MSA2. World J. Microbiol. Biotechnol. 28, 891–899.

Jones, N., Miller, J.H., 1992. *Jatropha curcas*: a multipurpose species for problematic sites. World Bank, Washington DC, USA. ASTAG technical papers-1 and, resources 1, p. 12.

Jongschapp, R.E.E., Corre, W.J., Bindraban, P.S., Brandenburg, W.A., 2007. Claims and facts on *Jatropha curcas* L.: global *Jatropha curcas* evaluation, breeding and propagation programme. Plant Research International report, Wageningen, p. 158.

King, A.J., He, W., Cuevas, J.A., Freudenberger, M., Ramiaramanana, D., Graha, I.A., 2009. Potential of *Jatropha curcas* as source of renewable oil and animal feed. J. Exp. Environ. Bot. 60, 2897–2905.

Kumar, A., Sharma, S., 2008. An evaluation of multipurpose oil seed crop for industrial uses (*Jatropha curcas* L.): a review. Ind. Crop. Prod. 28, 1–10.

Kumar, S., Parimallam, R., Arjunan, M.C., Vijayachandran, S.N., 2003. Variation in *Jatropha curcas* seed characteristics and germination. In: Hegde, N.G., Daniel, J.N., Dhar, S. (Eds.), Proceedings of the National Workshop on Jatropha and Other Perennial Oilseed Species, Pune, India. pp. 63–66.

Kumar, G.P., Yadav, S.K., Thawale, P.R., Singh, S.K., Juwarkar, A.A., 2008. Growth of *Jatropha curcas* on heavy metal contaminated soil amended with industrial wastes and *Azotobacter*—a greenhouse study. Bioresour. Technol. 99, 2078–2082.

Li, Z., Lin, B.L., Zhao, X., Sagisaka, M., Sagisaka, M., Shibazaki, R., 2010. System approach for evaluating the potential yield and plantation of *Jatropha curcas* L. on a global scale. Environ. Sci. Technol. 44, 2204–2209.

Liu, J., Chen, P., He, J., Deng, L., Wang, L., Lei, J., Rong, L., 2014. Extraction of oil from *Jatropha curcas* seeds by subcritical fluid extraction. Ind. Crop. Prod. 62, 235–241.

Liu, H., Wang, C., Chen, F., Shen, S., 2015. Proteomic analysis of oil bodies in mature *Jatropha curcas* seeds with different lipid content. J. Proteomics 113, 403–414.

Lopes, S.L., Fragoso, R., Duarte, E., Marques, P.A.S.S., 2015. Bioconversion of *Jatropha curcas* seed cake to hydrogen by a strain of *Enterobacter aerogenes*. Fuel 139, 715–719.

Lopez, O., Foidl, G., Foidl, N., 1997. Production of biogas from *J. curcas* fruitshells. In: Gubitz, G.M., Mittelbach, M., Trabi, M. (Eds.), Biofuels and Industrial Products from *Jatropha curcas*. Proceedings of the symposium "Jatropha 97", Managua, Nicaragua, 23-27 February. dbv-Verlag, Graz, pp. 118–122.

Luu, P.D., Truong, H.T., Luu, B.V., Pham, L.N., Imamura, K., Takenaka, N., Maeda, Y., 2014. Production of biodiesel from Vietnamese *Jatropha curcas* oil by a co-solvent method. Bioresour. Technol. 173, 309–316.

Maes, W.H., Achten, W.M.J., Reubens, B., Raes, D., Samson, R., Muys, B., 2009. Plant-water relationship and growth strategies of *Jatropha curcas* L. seedlings under different levels of drought stress. J. Arid Environ. 73, 877–882.

Maingi, R.N., 2010. The potential role of *Jatropha curcas* L. for environmental management and sustainable livelihoods in kibwezi, Kenya (M.Sc. thesis). Graduate School, Kenyatta University, Nairobi, Kenya.

Makkar, H.P.S., Becker, K., 2009. *Jatropha curcas*, a promising crop for the generation of bio-diesel and value added coproducts. Eur. J. Lipid Sci. Tech. 111, 773–787.

Mangaraj, S., Singh, R., 2006. Studies on some engineering properties of *Jatropha* for use as biodiesel. Bioenergy News 9, 18–20.

Mangkoedihardjo, S., Surahmaida, A., 2008. *Jatropha curcas* L. for phytoremediation of lead and cadmium polluted soils. World Appl. Sci. J. 4, 519–522.

Maravi, D.K., Mazumdar, P., Alam, S., Goud, V.V., Sahoo, L., 2015. Jatropha (*Jatropha curcas* L.). Methods Mol. Biol. 1224, 25–35.

Méndez, L., Rojas, J., Izaguirre, C., Contreras, B., Gómez, R., 2014. *Jatropha curcas* leaves analysis, reveals it as mineral source for low sodium diets. Food Chem. 165, 575–577.

Mergeai, G., 2008. Jatropha curcas: what sustainability? Tropicultura 26, 1.

Mujumdar, A.M., Misar, A.V., 2004. Anti-inflammatory activity of *Jatropha curcas* roots in mice and rats. J. Ethnopharmacol. 90, 11–15.

Munyao, C.M., Muisu, F., Mbego, J., Mburu, F., Peter Sirmah, P., 2013. Influence of land size on adoption of *Jatropha curcas* in Yatta District, Kenya. J. Nat. Sci. Res. 3 (4), 42–50.

Naik, D.V., Kumar, V., Prasad, B., Poddar, M.K., Behera, B., Bal, R., Khatri, O.P., Adhikari, D.K., Garg, M.O., 2015. Catalytic cracking of Jatropha-derived fast pyrolysis oils with VGO and their NMR characterization. RSC Adv. 5 (1), 398–409.

Negussie, A., Nacro, S., Achten, W.M.J., Norgrove, L., Kenis, M., Hadgu, K.M., Aynekulu, E., Hermy, M., Muys, B., 2015. Insufficient evidence of *Jatropha curcas* L. invasiveness: experimental observations in Burkina Faso, West Africa. BioEnergy Res. 8, 570–580.

Ogunwole, J.O., Patolia, J.S., Chaudhary, D.R., Gosh, A., Chikara, J., 2007. Improvement of quality of a degraded entisol with *Jatropha curcas* L. under Indian semi-arid conditions. In: Expert seminar on *Jatropha curcas* L. Agronomy and Genetics, Wageningen, The Netherlands, 26-28 March 2007. FACT Foundation.

Ogunwole, J.O., Chaudhary, D.R., Gosh, A., Dauda, C.K., Chikara, J., Patolia, S., 2008. Contribution of *Jatropha curcas* to soil quality improvement in a degraded Indian entisol. Acta Agric. Scand. B Soil Plant Sci. 58, 245–251.

Openshaw, K., 2000. A review of *Jatropha curcas*: an oil plant of unfulfilled promise. Biomass Bioenergy 19, 1–15.

Osoniyi, O., Onajobi, F., 2003. Coagulant and anticoagulant activities in *Jatropha curcas* latex. J. Ethnopharmacol. 89, 101–105.

Padma, N., 2007. Biotechnological interventions in *Jatropha* for biodiesel production. Curr. Sci. 93, 1347–1348.

Patel, D., Saraf, M., 2013. Influence of soil ameliorants and microflora on induction of antioxidant enzymes and growth promotion of *Jatropha curcas* under saline condition. Eur. J. Soil Biol. 55, 47–54.

Patil, V.K., Bhandare, P., Kulkarni, P.B., Naik, G.R., 2015. Progeny evaluation of *Jatropha curcas* and *Pongamia pinnata* with comparison to bioproductivity and biodiesel parameters. J. Forest. Res. 26, 137–142.

Pramanik, K., 2003. Properties and use of *Jatropha curcas* oil and diesel fuel blends in compression ignition engine. Renew. Energy 28, 239–248.

Pratt, J.H., Henry, E.M.T., Mbeza, H.F., Mlaka, E., Satali, L.B., 2002. Malawi agroforestry extension project marketing & enterprise program. Main report, Malawi Agroforestry Publication No. 47, pp. 44–46.

Rodríguez-Acosta, M., Sandoval-Ramírez, J., Zeferino-Díaz, R., 2010. Extraction and characterization of oils from three Mexican *Jatropha* species. J. Mex. Chem. Soc. 54 (2), 88–91.

Romijn, H.A., 2011. Land clearing and greenhouse gas emissions from Jatropha biofuels on African Miombo Woodlands. Energy Policy 39 (10), 5751–5762.

Roy, P.K., Datta, S., Nandi, S., Al Basir, F., 2014. Effect of mass transfer kinetics for maximum production of biodiesel from *Jatropha curcas* oil: a mathematical approach. Fuel 134, 39–44.

Salimon, J., Abdullah, R., 2008. Physicochemical properties of Malaysian *Jatropha curcas* seed oil. Sains Malays. 37 (4), 379–382.

Samy, R.P., Ignacimuthu, S., Sen, A., 1998. Screening of 34 Indian medicinal plants for antibacterial properties. J. Ethnopharmacol. 62, 173–181.

Sánchez, M., Bergamin, F., Peña, E., Martínez, M., Aracil, J., 2015. A comparative study of the production of esters from Jatropha oil using different short-chain alcohols: optimization and characterization. Fuel 143, 183–188.

Sánchez-Arreola, E., Martin-Torres, G., Lozada-Ramírez, J.D., Hernández, L.R., Bandala-González, E.R., Bach, H., 2015. Biodiesel production and de-oiled seed cake nutritional values of a Mexican edible *Jatropha curcas*. Renew. Energy 76, 143–147.

Saverys, S., Toussaint, A., Gueye, M., Defrise, L., Van Rattinghe, K., Baudoin, J.P., Terren, M., Jacquet de Haveskercke, P., Mergeai, G., 2008. Possible contributions of *Jatropha curcas* L. to rural poverty alleviation in Senegal: vision and facts. Tropicultura 26 (2), 125–126.

Sharma, G.D., Gupta, S.N., Khabiruddin, M., 1997. Cultivation of *Jatropha curcas* as a future source of hydrocarbon and other industrial products. In: Gubitz, G.M., Mittelbach, M., Trabi, M. (Eds.), Biofuels and Industrial Products from *Jatropha curcas*. Proceedings of the Symposium "Jatropha 97", Managua, Nicaragua, 23-27 February. dbv-Verlag, Graz, pp. 19–21.

Silva, E.N., Silveira, J.A.G., Ribeiro, R.V., Vieira, S.A., 2015. Photoprotective function of energy dissipation by thermal processes and photorespiratory mechanisms in *Jatropha curcas* plants during different intensities of drought and after recovery. Environ. Exp. Bot. 110, 36–45.

Slingerland, M., Schut, M., 2014. Jatropha developments in Mozambique: analysis of structural conditions influencing niche-regime interactions. Sustainability 6, 7541–7563. http://dx.doi.org/10.3390/su6117541.

Sreedevi, T.K., Wani, S.P., SrinivasaRao, Ch., Chaliganti, R., Reddy, R.L., 2009. Jatropha and Pongamia rainfed plantations on wastelands in India for improved livelihoods and protecting environment. In: Proceedings of the 6th International Biofuels Conference, New Delhi, India.

Srivastava, P., 2010. Evaluation of soil carbon sequestration potential of *Jatropha curcas* L. plantation growing in varying soils conditions. Ph.D. synopsis submitted to University of Lucknow, Lucknow.

Srivastava, P., Behera, S.K., Gupta, J., Jamil, S., Singh, N., Sharma, Y.K., 2011. Growth performance, variability in yield traits and oil content of selected accessions of *Jatropha curcas* L. growing in a large scale plantation site. Biomass Bioenergy 35, 3936–3942.

Srivastava, A., Jaidi, M., Kumar, S., Raj, S.K., 2015. Molecular identification of a new *Begomovirus* associated with leaf crumple disease of *Jatropha curcas* L. in India. Arch. Virol. 160, 617–619.

Staubmann, R., Foidl, G., Foidl, N., Gubitz, G.M., Lafferty, R.M., Valencia Arbizu, V.M., 1997. Production of biogas from *J. curcas* seed press cake. In: Gubitz, G.M., Mittelbach, M., Trabi, M. (Eds.), Biofuels and Industrial Products from *Jatropha curcas*. Proceedings of the Symposium "Jatropha 97", Managua, Nicaragua, 23-27 February. dbv-Verlag, Graz, pp. 19–21.

Sunil, N., Varaprasad, K.S., Sivaraj, N., Kumar, T.S., Abraham, B., Prasad, R.B.N., 2008. Assessing *Jatropha curcas* L. germplasm in situ—a case study. Biomass Bioenergy 32, 198–202.

Sunil, N., Sujatha, M., Kumar, V., Vanaja, M., Basha, S.D., Varaprasad, K.S., 2011. Correlating the phenotypic and molecular diversity in *Jatropha curcas* L.. Biomass Bioenergy 32, 1085–1096.

Tao, Y.-B., He, L.-L., Niu, L.-J., Xu, Z.-F., 2015. Isolation and characterization of an ubiquitin extension protein gene (JcUEP) promoter from *Jatropha curcas*. Planta 241, 823–836.

Tatikonda, L., Wani, S.P., Kannan, S., Beerelli, N., Sreedevi, T.K., Hoisington, D.A., Devib, P., Varshney, R.K., 2009. AFLP-based molecular characterization of an elite germplasm collection of *Jatropha curcas* L., a biofuel plant. Plant Sci. 76, 505–513.

Teo, S.H., Rashid, U., Taufiq-Yap, Y.H., 2014. Biodiesel production from crude *Jatropha curcas* oil using calcium based mixed oxide catalysts. Fuel 136, 244–252.

Teo, S.H., Taufiq-Yap, Y.H., Rashid, U., Islam, A., 2015. Hydrothermal effect on synthesis, characterization and catalytic properties of calcium methoxide for biodiesel production from crude *Jatropha curcas*. RSC Adv. 5 (6), 4266–4276.

Terren, M., Saverys, S., Jacquet de Haveskercke, P., Winandy, S., Mergeai, G., 2012. Attempted cultivation of *Jatropha curcas* L. in the lower Senegal River Valley: story of a failure. Tropicultura 30 (4), 204–208.

Tiwari, A.K., Kumar, A., Rehaman, H., 2007. Biodiesel production from *Jatropha* with high free fatty acids: an optimized process. Biomass Bioenergy 31, 569–575.

Tomar, N.S., Sharma, M., Agarwal, R.M., 2015. Phytochemical analysis of *Jatropha curcas* L. during different seasons and developmental stages and seedling growth of wheat (*Triticum aestivum* L.) as affected by extracts/leachates of *Jatropha curcas* L.. Physiol. Mol. Biol. Plants 21, 83–92.

Verma, K.C., Gaur, A.K., 2009. *Jatropha curcas* L.: substitute for conventional energy. World J. Agric. Sci. 5 (5), 552–556.

Wang, L., Gao, J., Qin, X., Shi, X., Luo, L., Zhang, G., Yu, H., Li, C., Hu, M., Liu, Q., Xu, Y., Chen, F., 2015. JcCBF2 gene from *Jatropha curcas* improves freezing tolerance of Arabidopsis thaliana during the early stage of stress. Mol. Biol. Rep. 42, 937–945.

Warra, A.A., 2012. Cosmetic potentials of physic nut (*Jatropha curcas* Linn.) seed oil. Am. J. Sci. Ind. Res. 2153-649X3, http://dx.doi.org/10.5251/ajsir.2012.3.6.358.366. Science Hub http://www.scihub.org/AJSIR.

Warra, A.A., Wawata, I.G., Umar, R.A., Gunu, S.Y., 2012. Soxhlet extraction, physicochemical analysis and cold process saponification of Nigerian *Jatropha curcas* L. seed oil. Can. J. Pure Appl. Sci. 6 (1), 1803–1807.

Wendimu, M.A., 2013. Jatropha potential on Marginal Land in Ethiopia: reality or myth? IFRO working paper, pp. 1–19.

Yadav, S.K., Juwarkar, A.A., Kumar, G.P., Thawale, P.R., Singh, S.K., Chakrabarti, T., 2009. Bioaccumulation and phyto-translocation of arsenic, chromium and zinc by *Jatropha curcas* L.: impact of dairy sludge and biofertilizer. Bioresour. Technol. 100, 4616–4622.

Zah, R., Gmuender, S., Muys, B., Achten, W.M.J., Norgrove, L., 2013. Can *Jatropha curcas* contribute to climate change mitigation? In: Jatropha Facts Series, Issue 2, ERA-ARD.

Zhuang, D., Jiang, D., Liu, D., Huang, Y., 2011. Assessment of bioenergy potential on marginal land in China. Renew. Sustain. Energy Rev. 15, 1050–1056.

POTENTIAL OF CASTOR BEAN (*RICINUS COMMUNIS* L.) FOR PHYTOREMEDIATION OF METALLIFEROUS WASTE ASSISTED BY PLANT GROWTH-PROMOTING BACTERIA: POSSIBLE COGENERATION OF ECONOMIC PRODUCTS

D. Annapurna[1], M. Rajkumar[2], M.N.V. Prasad[1]

University of Hyderabad, Hyderabad, Telangana, India[1]
Central University of Tamil Nadu, Thiruvarur, Tamil Nadu, India[2]

1 INTRODUCTION

Contamination of soil by heavy metals is an alarming environmental problem. Unplanned disposal of municipal solid waste (MSW) and industrial effluents have been implicated as the major source of heavy metal contamination. Soil contamination with toxic metals has become a serious problem posing potential health risks, as the toxic metals enter the food chain. With the growing interest in soil restoration, various physico-chemico-biological methods have been developed for treating heavy metal-contaminated sites. Among these methods, phytoremediation, the use of plants and beneficial microbes to remediate polluted soils, has been proposed as a promising innovative technology due to its ecological and economic significance (Ma et al., 2011a,b; Rajkumar et al., 2012). The plant- and microbial-based method is an emerging technology to remediate a broad range of organic and inorganic pollutants in the environment (Wenzel, 2009). There are different ways in which the plants can remove, degrade, immobilize, and reduce the concentration of toxic effects of metals in the soil. Plants and their rhizosperic organisms can be used for phytoremediation in different ways (Glick, 2010). The uptake and concentration of contaminants within the roots or above-ground portion of plants is termed as phytoaccumulation. Rhizoremediation is a specific type of phytoremediation involving the complex interaction of roots and associated microorganisms in the rhizospheric region. Phytostabilization aims

Bioremediation and Bioeconomy. http://dx.doi.org/10.1016/B978-0-12-802830-8.00008-3
Copyright © 2016 Elsevier Inc. All rights reserved.

to reduce contaminants in the soil through adsorption and accumulation by roots, adsorption onto roots, or precipitation within the rhizospheric region of plants to prevent further dispersal through erosion and leaching (Atabayeva et al., 2010; Compant et al., 2010; Conesa et al., 2012). It also helps in establishing vegetation on the surface of the contaminated soil and sediment, which accelerates natural attenuation. For soils polluted with inorganic contaminants such as metals, tolerant plants can be used on site to simply immobilize pollutants, thereby reducing its bioavailability. Plants can also facilitate biodegradation of organic pollutant by microbes in the rhizosphere, which is known as phytostimulation or biodegradation (Salt et al., 1998; Dietz and Schnoor, 2001).

Some plant species growing in metal-polluted soils were found to have the ability to tolerate/accumulate/detoxify high concentrations of pollutants without any impact on their growth and development. Plant-microbe-pollutant interactions play a large role in survival of plants and microbes in polluted soils and in most cases the rhizosperic microbes may strengthen the ability of plants to survive under stressed conditions. Plants growing in metal-polluted soils have been shown to structure rhizosperic microbial communities through influencing soil nutrient bioavailability and changing the composition of root exudates (Ma et al., 2011a; Rajkumar et al., 2012). Similarly, plants get benefit from the rhizosperic microbial populations through the recycling and solubilization of mineral nutrients, as well as the increased supply of metabolites including auxins, cytokinins, and gibberellins that stimulate plant growth (Dakora and Phillips, 2002). Moreover, the activities of soil microbial communities govern many processes in the rhizosphere such as metal mobilization and immobilization and impact plant fitness (e.g., plant growth-promoting rhizobacteria protect plant from pathogens). Thus alterations in the microbial consortia of rhizosphere soils greatly alter various processes in metal-polluted soils. These include recycling of nutrients, maintenance of soil structure, detoxification of pollutants, and control of plant pests and plant growth, etc. (Dakora and Phillips, 2002; Jones et al., 2003; Rajkumar et al., 2010; Figure 1).

Castor bean is reported as a suitable candidate for phytostabilization of cocontaminated sites by organics and inorganics (Romeiro et al., 2006; Olivares et al., 2013; Wang et al., 2013; Wu et al., 2012), heavy metals, and polycyclic aromatic hydrocarbons (PAHs). This process of phytostabilization is accomplished through phytoexclusion (use of plants with low metal uptake) (Andreazza et al., 2013), aided phytostabilization (improvement through soil amendments) (Bauddh and Singh, 2012a,b; 2015; De Abreu et al., 2012), hydraulic control (inhibition of pollutant leaching), and phytorestoration (phytostabilization with the help of native plants) (Pandey, 2013).

Recently, manipulation of rhizospheric processes of plants growing on metal-contaminated soils is considered as an important strategy to strengthen phytoremediation (Ananthi and Manikandan, 2013). Soil microorganisms, including the microbial consortia as well as the tissue-colonized symbiotic bacteria, are the integral part of the rhizosphere biota. It play a key role in the biogeochemical cycle, through various mechanisms involving recycling of plant minerals and nutrients in the environment (Rajkumar et al., 2012). Further studies confirmed that microbial consortia can affect the mobility and bioavailability of the trace metal in the soil. Pairing both of efficient plant growth promoting rhizobacteria (PGPR) with accumulator may be the best way for the restoration of vegetation on metal-contaminated land (Ma et al., 2011a,b). Improvement of the interaction between plants and beneficial rhizosphere microorganisms can enhance metal tolerance of the plants to heavy metals (Glick, 2010). For a successful restoration of municipal waste dumpsites with metalliferous waste, it is necessary to exploit such metal-tolerant plant-microbe partnerships. This knowledge will enable us to understand the ongoing biological rhizospheric processes and the interaction occurring at the interface of soil-root, underground.

Disposal of hazardous waste is considered as the major industrial problem. According to the Telangana State Pollution Control Board, it has been estimated that about 56,000 tons of industrial waste are being

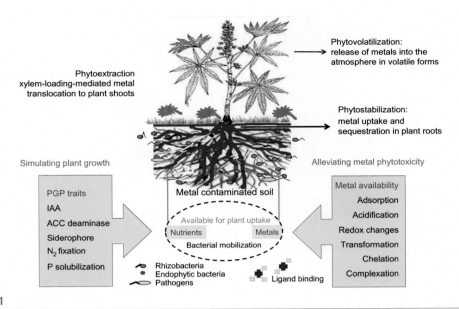

FIGURE 1

Plant-microbe interactions and rhizospheric processes are important for phytotechnologies mainly for stimulating plant growth and alleviating metal toxicity.

produced from about 400 industries located in and around Greater Hyderabad Municipal Corporation. The Pollution Control Board of the Telangana State has identified several dumpsites wherein the hazardous wastes is disposed into open spaces, diluted, and discharged into the water bodies and lakes without any prior treatment. Most of the contaminated land bearing industrial waste and MSW is known to have heavy metals. The most dominant plants growing on the industrial waste-contaminated area have been analyzed with basic ecological parameters. We found *Ricinus communis* as the most dominant plant species growing on the industrial waste-contaminated sites. Several other studies have also shown that *R. communis* grows spontaneously in metal-contaminated sites and has been proposed as a candidate for restoration of a wide range of contaminants (Zhuang et al., 2007; De Souza Costa et al., 2012; Bosiacki et al., 2013; Yi et al., 2014). It has been reported that *R. communis*, tolerant to various stresses, has proved to be the preferred phytoremediator when compared to *Brassica juncea*. Moreover, it grows fast and has been recommended for plantation on wasteland as it requires minimal inputs with little maintenance for its establishment (Ma et al., 2011c; Rajkumar and Freitas, 2008). Besides the use of its biomass as biofuel, animal feeds (if detoxified) and fertilizer, it can make contaminated soils productive (Figure 2).

Some investigations have been made with regard to ecophysiological response of *R. communis* under heavy metal stress and the role of plant growth-promoting bacteria on plant growth and phytoremediation in artificially metal-spiked soils (pot culture and greenhouse experiments). However, information on the composition of rhizosphere and endophytic microbial community associated with *R. communis* growing in soils that are polluted with heavy metals is rather scanty. Information on microbial community diversity in heavy metal-polluted soils can provide a new insight into plant-microbe interaction in polluted soils. Thus knowledge of new isolates and genetic information of metal resistance can be exploited for phytoremediation of metalliferous substrates using *R. communis*.

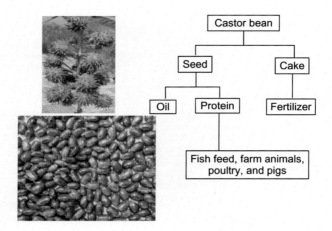

FIGURE 2

Phytoproducts from *Ricinus* potential phytostabilizer.

In the present study, we investigated the basic ecological parameters of vegetation growing in contaminated soils and the association of *R. communis* with microorganisms that would accelerate the process of phytoremediation. In this study with culture-dependent technique, we isolated and characterized endophytic and rhizosphere bacterial strains of *R. communis* growing in the heavily industrialized area of Greater Hyderabad Municipal Corporation, study sites are shown in Figure 3 and Figure 4.

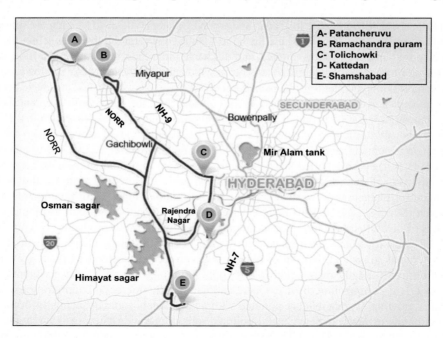

FIGURE 3

Study sites in Greater Hyderabad Municipal Corporation limits.

FIGURE 4

(a–d) *R. communis* (Castor bean) on contaminated periurban area of Greater Hyderabad Municipal Corporation. Please note the large leaf area (inset).

Our objectives were to characterize the microbial isolates of *R. communis* rhizosphere by partial 16S rRNA gene sequencing, metal mobilization and their capability to solubilize P, production of siderophores, indole acetic acid, and 1-aminocyclopropane-1-carboxylic acid (ACC) deaminase.

2 MATERIALS AND METHODS

2.1 STUDY SITES

Cover, density, and frequency are important aspects of plant community that can be measured by quadrant sampling. A quadrant delimits an area in which vegetation cover can be estimated, plants counted, or species listed. We established a permanent plot of $100\,m^2$ at the study site. We subdivided each plot into adjacent $10\,m \times 10\,m$ subplots. Plant specimens for all species were collected for taxonomic identification. We calculated relative density, relative dominance, and relative frequency of each plant species in order to estimate the importance value index of that area (Curtis and Mcintosh). Relative density is the density of one species as a percentage of total plant density.

Relative density = No. of individuals of species/total no. of individuals × 100
Relative frequency of one species as a percentage of total plant frequency
Relative frequency = Frequency of a species/sum frequency of all species × 100
Relative dominance = Dominance of a species/dominance of all species × 100
Important value index = Relative density + Relative dominance + Relative frequency

The study site of our investigation at the two sites was ~20 m². A total of eight soil samples were collected from each site from the upper 30-40 cm depth and immediately transported to the laboratory in polythene bags.

2.2 SOIL CHARACTERIZATION

Composite soil samples (each prepared from eight different randomly collected cores) were collected from the top layer (0-15 cm) at the sites. The soil was air dried and sieved to remove plant materials, soil macrofauna, and stones. The pH and EC was determined in soil/water (1:2.5; w/v) suspension with a pH meter and a conductivity meter, respectively. Organic matter was determined by loss on ignition (Cambardella and Elliott, 1992), organic carbon by rapid dichromate oxidation technique (Nelson and Sommers, 1996), available nitrogen by the alkaline potassium permanganate method (Subbiah and Asija, 1956), available phosphorus by Bray's method (Bray and Kurtz, 1966). The cation exchange capacity (CEC) was determined by extraction with 1 N sodium acetate, followed by washing with 95% of ethanol, leaching with 1 N ammonium acetate solution, and was measured by flame photometer (Jackson, 1973). Available micronutrients and total heavy metal in the soil were extracted with DTPA-Cacl$_2$ (soil to extractant ratio, 1:2) and analyzed by atomic absorption spectrophotometry (AAS) (Lindsay and Norvell, 1978).

2.3 ENUMERATION OF CULTURABLE SOIL BACTERIA

The bacterial strains were isolated from the rhizosphere and tissue interior of *R. communis* growing in metal-polluted soils. For the isolation of rhizosphere bacteria, harvested plants were shaken to remove the loosely attached soil. Soil adhering to the root was considered as the rhizosphere soil. About 1 g of soil samples were serially diluted using 25 mM phosphate buffer and spread over on Luria-Bertani medium (LB) amended with 50 mg of heavy metals. The plates were incubated at 37 °C for 48 h.

For the isolation of endophytic bacteria, root samples were washed with tap water followed by three rinses with deionized water. Healthy root tissues were sterilized by sequential immersion in 70% (v/v) ethanol for 1 min, and 3% sodium hypochlorite for 3 min and washed three times with sterile water to remove surface sterilization agents. In order to confirm the surface disinfection process was successful, sterility was checked by plating 100 microliter of final rinsed water on LB agar. After surface sterilization, the root tissue was cut and titrated in distilled water; appropriate dilutions were plated onto sucrose-minimal salts, low-phosphate agar medium (sucrose 1%; $(NH_4)_2SO_4$ 0.1%; K_2HPO_4 0.05%; $MgSO_4$ 0.05%; NaCl 0.01%; yeast extract 0.05%; $CaCO_3$ 0.05%; pH 7.2) amended with 50 mg of heavy metals. To isolate metal-resistant strains, the bacterial strains picked from the metal-resistant colonies were purified on the LB agar medium containing 50 mg L^{-1} of heavy metals and gradually taken to higher concentration (100-1000 mg L^{-1}) according to the procedure of Ma et al. (2011b) (Figure 5).

2.4 BACTERIAL GROWTH UNDER INCREASING METAL CONCENTRATION

Culture flasks containing 50 mL LB broth supplemented with different concentrations of Cu, Cd, Pb, and Zn varying from 0, 100, 200, and 400 ppm were inoculated with logarithmic-phase bacterial cultures. All the cultures including the controls were incubated at 27 ° C for 32 h at 200 rpm on shaker. The bacterial growth was monitored at fixed time intervals by means of optical density at 600 nm.

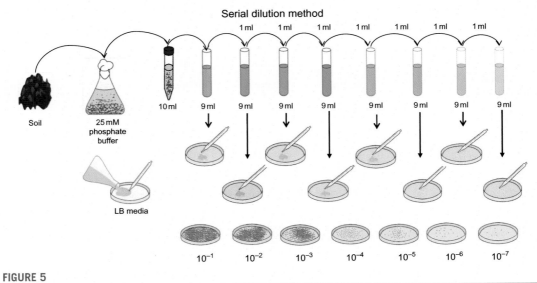

Enumeration of culturable soil bacteria.

2.5 BIOCHEMICAL CHARACTERIZATION OF BACTERIAL STRAINS

Preliminary identification of the bacterial strains was done based on the morphological, physiological, and biochemical characteristics according to the Bergey's Manual of Systematic Bacteriology vol. I (Krieg and Holt, 1984); colony morphology, gram-staining, motility, and biochemical tests: oxidase, catalase, indole, methyl red, Voges-Proskauer, citrate utilization, gelatin liquefaction, and starch hydrolysis tests (Figure 6).

2.6 MOLECULAR CHARACTERIZATION OF BACTERIAL STRAINS

Bacterial strains were grown in LB broth at 30 °C for 20 h. Cells were harvested after 20 h and genomic DNA was isolated, using Qiagen genomic DNA isolation kit. Amplification of 16S rDNA gene sequence was performed by polymerase chain reaction (PCR) using bacterial genome as template and universal bacterial primers pA (5′-GTTTGATCCTGGCTCAG-3′) and 1492r (5′-TACCTTGTTACGACTTCA-3′). The PCR mixture (25 μL) contained 1 μL template, 2.5 μL 10× Taq DNA polymerase buffer, 25 mM $MgCl_2$, 0.5 μL 10 mM dNTP; 10 pmol of each FP and RP 0.5 and 1 μL of 2.5 Unit Taq polymerase. The PCR hot start performed at 94 °C for 4 min, followed by 30 cycles of 94 for 1 min, 55 for 1 min, 72 for 1 min, 72 for 5 min followed by final extension at 15 °C for 5 min. Each amplification mixture (5 μL) was analyzed by agarose gel (1.5% w/v) electrophoresis in TAE buffer (0.4 M Tris acetate, 0.001 M EDTA) with 1 mg mL^{-1} ethidium bromide. The amplified DNA was purified from salts and primers using PCR purification kit according to the manufacturer's instruction. Sequencing of the purified PCR products was performed at Ocimum Biosolutions Pvt Ltd, Hyderabad. Partial 16S rRNA sequences obtained were matched against nucleotide sequences present in GenBank using the BLAST program (Altschul et al., 1997).

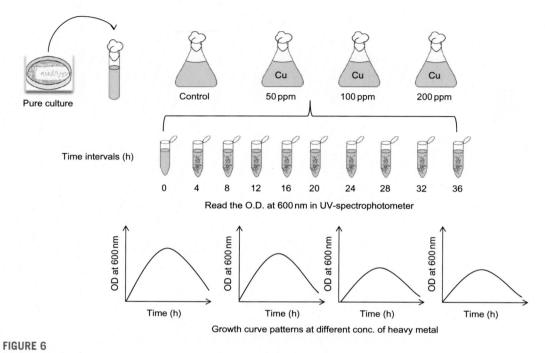

FIGURE 6

Bacterial growth under increasing heavy metal concentration.

3 CHARACTERIZATION OF PLANT GROWTH-PROMOTING BACTERIA
3.1 INDOLE ACETIC ACID PRODUCTION

Indole-3-acetic acid (IAA) is one of the best-characterized plant growth-promoting phytohormones. Tryptophan is generally considered to be the precursor of IAA. Bacteria associated with soil and rhizospheres have the ability to produce IAA; 80% of rhizospheric bacteria are known to produce IAA (Patten and Glick, 1996). Bacterial cultures were grown in LB medium supplemented without and with L-tryptophan (500 μg mL^{-1}). The liquid medium was inoculated with 24 h-old bacterial culture and incubated at 27 °C for 96 h at 175 rpm. The growth of bacteria was monitored at definite time intervals. Fully grown cultures were centrifuged at 10,000 rpm for 10 min. The IAA concentration in the culture supernatant was determined according to the method of Bric et al. (1991). To 2 mL of the supernatant obtained from the culture medium, 100 μL of 10 mM Ortho phosphoric acid and 4 mL of Salkowski reagent (1 mL of 0.5 M FeCl$_3$ 50 mL of 35% HClO$_4$) was added. The intensity of the pink color developed was measured after 25 min incubation at 530 nm. The IAA concentration in culture was determined using calibration curve of pure IAA as a standard (Figure 7).

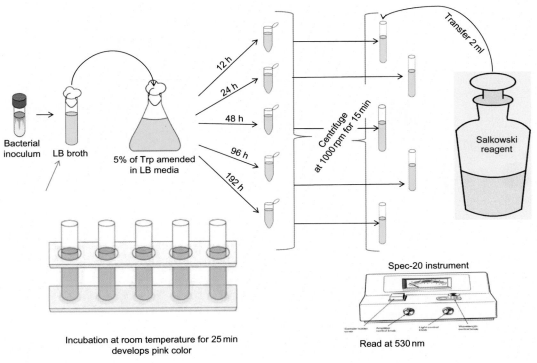

FIGURE 7

Protocol for the determination of IAA production by bacteria. Note: LB, Luria-Bertani medium.

3.2 PHOSPHATE SOLUBILIZATION

Phosphorus (P) is a major constituent in all living forms and is one of the three essential nutrients required for plant growth along with nitrogen and potassium. Phosphorus is the least mobile and unavailable macronutrient to the plants in most soil conditions although abundant in organic and inorganic forms. Solubilization of inorganic phosphate and mineralization of organic phosphate, which makes phosphorus available to the plants, are the two major roles played by PGPR (Rajkumar et al., 2012). The phosphate solubilizing ability of the bacterial isolates was quantitatively estimated by using National Botanical Research Institute's phosphate growth medium (NBRIP) medium with $1\,mL$ ($3 \times 10^7\,cells\,mL^{-1}$) of each bacterial inoculum and incubated at $27\,°C$ for 7 days at $175\,rpm$; sterile medium without bacterial inoculum was served as control. The solubilized phosphate concentration in culture supernatants was estimated by the method of Fiske and Subbarow (1925) (Figure 8).

3.3 METAL UPTAKE BY SOIL BACTERIA

Bacterial cultures were grown in $100\,mL$ of LB broth until the optical density (O.D.) reached 1.0 at $600\,nm$. Cells were then harvested by centrifugation at $8000\,rpm$ for $15\,min$ and the bacterial pellet was washed twice with sterile water. The harvested biomass was resuspended in $5\,mL$ of the respective

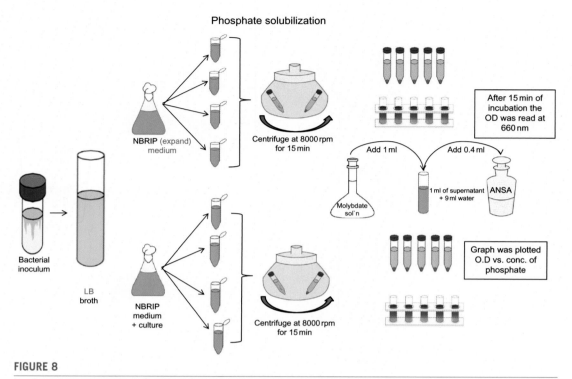

FIGURE 8

Protocol depicting phosphate solubilization by bacteria.

metal solution with varying concentration, 10, 20, 40, 60 mg L^{-1}. After incubation at room temperature for 10 h, the cells were harvested by centrifugation. The amount of residual metal present in the supernatant was measured by AAS (Figure 9).

4 RESULTS AND DISCUSSION

Researchers all over the world are searching for new plant species suitable to be used in phytoremediation. While selecting a species for both phytoremediation and biodiesel production, several factors have to be taken into account (Demirbas, 2009; Gui et al., 2008; Jayed et al., 2009). The species should be fast growing, high yield producing, with profuse root and shoot systems, tolerant to environmental stress condition, and also economically beneficial. *R. communis* has been used for phytoremediation of multi-metal-contaminated soils (Ma et al., 2011c; Rajkumar and Freitas, 2008). Their fast growth and the ability to produce a significant amount of biomass in a short period of time make them "trees of choice" for phytoremediation purposes. Moreover, since it is a widely distributed tree species in metal-polluted soils able to tolerate and accumulate high concentrations of heavy metals, it seemed very interesting to study the microbial diversity and the plant growth-promoting and metal-resistant potential of their associated microbial community.

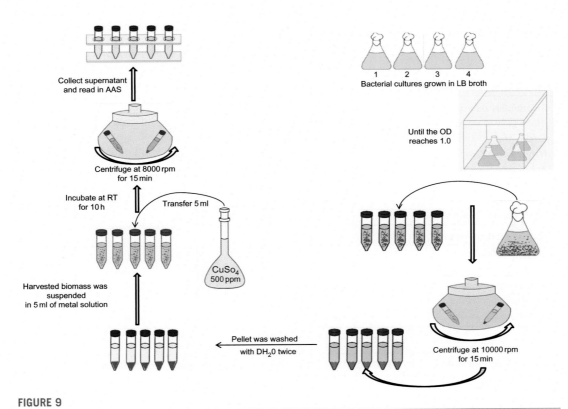

FIGURE 9

Biosorption of metal by bacteria. Note: AAS, atomic absorption spectrophotometry; LB, Luria-Bertani medium.

Investigating the distribution of the individual of plant species in small plots of the area, the relationship between the relative frequency, density, and dominance of the species has been studied. From Table 1, it is clear that the study sites were dominated by *R. communis* belonging to the family Euphorbiaceae. It has the highest number of individuals and contributed more than 40% of the total number of individuals on the plot. Other important and dominant species include *Amaranthus spinosus* and *Hyptis sueveoleus*, which belong to family Amaranthaceae and Lamiaceae, respectively. So we selected the rhizospheric soil of *R. communis* for our studies. Reference may be made to Table 2 for its phytoremediation and phytoproduct cogeneration.

4.1 ENUMERATION OF CULTURABLE SOIL BACTERIA

Cultivable bacteria were isolated from the rhizosphere and root interior of *R. communis* growing in metal-polluted regions (Table 3). The pH of soils was near neutral with a value of 6.74 and 6.12. Among the heavy metals, copper recorded the highest value. Similarly, the other heavy metals recorded were lead followed by cadmium in terms of mg metal kg^{-1} of dry soil. During the initial screening, seven Cu and four Pb-tolerant endophytes were isolated from the roots of L. *R. communis*

Table 1 Vegetation Sampling Analysis of the Study Site

Plant Species	Family	Relative Frequency	Relative Density	Relative Dominance	Important Value
Patancheru industrial area					
Amaranthus spinosus	Amaranthaceae	18.88	1.25	2.22	22.35
Alternanthera sessilis	Amaranthaceae	3.23	2.05	10.90	16.18
Cassia tora	Caesalpinaceae	3.55	3.17	3.26	9.98
Hyptis suaveolens	Lamiaceae	6.82	6.31	8.96	22.09
Achyranthes aspera	Amaranthceae	3.41	6.31	1.46	1.18
Ricinus communis	Euphorbiaceae	20.29	10.09	8.77	39.15
Trema orientalis	Ulmaceae	1.14	0.9	0.34	2.38
Parthenium hysterophorus	Asteraceae	1.14	0.9	0.36	2.40
Katedan industrial area					
Achranthus aspera	Amaranthaceae	2.27	1.8	0.54	4.61
Parthenium hysterophorus	Asteraceae	3.41	2.7	0.82	6.93
Ricinus communis	Euphorbiaceae	10.23	13.51	43.0	66.74
Amaranthus spinosus	Amaranthaceae	9.09	8.11	2.74	19.94
Alternanthera sessilis	Amaranthaceae	1.14	0.9	0.22	2.26

Table 2 *R. communis*: Phytoremediation Potential and Cogeneration of Economic Products

Shi et al. (2015)	Drought stress decreases cadmium accumulation in castor bean by altering root morphology
Kang et al. (2015)	Copper distribution in the root cells of castor seedlings in hydroponic culture
Bauddh and Singh, 2015	Bioaccumulation and partitioning of Cd in *as* influenced by organic and inorganic amendments
Bonanno (2014)	Biomonitoring potential of atmospheric pollution in urban areas
Olivares et al. (2013)	Mine tailings stabilization
Tyagi et al. (2013)	Phytochemical constituents under the influence of industrial effluent
Andreazza et al. (2013)	Potential phytoremediator for copper-contaminated soils
Wang et al. (2013)	Phytoextraction of metals and rhizoremediation of PAHS in cocontaminated soil by coplanting with other species
Hernández (2013)	Potential of castor bean for phytoremediation of mine tailings
Martins et al. (2013)	Leaf powder as a green adsorbent for the removal of heavy metals
Abreu et al. (2012)	Soil contaminated by heavy metals and boron and organic amendments
Bauddh and Singh (2012a,b)	Saline, drought, and Cd-contaminated soil
De Souza Costa et al. 2012	Cd- and Pb-contaminated soil
Varun et al. (2012)	Glass industry-contaminated substrates
Melo et al. (2012)	Arsenic accumulation grown on contaminated soils
De Souza Costa et al., 2012	Phytoremediation of Cd and Pb

Table 2 *R. communis*: **Phytoremediation Potential and Cogeneration of Economic Products—Cont'd**

Adhikari and Kumar (2012)	Phytoaccumulation and tolerance to nickel
Dos Santos et al. (2012)	Potential candidate for Pb-phytoextraction
Huang et al. (2011)	DDT and Cd cocontaminated soil
Nazir et al. (2011)	Heavy metal hyperaccumulator
Ye et al. (2010)	Arsenic speciation in phloem and xylem exudates of castor bean
Singh et al. (2010)	Heavy metal accumulation in fly ash pond
Coscione and Berton (2009)	Barium phytoextraction
Melo et al. (2009)	Accumulation of arsenic and nutrients by castor bean plants grown on an As-enriched nutrient solution
Niu et al. (2009)	Response of root and aerial biomass to phytoextraction of Cd and Pb
Shi and Cai (2009)	Cd tolerance and accumulation
Figueroa et al. (2008)	Phytochelatin production on silver mine waste
Liu et al. (2008)	Accumulation of Pb, Cu, and Zn
Rajkumar and Freitas (2008)	Plant growth-promoting bacteria in heavy metal-contaminated soil accelerates growth
Malarkodi et al. (2008)	Phytoextraction of nickel
Niu et al. (2007)	Evaluation of phytoextracting cadmium and lead
Zhi-Xin et al. (2007)	Phytoextraction of cd and pb in hydroponics
Romeiro et al. (2006)	Pb uptake and tolerance
Giordani et al. (2005)	Soil polluted by Ni
Stephan et al. (1994)	Phloem translocation of Fe, Cu, Mn, and Zn

Table 3 **Physicochemical Properties of the Rhizospheric Soil**

Soil sample	SS1	SS2
Texture	Sandy loam	Sandy loam
pH	6.74	6.12
CEC ($cmol\,kg^{-1}$)	4.14	3.03
Organic matter ($g\,kg^{-1}$)	10.10	8.55
Total N ($g\,kg^{-1}$)	0.93	1.09
Available P ($mg\,kg^{-1}$)	12.33	9.64
Available K ($mg\,kg^{-1}$)	17.66	25.11
Total Cu ($mg\,kg^{-1}$)	259	146
NH_4OAc ($mg\,kg^{-1}$)	71	33
Total Pb ($mg\,kg^{-1}$)	21	43
NH_4OAc ($mg\,kg^{-1}$)	1.1	1.2
Total Cd ($mg\,kg^{-1}$)	3.7	5.6
NH_4OAc ($mg\,kg^{-1}$)	0.3	0.4

whereas from the rhizosphere region, 32 metal-resistant bacterial strains were isolated. After secondary screening, the strains showing resistance up to and above $300 \, mg \, L^{-1}$ of the respective heavy metal were selected for studying metal tolerance mechanisms and plant growth-promoting properties. Among the isolates tested, two endophytes (strains SS1-E2 and SS1-E9) and four rhizosphere isolates (SS1-R2, SS1-R5, SS2-R4, and SS2-R9) showed maximum tolerance to Cu, Cd, and Pb, respectively. It is well known that microbes isolated from a metal-polluted environment often exhibit high resistance to metal stress since they have adapted to such stress conditions. All the strains were purified and maintained in trypticase soya broth agar slants at $4°C$ for further studies. The isolates were further characterized based on the morphological, physiological, and biochemical characteristics according to Bergey's Manual (Table 4). Although heavy metals exert their toxic effects on microbes and plants through various mechanisms, metal-tolerant bacteria can survive in rhizosphere soils and exert their beneficial effects on plants growing in polluted soils. Thus such beneficial bacterial strains can be isolated and selected for their potential application in the microbe-assisted phytoremediation of metal-polluted soils.

4.2 BACTERIAL GROWTH UNDER INCREASING METAL CONCENTRATION

The two endophytic strains, SS1-E2 and SS1-E9, were isolated from the roots of *R. communis* L. growing near the highly industrialized sites in Hyderabad. Strains SS1-E2 and SS1-E9 showed maximum tolerance up to $400 \, mg \, Cu \, L^{-1}$ and $600 \, mg \, Pb \, L^{-1}$, respectively. Thus the effect of different concentrations of Cu on SS1-E2 and of Pb on SS1-E9 growth was tested in liquid cultures. Growth pattern exhibited a variation in control compared to that of other concentrations used. In the case of SS1-E2, during

Table 4 Morphological and Biochemical Characteristics of the Isolates

	SS1-E2	SS1-E9	SS1-R2	SS1-R5	SS2-R4	SS2-R9
Gram's reaction	+	+	−	+	+	−
Shape	Rods	Rods	Rods	Rods	Rod	Cocci
Pigmentation	No	−	No	No	No	−
Oxidase test	+	+	+	−	+	−
Catalase test	+	+	+	+	+	−
Indole test	+	+	+	+	−	+
Methyl red test	+	−	−	−	+	+
Voges-Proskauer test	+	+	−	+	+	−
Citrate utilization	+	+	+	+	+	+
Nitrate reduction	−	−	+	+	+	−
H$_2$S production (TSI)	+	+	−	+	−	−
Gelatin hydrolysis	+	+	+	+	+	−
Starch hydrolysis	+	+	+	+	+	+
Urea hydrolysis	−	−	−	+	−	−
Glucose fermentation	+	−	+	−	−	+
Lactose fermentation	+	+	+	−	+	+

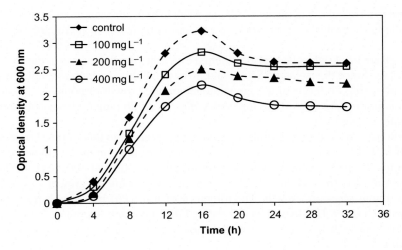

FIGURE 10

Effect of Cu on the growth pattern of SS1-E2.

the initial hours of incubation, maximum growth was observed in control followed by $100\,mg\,Cu\,L^{-1}$ compared to the concentrations 200, $400\,mg\,Cu\,L^{-1}$ (Figure 10).

Similarly, the growth curve pattern of SS1-E9 in liquid media with 400 or $600\,mg\,Pb\,L^{-1}$ showed a long lag phase of 12h indicating that the isolates could adapt to the stress environment, heavy metal, which was followed by a logarithmic phase of growth curve (Figure 11). It is well known that bacterial

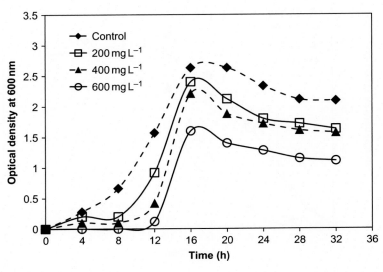

FIGURE 11

Effect of Pb on the growth pattern of SS1-E9.

strains isolated from metal-contaminated soils can adapt to high concentrations of metal through various strategies including physical sequestration, exclusion, complexation, and detoxification (Rajkumar et al., 2010).

The rhizosphere bacterial isolates also exhibited different growth patterns in the presence of respective heavy metals. Since the isolates SS1-R2 and SS1-R5 exhibited a high tolerance to Pb when cultivated under increasing Pb levels in the growth medium, the growth response of SS1-R2 and SS1-R5 in liquid media in the presence of varying concentrations of Pb ranging from 100 to 600 mg L^{-1} was assessed (Figures 12 and 13).

The growth pattern exhibited a variation in control compared to that of other concentrations used. In the case of SS1-R2, during the initial hours of incubation, maximum growth was observed in control followed by 200 mg L^{-1} compared to the concentrations (400 and 600 mg L^{-1}). The inhibition of growth was observed in the early hours, however, after a few hours the growth was resumed, entering into the logarithmic phase of growth. The tolerance of lead (Pb) by the two bacterial isolates was different from each other. However, SS1-R5 showed the maximum tolerance up to 400 mg L^{-1} of Cd. Compared with control, however, an overall decrease in growth of SS1-R5 was observed in the media amended with Pb.

Tolerance of Cd by the bacterial strains from the SS2 was tested based on their growth in the LB medium. The growth curves for SS2-R4 and SS2-R9 in the presence of varying concentrations of Cd ranging from 50 to 350 mg L^{-1} are shown in Figures 14 and 15.

The rhizosphere strain SS2-R4, exposed to different concentrations of Cd, showed a marked inhibition in growth. With the increase in concentration of metals, a progressive decrease in SS2-R4 growth was observed. On the other side, SS2-R9 exhibited a different growth pattern in control compared to

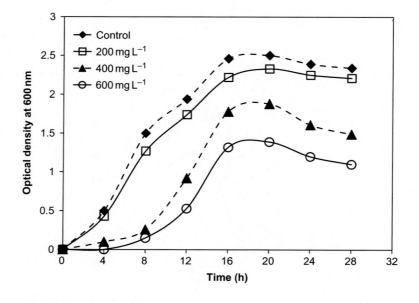

FIGURE 12

Effect of Pb on the growth pattern of SS1-R2.

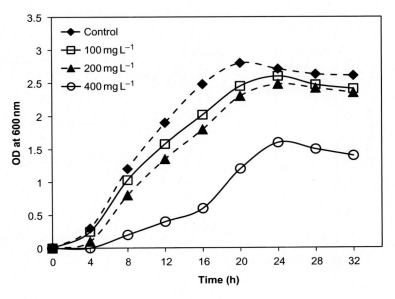

FIGURE 13

Effect of Pb on the growth pattern of SS1-R5.

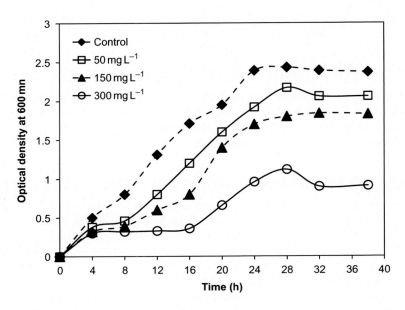

FIGURE 14

Effect of Cd on the growth of SS2-R4.

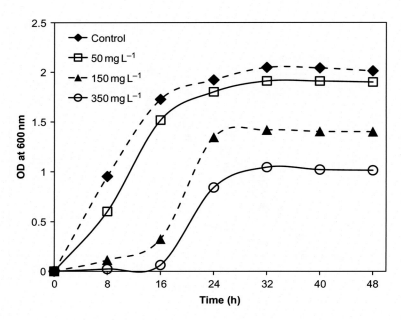

FIGURE 15

Effect of Cd on the growth of SS2-R9.

that of other concentrations used. Here, the isolate showed relatively short lag phase and enters into log phase within 4 h of incubation. But in the case of 350 mg Cd L^{-1}, it showed relatively long lag phase and thereafter entered into log phase and stationary phase.

4.3 PLANT GROWTH-PROMOTING TRAITS OF METAL-RESISTANT BACTERIA

Certain plant-associated microorganisms have the ability to promote host plant growth and alleviate metal stress by various mechanisms; namely, ACC deaminase, IAA, siderophores, and/or solubilization of P. The metal-resistant strains isolated from the rhizospheric root interior of *R. communis* were further assayed for a number of plant-promoting traits.

4.4 GROWTH AND ACC DEAMINASE ACTIVITY OF METAL-RESISTANT BACTERIA

The metal-tolerant isolates were tested for their ability to grow on Dworkin-Foster (DF) salts minimal medium with ACC. Among the rhizosphere and endophytic bacteria tested, the endophytes SS1-E2 and SS1-E9 grew in DF salts minimal medium with ACC as the sole source of nitrogen. However, maximum growth was observed in SS1-E2 compared to SS1-E9. In the absence of ACC, both the isolates showed limited growth (Figure 16).

It has been observed that increased accumulation of ACC in roots, caused by abiotic stress, can facilitate the colonization of the metal-tolerant bacteria in the rhizoplane. It is known that plants growing in the metal-contaminated soils accumulate ACC and thereby produce endogenous stress ethylene, which can reduce the root growth and consequently affect plant health. Under such conditions,

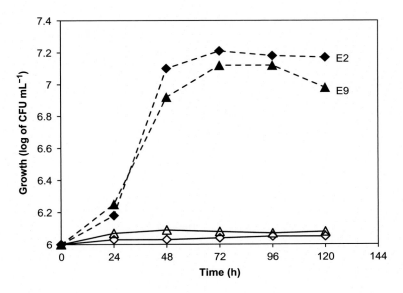

FIGURE 16

Growth of E2 and E9 on DF salts minimal medium.

certain rhizosphere bacteria play a significant role in plant growth through reducing stress ethylene concentration in root. The rhizosphere bacteria possessing the enzyme ACC metabolize the ethylene precursor, ACC, into ketobutyrate and ammonia (Glick et al., 2007). Similarly, when the plants exude more ACC to maintain the equilibrium between the rhizosphere soil and root interior, the stress ethylene production in plants decreases (Adams and Yang, 1976).

4.5 GROWTH AND INDOLE ACETIC ACID PRODUCTION

IAA production by endophytic bacterial strains was estimated in LB medium supplemented without and with L-tryptophan. Both the isolates produced IAA in culture when the medium was supplemented with L-tryptophan, which indicates that the endophytic isolates utilized tryptophan as a precursor for growth and IAA production. The estimation of IAA in culture filtrate at different time intervals showed a linear and time-dependent increase. In both cases, the growth and IAA production increased simultaneously, and a maximum IAA production was observed after 72 h of incubation. Similar results were also observed with the earlier observations of Garcia de Salamone et al. (2001) indicating that the induction of IAA production in stationary phase culture is probably due to delayed induction of a key enzyme in IAA biosynthetic pathway.

The analysis of IAA production by rhizosphere bacteria also showed a substantial amount of IAA production during the stationary phase of culture in LB medium supplemented with L-tryptophan. The growth and IAA production increased simultaneously and the maximum IAA production was observed by SS-R9 strain after 96 h of incubation. The production of IAA was dependent on the bacterial strain as well as on the concentration of tryptophan. Quantitatively, SS1-R9 produced significantly more IAA than SS1-R1. It produced 19.6 μg mL^{-1} whereas R2 produced nearly about 5 μg mL^{-1} of IAA, after 96 h of incubation (Figures 17 and 18).

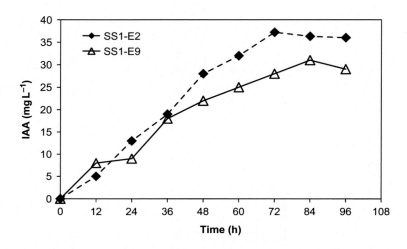

FIGURE 17

IAA production by endophyte bacterial strains.

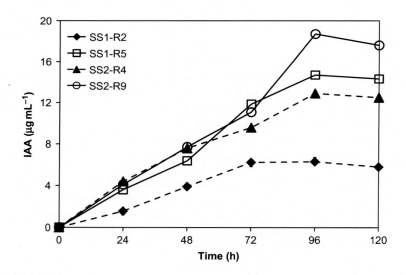

FIGURE 18

IAA production by rhizospheric isolates.

5 GROWTH AND PHOSPHATE SOLUBILIZATION

Quantitative estimation of phosphorus solubilization was carried out in NBRI medium for 144 h at 27 °C. Quantification of the phosphate was done at regular intervals of 24 h. The endophytic strains SS1-E2 and SS1-E9 utilized the insoluble calcium phosphate in the medium as a sole source of phosphate. Growth of the endophytes increased linearly up to 72 h incubation and

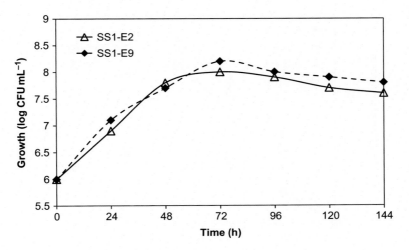

FIGURE 19

Growth by SS1-E2 and SS1-E9 in NBRIP medium.

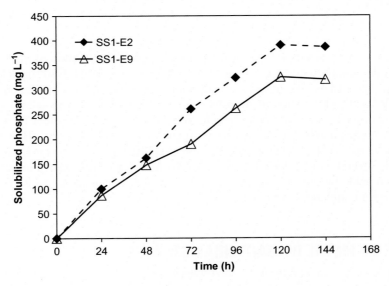

FIGURE 20

Phosphate solubilization by SS1-E2 and SS1-E9.

thereafter the stationary phase was reached. The maximum solubilization by SS1-E2 and SS1-E9 was achieved after 120 h of incubation (390.2 and 325 mg P L^{-1}). Further incubation up to 144 h did not improve the extent of solubilization (Figures 19–21). Phosphate solubilization potential of rhizosphere isolates SS1-R2, SS1-R5, SS2-R4, and SS2-R9 was monitored in NBRIP media for 6 days.

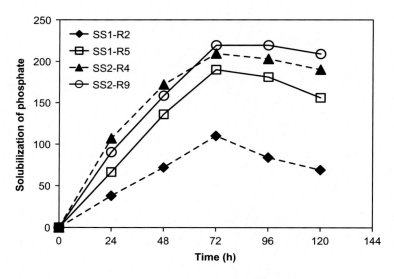

FIGURE 21

Phosphate solubilization by SS1-R2, SS1-R5, SS2-R4, SS2-R9 in NBRIP medium.

Among the isolates tested, SS2-R4 (value $mg\,P\,L^{-1}$) and SS1-R3 showed maximum phosphate solubilization potential after 120 h of incubation (SS2-R4-219; SS1-R3-209 $mg\,P\,L^{-1}$). Further incubation up to 120 h (data not shown), did not improve the extent of solubilization. However, the isolates SS1-R2 and SS1-R5 showed almost a similar pattern of phosphate solubilization. The final pH in the growth media was checked at regular intervals to find out whether the solubilization of phosphate is accompanied by the production of organic acid. In all cases, the pH of the medium was lowered. A decrease in pH indicates the production of acids, which may be the reason for phosphate solubilization. Several authors have reported that the decrease in pH clearly indicates the production of organic acids by the rhizosphere organisms, which is considered to be responsible for P solubilization (Cunningham and Kuiack, 1992).

6 UPTAKE OF METAL BY BACTERIAL STRAINS

The data expressing the capability of the rhizospheric bacteria isolated from two different locations, SS1 and SS2, to uptake heavy metals are shown in Figure 22.

It is clearly evident that the rhizosphere isolates exhibited different biosorption capacity toward Cd and Pb. The uptake of Cd and Pb by the four bacterial strains showed an increase in biosorption with increasing the initial concentration of metal. However, the maximum biosorption capacity for Cd was observed in the case of R5, and the highest quantity of Pb was adsorbed by R9. In the case of endophytic strains, both isolates showed minimum biosorption capacity for Cd and Pb (data not shown).

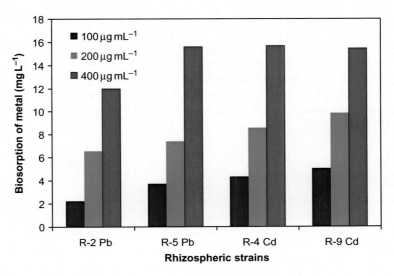

FIGURE 22

Metal uptake by the rhizospheric strains.

7 IDENTIFICATION OF BACTERIAL STRAINS

Assessment of plant growth-promoting parameters revealed the intrinsic ability of the rhizosphere and endophytic bacterial strains to produce IAA, utilizing ACC as the sole N source and solubilizing insoluble phosphate. All of the strains tested positive for IAA production and phosphate solubilization. However, except for rhizosphere bacteria, only the endophytes showed positive for ACC deaminase activity. Bacterial strains utilizing ACC as a sole source of nitrogen reduce stress ethylene production and play a major role in the amelioration of metal-induced impact on plant health. Previous studies have confirmed the efficiency of ACC utilizing PGPR to promote the root elongation and growth of plants under metal stress conditions. Hence, further studies including the inoculation effects of ACC utilizing endophytic strains SS1-E2 and SS1-E9 on the growth and phytoremediation potential of *R. communis* are under progress in order to test the usefulness of this novel isolate for future phytoremediation application. On the basis of morphological, physiological, and biochemical characteristics (Table 3) and comparative analysis of the sequence with the already available database showed that the strains SS1-E2 and SS1-E9 were close to the members of the genus *Enterobacter* and *Bacillus*. Partial sequence of 16S r DNA of E2 showed 99% homology with *Enterobacter* sp, and E9 showed 99% homology with *Bacillus* sp (Table 4).

8 CONCLUSION

Our study reveals that different bacterial communities living in association with rhizosphere soil and those colonized in the root tissues of *R. communis* growing in contaminated sites are able to withstand high levels of metal in the soil. A total of six potential bacterial isolates, two endophytes, and four rhizosphere bacteria exhibited tolerance to a range of metal ions that included Cu, Pb, and Cd. Increased tolerance to the heavy metals that was found among the isolates may be attributed to the high-level contamination in the soil. The isolated bacteria were found to have potential plant growth-promoting characteristics under metal-stressed conditions by producing IAA, ACC deaminase, and solubilizing phosphate. The metal-resistant plant growth-promoting bacteria isolated from the rhizosphere and tissues of *R. communis* serve as resource for experiments of seed inoculation to verify their ability to improve heavy metal phytoremediation process and protect plants from the effects of metals (Ahemad, 2015).

ACKNOWLEDGMENTS

D. Annapurna gratefully acknowledges the award of Junior Research scholarship by CSIR, New Delhi.

REFERENCES

Adams, D.O., Yang, S.F., 1976. Ethylene biosynthesis: identification of 1-amino cyclopropane carboxylic acid as an intermediate in the conversion of methionine to ethylene. Proc. Natl. Acad. Sci. U.S.A. 76, 170–174.

Adhikari, T., Kumar, A., 2012. Phytoaccumulation and tolerance of *Riccinus communis* L. to nickel. Int. J. Phytoremediat. 14 (5), 481–492.

Ahemad, M., 2015. Phosphate-solubilizing bacteria-assisted phytoremediation of metalliferous soils: a review. 3 Biotech 5, 111–121.

Altschul, S.F., Madden, T.L., Schaffer, A.A., Zhang, J., Zhang, Z., Miller, W., Lipman, D.J., 1997. Gapped BLAST and PSIBLAST: a new generation of protein database search programs. Nucleic Acid Res. 25, 3389–3402.

Ananthi, T.A.S., Manikandan, P.N.A., 2013. Potential of rhizobacteria for improving lead phytoextraction in *Ricinus communis*. Remediation 24 (1), 99–106.

Andreazza, R., Bortolon, L., Pieniz, S., Camargo, F.A.O., 2013. Use of high-yielding bioenergy plant castor bean (*Ricinus communis* L.) as a potential phytoremediator for copper-contaminated soils. Pedosphere 23 (5), 651–661.

Atabayeva, S., Sarsenbayev, B., Prasad, M.N.V., Jaime, A., da Silva, T., Kenzhebayeva, S., Usenbekov, B., Kirshibayev, Y., Asrandina, S., Beisenova, A., Danilova, A., Kotuhov, Y., 2010. Accumulation of trace metals in grasses of Kazakhstan: relevance to phytostabilization of mine waste and metal-smelting areas. Asian Australas. J. Plant Sci. Biotechnol. 4 (1), 91–97.

Bauddh, K., Singh, R.P., 2012a. Growth, tolerance efficiency and phytoremediation potential of *Ricinus communis* (L.) and *Brassica juncea* (L.) in salinity and drought affected cadmium contaminated soil. Ecotox. Environ. Safe. 85, 13–22.

Bauddh, K., Singh, R.P., 2012b. Cadmium tolerance and its phytoremediation by two oil yielding plants *Ricinus communis* (L.) and *Brassica juncea* (L.) from the contaminated soil. Int. J. Phytoremediat. 14 (8), 772–785.

Bauddh, K., Singh, R.P., 2015. Effects of organic and inorganic amendments on bio-accumulation and partitioning of Cd in *Brassica juncea* and *Ricinus communis*. Ecol. Eng. 74, 93–100.

Bonanno, G., 2014. *Ricinus communis* as an element biomonitor of atmospheric pollution in urban areas. Water Air Soil Pollut. 225 (2), 1852.

Bosiacki, M., Kleiber, T., Kaczmarek, J., 2013. Evaluation of suitability of *Amaranthus caudatus* L. and *Ricinus communis* L. in phytoextraction of cadmium and lead from contaminated substrates. Arch. Environ. Prot. 39 (3), 47–59.

Bray, R., Kurtz, L.T., 1966. Determination of total, organic and available forms of phosphorus in soil. Soil Sci. 59, 39–45.

Bric, J.M., Bostock, R.M., Silversone, S.E., 1991. Rapid in situ assay for indole acetic acid production by bacteria immobilization on a nitrocellulose membrane. Appl. Environ. Microbiol. 57, 535–538.

Cambardella, C.A., Elliott, E.T., 1992. Particulate soil organic matter. Changes across a grassland cultivation sequence. Soil Sci. Soc. Am. J. 56, 777–783.

Compant, S., Clement, C., Sessitch, A., 2010. Plant growth promoting bacteria in the rhizo- and endosphere of plants: their role, colonization, mechanisms involved and prospects for utilization. Soil Biol. Biochem. 42, 669–678.

Conesa, H.M., Evangelou, W.H., Robinson, B.H., Schulin, R., 2012. A critical review of current state of phytotechnologies to remediate soils: still a promising tool. Sci. World J. 2012, http://dx.doi.org/10.1100/2012/173829. Published online 4 January 2012.

Coscione, A.R., Berton, R.S., 2009. Barium extraction potential by mustard, sunflower and castor bean. Scientia Agricola 66, 59–63.

Cunningham, J.E., Kuiack, C., 1992. Production of citric and oxalic acids and solubilization of calcium phosphate by *Penicillium bilaii*. Appl. Environ. Microbiol. 52, 1451–1458.

Curtis, J.T., McIntosh, R.P., 1950. The interrelations of certain analytic and synthetic phytosociological characters. Ecology 31, 434–455.

Dakora, F.D., Phillips, D.A., 2002. Root exudates as mediators of mineral acquisition in low-nutrient environments. Plant Soil 245, 35–47.

De Abreu, C.A., Coscione, A.R., Pires, A.M., Paz-Ferreiro, J., 2012. Phytoremediation of a soil contaminated by heavy metals and boron using castor oil plants and organic matter amendments. J Geochem Explor 123, 3–7.

De Souza Costa, E.T., Guilherme, L.R.G., De Melo, É.E.C., Ribeiro, B.T., Dos Santos, B., Inácio, E., Da Costa Severiano, E., Faquin, V., Hale, B.A., 2012. Assessing the tolerance of castor bean to Cd and Pb for phytoremediation purposes. Biol. Trace Elem. Res. 145 (1), 93–100.

Demirbas, A., 2009. Progress and recent trends in biodiesel fuels. Energ. Convers. Manage. 50, 14–34.

Dietz, A.C., Schnoor, J.L., 2001. Advances in phytoremediation. Environ. Health Perspect. 109, 163–168.

Dos Santos, C.H., De Oliveira Garcia, A.L., Calonego, J.C., Spósito, T.H.N., Rigolin, I.M., 2012. Pb-phytoextraction potential by castor beans in soil contaminated ((Potencial de fitoextração de Pb por mamoneiras em solo contaminado)). Semina Cienc. Agrar. 33 (4), 1427–1433.

Figueroa, J.A.L., Wrobel, K., Afton, S., Caruso, J.A., Felix Gutierrez Corona, J., Wrobel, K., 2008. Effect of some heavy metals and soil humic substances on the phytochelatin production in wild plants from silver mine areas of Guanajuato, Mexico. Chemosphere 70 (11), 2084–2091.

Fiske, C.H., Subbarow, Y., 1925. A colorimetric determination of phosphorus. J. Biol. Chem. 66, 375–400.

Garcia de Salamone, I.E., Hynes, R.K., Nelson, L.N., 2001. Cytokinin production by plant growth promoting rhizobacteria and selected mutants. Can. J. Microbiol. 47, 103–113.

Giordani, C., Cecchi, S., Zanchi, C., 2005. Phytoremediation of soil polluted by nickel using agricultural crops. Environ. Manage. 36 (5), 675–681.

Glick, B.R., 2010. Using soil bacteria to facilitate phytoremediation. Biotechnol. Adv. 28, 367–374.

Glick, B.R., Todorovic, B., Czarny, J., Cheng, Z., Duan, J., McConkey, B., 2007. Promotion of plant growth by bacterial ACC deaminase. Crit. Rev. Plant Sci. 26, 227–242.

Gui, M.M., Lee, K.T., Bhatia, S., 2008. Feasibility of edible oil vs. non-edible oil vs. waste edible oil as biodiesel feedstock. Energy 33, 1646–1653.

Hernández, R.M., 2013. Potential of castor bean (*Ricinus communis* L.) for phytoremediation of mine tailings and oil production. J. Environ. Manage. 114, 316–323.

Huang, H., Yu, N., Wang, L., Gupta, D.K., He, Z., Wang, K., Zhu, Z., Yan, X., Li, T., Yang, X.-E., 2011. The phytoremediation potential of bioenergy crop *Ricinus communis* for DDTs and cadmium co-contaminated soil. Bioresour. Technol. 102 (23), 11034–11038.

Jackson, M.L., 1973. Soil Chemical Analysis. Prentice Hall of India, New Delhi.

Jayed, M.H., Masjuki, H.H., Saidur, R., Kalam, M.A., Jahirul, M.I., 2009. Environmental aspects and challenges of oilseed produced biodiesel in Southeast Asia. Renew. Sust. Energ. Rev. 13, 2452–2462.

Jones, D.L., Dennis, P.G., Owen, A.G., van Hees, P.A.W., 2003. Organic acid behavior in soils—misconceptions and knowledge gaps. Plant Soil 248, 31–41.

Kang, W., Bao, J., Zheng, J., Hu, H., Du, J., 2015. Distribution and chemical forms of copper in the root cells of castor seedlings and their tolerance to copper phytotoxicity in hydroponic culture. Environ. Sci. Pollut. R 22 (10), 7726–7734.

Krieg, R.N., Holt, J.G., 1984. Bergey's Manual of Systematic Bacteriology, vol. 1. Williams and Wilkins, Baltimore, USA, pp. 308–429.

Lindsay, W.L., Norvell, W.A., 1978. Development of a DTPA soil test for zinc, iron, manganese, and copper. Soil Sci. Soc. Amer. J. 42, 421–428.

Liu, X., Gao, Y., Khan, S., Duan, G., Chen, A., Ling, L., Zhao, L., Liu, Z., Wu, X., 2008. Accumulation of Pb, Cu, and Zn in native plants growing on contaminated sites and their potential accumulation capacity in Heqing, Yunnan. J. Environ. Sci. 20 (12), 1469–1474.

Ma, Y., Prasad, M.N.V., Rajkumar, M., Freitas, H., 2011a. Plant growth promoting rhizobacteria and endophytes accelerate phytoremediation of metalliferous soils. Biotechnol. Adv. 29, 248–258.

Ma, Y., Rajkumar, M., Luo, Y., Freitas, H., 2011b. Inoculation of endophytic bacteria on host and non-host plants—effects on plant growth and Ni uptake. J. Hazard. Mater. 196, 230–237.

Ma, Y., Rajkumar, M., Vicente, J.A.F., Freitas, H., 2011c. Inoculation of Ni-resistant plant growth promoting bacterium *Psychrobacter* sp. strain SRS8 for the improvement of nickel phytoextraction by energy crops. Int. J. Phytoremediation 13, 126–139.

Malarkodi, M., Krishnasamy, R., Chitdeshwari, T., 2008. Phytoextraction of nickel contaminated soil using castor phytoextractor. J. Plant Nutr. 31 (2), 219–229.

Martins, A.E., Pereira, M.S., Jorgetto, A.O., Martines Ma, U., Silva, R.I.V., Saeki, M.J., et al., 2013. The reactive surface of Castor leaf [*Ricinus communis* L.] powder as a green adsorbent for the removal of heavy metals from natural river water. Appl. Surf. Sci. 276, 24–30.

Melo, E.E.C., Costa, E.T.S., Guilherme, L.R.G., Faquin, V., Nascimento, C.W.A., 2009. Accumulation of arsenic and nutrients by castor bean plants grown on an as-enriched nutrient solution. J. Hazard. Mater. 168 (1), 479–483.

Melo, E.E.C., Guilherme, L.R.G., Nascimento, C.W.A., Penha, H.G.V., 2012. Availability and accumulation of arsenic in oilseeds grown in contaminated soils. Water Air Soil Pollut. 223 (1), 233–240.

Nazir, A., Malik, R.N., Ajaib, M., Khan, N., Siddiqui, M.F., 2011. Hyperaccumulators of heavy metals of industrial areas of Islamabad and Rawalpindi. Pak J Bot 43 (4), 1925–1933.

Nelson, D.W., Sommers, L.E., 1996. Total carbon, organic carbon, and organic matter. In: Page, A.L., et al. (Eds.), Methods of Soil Analysis, Part 2, second ed., In: Agronomy, vol. 9. American Society of Agronomy, Madison, WI, pp. 961–1010.

Niu, Z.-X., Sun, L.-N., Sun, T.-H., Li, Y.-S., Wang, H., 2007. Evaluation of phytoextracting cadmium and lead by sunflower, ricinus, alfalfa and mustard in hydroponic culture. J. Environ. Sci. 19 (8), 961–967.

Niu, Z., Sun, L., Sun, T., 2009. Response of root and aerial biomass to phytoextraction of Cd and Pb by sunflower, castor bean, alfalfa and mustard. Adv. Environ. Biol. 3 (3), 255–262.

Olivares, A.R., Carrillo-González, R., González-Chávez, M.D.C.A., Soto Hernández, R.M., 2013. Potential of castor bean (*Ricinus communis* L.) for phytoremediation of mine tailings and oil production. J. Environ. Manage. 114, 316–323.

Pandey, V.C., 2013. Suitability of *Ricinus communis* L. cultivation for phytoremediation of fly ash disposal sites. Ecol. Eng. 57, 336–341.

Patten, C.L., Glick, R.B., 1996. Bacterial biosynthesis of indole-3-acetic acid. Can. J. Microbiol. 42, 207–220.

POTENTIAL OF ORNAMENTAL PLANTS FOR PHYTOREMEDIATION OF HEAVY METALS AND INCOME GENERATION

W. Nakbanpote[1], O. Meesungnoen[1], M.N.V. Prasad[2]

Mahasarakham University, Maha Sarakham, Thailand[1]
University of Hyderabad, Hyderabad, Telangana, India[2]

1 INTRODUCTION

Soils and waters contaminated/polluted with toxic metals pose a major environmental problem that needs an effective and affordable technological solution. Phytoremediation is the process through which contaminated substrates are ameliorated by growing plants that have the ability to remove the contaminants. Phytoremediation of heavy metals includes processes such as phytostabilization, phytoextraction, phytovolatilization, and rhizofiltration. The key factor for successful phytoremediation is the identification of a plant that is tolerant and suitable for the specific area and conditions. Plant design for successful phytoremediation in a chosen contaminated area should not have an adverse effect on the local biodiversity. Other important aspects for the success of phytoremediation are economic benefits and by-product generation in order to convince local people and government of the advantages of phytoremediation. Therefore, ornamental plants are considered for phytoremediation and phytomanagement. Use of ornamental plants for remediation of a contaminated environment would also change the landscape for ecotourism. Some cut flowers such as marigold (*Tagetes* spp.) and water lily are regarded as model ornamental plants. Application of fertilizers, soil amendments, chemical chelate, microorganisms, pesticides, etc., for accelerating remediation and economic benefits should be balanced for the benefit of local people and environments.

2 CONTAMINATION OF HEAVY METALS AND PHYTOREMEDIATION

Essential heavy metals include iron (Fe), zinc (Zn), copper (Cu), manganese (Mn), and cobalt (Co). Cadmium (Cd), lead (Pb), and mercury (Hg), however, have no biological functions. An essential micronutrient can act as a toxin at high concentration. Heavy metals are not degraded biologically in soil. In soils, they occur as free metal ions; exchangeable metal ions; soluble metal complexes (sequestered to ligands); organically bound, precipitated, or insoluble compounds such as oxides, carbonates, and

Bioremediation and Bioeconomy. http://dx.doi.org/10.1016/B978-0-12-802830-8.00009-5
Copyright © 2016 Elsevier Inc. All rights reserved.

hydroxides; or they may form silicate (indigenous soil content) (Kabata-Pendias and Pendias, 2001). Anthropogenic activity leads to the accumulation of many harmful substances such as herbicides, radioactive elements, and heavy metals (Enger and Smith, 2010). Huge quantities of solid wastes (tailings) are generated during the beneficiation of precious ores, which are disposed of in nearby areas, creating vast barren land that can hardly support any vegetation. Agricultural soils and reservoirs in many parts of the world are slightly to moderately contaminated by heavy metal toxicity such as Cd, Cu, Zn, Ni, Co, Cr, Pb, and As. This could be due to long-term use of commercial mineral fertilizers, agrochemicals, compost, sewage sludge, and waste disposal (Marmiroli and Maestri, 2008).

The anthropogenic discharge of Cu into the environment is ubiquitous given its use as a biocide and in various metallurgical industries. Cu and Zn are also present in some sewage sludge, organic and inorganic fertilizers, and fungicides. Cr is commonly used in leather tanning agents, textile pigments and preservatives, antifouling paints, wood treatment/preservation, steel processing, aluminum alloys, electroplating, and microbial growth inhibition such as in cooling towers of power plants. The anthropogenic sources of Pb are mining activities, metallic smelting, coal combustion, automotive exhaust fumes, domestic utilization of Pb-based paints, land application of municipal wastewater and sludge, agricultural use of chemical fertilizers and insecticides, as well as waste disposal in landfills. One of the main sources of Cs is nuclear technology, which releases high amounts of radioactive wastes including Cs isotopes into the environment, causing great hazard to public health. ^{137}Cs is the most abundant anthropogenic radionuclide in the marine environment.

Arsenic has been used in various fields such as medicine, electronics, agriculture (pesticides, herbicides, insecticides, fertilizer, etc.), livestock (cattle and sheep dips), and as wood preservatives. Arsenic has both metallic and nonmetallic properties. It is now well recognized that consumption of As, even at low levels, leads to carcinogenesis. Cu and Zn may accumulate in plant tissues, causing many physiological and biochemical changes and growth reduction. Excessive amounts of Cd can cause various phytotoxic symptoms including chlorosis, growth inhibition, damage to root tips, and reduction in water and nutrient uptake. Toxic effects of Pb on plants include inhibition of photosynthesis, enzymatic activities, deficient mineral nutrition uptake and water imbalance, and inhibition of root elongation, which considerably reduces both the vegetative and reproductive growth of plants. Pb induces visible symptoms such as chlorosis, necrosis, and growth inhibition of roots and shoots, and changes in the branching pattern of roots. Cs is the scarcest of alkali metals and has no nutritional role in plants, although it can be absorbed by them easily, and it has a high mobility within plants. Hg is of great concern regarding the environment and is highly toxic, causing severe damage to the human central nervous system. Heavy metals have accumulated in the soil over decades because of the low mobility and binding to soil of Fe and Mn oxides and organic fractions, and bioavailability of heavy metals depends on soil conditions and plant factors (Greger, 1999).

Remediation of heavy metal-contaminated soil and water is an important concern. Various technologies are in place to clean up or reduce metals exposure from contact with metal-contaminated soil and water. Technologies for remediation of metal-contaminated soil include excavation, immobilization, vitrification, soil washing/flushing, precipitation, membrane filtration, adsorption, ion-exchange, permeable reactive barriers, biological treatment, and phytoremediation. Phytoremediation is a cost-effective technology that uses plants to degrade, assimilate, metabolize, or detoxify metal and organic chemical contamination (Organum and Bacon, 2006). Phytoremediation is one of the in situ procedures, environmentally friendly, reducing soil erosion, enriching soil organic matter leading to enhanced soil fertility, and is a sustainable technology for site restoration. It also helps in removing the carbon dioxide (CO_2)

from the air during the photosynthesis process. Furthermore, the use of plants to remediate contaminated soils and water could preserve the structure and biological functions of the environments.

Phytoremediation utilizes physical, chemical, and biological processes to remove, degrade, transform, or stabilize contaminants within soil and groundwater. Reference can be made to Ghosh and Singh (2005), Prasad and Prasad (2012), Raskin and Ensley (2000), for the mechanisms of heavy metal remediation.

Proper selection of plant species for phytoremediation plays an important role in the development of remediation methods (decontamination or stabilization), especially on metal-polluted soil. Hyperaccumulative plants for heavy metal have defense reactions and detoxification mechanisms. After exposure to heavy metals, some hyperaccumulative plants can change structurally and ultrastructurally and localize the heavy metals to subcellular compartments with little or no detectable metabolic activity (Barceló and Poschenrieder, 2004; Rattanapolsan et al., 2013; Mongkhonsin et al., 2011). Heavy metals induce oxidative damage in plants (Shaw et al., 2004). One of the consequences of metal accumulation is an increase in reactive oxygen species (ROS) contents, which is destructive if protective antioxidant mechanisms do not operate efficiently. Many antioxidant substances such as ascorbate and the enzymes ascorbate peroxidase (APX) and superoxide dismutase (SOD) are increased under heavy metal stress (Sytar et al., 2013). The enhancements of contents of carotenoids and polyamines have also been correlated with plant tolerance (Sytar et al., 2013). In addition, glutathione and phytochelatins have a role in conferring tolerance to heavy metal stress (Yadav, 2010; Panitlertumpai et al., 2013).

Another important criterion for the potential plants used for phytoremediation is the provision of economic benefits in addition to environmental remediation. In developing countries including Thailand, it is difficult to convince local people to grow metal-accumulating plants in their fields for removing pollutants unless financial incentive is offered. Phytomanagement studies should focus on combining phytoremediation with enhancing soil fertility, chemical chelate, microorganisms, and minimizing pesticides management, together with investigation of additional species for phytoremediation that may also enhance economic benefits.

3 ORNAMENTAL PLANTS FOR HEAVY METALS PHYTOREMEDIATION

Ornamental plants are grown for decorative purposes in gardens and landscape design projects for cut flowers and landscape beautification. Most commonly, ornamental garden plants are grown for the display of esthetic features including the flowers, leaves, scent, overall foliage texture, fruit, stem and bark, and esthetic form. Ornamental plants are separated into two groups of terrestrial plants and aquatic plants as shown in Figure 1. Terrestrial plants are important for flowers and foliage. Aquatic plants are divided into three groups: emergent, floating, and submergent plants (Figure 1). The principal factor controlling the distribution of aquatic plants is the depth and duration of flooding. Other factors may also control their distribution, abundance, and growth form, including nutrients, disturbance from waves, grazing, salinity, and contaminants. Ornamental plants also have utilitarian purposes such as production of perfume.

Ornamental plants commonly used in landscaping not only make the environment colorful but also remediate the contaminated environment. They will visually decorate the environment of metal-impacted areas. Since many of them are not edible plants, the risk to the food chain is reduced.

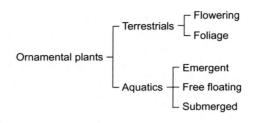

FIGURE 1

Ornamental plants classification: terrestrial and aquatic.

3.1 TERRESTRIAL ORNAMENTAL PLANTS FOR PHYTOREMEDIATION

The accumulation of heavy metals in plants depends on the metal(s) and their bioavailability in soils, species, and growth stages of plants, nutrients, seasons, and geological and environmental factors (Prasad and Freitas, 2003). Table 1 shows selected species of ornamental plants for heavy metal(s) accumulation studied in pot and field systems. The majority of ornamental plants concentrate heavy metal in roots; for example, *Nerium oleander* (Trigueros et al., 2012), *Quamoclit pinnata*, *Antirrhinum majus* (Cui et al., 2013), *Erica andevalensis* and *Erica australis* (Pérez-López et al., 2014), *Calendula officinalis* (Liu et al., 2008a), and *Tagetes patula* (Chaturvedi et al., 2014; Chintakovid et al., 2008). They can thus remedy contaminated soils to some extent through phytostablilization and at the same time beautify the environment and be sold for income. Growth stages of plants also have an effect on heavy metals accumulation in each plant's part. In the case of *Helianthus annuus* (sunflower), the adult plant showed a lower sensitivity to metal toxicity than young seedlings. The order of Cd content is roots > leaves > stems in young plants; stems > leaves > roots > flowers in adult plants (Nehnevajova et al., 2012). The imbalanced trace element concentrations in *T. erecta* (marigold) influence detoxification processes to heavy metals (Liu et al., 2011a,b).

The extraction of heavy metals by plants is usually limited by the availability of heavy metals in soils. Methods of increasing the availability of heavy metals in soils and their translocation to the plant shoots are vital to facilitate phytoextraction. Ethylenediaminetetraacetic acid (EDTA) and ethylene glycol-bis (2-aminoethyl) tetraacetic acid (EGTA) as chemical chelator could be applied to increase metals bioavailability (Tahmasbian and Safari Sinegani, 2014; Wei et al., 2012). However, enhanced phytoextraction by the application EDTA is not a suitable practice from economic and environmental standpoints. Sodium dodecyl sulfate as an important ionic surfactant was also used together with the synthetic chelators (EDTA, EGTA) (Wang and Liu, 2014; Liu et al., 2008a, 2009a,b, 2010). Citric acid application can increase Fe, Mn, Cu solubility, which indicates the possibility of using it to improve plant nutrition and mobility of metals (Tapia et al., 2013). To increase metals availability for plant uptake, the application of soil additive such as thioligands—for example, thiosulfate (TS)—and halogen salts such as potassium chloride (KCl), which are nontoxic chemicals commonly used in agriculture for their fertilizing properties (Cassina et al., 2012; Hao et al., 2012). However, caution must be taken with the metal leaching into deeper soil horizons. Tahmasbian and Safari Sinegani (2014) studied an ammonium combination of electrokinetic remediation and phytoremediation to decontaminate metal-polluted soil.

Phytohormones, especially cytokinins (CKs), can also play an important role in the processes related to metal stress. The exogenous application of CKs produces an alleviation of heavy metal toxicity

Table 1 Terrestrial Ornamental Plants for Phytoremediation and Phytostabilization

Botanical Name	Metals	Treatment	Results	References
Lonicera japonica	Cd	*System:* pot *Stress:* soil spiked with $CdCl_2$ (5–200 mg-Cd kg^{-1})	Synergistic interaction was found in accumulation and translocation between Cd and Fe, and a significantly negative correlation between Cd and Cu or Zn concentrations in plant tissues	Liu et al. (2009a)
Nerium oleander	Pb	*System:* hydroponic *Stress:* nutrient solution spiked with $Pb(NO_3)_2$ (20–100 µMPb)	BCF and TF were <1, which may be useful with regard to phytostabilizing Pb. Pb inhibited plant growth and increased malondialdehyde content in leaves	Trigueros et al. (2012)
Mirabilis jalapa	Cd	*System:* pot *Treatment:* EGTA or EDTA, ±SDS *Stress:* soil spiked with $CdCl_2$ (25 mg-Cd kg^{-1})	EGTA has better effectiveness and can bring lesser metal leaching risk than EDTA. Cd translocation ability under the EGTA was higher than the EDTA treatment. Single sodium dodecyl sulfate (SDS) treatment could not enhance the shoot Cd extraction	Wang and Liu (2014)
Quamoclit pennata, Antirrhinum majus, Celosia cristatapyramidalis	Pb	*System:* pot *Treatment:* soil spiked with $Pb(NO_3)_2$ (1000–5000 mg-Pb kg^{-1})	BCF and TF of these plants were <1. Only *C. cristatapyramidalis* could be identified as a Pb-accumulator	Cui et al. (2013)
Erica andevalensis, Erica australis	Al, As, Fe, Mn	*System:* field *Stress:* sulfide-mining waste	*Erica* plants can be considered as an Mn-accumulator and acid-, Al-, As-, Fe-, and Mn-tolerant. Both *Erica* species can be considered suitable for phytostabilization of metal(loid)-polluted sites of mine tailing	Pérez-López et al. (2014)
Chrysantemum maximum	Cd, Cu, Ni, Pb	*System:* soilless *Treatment:* *Glomus mosseae*, mycorrhizal fungus *Stress:* perlite exposed to Cd, Cu, and Ni (0.5–10 mg l^{-1}) and Pb (6–100 mg l^{-1})	Plants behaving as Pb-excluders. Cd, Cu, and Ni were accumulated in roots. Low accumulation in flowers was observed for Cd and Cu, concentration-dependent. Ni and Pb were not translocated to flowers. Mycorrhizal plants accumulated less Pb and Cu in both shoots and roots than nonmycorrhizal plants	González-Chávez and Carrillo-González (2013)

(Continued)

Table 1 Terrestrial Ornamental Plants for Phytoremediation and Phytostabilization—Cont'd

Botanical Name	Metals	Treatment	Results	References
Amaranthus hypochondriacus	Cd	*System*: pot *Treatment*: NPK fertilizer, repeated harvest *Stress*: soil spiked with $3CdSO_4$ ($5\,mg\text{-}Cd\,kg^{-1}$)	NPK increased dry biomass, resulting in a large increment of Cd accumulation. Repeated harvests had a significant effect on the plant biomass and thus on overall Cd removal. Plant growth time was found to significantly affect the amount of Cd extracted by the plants	Li et al. (2013)
Althaea rosea	Cd	*System*: pot *Treatment*: EGTA or EDTA and/or SDS *Stress*: soil spiked with $CdCl_2$ (30, $100\,mg\text{-}Cd\,kg^{-1}$)	EGTA and SDS increased plant biomass and promoted Cd accumulation, while EDTA was toxic to plant and ineffective in this regard	Liu et al. (2009b)
Calendula officinalis	Cd	*System*: pot *Treatment*: EDTA and EGTA, and/or SDS *Stress*: soil spiked with $CdCl_2$ (30, $100\,mg\text{-}Cd\,kg^{-1}$)	EDTA contamination led to retarded growth. EGTA and SDS, and EGTA alone, increased the total Cd accumulation	Liu et al. (2010)
Calendula officinalis, Althaea rosea	Cd	*System*: pot *Treatment*: EGTA and SDS *Stress*: soil spiked with $CdCl_2$ ($10\text{-}100\,mg\text{-}Cd\,kg^{-1}$)	EGTA and SDS could increase plant biomass and promote Cd accumulation. The two ornamental plants can be regarded as a potential Cd-hyperacumulator through applying the chemical agents	Liu et al. (2008a)
Calendula alata	Cs, Pb	*System*: hydroponic *Stress*: nutrient solution spiked with CsCl ($0.6\text{-}5\,mg\text{-}Cs\,l^{-1}$) or $Pb(C_2H_3O_2)_2$ ($0.6\text{-}5\,mg\text{-}Pb\,l^{-1}$)	*Calendula* had an extremely fast growth rate and could remediate Cs and Pb. The presence of Pb influenced uptake of Cs, and the effect is significantly inhibitory	Borghei et al. (2011)
Helianthus annuus	Cd	*System*: pot *Treatment*: chemical chelator, organic chelator, and/or electrical fields *Stress*: Cd-contaminated soil ($10.2\,mg\,kg^{-1}$)	EDTA as chemical chelator, cow manure extract and poultry manure extract as organic chelator, and electrical fields had no significant impacts on the dry weights of shoots and roots. Treatment with EDTA and electrical field (10 and 30 V) increased the Cd accumulated in shoot	Tahmasbian and Safari Sinegani (2014)

Plant	Metals	System/Stress/Treatment	Description	Reference
Helianthus annuus, Tagetes patula, Celocia cristata	Ca, Cr, Mn, Fe, Cu, Zn, Pb	*System:* pot *Stress:* contaminated soil	Sunflower, marigold, and cockscomb were found to be good accumulators of different elements. Concentration of different elements in different parts followed a trend of root > leaves > stem > flower	Chatterjee et al. (2012)
Helianthus annuus	Cr	*System:* pot *Treatment: Glomus intraradices* *Stress:* soil and sand (1:1) contaminated with $CrCl_3$ for Cr^{3+}, $Na_2Cr_2O_7$ for $Cr_2O_7^-$	Cr(III) and Cr(VI) depressed plant growth and decreased stomata conductance and net photosynthesis. Cr(VI) was more toxic than Cr(III). Arbuscular microrhiza (AM) fungus alleviated Cr toxicity and decreased Cr accumulation	Davies et al. (2002)
Helianthus annuus	Cd, Zn	*System:* pot *Treatment:* amendments *Stress:* Cd- and Zn-contaminated soil	Three amendments of swine manure salicylic acid (SA) and potassium chloride (KCl) increased height, flower diameter, and biomass. KCl increased Zn and Cd accumulations and suitable for environment and economy	Hao et al. (2012)
Helianthus annuus	Zn, Cd, Cu	*System:* pot *Stress:* metals-contaminated soil	Metals showed a decrease in shoot biomass and chlorophyll concentration. Increased ascorbate peroxidase (APOX) activity in adult indicated an elevated use of ascorbate after exposure to metal stress	Nehnevajova et al. (2012)
Helianthus annuus	Hg	*System:* pot *Treatment:* cytokinine and ammonium thiosulfate *Stress:* contaminated soil from petrochemical plant	Addition of cytokinine and ammonium thiosulfate increased Hg uptake and translocation	Cassina et al. (2012)
Helianthus annuus	Zn, Cd	*System:* pot *Treatment:* PGPB *Stress:* soil spiked with $CdCl_2$ (10-30mg-Cdkg^{-1}) and $ZnCl_2$ (100-1000mg-Znkg^{-1})	The PGPB strains *Ralstonia eutropha* and *Chrysiobacterium humi* reduced losses of weight in metal-exposed plants and induced changes in metal bioaccumulation. *C. humi* enhanced the short-term stabilization, lowering losses in plant biomass, and decreasing above-ground tissue contamination	Marques et al. (2013)

(Continued)

Table 1 Terrestrial Ornamental Plants for Phytoremediation and Phytostabilization—Cont'd

Botanical Name	Metals	Treatment	Results	References
Panicum maximum, Cosmos sulphureus, Tagetes erecta, Helianthus annuus	Cd	*System:* pot *Stress:* Cd-contaminated soil (50-400 mg kg^{-1})	Marigold showed higher potential, Cd accumulation in shoots was >100 mg kg^{-1}, TF>1 and BCF was >1 (Cd treatment of 50 mg kg^{-1}). Due to higher biomass and numerous roots, the total uptake of Cd by Guinea grass could be maximized	Rungruang et al. (2011)
Helianthus annus, Salvia splendens, Tagetes eracta	Cd	*System:* pot *Stress:* soil spiked with CdSO$_4$ (1-10 mg-Cd l^{-3})	The uptake of Cd by marigold, sunflower, and scarlet sage increased with the increase of Cd doses, while the yield of these plants decreased	Bosiacki (2008)
Ricinus communis, Tagetes erecta	Ni, Pb	*System:* pot *Treatment:* manure *Stress:* Ni- and Pb-contaminated soil	*R. communis* accumulated more Ni than *T. erecta.* Root of both crops contained higher Ni than aerial parts. Farmyard manure application enhanced Ni accumulation and reduced time requirement	Malarkodi et al. (2008)
Tagetes patula, Impatiens walleriana	Cd	*System:* pot *Treatment:* EDTA *Stress:* soil spiked with Cd(NO$_3$)$_2$ (20-80 mg-Cd kg^{-1})	Impatiens and marigold accumulated Cd at a rate of 200-1200 mg kg^{-1} in shoot, with BCFs and TFs of 8.5-15 and 1.7-2.6, respectively. EDTA application increased the translocation of Cd from roots to shoots	Wei et al. (2012)
Mirabilis jalapa, Impatiens balsamin, Tagetes erecta	Cr	*System:* pot *Stress:* Cr in tannery sludge	*M. jalapa* showed the strongest ability to tolerate and enrich Cr for phytoextraction, as Cr accumulation in roots > stems > leaves > inflorescence	Miao and Yan (2013)
Tagetes erecta, Chrysanthemum indicum, Gladiolus grandiflorus	Cd	*System:* pot *Stress:* soil spiked with CdCl$_2$ (5-100 mg-Cd kg^{-1})	Cd uptake increased with contents in soils and the maximum accumulation occurred in leaves. In view of higher biomass, Cd removal was the maximum with chrysanthemum > gladiolus > marigold, while gladiolus with highest tolerance and Cd-content holds potential to clean up the contaminated soil	Lal et al. (2008)
Impatiens balsamina, Calendula officinalis, Althaea rosea	Cd, Pb	*System:* pot and hydroponic *Stress:* soil spiked with CdCl$_2$ (10-100 mg-Cd kg^{-1}), Hoagland solution spiked with CdCl$_2$ (1-10 mg-Cd l^{-1}) plus Pb(NO$_3$)$_2$ (50, 100 mg-Pb l^{-1})	*C. officinalis* showed great tolerance to Cd and stronger ability to Cd accumulation. Cd in the roots was greater than in the shoots, a potential for phytostabilization. The relation between the Pb and Cd in solution showed highly significant negative correlation	Liu et al. (2008b)

Species	Metal	System/Stress/Treatment	Findings	Reference
Tagetes patula	Cd, Cu, Pb, and benzo[a]pyrene (B[a]P)	*System:* pot. *Stress:* soil, single B[a]P treatment (2–50 mg kg^{-1}), and co-contamination of B[a]P and CdCl$_2$ (20, 50 mg-Cd kg^{-1}), CuSO$_4$ (100, 500 mg-Cu kg^{-1}) or Pb(CH$_3$COO)$_2$ (1000, 3000 mg-Pb kg^{-1})	Marigold could be useful for phytoremediation of B[a]P and B[a]P-Cd-contaminated sites. Low concentration of B[a]P ($\leq$10 mg kg^{-1}) could facilitate plant growth, but Cd, Cu, and Pb inhibited effects on plant growth and B[a]P uptake and accumulation	Sun et al. (2011)
Tagetes patula	Fe	*System:* pot. *Stress:* Fe ore tailing	Marigold qualified well as a potential tool for phytostabilization and survived excess of heavy metals present in the Fe ore tailing	Chaturvedi et al. (2014)
Tagetes erecta	Cd, Pb	*System:* pot. *Stress:* soil spiked with Cd (1–10 mg l^{-1}) and Pb (100–1000 mg l^{-1})	Marigold accumulated Cd as leaves > stalks > inflorescences. Pb was found in stalks > leaves > inflorescences. Cd and Pb contents in plants depended on the increase concentrations	Bosiacki (2009)
Nugget marigold (triploid hybrid between *T. erecta* L. and *T. patula*)	As	*System:* pot and field. *Treatment:* P-fertilizer, 30–50 days. *Stress:* As-contaminated soil	P-fertilizer increased As uptake in flowering stage of nugget marigold. Arsenite (As(III)) and Arsenate (As(V)) were found in the roots, stems, and leaves	Chintakovid et al. (2008)
Tagetes erecta	Cu	*System:* pot. *Treatment:* *G. intraradices.* *Stress:* CuO (500–2500 mg-Cu kg^{-1}) spiked in perlite and peat moss (2:1)	Marigold colonized with AM accumulated more Cu in the roots as well as the whole plant. TF and BCF suggest the system marigold-mycorrhiza can potentially phytostabilize Cu in contaminated soils	Castillo et al. (2011)
Tagetes erecta	Cd	*Treatment:* *G. intraradices, G. constrictum,* and *G. mosseae.* *Stress:* soil and sand, spiked with CdCl$_2$ (0–50 mg-Cd kg^{-1})	AM fungi improved the capability of reactive oxygen species scavenging, and reduced Cd concentration in plants to alleviate marigold from Cd stress	Liu et al. (2011a,b)

in plants used for phytoremediation, especially in terms of biomass production. Moreover, the proven effects of CKs on stomatal opening and the consequent increase in transpiration efficiency could also produce an increase in pollutant uptake by plant (Cassina et al., 2012). Various amendments are required for plant growth. Some agronomic measures such as organic manures, chemical fertilizers, and plant hormones would have an effect on element uptake and plant growth. Cow manure extract and poultry manure extract, farmyard manure and poultry manure as organic chelator could be applied as amendments to increase bioavailability and phytoremediation (Malarkodi et al., 2008; Tahmasbian and Safari Sinegani, 2014).

The mutual symbiosis between arbuscular mycorrhizal (AM) fungi and roots of terrestrial plants have been reported to be direct physical links between soil and plant roots, increasing soil nutrient exploitation, and transfer of minerals to the roots (Smith and Read, 1997). The role of the symbiosis of the AM fungi (e.g., *Glomus intraradices*, *G. constrictrum*, and *G. mosseae*) with plant in the phytoaccumulation process could protect the plant from the toxicity induced by heavy metals, while at the same time improving the indices of phytoaccumulative yields. AM fungi can improve the capability of ROS scavenging to alleviate plants from heavy metals stress. The plant-mycorrhizal systems are not only being able to tolerate and accumulate high levels of heavy metals in harvestable parts, but also have a rapid growth rate and the potential to produce high biomass in the field. Therefore, the symbiotic system provides an attractive way to advance the phytostabilization and/or the phytoextraction of many terrestrial ornamental plants as shown in Table 1 (Castillo et al., 2011; Liu et al., 2011a,b; Davies et al., 2002; González-Chávez and Carrillo-González, 2013).

Plant growth-promoting bacteria (PGPB) capable of colonizing the plant root promote plant growth by various mechanisms: through the fixation of atmospheric nitrogen, utilization of 1-aminocyclopropane-1-carboxylic acid (ACC) as a sole nitrogen (N) source, production of siderophores and antipathogenic substances, production of plant growth regulators (phytohormones, such as auxins), and also through the transformation of nutrient elements. PGPB may help reducing the toxicity of heavy metals to plants in polluted environments. Marques et al. (2013) showed the PGPB strains *Ralstonia eutropha* and *Chrysiobacterium humi* reduced losses of weight in metal-exposed sunflower plants and induced changes in metal bioaccumulation and bioconcentration.

The goal is to evaluate the phytoremediation potential of plants with the assistance of different agricultural strategies. The bioavailable fraction of heavy metals in the soil can be reduced effectively in a short period of time and with the affordable costs of regular agronomic practices. For best practice, the recommendation is to maximize phytoextraction efficiency of plants by repeated harvests, harvesting at the squaring stage (soon after the flower begins to appear), and applying nitrogen-phosphorus-potassium (N-P-K) compounds or manure to increase metal accumulation and plant biomass, as a base application.

Tagetes patula (French marigold) is an ornamental plant that can be applied for both remediation and income in Thailand. *T. patula* is a beneficial plant for crop production. This plant can be used in combination with other species being used for the same purpose of protection as the pungent odor of these plants can keep insects away, and the vibrant color of the flowers will help in modifying the esthetics of the concerned area, which is badly needed for making the area more habitable. In addition, marigold secretes the chemical alpha-terthienyl from root tissue, which is an effective nematicide for root-knot nematodes and lesion nematodes control. Because of its allelopathic effect, marigold has been used as a companion plant or cover crop in cropping systems to protect crops against nematodes (Evenhuis et al., 2004). American (*T. erecta*) marigold, French (*T. patula*) marigolds, and nugget marigold, a triploid hybrid between *T. erecta* and *T. patula*, could also provide economic benefits to the remediators with large-scale plantation on wastelands, providing flowers to sell. Many researchers reported that *T. erecta* can be applied to soil contaminated with Cd, Cr, Ni, Pb, As, and Cu (Rungruang et al., 2011; Bosiacki, 2008; Malarkodi et al., 2008; Miao and yan, 2013; Lal et al., 2008). *T. patula* has potential for remediation of areas contaminated with Cd, Fe, Cu, Pb (Wei et al., 2012; San et al., 2011; Chaturvedi et al., 2014), and for the degradation of organic pollutants; for example, Reactive Blue 160 (Patil and Jadhav, 2013) and benzo[a]pyrene (B[a]P) (Sun et al., 2011). Nugget marigold has potential for As-contaminated areas (Chintakovid et al., 2008). Many researches showed heavy metals were mainly accumulated in leaves > stems > inflorescence (Chintakovid et al., 2008; Bosiacki, 2008; Lal et al., 2008). A symplastic pathway rather than an apoplastic bypass contributed greatly to root uptake, xylem loading, and translocation of heavy metals to the shoots. A hydroponic experiment of ^{108}Cd showed that the root to above-ground Cd translocation via phloem was an important and common physiological process as xylem determination of the Cd accumulation in stems and leaves of marigold seedlings (Qin et al., 2013). *T. patula* is a novel Cd accumulator able to tolerate Cd-induced toxicity by activation of its antioxidative defense system. Cd, which is known to stimulate the formation of free radicals, induced the antioxidative enzymatic activities of APX, SOD (CuSOD and CuZnSOD), and different isozymes of glutathione reductase in leaves (Liu et al., 2011a,b).

In conclusion, the advantages of marigold for phytoremediation are as follows: (1) It is not an edible crop, which will prevent accumulated heavy metals from entering the food chain; (2) it grows with high shoot biomass when exposed to heavy metals; (3) it has wide adaptability to different soils and climate conditions; and (4) it secretes phytochemicals from root tissues and therefore can serve as a nematicide for the control nematodes. Thus, using marigold may offer a new possibility for rendering agricultural soils or heavy metal-contaminated areas; for example: Cd contamination in Mae Sod District, Tak province; As contamination in Ron Phibun District, Nakhon Si Thammarat Province; and Pb contamination in Klity Creek in the Thungyai Naresuan Wildlife Sanctuary, Thong Pha Phum District, Kanchanaburi Province, in Thailand (Prasad and Nakbanpote, 2015).

Phytomanagement suggests that fast-growing trees with short rotation coppice systems like eucalyptus and willow should be grown to create green belts around contaminated land (Pulford and Watson, 2003). Figures 2 and 3 show some flowers and foliage leaves sold at a flower market in Bangkok, Thailand. Fern foliage is in great demand for floral bouquets. *Pityrogramma calomelanos* and *Pteris vittata*, indigenous hyperaccumulating fern species, are well acknowledged in literature for their As phytoremediation potential (Visoottiviseth et al., 2002; Cao et al., 2003). Although *Coleus blumei*, a garden plant, behaves as a nonaccumulator for Al, this plant can play an important role in the treatment of polluted water with available Al at pH below 5.0 (Panizza de León et al., 2011). *Chlorophytum comosum* (spider plant), *Amaranthus hypochondriacus* (amaranth), and *A. caudatus* had application value in the treatment of Cd-contaminated soils (Wang et al., 2012; Li et al., 2013; Bosiacki et al., 2013). *Gynura pseudochina*, a perennial herbal plant, has the potential to accumulate Cr (Mongkhonsin et al., 2011), Cd, and Zn (Panitlertumpai et al., 2013). *Panicum maximum* (Guinea grass) and *Cyperus rotundus* (nut grass) have properties that make them beneficial for covering the terrestrial area contaminated with Cd (Rungruang et al., 2011; Sao et al., 2007). Lemon-scented geraniums (*Pelargonium* sp. "Frensham" or scented geranium) accumulated large amounts of Cd, Pb, Ni, and Cu from soil in greenhouse experiments (Dan et al., 2000). Pellegrineschi et al. (1994) improved the ornamental quality of scented *Pelargonium* spp. This plant has a pleasant odor that adds scent to the toxic metal-contaminated soil.

FIGURE 2

Flowers and foliage for sale in a flower market in Bangkok, Thailand: (a) marigold blossoms, (b) many kinds of roses, and (c) foliage leaves for decoration and bouquets.

3.2 AQUATIC ORNAMENTAL PLANTS FOR PHYTOREMEDIATION

Freshwater as well as seawater resources are being contaminated by various toxic elements through anthropogenic activities and from natural sources. Remediation of the contaminated aquatic environment is as important as it is for terrestrial environment. Among the toxic substances found in water bodies, heavy metals deserve special attention. They are highly toxic at low doses, strongly persistent in the environment and living tissues, and easily transferred to the food chain. In addition, their monitoring

FIGURE 3

Popular flowers for garlands and flower arrangements in Thai style: (a) *Gomphrena globose* (bachelor's button), (b) *Calotropis gigantea* (crown flower), (c) *Rosa damascene* (rose), (d) *Jasminum officinale* (jasmine), (e, f) garlands and flower arrangements for special purposes, (g-j) ordinary garlands and flower arrangements for worship.

and removal is costly. Point and nonpoint sources of pollution are major causes of water-quality degradation. Concentration-based controls of discharge at point sources are insufficient for attaining water-quality targets. Nonpoint source pollution is difficult to control because it comes from many different sources and locations. The major sources of nonpoint pollution are agriculture and urban/rural activity. Agriculture is one of the largest consumers of chemical fertilizers, pesticide, and manure-based composts.

The most important role of plants in wetlands is that they increase the residence time of water, which means that they reduce the velocity and thereby increase the sedimentation of particles and associated pollutants. Aquatic plants have been recognized for controlling eutrophication by assimilation of nitrogen (N) and phosphorus (P). Plants' photosynthesis produce oxygen during day time. In addition, nutrients are removed simultaneously by roots of the aquatic plants and microbial communities.

Constructed wetlands are widely used for removal of heavy metals in treating wastewater. Sedimentation, sorption, coprecipitation, cation exchange, phytoaccumulation, and microbial activity are the main processes for the removal of heavy metals (Dhir, 2013). Iron toxicity is a serious problem in crop production in waterlogged soils. Iron, which is essential for plant growth, causes toxicity in plants at high concentration, in aerobic soils is found as insoluble Fe^{3+}, but a large amount of this element could be reduced to the more soluble Fe^{2+} in waterlogged soils characterized by anaerobic conditions and a low pH. In this case, wetland plants develop aerenchyma cells to transfer O_2 from aerial parts to the roots, inducing the precipitation of iron oxides or hydroxides on the root surfaces. In addition, iron oxide deposit on the roots (iron plaque) of many wetland plants have high capacity of functional groups on iron (hydr)oxides to sequester metals by adsorption and/or coprecipitation (Zhong et al., 2010).

Phytoremediation of heavy metals by means of constructed wetlands constitutes a low-cost, environmentally friendly alternative to conventional cleanup techniques. Furthermore, as metals accumulate mainly in roots, part of the biomass harvested from such wetland has many potential uses in nonfood industries. Some of these side products that could yield substantial economic benefits for affected communities are biogas and compost, fibers, and ornamental plants. Plants that have potential for phytoremediation in aquatic systems are categorized into three groups: aquatic ornamental plants, adapted terrestrial plants, and aquatic macrophytes.

3.2.1 Aquatic ornamental plants

Aquatic ornamental plants are known for high biomass production and tolerance to polluted environments. Some plants shows better performance in removing total N and P, chemical oxygen demand, biological oxygen demand, and heavy metals (Cr, Pb, Cd, Fe, Cu, Mn, etc.) from sewage. N and P have been identified as the two main nutrients needed to control for mitigating the serious situation of eutrophication. Li et al. (2015) showed that assimilation of *Oenanthe javanica* accounted for 28-34% of N reduction and 25.2-33.4% of P reduction. *Potamogeton crispus* accounted for 61.5% of N removal and 67.5% of P removal. *Zantedeschia aethiopica* significantly influenced the removal rate of N (Belmont and Metcalfe, 2003).

Aquatic ornamental plants have the dual benefits of water reclamation and flower production with financial returns. The aquatic plant species with ornamental flowers evaluated for heavy metal(s) removal and their high market values are shown in Table 2. *Iris* species could be widely used as phytoaccumulators to remediate Cd (Han et al., 2007), Cr, and Zn (Caldelas et al., 2012). The high biomass production and metal extraction capacity makes this species a good candidate for Cr rhizofiltration and Zn phytoextraction, as reflected by the level of exportation of each metal to leaves. The reduced exportation of Cr to leaves can be advantageous for flower production, yield of emerged parts, and human safety (Caldelas et al., 2012). *Zantedeschia aethiopica* (white calla lilies) have potential to remediate iron (Casierra-Posada et al., 2014).

Lotus plant (*Nelumbo nucifera*) can accumulate pollutants such as pesticides and heavy metals from sediment and water. Bioaccumulation of (−)-enantiomers of chiral polychlorinated biphenyls were in leaves > stems > roots of lotus plants (Dai et al., 2014). Hindu lotus (*Nelumbo nucifera*) had efficiency to treat total nitrogen and total phosphorus from zoo wastewater up to 58% and 38%, respectively (Jiang and Xinyuan, 1998).

Water lilies (*Nymphaea*) are planted in the mud, and its laminae, carried on long petioles, float on the water's surface. These plants are not only tolerant of high levels of toxic heavy metals but they can

Table 2 Aquatic Ornamental Plants for Phytoremediation

Botanical Name	Metal	Treatment	Results	References
Iris lactea and *Iris tectorum*	Cd	*System*: hydroponic *Stress*: Hoagland nutrient solution spiked with $CdCl_2$ (1–60 mg–Cd l^{-1})	Cd accumulated in the two iris were in shoots > roots. TEM showed Cd localizing in cell wall, cytoplasm, and inner surface of xylem in root tip	Han et al. (2007)
Iris lactea and *Iris tectorum*	Pb	*System*: hydroponic *Stress*: Hoagland nutrient solution spiked with $Pb(NO_3)_2$ (2–10 mg–Pb l^{-1})	Pb accumulated in the two iris were in roots > shoots. *I. lactea* was more tolerant to Pb than *I. tectorum*. Transmission electron microscope (TEM) showed Pb deposits were found along the plasma membrane of some root tip cells	Han et al. (2008)
Iris pseudacorus	Cr and Zn	*System*: hydroponic *Stress*: peat-perlite (50/50), $ZnCl_2$ (0.07–1.5 mM Zn) and $CrCl_3$ (0.04–0.8 mM Cr(III))	*Iris* showed a great capacity to tolerate and accumulate both Zn and Cr. The species was a good candidate for Cr rhizofiltration and Zn phytoextraction	Caldelas et al. (2012)
Iris pseudacorus	Pb	*System*: pot *Treatment*: $FeSO_4$ (100–500 mg–Fe kg^{-1}) *Stress*: clay from paddy field and vermiculite, spiked with $Pb(NO_3)_2$ (100–1000 mg–Pb kg^{-1})	Intermediate iron dose supply (100 mg kg^{-1}) generally enhanced Pb absorption and accumulation. Plant growth was inhibited by iron toxicity at high iron dose (500 mg kg^{-1})	Zhong et al. (2010)
Nymphaea	Cd	*System*: pot *Treatment*: heavy clay soil exposed to $Cd(NO_3)_2$ (50 mg–Cd l^{-1}) and/or $CaCl_2$ (500 mg–Ca l^{-1})	Maximum Cd and Ca accumulation were in the mature leaf lamina in daylight. The accumulation was inhibited by the herbicide 3-(3′,4′-dichlorophenyl)-1,1-dimethylure. Deposition and storage of heavy metals by epidermal glands represented a stage in the sequestration and detoxification of the metals	Lavid et al. (2001)
Nymphaea odorata	Pb	*System*: field sampling	About 28% of the Pb in shoots was accumulated directly from water, whereas there was no evidence of root uptake of Pb from sediments	Outridge (2000)

(Continued)

Table 2 Aquatic Ornamental Plants for Phytoremediation—Cont'd

Botanical Name	Metal	Treatment	Results	References
Nymphaea spontanea	Cr(VI)	*System:* hydroponics *Treatment:* $K_2Cr_2O_7$ (1-10 mg l^{-1}), binary metals Cr(VI) 2.5 mg l^{-1} and $Cu(NO_3)_2$ (Cu(II) 0.5 mg l^{-1})	Cr(VI) accumulation in water lilies follows the order: roots > leaves > petioles. Roots play an important role. The maturity of plant exerts a great effect on the removal and accumulation of Cr(VI). Removal of Cr(VI) was more efficient when the metal was present singly than in the presence of Cu(II). Significant toxicity effect was evident as shown in the reduction of chlorophyll, protein, and sugar contents in plants exposed to Cr(VI)	Choo et al. (2006)
Nuphar variegata	Cd	*System:* field, peatlands in the Algonquin Park-Haliburton region of Ontario	Yellow pond lily accumulated more Cd in its leaves from peatlands with low pH, low alkalinity, and low dissolved organic carbon. The petioles of Nuphar accumulate more Cd from the peatlands with low pH. The organic content of the sediment had no effect on *Nuphar* Cd levels	Thompson et al. (1997)
Zantedeschia aethiopica	Fe	*System:* hydroponic *Stress:* nutrient solution spiked with $FeSO_4$ (100 or 200 mg-Fe l^{-1})	White calla lily plants were moderately tolerant to excess iron toxicity, depending on the pH and moisture system. The plant would allow for the phytoremediation and rehabilitation of iron-contaminated wetlands	Casierra-Posada et al. (2014)

DOC, dissolved organic carbon.

also accumulate them in large amounts. The degree of metal uptake tended to increase with tissue age (Twining, 1993). Water lilies have been reported to accumulate heavy metals such as Cd, Pb, and Cr as shown in Table 2. Epidermal glands on the submerged surfaces of water lily leaf laminae, petioles, and the rhizome accumulate heavy metals. Typical co-crystallization of Cd and Ca accumulating in epidermal glands results in Mn depletion and sometimes causes depletion of other elements such as Fe, Pb, and Hg accumulating in laminar epidermal glands. Lavid et al. (2001) showed deposition and storage of Cd by water lily. Translocation of Cd was not observed in the plant. Cd is apparently immobilized in the epidermal glands in the early stages of heavy metal accumulation, and subsequently in other epidermal cells. This immobilization would explain the water lily's high tolerance to heavy metal. Photosynthetic energy or products may be involved in heavy metal uptake by the water lily epidermal glands, as metal uptake was lower in the dark, and green tissues in the leaves (laminae and petioles) accumulated a higher level of metals than the nongreen rhizome tissues.

Lotus (*Nelumbonaceae*) and water lily (*Nymphaeaceae*) are commonly found in ponds or wetland. Lotus is in high demand in markets because the flower is used for decoration at almost all festivals in Thailand. Commercial life of lotus and water lily in Thailand are shown in Figure 4. More 1 million

FIGURE 4

Commercial life of lotus (Nelumbonaceae) and water lily (Nymphaeaceae) in Thailand: (a) harvesting, (b) sorting, (c) and packing and marketing of lotus; (d) many colors of water lily, and (e) marketing as pots of their rhizome for planting and decoration.

lotus flowers per day are sold in Thai flower markets. Lotus requires a large amount of essential nutrients (e.g., nitrogen and phosphorus) from either fertilizer or compost for growth. Nutrients' mass balance are required in lotus ponds to effectively treat nonpoint sources (e.g., agricultural wastewater) of pollution discharged from agricultural lands (Seo et al., 2010).

The lotus (*Nelumbo nucifera*), also known as sacred lotus or Chinese arrowroot, is a valued aquatic plant grown and consumed throughout Asia. Almost every part of the lotus plant from stamens, pollen, flowers, stems, rhizomes to leaves can be consumed as food or used for various therapeutic purposes in India and China. Lotus has multiple uses besides food: the stems as fresh vegetables and fuel; rhizomes as fresh vegetables, canned food, dessert, and starch; seeds as dessert and medicine; flowers as religious ornaments; and several parts as raw materials to produce cosmetics. Scientifically, most parts of the lotus plant, such as the seed, rhizome, stamen, and leaves, have exhibited significant antioxidant activities (Leong et al., 2012). However, the lotus plant has metal-dominant hypertolerance as it adapts accumulation strategy to accumulate heavy metals. Leong et al. (2012) reported that the highest concentration of phenolic content, Pb and Cr, was found in the seeds of lotus plants grown in a former tin-mining pond. The concentration of Pb found exceeded the permissible limits of Codex Alimentarius. Although it was reported that the contents of heavy metals in lotus seeds from a mining area were no higher than in a reference area (Dingjian, 2012), and heavy metal in lotus roots were accumulated primarily independently from or partially negatively correlated with soil (Xiong et al., 2012, 2013), assessment of metal concentration should be a mandatory priority to utilization of contaminated areas for edible plant cultivation.

Wastewater is one of the most serious environmental problems in Thailand. The major source of water pollution in the country is domestic wastewater. According to the national policy concerning domestic wastewater management, the national target of untreated discharge to receiving water and environment should be less than 50% of generated domestic wastewater. Therefore, wastewater management at the local level and improvement of management capacity need to be supported (Simachaya, 2009). Floriculture activities in constructed wetlands could provide the economic benefits necessary in encouraging small communities to maintain a wastewater treatment system. It is feasible to treat domestic wastewater in small rural communities using constructed wetland. These systems are especially valuable for on-site wastewater treatment in developing countries because they involve simple technology, and the costs of construction and operation are low.

Nong (Lake) Han Kumphawapi in northeast Thailand is one of the largest natural freshwater lakes in the country. Kumphawapi is a shallow lake (<4 m water depth), but it has considerable seasonal fluctuations in water level (Chawchai et al., 2013). The lake's extensive, floating, herbaceous swamp vegetation has been described by Penny (1998, 1999), who noted a dominance of grasses (*Poaceae* including *Phragmites* sp.) and sedges (*Cyperaceae*), as well as *Eichhornia crassipes*, *Ipomoea aquatica*, *Ludwigia adscendens* (creeping water primrose), *Ludwigia octovalis*, *Nelumbo nucifera*, *Nymphaea lotus*, *Nymphoides indicum*, *Persicaria attenuata*, *Saccharum* spp., *Typha angustifolia*, and *Salvinia cucullata* (floating moss). Several fern taxa occur as epiphytic elements on the floating or partially rooted herbaceous substrate. The lake is used for water supply, flood control, waste disposal, fisheries, aquaculture, and tourism. Red Lotus Lake is the famous wetland (under the Ramsar Convention) and ecotourism place in Nong Han district, Udon Thani province, Thailand. Red lotus (*Nymphaea lotus* Linn. Varpubescens) is the dominant aquatic plant in Nong Han wetland. They overspread a large area, which makes this lake a beautiful place for ecotourism (Figure 5). A lot of tourists visit Nong Han and provide income for the local people. This is an important consideration to motivate them for phytoremediation.

FIGURE 5

Landscape for ecotourism: (a, b) Red lotus lake and ecotourism in Nong Han district, Udon Thani province; (c, d) marigold and ecotourism place in Pak Chong district, Nakhon Ratchasima Province.

3.2.2 Adapted terrestrial plants

Terrestrial plants can develop much longer, fibrous root systems covered with root hairs that create an extremely high surface area. Hydroponically cultivated roots of several terrestrial plants were discovered to be effective in absorbing, concentrating, or precipitating toxic metals from polluted effluents. This process was termed rhizofiltration. The stem cuttings of the terrestrial, ornamental plant, *Talinum triangulare* (waterleaf) in hydroponic medium can be definitely classified as a Cu hyperaccumulator, which indicates potential to support rhizofiltration (Rajkumar et al., 2009; Kumar and Prasad, 2010). *T. triangulare* (=*T. cuneifolium*) is a succulent shrub of about 60 cm height with cuneate to obovate leaves, flowers in terminal panicles, and purple-colored corolla. The plant flowers and fruits throughout the year. It is widely distributed in India, Arabia, and Africa. Cuttings are a ready means of propagation of these plants. *T. cuneifolium* was reported to accumulate high levels of Cu in its leaves. These plants showed absorption barriers at high soil Cu concentrations, indicating limits to uptake of the metal (Tiagi and Aery, 1986).

3.2.3 Aquatic macrophytes

Aquatic macrophytes can readily achieve biosorption and bioaccumulation of the soluble and bioavailable contaminants from water. A macrophyte is an aquatic plant that grows in or near water and is either emergent, submergent, or floating. Floating, aquatic, hyperaccumulating plants absorb or accumulate contaminants by its roots, while the submerged plants accumulate metals by their whole body (Rahman and Hasegawa, 2011). Benthic rooted macrophytes (both submerged and emergent) play an important role in metal bioavailability from sediments through rhizosphere exchanges and other carrier chelates. Macrophytes readily take up metals in their reduced form from sediments, which exist in anaerobic situations due to lack of oxygen, and oxidize them in the plant tissues, making them immobile and thus bioconcentrated in their tissues (Okurut et al., 1999). Potential aquatic macrophytes for metals phytoremediation are *Eichhornia crassipes* (water hyacinth), *Lemna* sp. (duckweed) and *Spirodela polyrhiza* (greater duckweed), *Azolla* sp. (water fern), *Salvinia* sp. (butterfly fern), *Pistia stratiotes* (water lettuce), *Nasturtium officinale* (watercess), *Elodea canadensis* (waterweed) and *E. acicularis* (needle spikerush), and *Hydrilla verticillata* (Esthwaite water weed), *Hydrocotyle umbellate* (pennywort), *Alternanthera philoxeroides* (alligator weed), and *Typha* spp. (bulrush or cattail) can remove various heavy metals from water. However, the efficiency of metal removal by these macrophytes depends on the size and root system of plants (Prasad, 2007; Figures 7 and 8).

Pistia stratiotes (water lettuce or water cabbage) (Araceae) is a monotypic genus with wide ecological amplitude (Šajna et al., 2007). It floats on the surface of the water, its roots hanging submerged beneath floating leaves. It is a common aquatic weed in eutrophic Indian water bodies. *Pistia stratiotes* mats degrade water quality by blocking the air-water interface, reducing oxygen levels in the water, and thus threatening aquatic life. It has been tested for metal remediation (Odjegba and Fasidi, 2007; Skinner et al., 2007), metal detoxification (Tewari et al., 2008; Upadhyay and Panda, 2009), and treatment of urban sewage (Zimmels et al., 2006), and has been found to ameliorate municipal sludge leachate with its metabolically active and metal detoxificatory enzymes (Tewari et al., 2008). Marchand et al. (2010) have estimated the efficiency removal (%) in constructed wetlands with plants grown in monocultures including *P. stratioites* (Table 3). A field study carried out by Lee et al. (1991)

Table 3 *Pistia*: **An Ornamental and Efficient Phytoremediator for a Variety of Contaminants**	
References	**Research Topic Invesigated**
Chowdhury and Mulligan (2011)	As biosorption
Rahman and Hasegawa (2011)	Phytoremediation of arsenic
Miretzky et al. (2010)	Cd (II) removal
Mufarrege et al. (2010)	Cr, Ni, Zn, and phosphorous removal
Marchand et al. (2010)	Metal and metalloid removal
Zhang et al. (2010)	Phytoremediation in engineered wetlands
Sanchez-Galvan et al. (2010)	Crude oil sorption
Tripathi et al. (2010)	Pharmacological activities, phytochemistry and diuretic, antidiabetic, antidermatophytic, antifungal, and antimicrobial properties
Upadhyay and Panda (2009)	Cu phytotoxicity
Hadad et al. (2009)	Ni and phosphorous sorption efficiencies
Espinoza-Quiñones et al. (2009)	Pb removal kinetics

Table 3 *Pistia*: **An Ornamental and Efficient Phytoremediator for a Variety of Contaminants—Cont'd**

References	Research Topic Invesigated
Leterme et al. (2009)	Nutritional value
Dhote and Dixit (2009)	Water quality improvement
Khan et al. (2009)	Removal of heavy metals from industrial wastewater
Bhakta and Munekage (2008)	Cd removal
Mishra and Tripathi (2008)	Concurrent removal and accumulation of heavy metals
Tewari et al. (2008)	Amelioration of municipal sludge and role of its metal detoxification enzymes
Noyo and George (2008)	Crude oil removal
Suñe et al. (2007)	Cd and Cr removal kinetics
Skinner et al. (2007)	Mercury (Hg) uptake
Odjegba and Fasidi (2007)	Phytoremediation
Nina et al. (2007)	Survival in a thermal stream
Miretzky et al. (2006)	Heavy metal removal by dead macrophytes
Henry-Silva et al. (2006)	Nutrional quality and apparent digestibility
Zimmels et al. (2006)	Urban sewage treatment
Maine et al. (2004)	Cd uptake
Miretzky et al. (2004)	Heavy metal removal
Volk et al. (2004)	Phytochemistry
Basu et al. (2003)	As reduction
Cardwell et al. (2002)	Metal accumulation
Venema (2001)	Fast spread
Keates et al. (2000)	Phytochemistry
den Hollander et al. (1999)	Survival strategy
Bini et al. (1999)	Water and sediment quality indicator
Abbasi and Ramasamy (1999)	Biomass estimation
Vymazal et al. (1998)	Wastewater treatment
Sen and Bhattacharyya (1994)	Ni (II) uptake
Patra and Panigrahi (1994)	Mercury (Hg) accumulation
Satyakala and Jamil (1992)	Cr toxicity
Lee et al. (1991)	Arsenic accumulation
Aliotta et al. (1991)	Allelochemicals (natural pesticides)
Abbasi et al. (1991)	Biogas production
Bassi et al. (1990)	Cr (VI) uptake
Dewald and Lounibos (1990)	Seasonal growth
Harley (1990)	Seed production and phenology
Dray and Center (1989)	Seed production
Sen et al. (1987)	Cr (VI) uptake
Mennema (1977)	Invasive weed
Thawil and Mercado (1975)	The life cycle of water lettuce
Hall and Okali (1974)	Phenology and productivity

reported that the average arsenic enrichment (bioaccumulation) factor of *P. stratiotes* was 8632 in roots and 2342 in leaf. It appears that arsenic translocation in *P. stratiotes* was slow and most of the arsenic was strongly adsorbed onto root surfaces from solution. This agrees with the earlier findings that arsenic compounds are less readily translocated through the root system of aquatic plants. In various research studies it has been noted that different components exhibited variable performance in removal of Cd from aquatic ecosystems, *Pistia stratiotes* being one of the efficient components (Bhakta and Munekage, 2008).

Plants with efficient phytofiltration of toxic metals are beneficial for cleanup of contaminated fish ponds by *P. stratiotes*. Bioconcentration of heavy metal by water weeds is described as the bioconcentration factor (BCF), which is the ratio of heavy metal accumulated by plants to that dissolved in the surrounding medium. For this, two BCFs were computed from the plant compartment concentrations as

$$BCF_R = \frac{CR_{root}}{C_{water}} \tag{1}$$

$$BCF_L = \frac{C_{leave}}{C_{water}} \tag{2}$$

The translocation of heavy metal from the roots to harvestable aerial part is generally expressed as the translocation factor (TF). It was calculated on a dry weight basis by dividing the heavy metal concentration in aerial parts by the heavy metal concentration in root. Based on the above two Eqs. (1) and (2), the TF can be expressed as:

$$\text{Translocation factor} \left(TF\right) = \frac{BCF_{leaf}}{BCF_{root}} \tag{3}$$

Many plants, particularly marshy plants and aquatic macrophytes, have shown the deposits of iron oxides or hydroxides on their root surface, which are called "iron plaque." There have been few detailed scientific investigations on the transfer of metals from the metal-accumulating macrophytes. Further, the transfer of metals to herbivores via metal-rich iron plaque on macrophyte roots needs critical investigation. This might be a significant method for assessing the food chain transfer and concomitant toxic functions.

Several of the wetland plants exhibiting iron plaque are expected to tolerate the metal-contaminated environment. Iron plaque acts as a barrier for the uptake of toxic metals that get adsorbed or immobilized by the plaque (Figure 6). The role of plaque in different species and the geochemical setting of the sediments needs detailed investigation (Zhong et al., 2010).

Typha species can be used to remediate both heavy metals and organic compounds. *Typha domingensis* have ability to absorb and accumulate Al, Fe, Zn, and Pb, the uptake ability for the roots toward the studied elements was in the following order: $Pb^{2+} > Fe^{3+} > Al^{3+} > Zn^{2+}$. The cattail was capable of accumulating the heavy metal ions preferentially from wastewater than from sediments (Hegazy et al., 2011). The accumulation of metal(s) in plant organs attained the highest values in roots, rhizomes, and old leaves. Rhizofiltration was found to be the best mechanism to explain *Typha* phytoremediation capability. In addition, *Typha angustifolia* (narrow-leaved cattail) was able to remediate reactive dyes under caustic conditions (Nilratnisakorn et al., 2007, 2009).

Eichhornia crassipes (water hyacinth) is qualified for use in wastewater by its absorption capacity of heavy metals and nutrients. Use of water hyacinths can help reduce eutrophication by removing ammonia, phosphorus, and nitrate from the municipal wastewater treatment plant effluence

FIGURE 6

(a-e) *Pistia* an aquatic ornamental in market. It is also an efficient phytoremediator for a variety of contaminants. *Pistia* roots are model experimental material for investigating the significance of plaque in biocgeochemistry of trace element (e) *Lemna*.

(Kutty et al., 2009). This plant represents an alternative for metals bioremediation in an aquatic system. However, water hyacinth may cause severe water management problems because of its huge vegetative reproduction and high growth rate. Therefore, the use of water hyacinth in phytoremediation technology should be considered carefully.

Elodea canadensis (waterweed) grows rapidly in favorable conditions and can choke shallow ponds, canals, and the margins of some slow-flowing rivers. *Nasturtium officinale* (watercress) is a popular vegetable in many countries, but should not be used as food from a contaminated site. Since watercresses are consumed as vegetables, these aquatic plants should be used with care in metals phytoremediation.

Alternanthera philoxeroides (alligator weed) is one of the most common aquatic weeds in contaminated/polluted ecosystems. This weed is native to South America and naturalized in India (Naqvi and Rizvi, 2000). Several Amaranthaceae produce large biomass and are suitable for environmental remediation and toxic metal cleanup (e.g., *Amaranthus retroflexus*) (Prasad, 2001). *Hydrocotyle umbellata* (pennywort) is an aquatic plant commonly found in many tropical countries. The plants grow very rapidly and serve an ornamental and decorative purpose. It is reported to remove trace metals from aquatic systems. Duckweed plants (*Lemna* sp. and *Spirodela polyrhiza*) have received the greatest attention for toxicity tests as they are common to many aquatic environments, including lakes, streams, and effluents.

Aquatic macrophytes and floating islands are commonly found in wetland ecosystems of aquatic ornamental plants as shown in Figures 7 and 8. Therefore, phytoremediation of heavy metals and other contaminants in an aquatic system should be constructed by growing ornamental plants and macrophytes in the system. In addition to phytoremediation purpose, the ornamental plants are esthetically appealing and the flowers provide income.

FIGURE 7

Aquatic macrophytes and floating islands commonly found in wetland ecosystem of aquatic ornamental plants: (a) floating island and *Salvinia cucullata* (floating moss), (b) floating island and *Phragmites* sp. (reed), (c) mass of grasses, reed, and cattail in floating island with *Ipomoea aquatica* (water morning glory) and *Ludwigia adscendens* (creeping water primrose), (d) *Utricularia aurea* (common bladderwort), growth regulator, soil amendments, pesticide, manure and fertilizer for commercial planting ornamental plants.

FIGURE 8

(a) *Eichhornia crassipes* (water hyacinth), (b) *Pistia stratiotes* (water lettuce) and *S. cucullata*, (c) *Lemma perpusilla* (duckweed), *Azolla pinnata* (water fern), and *P. stratiotes*, and (d) *A. pinnata*, *L. perpusilla*, and *Ceratophyllum demersum* (hornwort).

Peatlands low in alkalinity ($<100\,\mu\mathrm{eq}\,l^{-1}$ of Ca) have been classified as the most susceptible to the process of acidification. Therefore, the aquatic plants in these sensitive peatlands may begin to accumulate potentially toxic metals, such as Cd, with concurrent decreases in levels of essential elements. If so, then these effects may be transferred through the food chain. The effects of peatland parameters (pH, alkalinity, dissolved organic carbon, and sediment organic matter) on the accumulation of Cd in macrophytes is an important step in determining the variables that may lead to higher Cd in peatland vegetation and subsequently wildlife (Thompson et al., 1997).

Mine tailings rich in sulfides, such as pyrite, can form acid mine drainage (AMD) if they react with atmospheric oxygen and water, which may also promote the release of metals and As. To prevent AMD formation, mine tailings rich in sulfides may be saturated with water to reduce the penetration of atmospheric oxygen. An organic layer with plants on top of the mine tailings would consume oxygen, as would plant roots through respiration. Thus, phytostabilization on water-covered mine tailings may further reduce the oxygen penetration into the mine tailings and prevent the release of elevated levels of elements into the surroundings. Since some wetland plant species have been found with the latter property (e.g., *T. latifolia*, *Glyceria fluitans*, and *Phragmites australis*), it seems that wetland communities may easily establish on submerged mine tailings (Sheoran and Sheoran, 2006; Williams, 2002; Wood and Mcatamney, 1994; Woulds and Ngwenya, 2004; Ye et al., 2001).

Nymphaea spp. are esthetically appealing and potentially hyperaccumulators of nutrients and metals as they have extensive roots and provide a large surface area for biofilm formation, thus enhancing microbial activities. Nitrogen concentrations in water lily also generally increased in response to increased P loads. The high affinity of water lily for P, combined with their subsequent influence on N uptake, suggests that these components can play an important role in wetland nutrient cycling (Newman et al., 2004). Water lily (*Nymphae capensis*) could be used as a biomonitor of the discharge of acid waters from Reserve Creek, Australia, for example, which contain high concentrations of metals, in order to identify priority sites (hotspots) in three acid sulfate soil-impacted enrichments. Water lily lamina concentrations of a suite of metal(loid)s (Al and Fe) were significantly higher than in plants collected from an unpolluted "reference" drainage channel, thus validating the concept of using this species as a biomonitor (Stroud and Collins, 2014).

4 ORNAMENTAL PLANTS FOR PHYTOREMEDIATION, SUSTAINABILITY, AND BIOECONOMY

Floriculture was thought to be the best option for an alternative agricultural practice that would reduce the metal concentration levels of contaminated lands on the one hand while providing a steady source of income on the other, as the flowers are marketable as sold in the marketplace as well as providing landscape for tourism. In addition, many of them are not edible plants; the risk of entering metals into the human food chain is reduced. Cut flowers like the marigold (*Tagetes* spp.) and the water lily in the families of Nymphaeaceae and Nelumbonaceae can be model ornamental plants in both heavy metal phytoremediation and for their economic benefit when sold as offerings. However, application of growth regulator, soil amendments, pesticide, manure, and fertilizer for commercial ornamental plants (Figures 9 and 10) should be optimized for both income and remediation. In Thailand, natural wetlands and constructed wetlands are used for lotus cultivation, and several paddy fields have been converted for lotus cultivation (Figure 11).

Marigolds are popular today for planting in large areas or farms in Thailand. Some agriculturists manage entire processes of marigold flower production (seed maturing, planting, flower cutting, and direct selling to markets) and selling their marigold seeds by breeding their own varieties. Moreover, some marigold farms have become ecotourism sites such as the marigold farm in Pobpra District, Tak Province, near Mae Sot Province (this area is rich in Zn and other heavy metals, especially Cd contaminations). Chintakovid et al. (2008) studied the potential of nugget marigold for As phytoremediation in Ron Phibun District, Nakhon Si Thammarat Province, and found that this type of marigold accumulated high As concentration and grew well in As-contaminated fields. Chaturvedi et al. (2014)

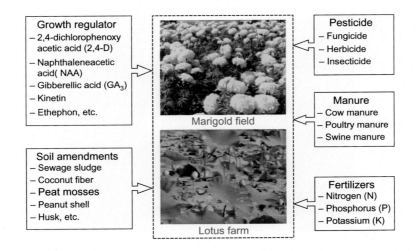

FIGURE 9

Marigold and lotus cultivation in contaminated environment.

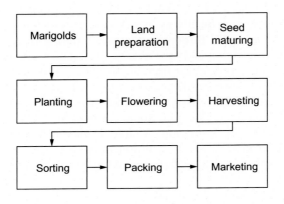

FIGURE 10

Processes for commercial planting of marigolds on contaminated soils.

found that the French marigold grew well in soil contaminated with iron ore tailings and could provide a phytostabilization of iron ore tailings.

Marigolds are easy to cultivate and transport. The species of marigolds is chosen based on marketing demand. There are various species of marigolds including *Tagetes erecta* (African marigold), *Tagetes patula* (French marigold), *Tagetes minuta* (wild marigold), *Tagetes lucida* (Mexican

FIGURE 11

Lotus replacing rice paddy.

mint marigold), and *Tagetes tenuifolia* (signet marigold). Figure 10 shows a flowchart of the marigold process from land preparation and planting through marketing. Land preparation is basic stage for well growing of plants. Seedlings, which have two or three true leaves, are ready for planting in the land. The marigold flowers are harvested during flowering stage. Then they are sorted by color and size. Packaging and time needed for transport are important parts of the process. In some areas marigold farms are far from the flower markets. In Bangkok about 200 million marigold flowers are sold in flower markets each year. The times of highest demand are Buddhist Sabbath days and Thai traditional festivals. On those days about a billion marigold flowers are sold in flower markets throughout Thailand. Some celebrations including weddings and funerals use mostly marigold flowers for decoration. A large display of marigold flowers for sale in the flower market is presented in Figure 2. In addition, popular flowers for garlands and flower arrangements in the Thai style for both special purposes and ordinary arrangements for worship are shown in Figure 3.

The planting of marigolds for commercial use requires special attention from seedling to harvesting. For growing seedlings, the recommended sowing method is in a plug tray. Good sowing media and growing support materials include peat mosses, coconut fiber, husk, and peanut shell. Soil amendments are applied in land preparation. Fertilizers including N and P are needed for strong and healthy seedlings. At the land preparation stage, fertilizers (N-P-K) are combined with soil, manures, and organic matters to produce exuberant growth. The amount of fertilizers per bed/plot depends on soil characteristics and properties. Pesticides are important for plant protection products, but some contain heavy metals such as copper, mercury, tin, etc. Plant growth regulators are needed for all stage of growth.

Landscapes for ecotourism are shown in Figure 5. Incomes form ecotourism and floral garden festivals are attractive tourists. Some parks and gardens have or are near wastewater treatment plant/

pond. Sewage sludge is useful for soil amendment because it contains a high amount of digested organic matter, N, and P. Nevertheless, the use of sewage sludge could contaminate with heavy metals, pathogenic microorganisms, methane production, and bad smells, and high nitrate could be toxic. Phytoremediation could solve these problems and planting ornamental plants also enhances esthetics. Moreover, a floral garden festival could provide income from tourists and also reduce the area needed for disposal of sewage sludge. In addition, wastelands are being considered for agriculture. Ornamental plants could be planted in urban areas for esthetics and to reduce air and soil pollution, and could be sustained by used water from houses. These considerations could motivate local people in and around contaminated or urban areas to cooperate for cleanup, esthetics, and also for income earning.

5 MANAGEMENT OF PHYTOREMEDIATING ORNAMENTAL BIOMASS

Once ornamental plants are used for contaminated soil remediation, they may be harvested and then incinerated together to recover the residual heavy metals. Alternatively, cut flowers that contain a limited amount of contaminants can be sold to households for their ornamental value. The other parts of the plants, however, which contain high content of heavy metal(s), should not be dispersedly discarded after death because the heavy metal contamination in soils would be diffused gradually from the discarded plants. Management and disposal of the huge amount of phytoremediating plants with high metals content is therefore an important concern. After harvest the metal-enriched biomass must be disposed of safely, which is a technical issue that remains partially unsolved. A variety of techniques under development include composting, compacting, pyrolysis, and biogas production (Ghosh and Singh, 2005).

Carbonization and incineration: Plants usually contain high moisture content, therefore, the biomass have to be dried before carbonization. In addition, the ash content of air-dried plants is too high to get a good fuel as an end-product. The high investment and technological level necessary also make carbonization. For hyperaccumulating plants, incineration is a suitable management technique.

Hydrolysis and fermentation: Liquid fuel, such as ethanol, may be produced from phytoremediating plants by hydrolysis together with fermentation, which would make the plants a good substrate. Some kind of pretreatment is necessary, therefore, to make the sugar more easily available for chemical hydrolysis. Pretreatment requires a relatively high temperature, strong acids, and pressurized reactors. Even it is economically feasible to produce fuel from phytoremediating plants, metals content in by-product sludge and the possibility of recontamination should be tested.

Anaerobic digestion and production of biogas: Anaerobic digestion is a biological process by which organic matter is degraded in the absence of oxygen, and biogas is produced as a by-product. Biogas production from phytoremediation biomass would be a viable, interesting, and environmentally sound idea for phytoremediating plants management. However, there are some limitations in biogas production from phytomass. The high lignin content can reduce the actual production, and the low bulk density could result in large voids with poor compaction and low feed rates. The mixture of phytomass and manures may produce more biogas. The use of phytoremediation plants in biogas production has good potential, but metals content and speciation in sludge from the anaerobic digestion should be investigated to prevent its redistribution in the environment.

FIGURE 12

Management of phytoremediating ornamental biomass: (a, b) mats from *Cyperus* spp., (c) tea from lotus petal, (d) composite biomaterials.

Other economic uses for phytoremediation include things like the making of mats from biomaterial from some aquatic macrophytes, such as *Cyperrus* spp. (Figure 12a and b); herbal tea prepared from petals of the lotus blossom (Figure 12c) and compost generated from the waste biomass (Figure 12d), which could serve both as compost and as growing media.

6 CONCLUSIONS

Phytoremediation is a technology that uses plants to degrade, assimilate, metabolize, or detoxify metal and organic chemical contamination. Important criteria for plants used for phytoremediation include the provision of economic benefits, harvesting management, and by-product utilization.

Growing ornamental plants based on their habitat and phytoremediation potential not only makes the environment colorful but also remediates the contaminants in terrestrial and aquatic environments. Many ornamental plants are not edible; therefore, the risk of entering metals into the food chain is reduced. Income from cut-flower sales and/or tourism may convince local people to remediate their contaminated lands and reservoirs. Management and disposal of phytoremediating plants with high contaminants are an important concern. However, ornamental plants will add a new dimension to the field of phytoremediation and phytomanagement of contaminated aquatic and terrestrial environments. Ornamental plants have the added advantage of enhancing the environment's esthetics besides cleaning up the environment and generating additional income, including additional employment opportunities (Figures 13 and 14).

FIGURE 13

Lotus: (a, b) lotus propagules and materials for its cultivation and propagation. (c) Flower market generates employment for packing and transport.

FIGURE 14

Ornamentals used in remediation generate employment for packing and transport.

ACKNOWLEDGMENTS

The authors gratefully acknowledge the receipt of financial support under the auspices of India-Thailand bilateral scientific cooperation ref. DST/INT/THAI/P-02/2012 dated 31-1-13. Also thanks to Pak Khlong Talat, (flower market), lotus farm, and marigold filed for permission of taking photographs. O. Meesungnoen would like to thank Science Achievement Scholarship of Thailand (SAST) (20/2555) for financial support.

REFERENCES

Abbasi, S.A., Ramasamy, E.V., 1999. Biotechnological Methods of Pollution Control. Orient Longman (Universities Press India Ltd.), Hyderabad. pp 168.

Abbasi, S.A., Nipaney, P.C., Panholzer, M.B., 1991. Biogas production from the aquatic weed pistia (*Pistia stratiotes*). Bioresour. Technol. 37, 211–214.

Aliotta, G., Monaco, P., Pinto, G., Pollio, A., Previtera, L., 1991. Potential allelochemicals from *Pistia stratiotes* L. J. Chem. Ecol. 17, 2223–2234.

Barceló, J., Poschenrieder, Ch., 2004. Structural and ultrastructural changes in heavy metal exposed plants. In: Prasad, M.N.V. (Ed.), Heavy Metal Stress in Plant, second ed. Springer-Verlag, Berlin Heidelberg, India, pp. 223–247.

Bassi, M., Corradi, M.G., Favali, M.A., 1990. Effect of chromium (VI) on two fresh water plants, *Lemna minor* and *Pistia stratiotes*: II. Biochemical and physiological observations. Cytobios 62, 101–109.

Basu, A., Kumar, S., Mukherjee, S., 2003. Arsenic reduction from aqueous environment by water lettuce (*Pistia stratiotes* L.). Indian J. Environ. Health 45, 143–150.

Belmont, M.A., Metcalfe, C.D., 2003. Feasibility of using ornamental plants (*Zantedeschia aethiopica*) in subsurface flow treatment wetlands to remove nitrogen, chemical oxygen demand and nonylphenol ethoxylate surfactants – a laboratory-scale study. Ecol. Eng. 21, 233–247.

Bhakta, J.N., Munekage, Y., 2008. Role of ecosystem components in Cd removal process of aquatic ecosystem. Ecol. Eng. 32, 274–280.

Bini, M.L., Thomaz, S.M., Murphy, K.J., Camargo, A.F.M., 1999. Aquaticmacrophyte distribution in relation to water and sediment condition in the Itaipu reservoir, Brazil. Hydrobiologia 415, 147–154.

Borghei, M., Arjmandi, R., Moogouei, R., 2011. Potential of *Calendula alata* for phytoremediation of stable cesium and lead from solutions. Environ. Monit. Assess. 181, 63–68.

Bosiacki, M., 2008. Accumulation of cadmium in selected species of ornamental plants. Acta Sci. Pol. Hortoru. 7, 21–31.

Bosiacki, M., 2009. Phytoextraction of cadmium and lead by selected cultivars of *Tagetes electa* L. Part II. Content of Cd and Pb in plants. Acta. Sci. Pol. Hortoru. 8, 15–26.

Bosiacki, M., Kleiber, T., Kaczmarek, J., 2013. Evaluation of suitability of *Amaranthus caudatus* L. and *Ricinus communis* L. in phytoextration of cadmium and lead from contaminated substrates. Arch. Environ. Prot. 39, 47–59.

Caldelas, C., Araus, J.L., Febrero, A., Bort, J., 2012. Accumulation and toxic effects of chromium and zinc in *Iris pseudacorus* L. Acta Physiol. Plant. 34, 1217–1228.

Cao, X., Ma, L.Q., Shiralipour, A., 2003. Effects of compost and phosphate amendments on arsenic mobility in soils and arsenic uptake by the hyperaccumulator, *Pteris vittata* L. Environ. Pollut. 126, 157–167.

Cardwell, A.J., Hawker, D.W., Greenway, M., 2002. Metal accumulation in aquatic macrophytes from southeast Queensland, Australia. Chemosphere 48, 653–663.

Casierra-Posada, F., Blanke, M.M., Guerrero-Guío, J., 2014. Iron tolerance in Calla Lilies (*Zantedeschia aethiopica*). Gesunde Pflanz. 66, 63–68.

Cassina, L., Tassi, E., Pedron, F., Petruzzelli, G., Ambrosini, P., Barbafieri, M., 2012. Using a plant hormone and a thioligand to improve phytoremediation of Hg-contaminated soil from a petrochemical plant. J. Hazard. Mater. 231–232, 36–42.

Castillo, O.S., Dasgupta-Schubert, N., Alvarado, C.J., Zaragoza, E.M., Villegas, H.J., 2011. The effect of the symbiosis between *Tagetes erecta* L. (marigold) and *Glomus intraradices* in the uptake of Copper(II) and its implications for phytoremediation. New Biotechnol. 29, 156–164.

Chatterjee, S., Singh, L., Chattopadhyay, B., Datta, S., Mukhopadhyay, S.K., 2012. A study on the waste metal remediation using floriculture at East Calcutta Wetlands, a Ramsar site in India. Environ. Monit. Assess. 184, 5139–5150.

Chaturvedi, N., Ahmed, M.J., Dhal, N.K., 2014. Effects of iron ore tailings on growth and physiological activities of *Tagetes patula* L. J. Soils Sediments 14, 721–730.

Chawchai, S., Chabangborn, A., Kylander, M., Löwemark, L., Mörth, C.-M., Blaauw, M., Klubseang, W., Reimer, P.J., Fritz, S.C., Wohlfarth, B., 2013. Lake Kumphawapi – an archive of Holocene palaeoenvironmental and palaeoclimatic changes in northeast Thailand. Quat. Sci. Rev. 68, 59–75.

Chintakovid, W., Visoottiviseth, P., Khokiattiwong, S., Lauengsuchonkul, S., 2008. Potential of the hybrid marigolds for arsenic phytoremediation and income generation of remediators in Ron Phibun District, Thailand. Chemosphere 70, 1532–1537.

Choo, T.P., Lee, C.K., Low, K.S., Hishamuddin, O., 2006. Accumulation of chromium (VI) from aqueous solutions using water lilies (*Nymphaea spontanea*). Chemosphere 62, 961–967.

Chowdhury, Md.R.I., Mulligan, C.N., 2011. Biosorption of arsenic from contaminated water by anaerobic biomass. J. Hazard. Mater. 190, 486–492.

Cui, S., Zhang, T., Zhao, S., Li, P., Zhou, Q., Zhang, Q., Han, Q., 2013. Evaluation of three ornamental plants for phytoremediation of Pb-contaminated soil. Int. J. Phytorem. 15, 299–306.

Dai, S., Wong, C.S., Qiu, J., Wang, M., Chai, T., Fan, L., Yang, S., 2014. Enantioselective accumulation of chiral polychlorinated biphenyls in lotus plant (*Nelumbonucifera* spp.). J. Hazard. Mater. 280, 612–618.

Dan, T.V., Raj, K.S., Saxena, P.K., 2000. Metal tolerance of scented geranium (*Pelargonium* sp. 'Frensham'): effects of cadmium and nickel on chlorophyll fluorescence kinetics. Int. J. Phytorem. 2, 91–104.

Davies Jr., F.T., Puryear, J.D., Newton, R.J., Egilla, J.N., Saraiva Grossi, J.A., 2002. Mycorrhizal fungi increase chromium uptake by sunflower plants: influence on tissue mineral concentration, growth, and gas exchange. J. Plant Nutr. 25, 2389–2407.

den Hollander, N.G., Schenk, I.W., Diouf, S., Kropff, M.J., Pieterse, A.H., 1999. Survival strategy of *Pistia stratiotes* L. in the Djoudj National Park in Senegal. Hydrobiologia 415, 21–27.

Dewald, L.B., Lounibos, L.P., 1990. Seasonal growth of *Pistia stratiotes* L. in South Florida. Aquat. Bot. 36, 263–275.

Dhir, B., 2013. Phytoremediation: Role of Aquatic Plants in Environmental Clean-Up. Springer India, New Delhi.

Dhote, S., Dixit, S., 2009. Water quality improvement through macrophytes – a review. Environ. Monit. Assess. 152, 149–153.

Dingjian, C., 2012. Contents of heavy metals in lotus seed from REEs mining area. J. Saudi Chem. Soc. 16, 175–176.

Dray, F.A., Center, T.D., 1989. Seed production by *Pistia stratiotes* L. (water lettuce) in the United States. Aquat. Bot. 33, 155–160.

Enger, E.D., Smith, B.F., 2010. Environmental Science: A Study of Interrelationships, twelfth ed. The McGraw-Hill Companies, Singapore.

Espinoza-Quiñones, F.R., Módenes, A.N., Costa, I.L., Palácio, S.M., Trigueros, D.E.G., 2009. Kinetics of lead bioaccumulation from a hydroponic medium by aquatic macrophytes *Pistia stratiotes* L. Water Air Soil Pollut. 203 (1–4), 29–37.

Evenhuis, A., Korthals, G.W., Molendijk, L.P.G., 2004. *Tagetes patula* as an effective catch crop for longterm control of *Pratylenchus penetrans*. Nematology 6, 877–881.

Ghosh, M., Singh, S.P., 2005. A review on phytoremediation of heavy metals and utilization of its by products. Appl. Ecol. Environ. Res. 3, 1–18.

González-Chávez, M.D.C.A., Carrillo-González, R., 2013. Tolerance of *Chrysantemum maximum* to heavy metals: the potential for its use in the revegetation of tailings heaps. J. Environ. Sci. (China) 25, 367–375.

Greger, M., 1999. Meter availability, uptake, transport and accumulation in plants. In: Prasad, M.N.V. (Ed.), Heavy Metal Stress in Plant, second ed., Springer-Verlag, Berlin Heidelberg, India, pp. 1–27.

Hadad, H.R., Pinciroli, Maine M.A., Mufarrege, M.M., 2009. Nickel and phosphorous sorption efficiencies, tissue accumulation kinetics and morphological effects on *Eichhornia crassipes*. Ecotoxicology 18 (5), 504–513.

Hall, J.B., Okali, D.U.U., 1974. Phenology and productivity of *Pistia stratiotes* on Volta Lake, Ghana. J. Appl. Ecol. 11, 709–726.

Han, Y.L., Yuan, H.Y., Huang, S.Z., Guo, Z., Xia, B., Gu, J., 2007. Cadmium tolerance and accumulation by two species of Iris. Ecotoxicology 16, 557–563.

Han, Y.L., Huang, S.Z., Gu, J.G., Qiu, S., Chen, J.M., 2008. Tolerance and accumulation of lead by species of *Iris* L. Ecotoxicology 17, 853–859.

Hao, X.Z., Zhou, D.M., Li, D.D., Jiang, P., 2012. Growth, cadmium and zinc accumulation of ornamental sunflower (*Helianthus annuus* L.) in contaminated soil with different amendments. Pedosphere 22, 631–639.

Harley, K.L.S., 1990. Production of viable seeds by water lettuce. *Pistia stratiotes* L, in Australia. Aquat. Bot. 36, 277–279.

Hegazy, A.K., Abdel-Ghani, N.T., Abdel-Ghani, G.A., 2011. Phytoremediation of industrial wastewater potentiality by *Typha domingensis*. Int. J. Environ. Sci. Technol. 8, 639–648.

Henry-Silva, G.G., Camargo, A.F.M., Pezzato, L.E., 2006. Apparent digestibility of aquatic macrophytes by Nile tilapia (*Oreochromis niloticus*) and water quality in relation nutrients concentrations. Rev. Bras. Zootec. 35, 641–647.

Jiang, Z., Xinyuan, Z., 1998. Treatment and utilization of wastewater in the Beijing Zoo by an aquatic macrophyte system. Ecol. Eng. 11, 101–110.

Kabata-Pendias, A., Pendias, H., 2001. Trace Elements in Soils and Plants, third ed. CRC Press, Boca Raton, FL.

Keates, S.E., Tarlyn, N.M., Loewus, F.A., Franceschi, V.R., 2000. L-Ascorbic acid and L-galactose are sources for oxalic acid and calcium oxalate in *Pistia stratiotes*. Phytochemistry 53, 433–440.

Khan, S., Ahmad, I., Shah, M.T., Rehman, S., Khaliq, A., 2009. Use of constructed wetland for the removal of heavy metals from industrial wastewater. J. Environ. Manag. 90, 3451–3457.

Kumar, A., Prasad, M.N.V., 2010. Propagation of *Talinum cuneifolium* L. (Portulacaceae), an ornamental plant and leafy vegetable, by stem cuttings. Floriculture Ornamental Biotech. 4 (SI1), 68–71.

Kutty, S.R.M., Ngatenah, S.N.I., Isa, N.H., Malakahmad, A., 2009. Nutrient removal from municipal wastewater treatment plant effluent using *Eichhornia crassipes*. World Acad. Sci. Eng. Technol. 3, 12–22.

Lal, K., Minhas, P.S., Shipra, Chaturvedi, R.K., Yadav, R.K., 2008. Extraction of cadmium and tolerance of three annual cut flowers on Cd-contaminated soils. Bioresour. Technol. 99, 1006–1011.

Lavid, N., Barkay, Z., Tel-Or, E., 2001. Accumulation of heavy metals in epidermal grands of waterlily (*Nymphaeaceae*). Planta 212, 313–322.

Lee, C.K., Low, K.S., Hew, N.S., 1991. Accumulation of arsenic by aquatic plants. Sci. Total Environ. 103, 215–227.

Leong, E.S., Tan, S., Chang, Y.P., 2012. Antioxidant properties and heavy metal content of lotus plant (*Nelumbo nucifera* gaertn) grown in ex-tin mining pond near Kampar, Malaysia. Food Sci. Technol. Res. 18, 461–465.

Leterme, P., Londoño, A.M., Muñoz, J.E., Súarez, J.S., Bedoya, C.A., Souffrant, W.B., Buldgen, A., 2009. Nutritional value of aquatic ferns (*Azolla filiculoides* Lam. and *Salvinia molesta* Mitchell) in pigs. Anim. Feed Sci. Technol. 149, 135–148.

Li, N., Li, Z., Fu, Q., Zhuang, P., Guo, B., Li, H., 2013. Agricultural technologies for enhancing the phytoremediation of cadmium-contaminated soil by *Amaranthus hypochondriacus* L. Water Air Soil Pollut. 224, 1673.

Li, J., Yang, X., Wang, Z., Shan, Y., Zheng, Z., 2015. Comparison of four aquatic plant treatment systems for nutrient removal from eutrophied water. Bioresour. Technol. 179, 1–7.

Liu, J.N., Zhou, Q.X., Sun, T., Ma, L.Q., Wang, S., 2008a. Identification and chemical enhancement of two ornamental plants for phytoremediation. Bull. Environ. Contam. Toxicol. 80, 260–265.

Liu, Jn, Zhou, Qx, Sun, T., Ma, L.Q., Wang, S., 2008b. Growth responses of three ornamental plants to Cd and Cd-Pb stress and their metal accumulation characteristics. J. Hazard. Mater. 151, 261–267.

Liu, Z., He, X., Chen, W., Yuan, F., Yan, K., Tao, D., 2009a. Accumulation and tolerance characteristics of cadmium in a potential hyperaccumulator-*Lonicera japonica* Thunb. J. Hazard. Mater. 169, 170–175.

Liu, J.N., Zhou, Q.X., Wang, S., Sun, T., 2009b. Cadmium tolerance and accumulation of *Althaea rosea* Cav. and its potential as a hyperaccumulator under chemical enhancement. Environ. Monit. Assess. 149, 419–427.

Liu, J., Zhou, Q., Wang, S., 2010. Evaluation of chemical enhancement on phytoremediation effect of Cd-contaminated soils with *Calendula officinalis* L. Int. J. Phytorem. 12, 503–515.

Liu, L.Z., Gong, Z.Q., Zhang, Y.L., Li, P.J., 2011a. Growth, cadmium accumulation and physiology of marigold (*Tagetes erecta* L.) as affected by arbuscular mycorrhizal fungi. Pedosphere 21, 319–327.

Liu, Y.T., Chen, Z.S., Hong, C.Y., 2011b. Cadmium-induced physiological response and antioxidant enzyme changes in the novel cadmium accumulator, *Tagetes patula*. J. Hazard. Mater. 189, 724–731.

Maine, M.A., Suné, N., Lagger, S.C., 2004. Chromium bioaccumulation: comparison of the capacity of two floating aquatic macrophytes. Water Res. 38, 1494–1501.

Malarkodi, M., Krishnasamy, R., Chitdeshwari, T., 2008. Phytoextraction of nickel contaminated soil using castor phytoextractor. J. Plant Nutr. 31, 219–229.

Marchand, L., Mench, M., Jacob, D.L., Otte, M.L., 2010. Metal and metalloid removal in constructed wetlands, with emphasis on the importance of plants and standardized measurements: a review. Environ. Pollut. 158, 3447–3461.

Marmiroli, N., Maestri, E., 2008. Health implications of trace elements in the environment and the food chain. In: Prasad, M.N.V. (Ed.), Trace Elements as Contaminants and Nutrients. John Willy & Sons, New Jersey, pp. 23–53.

Marques, A.P.G.C., Moreira, H., Franco, A.R., Rangel, A.O.S.S., Castro, P.M.L., 2013. Inoculating *Helianthus annuus* (sunflower) grown in zinc and cadmium contaminated soils with plant growth promoting bacteria – effects on phytoremediation strategies. Chemosphere 92, 74–83.

Mennema, J., 1977. Is waterlettuce (*Pistia stratiotes* L.) becoming a new aquatic weed in the Netherlands? Natura 74, 187–190.

Miao, Q., Yan, J., 2013. Comparison of three ornamental plants for phytoextraction potential of chromium removal from tannery sludge. J. Mater. Cycles Waste Manag. 15, 98–105.

Miretzky, P., Saralegui, A., Fernandez, A., 2004. Aquatic macrophytes potential for the simultaneous removal of heavy metals (Buenos Aires, Argentina). Chemosphere 57, 997–1005.

Miretzky, P., Saralegui, A., Cirelli, A.F., 2006. Simultaneous heavy metal removal mechanism by dead macrophytes. Chemosphere 62, 247–254.

Miretzky, P., Muñoz, C., Carrillo-Chavez, A., 2010. Cd (II) removal from aqueous solution by *Eleocharis acicularis* biomass, equilibrium and kinetic studies. Bioresour. Technol. 101, 2637–2642.

Mishra, V.K., Tripathi, B.D., 2008. Concurrent removal and accumulation of heavy metals by the three aquatic macrophytes. Bioresour. Technol. 99, 7091–7097.

Mongkhonsin, B., Nakbanpote, W., Nakai, I., Hokura, A., Jearanaikoon, N., 2011. Distribution and speciation of chromium accumulated in *Gynura pseudochina* (L.) DC. Environ. Exp. Bot. 47, 56–64.

Mufarrege, M.M., Hadad, H.R., Maine, M.A., 2010. Response of *Pistia stratiotes* to heavy metals (Cr, Ni, and Zn) and phosphorous. Arch. Environ. Contam. Toxicol. 58, 53–61.

Naqvi, S.M., Rizvi, S.A., 2000. Accumulation of chromium and copper in three different soils and bioaccumulation in an aquatic plant, *Alternanthera philoxeroides*. Bull. Environ. Contam. Toxicol. 65, 55–61.

Nehnevajova, E., Lyubenova, L., Herzig, R., Schröder, P., Schwitzguébel, J.P., Schmülling, T., 2012. Metal accumulation and response of antioxidant enzymes in seedlings and adult sunflower mutants with improved metal removal traits on a metal-contaminated soil. Environ. Exp. Bot. 76, 39–48.

Newman, S., McCormick, P.V., Miao, S.L., Laing, J.A., Kennedy, W.C., O'Dell, M.B., 2004. The effect of phosphorus enrichment on the nutrient status of a northern Everglades slough. Wetl. Ecol. Manag. 12, 63–79.

Nilratnisakorn, S., Thiravetyan, P., Nakbanpote, W., 2007. Synthetic reactive dye wastewater treatment by narrow-leaved cattails (*Typha angustifolia* Linn.): effects of dye, salinity and metals. Sci. Total Environ. 384, 67–76.

Nilratnisakorn, S., Thiravetyan, P., Nakbanpote, W., 2009. A constructed wetland model for synthetic reactive dye wastewater treatment by narrow-leaved cattails (*Typha angustifolia* Linn.). Water Sci. Technol. 60, 1565–1574.

Nina, S., Haler, M., Skornik, S., Kaligaric, M., 2007. Survival and expansion of *Pistia stratiotes* L. in a thermal stream in Slovenia. Aquat. Bot. 87, 75–79.

Noyo, E.E., George, E.O., 2008. Composition of water soluble fraction (WSF) of amukpe well-head crude oil before and after exposure to *Pistia stratiotes* L. Res. J. Appl. Sci. 3, 143–146.

Odjegba, V.J., Fasidi, I.O., 2007. Phytoremediation of heavy metals by *Eichhornia crassipes*. Environmentalist 27, 349–355.

Okurut, T.O., Rijs, G.B.J., Van Bruggen, J.J.A., 1999. Design and performance of experimental constructed wetlands in Unganda, planted with *Cyperus papyrus* and *Phragmites mauritianus*. Water Sci. Technol. 40, 265–271.

Organum, N., Bacon, F., 2006. Bioremediation technologies. In: Alvarez, P.J.J., Illman, W.A. (Eds.), Bioremediation and Natural Attenuation. John Wiley & Sons, New Jersey, pp. 351–455.

Outridge, P.M., 2000. Lead biogeochemistry in the littoral zones of south-central Ontario Lakes, Canada, after the elimination of gasoline lead additives. Water Air Soil Pollut. 118, 179–201.

Panitlertumpai, N., Nakbanpote, W., Sangdee, A., Thumanu, K., Nakai, I., Hokura, A., 2013. Zinc and/or cadmium accumulation in *Gynura pseudochina* (L.) DC. studied *in vitro* and the effect on crude protein. J. Mol. Struct. 1036, 279–291.

Panizza de León, A., González, R.C., González, M.B., Mier, M.V., Durán-Domínguez-de-Bazúa, C., 2011. Exploration of the ability of *Coleus blumei* to accumulate aluminum. Int. J. Phytorem. 13, 421–433.

Patil, A.V., Jadhav, J.P., 2013. Evaluation of phytoremediation potential of *Tagetes patula* L. for the degradation of textile dye Reactive Blue 160 and assessment of the toxicity of degraded metabolites by cytogenotoxicity. Chemosphere 92, 225–232.

Patra, R.R., Panigrahi, A.K., 1994. Changes in residual mercury accumulation and pigment contents in some aquatic plants, *Pistia* and *Hydrilla*, exposed to solid waste of a chloralkali industry. J. Environ. Biol. 15, 299–304.

Pellegrineschi, A., Damon, J.P., Valtorta, N., Paillard, N., Tepfer, D., 1994. Improvement of ornamental characters and fragrance production in lemon scented geranium through genetic transformation by *Agrobacterium rhizogenes*. Nat. Biotechnol. 12, 64–68.

Penny, D., 1998. Late Quaternary Palaeoenvironments in the Sakon Nakhon Basin, North-east Thailand. PhD thesis Monash University, Victoria, Australia 260 pp.

Penny, D., 1999. Palaeoenvironmental analysis of the Sakon Nakhon Basin, northeast Thailand: palynological perspectives on climate change and human occupation. Bull. Indo-Pacific Prehistory Assoc. 18, 139–149.

Pérez-López, R., Márquez-García, Bn., Abreu, M.M., Nieto, J.M., Córdoba, F., 2014. *Erica andevalensis* and *Erica australis* growing in the same extreme environments: phytostabilization potential of mining areas. Geoderma 230–231, 194–203.

Prasad, M.N.V., 2001. Bioremediation potential of Amaranthaceae. In: Leeson, A., Foote, E.A., Banks, M.K., Magar, V.S. (Eds.), Phytoremediation, Wetlands, and Sediments. 6th Int In Situ and On-Site Bioremediation Symposium. Battelle Press, Columbus, OH, pp. 165–172.

Prasad, M.N.V., 2007. Phytoremediation in India. In: Willey, N. (Ed.), Phytoremediation, Methods and Reviews. Humana Press, New Jersey.

Prasad, M.N.V., Freitas, H.M.O., 2003. Metal hyperaccumulation in plants – biodiversity prospecting for phytoremediation technology. Electron. J. Biotechnol. 6, 276–312.

Prasad, M.N.V., Nakbanpote, W., 2015. Integrated management of mine waste using biogeotechnologies focusing Thai mines. In: Thangavel, P., Sridevi, G. (Eds.), Environmental Sustainability: Role of Green Technologies. Springer, Berlin, pp. 229–249.

Prasad, M.N.V., Prasad, R., 2012. Nature's cure for cleanup of contaminated environment – a review of bioremediation strategies. Rev. Environ. Health 28, 181–189.

Pulford, I.D., Watson, C., 2003. Phytoremediation of trace metal-contaminated land by trees – a review. Environ. Int. 29, 529–540.

Qin, Q., Li, X., Wu, H., Zhang, Y., Feng, Q., Tai, P., 2013. Characterization of cadmium ([108]Cd) distribution and accumulation in *Tagetes erecta* L. Seedlings: effect of split-root and of remove-xylem/phloem. Chemosphere 93, 2284–2288.

Rahman, M.A., Hasegawa, H., 2011. Aquatic arsenic: phytoremediation using floating macrophytes. Chemosphere 83, 633–646.

Rajkumar, K., Sivakumar, S., Senthilkumar, P., Prabha, D., Subbhuraam, C.V., Song, Y.C., 2009. Effects of selected heavy metals (Pb, Cu, Ni, and Cd) in the aquatic medium on the restoration potential and accumulation in the stem cuttings of the terrestrial plant, *Talinum triangulare* Linn. Ecotoxicology 18, 952–960.

Raskin, I., Ensley, B.D., 2000. Phytoremediation of Toxic Metals: Using Plants to Clean Up the Environment. John Wiley & Sons, New York.

Rattanapolsan, L., Nakbanpote, W., Saensouke, P., 2013. Metals accumulation and leaf anatomy of *Murdannia spectabilis* growing in Zn/Cd contaminated soil. EnvironmentAsia 6, 71–82.

Rungruang, N., Babela, S., Parkpian, P., 2011. Screening of potential hyperaccumulator for cadmium from contaminated soil. Desalin. Water Treat. 32, 19–26.

Šajna, N., Haler, M., Škornik, S., Kaligarič, M., 2007. Survival and expansion of *Pistia stratiotes* L. in a thermal stream in Slovenia. Aquat. Bot. 87, 75–79.

Sanchez-Galvan, G., Mercado, F.J., Olguin, E.J., 2010. Sorption of crude oil by the non-living biomass of *Pistia stratiotes*. J. Biotechnol. 150S, S1–S576.

Sao, V., Nakbanpote, W., Thiravetyan, P., 2007. Cadmium accumulation by *Axonopus compressus* (Sw.) P. Beauv and *Cyperus rotundas* Linn growing in cadmium solution and cadmium-zinc contaminated soil. Songklanakarin J. Sci. Technol. 29, 881–892.

Satyakala, G., Jamil, K., 1992. Chromium induced biochemical changes in *Eichornia crassipes* (Mart) Solms and *Pistia stratiotes* L. Bull. Environ. Contam. Toxicol. 48, 921–992.

Sen, A.K., Bhattacharyya, M., 1994. Studies of uptake and toxic effects of Ni (II) on *Salvinia natans*. Water Air Soil Pollut. 78, 141–152.

Sen, A.K., Mondal, N.G., Mandal, S., 1987. Studies of uptake and toxic effects of Cr(VI) on *Pistia stratiotes*. Water Sci. Technol. 19, 119–127.

Seo, D.C., DeLaune, R.D., Han, M.J., Lee, Y.C., Bang, S.B., Oh, E.J., Chae, J.H., Kim, K.S., Park, J.H., Cho, J.S., 2010. Nutrient uptake and release in ponds under long-term and short-term lotus (*Nelumbo nucifera*) cultivation: influence of compost application. Ecol. Eng. 36, 1373–1382.

Shaw, B.P., Sahu, S.K., Mishra, R.K., 2004. Heavy metal induced oxidative damage in terrestrial plants. In: Prasad, M.N.V. (Ed.), Heavy Metal Stress in Plant, second ed. Springer-Verlag, Berlin Heidelberg, India, pp. 84–126.

Sheoran, A.S., Sheoran, V., 2006. Heavy metal removal mechanism of acid mine drainage in wetlands: a critical review. Miner. Eng. 19, 105–116.

Simachaya, W., 2009. Wastewater tariffs in Thailand. Ocean Coast. Manag. 52, 378–382.

Skinner, K., Wright, N., Porter-Goff, E., 2007. Mercury uptake and accumulation by four species of aquatic plants. Environ. Pollut. 145, 234–237.

Smith, S.E., Read, D.J., 1997. Mycorrhizal Symbiosis, second ed. Academic Press, San Diego and London.

Stroud, J.L., Collins, R.N., 2014. Improved detection of coastal acid sulfate soil hotspots through biomonitoring of metal(loid) accumulation in water lilies (*Nymphaea capensis*). Sci. Total Environ. 487, 500–505.

Sun, Y., Zhou, Q., Xu, Y., Wang, L., Liang, X., 2011. Phytoremediation for co-contaminated soils of benzo[a]pyrene (B[a]P) and heavy metals using ornamental plant *Tagetes patula*. J. Hazard. Mater. 2–3, 2075–2082.

Suñe, N., Sánchez, G., Caffaratti, S., Maine, M., 2007. Cadmium and chromium removal kinetics from solution by two aquatic macrophytes. Environ. Pollut. 145, 467–473.

Sytar, O., Kumar, A., Latowski, D., Kuczynska, P., Strzalka, K., Prasad, M.N.V., 2013. Heavy metal-induced oxidative damage, defense reactions, and detoxification mechanisms in plants. Acta Physiol. Plant. 35, 985–999.

Tahmasbian, I., Safari Sinegani, A.A., 2014. Chelate-assisted phytoextraction of cadmium from a mine soil by negatively charged sunflower. Int. J. Environ. Sci. Technol. 11, 695–702.

Tapia, Y., Eymar, E., Gárate, A., Masaguer, A., 2013. Effect of citric acid on metals mobility in pruning wastes and biosolids compost and metals uptake in *Atriplex halimus* and *Rosmarinus officinalis*. Environ. Monit. Assess. 185, 4221–4229.

Tewari, A., Singh, R., Singh, N.K., Rai, U.N., 2008. Amelioration of municipal sludge by *Pistia stratiotes* L.: role of antioxidant enzymes in detoxification of metals. Bioresour. Technol. 99, 8715–8721.

Thawil, B., Mercado, B.L., 1975. The life cycle of water lettuce (*Pistia stratiotes* L.). Philippine Weed Sci. Bull. 2, 11–15.

Thompson, E.S., Pick, F.R., Bendell-Young, L.I., 1997. The accumulation of cadmium by the yellow pond lily, *Nuphar variegatum*, in Ontario peatlands. Arch. Environ. Contam. Toxicol. 32, 161–165.

Tiagi, Y.D., Aery, N.C., 1986. Biogeochemical studies at the Khetri copper deposits of Rajasthan, India. J. Geochem. Explor. 26, 267–274.

Trigueros, D., Mingorance, M.D., Rossini Oliva, S., 2012. Evaluation of the ability of *Nerium oleander* L. to remediate Pb-contaminated soils. J. Geochem. Explor. 114, 126–133.

Tripathi, P., Kumar, R., Sharma, A.K., Mishra, A., Gupta, R., 2010. *Pistia stratiotes* (Jalkumbhi). Phcog. Rev. 4, 153–160. http://www.phcogrev.com/text.asp?2010/4/8/153/70909.

absorption, while simultaneously reducing their mobility in soil. Thus plants with phytostabilization potential can be of great value for the revegetation of mine tailings and contaminated soils (Antosiewicz et al., 2008; Mains et al., 2006). Furthermore, several amendments can be applied to the soil to promote the formation of insoluble contaminant forms (e.g., phosphate fertilizers, Fe oxide materials, organic materials, clay minerals) (Berti and Cunningham, 2000).

In contrast to phytostabilization, phytoextraction aims at removing the contaminants from the soil through absorption by roots followed by translocation and accumulation in the aerial parts (Figure 2). It is mainly applied to metals (Cd, Ni, Cu, Zn, Pb) but can also be used for other elements (Se, As) and organic compounds. Phytoextraction has been widely studied, mainly due to the potential for high efficiency and possible economic value (in metal recovery, energy production) (Prasad, 2004; Hernández-Allica et al., 2008; Pedron et al., 2009; Glass, 2000; Nascimento and Xing, 2006; Zhuang et al., 2007; Evangelou et al., 2015). Preferably, plants used in phytoextraction should possess, among others, the following characteristics (Ali et al., 2013; Hernández-Allica et al., 2008; Sakakibara et al., 2011; Shabani and Sayadi, 2012): (a) tolerance to high concentrations of metals; (b) the ability to

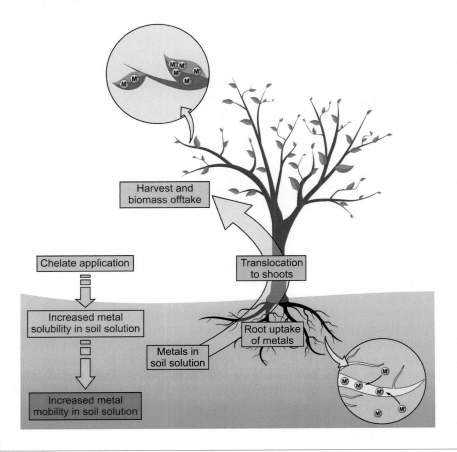

FIGURE 2

Processes involved in phytoextraction of metal-contaminated soil.

accumulate high concentrations in their aerial tissues; (c) rapid growth; (d) high biomass production; (e) profuse root system; (f) be easy to cultivate and harvest.

Phytoextraction preferentially uses hyperaccumulator plants, which have the ability to store high concentrations of specific metals in their aerial parts (0.01-1% dry weight (DW), depending on the metal) (Prasad, 2004; Hernández-Allica et al., 2008; Pedron et al., 2009; McGrath, 1998; Blaylock and Huang, 2000; Ma et al., 2001; McGrath and Zhao, 2003; Xie et al., 2009; Van der Ent et al., 2013). Although hyperaccumulators are phytoextractors par excellence, usually they are low biomass producers. Thus it is generally accepted that plants with a significant biomass production capacity can compensate their relatively lower metal accumulation capacity, to an extent where the amount of metal removed can be higher (Zhuang et al., 2007).

According to McGrath and Zhao (2003), phytoextraction efficiency is determined by the ability to hyperaccumulate metals and the biomass production. Therefore, if these factors influence the phytoextraction, they can be optimized to improve the phytoremediation process. One possibility is the addition of chemical agents into the soil in order to increase the bioavailability of metals and their root uptake (Varun et al., 2015; Pierzynski et al., 2002; Saifullah et al., 2009). This form of assisted phytoextraction, also known as induced phytoextraction, has shown great potential and has been widely studied.

Phytoextraction can only be considered effective if the accumulated contaminant is subsequently removed through harvesting. If most of the captured heavy metals are translocated to shoots, traditional farming methods can be used for harvesting. It is important to harvest the plants before leaf-fall or death and decomposition to ensure that contaminants do not disperse or return to the soil (Blaylock and Huang, 2000).

Phytoextraction potential can be estimated by calculation of bioconcentration factor (BCF) (or biological absorption coefficient) and translocation factor (TF) (Zhuang et al., 2007; Tu and Ma, 2004). The BCF, which is defined as the ratio of the total concentration of element in the harvested plant tissue (C_{plant}) to its concentration in the soil in which the plant was growing (C_{soil}), is calculated as follows:

$$BCF = \frac{C_{plant}}{C_{soil}}$$

TF, defined as the ratio of the total concentration of elements in the aerial parts of the plant (C_{shoot}) to the concentration in the root (C_{root}), is calculated as follows:

$$TF = \frac{C_{shoot}}{C_{root}}$$

Using both the BCF and the TF it is possible to assess the phytoextraction capacity of the plant. According to Díez Lázaro et al. (2006), a high root-to-shoot translocation (TF) of metals is a fundamental characteristic for a plant to be classified as effective in phytoextraction.

The commercial efficiency of phytoextraction can be estimated by the rate of metal accumulation and biomass production. Multiplying the rate of accumulation, metal (g)/plant tissue (kg), by the growth rate, plant tissue (kg)/hectare/year, gives the metal removal value: g/kg of metal per hectare and per year (Prasad, 2004; McGrath, 1998; Pierzynski et al., 2002; Liang et al., 2009). This rate of removal or extraction should reach several hundred, or at least 1 kg/ha/year, for the species to be commercially useful, and even then the remediation process may take from 15 to 20 years (Prasad, 2004).

After harvesting, the biomass disposal or processing is required (Figure 3). In the case of landfill deposition, thermal, physical, chemical, or microbiological processes can be used to reduce the volume/weight of biomass. In the case of incineration the energy produced represents an economic opportunity, and the ash can be further processed for extraction and recovery of metals with commercial value (phytomining) (Brooks et al., 1998a; Sheoran et al., 2009).

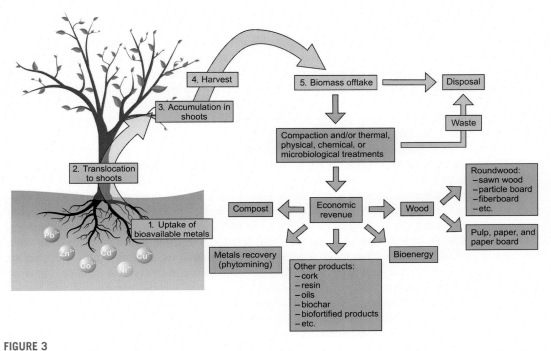

FIGURE 3

Biomass-processing opportunities involved in phytoextraction of metal-contaminated soil.

The commercial value of metals such as Ni, Zn, Cu, or Co may encourage the phytoremediation/phytomining processes. Economic results of phytomining can be greatly improved by using simple agricultural practices such as fertilization and weed control, as demonstrated by several studies using *Alyssum* species (Li et al., 2003; Bani et al., 2007, 2015).

Li et al. (2003) showed that with an addition of fertilizers and herbicides, among other agricultural management practices, *Alyssum murale* and *Alyssum corsicum* could reach a biomass yield of 20 t/ha and the consequent Ni phytoextraction production can be up to 400 kg Ni/ha. With this intensively managed phytoextraction, Li et al. (2003) estimated that the net value of an annual phytomining crop would be US $1749 net ha^{-1}. Later, Bani et al. (2015), after a 5-year field study, reported the maximum values of biomass production of *Alyssum murale* to the tune of 9.0 t/ha and Ni phytoextraction yield of 105 kg Ni/ha. These results, obtained with the use of fertilizers and weed control, allowed the authors to estimate that the net value of an annual phytomining crop would be US $1055 ha^{-1}.

Thus, future investigations on phytomining processes might focus on optimizing the plant cultivation processes via fertilization, the effects of pH variation, weed control, the addition of chelating agents, planting practices and harvest methods, research on hybridization or cloning to increase the metal hyperaccumulation potential, and biomass production, among others (Li et al., 2003; Bani et al., 2007, 2015; Morais et al., 2015).

According to Evangelou et al. (2015), in the bioenergy production the biomass may be used directly as heat or processed into gases or liquids (e.g., ethanol, biodiesel). Besides the bioenergy production, other potential economic products are wood, biochar, and biofortified products (Evangelou et al., 2015; Coates, 2005; Rockwood et al., 2006). However, these economic revenue opportunities should be analyzed carefully considering and evaluating all possible environmental and health risks.

Another interesting alternative comprises the usage of tree biomass (e.g., agricultural and forestry wastes) as adsorbents of heavy metals for treatment of industrial and domestic wastewaters. In this regard, a wide variety of native biomass have been tested as adsorbents, including several biomaterials derived from tree species, such as those of the genus *Quercus* (Prasad and Freitas, 2000), *Eucalyptus* (Chockalingam and Subramanian, 2009), *Sterculia* (Vinod et al., 2011), *Juglans* (Saqib et al., 2013), and *Pinus* (Saqib et al., 2013).

These phytoremediation techniques offer several advantages, but also some disadvantages, which should be considered when seeking to apply this technology. If low cost is an advantage, the time necessary to observe the results can be long. The pollutant concentration and the presence of other toxins should be within the tolerance limits of the plant to be used. Selecting plants with the efficiency for remediating varied contaminants simultaneously is not easy. These limitations and the possibility of these plants entering the food chain should be taken into account when applying this technology.

Some soils are so heavily contaminated that removal of metals using plants would take an unrealistic amount of time. On the other hand, conventional remediation technologies (physical, chemical, and thermal methods) can be prohibitive given the high economic costs. A typical practice is to choose drought-resistant fast-growing crops or fodder that can grow in metal-contaminated and nutrient-deficient soils. In this technique, contaminated soil is covered by vegetation tolerant to high concentrations of toxic elements, limiting the soil erosion and leaching of contaminants into groundwater (vegetation covers, vegetative caps, or phytocovers).

Thus, herbs (usually grasses), eventually shrubs or trees, established on landfills or tailings are used to minimize the infiltration of rainwater and contain the spread of pollutants. The roots increase soil aeration, promoting biodegradation, evaporation, and transpiration (Mendez and Maier, 2008; Williamson et al., 1982; Brooks et al., 1998b; Jorba and Vallejo, 2008; Prasad, 2015).

The difficulty of this technique is that tailings generally are not suitable for the development of plant roots. However, various investigations have been undertaken with the aim of developing processes of cultivation in tailings. For example, a technique in which an organic soil composed of sawdust, plant remains, and some NPK-fertilizers[1] is deposited on the surface was utilized by Hungarian agronomists (*Biological Reclamation Process, BRP*) (Oláh, 1988). With this technique, the researchers have proven that under normal overburden conditions, surface mined lands can be reclaimed to their pre-mine agricultural and forest productivity, without topsoil (Oláh, 1988). After only six months, were able to obtain 76 different plant species including cereals, shrubs, fruit trees, oaks and pines.

Therefore the addition of amendment, such as fertilizers and organic residues, is a common practice to facilitate revegetation of contaminated soils. The incorporation of organic residues (e.g., composts, manure, sewage sludge, biosolids) also provides a viable manner of recycling waste products (Kidd et al., 2015).

Vegetative covers are also often integrated with classical engineering technologies such as the enclosure, isolation, or encapsulation of hazardous wastes (Figure 4). In this case the hazardous wastes are isolated and contained in a landfill with a multilayer cover, encompassing vegetative cover, topsoil, protective cover layer, geocomposite, geotextile, and high-density polyethylene barrier (Prasad, 2015).

Portugal is an important producer of forestry products, with about 35% of the total land area forested equivalent to 3.15 million hectares (I.C.N.F., 2013). *Eucalyptus* spp., *Quercus suber*, and

---

[1] NPK-fertilizers are three-component fertilizers providing nitrogen (N), phosphorus (P), and potassium (K).

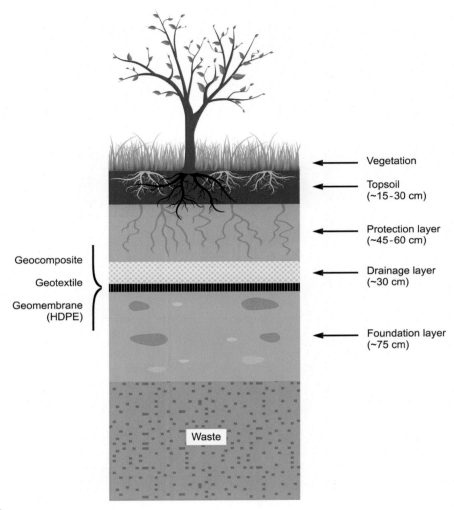

Geocomposite

Geotextile

Geomembrane
(HDPE)

Vegetation

Topsoil
(~15-30 cm)

Protection layer
(~45-60 cm)

Drainage layer
(~30 cm)

Foundation layer
(~75 cm)

Waste

FIGURE 4

Typical engineered phytocover integrated with the encapsulation technique.

Pinus pinaster tree species are the most representative forests in the Portuguese landscape, with 26% (~812,000 ha), 23% (~737,000 ha), and 23% (~714,000 ha) of forest territory occupation, respectively (I.C.N.F., 2013; Lopes and Cunha-e-Sá, 2014).

Portuguese forests contribute significantly to the employment and to the gross domestic product (GDP) of the country, providing a variety of services and industrial products (Lopes and Cunha-e-Sá, 2014; Louro et al., 2014). According to Lopes and Cunha-e-Sá (2014), these services include market services (e.g., timber exploitation, cork, resin, pine nuts) and nonmarket services (recreation, landscape, carbon sequestration, watershed protection, protection from soil erosion, and biodiversity) (Table 1). These authors also emphasize the increasing importance of biomass as a renewable resource for energy production.

Table 1 Forest Area Occupied by Tree Species in Portugal (I.C.N.F., 2013), Forest Services and Industrial Products and Overview of the Estimated Economic Value of Portuguese Forests Depending on Tree Species (Values per Hectare) (Lopes and Cunha-e-Sá, 2014)

| | Forest Area Occupied | Services and Products and Economic Importance | | | | | | |
| | | Market Services and Products[a] | | | Nonmarket Services[a] | | Total | |
			With Fire (€)	Without Fire (€)	With Fire (€)	Without Fire (€)	With Fire (€)	Without Fire (€)
Eucalyptus spp. (eucalyptus)	26%	*Timber* (pulp and paper industry)	166.32	202.84	128.8	192.05	295.13	394.89
Quercus suber (cork)	23%	Cork, timber (firewood), acorn	320.7	354.85	118.07	136.25	438.77	491.09
Pinus pinaster (maritime pine)	23%	*Timber* (firewood, furniture, construction, etc.), resin	157.69	192.31	134.56	191.39	292.25	383.69
Quercus rotundifolia (holm oak)	11%	*Timber* (firewood), acorn	44.87	49.8	185.83	190.81	230.7	240.61
Pinus pinea (stone pine)	6%	*Pine nuts*, timber, resin	283.52	302.25	144.88	157.75	428.4	460.01
Quercus spp.[b] (oak)	2%	*Timber* (furniture wood)	52.89	64.49	294.65	381.2	347.53	445.69
Acacia spp. (thorntree)	0.40%	-	-	-	115.31	138.42	115.31	138.42

[a]Estimates found by Lopes and Cunha-e-Sá (2014) considering market services (timber and nontimber forest products) and nonmarket services (recreation, carbon sequestration, watershed protection, protection from soil erosion, landscape, and biodiversity). Each type of services was considered in two scenarios: with fire risk and without fire risk.

[b]Quercus species excluding Quercus suber and Quercus rotundifolia.

3 STUDY AREAS

In recent decades many studies have been conducted in contaminated mining and industrial areas and in natural metalliferous soils (Antosiewicz et al., 2008; Zhuang et al., 2007; Williamson et al., 1982; Ernst, 1990; Robinson et al., 1997; Poschenrieder et al., 2001; Madejón et al., 2002; Kidd and Monterroso, 2005; Yanqun et al., 2005; Rodríguez et al., 2007; Saraswat and Rai, 2009; D'Souza et al., 2010; Lorestani et al., 2011; Varun et al., 2012; Maric et al., 2013; Zhang et al., 2014; Parraga-Aguado et al., 2014; Wójcik et al., 2014) in order to list and screen the indigenous species and evaluate their potential for phytoremediation of contaminated soils and ecological restoration. Also in Portugal, several studies have been performed to survey the indigenous plant species of diverse contaminated areas and evaluate their potential for environmental remediation (De Koe et al., 1991; Henriques and Fernandes, 1991; Alvarenga et al., 2004; Freitas et al., 2004a,b; Pratas et al., 2005, 2013; Díez Lázaro et al., 2006; Branquinho et al., 2007; Turnau et al., 2007; Abreu et al., 2008, 2012; Marques et al., 2009; Anawar et al., 2011; Valente et al., 2012; Favas et al., 2013; Gomes et al., 2014; Sousa et al., 2014).

In this chapter, important findings have been presented from several studies to evaluate the phyto-technological potential (phytoremediation, bioindication, biogeochemical prospecting) of trees grown on soils enriched with metals and metalloids in distinct abandoned mining areas of tin/tungsten (Sn/W), copper (Cu), and lead (Pb) of Portugal.

Related to Sn/W mines, the results obtained were from Sarzedas mine (central Portugal), Fragas do Cavalo mine (central Portugal), and Vale das Gatas mine (northern Portugal). As examples of Cu mines the results from São Domingos mine (southeast Portugal) and Borralhal mine (central Portugal) are presented. The Barbadalhos mine and the Sanheiro mine (central Portugal) are examples of abandoned Pb mines.

The Sarzedas study area includes two abandoned mines, Gatas and Santa, which form the denominated Sarzedas mine. The area is situated close to the Sarzedas village (Castelo Branco county, central Portugal). The Sarzedas mineralizations are emplaced in quartzous veins striking N60°E (Santa mine) and N20°W (Gatas mine), filling late Hercynian fractures, which cut the "Schist-Greywacke Complex" (Pratas et al., 2005). The acid rock veins cutting the complex are frequent, and these are identical to ante-Ordovician veins appearing in the area. These veins show indication of hydrothermal alteration, and they are mineralized by disseminated sulfides (Pratas et al., 2005). The main mineralization is of the vein type, and it is made of wolframite (ferberite), stibnite, pyrite, arsenopyrite, and more sparsely, by chalcopyrite, sphalerite, and galena (Pratas et al., 2005). Gold occurs in its native form. Soils are barely developed, and they are mainly mountain soils, especially cambisoils and lithosoils.

The Fragas do Cavalo mine is located close to the Oleiros village (Castelo Branco district, central Portugal). Mineralization occurs in quartz veins embedded in metasedimentary rocks of the Schist-Greywacke Complex in a unit consisting almost entirely of shales with a few thick levels of siltstones (Pratas et al., 2012). The mineralization consists mainly of wolframite, arsenopyrite, pyrite, and chalcopyrite (Pratas et al., 2012).

The Vale das Gatas mine is located in the district of Vila Real in northern Portugal. The geological units that emerge in this sector are metasedimentary rocks (Schist-Greywacke Complex), Hercynian granites and veinous rocks (mineralized and nonmineralized). The main mineralization of the veins is made of wolframite, cassiterite, scheelite, several sulfides (pyrite, chalcopyrite, sphalerite, galena, arsenopyrite, pyrrhotite, stannite, covellite, marcasite), silver (Ag), Pb, and bismuth (Bi) sulfosalts and native Bi were also present (Favas et al., 2013). The minerals that support mineralization are essentially quartz, fluorite, and muscovite. In this region, the characteristic soil units are Leptosols, Cambisols, and Anthrosols. The dominant soil unit of the studied area is the Leptosol type, which consists of soils whose principal characteristic is the presence of bedrock at 20 cm below the surface. These soils are divided into a distric (acid) subunit associated with the metasedimentary rocks and an umbric (acid and organic-rich) subunit associated with granitic rocks (Favas et al., 2013).

The São Domingos mine, located in southeast Portugal, is one of the historical mining centers, known for its activity since pre-Roman times, with extraction of Au, Ag, and mainly Cu (Freitas et al., 2004a; Gaspar, 1998). São Domingos mine belongs to the Iberian pyrite belt, with the outcropping area sequences formed by a unique vertical mass of cupriferous pyrite associated with zinc and lead sulfide. It is similar to many other volcanogenic massive sulfide mineralizations existing in this belt, from the genetic, morphological, and mineralogical point of view (Freitas et al., 2004a; Gaspar, 1998). The primary mineralization is made of massive sulfides (more than 85% of sulfides), essentially pyrite, sphalerite, chalcopyrite, galena, and sulfosalts (Gaspar, 1998).

The Borralhal mine is an abandoned Cu mine located in the Oleiros county (Castelo Branco district, central Portugal). The mineralizations are emplaced in quartzose veins striking N70°E that cut through the Schist-Greywacke Complex, and consist mainly of chalcopyrite, pyrite, bournonite, and other sulfosalts, with quartz-carbonate gangue (Pratas, 1996).

The Barbadalhos mine (also known as the Zorro mine) is an abandoned Pb mine in central Portugal, near the city of Coimbra. The mineralized quartz veins are situated in the Central Iberian Zone, near the contact between this geotectonic zone and the Ossa Morena Zone (Pratas et al., 2013). The mineralogy consists mainly of argentiferous galena and sphalerite. Chalcopyrite and arsenopyrite are also present in small amounts. The mineral supporting mineralization is essentially quartz, but calcite and some dolomite and ankerite are also present (Pratas et al., 2013). The soils of this area are poorly developed and predominantly acidic, showing variable thickness, between 10 and 80 cm and an overall clay texture with clay loam in the higher areas.

The Sanheiro mine (also known as the Sanguinheiro mine) is located in Penacova County, near the city of Coimbra (central Portugal). The mineralized quartz veins, striking N39°E, cut through the Schist-Greywacke Complex in the Central Iberian Zone, near the contact between this and the Ossa Morena Zone (Pratas, 1996). The mineralization consists mainly of galena and sphalerite with small amounts of pyrite, chalcopyrite, and marcasite in a gangue composed of quartz, barite, and siderite (Pratas, 1996; Fonseca and Martin, 1986).

4 METHODS

In the studied areas several line transects were made in mineralized and nonmineralized zones as well as in areas with tailings. Soils and plants were collected at 20 m intervals along the line transects (0, 20, 40 m, etc.) in circles of $\cong 2$ m radius.

At each location four random partial soil samples weighing 0.5 kg each were collected from 0 to 20 cm depth and mixed to obtain one composite sample to save time and costs. In the laboratory, soil samples were oven-dried at a constant temperature, manually homogenized and quartered. Two equivalent fractions were obtained from each quartered sample. One was used for the determination of pH and the other for chemical analysis. The samples for chemical analysis were sieved using a 2 mm mesh sieve to remove plant matter and subsequently screened to pass through a 250 μm screen.

Samples were also obtained from all species of plants whenever found growing within the 2 m radius of each sampling point. The plant-sampling methodology followed the orientations defined by Brooks (1983). The plant sample focused on the aerial parts, taking into consideration similar maturity of the plants and the proportionality of the different types of tissues, or the separation of different types of tissues (leaves and stems) in some species. The samples were separately collected into clean cellulose bags and brought to the laboratory on the same day. Although all the species present in the sample points have been sampled, this chapter presents only the results of the tree species.

The leaves/needles and stem samples of tree species followed the orientations defined by Özdemir (2005) and Sun et al. (2009). The samples were collected from trees of similar size and age. Leaves, needles, and stems were collected from outer branches of the middle canopy at south, west, east, and north directions and then were homogeneously mixed. When the vegetal material was sampled, different organs (stems and leaves/needles) were separated, and the tissues were separated by age (in the case of *Pinus pinaster*). Stems and needles (0-, 1-, 2-, 3-, and 4-year-old) were collected.

In the laboratory, the vegetal material was washed thoroughly, first in running water followed by distilled water, and then dried in a glasshouse. When dry, the material was milled into a homogenous powder.

The chemical analyses were performed at the Biogeochemistry Laboratory at the Department of Earth Sciences, University of Coimbra (Portugal) using the latest analytical methods. Soil pH was determined in water extract (1:2.5 v/v). The soil and plant samples were acid-digested for elemental analysis. Analytical methods included colorimetry for W, atomic absorption spectrophotometry (AAS, Perkin-Elmer, 2380) for Co, Cr, Cu, Fe, Mn, Ni, Pb, and Zn, and hydride generation system for As, Sb, and Sn. Data quality control was performed by inserting triplicate samples into each batch. Certified references materials (Virginia tobacco leaves CTA-VTL-2 and NIST 2709-San Joaquin Soil) were also used. The agreement between the certified reference values and the values determined by the analytical methods were in the range of 86–105%.

5 RESULTS AND DISCUSSION
5.1 TRACE METALS AND METALLOIDS IN SOIL

Concerning the tin/tungsten (Sn/W) mines, a summary of the pH and trace element data in soil from the Sarzedas, Fragas do Cavalo, and Vale das Gatas mines is shown in Figure 5. Among the elements present in the soils of Sarzedas mine area (Pratas et al., 2005), Ag, As, Pb, Sb, and W showed the most relevant anomalies. Low pH values observed near the mineralized area can be explained by the presence of sulfides in the mineralization. Therefore, soil pH was negatively correlated to mineralization (Pratas et al., 2005). High levels of sulfides, in particular pyrite and arsenopyrite, which are easily weathered, favored the dissolution of toxic elements, allowing higher dispersion and bioavailability.

In the soils of Fragas do Cavalo mine (Figure 5) stand out the anomalies of As, Pb, and W. These anomalies are clearly related to the mineralization of the area. However, these anomalies are limited to the mineralization vicinity, with small secondary dispersion aureoles (Pratas et al., 2012; Pratas, 1996).

Very high maximum values for Pb (6299 mg/kg), As (5770 mg/kg), and W (636 mg/kg) were observed at the Vale das Gatas mine (Figure 5). The Cu-Mn-W-As-Pb-Zn association, which reflects the presence of mineralized veins in the area, is inversely correlated with pH (Favas et al., 2013).

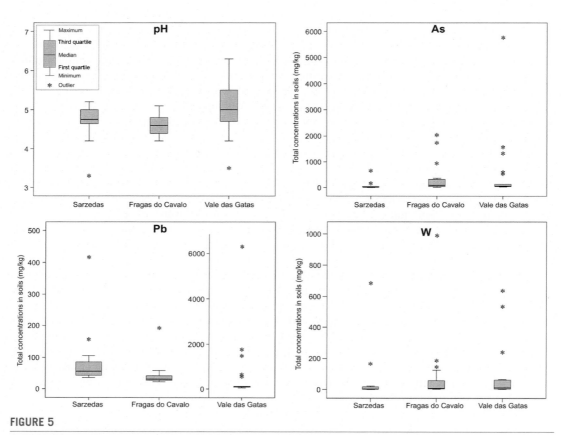

FIGURE 5

Arsenic, Pb, and W concentrations (mg/kg) and pH of soil samples of the Sarzedas mine ($n=24$), Fragas do Cavalo mine ($n=22$), and Vale das Gatas mine ($n=69$).

Related to the copper (Cu) mines, a summary of soil pH and trace element data from São Domingos mine and Borralhal mine is presented in Figure 6. High levels of Ag, As, Cu, Pb, and Zn were recorded in the soils of São Domingos mine area (Freitas et al., 2004a). Maximum concentration of As in these soils was very high, reaching 1291 mg/kg (see Figure 6). Copper concentration in these soils reached up to 1829 mg/kg (Figure 6) as a result of the former activities at the site (copper smelter). The concentration of Pb in the soil was also very high, 2694 mg/kg as the average value registered (Figure 6). The average Zn concentration in soils was of 218 mg/kg but it could reach 714 mg/kg (Freitas et al., 2004a), a level that can be extremely toxic for plants. Silver concentrations ranging from 2.5 to 16.6 mg/kg (Freitas et al., 2004a), cobalt, Cr, and Ni concentrations in soils were normally low, ranging from 20.1 to 54.3 mg/kg, from 5.1 to 84.6 mg/kg and from 27.2 to 52.9 mg/kg, respectively (Freitas et al., 2004a).

In the Borralhal mine area, high levels of Ag, As, Cu, Pb, and Sb were recorded in the soils (Pratas, 1996). These anomalies are clearly related to the mineralization of the area (chalcopyrite,

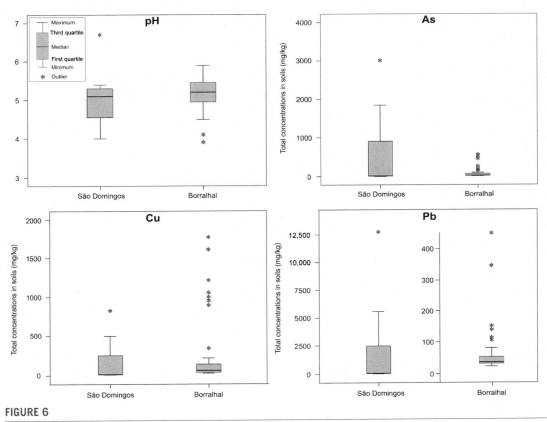

FIGURE 6

Arsenic, Cu, and Pb concentrations (mg/kg) and pH of soil samples of the São Domingos mine ($n=21$) and Borralhal mine ($n=43$).

pyrite, bournonite, and other sulfosalts). Arsenic, Cu, and Pb concentrations in these soils reached up to 555 mg/kg, 1773 mg/kg, and 451 mg/kg, respectively (Figure 6).

Concerning the lead (Pb) mines, a summary of the pH and trace element data in soils from Barbadalhos mine and Sanheiro mine is shown in Figure 7. High levels of Ag, As, Zn, and, mainly, Pb were recorded in the soils of Barbadalhos mine (Pratas et al., 2013). Lead concentration in these soils reached 9331 mg/kg while the average value was 928 mg/kg (Figure 7), obviously due to mining of galena at the site. In soils from mineralized zone, the mean Pb concentration (2380 mg/kg) was nearly nine times the threshold for industrial soils suggested by the Canadian Environmental Quality Guidelines (Canadian Council of Ministers for the Environment, 2006).

In Sanheiro mine there is a similar situation to that found in Barbadalhos mine. In Sanheiro mine were also recorded high levels of Ag, Zn, and, mainly, Pb (Figure 7). Lead concentration in these soils reached 11,314 mg/kg while the average value was 536 mg/kg.

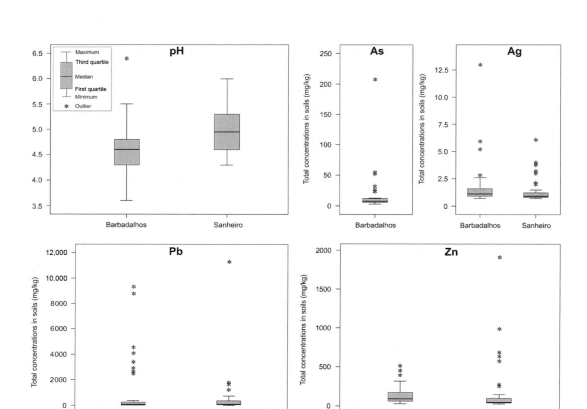

FIGURE 7

Arsenic, Ag, Pb, and Zn concentrations (mg/kg) and pH of soil samples of the Barbadalhos mine ($n=45$) and Sanheiro mine ($n=50$).

5.2 TRACE METALS AND METALLOIDS IN PLANTS

In the tree species of Sarzedas mine area, As was preferentially accumulated in the needles of *Pinus pinaster*. In fact, high accumulation of As was present in needles, and it increased in the older tissues (Figure 8). Therefore, these tissues are suited for recognizing the anomaly. This translocation is a common mechanism in plants to avoid toxicity in young leaves, as their metabolic activity is higher (Pratas et al., 2005).

With respect to Pb, it was observed that the accumulated concentrations are within the normal range in plants (Figure 9). The highest Pb concentrations were in *Quercus rotundifolia* tissues.

Conspicuous among the tree species that are capable of accumulating W was the *P. pinaster* (old needles) (Figure 10).

Considering other species of native flora (data not shown in this chapter), shrub and herbaceous species, it was found that (Favas et al., 2014; Pratas et al., 2005; Pratas, 1996): Arsenic is accumulated mostly in aerial tissues of *Digitalis purpurea*; species that are capable of accumulating Pb are *Halimium ocymoides*, *Digitalis purpurea*, and *Helichrysum stoechas*; W is accumulated by *D. purpurea*, *Cistus*

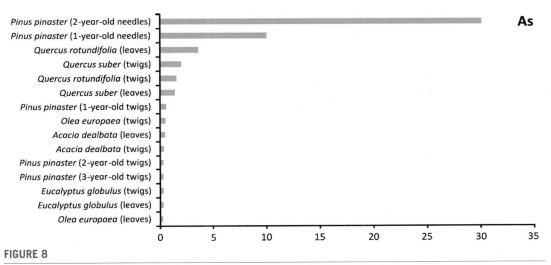

FIGURE 8

Accumulation of As (mg/kg DW) in tree species of the Sarzedas mine area.

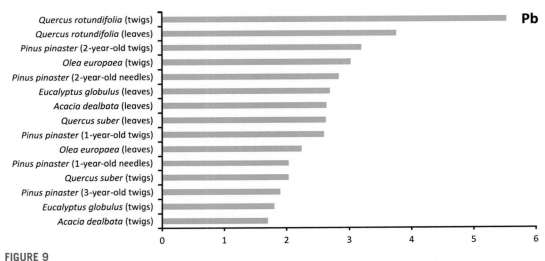

FIGURE 9

Accumulation of Pb (mg/kg DW) in tree species of the Sarzedas mine area.

ladanifer, Calluna vulgaris, and *H. stoechas. Digitalis purpurea* also accumulated substantial amount of Sb (Pratas et al., 2005), indicating its tolerance to this element, although the assimilation occurred at low concentrations in the soil.

Given the results obtained in the Sarzedas mine area, it was concluded that the tree species and/or their tissues best suited for metal(loid) bioindicating and/or with potential for mine restoration are by order of importance (Pratas et al., 2005): (1) As: old needles of *P. pinaster* and leaves of

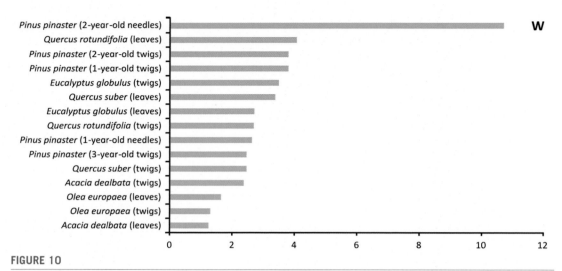

FIGURE 10

Accumulation of W (mg/kg DW) in tree species of the Sarzedas mine area.

Q. rotundifolia; (2) W: old stems and needles of *P. pinaster*, stems and leaves of *Q. rotundifolia*, and stems of *Eucalyptus globulus*; (3) Sb: stems of *P. pinaster*. Considering shrub and herbaceous, the species/tissues best suited are (Favas et al., 2014; Pratas et al., 2005): (1) As: aerial tissues of *D. purpurea*, *C. vulgaris*, *Chamaespartium tridentatum*; leaves of *C. ladanifer* and *Erica umbellate*; (2) W: *D. purpurea*, *C. tridentatum*; stem and leaves of *C. ladanifer* and *E. umbellate*; (3) Sb: *D. purpurea*, *E. umbellate*; stems of *C. ladanifer*, *C. vulgaris*, and *C. tridentatum*.

In the Fragas do Cavalo mine area, the concentrations of metal(loid)s in the tree species are relaively low compared with the other Sn/W studied mines. The highest As concentration was found in *Q. rotundifolia* (Figure 11) with values higher than normal levels in plants (Kabata-Pendias, 2010). Lead concentrations in all trees species (Figure 12) are within the normal range for plants

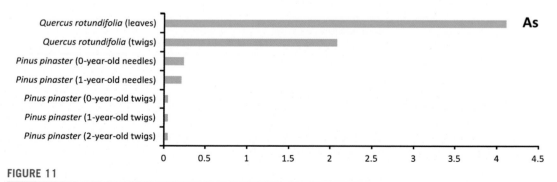

FIGURE 11

Accumulation of As (mg/kg DW) in tree species of the Fragas do Cavalo mine area.

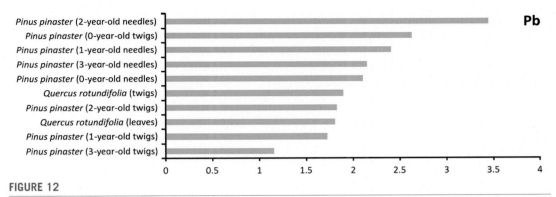

FIGURE 12

Accumulation of Pb (mg/kg DW) in tree species of the Fragas do Cavalo mine area.

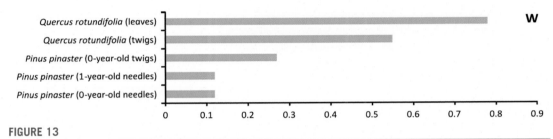

FIGURE 13

Accumulation of W (mg/kg DW) in tree species of the Fragas do Cavalo mine area.

(Kabata-Pendias, 2010). Tungsten concentrations in both leaves and twigs of *Q. rotundifolia* and in the twig shoots of *P. pinaster* ranged from 0.27 to 0.78 mg/kg DW (Figure 13), which are above the normal range for plants (Kabata-Pendias, 2010).

In the Vale das Gatas mine area the content variations in plant samples were, in general, strongly related to the content variations in soils. It has also been verified that in contaminated sites or tailings, the concentration of metals in plant tissues is higher due to the higher metal concentrations in the soil.

The *P. pinaster* trees growing on the tailings and contaminated soils of Vale das Gatas mine accumulated the studied elements in quantities greater than observed in plants of the representative areas of the local geochemical background. These values were also higher than those typically observed in this species.

In the *P. pinaster* samples from tailings and contaminated soil sites, the older needles (2 and 3 years old) show a tendency to accumulate higher concentrations of As, Pb, and W (Figures 14–16) and also Fe and Zn (Favas et al., 2013, 2014), while Ni and Cu were preferentially accumulated in young needles and stems (1 year old) (Favas et al., 2013, 2014).

Therefore, the results allowed the authors to conclude that the metal(loid) concentrations in plants depend as much on the plant organ as on its age, and in biogeochemical studies, it is important not to mix foliar and woody material in the same sample. The species showed a great variability in the

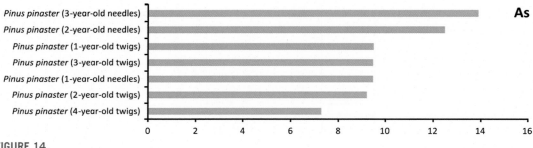

FIGURE 14

Accumulation of As (mg/kg DW) in *Pinus pinaster* tissues of the Vale das Gatas mine area.

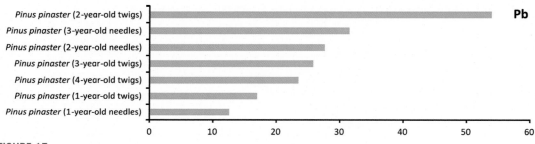

FIGURE 15

Accumulation of Pb (mg/kg DW) in *Pinus pinaster* tissues of the Vale das Gatas mine area.

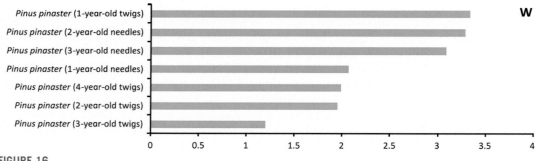

FIGURE 16

Accumulation of W (mg/kg DW) in *Pinus pinaster* tissues of the Vale das Gatas mine area.

accumulation behavior of As, Pb, and W, and also of Fe, Mn, Cu, Zn, and Ni (Favas et al., 2013) with the age of the organ. Thus, the older needles accumulated higher levels of As (Figure 14), Pb (Figure 15), W (Figure 16), and Fe and Zn (Favas et al., 2013, 2014). The 2-year-old stems may also be appropriate samples to detect higher levels of Fe, Zn, and Pb, while the 1-year-old needles and stems accumulated higher levels of Cu and Ni (Favas et al., 2013, 2014). The 2-year-old stems may also be appropriate samples to detect higher levels of Pb (Figure 15) and Fe and Zn (Favas et al., 2013, 2014).

Considering the shrub and herbaceous species (data not shown in this chapter), it was found that (Favas et al., 2002, 2014): The leaves of *Agrostis castellana* and *Holcus lanatus* reflect the Cu, Pb, and Ni pedogeochemical anomalies; the aerial parts of *Pteridium aquilinum* and *Juncus effusus* seem to be indicative of Zn anomalies in the soil; *H. lanatus* and *A. castellana* were the main accumulators of As, Cu, Fe, and Pb and were good accumulators of Zn. *Pteridium aquilinum* was a good accumulator of As, Pb, and Zn. *Juncus effusus* appeared to be a Zn accumulator.

In the tree samples from the São Domingos mine area, Cu concentrations in plant tissues ranged from 4.60 to 23.8 mg/kg DW (Figure 17). These Cu values are within the range considered normal for plants (Kabata-Pendias, 2010).

Arsenic concentrations in tree tissues ranged from 0.3 to 1.4 mg/kg DW (Figure 18). In fact, the maximum As concentration was recorded in herbaceous species (Freitas et al., 2004a): *Juncus conglomeratus, Thymus mastichina, J. effusus,* and *Scirpus holoschoenus.*

Lead concentrations in tree tissues ranged from 2.9 to 9.1 mg/kg DW (Figure 19), which are within the normal range for plants (Kabata-Pendias, 2010). On the other hand, Pb concentrations were rather high for

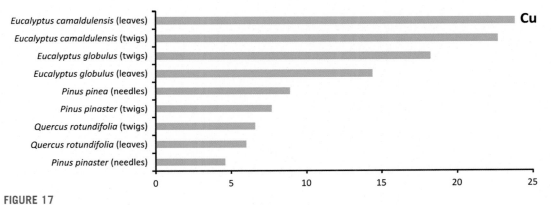

FIGURE 17

Accumulation of Cu (mg/kg DW) in tree species of the São Domingos mine area.

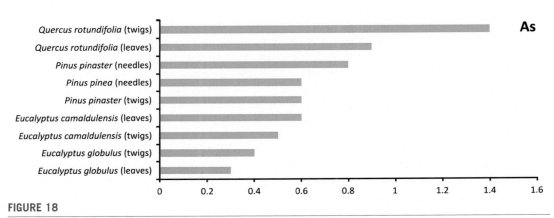

FIGURE 18

Accumulation of As (mg/kg DW) in tree species of the São Domingos mine area.

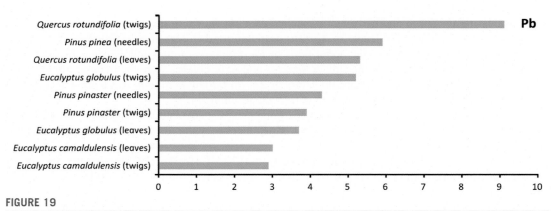

FIGURE 19

Accumulation of Pb (mg/kg DW) in tree species of the São Domingos mine area.

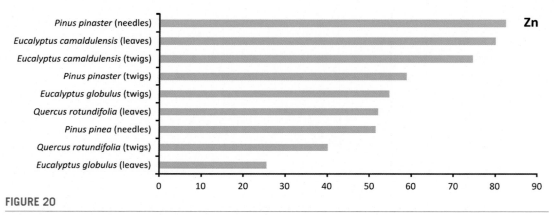

FIGURE 20

Accumulation of Zn (mg/kg DW) in tree species of the São Domingos mine area.

some herbaceous species, varying from 3.5 to 84.9 mg/kg DW (Freitas et al., 2004a). Semi-aquatic species sampled in the mining area, *J. conglomeratus* and *S. holoschoenus*, showed high accumulation of Pb in plant tissues. Lead above 20 mg/kg DW was found in leaves of two species of *Cistus*, typical Mediterranean shrubs known for their tolerance to drought and low nutrient availability (Freitas et al., 2004a).

Zinc concentrations in tree tissues are also within the normal range for plants (Kabata-Pendias, 2010) (Figure 20). Like other metals, the highest Zn concentrations were found in the herbaceous species *Cistus monspeliensis* and *Daphne gnidium* (Freitas et al., 2004a).

A few trees, *Eucalyptus*, *Quercus*, and *Pinus* species, were found in the contaminated area showing modest accumulation of different metal(loid)s in the above-ground tissues. However, due to their high biomass, they can be very effective for metal(loid)s phytoextraction and phytostabilization especially when established in the less contaminated soils on the peripheral zone of the study area (Freitas et al., 2004a).

In the tree species of Borralhal mine area, Cu was preferentially accumulated in the *Alnus glutinosa* tissues (both leaves and twigs) and in the twigs of *P. pinaster* (Figure 21). Copper concentrations reached 52.8 mg/kg in leaves of *A. glutinosa*. Arsenic concentrations in tree tissues ranged from

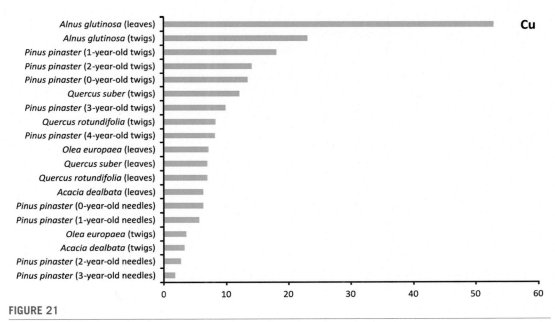

FIGURE 21

Accumulation of Cu (mg/kg DW) in tree species of the Borralhal mine area.

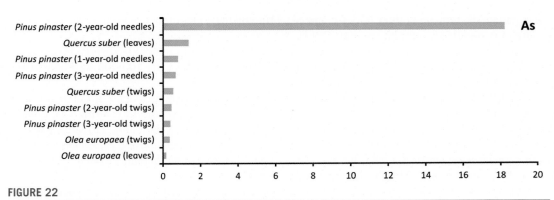

FIGURE 22

Accumulation of As (mg/kg DW) in tree species of the Borralhal mine area.

0.2 to 18 mg/kg DW (Figure 22). In fact, high accumulation of As was present in 2-year-old needles of *P. pinaster*. Lead concentrations in tree tissues ranged from 0.9 to 12 mg/kg DW, being higher in *P. pinaster* 3-year-old needles and 2-year-old twigs (Figure 23).

Samples from 7 tree species of the native flora of the Barbadalhos mine area were investigated, along with 18 types of shrubs, 17 herbs, 4 grasses, and 3 ferns.

Individual elements and species displayed different trends of accumulation. All plants collected along mineralized zone accumulated eight metals (Ag, Co, Cr, Cu, Fe, Ni, Pb, and Zn) but many plants from nonmineralized zone accumulated only five metals (Ag, Cu, Fe, Pb, and Zn). A few, however, did

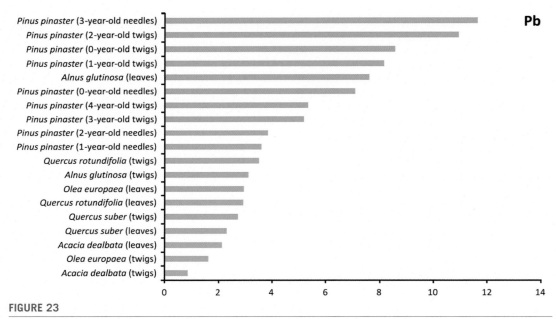

FIGURE 23

Accumulation of Pb (mg/kg DW) in tree species of the Borralhal mine area.

accumulate the remaining three (Co, Cr, and Ni), bringing the count of metals accumulated at par with those of mineralized zone (Pratas et al., 2013).

Most plants were seen to be tolerant to soil Pb concentrations. In mineralized zone, Pb concentrations in plants ranged from 1.11 to 548 mg/kg DW. This is far above the 100-400 mg Pb/kg content considered toxic for most plants (Alloway, 1990).

Significant accumulation of Pb was seen in these shrub and herbaceous species, listed in decreasing order (Pratas et al., 2013) (data not shown in this chapter): *Cistus salvifolius* (548 mg/kg), *Lonicera periclymenum* (318 mg/kg), *Anarrhinum bellidifolium*, *Phytolacca americana*, *Digitalis purpurea*, *Mentha suaveolens* (255-217 mg/kg). Pteridophytes like *Polystichum setiferum*, *Pteridium aquilinum*, and *Asplenium onopteris* also showed 117-251 mg/kg Pb in aerial parts. In plants from nonmineralized zone, Pb content was not significant, ranging from 0.94 to 11.6 mg/kg.

Though at first glance maximum Pb content (Figure 24) observed in trees like *Acacia dealbata* (84 mg/kg in leaves; 56.5 mg/kg in twigs), *Olea europaea* (62 mg/kg in twigs; 58 mg/kg in leaves), and *Quercus suber* (57.5 mg/kg in twigs) from mineralized zone is not very impressive compared to that of smaller plants mentioned above; nevertheless, these trees can be very effective due to their higher biomass. When combined with the hardy nature, biomass, and abundance of this species, the moderate accumulation indicates immense potential for phytoextraction of Pb in the area (Pratas et al., 2013).

In the future, however, any use of the *Acacia* species in revegetation actions should be avoided. These species are very aggressive invasive plants in the Portuguese ecosystems, with negative impacts on native biodiversity and community structure. Therefore, these species have been subject to containment actions and even to eradication actions.

Zinc concentrations in trees reached 140 mg/kg in leaves of *Q. suber* (Figure 25). However, concentrations reached 1020 mg/kg in *D. purpurea* and ranged from 262 to 887 mg/kg in *L. periclymenum*,

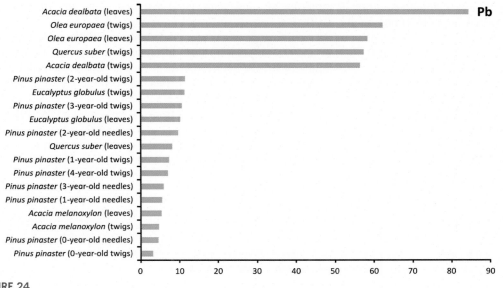

FIGURE 24

Accumulation of Pb (mg/kg DW) in tree species of the Barbadalhos mine area.

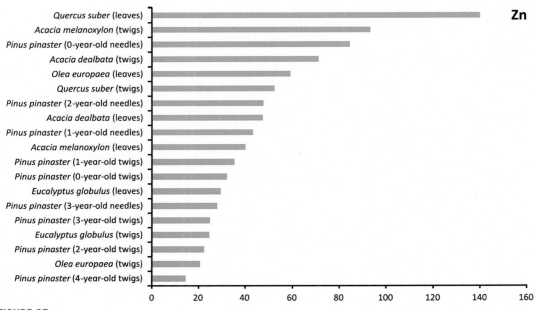

FIGURE 25

Accumulation of Zn (mg/kg DW) in tree species of the Barbadalhos mine area.

P. americana, Solanum nigrum, P. setiferum, M. suaveolens, Viola riviniana, and *A. bellidifolium,* listed in decreasing order (Pratas et al., 2013).

Metal budget within a plant under the influence of a contaminated rhizosphere may also vary based on the organ analyzed. In *P. pinaster* and *O. europaea* no such difference was observed for any metal. This was the case for most metals in the other five trees too, except for Fe, which was better concentrated in the leaves except in *E. globulus* and *A. melanoxylon.* Lead concentrations were about five times higher in *Q. suber* stems from mineralized zone (Pratas et al., 2013).

In the Sanheiro mine area, Pb concentrations in trees ranged from 2.9 to 37 mg/kg DW (Figure 26). Lead concentrations are pronounced in the species *E. globulus, Q. robur,* and *Q. suber,* reaching values higher than normal levels in plants (Kabata-Pendias, 2010). Zinc concentrations in tree tissues are within the normal range for plants (Kabata-Pendias, 2010), being higher in the species *Q. robur* and *P. pinaster* (Figure 27).

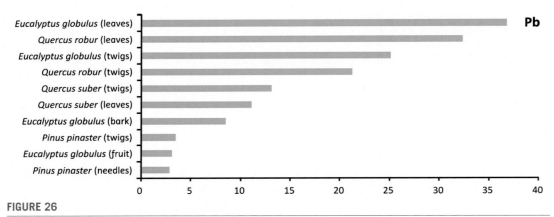

FIGURE 26

Accumulation of Pb (mg/kg DW) in tree species of the Sanheiro mine area.

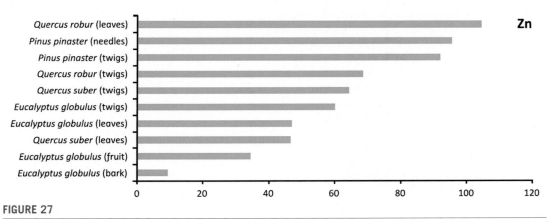

FIGURE 27

Accumulation of Zn (mg/kg DW) in tree species of the Sanheiro mine area.

6 CONCLUSIONS

Tree species growing on tailings and mine-contaminated soils show tolerance to imposed stress conditions (metal contamination and nutrient deficiency) and can fulfill the objectives of stabilization, pollution attenuation, and visual improvement.

A few trees, *Pinus*, *Eucalyptus*, *Quercus*, *Acacia*, *Alnus*, and *Olea* species, found in tailings and contaminated soils in abandoned mines of Portugal, show accumulation of different metals in the above-ground tissues. Hence these trees, combined with spontaneous flora of the herbaceous and arbustive strata, forming communities tolerant to toxic trace elements, play a major role in remediation of degraded mine soils. The species of the genera *Pinus*, *Eucalyptus*, and *Quercus* should be highlighted. These species are widely distributed throughout the territory and are very important from an economic point of view. Furthermore, these trees could grow and propagate in substrata with low nutrient conditions, which would be a great advantage in the revegetation of mine tailings.

The physicochemical properties of the metalliferous or metal-contaminated soils tend to inhibit soil-forming processes and plant growth. In addition to elevated metal(loid) concentrations, other adverse factors include absence of topsoil, erosion, drought, compaction, wide temperature fluctuations, absence of soil-forming fine materials, and shortage of essential nutrients (Pratas et al., 2005; Wong, 2003). Degraded soils of mines usually have low concentrations of important nutrients, like N, P, and K (Huenneke et al., 1990). Toxic metals can also adversely affect the number, diversity, and activity of soil organisms, inhibiting soil organic matter decomposition and N mineralization processes. The chemical form of the potential toxic metal, the presence of other chemicals that may aggravate or attenuate metal toxicity, the prevailing pH, and the poor nutrient status of contaminated soil affect the way in which plants respond to it. Substrate pH affects plant growth mainly through its effect on the solubility of chemicals, including toxic metals and nutrients.

It was also observed that despite lower accumulation, when compared to herbaceous and shrub species, trees can be very effective for metals phytoextraction and phytostabilization due to their higher biomass and bioproductivity (Freitas et al., 2004a; Pratas et al., 2013; Liu et al., 2013), especially when established in the less contaminated soils on the peripheral zone of the studied areas. Nevertheless, the role of the small and herbaceous plants and shrubs cannot be underestimated as their rapid growth rate, shorter life span, and fast successive generations can hasten the process of soil reclamation by providing repeated short-duration rapid phases or bursts of remediation. The timely removal of the plants' and shrubs' biomass from the site can make life easier for these candidate trees as the former have already played their part, and their timely disposal prevents the remediated metals from going back into the soil.

Metal toxicity issues do not generally arise in the case of native flora, considering that native plants become adapted over time to the locally elevated metal levels (Varun et al., 2012; D'Souza et al., 2013). Native plants may be better phytoremediators for contaminated lands than the known metal hyperaccumulators because these are generally slow growing with shallow root systems and low biomass. Plants tolerant to toxic metals and low nutrient status with a high rate of growth and biomass are the ideal species to remediate degraded soils and habitats like those around mines. The native flora displayed its ability to withstand high concentrations of heavy metals in the soil. Some species also displayed variable accumulation patterns for metals at different soil concentrations. This variation was also observed in different parts of the same plant, suggesting that full consideration of plant-soil interactions should be taken into account when choosing plant species for developing and utilizing methods such as phytoremediation.

The existing natural plant cover at abandoned mining sites can be increased manifold by wide-scale planting and maintenance of native species with higher metal accumulation potential for some years. Even dispersal of seeds obtained from plants on site is to be encouraged. Adding organic amendment is essential to facilitate the establishment and colonization of these "pioneer plants." They can eventually modify the man-made habitat and render it more suitable for subsequent plant communities. Allowing native species to remediate soils is an attractive proposition since native wild species do not require frequent irrigation, fertilization, and pesticide treatments, while simultaneously a plant community comparable to that existing in the vicinity can be established.

Therefore, mine restoration could benefit from a broader perspective including different groups of plant species as they can perform distinct functional roles in the remediation process. The use of leguminous plants, for example, may enrich the nutrient content, and the combined use of perennials and annuals can provide substantial inputs in terms of organic matter and nutrient recycling, thus contributing in distinct ways to the development of the soil (Freitas et al., 2004a; Hooper and Vitousek, 1997). This approach requires more information about plant communities growing on metal-contaminated soils in order to accurately determine their potential for remediation of polluted soils at abandoned mines. Ideal phytoremedial candidates can be screened out from the native flora, and after assessing their individual requirements, suitable conditions/amendments can be created to develop them as good competitors with enhanced growth and proliferation to their counterparts growing on the same metal-contaminated nutrient-depleted soils.

The spontaneous tree species of the studied mining areas display their ability to withstand high concentrations of heavy metal(loid)s in the soil. However, accumulation patterns of metal(loid)s in the plants tested varied. As metal concentrations in above-ground parts were maintained at low levels, metal tolerance in most cases may mainly depend on their metal-excluding ability. However, metal(loid) concentrations higher than toxic level in some species, indicate that internal detoxification, metal-tolerance mechanisms might also exist. Furthermore, the plants could grow and propagate in substrata with low nutrient conditions, which would be a great advantage in the revegetation of mine tailings. It was also observed that despite lower accumulation, trees of the studied regions can be very effective due to their higher biomass.

Therefore, based on the results, the aforementioned tree species can be useful for the purpose of mine restoration and minimization of mining impacts. This is accomplished by the partial removal of the bioavailable toxic elements, and/or by their retention to minimize the output of the toxic elements to the ecosystem. Especially when the metal concentrations in above-ground parts were maintained at low levels, the biomass can be used to obtain economic value. As examples, we can refer to the use of biomass as a renewable resource for energy production (bioenergy), or the timber extraction for wooden furniture and other manufactured wood products, or sawn wood and wood-processed products (e.g., particle board, fiberboard, and other panels) for carpentry and construction.

REFERENCES

Abreu, M.M., Tavares, M.T., Batista, M.J., 2008. Potential use of *Erica andevalensis* and *Erica australis* in phytoremediation of sulphide mine environments: São Domingos, Portugal. J. Geochem. Explor. 96, 210–222.

Abreu, M.M., Santos, E.S., Ferreira, M., Magalhães, M.C.F., 2012. *Cistus salviifolius* a promising species for mine wastes remediation. J. Geochem. Explor. 113, 86–93.

Ali, H., Khan, E., Sajad, M.A., 2013. Phytoremediation of heavy metals – concepts and applications. Chemosphere 91, 869–881.

Alloway, B.J., 1990. Heavy Metals in Soils. Blackie Academic & Professional, London.

Alvarenga, P.M., Araújo, M.F., Silva, J.A.L., 2004. Elemental uptake and root-leaves transfer in *Cistus ladanifer* L. growing in a contaminated pyrite mining area (Aljustrel-Portugal). Water Air Soil Pollut. 162, 81–96.

Anawar, H.M., Freitas, M.C., Canha, N., Santa Regina, I., 2011. Arsenic, antimony, and other trace element contamination in a mine tailings affected area and uptake by tolerant plant species. Environ. Geochem. Health 33, 353–362.

Antosiewicz, D.M., Escudě-Duran, C., Wierzbowska, E., Skłodowska, A., 2008. Indigenous plant species with the potential for the phytoremediation of arsenic and metals contaminated soil. Water Air Soil Pollut. 193, 197–210.

Bani, A., Echevarria, G., Sulçe, S., Morel, J.L., Mullai, A., 2007. In-situ phytoextraction of Ni by a native population of *Alyssum murale* on an ultramafic site (Albania). Plant Soil 293, 79–89.

Bani, A., Echevarria, G., Sulçe, S., Morel, J.L., 2015. Improving the agronomy of *Alyssum murale* for extensive phytomining: a five-year field study. Int. J. Phytoremediation 17, 117–127.

Berti, W.R., Cunningham, S.D., 2000. Phytostabilization of metals. In: Raskin, I., Ensley, B.D. (Eds.), Phytoremediation of Toxic Metals. Using Plants to Clean Up the Environment. John Wiley & Sons, Inc, New York, pp. 71–88.

Blaylock, M.J., Huang, J.W., 2000. Phytoextraction of metals. In: Raskin, I., Ensley, B.D. (Eds.), Phytoremediation of Toxic Metals. Using Plants to Clean Up the Environment. John Wiley & Sons, Inc, New York, pp. 53–70.

Branquinho, C., Serrano, H.C., Pinto, M.J., Martins-Loução, M.A., 2007. Revisiting the plant hyperaccumulation criteria to rare plants and earth abundant elements. Environ. Pollut. 146, 437–443.

Brooks, R., 1983. Biological Methods of Prospecting for Minerals. Wiley, New York.

Brooks, R.R., Chambers, M.F., Nicks, L.J., Robinson, B.H., 1998a. Phytomining. Trends Plant Sci. 3, 359–362.

Brooks, R.R., Chiarucci, A., Jaffré, T., 1998b. Revegetation and stabilization of mine dumps and other degraded terrain. In: Brooks, R.R. (Ed.), Plants that Hyperaccumulate Heavy Metals: Their Role in Phytoremediation, Microbiology, Archaeology, Mineral Exploration and Phytomining. CAB International, New York, pp. 227–247.

Canadian Council of Ministers for the Environment, 2006. Canadian Soil Quality Guidelines for the Protection of Environmental and Human Health. update 6.02, Publication no. 1299.

Chaney, R.L., Malik, M., Li, Y.M., Brown, S.L., Angle, J.S., Baker, A.J.M., 1997. Phytoremediation of soil metals. Curr. Opin. Biotechnol. 8, 279–284.

Chockalingam, E., Subramanian, S., 2009. Utility of *Eucalyptus tereticornis* (Smith) bark and *Desulfotomaculum nigrificans* for the remediation of acid mine drainage. Bioresour. Technol. 100, 615–621.

Coates, W., 2005. Tree species selection for a mine tailings bioremediation project in Peru. Biomass Bioenergy 28, 418–423.

D'Souza, R., Varun, M., Masih, J., Paul, M.S., 2010. Identification of *Calotropis procera* L. as a potential phytoaccumulator of heavy metals from contaminated soils in Urban North Central India. J. Hazard. Mater. 184, 457–464.

D'Souza, R., Varun, M., Pratas, J., Paul, M.S., 2013. Spatial distribution of heavy metals in soil and flora associated with the glass industry in North Central India: implications for phytoremediation. Soil Sediment Contam. 22, 1–20.

De Koe, T., Beek, M., Haarsma, M., Ernst, W., 1991. Heavy metals and arsenic grasses and soils of mine spoils in North East Portugal, with particular reference to some Portuguese goldmines. In: Nath, B. (Ed.), Environmental Pollution. Proceedings of the International Conference on Environmental Pollution, ICEP-1, vol. 1, pp. 373–380.

Dickinson, N.M., Baker, A.J.M., Doronila, A., Laidlaw, S., Reeves, R.D., 2009. Phytoremediation of inorganics: realism and synergies. Int. J. Phytoremediation 11, 97–114.

Díez Lázaro, J., Kidd, P.S., Monterroso Martínez, C., 2006. A phytogeochemical study of the Trás-os-Montes region (NE Portugal): possible species for plant-based soil remediation technologies. Sci. Total Environ. 354, 265–277.

Domínguez, M.T., Madrid, F., Marañón, T., Murillo, J.M., 2009. Cadmium availability in soil and retention in oak roots: potential for phytostabilization. Chemosphere 76, 480–486.

Ensley, B.D., 2000. Rationale for use of phytoremediation. In: Raskin, I., Ensley, B.D. (Eds.), Phytoremediation of Toxic Metals. Using Plants to Clean Up the Environment. John Wiley & Sons, Inc, New York, pp. 3–11.

Ernst, W., 1990. Mine vegetation in Europe. In: Shaw, A.J. (Ed.), Heavy Metal Tolerance in Plants: Evolutionary Aspects. CRC Press, Boca Raton, Florida, pp. 21–37.

Evangelou, M.W.H., Papazoglou, E.G., Robinson, B.H., Schulin, R., 2015. Phytomanagement: phytoremediation and the production of biomass for economic revenue on contaminated land. In: Ansari, A.A., Gill, S.S., Gill, R., Lanza, G.R., Newman, L. (Eds.), Phytoremediation: Management of Environmental Contaminants, vol. 1. Springer, Cham, Heidelberg, New York, Dordrecht, London, pp. 115–132.

Evanko, C.R., Dzombak, D.A., 1997. Remediation of metals-contaminated soils and groundwater. Technology evaluation report. GWRTAC – Ground-Water Remediation Technologies Analysis Center, Pittsburgh.

Favas, P.J.C., Pratas, J.A.M.S., Conde, L.E.N., 2002. Bioaccumulation of heavy metals in plants that colonize tailings (Vale das Gatas mine-Vila Real-North of Portugal). In: Prego, R., Duarte, A., Panteleitchouk, A., Santos, T.R. (Eds.), Studies on Environmental Contamination in Iberian Peninsula. Instituto Piaget, Lisbon, pp. 259–272 (in Portuguese).

Favas, P.J.C., Pratas, J., Prasad, M.N.V., 2013. Temporal variation in the arsenic and metal accumulation in the maritime pine tree grown on contaminated soils. Int. J. Environ. Sci. Technol. 10, 809–826.

Favas, P.J.C., Pratas, J., Varun, M., D'Souza, R., Paul, M.S., 2014. Phytoremediation of soils contaminated with metals and metalloids at mining areas: potential of native flora. In: Hernández-Soriano, M.C. (Ed.), Environmental Risk Assessment of Soil Contamination. Intech, Rijeka, pp. 485–517.

Fonseca, E.C., Martin, H., 1986. The selective extraction of Pb and Zn in selected mineral and soil samples, application in geochemical exploration (Portugal). J. Geochem. Explor. 26, 231–248.

Freitas, H., Prasad, M.N.V., Pratas, J., 2004a. Plant community tolerant to trace elements growing on the degraded soils of São Domingos mine in the south east of Portugal: environmental implications. Environ. Int. 30, 65–72.

Freitas, H., Prasad, M.N.V., Pratas, J., 2004b. Analysis of serpentinophytes from north-east of Portugal for trace metal accumulation – relevance to the management of mine environment. Chemosphere 54, 1625–1642.

Gaspar, O.C., 1998. História da mineração dos depósitos de sulfuretos maciços vulcanogénicos da Faixa Piritosa Portuguesa. Bol. Minas 35, 401–414.

Glass, D.J., 2000. Economic potential of phytoremediation. In: Raskin, I., Ensley, B.D. (Eds.), Phytoremediation of Toxic Metals. Using Plants to Clean Up the Environment. John Wiley & Sons, Inc, New York, pp. 15–31.

Gomes, P., Valente, T., Pamplona, J., Braga, M.A.S., Pissarra, J., Gil, J.A.G., Torre, N.L., 2014. Metal uptake by native plants and revegetation potential of mining sulfide-rich waste-dumps. Int. J. Phytoremediation 16, 1087–1103.

Gómez Orea, D., 2004. Recuperación de Espacios Degradados. Ediciones Mundi-Prensa, Madrid, Barcelona, México.

Henriques, F.S., Fernandes, J.C., 1991. Metal uptake and distribution in rush (*Juncus conglomeratus* L.) plants growing in pyrite mine tailings at Lousal, Portugal. Sci. Total Environ. 102, 253–260.

Hernández-Allica, J., Becerril, J.M., Garbisu, C., 2008. Assessment of the phytoextraction potential of high biomass crop plants. Environ. Pollut. 152, 32–40.

Hooper, D.U., Vitousek, P.M., 1997. The effects of plant composition and diversity on ecosystem processes. Science 277, 1302–1305.

Huenneke, L.F., Hamburg, S.P., Koide, R., Mooney, H.A., Vitousek, P.M., 1990. Effects of soil resources on plant invasion and community structure in Californian serpentine grassland. Ecology 71, 478–491.

I.C.N.F., 2013. IFN6 – Áreas dos Usos do Solo e das Espécies Florestais de Portugal Continental. Resultados Preliminares. Instituto da Conservação da Natureza e das Florestas, Lisboa.

Jorba, M., Vallejo, R., 2008. La restauración ecológica de canteras: un caso con aplicación de enmiendas orgánicas y riegos. Ecosistemas 17, 119–132.

Kabata-Pendias, A., 2010. Trace Elements in Soils and Plants, fourth ed. CRC Press, Boca Raton.

Kidd, P.S., Monterroso, C., 2005. Metal extraction by *Alyssum serpyllifolium* ssp. *lusitanicum* on mine-spoil soils from Spain. Sci. Total Environ. 336, 1–11.

Kidd, P., Mench, M., Álvarez-López, V., Bert, V., Dimitriou, I., Friesl-Hanl, W., Herzig, R., Janssen, J.O., Kolbas, A., Müller, I., Neu, S., Renella, G., Ruttens, A., Vangronsveld, J., Puschenreiter, M., 2015. Agronomic practices for improving gentle remediation of trace element-contaminated soils. Int. J. Phytoremediation. http://dx.doi.org/10.1080/15226514.2014.1003788.

Li, Y.-M., Chaney, R., Brewer, E., Roseberg, R., Angle, J.S., Baker, A., Reeves, R., Nelkin, J., 2003. Development of a technology for commercial phytoextraction of nickel: economic and technical considerations. Plant Soil 249, 107–115.

Liang, H.M., Lin, T.H., Chiou, J.M., Yeh, K.C., 2009. Model evaluation of the phytoextraction potential of heavy metal hyperaccumulators and non-hyperaccumulators. Environ. Pollut. 157, 1945–1952.

Liu, W., Ni, J., Zhou, Q., 2013. Uptake of heavy metals by trees: prospects for phytoremediation. Mater. Sci. Forum 743–744, 768–781.

Lopes, A.M., Cunha-e-Sá, M.A., 2014. The economic value of Portuguese forests – the effect of tree species on valuation of forest ecosystems. In: VI Congress of the Spanish-Portuguese Association of Resource and Environmental Economics (AERNA), Girona, Catalonia, Spain, pp. 29. https://editorialexpress.com/cgi-bin/conference/download.cgi?db_name=AERNA2014&paper_id=65. Accessed on January 2015.

Lorestani, B., Cheraghi, M., Yousefi, N., 2011. Phytoremediation potential of native plants growing on a heavy metals contaminated soil of copper mine in Iran. World Acad. Sci. Eng. Technol. 53, 377–382.

Louro, G., Monteiro, M., Constantino, L., Rego, F., 2014. The Portuguese forest based chains: sector analyses. In: Reboredo, F. (Ed.), Forest Context and Policies in Portugal. Present and Future Challenges. World Forests, vol. 19. Springer, Cham, Heidelberg, New York, Dordrecht, London, pp. 39–65.

Ma, L.Q., Komar, K.M., Tu, C., Zhang, W., Cai, Y., Kennelley, E.D., 2001. A fern that hyperaccumulates arsenic. Nature 409, 579.

Madejón, P., Murillo, J.M., Marañon, T., Cabrera, F., López, R., 2002. Bioaccumulation of As, Cd, Cu, Fe and Pb in wild grasses affected by the Aznalcóllar mine spill (SW Spain). Sci. Total Environ. 290, 105–120.

Mains, D., Craw, D., Rufaut, C., Smith, C., 2006. Phytostabilization of gold mine tailings from New Zealand. Part 2: experimental evaluation of arsenic mobilization during revegetation. Int. J. Phytoremediation 8, 163–183.

Maric, M., Antonijevic, M., Alagic, S., 2013. The investigation of the possibility for using some wild and cultivated plants as hyperaccumulators of heavy metals from contaminated soil. Environ. Sci. Pollut. Res. 20, 1181–1188.

Marques, A.P.G.C., Moreira, H., Rangel, A.O.S.S., Castro, P.M.L., 2009. Arsenic, lead and nickel accumulation in *Rubus ulmifolius* growing in contaminated soil in Portugal. J. Hazard. Mater. 165, 174–179.

McGrath, S.P., 1998. Phytoextraction for soil remediation. In: Brooks, R.R. (Ed.), Plants that Hyperaccumulate Heavy Metals: Their Role in Phytoremediation, Microbiology, Archaeology, Mineral Exploration and Phytomining. CAB International, New York, pp. 261–287.

McGrath, S.P., Zhao, F.J., 2003. Phytoextraction of metals and metalloids from contaminated soils. Curr. Opin. Biotechnol. 14, 277–282.

Mendez, M.O., Maier, R.M., 2008. Phytoremediation of mine tailings in temperate and arid environments. Rev. Environ. Sci. Biotechnol. 7, 47–59.

Morais, I., Campos, J.S., Favas, P.J.C., Pratas, J., Pita, F., Prasad, M.N.V., 2015. Nickel accumulation by *Alyssum serpyllifolium* subsp. *lusitanicum* (Brassicaceae) from serpentine soils of Bragança and Morais (Portugal) ultramafic massifs: plant-soil relationships and prospects for phytomining. Aust. J. Bot. 63 (2), 17–30.

Nascimento, C.W.A., Xing, B., 2006. Phytoextraction: a review on enhanced metal availability and plant accumulation. Sci. Agric. 63, 299–311.

Oláh, P., 1988. Biological Reclamation Process. Canadian Reclamation 3, Canadian Land Reclamation Association, Guelph, Ontario, Canada.

Özdemir, Z., 2005. *Pinus brutia* as a biogeochemical medium to detect iron and zinc in soil analysis, chromite deposits of the area Mersin, Turkey. Chem. Erde-Geochem. 65, 79–88.

Parraga-Aguado, I., Querejeta, J.-I., González-Alcaraz, M.-N., Jiménez-Cárceles, F.J., Conesa, H.M., 2014. Usefulness of pioneer vegetation for the phytomanagement of metal(loid)s enriched tailings: grasses vs. shrubs vs. trees. J. Environ. Manag. 133, 51–58.

Pedron, F., Petruzzelli, G., Barbafieri, M., Tassi, E., 2009. Strategies to use phytoextraction in very acidic soil contaminated by heavy metals. Chemosphere 75, 808–814.

Pierzynski, G., Kulakow, P., Erickson, L., Jackson, L., 2002. Plant system technologies for environmental management of metals in soils: educational materials. J. Nat. Resour. Life Sci. Educ. 31, 31–37.

Poschenrieder, C., Bech, J., Llugany, M., Pace, A., Fenes, E., Barceló, J., 2001. Copper in plant species in a copper gradient in Catalonia (North East Spain) and their potential for phytoremediation. Plant Soil 230, 247–256.

Prasad, M.N.V., 2004. Phytoremediation of metals and radionuclides in the environment: the case for natural hyperaccumulators, metal transporters, soil-amending chelators and transgenic plants. In: Prasad, M.N.V. (Ed.), Heavy Metal Stress in Plants: From Biomolecules to Ecosystems, second ed. Springer, Berlin, pp. 345–391.

Prasad, M.N.V., 2015. Engineered phyto-covers as natural caps for containment of hazardous mine and municipal solid waste dump sites – possible energy sources. In: Öztürk, M., Ashraf, M., Aksoy, A., Ahmad, M.S.A. (Eds.), Phytoremediation for Green Energy. Springer, Dordrecht, Heidelberg, New York, London, pp. 55–68.

Prasad, M.N.V., Freitas, H., 2000. Removal of toxic metals from solution by leaf, stem and root phytomass of *Quercus ilex* L. (holly oak). Environ. Pollut. 110, 277–283.

Prasad, M.N.V., Freitas, H.M.O., 2003. Metal hyperaccumulation in plants – biodiversity prospecting for phytoremediation technology. Electron. J. Biotechnol. 6, 285–321.

Pratas, J.A.M.S., 1996. Biogeochemical Prospecting Applications. Selection of bioindicator species in some mining areas of Portugal. PhD Thesis, University of Coimbra (in Portuguese).

Pratas, J., Prasad, M.N.V., Freitas, H., Conde, L., 2005. Plants growing in abandoned mines of Portugal are useful for biogeochemical exploration of arsenic, antimony, tungsten and mine reclamation. J. Geochem. Explor. 85, 99–107.

Pratas, J., Favas, P.J.C., Conde, L., 2012. Metalotolerant plant species and their potential in biogeochemical prospecting and environmental restoration (Fragas do Cavalo old mine, Oleiros, Central Portugal). In: Henriques, M.H., Andrade, A.I., Lopes, F.C., Pena dos Reis, R., Quinta-Ferreira, M., Barata, M.T. (Eds.), To Develop the Earth. Imprensa da Universidade de Coimbra, Coimbra, pp. 301–310 (in Portuguese).

Pratas, J., Favas, P.J.C., D'Souza, R., Varun, M., Paul, M.S., 2013. Phytoremedial assessment of flora tolerant to heavy metals in the contaminated soils of an abandoned Pb mine in Central Portugal. Chemosphere 90, 2216–2225.

Robinson, B.H., Chiarucci, A., Brooks, R.R., Petit, D., Kirkman, J.H., Gregg, P.E.H., DeDominicis, V., 1997. The nickel hyperaccumulator plant *Alyssum bertolonii* as a potential agent for phytoremediation and phytomining of nickel. J. Geochem. Explor. 59, 75–86.

Rockwood, D.L., Carter, D.R., Langholtz, M.H., Stricker, J.A., 2006. *Eucalyptus* and *Populus* short rotation woody crops for phosphate mined lands in Florida USA. Biomass Bioenergy 30, 728–734.

Rodríguez, N., Amils, R., Jiménez-Ballesta, R., Rufo, L., De la Fuente, V., 2007. Heavy metal content in *Erica andevalensis*: an endemic plant from the extreme acidic environment of Tinto River and its soils. Arid Land Res. Manag. 21, 51–65.

Saifullah, Meers, E., Qadir, M., Caritat, P., Tack, F.M.G., Laing, G.D., Zia, M.H., 2009. EDTA-assisted Pb phytoextraction. Chemosphere 74, 1279–1291.

Sakakibara, M., Ohmori, Y., Ha, N.T.H., Sano, S., Sera, K., 2011. Phytoremediation of heavy metal contaminated water and sediment by *Eleocharis acicularis*. Clean: Soil, Air, Water 39, 735–741.

Saqib, A.N.S., Waseem, A., Khan, A.F., Mahmood, Q., Khan, A., Habib, A., Khan, A.R., 2013. Arsenic bioremediation by low cost materials derived from blue pine (*Pinus wallichiana*) and walnut (*Juglans regia*). Ecol. Eng. 51, 88–94.

Saraswat, S., Rai, J.P.N., 2009. Phytoextraction potential of six plant species grown in multimetal contaminated soil. Chem. Ecol. 25, 1–11.

Shabani, N., Sayadi, M.H., 2012. Evaluation of heavy metals accumulation by two emergent macrophytes from the polluted soil: an experimental study. Environmentalist 32, 91–98.

Sheoran, V., Sheoran, A.S., Poonia, P., 2009. Phytomining: a review. Miner. Eng. 22, 1007–1019.

Sousa, N.R., Ramos, M.A., Marques, A.P.G.C., Castro, P.M.L., 2014. A genotype dependent-response to cadmium contamination in soil is displayed by *Pinus pinaster* in symbiosis with different mycorrhizal fungi. Appl. Soil Ecol. 76, 7–13.

Sun, F.F., Wen, D.Z., Kuang, Y.W., Li, J., Zhang, J.G., 2009. Concentrations of sulphur and heavy metals in needles and rooting soils of Masson pine (*Pinus massoniana* L.) trees growing along an urban–rural gradient in Guangzhou. China. Environ. Monit. Assess. 154, 263–274.

Tu, S., Ma, L.Q., 2004. Comparison of arsenic and phosphate uptake and distribution in arsenic hyperaccumulating and nonhyperaccumulating fern. J. Plant Nutr. 27, 1227–1242.

Turnau, K., Henriques, F.S., Anielska, T., Renker, C., Buscot, F., 2007. Metal uptake and detoxification mechanisms in *Erica andevalensis* growing in a pyrite mine tailing. Environ. Exp. Bot. 61, 117–123.

Valente, T., Gomes, P., Pamplona, J., Torre, M.L., 2012. Natural stabilization of mine waste-dumps – evolution of the vegetation cover in distinctive geochemical and mineralogical environments. J. Geochem. Explor. 123, 152–161.

Van der Ent, A., Baker, A.J.M., Reeves, R.D., Pollard, A.J., Schat, H., 2013. Hyperaccumulators of metal and metalloid trace elements: facts and fiction. Plant Soil 362, 319–334.

Varun, M., D'Souza, R., Pratas, J., Paul, M.S., 2012. Metal contamination of soils and plants associated with the glass industry in North-Central India: prospects of phytoremediation. Environ. Sci. Pollut. Res. 19, 269–281.

Varun, M., D'Souza, R., Favas, P.J.C., Pratas, J., Paul, M.S., 2015. Utilization and supplementation of phytoextraction potential of some terrestrial plants in metal-contaminated soils. In: Ansari, A.A., Gill, S.S., Gill, R., Lanza, G.R., Newman, L. (Eds.), Phytoremediation: Management of Environmental Contaminants, vol. 1, Springer, Cham, Heidelberg, New York, Dordrecht, London, pp. 177–200.

Vinod, V.T.P., Sashidhar, R.B., Sivaprasad, N., Sarma, V.U.M., Satyanarayana, N., Kumaresan, R., Rao, T.N., Raviprasad, P., 2011. Bioremediation of mercury (II) from aqueous solution by gum karaya (*Sterculia urens*): a natural hydrocolloid. Desalination 272, 270–277.

Williamson, N.A., Johnson, M.S., Bradshaw, A.D., 1982. Mine Wastes Reclamation. The Establishment of Vegetation on Metal Mine Wastes. Mining Journal Books, London.

Wójcik, M., Sugier, P., Siebielec, G., 2014. Metal accumulation strategies in plants spontaneously inhabiting Zn-Pb waste deposits. Sci. Total Environ. 487, 313–322.

Wong, M.H., 2003. Ecological restoration of mine degraded soils with emphasis on metal contaminated soils. Chemosphere 50, 775–780.

Xie, Q.E., Yan, X.L., Liao, X.Y., Li, X., 2009. The arsenic hyperaccumulator fern *Pteris vittata* L. Environ. Sci. Technol. 43, 8488–8495.

Yanqun, Z., Yuan, L., Jianjun, C., Haiyan, C., Li, Q., Schvartz, C., 2005. Hyperaccumulation of Pb, Zn and Cd in herbaceous grown on lead-zinc mining area in Yunnan, China. Environ. Int. 31, 755–762.

Zhang, Z., Sugawara, K., Hatayama, M., Huang, Y., Inoue, C., 2014. Screening of As-accumulating plants using a foliar application and a native accumulation of As. Int. J. Phytoremediation 16, 257–266.

Zhuang, P., Yang, Q.W., Wang, H.B., Shu, W.S., 2007. Phytoextraction of heavy metals by eight plant species in the field. Water Air Soil Pollut. 184, 235–242.

RICE PADDIES FOR TRACE ELEMENT CLEANUP: BIOECONOMIC PERSPECTIVES

11

A. Sebastian, M.N.V. Prasad

University of Hyderabad, Hyderabad, Telangana, India

1 INTRODUCTION

The rice plant is a high-biomass-producing cereal. Even though this crop follows maize and sugar cane in worldwide cultivation, rice grain contributes more than one-fifth of the daily caloric intake of the human population (Bhullar and Gruissem, 2013). Cultivation of rice can be done in diverse climatic conditions and is cost effective (Felkner et al., 2009). Approximately 162.3 million hectares of land around the world are dedicated for rice cultivation, producing more than 738.1 million tons of rice grain. Asian countries such as China and India produce more than 50% of rice around the world. India, Thailand, and Vietnam are the chief countries involved in rice trade. These countries export more than 25 million tons of rice annually (Suwannaporn and Lineman, 2008). India exports mainly basmati grain whereas jasmine rice is the main export commodity from Thailand and Vietnam. Major rice-importing countries belong to Africa and the Persian Gulf. Crop productivity is limited in rice-exporting countries because of lack of advancement in farming technology. It has been said that avoidance of postharvest loss with proper infrastructure development in India alone could feed 100 million people for over a year (Basavaraja et al., 2007). But the advent of system intensification of rice has resulted in higher yield from rice paddies (Chowdhury et al., 2014).

Management and irrigation practices make rice cultivation unique. Rice cultivation may be either on wet or dry land. Field management practices in wetland cultivation include submergence, puddling, and addition of nutrients in the form of manure and fertilizers (Kögel-Knabner et al., 2010). These practices bring changes in solubility of mineral nutrients, organic matter content, and dissolved oxygen. Prevalence of anoxia in the rice paddies raises environmental issues such as emission of greenhouse gases, especially methane (Huang et al., 2004). Another matter of concern during rice cultivation is trace element (TEs) contamination. It is reported that rice grains harvested from Asian countries are contaminated with TEs such as Cd above the international standard (Simmons et al., 2005; Bian et al., 2014; Zhou et al., 2014). Hence toxic TEs contamination in paddy soil becomes a major topic of health concern. It was the mismanagement of mine wastewater release and usage of TE-contaminated fertilizer that caused toxic TEs contamination of paddy soils (Sebastian and Prasad, 2014a). The capacity of rice plants to accumulate toxic TEs has been found to vary, depending on the type of metal as well as physiochemical conditions of the soil. Metal accumulation capacity, habitat range, and germplasm collections adapted to varying climatic conditions makes the rice plant ideal for phytoremediation

Bioremediation and Bioeconomy. http://dx.doi.org/10.1016/B978-0-12-802830-8.00011-3
Copyright © 2016 Elsevier Inc. All rights reserved.

(Bosetti et al., 2011). Low labor cost compared with other cereal crops, well-developed mechanization for cultivation practices, ambient biomass productivity, and plentiful by-products from the rice industry help to formulate commercial bioremediation strategies using rice paddies.

2 RICE PLANT AS A PHYTOREMEDIATION CROP

Phytoremediation crops are characterized by substantial metal tolerance with high biomass yield (Rugh, 2004). Apart from these factors, extensive root systems that provide more capacity to absorb large amounts of water from the soil are also important in phytoremediation perspective. Even though many herbs and shrubs pose metal accumulation characteristics, they require enormous time to grow and require specific environmental conditions. Most of the plants can accumulate about 100 ppm zinc and 1 ppm Cd in shoots, whereas metal hyperaccumulator plants accumulate these metals up to 30,000 and 1500 ppm, respectively (Rascio and Navari-Izzo, 2011). On the other hand, metal accumulation characteristics can be classified into natural hyperaccumulation or assisted hyperaccumulation; in the latter case, accumulation is promoted with the help of external agents such as soil amendments or microbes (Rajkumar et al., 2012). Rice plants possess many of the beneficial aspects of the phytoremediation characteristics mentioned earlier that make this plant a potential phytoremediation crop (Figure 1).

One of the major drawbacks of modern rice cultivation for food is the accumulation of toxic TEs in rice grain (Sebastian and Prasad, 2014a). This provides clear evidence of rice plants' ability to accumulate a considerable amount of heavy metals. Selected studies on rice plants that indicate toxic TEs accumulation capacity are summarized in Table 1. Major features that make rice plants suitable for commercial bioremediation are their short life cycle, adaptability to grow well in diverse environmental conditions, minimum growth requirements, and cost effectiveness. Typically rice plants complete their life cycle in 3-4 months (Li and Cui, 2014). This means four harvests a year and removal of large quantity of biomass contaminated with metals from the field. For example, a rice variety that accumulates $10 \mu g\,g^{-1}$ tissue and produces minimum biomass of 15 tons ha^{-1} is able to remove 0.15 kg metal per hectare during harvest and approximately one-half kg of metal per hectare in a year, respectively. Typically 40-60 kg of

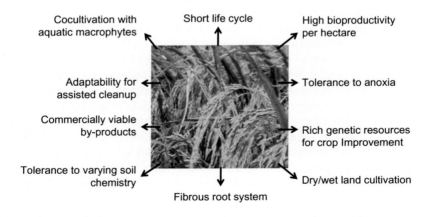

FIGURE 1

Features of rice plant as phytoremediation crop.

Table 1 Selected Studies on Toxic Trace Element Accumulation in Rice Plants

Metal	Type of Study/Duration	Metal Added	Accumulation	Reference
Cadmium	Field study/90 days	1.2 mg Cd kg⁻¹	0.29 mg kg⁻¹ rice	Wang et al. (2011)
	Hydrophonics/10 days	0.18 μM Cd L⁻¹	40 μg⁻¹ shoot	Uraguchi et al. (2009)
	Greenhouse/180 days	6.0 mg Cd dm⁻³	180 mg kg⁻¹ plant	Pereira et al. (2011)
	Pot culture/90 days	100 mg Cd kg⁻¹	2.86 mg kg⁻¹ rice	Liu et al. (2005)
	Pot culture/45 days	60 mg kg⁻¹	23.4 μg g⁻¹ root	Vijayarengan (2011)
Lead	Greenhouse/700 days	16 mg kg⁻¹	400 mg g⁻¹ root	Panichpat and Srinives (2009)
	Pot culture/40 days	2 g kg⁻¹	14 mg g⁻¹ root	Ma et al. (2012)
	Sand culture/104 days	1 mM kg⁻¹	70 μg g⁻¹ leaf	Chatterjee et al. (2004)
	Pot culture/60 days	300 mg kg⁻¹	35 mg g⁻¹ leaf	Yizong et al. (2009)
Chromium	Hydroponics/10 days	1 mg L⁻¹	5.7 mg kg⁻¹ leaf	Zong et al. (2010)
	Pot culture/grain filling	400 mg kg⁻¹	137 mg kg⁻¹ root	Zhu et al. (2010)
	Field study/harvest	6.22 μg g⁻¹	0.65 μg g⁻¹ rice	Schuhmacher et al. (1994)
	Field study/harvest	48 mg kg⁻¹	5.6 mg kg⁻¹ straw	Bhattacharyya et al. (2005)
Arsenic	Field study/harvest	3.30 mg kg⁻¹	4.95 mg kg⁻¹ shoot	Alam and Rahman (2003)
	Field study/harvest	7.64 mg kg⁻¹	30.42 mg kg⁻¹ root	Liu et al. (2007)
	Field study/harvest	14.09 mg kg⁻¹	28.63 mg kg⁻¹ root	Bhattacharya et al. (2010a)
	Field study/harvest	5.85 mg kg⁻¹	1.65 mg kg⁻¹ straw	Bhattacharya et al. (2010b)
Mercury	Field study/120 days	29 mg kg⁻¹	3.2 mg kg⁻¹ root	Meng et al. (2012)
	Pot culture/harvest	4 mg kg⁻¹	170 μg kg⁻¹ grain	Liu et al. (2013)
	Pot culture/150 days	46 μg kg⁻¹	14 μg g⁻¹ root	Wang et al. (2014)

rice seed, which is worth about US $30, is sufficient for sowing a hectare of land. Apart from this, availability of ambient nutrient reserve in the seeds helps to germinate rice seeds more efficiently compared with other phytoremediation crops such as *Brassica juncea* which has a smaller reserve of nutrients in seeds, thus limiting germination events. Attractive rice paddies also meet aesthetic demands such as incorporating greenery in metal-contaminated sites, which are usually barren lands. Another noticeable feature of rice plants is the ability to grow in both waterlogged and dry land soils. All these characteristics are in the favor of remediation of toxic TE-contaminated sites using rice plants. The extent to which rice plants accumulate toxic TEs varies with the variety of rice and metabolic status of the roots (El-Habet et al., 2014). TEs pollution from anthropogenic activities is often restricted to 25 cm soil depth (Ramakrishnaiaha and Somashekar, 2010). Interestingly, the root system of the rice plant is located at this depth and indicates maximum accumulation potential of toxic TEs. Wetland soils, which are often near mine tailings, trap toxic TEs (Owen and Otton, 1995). The ability of rice plants to grow well in wetland conditions makes this plant a phytoremediation agent of mine leachate. It is clear that rice plants grow well in the presence of Cd at the rate of $20 \mu g \, g^{-1}$ of soil (see Table 1). This concentration often meets the requirement of phytoremediation of metal-contaminated agricultural soils using rice plants where the major input of Cd is rock phosphate fertilizer. It must be noted that absorption ability of rice roots is determined by the presence of oxygen and physiochemical properties of the soil. Prevalence of anoxia reduces liable metals in the solution (Violante et al., 2010). Wetland-grown rice plants overcome anoxia by radical oxygen loss that mobilizes metals from the soil (Meia et al., 2009). Thus adaptability to anoxia along with ability to grow well in both dry and wetland conditions makes rice plants suitable for phytoremediation.

Rice plants belong to the family Poaceae, members of which are characterized by a fibrous adventitious root system that helps to pump water more efficiently from the soil, an important characteristic for meeting the primary requirement of a metal accumulator plant. Fibrous adventitious roots of rice plants also help to cope with nutrient deficiency. Extensive fibrous root systems firmly cling to soil particles, preventing paddy soil erosion. Thus the morphological features of roots of rice plants help to prevent outflow of TEs from the contaminated soil through soil erosion. Plants with higher root biomass tend to accumulate more TEs (Fu-Zhong et al., 2011). Thus cultivation strategies for the enhancement of root biomass trigger a higher rate of mobilization of toxic TEs from soil to rice plants. Available information on rice roots from germplasm collection also enables selection of an ideal rice variety for the remediation of particular metal-contaminated sites. Cadmium-tolerant rice cultivar tends to accumulate more Cd in the root compared with Cd-sensitive cultivar (Uraguchi et al., 2009). All these findings indicate the characteristics of roots could be the prime reason that makes the rice plant promising for soil remediation.

The shoot system of rice also poses promising metal accumulation characteristics. Bt transgenic rice is reported to accumulate Pb up to $1000 \, mg \, kg^{-1}$ of straw and indicates the potential for rice leaves to accumulate toxic TEs (Wang et al., 2009). Metals such as Cr have been found to accumulate more in shoots compared with roots. Studies conducted on typical rice-growing regions indicated that Mn accumulated more in shoot than in root (Sasaki et al., 2011). Bioaccumulation factor for TEs in rice plants was in the decreasing order of $Zn > Mn > Cd > Cu > Cr > Pb$ (Satpathy et al., 2014). Studies with plant hormones indicated that supplement of abscisic acid and gibberellic acid restricts translocation of toxic TEs such as Ni and Cd (Rubio et al., 1994). Translocation is among the critical factors that make plants suitable as a phytoremediation crop. Increase of Cd in the shoot was found more in a Cd-susceptible variety than in a Cd-tolerant variety (Uraguchi et al., 2009). Higher degree of translocation of Cr and Zn was also observed in rice plants (Payus and Talip, 2014). It has been reported that translocation coefficient is in the decreasing order of Cd > As > Pb in rice plants that were grown in the Changjiang Delta region (Yuan et al., 2012). Arsenic also has shown a tendency to concentrate more in rice grain than in other parts of the plant. Studies on differential metal accumulation of rice plant during irrigation with water of Ramgarh Lake, India, indicate that metal accumulation pattern in the rice root follows decreasing order of As > Pb > Mn > Cd > Cr > Zn > Cu > Hg whereas in the leaf the decreasing order is As > Cd > Pb > Mn > Cr > Zn > Cu > Hg (Singh et al., 2011). Apart from these, root-to-leaf translocation of Cd was found to retard by fertilization with ammonium-based fertilizers (Sebastian and Prasad, 2014b). Most often metal accumulation factor in rice was described in the decreasing order root > shoot > leaf > grain. The studies mentioned have found that rice plant parts differentially accumulate metals based on their specific chemical nature.

Metal accumulation in rice also varies with rice varieties. Germplasm of rice plant is rich with approximately 108,256 varieties (Yadav et al., 2013). Based on origin, Indian rice varieties cover the largest number of accessions, approximately 16,013, and are followed by Lao People's Democratic Republic having 15,280 accessions at the International Rice Gene Bank Collection at International Rice Research Institute (IRRI). Most of the studies on heavy metal accumulation were carried out on Japonica varieties because of issues of metal contamination in regions where these cultivars are being cultivated. Diverse Japonica rice germplasm had shown natural variation of about 13- to 23-fold difference in grain Cd concentration (Arao and Ae, 2003; Ueno et al., 2009). It is the higher root activity, high shoot-to-root ratio, and water consumption that made rice plants accumulate more heavy metals (Ishikawa et al., 2010). Quantitative traits of Cd accumulation in rice plants are found to locate at chromosome 2, 7, 3, 4, 6, 8, 5, 11, and 10 (Ueno et al., 2009; Ishikawa et al., 2010; Xue et al., 2009). These traits are found to be similar in lowland as well as upland conditions and hence open a way to improve rice plants for metal accumulation. It has been reported that genotype variation affects As accumulation in rice irrespective of the nature of the cultivation practice; that is, dry or wetland cultivation (Wu et al., 2011). All these findings indicate that natural metal accumulation capacities of rice plants are stable irrespective of cultivation practices. These studies also pointed out stability of rice plants in metal accumulation characteristics irrespective of varying edaphic factors such as soils, metals, pH, organic carbon, cation exchange capacity (CEC), and ecological niches. It can be summarized that minimum growth requirement, fast growth rate, and genetic plasticity make rice plants ideal for phytoremediation.

3 ASSISTED CLEANUP USING RICE PLANTS

Nutritional requirements of rice plants are well studied, and this allows feeding of rice plants for enhanced metal accumulation. Fertilization often leads to higher biomass productivity in rice paddies (Zhang et al., 2012). Nitrogen fertilization is crucial during rice cultivation for enhancement of yield. Ammonium as N source is favored during wetland cultivation compared with nitrate because of problems with denitrification and leaching of nitrate (Smolders et al., 2010). Dissolved ammonium binds tightly with soil and hence prevents leaching of N. Organic manure as well as synthetic fertilizers such as urea is a practical solution for improvement of rice biomass. Humphreys et al. (1987) reported that enhanced agronomic efficiency (56 kg grain per kg applied N) of surface-applied N fertilizer was greatest where N application was prior to permanent flooding in a combine-sown rice crop (Humphreys et al., 1987). But agronomic efficiency was 8.2 kg grain per kg N when N applied at the time of sowing.

Mechanical disturbance of soil during ground preparation and combine sowing increased N mineralization of high clay soils (Craswell and Waring, 1972). N mineralization leads to increased nitrate levels in soils and cause N loss during permanent flooding because of denitrification. It must be noted that nitrogen fertilization of rice plants is often practiced during the tillering stage to increase the number of panicles. But from the phytoremediation point of view, fertilizing can be done before the tillering stage, which increases plant biomass for metal sequestration. The increase in biomass often leads to tissue dilution of the metals and hence lessens the toxicity (Wright et al., 1998). It has been reported that nitrogen fertilization leads to enhanced accumulation of Cd in rice plants while limiting Cd translocation (Sebastian and Prasad, 2014b). On the other hand, increase of nicotianamine during nitrogen fertilization leads to more accumulation of Zn, Fe, and Cu in the aerial plant part of cereals (Barunawati et al., 2013). Typical rice fertilization strategy is based on nitrogen, phosphorous, and potassium, which are essential to produce high rice plant biomass. Application of ammonium-based nitrogen fertilizers leads to accumulation of more amounts of toxic TEs in rice plants because of soil acidification (Casova et al., 2009; Eriksson, 1990). Hence application of ammonium fertilizer helps in the release of plant-available Cd in soil plus more accumulation of metal. Monoammonium phosphates are also potential fertilizers that help to enhance Cd accumulation in rice because of soil acidification capability, which enhances metal accumulation (McLaughlin et al., 1995). Since more than 70% of arable soils are acidic, chance of removal of metal from the field with rice plants is very high due to radical oxygen loss that prevents metal toxicity by formation of metal plaque on the surface of roots (Chlopecka and Adriano, 1997). Application of S and Ca leads to formation of insoluble TEs complexes in soil solution. So in order to avoid metal precipitation, application of S and Ca must be restricted. Chlorinated water could be another approach that enhances metal solubility in the field to extract more metals. The well-developed nutrient management practices such as site-specific nutrient management (SSNM) and a system of rice intensification that enhances rice biomass productivity also point to a higher possibility of improvement of metal accumulation in rice plants by meeting nutrient requirements (Johnston et al., 2009).

Organic manure application-accelerated plant growth is a promising strategy of assisted cleanup because these amendments are eco-friendly as well as economically feasible. Organic manure is applied before planting rice plants. This approach usually helps to enhance solubility of metals in the soil along with increase of metal chelator pool. Application of organic matter improves soil properties like CEC, aeration, water-holding capacity, and the amount of plant nutrients, all of which influence Cd accumulation in plants (Adeniyan et al., 2011). Incorporation of organic matter into soil for enhanced growth of rice plants requires attention because of the dynamics of N content in flooded conditions. Mineralization of organic N retarded under flooded condition because of anoxia (Ehrenfelda and Yu, 2012). But the draining of the soil ensures aerobic condition that leads to NH_4^+ conversion to NO_3^- via nitrification processes, and thus organic matter breakdown is accelerated. Hence draining of the field ensures maximum mineralization of organic material added. This indicates that intermittent draining of the field supports the growth of rice plants because of more N availability, which results in higher biomass productivity. Germination and growth of rice seedlings are accelerated in fields rich in organic matter. Nutrient uptake studies in rice fields indicate that the crop acquires more than half the nitrogen (N) and micronutrients from mineralized organic matter such as crop residues (Myint et al., 2010). Livestock wastes in paddy soils help to replenish organic matter, nutrients, and metal complexing agents such as humic acids (Antil and Singh, 2007). The higher amount of nutrients withheld in bioorganic waste help the plant to combat with toxic TEs (Park et al., 2011). Studies with application of vermicompost indicate that rice plants accumulate more Cd during vermicompost application (Sebastian and Prasad, 2013). The enhancement of root growth during vermicompost amendment application in turn favored Cd accumulation. Cd tolerance achieved with the addition of farmyard manures in this study also points to the potential of these manures to support establishment of rice seedlings in the presence of toxic TEs. In short, organic manure helps rice plants to combat the toxicity of TEs during the early growth period of rice plants.

Microbe-assisted metal cleanup with rice plants also has tremendous potential. It has been reported that microbial methylation contributes to accumulation of methylated As in rice grains (Zheng et al., 2013). Microbial oxidation of As (III) to As (V) and methylation of As are the As detoxification mechanisms operating in the soil that make As unavailable for plant uptake. Increase of Fe-reducing bacteria in the rhizosphere of rice plant is supposed to increase As uptake by rice. But it must be noted that decrease of nitrate-dependent, ferrous ion reducing bacteria in the soil mobilizes As in the soil and hence enhances As uptake (Burton et al., 2008). Iron plaque formation in rice roots assisted with aerobes in the soil act as a high-affinity adsorbing surface for a number of TEs. Iron plaque not only acts as adsorbing surface but also supports growth of rice plants by protecting roots from metal toxicity. This ensures enhanced biomass production including that of roots and results in more removal of metals from the soil during a harvest that includes roots. This also points to the importance of removal of rice roots that accumulate metals during a phytoremediation program, since anaerobic decomposition of rice roots will release the bound metal again into soil. *Azotobacter* is a noticeable bacterium that improves rice plant growth (Wani et al., 2013). Recommended application of this bacterium is 0.5 kg per hectare for promoting plant growth. These bacteria enhance plant growth through increase of N content in the soil. Other than bacteria, arbuscular mycorrhizal fungi are reported to form symbiosis with rice plants. Colonization of rice roots with this group of fungi is supposed to downregulate Si transporters that mediate As uptake, which in turn reduces As accumulation in the rice plants (Yeasmin et al., 2007). Mychorrhizal colonization was also reported to reduce uptake of Cd by creating a barrier before Cd entry into roots (Khan et al., 2000). In short, adaptability to varying nutrient management practices and ability to form symbiosis with microbes promise assisted cleanup using rice plants.

4 RICE INDUSTRY BY-PRODUCTS FOR BIOREMEDIATION

Rice cultivation and related industry offer various by-products. By-products obtained even after phytoremediation can be utilized for commercial purposes (Figure 2). Straw, husk, bran, germ, and broken rice are the chief by-products of the rice industry (Esa et al., 2013). One ton of rice grain production release approximately 220 kg straw and husk.

Rice straw acts as potential absorbent of TEs such as Fe, Mn, Zn, Pb, and Cd (Nawar et al., 2013). Metal absorption capacity of straw was found to depend on the type of metal, pH, metal concentration, and contact time. More percentage of absorption with increase in absorbent as well as contact time indicates that rice straw acts as a potential metal biosorbent. Commercial ion exchange resins made up of petroleum products are not biodegradable. Hence these resins have more negative environmental impact. Apart from this, efficiency of these resins is low because of less availability of functional groups. Biosorbents with high ion exchange capacity prepared from rice straw efficiently removed toxic TEs from wastewater (Rungrodnimitchai, 2014). These lignocellulosic biosorbents are biodegradable, and hence disposal is convenient. It is also notable that sodium hydroxide treatment helps in high-efficiency removal of lignin in rice straw and hence eases the preparation of cellulose phosphate from straw, which is an efficient biosorbent. Grafting of dimethylaminoethyl methacrylate (DMAEM) on rice straw

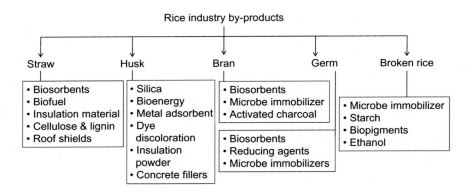

FIGURE 2

Rice paddy by-products applications.

with the help of potassium permanganate/nitric acid redox method was also reported to act as an efficient metal absorbent (Mostafa et al., 2012).

Rice husk is the main residue during polishing of rice grain. Rice husk has shown higher rate of adsorption of Pb and Zn (Elham et al., 2010). It is also a potential adsorbent of As (V), and the adsorption rate is found to increase with chemical treatment (Lata and Samadder, 2014). Its higher rate of metal adsorption capacity makes husk a potential amendment that immobilizes toxic TEs in the soil. Cellulose phosphate prepared with rice husk has shown 100% removal of divalent Cd from the solution (Athinarayanan et al., 2014). The ability to produce higher surface area with help of mechanical methods makes husk a more attractive material for biosorption. Rice husk is rich in cellulose and hemicellulose, which accounts for more than 50% bulk of the material. These sugar fibers make husk insoluble in water; hence husk can be efficiently used for wastewater treatment. Apart from cellulose, silica in the rice husk also helps in adsorption (Awizar et al., 2013). Husk can be processed as a biosorbent by washing it with tap water or chemical treatment. Treatment with epichlorohydrin, tartaric acid, orthophosphoric acid, and polyaniline are reported to increase the metal adsorption capacity of husk (Lata and Samadder, 2014). Briefly, metal absorption capacity as well as easier manipulation of ion exchange properties via chemical treatments makes rice husk an efficient metal-binding agent.

Rice husk can also be utilized for production of silica, activated carbon, tetramethoxysilane, insulation materials, concrete fillers, oils, absorbents, and insulation powder for steel mills (Thiravetyan, 2012). Rice husk contains amorphous silica which is being used in low permeability concrete for use in bridges, marine environments, and nuclear power plants. Rice husk ash has more than 90% amorphous silica that could be used as a substitute for silica fumes used in industry. Silica is prepared from the rice husk ash by application of high temperature. Digestion of the rice husk ash with caustic leads to extraction of silica as sodium silicate in the solution. Sodium is precipitated by passage of carbon dioxide, and the filtrate-containing silica is dried and packaged. Through this method 1 t of silica can be produced from 1.6 t of husk ash and the estimated selling cost of silica is US $1 kg^{-1} silica. Apart from the industrial applications, rice husk is also a source of energy. Husk produced from rice industry around the world has an energy content of about 14 GJ/ton, which is equivalent to energy produced from 1 billion barrels of oil per year. The usage of husk as an energy source would have an economic advantage of approximately US $7 billion year^{-1}.

Rice bran can be used for metal removal process because of its higher capacity of sorption (Montanher et al., 2005). These residues are also characterized by abundance of water-soluble vitamins, carbohydrates, potassium, nitrogen, and phosphorus. The presence of water-soluble components makes higher affinity contact of water with rice bran; hence bran has high affinity to bind dissolved metals. Being rich in nutrients, rice bran also acts as immobilizer of microbes. *Rhizopus arrhizus* immobilized in rice bran has shown more Ni uptake compared with pure rice bran used for biosorption (Gurel et al., 2010). Application of activated charcoal is not economically feasible because of low availability of the material and higher production cost. But removal of metals using activated charcoal is technically easier. Rice bran can be utilized to produce activated charcoal, which is economically feasible. It is reported that carbon produced from rice bran efficiently binds with hexavalent Cr (Ranjan and Hasan, 2010). Abundance of functional groups makes rice bran an important adsorbent where chemical conversion of elements is frequent. This allows reduction of metallic ions such as conversion of Cr^{6+} to Cr^{3+}. The ability to support chemical changes explains high-efficiency removal of Cd and Pb with rice bran compared to other green materials being used for metal removal such as pine sawdust.

Rice germ is often removed with rice bran. It is rich in polyunsaturated fatty acids, and the presence of these fatty acids enhances the metal-reducing power of rice germ (Nagendraprasad et al., 2011). Thus any interaction of metals with rice germ often ends up in reduction of metals. Germ oil can also be extracted and used for energy purposes. Rice germ is also rich in lectins, which help microbial immobilization in rice bran (Zhao-Wen et al., 1984). It is also important to note that rice bran together with rice germ holds the major portion of metal accumulated in rice grain. Hence removal of rice bran and the resulting white rice is important from the perspective of human health.

Broken rice is another notable by-product from the rice industry which contain proteases and starch (Gohel and Duan, 2012). This shapeless rice can be utilized for microbial immobilization. Many of the species of *Aspergillus* are well known to grow well in broken rice (Paranthaman et al., 2009). Biopigment production is another process utilizing broken rice. Culturing with fungus *Monascus ruber* MTCC232 has been found to produce pigments of red, orange, and yellow colors (Dikshit and Tallapragada, 2011). Non cook process based ethanol production potential is also high with broken rice. In this method, starch-hydrolyzing enzymes are used for conversion of starch to simple sugars, which can be converted into ethanol by fermentation.

5 SUSTAINABLE BIOREMEDIATION PROSPECTS OF RICE PADDIES

Rice paddies, being agro-wetland ecosystems, are ideal places for growing various aquatic macrophytes and algae that enhance biomass productivity (Linke et al., 2014). Besides this, remediation of paddy fields also can be achieved with the help of these macrophytes or algae (Figure 3). Aquatic macrophytes are well known for rhizofiltration, whereas algal biomass is reported to work as biosorbing agents (Rai, 2009). Biomass produced from paddy fields can be utilized for production of biosorbents or compost. Both these products have tremendous scope in remediation of metal-contaminated sites. Aquatic macrophyte cocultivation benefits in rice paddies are well studied with *Azolla* (Figures 4–6) (Table 2) (Bocchi and Malgioglio, 2010). This aquatic plant helps to enhance biomass yield from the rice paddies as well as nitrogen content in the field. This makes the genus *Azolla* an important component in rice cultivation for boosting the yield and biomass. *Azolla caroliniana* is reported to have capacity for metal hyperaccumulation. This fern has been found to accumulate significant level

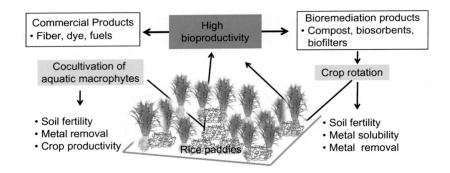

FIGURE 3

Benefits of incorporation of sustainable agriculture practices for rice cultivation to boost bioeconomy.

FIGURE 4

Azolla (nitrogen fixer) culture preparation: (a) Construction of 1×2 m pit using clay bricks; (b) silpaulin lining of the constructed pit; (c) preparation of sand and manure; (d) spread of sand and manure in silpaulin-lined pit.

FIGURE 5

(a) Cow dung slurry mixture with sand and manure in silpaulin-lined pit; (b) *Azolla* source culture for mass multiplication; (c) introduction of *Azolla* culture to the constructed pit; (d) pit with fully grown *Azolla* ready for spread in rice paddies.

of metals, especially Cr (III) (Bennicelli et al., 2004). The usage of *Azolla* is reported to suitable at the level of contamination of 1-20 ppm metals in the solution. It is also reported that growth of aquatic macrophytes such as *Eichhornia crassipes*, *Lemna minor*, and *Pistia stratiotes* in the lowland fields act as an efficient rhizofiltration agent (Karkhanis et al., 2005). *Elatine hexandra*, *Althenia filiformis*, and *Monita rivularis* in the flooded rice fields significantly remove Cd. These macrophytes are also a potential remediating agent of salinity, which is a serious agricultural problem. Algal biomass in the flooded plain is also found to be effective in metal removal (Reniger, 1977). Successive cropping and harvesting of algal biomass reduces levels of toxic TEs in the paddy soil even though the process requires considerable effort to collect the algal biomass (Vandenhove et al., 2001). Incorporation of blue-green algae in the field not only removes metal from the field but also adds to nitrogen economy.

Aquatic plants obtained from the field can be used for manufacture of biofilters where biomass is packed in to filters. These filters are efficient in the removal of metal ions, especially from water bodies. *Azolla* binding filters are reported to bind metals efficiently (Pb, 10%; Cr, 3%; U, 3.5%; Cr^{3+}, 9%; Ni, 3%) (Sachdeva and Sharma, 2012). Metal uptake capacity of algal biomass also can be improved by

FIGURE 6

(a–d) *Azolla* introduction to rice paddies.

Table 2 *Azolla* **Culture Preparation: Expenditure**

Expenditure Per Tank/Pit of Size 1×2 m	Per Year in INR (Indian rupees)
88 kg cow dung (Rs. 1 for kg cow dung)	88
250 g magnesium sulfate	10
88 g muriate of potash	10
1733 g single super phosphate	18
2 kg mineral mixture	60
4 m² silpaulin sheet (durable for 3 years)	200
Bricks: 30 (durable for 3 years)	300
Labor: 12 laborers per year @ Rs. 300 day⁻¹	3600
Total expenditure per year	4286

1.5 kg per tank per day for 350 days = 525 kg
Compost: 100 kg per tank cost = Rs. 300
Expenditure
Azolla production cost per kg = Rs. 8.16

chemical treatments with CaCl$_2$. It is well known that dead algal biomass is more efficient in removing metals. Thus harvested algal biomass from rice paddies has great scope in commercial production of biosorbents. Residual algal-bacterial biomass in a symbiotic relationship can be induced, and this approach has been found to enhance metal biosorption as well as biodegradation of organic pollutants in photo bioreactor (Munoz et al., 2006). Immobilized or granulated algal biomass filled column is another promising technique that help sorption of metals. The use of algae turf scrubbers is a promising technique for metal ion removal using algal biomass (Mulbry et al., 2008). But economic gain from these mechanical systems is low because of the expenditure needed for meeting specific growth requirements. Constructed rice paddies with algae can be a good alternative for algae turf scrubbers due to the economic gain from by-products from rice plants and algae.

Rainfed, lowland rice cultivation areas where one crop of rice practiced are ideal for crop rotation with legume (Seng et al., 2008). Crop rotation with legume not only supports biomass productivity but also enhances fertility of the paddy soil (Graham and Vance, 2003). Notable aspects of legumes are short maturation period (55-90 days), drought tolerance, and ability to fix nitrogen, which improves nitrogen content in the soil. Rotation of legume with rice plants helps to intensify the land use along with increase of crop productivity per area per year. Cultivation of *Sesbania*, which have a short maturation period (approximately 60 days), helps green manure production. Cultivation of mung bean or cow pea not only adds to nutrient improvement of the field but also provides economic gain to the farmer. Soil acidification during legume-mediated nitrogen fixation is also beneficial for phytoremediation of paddy fields because soil acidity renders metal more soluble for plant uptake. For example, growth of the oilseed rape has been found to enhance Cd accumulation in rice plants (Yu et al., 2014). This indicates the promising role of legume in mobilization of metals where subsequent removal of metals is achieved by growth of rice plants. Legume growth also contributes to produce phytoremediation products. It is reported that application of rapeseed cake together with phytoextraction using *Sedum plumbizincicola* reduces Cd uptake in rice plants (Shen et al., 2010). Legume cotyledon is enriched with phytic acid. The compound is rich in phosphate and shows high-affinity binding with metal ions. This indicates potential use of legume cotyledons for usage in biofilter preparation, which helps to biosorb metal ions.

Compost preparation is another promising approach that helps to utilize biomass produced from rice paddies. Composts improve soil properties including aeration and porosity (Pagliai et al., 2004). Increase of both these properties helps to enhance root growth. Vermicompost can be produced from rice straw, rice husk, and rice bran. Addition of vermicompost to soil enhances root growth, which leads to Cd accumulation in rice seedlings (Sebastian and Prasad, 2013). This indicates that addition of compost to soil improves the phytoextraction process. Commercially important intercrops that produce fiber, dye, biofuels, etc., also can be cultivated in rice paddies.

To summarize, rice paddies offer great opportunities for enhancement of biomass, and the biomass obtained can be converted into commercially important phytoremediation agents. Thus it is clear that rice paddies also contribute to bioeconomy through biomass and commercially important by-products of cocultivated or crop-rotated plants that help in sustainable agriculture.

6 OUTLOOK

Rice plants pose all the essential characteristics needed for metal accumulation. Root growth is the most important factor for controlling metal tolerance in plants. The available information from rice germplasm with regard to root characteristics and the information about higher root biomass producing rice gene such as OsEXPA8 can be utilized for development of metal hyperaccumulator rice plants.

There is tremendous scope for assisted metal accumulation using rice plants. Chemical fertilizers to be used during rice cultivation must be screened for enhanced metal accumulation. It is also important to utilize the by-products from rice industry for bioremediation process. Cocultivation as well as crop rotation opportunities in rice paddies must be studied for a successful sustainable bioremediation that also provides economic gain.

REFERENCES

Adeniyan, O.N., Ojo, A.O., Akinbode, O.A., Adediran, J.A., 2011. Comparative study of different organic manures and NPK fertilizer for improvement of soil chemical properties and dry matter yield of maize in two different soils. J. Soil Sci. Environ. Manage. 2 (1), 9–13.

Alam, M.Z., Rahman, M., 2003. Accumulation of arsenic in rice plant from arsenic contaminated irrigation water and effect on nutrient content. In: Fate of Arsenic in the Environment. Presented in the BUET-UNU Symposium, Dhaka, Bangladesh, 5–6 February. pp. 131–135.

Antil, R.S., Singh, M., 2007. Effects of organic manures and fertilizers on organic matter and nutrients status of the soil. Arch. Agron. Soil Sci. 53, 519–528.

Arao, T., Ae, N., 2003. Genotypic variations in cadmium levels of rice grain. Soil Sci. Plant Nutr. 287, 223–233.

Athinarayanan, J., Periasamy, V.S., Alshatwi, A.A., 2014. Biogenic silica-metal phosphate (metal=Ca, Fe or Zn) nanocomposites: fabrication from rice husk and their biomedical applications. J. Mater. Sci. Mater. Med. 25 (7), 1637–1644.

Awizar, D.A., Othman, N.K., Jalar, A., Daud, A.R., Rahman, I.A., Al-hardan, N.H., 2013. Nanosilicate extraction from rice husk ash as green corrosion inhibitor. Int. J. Electrochem. Sci. 8, 1759–1769.

Barunawati, N., Giehl, R.F.H., Bauer, B., Wirén, N.V., 2013. The influence of inorganic nitrogen fertilizer forms on micronutrient retranslocation and accumulation in grains of winter wheat. Front. Plant Sci. 4, 1–11.

Basavaraja, H., Mahajanashetti, S.B., Udagatti, N.C., 2007. Economic analysis of post-harvest losses in food grains in India: a case study of Karnataka. Agric. Econ. Res. Rev. 20, 117–126.

Bennicelli, R., Stepniewska, Z., Banach, A., Szajnocha, K., Ostrowski, J., 2004. The ability of *Azolla caroliniana* to remove heavy metals (Hg(II), Cr(III), Cr(VI)) from municipal waste water. Chemosphere 55, 141–146.

Bhattacharya, P., Samal, A.C., Majumdar, J., Sant, S.C., 2010a. Uptake of arsenic in rice plant varieties cultivated with arsenic rich groundwater. EnvironmentAsia 3 (2), 34–37.

Bhattacharya, P., Samal, A.C., Majumdar, J., Santra, S.C., 2010b. Accumulation of arsenic and its distribution in rice plant (*Oryza sativa* L.) in Gangetic West Bengal, India. Paddy Water Environ. 8, 63–67.

Bhattacharyya, P., Chakraborty, A., Chakrabarti, K., Tripathy, S., Powell, M.A., 2005. Chromium uptake by rice and accumulation in soil amended with municipal solid waste compost. Chemosphere 60, 1481–1486.

Bhullar, N.K., Gruissem, W., 2013. Nutritional enhancement of rice for human health: the contribution of biotechnology. Biotechnol. Adv. 31, 50–57.

Bian, R., Joseph, S., Cui, L., Pan, G., Li, L., Liu, X., Zhang, A., Rutlidge, H., Wong, S., Chia, C., Marjo, C., Gong, B., Munroe, P., Donne, S., 2014. A three-year experiment confirms continuous immobilization of cadmium and lead in contaminated paddy field with biochar amendment. J. Hazard. Mater. 272, 121–128.

Bocchi, S., Malgioglio, A., 2010. Azolla-Anabaena as a biofertilizer for rice paddy fields in the Po Valley, a temperate rice area in Northern Italy. Int. J. Agron. 2010, 1–5.

Bosetti, F., Zucchi, M.I., Pinheiro, J.B., 2011. Molecular and morphological diversity in Japanese rice germplasm. Plant Genet. Resour. 9, 229–232.

Burton, E.D., Bush, R.T., Leigh, A., Sullivan, L.A., Johnston, S.G., Hocking, R.K., 2008. Mobility of arsenic and selected metals during re-flooding of iron- and organic-rich acid-sulfate soil. Chem. Geol. 258, 64–73.

Casova, K., Cerny, J., Szakova, J., Balik, J., Tlustos, P., 2009. Cadmium balance in soils under different fertilization managements including sewage sludge application. Plant Soil Environ. 55 (8), 353–361.

Chatterjee, C., Dube, B.K., Sinha, P., Srivastava, P., 2004. Detrimental effects of lead phytotoxicity on growth, yield, and metabolism of rice. Commun. Soil Sci. Plant Anal. 35, 255–265.

Chlopecka, A., Adriano, D.C., 1997. Influence of zeolite, apatite and Fe-oxide on Cd and Pb uptake by crops. Sci. Total Environ. 207, 195–206.

Chowdhury, M.D.R., Kumar, V., Sattar, A., Brahmachari, K., 2014. Studies on the water use efficiency and nutrient uptake by rice under system of intensification. Bioscan 9 (1), 85–88.

Craswell, E.T., Waring, S.A., 1972. Effect of grinding on the decomposition of soil organic matter. I. The mineralization of organic nitrogen in relation to soil type. Soil Biol. Biochem. 4, 427–433.

Dikshit, R., Tallapragada, P., 2011. Monascus purpureus: a potential source for natural pigment production. J. Microbiol. Biotechnol. Res. 1 (4), 164–174.

Ehrenfelda, J.G., Yu, S., 2012. Patterns of nitrogen mineralization in wetlands of the New Jersey pinelands along a shallow water table gradient. Am. Midl. Nat. 167 (2), 322–335.

El-Habet, H.B., Naeem, E.S., Abel-Meeged, T.M., Sedeek, S., 2014. Evaluation of genotypic variation in lead and cadmium accumulation of rice (*Oryza sativa*) in different water conditions in Egypt. Int. J. Plant Soil Sci. 3, 911–933.

Elham, A., Hossein, T., Mahnoosh, H., 2010. Removal of Zn(II) and Pb (II) ions using rice husk in food industrial wastewater. J. Appl. Sci. Environ. Manag. 14 (4), 159–162.

Eriksson, J.E., 1990. Effects of nitrogen-containing fertilizers on solubility and plant uptake of cadmium. Water Air Soil Pollut. 49, 355–368.

Esa, N.M., Ling, T.B., Peng, L.S., 2013. By-products of rice processing: an overview of health benefits and applications. J. Rice Res. 1, 1–11.

Felkner, J., Tazhibayeva, K., Townsend, R., 2009. Impact of climate change on rice production in Thailand. Am. Econ. Rev. 99, 205–210.

Fu-Zhong, W., Wan-Qin, Y., Zhang, J., Li-Qiang, Z., 2011. Growth responses and metal accumulation in an ornamental plant (*Osmanthus fragrans var. thunbergii*) submitted to different Cd levels. ISRN Ecol. 2011, 1–7.

Gohel, V., Duan, G., 2012. No-cook process for ethanol production using Indian broken rice and pearl millet. Int. J. Microbiol. 2012, 1–9.

Graham, P.H., Vance, C.P., 2003. Legumes: importance and constraints to greater use. Plant Physiol. 131, 872–877.

Gurel, L., Senturk, I., Bahadir, T., Buyukgungor, H., 2010. Treatment of nickel plating industrial wastewater by fungus immobilized onto rice bran. J. Microb. Biochem. Technol. 2, 34–37.

Huang, Y., Zhang, W., Zheng, X., Li, J., Yu, Y., 2004. Modeling methane emission from rice paddies with various agricultural practices. J. Geophys. Res. Atmos. 109, 1–12.

Humphreys, E., Chalk, P.M., Muirhead, W.A., Melhuish, F.M., White, R.J.G., 1987. Effects of time of urea application on combine-sown Calrose rice in South-east Australia. III. Fertiliser nitrogen recovery, efficiency of fertilization and soil nitrogen supply. Aust. J. Agric. Res. 38, 129–138.

Ishikawa, S., Abe, T., Kuramata, M., Yamaguchi, M.O.T., Yamamoto, T., Yano, M., 2010. Major quantitative trait locus for increasing cadmium specific concentration in rice grain is located on the short arm of chromosome 7. J. Exp. Bot. 61 (3), 923–934.

Johnston, A.M., Khurana, H.S., Majumdar, K., Satyanarayana, T., 2009. Site-specific nutrient management— concept, current research and future challenges in Indian agriculture. J. Indian Soc. Soil Sci. 57, 1–10.

Karkhanis, M., Jadia, C.D., Fulekar, M.H., 2005. Rhizofilteration of metals from coal ash leachate. Asian J. Water Environ. Pollut. 3 (1), 91–94.

Khan, A.G., Kuek, C., Chaudhry, T.M., Khoo, C.S., Hayes, W.J., 2000. Role of plants, mycorrhizae and phytochelators in heavy metal contaminated land remediation. Chemosphere 41, 197–207.

Kögel-Knabner, I., Amelung, W., Cao, Z., Fiedler, S., Frenzel, P., Jahn, R., Kalbitz, K., Kölbl, A., Schloter, M., 2010. Biogeochemistry of paddy soils. Geoderma 157 (1–2), 1–14.

Lata, S., Samadder, S.R., 2014. Removal of heavy metals using rice husk: a review. Int. J. Environ. Res. Dev. 4, 165–170.

Li, W., Cui, X., 2014. Focus on rice: towards better understanding of the life cycle of crop plants. Mol. Plant 7 (6), 931–933.

Linke, M.G., Godoy, R.S., Rolon, A.S., Maltchik, L., 2014. Can organic rice crops help conserve aquatic plants in southern Brazil wetlands? Appl. Veg. Sci. 17, 346–355.

Liu, J., Zhu, Q., Zhang, Z., Xu, J., Yang, J., Wong, M.H., 2005. Variations in cadmium accumulation among rice cultivars and types and the selection of cultivars for reducing cadmium in the diet. J. Sci. Food Agric. 85, 147–153.

Liu, W.X., Shen, L.F., Liu, J.W., Wang, Y.W., Li, S.R., 2007. Uptake of toxic heavy metals by rice (*Oryza sativa* L.) cultivated in the agricultural soil near Zhengzhou city, People's Republic of China. Bull. Environ. Contam. Toxicol. 79, 209–221.

Liu, C., Wu, C., Rafiq, M.T., Aziz, R., Hou, D., Ding, Z., Lin, Z., Lou, L., Feng, Y., Li, T., Yang, X., 2013. Accumulation of mercury in rice grain and cabbage grown on representative Chinese soils. J. Zhejiang Univ. (Sci.) 14, 1144–1151.

Ma, X., Liu, J., Wan, M., 2012. Differences between rice cultivars in iron plaque formation on roots and plant lead tolerance. Adv. J. Food Sci. Technol. 5 (2), 160–163.

McLaughlin, M.J., Maier, N.A., Freeman, K., Tiller, K.G., Williams, C.M.J., Smart, M.K., 1995. Effect of potassic and phosphatic fertilizer type, fertilizer Cd concentration and zinc rate on cadmium uptake by potatoes. Fert. Res. 40, 63–70.

Meia, X.Q., Yea, Z.H., Wong, M.H., 2009. The relationship of root porosity and radial oxygen loss on arsenic tolerance and uptake in rice grains and straw. Environ. Pollut. 157, 2550–2557.

Meng, B.O., Feng, X., Qiu, G., Wang, D., Liang, P., Li, P., Shang, L., 2012. Inorganic mercury accumulation in rice (*Oryza sativa* L.). Environ. Toxicol. Chem. 31, 2093–2098.

Montanher, S.F., Oliveira, E.A., Rollemberg, M.C., 2005. Removal of metal ions from aqueous solutions by sorption onto rice bran. J. Hazard. Mater. 117, 207–211.

Mostafa, K.M., Samarkandy, A.R., El-Sanabary, A.W., 2012. Harnessing of chemically modified rice straw plant waste as unique adsorbent for reducing organic and inorganic pollutants. Int. J. Org. Chem. 2, 143–151.

Mulbry, W., Kondrad, S., Pizarro, C., Kebede-Westhead, E., 2008. Treatment of dairy manure effluent using freshwater algae: algal productivity and recovery of manure nutrients using pilot-scale algal turf scrubbers. Bioresour. Technol. 99, 8137–8142.

Munoz, R., Alvarez, M.T., Munoz, A., Terrazas, E., Guieysse, B., Mattiasson, B., 2006. Sequential removal of heavy metals ions and organic pollutants using an algal-bacterial consortium. Chemosphere 63, 903–911.

Myint, A.K., Yamakawa, T., Kajihara, Y., Zenmyo, T., 2010. Application of different organic and mineral fertilizers on the growth, yield and nutrient accumulation of rice in a Japanese ordinary paddy field. Sci. World J. 5 (2), 47–54.

Nagendraprasad, M.N., Sanjay, K.R., Khatokar, M.S., Vismaya, M.N., Swamy, S.N., 2011. Health benefits of rice bran—a review. J. Nutr. Food Sci. 1, 1–7.

Nawar, N., Ebrahim, M., Sami, E., 2013. Removal of heavy metals Fe^{3+}, Mn^{2+}, Zn^{2+}, Pb^{2+} and Cd^{2+} from wastewater by using rice straw as low cost adsorbent. AJIS 2, 85–95.

Owen, D.E., Otton, J.K., 1995. Mountain wetlands: efficient uranium filters—potential impacts. Ecol. Eng. 5, 77–93.

Pagliai, M., Vignozzi, N., Pellegrini, S., 2004. Soil structure and the effect of management practices. Soil Tillage Res. 79, 131–143.

Panichpat, T., Srinives, P., 2009. Partitioning of lead accumulation in rice plants. Thai J. Agric. Sci. 42 (1), 35–40.

Paranthaman, R., Alagusundaram, K., Indhumathi, J., 2009. Production of protease from rice mill wastes by *Aspergillus niger* in solid state fermentation. World J. Agric. Sci. 5 (3), 308–312.

Park, J.H., Lamb, D., Paneerselvam, P., Choppala, G., Bolan, N.S., Chung, J.W., 2011. Role of organic amendments on enhanced bioremediation of heavy metal(loid) contaminated soils. J. Hazard. Mater. 183 (3), 549–574.

Payus, C., Talip, A.F.A., 2014. Assessment of heavy metals accumulation in paddy rice (*Oryza sativa* L.). Afr. J. Agric. Res. 9 (41), 3082–3090.

Pereira, B.F.F., Rozane, D.E., Araújo, R.S., Barth, G., Queiroz, R.J.B., Nogueira, T.A.R., Moraes, M.F., Cabral, C.P., Boaretto, A.E., Malavolta, E., 2011. Cadmium availability and accumulation by lettuce and rice. Braz. J. Soil Sci. 35, 645–646.

Rai, P.K., 2009. Heavy metal phytoremediation from aquatic ecosystems with special reference to macrophytes. Crit. Rev. Environ. Sci. Technol. 39, 697–753.

Rajkumar, M., Sandhya, S., Prasad, M.N.V., Freitas, H., 2012. Perspectives of plant-associated microbes in heavy metal phytoremediation. Biotechnol. Adv. 30 (6), 1562–1574.

Ramakrishnaiaha, H., Somashekar, R.K., 2010. Heavy metal contamination in roadside soil and their mobility in relations to pH and organic carbon. Soil Sediment Contam. Int. J. 11, 643–654.

Ranjan, D., Hasan, S.H., 2010. Rice bran carbon: an alternative to commercial activated carbon for the removal of hexavalent chromium from aqueous solution. Bioresources 5, 1661–1664.

Rascio, N., Navari-Izzo, F., 2011. Heavy metal hyper accumulating plants: how and why do they do it? And what makes them so interesting? Plant Sci. 180, 169–181.

Reniger, P., 1977. Concentration of cadmium in aquatic plants and algal mass in flooded rice culture. Environ. Pollut. 14, 297–302.

Rubio, M.I., Escrig, C., Martinez-Cortina, F., Lopez-Benet, J., Sanz, A., 1994. Cadmium and nickel accumulation in rice plants. Effects on mineral nutrition and possible interactions of abscisic and gibberellic acids. Plant Growth Regul. 14, 151–157.

Rugh, C.L., 2004. Genetically engineered phytoremediation: one man's trash is another man's transgene. Trends Biotechnol. 22, 496–498.

Rungrodnimitchai, R., 2014. Rapid preparation of biosorbents with high ion exchange capacity from rice straw and bagasse for removal of heavy metals. Sci. World J. 2014, 1–9.

Sachdeva, S., Sharma, A., 2012. Azolla: role in phytoremediation of heavy metals. Proceeding of the National Conference "Science in Media 2012" Organized by YMCA University of Science and Technology, Faridabad, Haryana (India). IJMRS Int. J. Eng. Sci., ISSN (Online), 2277–9698.

Sasaki, A., Yamaji, N., Ma, J.F., 2011. OsYSL6 is involved in the detoxification of excess manganese in rice. Plant Physiol. 157, 1832–1840.

Satpathy, D., Reddy, M.V., Dhal, S.P., 2014. Risk assessment of heavy metal contamination in paddy soil, plants, and grains (*Oryza sativa* L.) at the east coast of India. Biomed Res. Int. 2014, 1–11 (Article ID 545473).

Schuhmacher, M., Domingo, J.L., Llobet, J.M., Corbella, J., 1994. Cadmium, chromium, copper, and zinc in rice and rice field soil from southern Catalonia, Spain. Bull. Environ. Contam. Toxicol. 53, 54–60.

Sebastian, A., Prasad, M.N.V., 2013. Cadmium accumulation retard activity of functional components of photo assimilation and growth of rice cultivars amended with vermicompost. Int. J. Phytoremediation 15, 965–978.

Sebastian, A., Prasad, M.N.V., 2014a. Cadmium minimization in rice. A review. Agron. Sustain. Dev. 34, 155–173.

Sebastian, A., Prasad, M.N.V., 2014b. Photosynthesis-mediated decrease in cadmium translocation protects shoot growth of *Oryza sativa* seedlings up on ammonium phosphate—sulfur fertilization. Environ. Sci. Pollut. Res. 21, 986–997.

Seng, V., Eastick, R., Fukai, S., Ouk, M., Men, S., Chan, S.Y., Nget, S., 2008. Crop diversification in lowland rice cropping systems in Cambodia: effect of soil type on legume production Global Issues. Paddock Action. Proceedings of 14th Agronomy Conference 2008, 21–25 September 2008, Adelaide, South Australia.

Shen, L.B., Wu, L.H., Tan, W.N., Han, X.R., Luo, Y.M., Ouyang, Y.N., Jin, Q.Y., Jiang, Y.G., 2010. Effects of Sedum plumbizincicola—*Oryza sativa* rotation and phosphate amendment on Cd and Zn uptake by *O. sativa*. Ying Yong Sheng Tai Xue Bao 21 (11), 2952–2958.

Simmons, R.W., Pongsakul, P., Saiyasitpanich, D., Klinphoklap, S., 2005. Elevated levels of Cd and zinc in paddy soils and elevated levels of Cd in rice grain downstream of a zinc mineralized area in Thailand: implications for public health. Environ. Geochem. Health 27, 501–511.

Singh, J., Upadhyay, S.K., Pathak, R.K., Gupta, V., 2011. Accumulation of heavy metals in soil and paddy crop (*Oryza sativa*), irrigated with water of Ramgarh Lake, Gorakhpur, UP, India. Toxicol. Environ. Chem. 93 (3), 462–473.

Smolders, A.J.P., Lucassen, E.C.H.E.T., Bobbink, R., Roelofs, J.G.M., Lamers, L.P.M., 2010. How nitrate leaching from agricultural lands provokes phosphate eutrophication in groundwater fed wetlands: the sulphur bridge. Biogeochemistry 98, 1–7.

Suwannaporn, P., Lineman, A., 2008. Rice-eating quality among consumers in different rice grain preference countries. J. Sens. Stud. 23, 1–13.

Thiravetyan, P., 2012. Application of rice husk and bagasse ash as adsorbents in water treatment. In: Bhatnagar, A. (Ed.), Application of Adsorbents for Water Pollution Control. Bentham Science Publishers, Sharjah, pp. 432–454.

Ueno, D., Koyama, E., Kono, I.O.T., Yano, M., Ma, J.F., 2009. Identification of a novel major quantitative trait locus controlling distribution of Cd between roots and shoots in rice. Plant Cell Physiol. 50 (12), 2223–2233.

Uraguchi, S., Mori, S., Kuramata, M., Kawasaki, A., Arao, T., Ishikawa, S., 2009. Root-to-shoot Cd translocation via the xylem is the major process determining shoot and grain cadmium accumulation in rice. J. Exp. Bot. 60 (9), 2677–2688.

Vandenhove, H., Van Hees, M., Van Winkel, S., 2001. Feasibility of phytoextraction to clean up low-level uranium-contaminated soil. Int. J. Phytoremediation 3, 301–320.

Vijayarengan, P., 2011. Mineral nutrient variations in rice (*Oryza sativa*) after treatment with exogenous cadmium. Int. J. Environ. Biol. 2 (3), 147–152.

Violante, A., Cozzolino, Perelomov, L., Caporale, A.G., Pigna, M., 2010. Mobility and bioavailability of heavy metals and metalloids in soil environments. J.Soil Sci. Plant Nutr. 10 (3), 268–292.

Wang, H., Huang, J., Ye, Q., Wu, D., Chen, Z., 2009. Modified accumulation of selected heavy metals in Bt transgenic rice. J. Environ. Sci. (China) 21 (11), 1607–1612.

Wang, M.Y., Chen, A.K., Wong, M.H., Qiu, R.L., Cheng, H., Ye, Z.H., 2011. Cadmium accumulation in and tolerance of rice (*Oryza sativa* L.) varieties with different rates of radial oxygen loss. Environ. Pollut. 159, 1730–1736.

Wang, X., Ye, Z., Li, B., Huang, L., Meng, M., Shi, J., Jiang, G., 2014. Growing rice aerobically markedly decreases mercury accumulation by reducing both Hg bioavailability and the production of MeHg. Environ. Sci. Technol. 48, 1878–1885.

Wani, S., Chand, S., Ali, T., 2013. Potential use of azotobacter chroococcum in crop production: an overview. Curr. Agric. Res. J. 1, 35–38.

Wright, D.P., Scholes, J.D., Read, D.J., 1998. Effects of VA mycorrhizal colonization on photosynthesis and biomass production of *Trifolium repens* L. Plant Cell Environ. 21 (2), 209–216.

Wu, Z., Ren, H., McGrath, S.P., Wu, P., Zhao, F.J., 2011. Investigating the contribution of the phosphate transport pathway to arsenic accumulation in rice. Plant Physiol. 157 (1), 498–508.

Xue, D., Chen, M., Zhang, G., 2009. Mapping of QTLs associated with cadmium tolerance and accumulation during seedling stage in rice (*Oryza sativa* L.). Euphytica 165, 587–596.

Yadav, S., Singh, A., Singh, M.R., Goel, N., Vinod, K.K., Mohapatra, T., Singh, A.K., 2013. Assessment of genetic diversity in Indian rice germplasm (*Oryza sativa* L.): use of random versus trait-linked microsatellite markers. J. Genet. 92, 545–557.

Yeasmin, T., Zaman, P., Rahman, A., Absar, N., Khanum, N.S., 2007. Arbuscular mycorrhizal fungus inoculums production in rice plants. Afr. J. Agric. Res. 4, 463–467.

Yizong, H., Ying, H., Yunxia, L., 2009. Heavy metal accumulation in iron plaque and growth of rice plants upon exposure to single and combined contamination by copper, cadmium and lead. Acta Ecol. Sin. 29, 320–326.

Yu, L., Zhu, J., Huang, Q., Su, D., Jiang, R., Li, H., 2014. Application of a rotation system to oilseed rape and rice fields in Cd-contaminated agricultural land to ensure food safety. Ecotoxicol. Environ. Saf. 108, 287–293.

Yuan, X., Li, T., Li, J., 2012. The influence of paddy soil components and As, Pb and Cd speciation on their uptake by rice in the Changjiang delta region. In: Asia Pacific Conference on Environmental Science and Technology. In: Advances in Biomedical Engineering, vol. 6, pp. 147–152.

Zhang, Y., Zhang, G., Xiao, N., Wang, L., Fu, Y., Sun, Z., Fang, R., Chen, X., 2012. The rice 'nutrition response and root growth' (NRR) gene regulates heading date. Mol. Plant 6 (2), 585–588.

Zhao-Wen, S., Sun, C., Zhu, Z., Xi-Hua, T., Rui-Juan, S., 1984. Purification and properties of rice germ lectin. Can. J. Biochem. Cell Biol. 62 (10), 1027–1032.

Zheng, R., Sun, G.X., Zhu, Y.G., 2013. Effects of microbial processes on the fate of arsenic in paddy soil. Chin. Sci. Bull. 58, 186–193.

Zhou, H., Zhou, X., Zeng, M., Liao, B.H., Liu, L., Yang, W.T., Wu, Y.M., Qiu, Q.Y., Wang, Y.J., 2014. Effects of combined amendments on heavy metal accumulation in rice (*Oryza sativa* L.) planted on contaminated paddy soil. Ecotoxicol. Environ. Saf. 101, 226–232.

Zhu, X., Lin, L., Zhang, Q., Liu, Q., Ma, X., Ye, L., Sha, J., 2010. Effects of zinc and chromium stresses on heavy metal accumulation of rice roots at different growth stages of rice plants. In: iCBBE Proceedings. icbbe. org/2011/Proceeding2010.aspx.

Zong, H.Y., Ying, H., Yun Xia, L., 2010. Combined effects of chromium and arsenic on rice seedlings (*Oryza sativa* L.) growth in a solution culture supplied with or without P fertilizer. Sci. China Life Sci. 53 (12), 1459–1466.

CULTIVATION OF SWEET SORGHUM ON HEAVY METAL-CONTAMINATED SOILS BY PHYTOREMEDIATION APPROACH FOR PRODUCTION OF BIOETHANOL

A. Sathya, V. Kanaganahalli, P. Srinivas Rao, S. Gopalakrishnan

International Crops Research Institute for the Semi-Arid Tropics (ICRISAT), Patancheru, Telangana, India

1 INTRODUCTION

The term *heavy metal* (HM) has a wide range of meanings, and there has been no consistent definition by any authoritative body such as International Union of Pure and Applied Chemistry (IUPAC) over the past 60 years (Duffus, 2002). But over the past 2 decades, this term has been used by numerous publications and legislations for indicating a group name for metals or semimetals that cause human, phyto, animal, and also ecotoxicity. Though the imprecise term is defined by several researchers at various levels including density, atomic number, atomic weight, chemical properties, and toxicity, there is no connectivity between these properties. Since this chapter deals with bioremediation aspects, HMs causing human and ecotoxic effects were considered further. Three kinds of HMs are of concern, including toxic metals (Hg, Cr, Pb, Zn, Cu, Ni, Cd, As, Co, Sn, etc.), precious metals (Pd, Pt, Ag, Au, Ru, etc.), and radionuclides (U, Th, Ra, Am, etc.) (Wang and Chen, 2006).

The stability and nondegradability of metals mean higher exposure of HMs to humans and animals, and numerous reports are available for the related health and ecological issues (Caussy et al., 2003; Li et al., 1995). Toxic effects of HMs may be chronic or acute, which depends on the route of transfer and reactive forms. For instance, for Cd, all forms are toxic; for Pb, organic forms are highly toxic; for As, inorganic arsenate [As(+5)] or [As(+3)] is highly toxic; for Hg and Hg(II), organomercurials, mainly methylmercury (Mudgal et al., 2010). Toxic effects of these major HMs are periodically reviewed by many researchers (Burbacher et al., 1990; Duruibe et al., 2007; Wongsasuluk et al., 2014; Zatta, 2001). Representatives of HM-related health issues and their tolerable limits are summarized in Table 1.

The HMs are sourced from both natural and anthropogenic activities (Chopra et al., 2009; Li et al., 2009). Parent rocks are the natural contributors, and their HM content is usually found to be low

Bioremediation and Bioeconomy. http://dx.doi.org/10.1016/B978-0-12-802830-8.00012-5
Copyright © 2016 Elsevier Inc. All rights reserved.

Table 1 Effect of HMs on Humans

HMs	Effect on Human Health	JECFA Tolerable Limits (x/kg bw/day)
As	Bronchitis, dermatitis, poisoning	2.1 µg
Cd	Renal dysfunction, bone defects, blood pressure, bronchitis, cancer	25 µg
Pb	Mental retardation in children, developmental delay, fatal infant encephalopathy, congenital paralysis, sensor neural deafness, epilepsy, acute/chronic damage of CNS, liver, kidney, and GI tract	0.025 µg
Hg	Tremors, gingivitis, psychological changes, acrodynia, spontaneous abortion, protoplasm poisoning, CNS damage	4 µg
Zn	CNS damage, corrosive effects on skin	0.3-1 mg
Cr	CNS damage, fatigue, irritability	–
Cu	Anemia, liver and kidney damage, stomach and intestinal irritation	0.5 mg

JECFA, Joint FAO/WHO Expert Committee on Food Additives.
Source: JECFA (1982, 2011, 2013).

depending on the parent rock composition. Various anthropogenic activities transfer the HMs through air, water, and soil, the major transmitters of any kind of pollutants. Such routes and sources are (1) air, which has mining, smelting, and refining of fossil fuels; smoke from production units of metallic goods; and vehicular exhaust; (2) water, having domestic and industrial sewage and effluents, thermal power plants, and atmospheric fallout; and (3) soil having agricultural and animal wastes, municipal and industrial sewage, coal ashes, fertilizers, discarded metal goods, and atmospheric fallout. Anthropogenic sources are the maximum contributors for metals rather than natural sources (Nriagu and Pacyna, 1988). A recent study by Millward and Turner (2001) also states that, anthropogenic factors are the major metal contributors as they alter the natural biogeochemical cycles.

According to the Environmental Protection Agency (EPA) report, the United States has more than 40,000 contaminated sites as of May 2004. In addition, 100,000 ha of cropland, 55,000 ha of pasture, and 50,000 ha of forest have been lost by HM contamination and demands for reclamation (McGrath et al., 2001; Ragnarsdottir and Hawkins, 2005). In Europe, around 2 million sites were contaminated with HMs, cyanide, mineral oil and chlorinated hydrocarbons (EEA, 2005). In developing countries, particularly in India, China, Pakistan, and Bangladesh, HM pollution occurs by the release of untreated industrial effluents into the surface drains, which further spreads to agricultural lands. Untreated effluent is sometimes used for irrigation due to water scarcity, where it acts as an HM source for agricultural croplands (Ragnarsdottir and Hawkins, 2005). In addition, most agricultural land has been used for construction purposes. All these factors together have led to shrinkage of healthy agricultural cropland. The increasing demand for lands has forced farmers to use contaminated sites for crop cultivation. This chapter deals with remediation of HM-contaminated sites, specifically by phytoremediation, and the role of phytoremediation in improving the economy.

2 REMEDIATION MEASURES FOR HMs

Many remediation techniques involving physical (soil replacement, thermal desorption) and chemical (leaching and fixation) methods, along with use of a broad range of chemical additives, are available for HM removal from soil (Bricka et al., 1993; Yao et al., 2012). Due to the difficulties involving scale-up processing, adaptability, site conditions, low efficiency, loss of soil structure and fertility, metal specificity, and cost-effectiveness, they have not been promoted on a large scale. Figures 1 and 2 depict various physical and chemical remediation measures with their pros and cons.

The phytotoxic effects of HMs include reduction in plant growth and protein content, loss of mineral homeostasis, and hence loss in yield and crop quality (Chibuike and Obiora, 2014). Yet some plant species are able to tolerate the negative effects of HM in addition to having the ability to extract in their tissues. This paves a way for the development of a technology called phytoremediation, also known as botanical bioremediation or green remediation, which involves the use of plant species for the extraction, removal, sequestration, detoxification, immobilization of toxic substances through various mechanisms. This also overcomes the negative effects associated with the physical and chemical remediation methods mentioned earlier. Phytoremediation can be applied for different matrixes such as soil, water, and sediment. All these aspects have been summarized by many reviews from past decades to the current scenario (Chaney et al., 1997; Gomes, 2012; Moffat, 1995; Salt et al., 1977; Sharma and Pandey, 2014).

Besides the chemical states of HMs, there are different physical forms: (a) dissolved (in soil solution), (b) exchangeable (organic and inorganic components), (c) as structural components of the lattices of soil minerals, and (d) as insoluble precipitates with other soil components. The former two forms are available to the plants. The latter two forms remain for a long term (Aydinalp and Marinova, 2003). Depending on the physical and chemical states of HMs, phytoremediation in soil occurs through any of the following modes: phytoextraction, phytostabilization, and phytovolatilization (Sharma and Pandey, 2014), which is depicted in Figure 3.

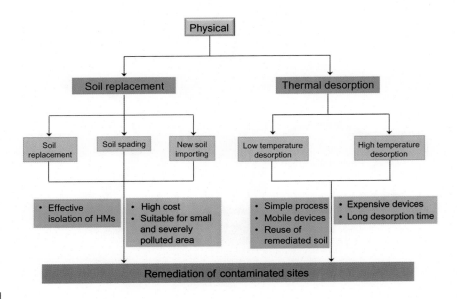

FIGURE 1

Physical remediation measures for HM removal. Strategies/technologies involved for HM remediation are in blue shapes. Positive impacts and risk factors associated with the strategies have been indicated in green and orange shapes respectively.

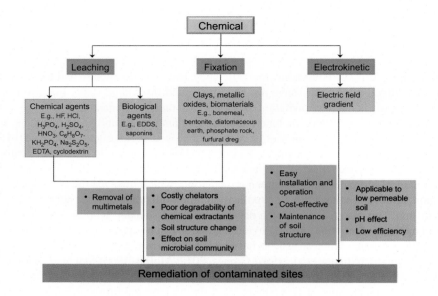

FIGURE 2

Chemical remediation measures for HM removal. Strategies/chemicals involved for HM remediation are in blue shapes. Positive impacts and risk factors associated with the strategies have been indicated in green and orange shapes respectively.

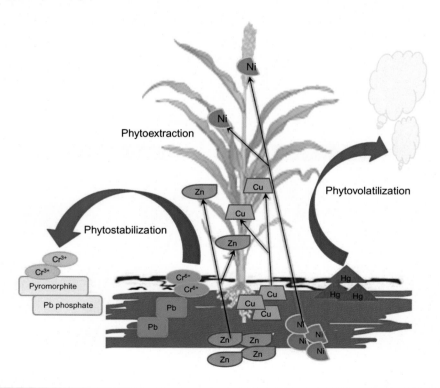

FIGURE 3

HM removal mechanisms in soil by phytoremediation.

Phytoextraction: Also known as phytoaccumulation, in phytoextraction the HMs in soil are transferred to the above-ground biomass such as the shoots and leaves with the aid of roots, and through the process of absorption, concentration, and precipitation. Most of the HMs including Ni (Kukier and Chaney, 2004), Cu (Delorme et al., 2001), and Zn (Sun et al., 2010) have been extracted by this mode.

Phytostabilization: In phytostabilization, toxic forms of HMs are changed into nontoxic or less toxic forms and reduce the bioavailability. The HMs are absorbed or accumulated in roots or precipitated into the root zone, either naturally or by chemical amendments. Since this process is confined to the rhizosphere, migration of HMs can be inhibited, which helps in conserving ground- and surface water and reduces bioavailability of metal into the food chain. Many HMs including As, Pb, Cd, Cr, Cu, and Zn can be stabilized by this mode (Brennan and Shelley, 1999). Still, it has some drawbacks such as contaminant remaining in soil and requirement for extensive fertilizers and soil amendments application (Sharma and Pandey, 2014).

Phytovolatilization: Phytovolatilization involves the transformation of HMs into volatile forms through leaves, which are further transpired into atmosphere. Hg is the major HM, phytoremediated by phytovolatilization (Rugh et al., 1996). But the negative impact associated with Hg is that recycling by precipitation and redeposition of atmospheric Hg into lakes and oceans leads to production of methylmercury.

A strengths, weaknesses, opportunities, and threats analysis done on phytoremediation by Gomes (2012) revealed that phytoremediation has its own pros and cons as do physical and chemical remediation measures. Major weaknesses observed were (i) plant selection with requirement of multitraits such as fast growth, high biomass, deep roots, and easy harvesting; (ii) slow process; and (iii) limited practical experience. Though phytoremediation has these weaknesses, it has many supporting strengths such as (i) high public acceptance; (ii) maintenance of soil biological components; (iii) environmental benefits such as control of soil erosion, carbon sequestration, and creation of wildlife habitat; (iv) generation of recyclable metal-rich plant tissues; (v) socioeconomic benefits via local labor employment and buildup of value-added industry; (vi) cost-effectiveness; and (vii) sustainability.

3 PHYTOREMEDIATION OF HMs: HYPERACCUMULATORS

Plant selection is a crucial first step for phytoremediation. It is already known that the elements present in the soil will be reflected to some extent in the plants, this extent represented by concentration, a platform for differentiating and selecting a plant for phytoremediation purpose. The term *hyperaccumulators* was initially used by Jaffré et al. (1976) to indicate higher metal uptake potential (25% on Dry Matter (DM)) of *Sebertia acuminata*. It was further used by Brooks et al. (1977) for describing plants whose dried tissues have $>1000 \mu g\, g^{-1}$ Ni; they proposed this is the discriminatory concentration threshold for differentiating normal plants and hyperaccumulating plants. Jaffré (1980) has refined the nomenclature by using the terms *hypermanganesophores* and *hypernickelophores* to describe metal extraction specificity of the plants. But the public exposure toward the use of hyperaccumulating plants for HM removal has increased in later years (Chaney, 1983; Anonymous, 1990). The necessary features that distinguish hyperaccumulators from nonhyperaccumulating plant species include higher rate of (i) HM uptake, (ii) root-to-shoot translocation, and (iii) detoxification and sequestration of HM (Rascio and Navari-Izzo, 2011).

Baker and Brooks (1989) reviewed the distribution of terrestrial plants with hyperaccumulating potential and found 145 hyperaccumulators for Ni with a distribution in 6 suborders, 17 orders, and 22 families including herbs, shrubs, and trees. This indicates that hyperaccumulators are not closely related, but they possess the common feature of growth on metalliferous soils without phytotoxic effects (Rascio and Navari-Izzo, 2011). A recent review by van der Ent et al. (2012) summarized several criteria used for hyperaccumulation threshold and suggested concentration criteria for different metals based on critical evaluation of numerous hyperaccumulation reports: Cd, Se, and Tl—100 μg g^{-1}; Co, Cu, and Cr—300 μg g^{-1}; Ni, Pb, and As—1000 μg g^{-1}; Zn—3000 μg g^{-1}; Mn—10,000 μg g^{-1}. This evaluation also identified more than 500 plant taxa that can be categorized as hyperaccumulators for one or more elements including Asteraceae, Brassicaceae, Caryophyllaceae, Cunouniaceae, Cyperaceae, Euphorbiaceae, Fabaceae, Flacourtiaceae, Lamiaceae, Poaceae, and Violaceae. The most-researched studies of hyperaccumulation model systems include *Thlaspi* sp., (Delorme et al., 2001; Idris et al., 2004), *Brassica* sp., (Quartacci et al., 2014; Wu et al., 2013), and *Alyssum* sp. (Barzanti et al., 2011; Bayramoglu et al., 2012). But features found in these plant families such as slow growth, shallow root system, small biomass, and unknown agronomic potential of hyperaccumulators have made it necessary to find alternative plant species for phytoremediation.

4 PHYTOREMEDIATION OF HMs: ENERGY CROPS

Various research groups differ on plant selection for phytoremediation. Many research groups have suggested that the use of hyperaccumulators for HM remediation is of prime importance rather than biomass (Chaney et al., 1997; van der Ent et al., 2012). However, the success of phytoremediation depends not only on the complete removal of toxic substances but also on the generation of valuable biomass including timber, bioenergy, feedstock for pyrolysis, and biofortified products or ecologically important species in order to demonstrate cost-effectiveness (Conesa et al., 2012; van der Lelie et al., 2001). Report of Meers et al. (2010) strongly supports the use of bioenergy crops for HM phytoremediation. Maize was tested under field conditions in Flanders, Belgium, in soil contaminated with HMs, Pb, Cd, Zn, and As by historic smelter activities. They stated that cultivation of energy maize in this region could result in the production of 30,000-42,000 kWh including electrical and thermal renewable energy per hectare. This could replace a coal-fed power plant with the reduction of up to 21 tons ha^{-1} year^{-1} CO$_2$ along with the HM removal.

Energy crops fall into two categories: annuals: sweet sorghum and fiber sorghum, kenaf, and rapeseed; and perennials, a category further subdivided into (a) agricultural: wheat, sugar beet, cardoon, reeds, miscanthus, switchgrass, and canary reed grass; and (b) forest: willows, poplars, eucalyptus, and black locust (Simpson et al., 2009). High biomass crops with HM tolerance such as Indian mustard, oat, maize, barley, sunflower, ryegrass, fast-growing willow, and poplars have been studied (Komárek et al., 2007; Meers et al., 2005; Shen et al., 2002; Vervaeke et al., 2003). Table 2 summarizes the HM removal efficiency of several agricultural crops (Zhuang et al., 2009). In conclusion, knowledge of energy crop cultivation under contaminated conditions will provide new avenues for bioeconomy and also for reclamation of contaminated soils.

5 PHYTOREMEDIATION OF HMs BY SUGAR CROPS: SWEET SORGHUM

Among the energy crops, sugar crops (sugarcane, sugar beet, and sweet sorghum) with HM remediation capacity have a great impact on the bioenergy-bioethanol (Yadav et al., 2011). Reports of Rayment et al. (2002), Yadav et al. (2010), and Jain et al. (2010) on sugarcane reveals their HM accumulation

Table 2 Comparison of HM Removal Efficiency by Agricultural Crops

Crop	Binomial Name	Biomass (tons ha^{-1})	Heavy Metal Uptake (kg ha^{-1})				References
			Pb	Cd	Zn	Cu	
Sorghum	*Sorghum bicolor*	25.8 (dw)	0.35	0.052	1.44	0.24	Zhuang et al. (2009)
Sunflower	*Helianthus annuus*	22.1 (dw)	0.38	0.006	1.22	0.64	Marchiol et al. (2007)
		–	0.091	0.002	0.41	0.12	Marchiol et al. (2007)
		24 (dw)	0.016	–	2.14	–	Madejón et al. (2003)
Indian mustard	*Brassica juncea*	7.3 (dw)	–	0.007	0.89	0.15	Keller et al. (2003)
Tobacco	*Nicotiana tabacum*	12.6 (dw)	–	0.042	1.83	0.47	
Alfalfa	*Medicago sativa*	45.9 (ww)	0.115	0.013	0.438	0.124	Ciura et al. (2005)
Maize	*Zea mays*	92.7 (ww)	0.042	0.0093	0.45	0.096	
Barley	*Hordeum vulgare*	5.84 (dw)	0.035	0.014	1.07	0.057	Soriano and Fereres (2003)

dw, dry weight; ww, wet weight.
Source: Zhuang et al. (2009).

potential for Cd, Zn, and Hg. Sugarcane grown near the municipal landfill site and medical waste treatment system in Brazil is also found to accumulate the HM such as Cd, Cr, Cu, Hg, Mn, Pb, and Zn (Segura-Muñoz et al., 2006). A study on HM content of vegetables in Pakistan revealed that sugar beet has the potential to accumulate Cd, Pb, As, and Hg but at the safest and legally permissible level (Abbas et al., 2010).

Sorghum (*Sorghum bicolor* L. Moench), a C4 annual grass valued for food, feed, fiber, and feedstock is known by many names such as jowar (India), kaoliang (China), great millet and guinea corn (west Africa), kafir corn (south Africa), mtama (north Africa), dura (Sudan), and milo or milo-maize (Unites States) (Purseglove, 1972). It is also known as "sugarcane of the desert" and "camel among crop" for its hardiness in drought. It can be grown in tropical, subtropical, temperate, and semiarid regions, and also in poor-quality soils (Sanderson et al., 1992). It has many salient features such as rapid growth, high sugar content (10-15%), higher biomass, wider adaptability to harsh agroclimatic conditions, and metal-absorbing property (Zhuang et al., 2009; Rao et al., 2009). Sweet sorghum is type of *S. bicolor*, generally cultivated for syrup and also forage and feed. It has sweet juicy stalks and higher quantity of sugars (both glucose and fructose) than grain sorghum; hence, it is called sweet sorghum. So, it can serve as the best candidate for playing dual role in phytoremediation and bioeconomy via bioethanol "sweet fuel" production and also related coproduct generation (Rao et al., 2012). The sweet sorghum varieties developed by International Crops Research Institute for the Semi-Arid Tropics (ICRISAT) and the related research work done on biofuel will be discussed later in this chapter.

The first hint of the HM absorption of sorghum was shown by An (2004) during an ecotoxic assessment, when he found sorghum can accumulate more Cd than cucumber, wheat, and sweet corn. Subsequent proof was given by Kaplan et al. (2005) during the analysis on sulfur-containing waste as a soil amendment under pot trials. It has been observed that, Ni, Cr, and Co accumulation occurs in roots, whereas Cd accumulation occurs in straw of sorghum. This indicates the efficiency of sorghum fibrous roots in HM absorption. A similar trend was observed on a short-term study on hydroponically grown *S. bicolor* for Pb, Cd, and Zn removal (Hernández-Allica et al., 2008; Soudek et al., 2012). In a 3-month microcosm study with artificially polluted soil containing Cd and Zn, sorghum was found to have 122 mg Cd kg^{-1} dry weight (DW) in shoot. This is higher than the threshold value of 100 mg kg^{-1} DW set for hyperaccumulators. From the various soil physicochemical and biological properties evaluated, it is noted that, phytoremediation with sorghum is able to recover the soil function, though the experimental soil has more phytotoxicity than the control treatments (Epelde et al., 2009).

The first *in situ* phytoremediation pilot plant was established during 2005 in Torviscosa, Italy, where Marchiol et al. (2007) designed a study to evaluate the phytoremediation effect of high biomass crops including *S. bicolor* and *H. annuus*. The experimental site possesses multimetal contaminants including As, Cd, Cu, Zn, and Fe, and it is under the national priority list of polluted sites in Italy. Limited metal extraction is observed in both the plants; however, the experimental design did not involve any practices such as soil amendments for enhancing metal bioavailability. It is understood that the factors considered for successful agriculture should also be considered for successful phytoremediation.

A study by Zhuang et al. (2009) is the first study conducted on HM remediation by sweet sorghum in a field at Lechang city, China, which has been contaminated with Pb, Cd, Zn, and Cu via atmospheric emissions and surface irrigation with mining wastewaters. The study was designed to evaluate the sweet sorghum varieties Keller, Rio, and Mary (bred for ethanol production in the United States) alone and in combination with soil amendments such as $(NH_4)_2SO_4$, NH_4NO_3, and ethylenediaminetetraacetic acid (EDTA). It was observed that there was no difference between the cultivars and biomass yield, whereas HM extraction efficiency was in the order of Keller > Mary > Rio. Among the soil amendments, EDTA promotes Pb accumulation, whereas $(NH_4)_2SO_4$ and NH_4NO_3 promote Zn and Cd accumulation. So, this assisted phytoextraction with facilitated agronomic practices serves as a sustainable remediation measure.

6 SWEET SORGHUM: A FEEDSTOCK FOR "SWEET FUEL" BIOETHANOL

Every year, fossil fuel sources are getting depleted, and they are anticipated to run out within the next 40-50 years. In addition, the consequences of fossil fuels such as global warming, acid rain, and urban smog have necessitated the shift to renewable energy sources such as biofuels, which are less harmful to the environment and also sustainable. Among the biofuels, ethanol is one of the prime alternatives. Though it has 68% lower equivalent energy than petroleum fuel, it is gaining in importance due to its complete combustion and release of less toxic by-products than other alcoholic and fossil fuels. Another contributing factor is its production from a broad range of feedstocks such as sugar (sugarcane, sugar beet, sweet sorghum), starch (corn, cassava), and lignocellulosic (agri-by-products: corn stover and fiber, wheat and barley straws, sugarcane bagasse, seed cake; woody biomass: hardwood and softwood; energy crops: switchgrass, poplar, banagrass, miscanthus, etc.) (Minteer, 2006; Vohra et al., 2014). All these factors will make ethanol the "fuel of the future," to use Henry Ford's phrase coined during his preparation of the Model T Ford. He designed the model to run on either gasoline or ethanol, with the vision of building a vehicle that was affordable for the working family and powered by a fuel that would boost the rural farm economy (Kovarik, 1998).

During 2009-2010, the world ethanol production was about 100 billion liters with consumption rates of 68% for fuel, 21% for industrial, and 11% for potable (Lichts, 2010). For fuel substitution, it is used with gasoline as either E15 (15:85%, ethanol vs. gasoline) or E85 (85:15%, ethanol vs. gasoline). Each country has set its own regulations for blending ethanol and even has set target requirements for the future (Cheng and Timilsin, 2011). In India in 2003, the Government of India (GOI) mandated the use of 5% ethanol blend in gasoline through its ambitious Ethanol Blending Program (EBP). Since 2003, the trade balance for ethanol has been generally negative. The balance has tapered down, however, from its peak of $140 million in 2005 to $11 million in 2012, indicating a gradual rise in export of ethanol and other spirits. In order to promote biofuels as an alternative energy source, the GOI in 2009 announced a comprehensive National Policy on Biofuels formulated by the Ministry of New and Renewable Energy (MNRE), calling for blending at least 20% of biofuels with diesel (biodiesel) and petrol (bioethanol) by 2017. The policies are designed to facilitate and bring about optimal development and utilization of indigenous biomass feedstock for biofuel production (Basavaraj et al., 2012).

Starch-based feedstocks contribute for higher ethanol production than sugar-based feedstocks (60 vs. 40%). But the use of starch based feedstock is limited due to higher energy requirement for its saccharification process (Vohra et al., 2014; Mussatto et al., 2010). Among the feedstocks, sugarcane is the major stock in tropical areas such as Brazil, Colombia, and India, whereas corn is the major stock in the United States, the European Union, and China (Cheng and Timilsin, 2011). Increased ethanol demand, decreased feedstock production, lack of clear technology, and the question of food/feed versus fuel have created the need for alternative feedstocks (Vohra et al., 2014).

Sweet sorghum has come to play a key role for this interlinked issue, and ethanol production from sweet sorghum is not a new process since it has 3 decades of history by various technologies (Christakopoulos et al., 1993; Kargi and Curme, 1985; Kargi et al., 1985; Lezinou et al., 1994; Mamma et al., 1995). However, it has received renewed attention for its beneficial characteristics, such as ethanol production at lower cost over sugarcane and sugar beet along with several by-products that enhance farmers' economy (Tables 3 and 4). In addition, the technology for alcoholic fermentation from sugar- and sucrose-containing feedstock is a well-known and mastered process. Some varieties of sweet sorghum have a significant sucrose content (500 gallons syrup per hectare), a major feature required for ethanol. In sweet sorghum, the sugar is stored in the main stalk, which can be recovered by pressing the stalks through rollers (similar to sugarcane processing). This yields about 20 gallons of ethanol per ton of stalks (Kojima and Johnson, 2005). Approximately, 50-85 tons ha^{-1} of sweet sorghum stalks yields 39.7-42.5 tons ha^{-1} of juice, which after fermentation produces 3450-4132 L ha^{-1} ethanol (Serna-Saldívar et al., 2012). Other studies have shown similar ethanol production levels: 3296 L ha^{-1} (Kim and Day, 2011) and 4750-5220 L ha^{-1} (Wu et al., 2010). In addition, sweet sorghum can rule out the food/feed versus fuel question due to the farmer's benefit from sorghum grains, after the stalk is harvested for juice (Kojima and Johnson, 2005). Besides this, the pressed stalk, called sweet sorghum bagasse, has several avenues in improving rural economy via ruminant/poultry feed and as raw material for biofertilizer production, paper making, and co-product generation including power (Rao et al., 2012).

Table 3 Favorable Traits of Sweet Sorghum

As Crop	As Ethanol Source	As Bagasse	As Raw Material for Industrial Products
Short duration (3-4 months)	Eco-friendly processing	High biological value	Paper and pulp making
C4 dryland crop	Less sulfur	Rich in micronutrients	Butanol, lactic acid, acetic acid production
Good tolerance to biotic and abiotic constraints	High octane ring	Ruminant/poultry feed	Alcoholic and nonalcoholic beverage production
Meets fodder and food needs	Automobile friendly (up to 25% of ethanol-petrol mixture without engine modification)	Power generation	Coproduct generation: dry ice, fuel oil, and methane
Non-invasive species		Biocompost	
Low soil N_2O and CO_2 emission		Good for silage making	
Seed propagated			

Source: Rao et al. (2012).

Table 4 Comparison Between Sugarcane, Sugar Beet, and Sweet Sorghum on Agronomic Traits and Ethanol Production Parameters

Traits	Sugarcane	Sugar Beet	Sweet Sorghum
Crop duration	About 7 months	About 5-6 months	About 4 months
Growing season	Only one season	Only one season	Temperate: 1 season; Tropical: 2/3 seasons
Soil requirement	Grows well in drain soil	Grows well in sandy loam; also tolerates alkalinity	All types of drained soil
Water management ($m^3 h^{-1}$)	36,000	18,000	4000-8000
Crop management	Requires good management	Greater fertilizer requirement; requires moderate management	Little fertilizer required; less pest and disease complex; easy management
Yield per ha (tons)	70-80	30-40	54-69
Sugar content on weight basis (%)	10-12	15-18	7-12
Sugar yield (tons ha^{-1})	7-8	5-6	6-8
Ethanol production from juice ($L ha^{-1}$)	3000-5000	5000-6000	3000
Harvesting	Mechanically harvested	Very simple; normally manual	Very simple; both manual and mechanical harvest

Source: Almodares and Hadi (2009) and Rao et al. (2009).

7 MICROBE-ASSISTED PHYTOREMEDIATION

Since phytoremediation is a time-consuming process, it can be enhanced by a plant microbe-mediated approach, popularly referred as biophytoremediation, because many microorganisms have been reported for their HM tolerance and removal of a broad spectrum of metal species. Bacterial cells (approximately 1.0-1.5 mm^3) have an extremely higher surface area-to-volume ratio, which influences HM absorption, than inorganic soil components by a metabolism-independent, passive/metabolism-dependent active process (Ledin et al., 1996). The major microbial groups—bacteria, fungi, yeast, and algae—play a better role in bioremediation by their biosorption properties. Some of the tested species includes *Bacillus subtilis*, *Rhizopus arrhizus*, *Saccharomyces cerevisiae*, and *Scytonema hofmanni* (Vijayaraghavan and Yun, 2008). Though the microbial biosorbents are cheaper and more effective alternatives for HM removal, effectiveness can be attained mainly in aqueous solutions (Joshi and Juwarkar, 2009; Kapoor et al., 1999; Wang and Chen, 2009).

Plant growth-promoting (PGP) microbes, a group of microbes found in rhizosphere or in association with the roots or other plant parts as endophytes, can play a key role in this scenario (Glick et al., 1999). The higher bacterial biomass in the rhizosphere occurs because of the nutrient release (especially small molecules such as amino acids, sugars, and organic acids) from the roots. In turn the PGP bacteria will support for plants by various direct (nitrogen fixation, phosphate solubilization, iron chelation, and phytohormone production) or indirect (suppression of plant pathogenic organisms, induction of resistance in host plants against plant pathogens and abiotic stresses) mechanisms (Penrose and Glick, 2001).

A wide range of PGP microbes is able to alleviate HM stress in soil by enhancing the HM uptake of plants as well as by increasing plant growth. Some of them are *Rhizobium, Pseudomonas, Agrobacterium, Burkholderia, Azospirillum, Bacillus, Azotobacter, Serratia, Alcaligenes,* and *Arthrobacter* (Carlot et al., 2002; Glick, 2003). A detailed review of PGP bacteria and their role in phytoremediation has been provided by many researchers (Ma et al., 2011; Rajkumar et al., 2012) and also demonstrated experimentally on various agricultural crops such as *Zea mays, Vigna mungo,* and *H. annuus* (Ganesan, 2008; Jiang et al., 2008; Rajkumar et al., 2008) and hyperaccumulating crops such as *Alyssum murale* and *Salix caprea* (Abou-Shanab et al., 2008; Kuffner et al., 2008). The following mechanisms were suggested for HM removal by microbes.

Siderophores: The low-molecular mass (400-1000 Da) compounds with high-association constants for complexing Fe and also other metals Al, Cd, Cu, Ga, In, Pb, and Zn (Rajkumar et al., 2010). Production of siderophores by PGP *Pseudomonas* spp., its metal complexes formation with high solubility, and the resulting higher HM uptake by plant has been reported (Wu et al., 2006a,b).

Organic acids: Low-molecular-weight organic acids of PGP microbes bind to metal ions in soil solution and increase the metal bioavailability to plants. However, the stability of the ligand:metal complexes is dependent on several factors such as the nature of organic acids (number of carboxylic groups and their position), the binding form of the HMs, and pH of soil solution (Saravanan et al., 2007).

Biosurfactants: The amphiphilic molecules with nonpolar tail and polar/ionic head, produced by microbes, are able to form complexes with HMs at the soil interface, desorbing metals from soil matrix and thus increasing metal solubility/bioavailability in the soil solution. Several studies have demonstrated the role of microbial biosurfactants in facilitating the release of adsorbed HMs and in enhancing the phytoextraction potential of plants. Still, documentation under field conditions is lacking (Basak and Das, 2014; Franzetti et al., 2014).

Polymeric substances and glycoprotein: Extracellular polymeric substances, mucopolysaccharides, and proteins produced by plant-associated microbes can make complexes with HMs and decrease their mobility in soils. Glomalin, an insoluble glycoprotein produced by arbuscular mycorrhizal fungi, proved its ability to form complexes with HMs Cu, Pb, and Cd (Bano and Ashfaq, 2013; Foster et al., 2000; Liu et al., 2001).

Metal reduction and oxidization: Mobility of HMs can be enhanced through oxidation and reduction reactions of plant-associated Fe- and S- oxidizing bacteria. Use of Fe-reducing bacteria and the Fe/S oxidizing bacteria together showed significantly increased mobility of Cu, Cd, Hg, and Zn, which might be due to coupled and synergistic metabolism of oxidizing and reducing microbes (Beolchini et al., 2009; Shi et al., 2011; Wani et al., 2007).

Biosorption: The plant-associated microbes may contribute to plant metal uptake through biosorption (microbial adsorption of soluble/insoluble organic/inorganic metals). Among the microbes, mycorrhizal fungi are key partners. The large surface area, cell wall (chitin, extracellular slime, etc.) and intracellular compounds (metallothioneins, P-rich amorphic material) of fungi endow them with a strong capacity for HM absorption from soil (He and Chen, 2014; Volesky and Holan, 1995).

The other interesting mechanism by which PGP microbes can alleviate HM stress on plant growth is through the production of enzyme 1-aminocyclopropane-1-carboxylate (ACC) deaminase, which controls the production of ethylene, a stress hormone (Glick, 2014). Similarly production of indole-3-acetic acid by rhizobacteria can enhance HM uptake by plant (Zaidi et al., 2006). The role of endophytic bacteria interacting with their host plant is of significance in the process of phytoremediation. Under

HM stress, the endophytes with the ability of stress tolerance can alleviate the stress and allocate the metals to the plant shoot (Ma et al., 2011; Weyens et al., 2009).

Few reports are available on the assistance of PGP microbes in HM remediation by sweet sorghum. The report of Duponnois et al. (2006) documents that the inoculation of Cd-tolerant *Pseudomonas* strains, mainly *P. monteilii*, showed significantly improved Cd uptake by sorghum plants under glasshouse conditions. Measurement of catabolic potentials on 16 substrates showed that pseudomonad strains presented a higher use of ketoglutaric and hydroxybutyric acids, as opposed to fumaric acid in control soil samples. It is suggested that fluorescent pseudomonads could act on the effect of small organic acids on phytoextraction of HMs from soil. Subsequently, Abou-Shanab et al. (2008) examined the ability of four bacterial isolates (*B. subtilis*, *Bacillus pumilus*, *Pseudomonas pseudoalcaligenes*, and *Brevibacterium halotolerans*) on HM removal capacity of sorghum. Sorghum roots accumulated higher concentration of Cr followed by metals Pb, Zn, and Cu. A comparative analysis was done on phytore-mediation efficiency of sweet sorghum, *Phytolacca acinosa* and *Solanum nigrum*, with the inoculation of PGP endophyte *Bacillus* sp. SLS18 on Mn- and Cd-amended soils. Sweet sorghum was found to have higher metal absorption (Mn vs. Cd; 65% vs. 40%) than *P. acinosa* (Mn vs. Cd; 55% vs. 31%) and *S. nigrum* (Mn vs. Cd; 18% vs. 25%). The effect of this remediation process on biomass was also observed in the order of sweet sorghum > *P. acinosa* > *S. nigrum*.

8 WORK AT ICRISAT

ICRISAT has developed several improved hybrid parental lines of sweet sorghum with high stalk sugar content that are currently being tested in pilot studies for sweet sorghum-based ethanol production in India, the Philippines, Mali, and Mozambique. Concerted research efforts under National Agricultural Research System (NARS) have led to the development and release of cultivars like SSV 84, CSV 19SS, CSH 22SS, and CSV 24SS for all India cultivation with productivity ranging from 40 to 50 tons ha^{-1} (Vinutha et al., 2014). Trial data over 3 years (2005-2007) and six seasons indicated that there is no reduction in grain yield while improving the sugar yield. Sugar yield and associated traits have greater genotype × environment interaction; therefore, it is prudent to breed for season-specific hybrids.

ICRISAT launched a global **BioPower** initiative in 2007 to find ways to empower the dryland poor to benefit from emerging opportunities in renewable energies. This involves the collaborative partnership of NARS, particularly India, the Philippines, Mali, and private sector partners in Brazil, the United States, Germany, and Mexico. ICRISAT focuses on hybrids parent development to produce cultivars withstanding biotic and abiotic stresses thereby strengthening sweet sorghum value chains and their impact. The ICRISAT has made the first attempt in India to evaluate and identify useful high biomass producing sweet sorghum germplasm from world collections. The sweet sorghum program at ICRISAT mainly focuses on developing primarily hybrid parents adapted to rainy and post-rainy seasons due to the highly significant interaction of genotype by environment (G × E). However about 100 sweet sorghum varieties and restorer lines and 50 improved hybrids were identified. ICSV 93046, ICSV 25274, ICSV 25280, and ICSSH 58 were identified for release owing to their superior performance in All India Coordinated Sorghum Improvement Project (AICSIP) multilocation trials during 2008-2012 (Rao et al., 2013). Sweet sorghum improvement aims for simultaneous improvement of stalk sugar traits such as total soluble sugars or (brix %), green stalk yield, juice quantity, girth of the stalk, and grain yield. Conventional breeding approaches are practiced for an increase in sucrose yield; R lines showed a brix percentage of 12-24% in the rainy season and 9-19% in the post-rainy season. In total, 600 A/B pairs were screened at ICRISAT and the brix percentage ranged from 10% to 15% in the rainy season and 8% to 13% in the post-rainy season (Rao et al., 2009). Sweet sorghum bagasse is highly palatable and intake by livestock is more than normal sorghum stover (Blummel et al., 2009).

Some insect- and pest-resistant materials have been developed at ICRISAT, such as ICSR 93034 and ICSV 700. ICSV 93046 (ICSV 700 × ICSV 708) is a promising sweet sorghum variety tolerant to shoot fly, stem borer, and leaf diseases; it also displays stay-green stems and leaves even after physiological maturity and has good grain (3.4-4.1 tons ha^{-1}) and biomass yield. Another hybrid, ICSSH 72, shows excellent fodder quality in the rainy season and is resistant to leaf diseases. SPV 422 also exhibits resistance to leaf diseases and other hybrids developed at ICRISAT, India; for example, ICSSH 21 (ICSA 38 × NTJ 2) and ICSSH 58 (ICSA 731 × ICSV 93046) are under advance testing stages. ICSSH 30 variety shows superior grain yields in both rainy and post-rainy seasons whereas ICSSH 39 and 28 are best for sugar yield. ICSSH 24 variety is supposed to be best suited for the rainy season (Vinutha et al., 2014). Some of the varieties and hybrids developed from ICRISAT are given in Table 5.

9 WORK AT INDIAN NARS

Concerted research efforts at AICSIP centers have resulted in the identification of several promising sweet sorghum varieties such as SSV 96, GSSV 148, SR 350-3, SSV 74, HES 13, HES 4, SSV 119, and SSV 12611 for total soluble solids (TSS%) and juice yield during 1991-1992 trials, GSSV 148 for cane sugar during 1993-1994 trials, NSS 104 and HES 4 for green cane yield, juice yield, juice extraction, and total sugar content during 1999-2000 trials, and RSSV 48 for better alcohol yield during 2001-2002. An evaluation of 11 promising sweet sorghum varieties bred at different AICSIP centers indicated superiority of the varieties NSSV 255 and RSSV 56 for green cane yield, juice yield, juice extractability, commercial cane sugar (CCS) yield (q ha^{-1}), and percent nonreducing sugars over the rest of the varieties. The varieties RSSV 79, PKV809, NSSV 256, and NSSV 6 excelled the check with superior performance for green cane yield, juice yield, juice extractability, CCS yield, and total sugars (Reddy et al., 2007).

The Rusni distillery, established in 2007 near Sangareddy in the Medak district of Telangana, India, was the first sweet sorghum distillery amenable to use multiple feedstocks for transport-grade ethanol production. It generated 99.4% of fuel ethanol with a total capacity of 40 kiloliters per day (KLPD). It also produced 96% extra neutral alcohol (ENA) and 99.8% pharma alcohol from agro-based raw materials such as sweet sorghum juice, molded grains, broken rice, cassava, and rotten fruits. ICRISAT has incubated sweet sorghum ethanol production in partnership with Rusni Distilleries through its Agri-Business Incubator. A pilot-scale sweet sorghum distillery of 30 KLPD capacity was established in 2009 at Nanded, Maharashtra. It used commercially grown sweet sorghum cultivars such as CSH 22SS, ICSV 93046, sugargrace, JK Recova, and RSSV 9 in the 25 km radius of the distillery to produce transport-grade ethanol and ENA during 2009-2010. However, it could not continue operations due to the low mandated ethanol price. The Cabinet Committee of Economic Affairs (CCEA) of GOI on November 22, 2012, recommended 5% mandatory blending of ethanol with gasoline (Aradhey and Lagos, 2013). The government's current target of 5% blending of ethanol in gasoline has been partially successful in years of surplus sugar production and unfulfilled when sugar production declines.

Table 5 Sweet sorghum Varieties and Hybrids Developed from ICRISAT						
Varieties				**Hybrids**		
ICSV 93046	ICSSV 10032	ICSSV 10059	ICSSV 12021	ICSSH 1	ICSSH 28	ICSSH 55
ICSSV 10001	ICSSV 10033	ICSSV 10060	ICSSV 12022	ICSSH 2	ICSSH 29	ICSSH 56
ICSSV 10005	ICSSV 10034	ICSSV 10061	ICSSV 12023	ICSSH 3	ICSSH 30	ICSSH 57
ICSSV 10006	ICSSV 10035	ICSSV 10062	ICSSV 12024	ICSSH 4	ICSSH 31	ICSSH 58
ICSSV 10007	ICSSV 10036	ICSSV 10063	ICSSV 12025	ICSSH 5	ICSSH 32	ICSSH 59
ICSSV 10008	ICSSV 10037	ICSSV 10064	ICSSV 12026	ICSSH 6	ICSSH 33	ICSSH 60
ICSSV 10009	ICSSV 10038	ICSSV 10065	ICSSV 12027	ICSSH 7	ICSSH 34	ICSSH 61
ICSSV 10010	ICSSV 10039	ICSSV 10066	ICSSV 14001	ICSSH 8	ICSSH 35	ICSSH 62
ICSSV 10011	ICSSV 10040	ICSSV 10067	ICSSV 14002	ICSSH 9	ICSSH 36	ICSSH 63
ICSSV 10012	ICSSV 10041	ICSSV 10068	ICSSV 14003	ICSSH 10	ICSSH 37	ICSSH 64
ICSSV 10013	ICSSV 10042	ICSSV 10069	ICSSV 14004	ICSSH 11	ICSSH 38	ICSSH 65
ICSSV 10014	ICSSV 10043	ICSSV 12005	ICSSV 14005	ICSSH 12	ICSSH 39	ICSSH 66
ICSSV 10015	ICSSV 10044	ICSSV 12006	ICSSV 12020	ICSSH 13	ICSSH 40	ICSSH 67
ICSSV 10016	ICSSV 10045	ICSSV 12007	ICSSV 12021	ICSSH 14	ICSSH 41	ICSSH 68
ICSSV 10017	ICSSV 10046	ICSSV 12008	ICSSV 12022	ICSSH 15	ICSSH 42	ICSSH 69
ICSSV 10018	ICSSV 10047	ICSSV 12009	ICSSV 12023	ICSSH 16	ICSSH 43	ICSSH 70
ICSSV 10019	ICSSV 10048	ICSSV 12010	ICSSV 12024	ICSSH 17	ICSSH 44	ICSSH 71
ICSSV 10021	ICSSV 10049	ICSSV 12011	ICSSV 12025	ICSSH 18	ICSSH 45	ICSSH 72
ICSSV 10022	ICSSV 10050	ICSSV 12012	ICSSV 12026	ICSSH 19	ICSSH 46	ICSSH 73
ICSSV 10024	ICSSV 10051	ICSSV 12013	ICSSV 12027	ICSSH 20	ICSSH 47	ICSSH 74
ICSSV 10025	ICSSV 10052	ICSSV 12014	ICSSV 14001	ICSSH 21	ICSSH 48	ICSSH 75
ICSSV 10026	ICSSV 10053	ICSSV 12015	ICSSV 14002	ICSSH 22	ICSSH 49	ICSSH 76
ICSSV 10027	ICSSV 10054	ICSSV 12016	ICSSV 14003	ICSSH 23	ICSSH 50	ICSSH 77
ICSSV 10028	ICSSV 10055	ICSSV 12017	ICSSV 14004	ICSSH 24	ICSSH 51	ICSSH 78
ICSSV 10029	ICSSV 10056	ICSSV 12018	ICSSV 14005	ICSSH 25	ICSSH 52	
ICSSV 10030	ICSSV 10057	ICSSV 12019		ICSSH 26	ICSSH 53	
ICSSV 10031	ICSSV 10058	ICSSV 12020		ICSSH 27	ICSSH 54	

The interim price of US 0.441^{-1} would no longer hold as the price would now be decided by market forces. It is expected this decision will have a positive effect on forthcoming distilleries in India.

10 WORK IN OTHER COUNTRIES

Other countries involved in sweet sorghum research and development are the United States, Brazil, Colombia, Haiti, Argentina, Italy, Germany, Hungary, France, China, the Philippines, and Indonesia. Nonfood crops and materials such as cassava and sweet sorghum are the priority choice for biofuel ethanol production in China. Sorghum Research Institute (SRI) of Liaoning Academy of Agricultural Sciences (LAAS) is the lead organization involved in sweet sorghum research in China since the 1980s. So far 17 promising sweet sorghum hybrids were released nationally. A few industries such as ZTE energy company limited

(ZTE, Inner Mongolia), Fuxin Green BioEnergy Corporation (FGBE), Xinjiang Santai Distillery, Liaoning Guofu Bioenergy Development Company Limited, Binzhou Guanghua Biology Energy Company Ltd, Jiangxi Qishengyuan Agri-Biology Science and Technology Company Ltd, Jilin Fuel Alcohol Company Limited, and Heilongjiang Huachuan Siyi Bio-fuel Ethanol Company Ltd conducted either large-scale sweet sorghum processing trials or are at the commercialization stage (Reddy et al., 2011).

In the Philippines, sweet sorghum has been proven to be a technically and economically viable alternative feedstock for bioethanol production. The plantation, agronomic performance, and actual bioethanol production of sweet sorghum have been evaluated on different plantation sites nationwide. A hectare of sweet sorghum plantation can potentially provide farmers with an annual net income of US $1860.47 at a stalk-selling price of US $22 and grain price of US $0.30. The San Carlos Bioenergy Inc. (SCBI) became the first commercial distillery to process sweet sorghum bioethanol in Southeast Asia under the Department of Agriculture (DA) and produced 14,000 l of fuel-grade ethanol in 2012 (Demafelis et al., 2013). The Ecofuels 300 KLPD distillery at San Mariano, Isabela, is planning to use sugarcane and sweet sorghum as feedstocks for ethanol production commercially. The sweet sorghum growers are enthusiastic as the ratoon (new shoot) yields are about 20-25% higher than that of plant crop. The Bapamin enterprises based in Batac have been successfully marketing vinegar and hand sanitizer made from sweet sorghum since 2009 (Reddy et al., 2011).

In the United States, a sweet sorghum distillery is under construction in South Florida by South Eastern Biofuels Ltd. In Brazil, large-scale sweet sorghum pilot trials are being conducted in the last 3 years by Ceres Inc, Chromatin Inc, Advanta Inc, Dow Agro Sciences, as well as Empresa Brasileira de Pesquisa Agropecuária (EMBRAPA) in the areas of sugarcane renovation to commercialize sweet sorghum (Nass et al., 2007). The Government of Brazil has identified 1.8 m ha for sweet sorghum plantation to augment fuel-grade ethanol production. In some African countries like Mozambique, Kenya, South Africa, and Ethiopia, sweet sorghum adaptation trials are being conducted in pilot scale to assess feasibility of sweet sorghum for biofuel production.

11 FUTURE OUTLOOK

The phytoremediation potential of sweet sorghum has not been studied extensively enough under field conditions. The published data has come from trials using pot conditions or microcosm experiments, which are not adequate for future cleanup of contaminated areas. In an economic viability assessment done by Basavaraj et al. (2013) on ethanol production from sweet sorghum, net present value (NPV), the indicator of economic viability assessment, is found to be negative. So, it may be difficult for the industry to take off under the current scenario of fluctuating ethanol price, feedstock price, and ethanol recovery rate. A well-developed technology for phytoremediation and ethanol production involving sweet sorghum along with well-defined policy support is crucial to meet future blending requirements and also to improve rural while adhering to greenfield biorefinery approach.

REFERENCES

Abbas, M., Parveen, Z., Iqbal, M., Riazuddin, I.S., Ahmed, M., Bhutto, R., 2010. Monitoring of toxic metals (cadmium, lead, arsenic and mercury) in vegetables of Sindh, Pakistan. Kathmandu Univ. J. Sci. Eng. Technol. 6, 60–65.

Abou-Shanab, R.A., Ghanem, K., Ghanem, N., Al-Kolaibe, A., 2008. The role of bacteria on heavy-metal extraction and uptake by plants growing on multi-metal-contaminated soils. World J. Microbiol. Biotechnol. 24, 253–262.

Almodares, A., Hadi, M.R., 2009. Production of bioethanol from sweet sorghum: a review. Afr. J. Agric. Res. 4 (9), 772–780.

An, Y.J., 2004. Soil ecotoxicity assessment using cadmium sensitive plants. Environ. Pollut. 127 (1), 21–26.

Anonymous, 1990. NEA dumps on science art. Science 250, 1515.

Aradhey, A., Lagos, J., 2013. India – Biofuel Annual. *GAIN Report*. Last accessed at http://gain.fas.usda.gov/Recent%20GAIN%20Publications/Biofuels%20Annual_New%20Delhi_India_8-13-2013.pdf.

Aydinalp, C., Marinova, S., 2003. Distribution and forms of heavy metals in some agricultural soils. Pol. J. Environ. Stud. 12 (5), 629–633.

Baker, A.J.M., Brooks, R.R., 1989. Terrestrial higher plants which hyperaccumulate metallic elements—a review of their distribution, ecology and phytochemistry. Biorecovery 1, 81–126.

Bano, S.A., Ashfaq, D., 2013. Role of mycorrhiza to reduce heavy metal stress. Nat. Sci. 5 (12A), 16–20.

Barzanti, R., Colzi, I., Arnetoli, M., Gallo, A., Pignattelli, S., Gabbrielli, R., Gonnelli, C., 2011. Cadmium phytoextraction potential of different *Alyssum* species. J. Hazard. Mater. 196, 66–72.

Basak, G., Das, N., 2014. Characterization of sophorolipid biosurfactant produced by *Cryptococcus* sp. VITGBN2 and its application on Zn(II) removal from electroplating wastewater. J. Environ. Biol. 35, 1087–1094.

Basavaraj, G., Rao, P.P., Basu, K., Reddy, Ch.R., Kumar, A.A., Rao, P.S., Reddy, B.V.S., 2012. A review of the national biofuel policy in India: a critique of the need to promote alternative feedstocks. Working Paper Series no. 34. International Crops Research Institute for the Semi-Arid Tropics, Andhra Pradesh, India.

Basavaraj, G., Rao, P.P., Basu, K., Reddy, Ch.R., Kumar, A.A., Rao, P.S., Reddy, B.V.S., 2013. Assessing viability of bio-ethanol production from sweet sorghum in India. Energy Policy 56, 501–508.

Bayramoglu, G., Arica, M.Y., Adiguzel, N., 2012. Removal of Ni(II) and Cu(II) ions using native and acid treated Ni-hyperaccumulator plant *Alyssum discolor* from Turkish serpentine soil. Chemosphere 89 (3), 302–309.

Beolchini, F., Anno, D.A., Propris, L.D., Ubaldini, S., Cerrone, F., Danovaro, R., 2009. Auto and heterotrophic acidophilic bacteria enhance the bioremediation efficiency of sediments contaminated by heavy metals. Chemosphere 74, 1321–1326.

Blummel, M., Rao, S.S., Palaniswami, S., Shah, L., Reddy, B.V., 2009. Evaluation of sweet sorghum (*Sorghum bicolor* L. Moench) used for bio-ethonol production in the context of optimizing whole plant utilization. Anim. Nutr. Feed. Technol. 9 (1), 1–10.

Brennan, M.A., Shelley, M.L., 1999. A model of the uptake, translocation, and accumulation of lead (Pb) by maize for the purpose of phytoextraction. Ecol. Eng. 12, 271–297.

Bricka, R.M., Williford, C.W., Jones, L.W., 1993. Technology assessment of currently available and developmental techniques for heavy metals – contaminated soils treatment. *Technical Report, IRRP-93-4*, pp. 1–3. U.S. Army Engineer Waterways Experiment Station, USA.

Brooks, R.R., Lee, J., Reeves, R.D., Jaffre, T., 1977. Detection of nickeliferous rocks by analysis of herbarium specimens of indicator plants. J. Geochem. Explor. 7, 49–77.

Burbacher, T.M., Rodier, P.M., Weiss, B., 1990. Methylmercury developmental neurotoxicity: a comparison of effects in humans and animals. Neurotoxicol. Teratol. 12 (3), 191–202.

Carlot, M., Giacomini, A., Casella, S., 2002. Aspects of plant-microbe interactions in heavy metal polluted soil. Acta Biotechnol. 22, 13–20.

Caussy, D., Gochfeld, M., Gurzau, E., Neagu, C., 2003. Lessons from case studies of metals: investigating exposure, bioavailability, and risk. Ecotoxicol. Environ. Saf. 54, 45–51.

Chaney, R.L., 1983. Plant uptake of inorganic waste constituents. In: Parr, J.F., Marsh, P.D., Kla, J.M. (Eds.), Land Treatment of Hazardous Wastes. Noyes Data Corporation, Park Ridge, NJ, pp. 50–76.

Chaney, R.L., Malik, M., Li, Y.M., Brown, S.L., Brewer, E.P., Angle, J.S., Baker, A.J.M., 1997. Phytoremediation of soil metals. Curr. Opin. Biotechnol. 8, 279–284.

Cheng, G.R., Timilsin, A., 2011. Status and barriers of advanced biofuel technologies: a review. Renew. Energy 36, 3541–3549.

Chibuike, G.U., Obiora, S.C., 2014. Heavy metal polluted soils: effect on plants and bioremediation methods. Appl. Environ. Soil Sci. 2014, 12. Article ID 752708.

Chopra, A.K., Pathak, C., Prasad, G., 2009. Scenario of heavy metal contamination in agricultural soil and its management. J. Appl. Nat. Sci. 1 (1), 99–108.

Christakopoulos, P., Li, L.W., Kekos, F., Macris, B.J., 1993. Direct conversion of sorghum carbohydrates to ethanol by a mixed microbial culture. Bioresour. Technol. 45, 89–92.

Ciura, J., Poniedzialek, M., Sekara, A., Jedrszczyk, E., 2005. The possibility of using crops as metal phytoremediation. Pol. J. Environ. Stud. 14 (1), 17–22.

Conesa, H.M., Evangelou, M.W.H., Robinson, B.H., Schulin, R., 2012. A critical view of current state of phytotechnologies to remediate soils: still a promising tool? Sci. World J. 2012, 173829. http://dx.doi.org/10.1100/2012/173829.

Delorme, T.A., Gagliardi, J.V., Angle, J.S., Chaney, R.L., 2001. Influence of the zinc hyperaccumulator *Thlaspi caerulescens* J. & C. Presl. and the nonmetal accumulator *Trifolium pratense* L. on soil microbial population. Can. J. Microbiol. 47, 773–776.

Demafelis, R.B., El Jirie, N.B., Hourani, K.A., Tongko, B., 2013. Potential of bioethanol production from sweet sorghum in the Philippines: an income analyses for farmers and distilleries. Sugar Tech. 15 (3), 225–231.

Duffus, J.H., 2002. Heavy metals—a meaningless term? Pure Appl. Chem. 74 (5), 793–807.

Duponnois, R., Kisa, M., Assigbetse, K., Prin, Y., Thioulouse, J., Issartel, M., Moulin, P., Lepage, M., 2006. Fluorescent pseudomonads occurring in *Macrotermes subhyalinus* mound structures decrease Cd toxicity and improve its accumulation in sorghum plants. Sci. Total Environ. 370, 391–400.

Duruibe, J.O., Ogwuegbu, M.O.C., Egwurugwu, J.N., 2007. Heavy metal pollution and human biotoxic effects. Int. J. Phys. Sci. 2 (5), 112–118.

EEA, 2005. The European Environment – State and Outlook 2005. European Environment Agency, Copenhagen.

Epelde, L., Mijangos, I., Becerril, J.M., Garbisu, C., 2009. Soil microbial community as bioindicator of the recovery of soil functioning derived from metal phytoextraction with sorghum. Soil Biol. Biochem. 41, 1788–1794.

Foster, L.J.R., Moy, Y.P., Rogers, P.L., 2000. Metal binding capabilities of *Rhizobium etli* and its extracellular polymeric substances. Biotechnol. Lett. 22, 1757–1760.

Franzetti, A., Gandolfi, I., Fracchia, L., Van Hamme, J., Gkorezis, P., Roger, M., Ibrahim, B.M., 2014. Biosurfactant use in heavy metal removal from industrial effluents and contaminated sites. In: Kosaric, N., Sukan, F.V. (Eds.), Biosurfactants: Production and Utilization—Processes, Technologies and Economics. CRC Press, Boca Raton, FL, pp. 361–369.

Ganesan, V., 2008. Rhizoremediation of cadmium soil using a cadmium-resistant plant growth-promoting rhizopseudomonad. Curr. Microbiol. 56, 403–407.

Glick, B.R., 2003. Phytoremediation: synergistic use of plants and bacteria to clean up the environment. Biotechnol. Adv. 21, 383–393.

Glick, B.R., 2014. Bacteria with ACC deaminase can promote plant growth and help to feed the world. Microbiol. Res. 169 (1), 30–39. http://dx.doi.org/10.1016/j.micres.2013.09.009.

Glick, B.R., Patten, C.L., Holguin, G., Penrose, D.M., 1999. Biochemical and Genetic Mechanisms Used by Plant Growth Promoting Bacteria. Imperial College Press, London.

Gomes, H.I., 2012. Phytoremediation for bioenergy: challenges and opportunities. Environ. Technol. Rev. 1 (1), 59–66.

He, J., Chen, J.P., 2014. A comprehensive review on biosorption of heavy metals by algal biomass: materials, performances, chemistry, and modeling simulation tools. Bioresour. Technol. 160, 67–78.

Hernández-Allica, J., Becerril, J.M., Garbisu, C., 2008. Assessment of the phytoextraction potential of high biomass crop plants. Environ. Pollut. 152, 32–40.

Idris, R., Trifonova, R., Puschenreiter, M., Wenze, W.W., Sessitsch, A., 2004. Bacterial communities associated with flowering plants of the Ni hyperaccumulator *Thlaspi goesingense*. Appl. Environ. Microbiol. 70, 2667–2677.

Jaffré, T., 1980. Étude Écologique du Peuplement Végétal des Sols Dérivés de Roches Ultrabasiques en Nouvelle Calédonie. ORSTOM, Paris. pp. 273.

Jaffré, T., Brooks, R.R., Lee, J., Reeves, R.D., 1976. *Sebertia acuminata*: a hyperaccumulator of nickel from New Caledonia. Science 193, 579–580.

Jain, R., Srivastava, S., Solomon, S., Shrivastava, A.K., Chandra, A., 2010. Impact of excess zinc on growth parameters, cell division, nutrient accumulation, photosynthetic pigments and oxidative stress of sugar cane. Acta Physiol. Plant. 32, 979–986.

JECFA, 1982. Evaluation of certain food additives and contaminants (Twenty-sixth report of the Joint FAO/WHO Expert Committee on Food Additives). WHO Technical Report Series, No. 683.

JECFA, 2011. Evaluation of certain contaminants in food (Seventy-second report of the Joint FAO/WHO Expert Committee on Food Additives). WHO Technical Report Series, No. 959.

JECFA, 2013. Evaluation of certain food additives and contaminants (Seventy-seventh report of the Joint FAO/WHO Expert Committee on Food Additives). WHO Technical Report Series, No. 983.

Jiang, C.Y., Sheng, X.F., Qian, M., Wang, Q.Y., 2008. Isolation and characterization of a heavy metal resistant *Burkholderia* sp. from heavy metal-contaminated paddy field soil and its potential in promoting plant growth and heavy metal accumulation in metal polluted soil. Chemosphere 72, 157–164.

Joshi, P., Juwarkar, A., 2009. In vivo studies to elucidate the role of extracellular polymeric substances from *Azotobacter* in immobilization of heavy metals. Environ. Sci. Technol. 43, 5884–5889.

Kaplan, M., Orman, S., Kadar, I., Koncz, J., 2005. Heavy metal accumulation in calcareous soil and sorghum plants after addition of sulphur-containing waste as a soil amendment in Turkey. Agric. Ecosyst. Environ. 111, 41–46.

Kapoor, A., Viraraghavan, T., Cullimore, R.D., 1999. Removal of heavy metals using the fungus *Aspergillus niger*. Bioresour. Technol. 70, 95–104.

Kargi, F., Curme, J.A., 1985. Solid state fermentation of sweet sorghum to ethanol in a rotary-drum fermentor. Biotechnol. Bioeng. 27, 1122–1125.

Kargi, F., Curme, J.A., Sheehan, J.J., 1985. Solid state fermentation of sweet sorghum to ethanol. Biotechnol. Bioeng. 27, 34–40.

Keller, C., Hammer, D., Kayser, A., Richner, W., Brodbeck, M., Sennhauser, M., 2003. Root development and heavy metal phytoextraction efficiency: comparison of different plant species in the field. Plant Soil 249 (1), 67–81.

Kim, M., Day, D.F., 2011. Composition of sugar cane, energy cane, and sweet sorghum suitable for ethanol production at Louisiana sugar mills. J. Ind. Microbiol. Biotechnol. 38 (7), 803–807.

Kojima, M., Johnson, T., 2005. Potential for biofuels for transport in developing countries. Energy Sector Management Assistance Programme report. The International Bank for Reconstruction and Development/ The World Bank, Washington, DC.

Komárek, M., Tlustoš, P., Szákova, J., Richner, W., Brodbeck, M., Sennhauser, M., 2007. The use of maize and poplar in chelant-enhanced phytoextraction of lead from contaminated agricultural soils. Chemosphere 67 (4), 640–651.

Kovarik, B., 1998. Henry Ford, Charles Kettering and the fuel of the future. Automot. Hist. Rev. Spring 32, 7–27.

Kuffner, M., Puschenreiter, M., Wieshammer, G., Gorfer, M., Sessitsch, A., 2008. Rhizosphere bacteria affect growth and metal uptake of heavy metal accumulating willows. Plant Soil 304, 35–44.

Kukier, U., Chaney, R.L., 2004. In situ remediation of nickel phytotoxicity for different plant species. J. Plant Nutr. 27, 465–495.

Ledin, M., Krantz-Rulcker, C., Allard, B., 1996. Zn, Cd and Hg accumulation by microorganisms, organic and inorganic soil components in multicompartment system. Soil Biol. Biochem. 28, 791–799.

Lezinou, V., Christakopoulos, P., Kekos, D., Macris, B.J., 1994. Simultaneous saccharification and fermentation of sweet sorghum polysaccharides to ethanol in a fed-batch process. Biotechnol. Lett. 16, 983–988.

Li, X., Coles, B.J., Ramsey, M.H., Thornton, I., 1995. Chemical partitioning of the new National Institute of Standards and Technology standard reference materials (SRM 2709–2711) by sequential extraction using inductively coupled plasma atomic emission spectrometry. Analyst 120, 1415–1419.

Li, J.L., He, M., Han, W., Gu, Y.F., 2009. Analysis and assessment on heavy metal sources in the coastal soils developed from alluvial deposits using multivariate statistical methods. J. Hazard. Mater. 164, 976–981.

Licht, F.O., 2010. World Ethanol and Biofuels Report. 8 (16), 16–21.

Liu, Y., Lam, M.C., Fang, H.H.P., 2001. Adsorption of heavy metals by EPS of activated sludge. Water Sci. Technol. 43, 59–66.

Ma, Y., Prasad, M.N.V., Rajkumar, M., Freitas, H., 2011. Plant growth promoting rhizobacteria and endophytes accelerate phytoremediation of metalliferous soils. Biotechnol. Adv. 29, 248–258.

Madejón, P., Murillo, J.M., Marañón, T., Cabrera, F., Soriano, M.A., 2003. Trace element and nutrient accumulation in sunflower plants two years after the Aznalcóllar spill. Sci. Total Environ. 307 (1–3), 239–257.

Mamma, D., Christakopoulos, P., Koullas, D., Macris, K.B.J., Koukios, E., 1995. An alternative approach to the bioconversion of sweet sorghum carbohydrates to ethanol. Biomass Bioenergy 8 (2), 99–103.

Marchiol, L., Fellet, G., Perosa, D., Zerbi, G., 2007. Removal of trace metals by *Sorghum bicolor* and *Helianthus annuus* in a site polluted by industrial wastes: a field experience. Plant Physiol. Biochem. 45, 379–387.

McGrath, S.P., Zhao, F.J., Lombi, E., 2001. Plant and rhizosphere process involved in phytoremediation of metal-contaminated soils. Plant Soil 232, 207–214.

Meers, E., Ruttens, A., Hopgood, M., Lesage, E., Tack, F.M.G., 2005. Potential of *Brassica rapa*, *Cannabis sativa*, *Helianthus annuus* and *Zea mays* for phytoextraction of heavy metals from calcareous dredged sediment derived soils. Chemosphere 61 (4), 561–572.

Meers, E., Slycken, S.V., Adriaensen, K., Ruttens, A., Vangronsveld, J., Laing, G.D., Witters, N., Thewys, T., Tack, F.M.G., 2010. The use of bio-energy crops (*Zea mays*) for phytoattenuation of heavy metals on moderately contaminated soils: a field experiment. Chemosphere 78, 35–41.

Millward, G.E., Turner, A., 2001. Metal pollution. In: Steele, J.H., Turekian, K.K., Thorpe, S.A. (Eds.), Encyclopedia of Ocean Sciences. Academic Press, San Diego, CA, pp. 1730–1737.

Minteer, S., 2006. Alcoholic Fuels. Taylor & Francis, New York.

Moffat, A., 1995. Plants proving their worth in toxic metal cleanup. Science 269, 302–303.

Mudgal, V., Madaan, N., Mudgal, A., Singh, R.B., Mishra, S., 2010. Effect of toxic metals on human health. Open Nutraceuticals J. 3, 94–99.

Mussatto, S.I., Dragone, G., Guimarães, P.M.R., Paulo, J., Silva, A., Carneiro, L.M., Roberto, I.C., Vicente, A., Domingues, L., Teixeira, J.A., 2010. Technological trends, global market, and challenges of bio-ethanol production. Biotechnol. Adv. 28, 817–830.

Nass, L.L., Pereira, P.A.A., Ellis, D., 2007. Biofuels in Brazil: an overview. Crop Sci. 47 (6), 2228–2237.

Nriagu, J.O., Pacyna, J.M., 1988. Quantitative assessment of worldwide contamination of air, water and soils by trace metals. Nature 333, 134–139.

Penrose, D.M., Glick, B.R., 2001. Levels of 1-aminocyclopropane-1-carboxylic acid (ACC) in exudates and extracts of canola seeds treated with plant growth-promoting bacteria. Can. J. Microbiol. 47 (4), 368–372.

Purseglove, J.W., 1972. Tropical Crops: Monocotyledons. Longman Group Limited, London. pp. 334.

Quartacci, M.F., Micaelli, F., Sgherri, C., 2014. *Brassica carinata* planting pattern influences phytoextraction of metals from a multiple contaminated soil. Agrochimica 58 (1).

Ragnarsdottir, K.V., Hawkins, D., 2005. Trace metals in soils and their relationship with scrapie occurrence. Geochim. Cosmochim. Acta 69, A194–A196.

Rajkumar, M., Ma, Y., Freitas, H., 2008. Characterization of metal-resistant plant-growth promoting *Bacillus weihenstephanensis* isolated from serpentine soil in Portugal. J. Basic Microbiol. 48, 1–9.

Rajkumar, M., Ae, N., Prasad, M.N.V., Freitas, H., 2010. Potential of siderophore-producing bacteria for improving heavy metal phytoextraction. Trends Biotechnol. 28, 142–149.

Rajkumar, M., Sandhya, S., Prasad, M.N.V., Freitas, H., 2012. Perspectives of plant-associated microbes in heavy metal phytoremediation. Biotechnol. Adv. 30, 1562–1574.

Rao, P.S., Rao, S.S., Seetharama, N., Umakath, A.V., Reddy, P.S., Reddy, B.V.S., Gowda, C.L.L., 2009. Sweet Sorghum for Biofuel and Strategies for its Improvement. International Crops Research Institute for the Semi-Arid Tropics, Andhra Pradesh, India.

Rao, P.S., Reddy, B.V.S., Reddy, R.Ch., Blümmel, M., Kumar, A.A., Rao, P.P., Basavaraj, G., 2012. Utilizing co-products of the sweet sorghum based biofuel industry as livestock feed in decentralized systems. In: Makkar, H.P.S. (Ed.), Biofuel Co-Products as Livestock Feed: Opportunities and Challenges. FAO, Rome, Italy, pp. 229–242.

Rao, P.S., Kumar, C.G., Reddy, B.V., 2013. Sweet sorghum: from theory to practice. In: Rao, P.S., Kumar, C.G. (Eds.), Characterization of Improved Sweet Sorghum Cultivars. Springer, India, pp. 1–15.

Rascio, N., Navari-Izzo, F., 2011. Heavy metal hyperaccumulating plants: how and why do they do it? And what makes them so interesting? Plant Sci. 180 (2), 169–181.

Rayment, G.E., Jeffrey, A.J., Barry, G.A., 2002. Heavy metals in Australian sugar cane. Commun. Soil Sci. Plant Anal. 33, 3203–3212.

Reddy, B.V., Ashok Kumar, A., Ramesh, S., 2007. Sweet sorghum: a water saving bio-energy crop. In: International Conference on Linkages Between Energy and Water Management for Agriculture in Developing Countries, January 29–30, 2007. IWMI, ICRISAT, Hyderabad, India.

Reddy, B.V.S., Layaoen, H., Dar, W.D., Rao, P.S., Eusebio, J.E., 2011. Sweet Sorghum in the Philippines: Status and Future. International Crops Research Institute for the Semi-Arid Tropics, Andhra Pradesh, India.

Rugh, C.L., Wilde, H.D., Stack, N.M., Thompson, D.M., Summers, A.O., Meagher, R.B., 1996. Mercuric ion reduction and resistance in transgenic *Arabidopsis thaliana* plants expressing a modified bacterial *merA* gene. Proc. Natl. Acad. Sci. 93, 3182–3187.

Salt, D.E., Blaylock, M., Kumar, N.P.B.A., Dushenkov, V., Ensley, B.D., Chet, I., Raskin, I., 1977. Phytoremediation: a novel strategy for the removal of toxic metals from the environment using plants. Nat. Biotechnol. 13, 468–475.

Sanderson, M.A., Jones, R.M., Ward, J., Wolfe, R., 1992. Silage sorghum performance trial at Stephen-ville. Forage Research in Texas. *Report PR-5018*. Texas Agricultural Experimental Station, Stephenville, USA.

Saravanan, V.S., Madhaiyan, M., Thangaraju, M., 2007. Solubilization of zinc compounds by the diazotrophic, plant growth promoting bacterium *Gluconacetobacter diazotrophicus*. Chemosphere 66, 1794–1798.

Segura-Muñoz, S.I., da Silva, O.A., Nikaido, M., Trevilato, T.M., Bocio, A., Takayanagui, A.M., Domingo, J.L., 2006. Metal levels in sugar cane (*Saccharum* spp.) samples from an area under the influence of a municipal landfill and a medical waste treatment system in Brazil. Environ. Int. 32 (1), 52–57.

Serna-Saldívar, S.O., Chuck Hernandez, C., Perez Carrillo, E., Heredia-Olea, E., 2012. Sorghum as a multifunctional crop for the production of fuel ethanol: current status and future trends. In: Lima, M.A.P. (Ed.), Bioethanol. InTech, China, pp. 51–74.

Sharma, P., Pandey, S., 2014. Status of phytoremediation in world scenario. Int. J. Environ. Biorem. Biodegrad. 2 (4), 178–191.

Shen, Z.G., Li, X.D., Wang, C.C., Chen, H.M., Chua, H., 2002. Lead phytoextraction from contaminated soil with high-biomass plant species. J. Environ. Qual. 31 (6), 1893–1900.

Shi, J.Y., Lin, H.R., Yuan, X.F., Chen, X.C., Shen, C.F., Chen, Y.X., 2011. Enhancement of copper availability and microbial community changes in rice rhizospheres affected by sulfur. Molecules 16, 1409–1417.

Simpson, J.A., Picchi, G., Gordon, A.M., Thevathasan, N.V., Stanturf, J., Nicholas, I., 2009. Short Rotation Crops for Bioenergy Systems. Environmental Benefits Associated with Short-Rotation Woody Crops. IEA, Paris. Technical Review No. 3, IEA Bioenergy.

Soriano, M.A., Fereres, E., 2003. Use of crops for in situ phytoremediation of polluted soils following a toxic flood from a mine spill. Plant Soil 256 (2), 253–264.

Soudek, P., Petrová, S., Vaněk, T., 2012. Phytostabilization or accumulation of heavy metals by using of energy crop *Sorghum* sp. Int. Proc. Chem. Biol. Environ. Eng. 46, 25–29.

Sun, L.N., Zhang, Y.F., He, L.Y., Chen, Z.J., Wang, Q.Y., Qian, M., Sheng, X.F., 2010. Genetic diversity and characterization of heavy metal-resistant-endophytic bacteria from two copper-tolerant plant species on copper mine wasteland. Bioresour. Technol. 101, 501–509.

van der Ent, A., Baker, A.J.M., Reeves, R.D., Pollard, A.J., Schat, H., 2012. Hyperaccumulators of metal and metalloid trace elements: facts and fiction. Plant Soil 362 (1-2), 319–334. http://dx.doi.org/10.1007/s11104-012-1287-3.

van der Lelie, D., Schwitzguébel, J.P., Glass, D.J., Gronsveld, J.V., Baker, A., 2001. Assessing phytoremediation's progress. Environ. Sci. Technol. 35 (21), 447–452.

Vervaeke, P., Luyssaert, S., Mertens, J., Meers, E., Tack, F.M.G., Lust, N., 2003. Phytoremediation prospects of willow stands on contaminated sediment: a field trial. Environ. Pollut. 126 (2), 275–282.

Vijayaraghavan, K., Yun, Y.S., 2008. Bacterial biosorbents and biosorption. Biotechnol. Adv. 26, 266–291.

Vinutha, K.S., Rayaprolu, L., Yadagiri, K., Umakanth, A.V., Patil, J.V., Rao, P.S., 2014. Sweet sorghum research and development in India: status and prospects. Sugar Tech. 16 (2), 133–143.

Vohra, M., Manwar, J., Manmode, R., Padgilwar, S., Patil, S., 2014. Bioethanol production: feedstock and current technologies. J. Environ. Chem. Eng. 2, 573–584.

Volesky, B., Holan, Z.R., 1995. Biosorption of heavy metals. Biotechnol. Prog. 11, 235–250.

Wang, J.L., Chen, C., 2006. Biosorption of heavy metals by *Saccharomyces cerevisiae*: a review. Biotechnol. Adv. 24, 427–451.

Wang, J., Chen, C., 2009. Biosorbents for heavy metals removal and their future. Biotechnol. Adv. 27, 195–226.

Wani, P.A., Khan, M.S., Zaidi, A., 2007. Chromium reduction, plant growth-promoting potentials, and metal solubilization by *Bacillus* sp. isolated from alluvial soil. Curr. Microbiol. 54, 237–243.

Weyens, N., van der Lelie, D., Taghavi, S., Vangronsveld, J., 2009. Phytoremediation: plant endophyte partnerships take the challenge. Curr. Opin. Biotechnol. 20, 248–254.

Wongsasuluk, P., Chotpantarat, S., Siriwong, W., Robson, M., 2014. Heavy metal contamination and human health risk assessment in drinking water from shallow groundwater wells in an agricultural area in Ubon Ratchathani province, Thailand. Environ. Geochem. Health 36 (1), 169–182.

Wu, C.H., Wood, T.K., Mulchandani, A., Chen, W., 2006a. Engineering plant-microbe symbiosis for rhizoremediation of heavy metals. Appl. Environ. Microbiol. 72, 1129–1134.

Wu, S.C., Cheung, K.C., Luo, Y.M., Wong, M.H., 2006b. Effects of inoculation of plant growth promoting rhizobacteria on metal uptake by *Brassica juncea*. Environ. Pollut. 140, 124–135.

Wu, X., Staggenborg, S., Propheter, J.L., Rooney, W.L., Yu, J., Wang, D., 2010. Features of sweet sorghum juice and their performance in ethanol fermentation. Ind. Crop. Prod. 31 (1), 164–170.

Wu, Z., McGrouther, K., Chen, D., Wu, W., Wang, H., 2013. Subcellular distribution of metals within *Brassica chinensis* L. in response to elevated lead and chromium stress. J. Agric. Food Chem. 61 (20), 4715–4722.

Yadav, D.V., Jain, R., Rai, R.K., 2010. Impact of heavy metals on sugar cane. In: Sherameti, I., Varma, A. (Eds.), Soil Heavy Metals. Springer, Berlin Heidelberg, pp. 339–367.

Yadav, D.V., Jain, R., Rai, R.K., 2011. Detoxification of heavy metals from soils through sugar crops. In: Sherameti, I., Varma, A. (Eds.), Detoxification of Heavy Metals. Springer, Berlin Heidelberg, pp. 389–405.

Yao, Z., Li, J., Xie, H., Yu, C., 2012. Review on remediation technologies of soil contaminated by heavy metals. Procedia Environ. Sci. 16, 722–729.

Zaidi, S., Usmani, S., Singh, B.R., Musarrat, J., 2006. Significance of *Bacillus subtilis* strain SJ-101 as a bioinoculant for concurrent plant growth promotion and nickel accumulation in *Brassica juncea*. Chemosphere 64, 991–997.

Zatta, P., 2001. Metals and the brain: from neurochemistry to neurodegeneration. Brain Res. Bull. 55 (2), 123–124.

Zhuang, P., Shu, W., Li, Z., Liao, B., Li, J., Shao, J., 2009. Removal of metals by sorghum plants from contaminated land. J. Environ. Sci. 21, 1432–1437.

BROWNFIELD DEVELOPMENT FOR SMART BIOECONOMY

MULBERRY AND VETIVER FOR PHYTOSTABILIZATION OF MINE OVERBURDEN: COGENERATION OF ECONOMIC PRODUCTS

M.N.V. Prasad[1], W. Nakbanpote[2], C. Phadermrod[3], D. Rose[1], S. Suthari[1]

University of Hyderabad, Hyderabad, Telangana, India[1]
Mahasarakham University, Maha Sarakham, Thailand[2]
Padaeng Industry Public Co. Ltd. (Mae Sot Office), Tak, Thailand[3]

1 INTRODUCTION

India has rich sources of minerals and coal (Figure 1). The Jharkhand state has many mines (Figure 2). The mine waste is a source of toxic metal leachates resulting in acid mine drainage (AMD), ultimately leading to environmental degradation and human health risk (Prasad and Jeeva, 2009). Mine waste is generally unfit for vegetation due to a lack of essential nutrients, drought, extreme pH, and lack of plant health- and growth-promoting microbial consortia (Gonzalez-sangregorio et al., 1991; Wong et al., 1998; Tordoff et al., 2000; Wong, 2003; Freitas et al., 2004; Mendez et al., 2007; Juwarkar et al., 2009; Conesa and Faz, 2011; Rajkumar et al., 2012).

A possible solution for such mine waste and coal mine overburden is phytostabilization (Pandey, 2012, 2013; Prasad, 2007; Maiti, 2007; Prasad and Prasad, 2012). Establishment of vegetation cover with desirable and economically important species would decrease the spread of contamination through wind and water erosion, and decrease the leaching of heavy metals and AMD into the groundwater. Vegetation, possibly aided by microbiota or soil amendments, promotes soil development, nutrient recycling, and the development of microbial communities. It is a non-invasive, cost-efficient method to reclaim mine waste areas and reduce risks of heavy metal pollution and AMD. Recultivation of brownfields has been successful both in temperate and tropical environments. Mycorrhizae and plant growth-promoting bacteria have been researched extensively to aid plants in overcoming these hostile situations (Leung et al., 2013). Mycorrhizae and growth-promoting bacteria can also decrease the amount of soil amendments needed, lowering costs.

2 ENVIRONMENTAL ISSUES ASSOCIATED WITH MINE OVERBURDENS

Mining, especially opencast mining, is an intensive process. Unless planned properly it can result in enormous environmental degradation of the mined area. The environmental effects of mine waste include

Bioremediation and Bioeconomy. http://dx.doi.org/10.1016/B978-0-12-802830-8.00013-7
Copyright © 2016 Elsevier Inc. All rights reserved.

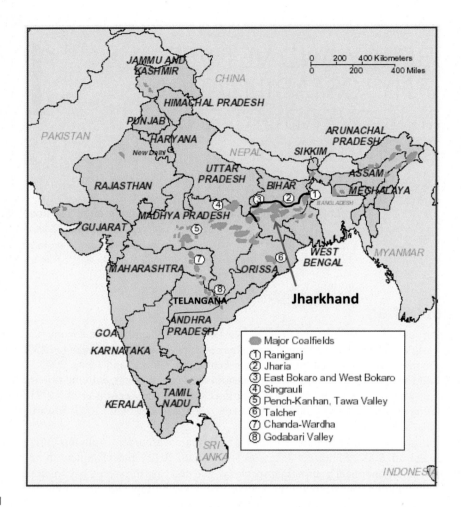

FIGURE 1

Major coal fields of India. Please note the arrow indicates the mine-rich Jharkhand State.

(a) Release of toxic metals
(b) Damage to heritage
(c) Air, water, soil pollution
(d) Sulfide minerals and acid drainage

The physical and chemical characteristics of the land also undergo a drastic change due to the ecosystem degradation.

Metal mobility is influenced by many processes driven by the plant, mycorrhizae, bacteria, and soil status. Phytostabilization has great potential as a cost-efficient and non-invasive method for the reclamation of mine waste areas, but more research is needed on the effectiveness of revegetation on risk reduction and to determine which processes contribute most to positive effects.

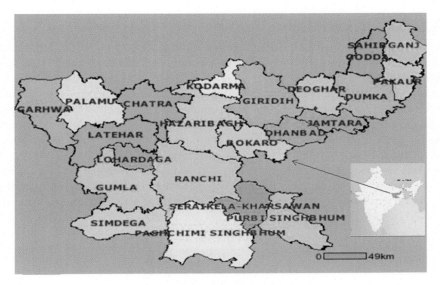

FIGURE 2

Study area location: West Bokaro Coalfield, Jharkhand.

Through surface runoff, eolian (wind) dispersion, and leaching, heavy metals can spread for tens of kilometers in the environment (Salomons, 1995; Castro-Larragoitia et al., 1997; González and González-Chávez, 2006; Nikolić et al., 2011).

Surface runoff and leaching may pollute the groundwater, while the dust spread by the wind settles on vegetation, including agricultural crops, and enters the food chain when consumed. Both acidity and heavy metals can have a severe impact on ecological and human health (Jennings et al., 2008; Tayfur and Baba, 2011).

Traditional methods to treat mine waste include soil washing, capping, and storing the waste in exhausted open pit mines (Mendez and Maier, 2008; Marques et al., 2009). A possible non-invasive remediation measure to improve mine waste and other polluted areas, is phytoremediation. Phytoremediation is the use of plants and associated microorganisms to remove, immobilize, or degrade harmful environmental contaminants.

3 RISKS ASSOCIATED WITH COAL MINING OVERBURDEN POLLUTION

AMD is the largest source of environmental problems caused by the mining industry. AMD is the result of tailings and overburden being exposed to air and water. The oxidation of pyrite and other sulfide minerals in the presence of oxidizing bacteria results in the production of acids. This oxidation process occurs slowly in all soils. Tailings and overburden, with their small grain size and thus greater surface exposure, are more prone to generating AMD. The oxidation process is accelerated beyond the natural buffering capacity of the host rock and water resources (Jennings et al., 2008). Mine tailings range from highly acidic (pH 2) to alkaline (pH 9), depending on the carbonate content and acid-generating potential.

4 COAL MINE OVERBURDENS

An *overburden* may be defined as the area that lies above a zone of scientific or economic interest. During mining, the term is usually associated with the rock, soil, and biota found lying above the coal seams (Figure 3).

Before beginning any mining activity, if appropriate measures are taken, the overburdens may be protected from being contaminated by toxic compounds and thus used for reclamation purposes. For this purpose, the topsoil and subsoils removed before mining carefully and conserved. The overburdens are removed from the location using shovels and draglines and deposited in another location, but care is taken to contain them so that they do not erode into close-lying water sources.

Interburdens are the areas found between two coal seams and are usually contaminated by the explosives used during blasting and acid runoff. They generally consist of shale and sandstone and are numbered from zero onward from bottom to top. Once the mining activity is complete, ideally, the overburdens should be used for reafforestation programs, but the loss of the topsoil, higher concentration

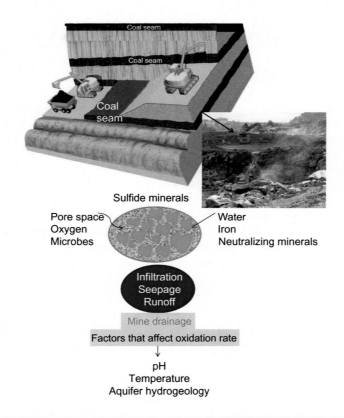

FIGURE 3

Coal mining generates a very large amount of waste including acid mine drainage. Factors affecting mine waste are briefly presented.

of inorganic elements, and the presence of organic pollutants such as those present in the bombs used for blasting make reafforestation a difficult endeavor. Another peculiarity of certain overburdens is that they tend to catch fire during summer months. A probable cause could be the production of combustible gases such as methane, ethane, etc.

4.1 ENVIRONMENTAL CONCERNS

One ton of coal leaves about 25 tons of overburden. Pyrite is contained in overburden; when pyrite is exposed to water and air, it forms dangerous substances like sulfuric acid and iron hydroxide. When this mixes with water, it forms AMD, which is highly destructive to the environment (Figure 3).

The overburden of waste and uneconomic mineralized rock is required to be removed to mine the useful mineral resource in a surface mining operation. In this process, a dump is formed by casting the waste material and dumping it nearby. The dump so formed is known as mine waste dump. Waste dump may be classified as internal and external dump. External dump is created outside the pit whereas internal dump is created inside at the back of the mining area.

Dump configuration is based on height of the dump, volume, slope angle, slope and degree of confinement, foundation conditions, dump material properties, dumping method, dumping rate, seismicity and dynamic stability, topography, dump drainage condition, bulk gradation, plasticity of fines (soil science parameter), index properties and classification, hydraulic conductivity, consolidation, strength, mineralogy and soil chemistry, *in situ* density, and compaction.

Tailing is the product discarded after mining and processing to remove the economic products. Processing may range from simple mechanical sorting to crushing and grinding followed by physical or chemical processing. It contains all other constituents of the ore except the majority of the extracted metal. It may also contain heavy metals and other substances at different concentration levels that can be toxic to biota in the environment.

Phytostabilization plant is described as the establishment of vegetation cover with desirable and economically important plant on the overburdens produced in mining activities. The overburdens often consist of sandstone, shale and separate two coal seams from each other as interburden. In a coal mine, the regular arrangement of the coal seams is seen to be that in levels numbered from zero to 'n' going from bottom to top. Each coal seam alternates with an interburden and hence all mining activities give rise to a substantial amount of overburden when blasting, drilling, and fracturing activities are carried out. The overburden produced may be stored and used to fill the quarry back up once all mining activity has been completed (Figure 4).

5 MULBERRY CULTIVATION ON COAL MINE OVERBURDEN IN INDIA

Study site: Lying along NH-23 on the road connecting Ranchi to Hazaribagh, a little outside the district capital of Ramgarh, is the town of West Bokaro. The location coordinates 23°48′N, and 85°45′E. As discussed earlier, mining may be carried out via surface mining or underground methods. Mines like the one in the West Bokaro coalfields are the kinds in which most coal is present near the surface, making surface mining a viable option. Before mining is begun, the area is cleared off all vegetation and the land is dug up using heavy machinery like draglines, trucks, wheel loader, scoop, etc.

FIGURE 4

West Bokaro coalfield mine waste dump: (a) the mine waste site, (b) revegetation of the mine waste dump, (c) phytostabilization by vetiver, (d) phytostabilization by native vegetation.

Cultivation of mulberry on the overburdens has been seen to be economically viable. *Morus alba* L. of Moraceae, the plant on which the larvae of the silkworm feed, takes up heavy metals and sequesters them within their idioblasts. The sequestration process causes no harm to the plant, which thrives under appropriate cultivation conditions. The leaves are then plucked off and fed to the larvae, leading to the production of silk (Figures 5–7), thus making this reclamation strategy an ideal cottage industry practice as well.

While industrial, especially mining activities, are a must and need to be carried out for economic growth and social security, their associated drawbacks such as air, soil, and water pollution, loss of biodiversity, shrinkage of forests, heavy metal contamination, etc., can no longer be overlooked and need immediate attention.

FIGURE 5

Phytostabilization of coal mine waste dump with Mulberry cultivation: (a) mulberry in foreground with local vegetation behind, (b) mulberry bushes.

Using the approach of natural attenuation, in which natural processes are used to reduce the effects of toxic contaminants present in a given system, in this case the plants—locally growing plant species *Shorea*, *Prosopis*, etc.—functioning as hyperaccumulators are grown along with other exotic species like mulberry and *Senna*, which also have the ability to hyperaccumulate the heavy metals present in the system, in this case the overburdens (waste generated during mining activities), thus leading to their reclamation and restoration.

Special emphasis is laid on plants that can also lead to some sort of economic activity, such as *M. alba*, which finds application in the cottage industry of sericulture and therefore is of both environmental and economic advantage by benefitting reclamation activities as well as providing the locals economic security and empowerment.

Though refilling the quarries with the overburdens is an ideal practice, it has been seen that the activity does not solve anything except maybe clearing space. The original vegetation cover having been lost, the overburdens are often loose and dry. With wind and rains they often erode into the nearby water sources, a must for refineries, potentially contaminating kilometers of area lying downstream. This is why it is extremely important to conserve the topsoil before carrying out mining activity in any area. Most importantly, the topsoil hosts soil microorganisms that are absolutely essential for plants to grow and a vegetative cover to be established.

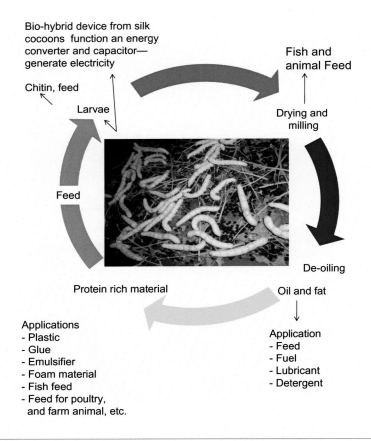

Bio-hybrid device from silk cocoons function an energy converter and capacitor—generate electricity

Chitin, feed

Larvae

Feed

Protein rich material

Applications
- Plastic
- Glue
- Emulsifier
- Foam material
- Fish feed
- Feed for poultry, and farm animal, etc.

Fish and animal Feed

Drying and milling

De-oiling

Oil and fat

Application
- Feed
- Fuel
- Lubricant
- Detergent

FIGURE 6

Bioeconomics of silkworm larvae cultured with mulberry leaves. Silk worm larvae thus reared is a resource for a wide variety of products.

Plants are imperative in carrying out the reclamation activity on overburdens for reasons discussed earlier. As an efficient and cost-effective method, especially in areas that do not require much maintenance, the approach has been to investigate the existing plant species and analyze the ones best suited to growing in such contaminated sites. Since the native species are already adapted to the local area and climate, efforts are being made to identify plants that can act as hyperaccumulators, along with established exotic species, of the contaminants present in the overburdens and help in establishing a primary vegetative cover.

Insects are fast growing with less need for feed compared to higher animals and produce highly invaluable biomass. Use of insect larvae for the production of feed and as value additions is gaining considerable importance in the developed world. Growing demand for arable land is increasing. Since, insects are enriched with nutrients, vitamins, rearing insects on phytomass harvested and produced from mine overburden is a usable feed for insect larvae. Insects can feed on phytomass produced from mine overburden. *Morus* is a classic example with capacity for metal accumulation (Singh and Behal, 2002; Ashfaq et al., 2009, 2010; Katayama et al., 2013): Silkworm larvae thus produced using *Morus* phytomass can be further utilized (Figures 5–7) in addition to silk producing cottage industry which supports rural employment and revenue generation. The excrement of silkworm enriches the soil for better yield

FIGURE 7

Silk work culture and silk production. (a) Mulberry is the best feed for rearing silk worms. (b) Silk worm larvae. (c) Large scale production of silk worm in as cottage industry in rural India. (d,e) Silk producing coccons. (g) Silk yarn. The waste from silkworm farms is used by farmers as biofertilizer for crops.

and the farmers have been utilizing it as biomanure for their crops. In India, Jharkhand State has many coal mines and incidentally silkworm rearing and silk production tops the list in India (Figures 5–7).

6 PHYTOSTABILIZATION OF MINE WASTE BY VETIVER GRASS

The heavy metals contamination of the environment by soil erosion in mining industry areas attracts worldwide concern. Phytoremediation is an affordable and sustainable technology that is most useful when contaminants are within the root zone of the plants (top 1-2 m of soil) (Juwarkar and Jambhulkar, 2008;

Ali et al., 2013). Phytostabilization and phytoextraction are two techniques that combine as *in situ* phytoremediation for metals. Revegetation methods are thought to be the most practical and economical method for rehabilitation of the mine wastelands for long-term stability of the land surface. A good vegetation cover is beneficial in the restoration of contaminated land and results in enhanced amenity values as well as in prevention of surface soil erosion (Baker et al., 1994; Truong, 1999). However, major limiting factors for plant establishment on mine tailings are toxicity of heavy metals and nutrient deficiencies (Bradshaw, 1987; Pichtel and Salt, 1998).

Vetiver grass (*Vetiveria zizanioides* (L.) Nash), recently reclassified as *Chrysopogon zizanioides* (L.) Roberty, belongs to the Poaceae family as of maize, sorghum, sugarcane, and lemongrass. Vetiver grass, called "wonder grass," "miracle grass," and "magic grass" in various parts of the world and popularly known as "khus", is a plant with considerable promise for soil and water conservation as well as phytoremediation applications (Table 2), because of its unique morphological and physiological characteristics (Greenfield, 1989; Truong et al., 1995; Chomchalow, 2001; Xia, 2004; Chaplot, 2014c (Table 2)). Vetiver is a tall (3-4 m), massive, tufted, perennial, scented grass with a straight stem, long narrow leaves, and a lacework root system that is abundant, complex, and extensive can penetrate to the deeper layers of the soil (Truong, 2000; Pichai et al., 2001). Vetiver is a versatile, hardy plant that is fast growing and can survive in a harsh environment. It also has been found to be highly tolerant to extreme soil conditions including high metal concentration (Knoll, 1997; Truong and Baker, 1998). The grass can grow well in soil contaminated with multiple elements (Roongtanakiat and Chairoj, 2001; Pang et al., 2003; Roongtanakiat et al., 2008). No obvious symptoms have been observed under adverse soil conditions such as alkalinity (pH 11), salinity (47.2 mS cm^{-1}), Mg (2400 mg kg^{-1}), As (100-250 mg kg^{-1}), Cd (mg kg^{-1}), Cu (174 mg kg^{-1}), Cr (200-600 mg kg^{-1}), Ni (50-100 mg kg^{-1}), Hg (>6 mg kg^{-1}), Pb (>1500 mg kg^{-1}), Se (>74 mg kg^{-1}), and Zn (>750 mg kg^{-1}) (Truong, 2000). In addition, vetiver grass has been shown to be highly tolerant to adverse climatic conditions such as frost, heat, wave, temperature (5-55°C), drought, flood and inundation, edaphic conditions, as well as to elevated levels of heavy metals, herbicides, pesticides, and showed high efficiency in absorbing dissolved nitrogen, phosphorus, and sulfate (Yang et al., 2003).

In term of phytostabilizing metal-contaminated sites, a lower metal concentration in shoot is preferred, in order to prevent metal from entering the ecosystem through the food chain. In phytostabilization, plants stabilize the pollutants in soils, reducing their mobility into the soil and therefore the risk of leaching into groundwater. Therefore, the extensive root system of vetiver grass has good potential to stabilize tailings' particles and avoid erosion (Truong, 1999). Vetiver has been found to accumulate most of the heavy metals in its roots, making this species more suitable for stabilizing mine tailings and the danger of transferring toxic metals to grazing animals minimal (Yang et al., 2003). Vetiver grass is highly tolerant to hostile soil conditions and widely used as a natural, effective, and low-cost alternative means to vegetate the heavy metal-contaminated land (Truong, 1996; Flathman and Lanza, 1998). In addition, vetiver is highly successful in the rehabilitation of old quarries and mines and is able to stabilize the erodible surface first so other species can colonize the areas later (Truong, 2000). Vetiver grass has been successfully used to stabilize mining overburden and highly saline, sodic, and alkaline tailings of gold mines (Randloff et al., 1995). In South Africa, it has been used effectively to stabilize waste and slime dams from platinum and gold mines (Knoll, 1997). Vetiver system should provide a powerful phytoremedial tool for the attenuation of the mercury pollution problem in Yolo and Lake counties by trapping and containing both the air -and water-borne insoluble mercury (Hg) at sources, and by reducing the soluble fraction in AMD (Truong, 2000). In Australia, *Chrysopogon zizanioides* has been successfully used to stabilize mining overburden, coal and gold mine tailings (Truong and

Baker, 1996). Vetiver grass was used to stabilize a landfill and industrial waste site contaminated with heavy metals such as As, Cd, Cr, Ni, Cu, Pb, and Hg (Truong and Baker, 1998). In China, vetiver grass was planted on a large scale for pollution control and mine tailings stabilization (Chen, 2000). In India, vetiver showed itself an excellent source for remediation and restoration of coal fly ash, containing heavy metals (Zn, Pb, Cu, Ni, and As) and acidic/alkaline (pH) (Verma et al., 2014; Ghosh et al., 2015).

Heavy metals tolerance properties of vetiver involved different mechanisms to detoxify active oxygen species that existed in different parts of the plant. Proportions of lead and zinc (Pb/Zn) tailings greatly inhibited the leaf growth, dry matter accumulation, and photosynthesis, but stimulated the accumulation of proline and abscisic acid, and enhanced the activities of superoxide dismutase, peroxidase, and catalase (Pang et al., 2003). In addition, application of vetiver for phytoremediation depends on various factors such as physical and chemical properties of growth media as well as agronomic practice. Nitrogen fertilizer application could greatly alleviate the adverse effects of high proportions of Pb/Zn tailings on vetiver grass growth (Pang et al., 2003). Organic matter application significantly improved soil physical characteristics and nutrient availability, while a chelating agent could modify metal bioavailability, plant uptake, and translocation (Walker et al., 2003). Domestic refuse alone and/or in combination and artificial fertilizer significantly improved the survival rates and growth of vetiver on the lead (Pb) and zinc (Zn) tailings (Yang et al., 2003). For phytoremediation of tailings, therefore, soil chemical fertilizer and amendment materials such as organic matter and chelating agents are often required to establish successful vegetation and metal uptake. The grass grew better with Osmocote in comparison to plants fertilized with 10-10-10 (NPK-Nitrogen-Phosphorus-Potassium) fertilizer, and maximum Pb levels were observed in root tissues (Wilde et al., 2005). In the case of iron (Fe) ore mining rehabilitation, the combination of soil amendment materials, especially diethylene triamene penta acetic acid (DTPA) and compost, was more effective than sole use of chelating agents and compost in enhancing vetiver growth and nutrient and heavy metals uptake (Roongtanakiat et al., 2008). Ethylene diamine tetra acetic acid (EDTA) could enhance vetiver growth on zinc mine soil, collected from Tak Province in Thailand, to uptake Zn, Mn, and Cu (but not Fe), while DTPA increased the phytoavailable metals, but not their uptake (Roongtanakiat et al., 2009). The translocation ratio of Pb from vetiver roots to shoots was significantly increased with the application of EDTA (Chen et al., 2004). In addition, phytoremediation properties of vetiver differed by ecotype and metal. Roongtanakiat and Chairoj (2001) planted three vetiver grass ecotypes—Kamphaeng Phet (*Chrysopogon zizanioides*), Ratchaburi (*Chrysopogon nemoralis*), and Surat Thani (*Chrysopogon zizanioides*)—in soil supplemented with different amounts of manganese (Mn), zinc (Zn), copper (Cu), cadmium (Cd), and lead (Pb). The roots of Ratchaburi ecotype could also absorb significantly higher amounts of Zn, Cd, and Pb than those of Surat Thani and Kamphaeng Phet ecotypes. For Mn and Cu, Ratchaburi and Surat Thani ecotypes could uptake more than those of Kamphaeng Phet. In addition, Ratchaburi ecotype is growing in Zn-, Cd-, and Pb-contaminated soil, which were collected near an operational zinc mine in Mae Sot District, Tak Province, accumulated more heavy metals in their roots than shoots (Roongtanakiat and Sanoh, 2011). Chantachon et al. (2004) also showed that *Chrysopogon zizanioides* could uptake more lead from soil than *Chrysopogon nemoralis* and that the phytomass was decreased when the lead concentration was increased. In comparison, *Chrysopogon zizanioides* showed greater phytomass than *Chrysopogon nemoralis*. In addition, Antiochia et al. (2007) showed the vetiver plant can be considered a quite good hyperaccumulator only for Pb and Zn, but the grass had a poor efficiency for Cr and Cu uptake.

In the case of wastewater treatment, vetiver grass has the ability to uptake heavy metals from industrial wastewater such as from a battery manufacturing plant, an electric lamp plant, and an ink manufacturing facility. However, the concentration of heavy metals in wastewater played an important role in vetiver growth and metal accumulation. Vetiver absorbed a substantial amount of Cd, Hg, and Pb in wastewater (Sripen, 1996). Arsenic-contaminated water could be cleaned up by a planted pond consisting of "red earth": rice husks and gravel planted with vetiver and *Colocasia esculenta* (Johne et al., 2008). In addition, the efficiency of vetiver grass depended on the ecotype (Roongtanakiat et al., 2007). Vetiver grass is a good candidate for remediation of wastewaters by constructed wetland technology. The grass reduced total dissolved soilds, chemical oxygen demand, calcium, chlorides, total nitrogen, and sulfates in different wastewater of sewage farm (Dhanya and Jaya, 2013). Singh et al. (2008) studied the effective removal of phenol (200-1000 mg l^{-1}) by both *in vitro* and *in vivo* plants of vetiver grass. The grass could effectively remove phenol and develop resistance to phenol-induced reactive oxygen species through a biochemical mechanism. Although plant growth was reduced in the presence of phenol (200 mg l^{-1}), the results of the reuse study indicated the possibility of plants getting adapted to phenol without any decline in potential for phenol remediation. Ho et al. (2013) demonstrated the potential to improve phytoremediation of aromatic pollutants (toluene) by inoculating *Achromobacter xylosoxidans* F3B, a functional endophytic bacterial strain, into vetiver grass. The strain F3B protected plants against toluene stress and maintained chlorophyll content of leaves, with a 30% reduction of evapotranspiration through vetiver leaves. In addition, the endophytic bacteria grew well in vetiver roots without greatly interfering with the diversity of native endophytes.

7 VETIVER ECOTYPE AND PROPAGATION IN THAILAND

Vetiver is reportedly indigenous to at least 70 tropical or subtropical countries (National Resource Council, 1995), most notably Thailand. There are at least 12 known species of vetiver spread over the worldwide, and the native species propagates by both sexual and asexual reproduction. Vetiver can live some 50 years (National Resource Council, 1995), which makes this species an efficient, low-cost, and longterm remediation option for phytoremediation. There are two species of vetiver in Thailand, *Chrysopogon nemoralis* (Balansa) Holttum and *Chrysopogon zizanioides* (L.) Roberty. Both species have distinct ecological characteristics that make them adapt to different habitats. They are commonly found in all regions of Thailand, and there are many ecotypes. Thai vetiver ecotypes have been named after the provinces where they were first found. The Department of Land Development has performed a comparative study of 28 vetiver ecotypes. Of these, 11 of *Chrysopogon nemoralis* and 17 of *Chrysopogon zizanioides* ecotypes. Some ecotypes have proven suitable to grow in various soil types (Table 1; ORDPB, 2000).

Table 1 Suitability of Vetiver Ecotypes in Various Soil Types Roongtanakiat (2009)

Soil Type	*Chrysopogon nemoralis*	*Chrysopogon zizanioides*
Sandy soil	Nakhon Sawan, Kamphaeng Phet 1, Roi Et, Ratchaburi	Kamphaeng Phet 2, Songkhla 3
Clay loam soil	Loei, Nakhon Sawan, Ratchaburi, Kamphaeng Phet 1, Prachuap khiri Khan	Surat Thani, Songkhla 3
Leterite soil	Prachuap Khiri Khan, Loei	Kamphaeng Phet 2, Songkhla 3, Surat Thani, Sri Lanka

The main goals in vetiver propagation are quality planting material, low cost, hardiness, and being easy to transport. Vetiver in cultivation rarely produces seeds. Thus only asexual reproduction will be propagated from tiller, slip, cutting, culm, clump, and ratoon. Although vetiver has been cultivated for a long time, nothing has been done with respect to its propagation techniques until recently. Only two methods have been employed: a traditional method of planting bare-root tillers in the field, and a rather new method of planting tillers in a polyethylene bag (commonly referred to as polybag) for a period of time and then transplanting them in the field. Other planting methods include tissue-cultured plantlets in which the explants are obtained from young shoot or young inflorescence; open hardening, in which plantlets are taken out from the bottle, then grown for further multiplication in the polybags, in the nursery beds, or in the field; or growth in strips, dibbling tubes, or nursery blocks prior to field planting (Chomchalow, 2000).

Vetiver is easy to propagate at a relatively low cost. The following goals are the criteria for the manager of the project to follow in choosing appropriate techniques for vetiver propagation (Chomchalow, 2000, 2014).

Quality planting material: One of the major goals in propagating vetiver is to produce only high-quality planting material. Only high-quality planting material should be used in transplanting in the field. High quality includes healthy and vigorous growth.

Low cost: Efficient nursery management will reduce the extra cost of propagating vetiver. The returns to production for a given input will often more than pay for vetiver needs such as water, nutrients, light, etc. It also needs subdued sunlight for its early growth in the nursery, and brighter sunlight at a later stage, especially during the hardening period.

Hardiness: By its nature, vetiver is a tough plant. However, during its early stage of growth as propagating material grown in any form of containers, it is rather weak, especially when subject to long transportation through rugged terrain in the hot sun before being transplanted. A period of hardening or acclimatization (i.e., exposure to sunlight) prior to field planting is needed.

Being easy to transport: Containerized-planting material is considered most practical. Trays to hold polybags or other containerized planting materials should also be lightweight, small volume, and easy to handle. A one-way transport, like the use of strip planting, biodegradable nursery blocks, dibbling tubes (tubes removed and retained), etc., has the advantage in that no containers and other materials are to be collected and returned to the nursery for reuse or throwing away.

In addition, simplicity, low cost, and low maintenance are the main advantages of vetiver system technology (VST) over chemical and engineering methods for contaminated land treatments (Truong et al., 2008).

Simplicity: Application of the vetiver system is rather simple compared with other conventional methods. In addition to an appropriate initial design, it only requires standard land preparation for planting and weed control in the establishment phase (Table 2).

Low cost: Application of the vetiver system in contaminated land treatment costs a fraction of conventional methods such as chemical or mechanical treatment. Most of the cost lies in the planting material, with small amounts in fertilizer, herbicides, and planting labor.

Minimal maintenance: When properly established, the VST requires practically no maintenance to keep it functioning. Harvesting two or three times a year to export nutrients and to remove top growth for other usages is all that is needed. This is in sharp contrast to other systems that need regular costly maintenance and a skilled operator, often an engineer, to operate efficiently.

Table 2 Application of vetiver grass for phytoremediation

References	Research Topics
Abaga et al. (2014a)	Is vetiver grass of interest for the remediation of Cu and Cd to protect marketing gardens in Burkina Faso?
Abaga et al. (2014b)	Phytoremediation of endosulfan in cotton-cultivated soils
Chaplot (2014a)	Useful for controlling run-off, soil loss, and productivity
Chaplot (2014b)	Erosion control and phytostabilization of overburden
Falola et al. (2014)	Nutritional and antinutritional components of vetiver grass
Filippi (2014)	Nor-sesquiterpenes
Ghotbizadeh and Sepaskhah (2014)	Cultivation of vetiver with saline water
Herrera (2014)	Vetiver system for soil erosion control and slope protection
Jayashree et al. (2014a)	Antitermite properties of root and leaf powder of vetiver
Jayashree et al. (2014b)	Influence of rooting media on vetiver
Jotisankasa et al. (2014)	Infiltration and stability of soil slope with vetiver grass subjected to rainfall from numerical modeling
Juntuek et al. (2014)	Effect of vetiver grass fiber on soil burial degradation of natural rubber and polylactic acid (PLA) composites
Kadarohman et al. (2014)	Quality and chemical composition of organic and nonorganic vetiver oil
Liu and Shi (2014)	Fast detection of heavy metal lead (Pb) in vetiver grass leaves using near-infrared spectroscopy
Maignana et al. (2014)	Evaluation of antiepileptic activity of *Vetiveria zizanioides* oil in mice
Manh et al. (2014)	Treatment of wastewater from slaughterhouse by biodigester and *V. zizanioides* L.
Oku et al. (2014)	Green structure for soil and water conservation on cultivated steep land
Pazoki et al. (2014)	Attenuation of municipal landfill leachate through land treatment
Prasad et al. (2014)	Heavy metals affect yield, essential oil compound, and rhizosphere microflora of vetiver
Quang et al. (2014)	Soil conservation methods in the uplands of Vietnam: an agent-based modeling approach
Rahardjo et al. (2014)	Performance of an instrumented slope covered with shrubs and deep-rooted grass
Samuel et al. (2014)	Performance evaluation of a dual-flow recharge filter for improving groundwater quality
Sciarrone et al. (2014)	Rapid isolation of high solute amounts using an online chromatography
Singh et al. (2014a)	Efficient C sequestration and benefits of medicinal vetiver cropping in tropical regions
Singh et al. (2014b)	Molecular diversity and SSR transferability studies in vetiver grass *V. zizanioides* (L.) Nash
Sinha et al. (2014)	Evaluation of toxicity of essential oils palmarosa, citronella, lemongrass, and vetiver in human lymphocytes
Vinayagamoorthy and Rajeswari (2014)	Mechanical performance studies on *V. zizanioides*/jute/glass fiber-reinforced hybrid polymeric composites
Yaseen et al. (2014)	Growth, yield, and economics of vetiver *V. zizanioides* (L.) Nash under intercropping system

Table 2 Application of vetiver grass for phytoremediation—cont'd

References	Research Topics
Ye et al. (2014a)	Vetiver grass for the recovery of organochlorine pesticides and heavy metals from a pesticide factory site
Ye et al. (2014b)	Extraction of PBDEs/PCBs/PAHs and heavy metals from an electronic waste using *Vetiver* grass phytoremediation
Zhang et al. (2014a)	Accumulation of chromium in *V. zizanioides* assisted by earthworm *Eisenia foelide* in contaminated soil
Zhang et al. (2014b)	Application of *V. zizanioides* assisted by different species of earthworm in chromium-contaminated soil remediation
Zhang et al. (2014c)	Effect of cadmium on growth, photosynthesis, mineral nutrition, and metal accumulation by vetiver grass
Zhuang et al. (2014)	Arsenic uptake from groundwater by fulvic acid and straw water amendments
Akhzari and Bidgoli (2013)	Tolerant to drought and salinity stresses
Altenor et al. (2013)	Pilot-scale synthesis of activated carbons from vetiver roots and sugar cane bagasse
Bharti and Banerjee (2013)	Bioassay analysis of efficacy of phytoremediation in decontamination of coal mine effluent
Boonyanuphap (2013)	Rehabilitation measures for landslide-damaged mountainous agricultural lands in lower Northern Thailand
Falola et al. (2013)	Silage quality and forage acceptability ensiled with cassava peels by goat
Kadarohman et al. (2013)	Biolarvicidal activity of vetiver
Kantawanichkul et al. (2013)	Treatment of domestic wastewater by vertical flow constructed wetland planted with vetiver
Maneecharoen et al. (2013)	Ecological erosion control by limited life geotextiles and grasses such as vetiver and ruzi
Oshunsanya (2013)	Crop yields as influenced by land preparation methods established within vetiver grass alleys for sustainable agriculture in Southwest Nigeria
Pu et al. (2013)	Characteristics of soil comprehensive antierodibility under sloped cropland with hedgerows
Sangeeta et al. (2013)	Use of Reclaimed Mine Soil Index for screening of tree species for reclamation of coal mine-degraded land
Sawangsuriya et al. (2013)	Comparison of erosion susceptibility and slope stability of repaired highway embankment
Sujaritjun et al. (2013)	Mechanical property of surface modified natural fiber reinforced PLA biocomposites
Boonying et al. (2012) and Nuthong et al. (2011)	Production of PLA composite
Chen et al. (2012)	The phytoattenuation of the soil metal contamination: the effects of plant growth regulators GA3 and IAA by vetiver and sunflower
Chou et al. (2012)	Study of the chemical composition, antioxidant activity, and anti-inflammatory activity of essential oil from *Vetiveria zizanioides*
Donjadee and Chinnarasri (2012)	Effects of rainfall intensity and slope gradient on the application of vetiver grass mulch in soil and water conservation
Holanda et al. (2012)	Comparison of different containers in the production of seedlings of vetiver grass for erosion control

(Continued)

Table 2 Application of vetiver grass for phytoremediation—cont'd

References	Research Topics
Jampeetong et al. (2012)	Effects of inorganic nitrogen forms on growth, morphology, nitrogen uptake, and nutrient allocation in *V. zizanioides*
Kuang and Zhang (2012)	Growth and physiobiochemical responses of *V. zizanioides* to solidification/stabilization products made from dredged sediment
Lima et al. (2012)	Phytochemical screening, antinociceptive, and anti-inflammatory activities of *Chrysopogon zizanioides* essential oil
Mallavarapu et al. (2012)	Constituents of South Indian vetiver oils
Manzoor et al. (2012)	Chemical composition and antitermitic activity of essential oils of *V. zizanioides* against *Heterotermes indicola* (Wasmann) from Pakistan
Nyomora et al. (2012)	Establishment and growth of vetiver grass exposed to landfill leachate
Rodkvamtook et al. (2012)	Efficacy of plant essential oils for the repellents against chiggers *Leptotrombidium imphalum* vector of scrub typhus
Roongtanakiat et al. (2012)	Radiosensitivity of vetiver to acute and chronic gamma irradiation
Santos et al. (2012)	*In vitro* propagation and conservation of vetiver grass (Propagação e conservação *in vitro* de Vetiver).
Sukwat et al. (2012)	Development of prototype of young Buddhist environmental education
Sutapun et al. (2012)	Effect of heat treatment on chemical structure of a bio-filler from vetiver grass
Vollú et al. (2012)	Molecular diversity of nitrogen-fixing bacteria associated with *Chrysopogon zizanioides* (L.) Roberty (Vetiver), an essential oil producer plant
Wu et al. (2012)	Root exudates of wetland plants influenced by nutrient status and types of plant cultivation
Bolan et al. (2011)	Suitable candidate for phytostabilization contaminant containment
Sadunishvili et al. (2009)	Influence of hydrocarbons on plant cell ultrastructure and main metabolic enzymes
Sarah et al. (2009a)	*Jatropha curcas*: a potential crop for phytoremediation of coal fly ash
Sarah et al. (2009b)	Fly ash-trapping and metal-accumulating capacity of plants: implication for green belt around thermal power plants
Li-ping et al. (2009)	Fertilizing reclamation of arbuscular mycorrhizal fungi on coal mine complex substrate
Dhillon et al. (2008)	Evaluation of different agroforestry tree species for their suitability in the phytoremediation of seleniferous soils
Rentz et al. (2005)	Benzo[a]pyrene co-metabolism in the presence of plant root extracts and exudates
Pulford and Watson (2003)	Phytoremediation of heavy metal-contaminated lands by trees: a review
Dutta and Agrawal (2001)	Litterfall, litter decomposition, and nutrient release in five exotic plant species planted on coal mine spoils
Hegde and Fletcher (1996)	Influence of plant growth stage and season on the release of root phenolics by mulberry for phytoremediation technology
Chibrik and Salamatova (1985)	Mycosymbiotrophism in cultivated phytocenoses of the Korkino open coal mine
Brenner (1984)	Restoration of natural ecosystems on surface coal mine lands in the northeastern United States

Note: Vetiver grass = *Chrysopogon zizanioides* (L.) Roberty (syn. *Vetiveria zizanioides* (L.) Nash).

8 PHYTOSTABILIZATION OF ZINC MINE WASTE BY VETIVER GRASS IN THAILAND

In Thailand, King Bhumibol Adulyadej, Rama IX, graciously provided initiative for utilization of vetiver on June 22, 1991. This provided for utilization of vetiver for soil and water conservation with a wide range of cultivation patterns. Although vetiver is not a hyperaccumulator, it has been found to be highly tolerant to extremely adverse soil conditions. Therefore, vetiver could be used for rehabilitation of mine tailings, which are often extremely acidic or alkaline, high in heavy metals and low in plant nutrients. Planting vetiver in rows or contour on the side slopes could prevent contamination of heavy metals in adjacent areas polluted by wind and water erosion, leachate, and run-off problems that are often the causes of off-site contamination (Truong, 1996, 1999; Flathman and Lanza, 1998; Yang et al., 2003). The vetiver shoots should be cut regularly in order to stimulate their growth into thick clumps. Then heavy metals would be uptaken and translocated to the new shoots while contaminant level should be reduced gradually. Vetiver shoots and roots can be safely disposed of away from contaminated sites or be used as value-added material such as green fuel, handicraft product, roofing, and mulching material.

Padaeng Industry Public Company Limited (PDI) was established on 10 April 1981, by Thai public (Ministry of Finance) and private investors and a private company from Belgium (Vieille Montagne). The company engaged in mining and refining business with the objective of producing zinc metal and zinc alloys to serve its customers. The company's mine is located in the Mae Sot District and its refinery in Muang District of Tak Province (Figure 8), the roaster plant in Rayong Province, and the head office in Bangkok. The company was granted the first mining lease in 1982. Mining operation was started in 1984 after mining preparation and finishing of the refinery plant construction. The mining leases and related activity areas covered about 332 hectares. The original surface of zinc ore deposit was on the top of a hill and the hill slope without soil covering. Some parts of the area covered by leases and permits have not been used. They have been left as natural forest and to be a buffer zone to the surrounding areas. The area for rehabilitation is about 167 hectares. Rehabilitation is being done as soon as possible after finishing each part of the mining and related activities. PDI has been taking serious action by coordinating with the Office of the Royal Development Project Board and Huay Sai Royal Development Study Centre in Phetchaburi Province to bring vetiver for growing on bare soil surface to prevent erosion and protect soil moisture. It will be suitable for the pioneer plantation before planting other trees, then the bare land will become real forest (Figure 9).

Rehabilitation was completed in places where mining and related activities have finished; namely, at the mine pit edge, waste collecting areas, and others. The first step for rehabilitation is land preparation, which is done by purchasing fertilized soil from outside to cap bare soil at least 30 cm thick before planting. The company provides a budget of 50 million Thai Baht for fertilized soil due to unsuitable soil within the project area. In the second step, the bare soil is covered principally by vetiver with the addition of ruzi grass (*Brachiaria ruziziensis* Germ. & C.M. Evrard). This step is intended for soil rehabilitation, adding more organic matter into the soil, preventing soil erosion, reducing velocity of running water on the surface, and protecting moisture in the soil as long as possible. Finally, trees of local species, such as teak, ironwood, Siamese sal, local cork tree, orchid tree, etc., are planted in principal. Additional fast-growing trees, such as eucalyptus, Royal Poinciana (*Delonix*), and *Leucaena* are also planted. Legumes have been important crop plants with the benefit of improving soil quality (Vandermeer, 1989). For long-term remediation, herbaceous legumes can be used as a pioneer species to solve the problem of nitrogen deficiencies in mining wastelands because of N_2-fixing ability of their microbial colonization (Bradshaw, 1987; Archer et al., 1988).

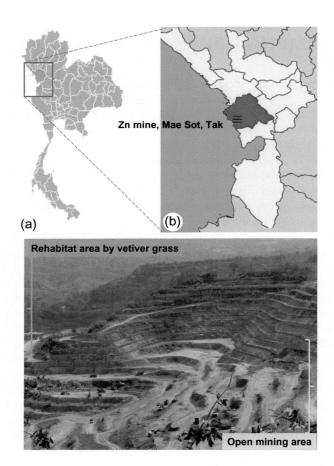

FIGURE 8

Padaeng zinc mine in Mae Sot, Tak Province, Thailand: (a) map showing area in Thailand, (b) map showing area in province, (c) zinc mining area terraces to be cultivated with vetiver.

Pinthong et al. (1998) also studied the capability of vetiver hedgerow in decontamination of agrochemical residues. All together, about 97,600 trees were planted. Plantation is done shortly before and at the start of the rainy season, then there is rainwater to enable them to grow enough to survive the summer heat.

The company started growing vetiver for conserving soil and water on 15 August 2003, and has grown vetiver each year since then. An experiment was conducted at Padaeng mine in order to compare development of two *Chrysopogon nemoralis* ecotypes, Nakhon Sawan and Prachuap Khiri Khan, and two *Chrysopogon zizanioides* ecotypes, Kamphaeng Phet 2 and Surat Thani, grown in zinc mining area. It was found that both *Chrysopogon zizanioides* ecotypes gave better growth performance than of *Chrysopogon nemoralis*, while Kamphaeng Phet 2 gave the highest plant height and shoot dry weight (Roongtanakiat, 2009). Therefore, the species of Kamphaeng Phet 2 was the most appropriate for the climate and the soil of Padaeng mine. A total of 19.17 million of vetiver sprouts were grown. An area of 103 hectares has been rehabilitated till the end of 2014 from total 166 hectares (or 62%). Therefore, Padaeng mine is one of the big mines of Thailand where the vetiver has been growing as big mountain

FIGURE 9

Succession of vetiver grass grown on mine waste: (a) 2006, (b) 2007, (c) 2008, (d) 2014. Please note well-established vetiver grass.

covering with vetiver (Figure 9). The company also invites members of government agencies and of the community, especially groups of teachers and students from local schools, colleges, and university, to join in the activity and to learn about the usefulness of vetiver. Each year almost a thousand people participate (Figure 10). After that vetiver was grown as underground walls to cover bare soil to prevent topsoil erosion and trees of local species were planted. The slope of each bench is 35° whereas the overall slope is 27°. As the overburden was piled from the ground level to the top of hill, the rehabilitation

FIGURE 10

Schoolchildren taking part in massive vetiver plantation on barren mine waste slopes: (a and b) climbing the slopes, (c and d) planting shoots.

can be made starting from the lower level and working towards the top. No need to wait till finishing the whole overburden piling in each place.

For phytoremediation of zinc mine tailings, application of organic matter and chelating agents have been researched to establish successful vegetation and metal uptake.

8.1 FERTILIZER AND SOIL AMENDMENTS

Primary nutrients are needed in large quantities for plant growth. The Land Development Department (LDD, 1994) reported that concentrations of N, P, and K in vetiver shoot were 2.5%, 0.17%, and 1.5%, respectively. The vetiver grown in zinc mine soil, which has lower fertility than iron ore tailings, had lower concentrations of primary nutrients in shoot of 2.12-2.55%, 0.44-0.50%, and 1.26-1.40%, respectively (Roongtanakiat, 2009). Application of organic fertilizer can increase vetiver yield and may reduce toxicity of heavy metal (Yong et al., 1992). The influence of organic and inorganic fertilizers on the growth of vetiver grown in zinc mine soils has been compared in pot experiment. In the case of vetiver grown in zinc mine soil, the use of compost elevated vetiver biomass while the inorganic fertilizer decreased vetiver growth, which gave biomass significantly different to those in control and compost treatments. For vetiver cultivation on deteriorated land with low fertility, the LDD recommended filling the bottom of the plant holes with manure or compost. Once the tillers start to sprout, the 15-15-15 inorganic fertilizer should be added to accelerate growth at the rate of 25 kg/rai (0.4 acre), along the contour (ORDPB, 2000).

Besides increasing organic matter and nutrient content in soil, application of organic amendments like compost to mine tailings is known to increase water-holding capacity, cation exchange capacity, and to improve the structure of mine tailings by forming stable aggregates (Stevenson and Cole, 1999; Ye et al., 2000). These amendments also mitigate the toxicity of heavy metals and plant failure to grow in their absence (Brown et al., 2003). A field experiment performed at PDI revealed that application of compost could significantly increase growth and shoot dry weight of vetiver; however, there was no significant difference between 47 ton/rai and 8 ton/rai applications. Hence, the application 4 ton/rai of compost was suggested for vetiver plantation in this area, as recommended by LDD (1998).

8.2 CHELATING AGENT APPLICATION

Since plant uptake requires metals in an environmentally mobile form, the negative charge of various soil particles tends to attract and bind heavy metals that are cations and prevent them from becoming soluble and diffuse to root surface. Using chelating agents such as EDTA, DTPA, nitrilotriacetic acid, and cyclo hexane diamine tetra acetic acid have been developed to overcome these problems (Huang and Cunningham, 1996; Robinson et al., 1999; Cooper et al., 1999). However, the effects of chelating agents on growth performance and heavy metal uptake can differ among chelating agents, heavy metals, and soils. In the zinc mine soil, the combination of DTPA and compost application actually reduced growth of vetiver in both height and biomass. The EDTA could enhance concentration and uptake of Zn, Mn, and Cu but not Fe while use of DTPA increased the mentioned heavy metal concentrations but not uptakes. These studies also revealed that application of only compost to zinc mine soil did not affect heavy metal uptakes by vetiver (Roongtanakiat, 2009).

8.3 TRANSLOCATION OF HEAVY METALS IN VETIVER

Distribution of heavy metals in vetiver plant can be divided into three groups: (1) very little of the As, Cd, Cr, and Hg absorbed, were translocated to the shoots (1-5%); (2) a moderate proportion of Cu, Pb, Ni, and Se was translocated (16-33%); (3) Zn was almost evenly distributed between shoot and root (40%) (Truong, 1999). However, numerous investigators concluded that vetiver root accumulated higher heavy metal concentrations than shoot, and they concluded that vetiver is a nonhyperaccumulator plant (Greenfield, 1989; Truong, 1999; Roongtanakiat, 2006). When vetiver plants were more mature, they could not concentrate higher heavy metal in the shoot. The shoot heavy metal concentration decreased, possibly due to the dilution effect of increasing biomass, while the root heavy metal concentrations increased (Roongtanakiat and Chairoj, 2001). The vetiver grown on iron tailings and zinc mine soils could translocate higher quantities of heavy metal from root to shoot with translocation factors of 0.55-0.86 and 0.50-0.89, respectively. Although soil amendments can enhance some metal translocations, the compost and chelating agents did not affect the Zn translocation of vetiver grown in both mine soils (Alloway, 1995).

9 ECONOMIC PRODUCTS FROM VETIVER

The vetiver grass is a very cost-effective, environmentally friendly, and practical phytoremedial tool for the control and attenuation of heavy metal pollution when appropriately applied (Zhuang and Hazelton, 2014). Vetiver at the age of 3-4 months requires leaf pruning to stimulate and maintain hedgerows and fertility of the prolific and efficient root system. Disposal of those plant tissues after environmental restoration applications can be a serious consideration due to the associated cost and handling concerns. Vetiver poses several alternative income-generating options. All parts of vetiver—roots, trunk, and leaves—can be beneficial, as: forage, mushroom substrate, mulch, cover crop, toxic substance absorption, wastewater treatment, pest and weed control, aromatic oils, herbs, handicrafts, paper, and building material. In addition, vetiver grass can be used as a biofuel (both as a direct fuel and its potential conversion to cellulosic ethanol) and has overall potential relating to climate change and the sequestering of atmospheric carbon. The promotion of vetiver grass as an income-generating crop and its development from merely being an agricultural waste to economical partial substitutes or raw materials would reduce environmental deterioration and deforestation (Grimshaw, 1997; Niityoungskul and Hengsadeekul, 2002).

Carbon sequestration: Vetiver has to be one of the world's best carbon sequesters. The anatomical study revealed that vetiver ecotypes have distinct characteristics of hydrophytes; that is, having large cellular pores that enable the storage of water and air, as well as circulating water and air in the vetiver leaf. Vetiver can sequestrate higher amount of carbon in the areas having high concentration of clay particles. Carbon and nitrogen were all accumulated in the roots, shoots, stem, and leaves, respectively. Thammathavorn and Khanema (2010) studied and compared the potential of vetiver in carbon sequestration. Ratchaburi ecotype has increased carbon at the highest rate of 11 tons ha^{-1} year^{-1}, while the lowest was the Loei ecotype at 4 tons ha^{-1} year^{-1}. Research by CIAT (International Center for Tropical Agriculture-Cali, Colombia) has shown that deep-rooted tropical grasses in South America can sequester 100-500 tons of carbon per ha/year. In addition, carbon must be transformed to mineralized carbon aided by microbial activity such as rhizobacteria and mycorrhizae in the soil associated with the root zones. This association is one of the reasons vetiver hedgerows produce such a high amount of phytomass on the one hand and such high amounts of carbon added to the sequestered soil carbon pool on the other (Mendez et al., 2007).

Phytoremediation: Vetiver is a C$_4$ plant with a high efficiency in converting solar radiation to phytomass (Mucciarelli et al., 1998). Its phytomass production (dry weight) usually ranges from 20 to 40 tons ha^{-1} year^{-1}. Owing to its high phytomass, vetiver has great potential for phytoremediation even though vetiver is not a hyperaccumulator. Moreover, its unique characteristic of long, deep roots allows it to penetrate to clean up the deep soil layer. Vetiver's ability to accumulate the highest amounts of heavy metals in roots prompted the choice of this species as more suitable for phytostabilizing mine tailings. Also, the danger of transferring toxic metals to the food chain was minimal (Yang et al., 2003).

Construction materials: Vetiver can be transformed from simply being an agricultural material for soil conservation into an industrial, low-cost construction material. Using vetiver grass ash as cement replacement material, vetiver fiber-clay composite for construction, and vetiver pulp as fiber reinforcement. Hengsakeekul and Nimityongskul (2004) utilized vetiver grass and clay to construct a paddy storage silo for demonstration at the Royal Chitralada project. The cylindrical silo had a diameter and height of 3 m and capacity of 10 tons. The quality of paddy stored in the vetiver-clay silo was unchanged for a period of 6 months. The Association of Siamese Architects decided to build its new center in order to support and provide service to the users. The creation of design from environmentally manufactured materials stimulated the demand of the market for the use of alternative materials. The actual structure was assembled from vetiver board (VB) and rubber inserted with wood to enhance the structure's durability and strength. The VB covered the inner structure, giving the exterior a different visual effect. In addition to its use in the walls, ceiling and floors of rooms, VB has also been used in making furniture such as cupboards, tables, and chairs (Anon, 2009).

Handicrafts: Handicrafts made from vetiver benefit both those who make the handicrafts and those who produce the raw material. Special programs in Thailand and Venezuela have used specially treated vetiver leaves for material for the high-end handicraft market. The Vetiver Network International in cooperation with the Royal Development Projects Board has supported the training of Indians and Chinese in vetiver handicrafts by Thai experts. Zehra Tyabji and Rashmi Ranade, from women weavers, received training and are now available to run training workshops in India. In India, traditionally vetiver roots have been used for handicrafts, but the proper treatment and use of vetiver leaves results in a much more finely crafted product that can be sold in top-end markets where there is a growing demand for "natural" products (Grimshaw, 1997).

Media for mushroom: Vetiver (*Chrysopogon zizanioides*) showed the opportunity for substitution of sawdust with cultivar on the media in mycelia growth of *Ganoderma lucidum* in plastic bags. With the ratio of sawdust and vetiver leaves at 100:0 and 20:80, when supplemented with $CaCO_3$ 0.5%, $MgSO_4 \cdot 7H_2O$ 0.2%, and fine rice barn 6%, gave the biological efficiency (%BE = fresh weight of mushroom $\times$ 100/dry weight of substrates) of 42.72% and 40.71%, respectively (Sornprasert and Aroonsrimorakot, 2014).

Aromatic oil: Spongy root mass of certain cultivars of *Chrysopogon zizanioides* contains trace amounts of essential or volatile oil, known as "vetiver oil" or "khus oil." The grass has been cultivated for its roots, which contain an essential oil used extensively in perfumery, cosmetics, and for biomedical utilization (Lavania, 1998; Chomchalow, 2000, 2001). In addition, vetiver root aromatic oil has been used as a repellent for the control of termites (Zhu et al., 2001). Active aromatic compounds of *Chrysopogon zizanioides* are benzoic acid, vetiverol, furfurol A and B, vetivone, vetivene, vetivenyl, vetivenate. The aromatic oil extracts are used as antiseptic, aphrodisiac, cicatrisant, nervine, sedative, tonic, sedative, and vulnerary (Anderson, 1970; Massardo et al., 2006; Champagnat et al., 2008). Vetiver oil production is closely related to plant metabolism. Oil yield and composition in vetiver can be influenced by several factors such as geographical origin, cultivar, cultivation methods, soil fertilization, the presence of microorganisms, and environmental parameters (Dethier et al., 1997; Adams et al., 2003; Pripdeevech et al., 2006). Low temperature (such as in winter) causes a decrease in plant metabolic activities and hence a decrease in oil production (Massardo et al., 2006). In addition, aromatic grasses produce more essential oil in stress conditions (Farooqi et al., 1999). Aromatic grasses are inedible, industrial and commercial crops, perennial in nature with multiple harvests and tolerant to stress conditions (pH variability, heavy metal toxicity, and drought) (Gupta et al., 2013). Most importantly, essential oil is extracted through hydrodistillation and free from the risk of toxic heavy metals (Figure 11) (Dowthwaite and Rajani, 2000; Zheljazkov et al., 2006;

FIGURE 11

Vetiver grass and some vetiver products: (a) vetiver grass grown in rows on mine waste, (b) vetiver in pot cultivation, (c and d) vetiver root powder, (e) distillation of roots for perfume. Major vetiver oil-producing countries are Haiti, Indonesia (Java), China, India, Brazil, Japan. Major consumers of vetiver oil are the United States, Europe (France), India, Japan. Major uses of vetiver oil are perfumery (perfume, blending, fixative), flavors, cosmetics, masticatories.

Khajanchi et al., 2013). Rotkittikhun et al. (2010) reported that lead (Pb) could increase the oil production of *Chrysopogon zizanioides*, although lead toxicity decreased the length of shoots and roots of the grass. The high number of total constituents of vetiver oil was found in plants grown in soil spiked with Pb (1000 mg Pb kg^{-1}). The predominant compound was khusimol (10.7-18.1%), followed by (E)-isovalencenol (10.3-15.6%).

Biofuel: Vetiver is well placed as a potential furnace feedstock, and has the added advantage that it can be grown on marginal lands to feed local steam-based generating plants creating energy for nearby communities. When grown as a field crop at about 100,000 plants per ha (0.3×0.3 m spacing), vetiver can produce up to 40-100 tons of dry biomass on soils of reasonable depth and fertility. Vetiver has an energy value of 7000 BTU/lb compared to petroleum $-18,000$; coal 12-13,000, dry wood 8500, and sugarcane bagasse 4000. These non-vetiver phytomass sources are used as feedstock to generate electricity (Tulachan et al., 2014). The advantage of vetiver feedstock over other fuels is that it is clean, entirely renewable, it does not emit noxious chemicals when burned, and in the burning process does not add net carbon amounts to the atmosphere (Grimshaw, 1997).

10 CONCLUSIONS

With vetiver phytoremediation, the massive and deep root system of vetiver can prevent leaching and run-off of heavy metals to nearby areas and groundwater by immobilizing and stabilizing heavy metals. In addition, this grass can be an excellent pioneer plant to conserve water and improve soil quality. For a nonhyperaccumulator plant like vetiver, improving phytomass and propagation are necessary for high efficiency of phytoremediation. Application of organic fertilizer can increase vetiver yield and may reduce toxicity of heavy metal. Once the vetiver is fully grown, the aerial growth should be harvested periodically to accelerate new growth. An important advantage of harvested vetiver is that it is not considered hazardous waste, unlike hyperaccumulator residual. It can be used safely for bioenergy production, compost, or even as material for handicrafts. Vetiver grass is not a weed; it is not invasive. The cultivar of vetiver is easily asexually reproduced using simple agricultural practices by root subdivision; each slip normally consists of 2-3 tillers. For long-term remediation, metal-tolerant species are commonly used for revegetation of mine tailings, and herbaceous legumes can be used as the pioneer species to solve the problem of nitrogen deficiencies.

ACKNOWLEDGMENTS

Thanks are due to those at TATA Steel, West Bokaro Coalfields, for permission to conduct fieldwork with necessary logistics. Special thanks to Col. (Retd.) Bhavani and Mr. Roshan Kumar for guidance and help in fieldwork. Thanks are also due to Padaeng Industry Public Co. Ltd. (Mae Sot Office, Tak Province, Thailand), for permission to conduct this phytostabilization study. Financial support under the auspices of India-Thailand bilateral scientific cooperation project "Phytomanagement of mine spoils" (ref. DST/INT/THAI/P-02/2012 dated 31st January 2013) is gratefully acknowledged. Dr. Sateesh Suthari is a recipient of the Start-Up Research Grant (Young Scientists) SERB/LS-293/2014 by the Science and Engineering Research Board (a statutory body under Department of Science & Technology, Government of India) on "Plant taxonomic surveillance and survey of contaminated and polluted ecosystems in Peri-Urban Hyderabad: A randomized crossover study of populations and communities." The authors thank Mr. A. Sampath, Research Scholar, Kakatiya University, Warangal, for providing silkworm photographs (Figure 7d–g).

REFERENCES

Abaga, N.O.Z., Dousset, S., Mbengue, S., Munier-Lamy, C., 2014a. Is vetiver grass of interest for the remediation of Cu and Cd to protect marketing gardens in Burkina Faso? Chemosphere 113, 42–47.

Abaga, N.O.Z., Dousset, S., Munier-Lamy, C., Billet, D., 2014b. Effectiveness of vetiver grass (*Vetiveria zizanioides* (L.) Nash) for phytoremediation of endosulfan in two cotton soils from Burkina Faso. Int. J. Phytorem. 16 (1), 95–108.

Adams, R.P., Pandey, R.N., Dafforn, M.R., James, S.A., 2003. Vetiver DNA-finerprinted cultivars: effects of environment on growth, oil yields and composition. J. Essent. Oil Res. 15, 363–371.

Akhzari, D., Bidgoli, R.D., 2013. Effect of drought and salinity stresses on growth of vetiver grass (*Vetiveria zizanioides* Stapf). World Appl. Sci. J. 24 (3), 390–394.

Ali, H., Khan, E., Sajad, M.A., 2013. Phytoremediation of heavy metals—concepts and applications. Chemosphere 91 (7), 869–881.

Alloway, B.J., 1995. Heavy Metals in Soils. Blackie Academic & Professional, London, UK.

Altenor, S., Ncibi, M.C., Brehm, N., Emmanuel, E., Gaspard, S., 2013. Pilot-scale synthesis of activated carbons from Vetiver roots and sugar cane bagasse. Waste Biomass Valorization 4 (3), 485–495.

Anderson, N.H., 1970. Biogenetic implications of the antipodal sesquiterpenes of vetiver oil. Phytochemistry 9, 145–151.

Anon, 2009. Vetiver grass board decorates ASA Centre in Bangkok. Vetiverim 49, 4.

Antiochia, R., Campanella, L., Ghezzi, P., Movassaghi, K., 2007. The use of vetiver for remediation of heavy metal soil contamination. Anal. Bioanal. Chem. 388, 947–956.

Archer, I.M., Marshman, N.A., Salomons, W., 1988. Development of a revegetation programme from copper and sulphide-bearing mine waste in the humid tropic. In: Solomons, W., Forstner, U. (Eds.), Environmental Management of Solid Waste. Overseas Typographers, Makati, pp. 166–184.

Ashfaq, M., Ali, S., Hanif, M.A., 2009. Bioaccumulation of cobalt in silkworm *Bombyx mori* L. in relation to mulberry, soil and wastewater metal concentrations. Process Biochem. 44, 1179–1184.

Ashfaq, M., Afzal, W., Hanif, M.A., 2010. Effect of ZnII deposition in soil on mulberry-silk worm food chain. Afr. J. Biotechnol. 9 (11), 1665–1672.

Baker, A.J.M., McGrath, S.P., Sidoli, C.M.D., Reeves, R.D., 1994. The possibility of *in situ* heavy metal decontamination of polluted soils using crops of metal-accumulating plants. Resour. Conserv. Recycl. 11, 41–49.

Bharti, S., Banerjee, T.K., 2013. Bioassay analysis of efficacy of phytoremediation in decontamination of coal mine effluent. Ecotoxicol. Environ. Saf. 92, 312–319. http://dx.doi.org/10.1016/j.ecoenv.2013.03.004.

Bolan, N.S., Park, J.H., Robinson, B., Naidu, R., Huh, K.Y., 2011. Phytostabilization: a green approach to contaminant containment. Chapt. IV. In: Sparks, D.L. (Ed.), Advances in Agronomy, vol. 112. Academic Press, New York, pp. 145–204.

Boonyanuphap, J., 2013. Cost-benefit analysis of vetiver system-based rehabilitation measures for landslide-damaged mountainous agricultural lands in the lower Northern Thailand. Nat. Hazard. 69 (1), 599–629.

Boonying, S., Sutapun, W., Supakarn, N., Ruksakulpiwat, Y., 2012. Crystallization behavior of vetiver grass fiber-polylactic acid composite. Adv. Mater. Res. 410, 55–58.

Bradshaw, D., 1987. Reclamation of land and ecology of ecosystem. In: William, R.J., Gilpin, M.E., Aber, J.D. (Eds.), Restoration Ecology. Cambridge University Press, Cambridge, pp. 53–74.

Brenner, F.J., 1984. Restoration of natural ecosystems on surface coal mine lands in the northeastern United States. Stud. Environ. Sci. 25, 211–225.

Brown, S.L., Henry, C.L., Chaney, R., Compton, H., Devolder, P.S., 2003. Using municipal biosolids in combination with other residuals to restore metal-contaminated mining areas. Plant Soil 249, 203–215.

Castro-Larragoitia, J., Kramar, U., Puchelt, H., 1997. 200 years of mining activities at La Paz, San Luis Potosi, Mexico—consequences for environment and geochemical exploration. J. Geochem. Expl. 58, 81–91.

Champagnat, P., Heitz, A., Carnat, A., Fraisse, D., Carnat, A., Lamaison, J., 2008. Flavonoids from *Vetiveria zizanioides* and *Vetiveria nigritana* (Poaceae). Biochem. Syst. Ecol. 36, 68–70.

Chantachon, S., Kruatrachue, M., Pokethitiyook, P., Upatham, S., Tantanasarit, S., Soonthornsarathool, V., 2004. Phytoextraction and accumulation of lead from contaminated soil by vetiver grass: laboratory and simulated field study. Water Air Soil Pollut. 154, 37–55.

Chaplot, P.C., 2014a. Effect of *in situ* moisture conservation practices on run-off, soil loss and productivity of pearl millet (*Pennisetum glaucum*). Ann. Biol. 30 (2), 246–248.

Chaplot, P.C., 2014b. Effect of moisture conservation measures on runoff, soil loss and grass productivity in non-arable land. Ann. Biol. 30 (3), 495–497.

Chaplot, P.C., 2014c. Suitable techniques for establishment of vetiver grass (*Vetiveria zizanioides*) in Rajasthan. Ann. Agri Bio Res. 19 (2), 224–226.

Chen, H., 2000. Chemical method and phytoremediation of soil contaminated with heavy metals. Chemosphere 41, 229–234.

Chen, K.F., Yeh, T.Y., Hsu, Y.H., Chen, C.W., 2012. The phytoattenuation of the soil metal contamination: the effects of plant growth regulators (GA3 and IAA) by employing wetland macrophyte vetiver and energy plant sunflower. Desalin. Water Treat. 45 (1–3), 144–152.

Chen, Y., Shen, Z., Li, X., 2004. The use of vetiver grass (*Vetiveria zizanioides*) in the phytoremediation of soils contaminated with heavy metals. Appl. Geochem. 19, 1553–1565.

Chibrik, T.S., Salamatova, N.A., 1985. Mycosymbiotrophism in cultivated phytocenoses of the Korkino open coal mine. In: Plants in the Industrial Environment. Ural State University, Sverdlovsk, Russia, pp. 54–69.

Chomchalow, N., 2000. Technique of vetiver propagation with special reference to Thailand, Pacific Rim Vetiver Network, Techical Pollution No.2000/1. Pacific Rim. Office of the Royal Development Project Board, Bangkok, Thailand.

Chomchalow, N., 2001. The utilization of vetiver as medicinal and aromatic plants with special reference to Thailand, Pacific Rim Vetiver Network, Techical Pollution No.2001/1. Pacific Rim. Office of the Royal Development Project Board, Bangkok, Thailand.

Chomchalow, N., 2014. Newly discovered plants that bear the names of HRH Princess Maha Chakri Sirindhorn. Acta Hortic. 1025, 115–122.

Chou, S.T., Lai, C.P., Lin, C.C., Shih, Y., 2012. Study of the chemical composition, antioxidant activity and anti-inflammatory activity of essential oil from *Vetiveria zizanioides*. Food Chem. 134 (1), 262–268.

Conesa, H.M., Faz, Á., 2011. Metal uptake by spontaneous vegetation in acidic mine tailings from a semiarid area in South Spain: Implications for revegetation and land management. Water Air Soil Pollution 215, 221–227.

Cooper, E.M., Sims, J.T., Cunningham, S.D., Huang, J.W., Berti, W.R., 1999. Chelate-assisted phytoextraction of lead from contaminated soils. J. Environ. Qual. 28, 1709–1719.

Dethier, M., Sakubu, S., Ciza, A., Cordier, Y., Menut, C., Lamaty, G., 1997. Aromatic plants of tropical central Africa. XXVIII. Influence of cultural treatment and harvest time on vetiver oil quality in Burundi. J. Essent. Oil Res. 9, 447–451.

Dhanya, G., Jaya, D.S., 2013. Pollutant removal in wastewater by vetiver grass in constructed wetland system. Int. J. Eng. Res. Technol. 2, 1361–1368.

Dhillon, K.S., Dhillon, S.K., Thind, H.S., 2008. Evaluation of different agroforestry tree species for their suitability in the phytoremediation of seleniferous soils. Soil Use Manage. 24 (2), 208–216.

Donjadee, S., Chinnarasri, C., 2012. Effects of rainfall intensity and slope gradient on the application of vetiver grass mulch in soil and water conservation. Int. J. Sediment Res. 27 (2), 168–177.

Dowthwaite, S.V., Rajani, S., 2000. Vetiver: perfumer's liquid gold. In: Proc. ICV-2 held in Cha-am, Phethchaburi, Thailand, 18–22 January, pp. 478–481.

Dutta, R.K., Agrawal, M., 2001. Litterfall, litter decomposition and nutrient release in five exotic plant species planted on coal mine spoils. Pedobiologia 45, 298–312.

Falola, O.O., Alasa, M.C., Amuda, A.J., Babayemi, O.J., 2014. Nutritional and antinutritional components of vetiver grass (*Chrysopogon zizanioides* (L.) Roberty) at different stages of growth. Pakistan J. Nutr. 12 (11), 957–959.

Falola, O.O., Alasa, M.C., Babayemi, O.J., 2013. Assessment of silage quality and forage acceptability of vetiver grass (*Chrysopogon zizanioides* L. Roberty) ensiled with cassava peels by wad goat. Pakistan J. Nutr. 12 (6), 529–533.

Farooqi, A.H.A., Sangwan, N.S., Sangwan, R.S., 1999. Effect of different photoperiodic regimes on growth: flowering and essential oil in *Mentha* species. Plant Growth Regul. 29, 181–187.

Filippi, J.J., 2014. Norsesquiterpenes as markers of overheating in Indonesian vetiver oil. Flavour Frag. J. 29 (3), 137–142.

Flathman, P.E., Lanza, G.R., 1998. Phytoremediation: current view on emerging green technology. J. Soil Contam. 7, 415–432.

Freitas, H., Prasad, M.N.V., Pratas, J., 2004. Plant community tolerant to trace elements growing on the degraded soils of Sao Domingos mine in the south east of Portugal: environmental implications. Environ. Int. 30 (1), 65–72.

Ghosh, M., Paul, J., Jana, A., De, A., Mukherjee, A., 2015. Use of the grass, *Vetiveria zizanioides* (L.) Nash for detoxification and phytoremediation of soils contaminated with fly ash from thermal power plants. Ecol. Eng. 74, 258–265.

Ghotbizadeh, M., Sepaskhah, A.R., 2014. Effect of irrigation interval and water salinity on growth of vetiver (*Vetiveria zizanioides*). Int. J. Plant Prod. 9 (1), 17–38.

González, R.C., González-Chávez, M.C.A., 2006. Metal accumulation in wild plants surrounding mining wastes. Environ. Pollut. 144 (1), 84–92.

Gonzalez-Sangregorio, M.V., Trasar-Cepeda, M.C., Leiros, M.C., Gil-Sotres, F., Guitian-Ojea, F., 1991. Early stages of lignite mine soil genesis—changes in biochemical properties. Soil Biol. Biochem. 23 (6), 589–595.

Greenfield, J.C., 1989. Vetiver Grass: The Idea Plant for Vegetative Soil and Moisture Conservation. ASTAG-The World Bank, Washington, DC.

Grimshaw, R.G., 1997. Vetiver grass technology and its application in China. Vetiver Newslett. 1, 4–6.

Gupta, A.K., Verma, S.K., Khan, K., Verma, R.K., 2013. Phytoremediation using aromatic plants: a sustainable approach for remediation of heavy metals polluted sites. Environ. Sci. Technol. 47, 10115–10116.

Hegde, R.S., Fletcher, J.S., 1996. Influence of plant growth stage and season on the release of root phenolics by mulberry as related to development of phytoremediation technology. Chemosphere 32, 2471–2479.

Hengsakeekul, T., Nimityongskul, P., 2004. Construction of paddy storage silo using vetiver grass and clay. AU J. Technol. 7, 120–128.

Herrera, L.E.C., 2014. Combination of vetiver system and erosion control blankets as a solution to slope protection. Environmental Connection Conference 1, 151–161.

Ho, Y.-N., Hsieh, J.-L., Huang, C.-C., 2013. Construction of a plant-microbe phytoremediation system: combination of vetiver grass with a functional endophytic bacterium, *Achromobacter xylosoxidans* F3B, for aromatic pollutants removal. Bioresour. Technol. 14, 43–47.

Holanda, F.S.R., Filho, R.N.A., Lima, J.C.B., Rocha, I.P., 2012. Comparison of different containers in the production of seedlings of vetiver grass for erosion control. Revista Brasileirade Ciencias Agrarias 7 (3), 440–445.

Huang, J.E., Cunningham, S.D., 1996. Lead phytoextraction: species variation in lead uptake and translocation. New Phytol. 134, 75–84.

Jampeetong, A., Brix, H., Kantawanichkul, S., 2012. Effects of inorganic nitrogen forms on growth, morphology, nitrogen uptake capacity and nutrient allocation of four tropical aquatic macrophytes (*Salvinia cucullata, Ipomoea aquatica, Cyperus involucratus*, and *Vetiveria zizanioides*). Aquat. Bot. 97 (1), 10–16.

Jayashree, S., Rathinamala, J., Lakshmanaperumalsamy, P., 2014a. Anti-termite properties of root and leaf powder of vetiver grass. J. Environ. Biol. 35 (1), 193–196.

Jayashree, S., Rathinamala, J., Turan, M., Lakshmanaperumalsamy, P., 2014b. Influence of rooting media on *Chrysopogon zizanioides* (L.) Roberty. J. Plant Nutr. 37 (7), 965–978.

Jennings, S.R., Neuman, D.R., Blicker, P.S., 2008. Acid Mine Drainage and Effects on Fish Health and Ecology: A Review. Reclamation Research Group Publication, Bozeman, MT.

Johne, S., Watzke, R., Visoottiviseht, P., 2008. Cleanup of arsenic-contaminated waters by means of soil feilters planted with vetiver and other plants. Vetiverim 44, 11–12.

Jotisankasa, A., Mairaing, W., Tansamrit, S., 2014. Infiltration and stability of soil slope with vetiver grass subjected to rainfall from numerical modeling. Unsaturated Soils: Research and Applications. In: Proc. 6th International Conference on Unsaturated Soils, UNSAT 2, pp. 1241–1247.

Juntuek, P., Chumsamrong, P., Ruksakulpiwat, Y., Ruksakulpiwat, C., 2014. Effect of vetiver grass fiber on soil burial degradation of natural rubber and polylactic acid composites. Int. Polym. Proc. 29 (3), 379–388.

Juwarkar, A.A., Jambhulkar, P.H., 2008. Phytoremediation of coal mine spoil dump through integrated biotechnological approach. Bioresour. Technol. 99, 4732–4741.

Juwarkar, A.A., Yadav, S.K., Thawale, P.R., Kumar, P., Singh, S.K., Chakrabarti, T., 2009. Developmental strategies for sustainable ecosystem on mine spoil dumps: a case of study. Environ. Monit. Assess. 157 (1–4), 471–481.

Kadarohman, A., Eko, R.S., Dwiyanti, G., Lailatul, L.K., Kadarusman, E., Ahmad Nur, F., 2014. Quality and chemical composition of organic and non-organic vetiver oil. Indo. J. Chem. 14 (1), 43–50.

Kadarohman, A., Sardjono, R.E., Aisyah, S., Khumaisah, L.L., 2013. Biolarvicidal of vetiver oil and ethanol extract of vetiver root distillation waste (*Vetiveria zizanioides*) effectiveness toward *Aedes aegypti*, Culex sp., and *Anopheles sundaicus*. J. Essent. Oil Bear. Plants 16 (6), 749–762.

Kantawanichkul, S., Sattayapanich, S., Van Dien, F., 2013. Treatment of domestic wastewater by vertical flow constructed wetland planted with umbrella sedge and vetiver grass. Water Sci. Technol. 68 (6), 1345–1351.

Katayama, H., Banba, N., Sugimura, Y., Tatsumi, M., Kusakari, S., Oyama, H., Nakahira, A., 2013. Subcellular compartmentation of strontium and zinc in mulberry idioblasts in relation to phytoremediation potential. Environ. Exp. Bot. 85, 30–35. ISSN: 0098-8472.

Khajanchi, L., Yadava, R.K., Kaurb, R., Bundelaa, D.S., Khana, M.I., Chaudharya, M., Meenaa, R.L., Dara, S.R., Singha, G., 2013. Productivity essential oil yield, and heavy metal accumulation in lemon grass (*Cymbopogon flexuosus*) under varied wastewater-groundwater irrigation regimes. Ind. Crop. Prod. 45, 270–278.

Knoll, C., 1997. Rehabilitation with vetiver. African Mining 2, 43–48.

Kuang, C., Zhang, T., 2012. Growth and physio-biochemical responses of *Vetiveria zizanioides* to solidification/stabilization products made from dredged sediment. Adv. Mater. Res. 518–523, 3375–3386.

Lavania, U.C., 1988. Enhanced productivity of the essential oil in the artificial autopolyploid of vetiver (*Vetiveria zizanioides* (L.) Nash). Euphytica 38, 271–276.

LDD, 1994. Vetiver Grass. Land Development Department/Ministry of Agriculture and Cooperatives, Bangkok, Thailand.

LDD, 1998. Vetiver Grass Overview. Land Development Department/Ministry of Agriculture and Cooperatives, Bangkok, Thailand.

Leung, H.-M., Wang, Z.-W., Ye, Z.-H., Yung, K.-L., Peng, X.-L., Cheung, K.-C., 2013. Interactions between arbuscular mycorrhizae and plants in phytoremediation of metal-contaminated soils: a review. Pedosphere 23 (5), 549–563.

Lima, G.M., Quintans-Júnior, L.J., Thomazzi, S.M., Almeida, E.M.S.A., Melo, M.S., Serafini, M.R., Cavalcanti, S.C.H., Gelain, D.P., Santos, J.P.A., Blank, A.F., Alves, P.B., Oliveira Neta, P.M., Lima, J.T., Rocha, R.F., Moreira, J.C.F., Araújo, A.A.S., 2012. Phytochemical screening, antinociceptive and anti-inflammatory activities of *Chrysopogon zizanioides* essential oil. Braz. J. Pharmacog. 22 (2), 443–450.

Li-ping, W., Kui-mei, Q., Shi-long, H., Bo, F., 2009. Fertilising reclamation of arbuscular mycorrhizal fungi on coal mine complex substrate. Procedia Earth Planet. Sci. 1, 1101–1106.

Liu, Y., Shi, Y., 2014. Fast detection of heavy metal lead (Pb) in vetiver grass leaves using near infrared spectroscopy. Nongye Jixie Xuebao/Trans. Chin. Soc. Agr. Mach. 45 (3), 232–236.

Maignana, K.R., Saradha, R.A., Arunkumar, R., Lakshmipathy, P.R., Madhavi, E., Sobita, D.T., 2014. Evaluation of antiepileptic activity of *Vetiveria zizanioides* oil in mice. Int. J. Pharm. Sci. Rev. Res. 25 (2), 248–251. art. no. 47.

Maiti, S.K., 2007. Bioreclamation of coalmine overburden dumps—with special empasis on micronutrients and heavy metals accumulation in tree species. Environ. Monit. Assess. 125, 111–122.

Mallavarapu, G.R., Syamasundar, K.V., Rameshc, S., Rajeswara Rao, B.R., 2012. Constituents of South Indian vetiver oils. Nat. Prod. Commun. 7 (2), 223–225.

Maneecharoen, J., Htwe, W., Bergado, D.T., Baral, P., 2013. Ecological erosion control by limited life geotextiles (LLGs) as well as with vetiver and ruzi grasses. Indian Geotech. J. 43 (4), 388–406.

Manh, L.H., Dung, N.N.X., Van Am, L., Le Minh, B.T., 2014. Treatment of wastewater from slaughterhouse by biodigester and *Vetiveria zizanioides* L. Livest. Res. Rural Dev. 26 (4), Article #68. http://www.lrrd.org/lrrd26/4/manh26068.htm

Manzoor, F., Naz, N., Malik, S.A., Siddiqui, B.S., Syed, A., Perwaiz, S., 2012. Chemical analysis and comparison of antitermitic activity of essential oils of neem (*Azadirachta indica*), vetiver (*Vetiveria zizanioides*) and mint *(Mentha arvensis)* against *Heterotermes indicola* (Wasmann) from Pakistan. Asian J. Chem. 24 (5), 2069–2072.

Marques, A.P.G.C., Rangel, A.O.S.S., Castro, P.M.L., 2009. Remediation of heavy metal contaminated soils: phytoremediation as a potentially promising clean-up technology. Crit. Rev. Environ. Sci. Technol. 39 (8), 622–654.

Massardo, D., Senatore, F., Alifano, P., Giudice, L.D., Pontieri, P., 2006. Vetiver oil production correlates with early root growth. Biochem. Syst. Ecol. 34, 376–382.

Mendez, M.O., Glenn, E.P., Maier, R.M., 2007. Phytostabilization potential of quailbush for mine tailings: growth, metal accumulation, and microbial community changes. J. Environ. Qual. 36 (1), 245–253.

Mendez, M.O., Maier, R.M., 2008. Phytostabilization of mine tailings in arid and semiarid environments—an emerging remediation technology. Environ. Health Perspect. 116 (3), 278.

Mucciarelli, M., Bertea, C.M., Cozzo, M., Scannerini, S., Gallino, M., Scannerini, S., 1998. *Vetiveria zizanioides* as a tool for environmental engineering. Acta Hortic. 457, 261–269.

National Resource Council, 1995. Vetiver Grass: A Thin Green Line Against Erosion. National Academy Press, Washington, DC.

Niityoungskul, P., Hengsadeekul, T., 2002. The construction of vetiver-clay composite storage bin. In: Summary Report of the Royal Project Foundation for 2002 on Research and Development Project on Vetiver Grass. The Royal Project Foundation, Chiang Mai, Thailand, pp. 31–39.

Nikolić, D., Milošević, N., Živković, Ž., Mihajlović, I., Kovačević, R., Petrović, N., 2011. Multi-criteria analysis of soil pollution by heavy metals in the vicinity of the copper smelting plant in Bor (Serbia). J. Serb. Chem. Soc. 76 (4), 625–641.

Nuthong, W., Uawongsuwan, P., Pivsa-Art, W., Hamada, H., 2013. Impact property of flexible epoxy treated natural fiber reinforced PLA composites. Energy Procedia 34, 839–847.

Nyomora, A.M.S., Njau, K.N., Mligo, L., 2012. Establishment and growth of vetiver grass exposed to landfill leachate. J. Solid Waste Technol. Manag. 38 (2), 82–91.

Oku, E., Aiyelari, A., Truong, P., 2014. Green structure for soil and water conservation on cultivated steep land. Kasetsart J. (Nat. Sci.) 48 (2), 167–174.

ORDPB, 2000. Factual Tips About Vetiver Grass. Office of the Royal Development Projects Board, Bangkok, Thailand.

Oshunsanya, S.O., 2013. Crop yields as influenced by land preparation methods established within vetiver grass alleys for sustainable agriculture in Southwest Nigeria. Agroecol. Sust. Food Syst. 37 (5), 578–591.

Pandey, V.C., 2012. Invasive species based efficient green technology for phytoremediation of fly ash deposits. J. Geochem. Explor. 123, 13–18.

Pandey, V.C., 2013. Suitability of *Ricinus communis* L. for phytoremediation fly ash disposal sites. Ecol. Eng. 57, 336–341.

Pang, J., Chan, G.S.Y., Zhang, J., Liang, J., Wong, M.H., 2003. Physiological aspects of vetiver grass for rehabilitation in abandoned metalliferous mine wastes. Chemosphere 52, 1559–1570.

Pazoki, M., Abdoli, M.A., Karbassi, A., Mehrdadi, N., Yaghmaeian, K., 2014. Attenuation of municipal landfill leachate through land treatment. J. Environ. Health Sci. Eng. 12 (1). art. no. 12.

Pichai, N.M.R., Samjiamjiaras, R., Thammanoon, H., 2001. The wonders of a grass, vetiver and its multifold applications. Asian Infrastruct. Res. Rev. 3, 1–4.

Pichtel, J., Salt, C.A., 1998. Vegetative growth and trace metal accumulation on metalliferous wastes. J. Environ. Qual. 27, 618–642.

Pinthong, J., Inpithuksa, S., Ramlee, A., 1998. The capability of vetiver hedgerow in decontamination of agrochemical residues; a case study on the production of cabbage at Non Hoi Development Center. In: Proc. First International Conference on Vetiver, Chiang Rai, Thailand, 4–8 February 1996, pp. 91–98.

Prasad, A., Chand, S., Kumar, S., Chattopadhyay, A., Patra, D.D., 2014. Heavy metals affect yield, essential oil compound, and rhizosphere microflora of vetiver (*Vetiveria zizanioides* Linn. Nash) grass. Communications in Soil Science and Plant Analysis 45 (11), 1511–1522. http://dx.doi.org/10.1080/00103624.2014.904334.

Prasad, M.N.V., 2007. Phytoremediation in India. In: Willey, N. (Ed.), Phytoremediaiton—Methods and Reviews. Humana Press, Totowa, NJ, pp. 435–454.

Prasad, M.N.V., Jeeva, S., 2009. Coal mining and its leachate are potential threats to *Nepenthes khasiana* Nepenthaceae that preys on insects—an endemic plant in north eastern India. BioDiCon 23, 29–33.

Prasad, M.N.V., Prasad, R., 2012. Nature's cure for cleanup of contaminated environment—a review of bioremediation strategies. Rev. Environ. Health 28, 181–189.

Pripdeevech, P., Wongpornchai, S., Promsiri, A., 2006. Highly volatile constituents of *Vetiveria zizanioides* roots grown under different cultivation conditions. Molecules 11, 817–826.

Pu, Y., Xie, D., Lin, C., Wei, C., 2013. Characteristics of soil comprehensive anti-erodibility under sloped cropland with hedgerows. Nongye Gongcheng Xuebao/Trans. Chin. Soc. Agr. Eng. 29 (18), 125–135.

Pulford, I.D., Watson, C., 2003. Phytoremediation of heavy metal contaminated lands by trees: a review. Environ. Int. 29, 529–540.

Quang, D.V., Schreinemachers, P., Berger, T., 2014. Ex-ante assessment of soil conservation methods in the uplands of Vietnam: an agent-based modeling approach. Agric. Syst. 123, 108–119.

Rahardjo, H., Satyanaga, A., Leong, E.C., Santoso, V.A., Ng, Y.S., 2014. Performance of an instrumented slope covered with shrubs and deep-rooted grass. Soils Found. 54 (3), 417–425.

Rajkumar, M., Sandhya, S., Prasad, M.N.V., Freitas, H., 2012. Perspectives of plant associated microbes in heavy metal phytoremediation. Biotechnol. Adv. 30, 1562–1574.

Randloff, B., Walsh, K., Melzer, A., 1995. Direct revegetation of coal tailing at BHP, Saraji Mine. In: Proc. 12th Annual Australian Mining Council Environment Workshop, Darwin, Australia, 17–19 June 1995, pp. 849–854.

Rentz, J.A., Alvarez, P.J.J., Schnoor, J.L., 2005. Benzo[a]pyrene co-metabolism in the presence of plant root extracts and exudates. Environ. Pollut. 136 (3), 477–484.

Robinson, B.H., Brooks, R.R., Clothier, B.E., 1999. Soil amendments affecting nickel and cobalt uptake by Berkheya coddii: potential use for phytomining and phytoremediation. Ann. Bot. 84, 689–694.

Rodkvamtook, W., Prasartvit, A., Jatisatienr, C., Jatisatienr, A., Gaywee, J., Eamsobhana, P., 2012. Efficacy of plant essential oils for the repellents against chiggers (*Leptotrombidium imphalum*) vector of scrub typhus. J. Med. Assoc. Thai. 95 (Suppl. 5), S103–S106.

Roongtanakiat, N., 2006. Vetiver in Thailand: general aspects and basic studies. KU Sci. J. 24, 13–19.

Roongtanakiat, N., 2009. Vetiver phytoremediation for heavy metal decontamination, Pacific Rim Vetiver Network, Techical Pollution No.2009/1. Pacific Rim. Office of the Royal Development Project Board, Bangkok, Thailand.

Roongtanakiat, N., Chairoj, P., 2001. Vetiver grass for the remediation of soil contaminated with heavy metals. Kasetsart J. Nat. Sci. 35, 433–440.

Roongtanakiat, N., Jompuk, P., Rattanawongwiboon, T., Puingam, R., 2012. Radiosensitivity of vetiver to acute and chronic gamma irradiation. Kasetsart J. Nat. Sci. 46 (3), 383–393.

Roongtanakiat, N., Osotsapar, Y., Yindiram, C., 2008. Effects of soil amendment on growth and heavy metals content in vetiver grown on iron ore tailings. Kasetsart J. Nat. Sci. 42, 397–406.

Roongtanakiat, N., Osotsapar, Y., Yindiram, C., 2009. Influence of heavy metals and soil amendments on vetiver (*Chrysopogon zizanioides*) grown in zinc mine soil. Kasetsart J. Nat. Sci. 43, 37–49.

Roongtanakiat, N., Sanoh, S., 2011. Phytoextraction of zinc, cadmium and lead from contaminated soil by vetiver grass. Kasetsart J. Nat. Sci. 45, 603–612.

Roongtanakiat, N., Tangruangkiat, S., Meesat, R., 2007. Utilization of vetiver grass (*Vetiveria zizanioides*) for removal of heavy metals from industrial wastewaters. ScienceAsia 33, 397–403.

Rotkittikhun, P., Kruatrachue, M., Pokethitiyook, P., Baker, A.J.M., 2010. Tolerance and accumulation of lead in *Vetiveria zizanioides* and its effect on oil production. J. Environ. Biol. 31, 329–334.

Sadunishvili, T., Kvesitadze, E., Betsiashvili, M., Kuprava, N., Zaalishvil, G., Kvesitadze, G., 2009. Influence of hydrocarbons on plant cell ultrastructure and main metabolic enzymes. World Academy of Science, Engineering and Technology 57, 271–276.

Salomons, W., 1995. Environmental impact of metals derived from mining activities: processes, predictions, prevention. J. Geochem. Explor. 52 (1–2), 5–23.

Samuel, M.P., Senthilvel, S., Mathew, A.C., 2014. Performance evaluation of a dual-flow recharge filter for improving groundwater quality. Water Environ. Res. 86 (7), 615–625.

Sangeeta, M., Maiti, S.K., Masto, R.E., 2013. Use of Reclaimed Mine Soil Index (RMSI) for screening of tree species for reclamation of coal mine degraded land. Ecol. Eng. 57, 133–142.

Santos, T.C., Arrigoni-Blank, M.F., Blank, A.F., 2012. *In vitro* propagation and conservation of vetiver grass (Propagação e conservação *in vitro* de Vetiver). Hortic. Bras. 30 (3), 507–513.

Sarah, J., Abhilash, P.C., Nandita, S., Sharma, P.N., 2009a. *Jatropha curcas*: a potential crop for phytoremediation of coal fly ash. J. Hazard. Mater. 172, 269–275.

Sarah, J., Abhilash, P.C., Singh, A., Singh, N., Behl, H.M., 2009b. Fly ash trapping and metal accumulating capacity of plants: implication for green belt around thermal power plants. Landscape Urban Plan. 92, 136–147.

Sawangsuriya, A., Jotisankasa, A., Sukolrat, J., Dechasakulsom, M., Mahatumrongchai, V., Milindalekha, P., Anuvechsirikiat, S., 2013. Comparison of erosion susceptibility and slope stability of repaired highway embankment. Geotechnical Special Publication (231 GSP), pp. 1912–1921.

Sciarrone, D., Pantò, S., Tranchida, P.Q., Dugo, P., Mondello, L., 2014. Rapid isolation of high solute amounts using an online four-dimensional preparative system: normal phase-liquid chromatography coupled to methyl siloxane-ionic liquid-wax phase gas chromatography. Anal. Chem. 86 (9), 4295–4301.

Singh, M., Guleria, N., Prakasa Rao, E.V.S., Goswami, P., 2014a. Efficient C sequestration and benefits of medicinal vetiver cropping in tropical regions. Agron. Sustain. Dev. 34 (3), 603–607.

Singh, R., Narzary, D., Bhardwaj, J., Singh, A.K., Kumar, S., Kumar, A., 2014b. Molecular diversity and SSR transferability studies in vetiver grass (*Vetiveria zizanioides* (L.) Nash). Ind. Crop. Prod. 53, 187–198.

Singh, V.K., Behal, K.K., 2002. Effect of coal ash amended soil on growth of mulberry plant (*Morus alba*). J. Ecophysiol. Occup. Health 2 (3&4), 243–254.

Singh, S., Melo, J.S., Eapen, S., D'Souza, S.F., 2008. Potential of vetiver (*Vetiveria zizanioides* (L.) Nash) for phytoremediation of phenol. Ecotoxicol. Environ. Saf. 71, 671–676.

Sinha, S., Jothiramajayam, M., Ghosh, M., Mukherjee, A., 2014. Evaluation of toxicity of essential oils palmarosa, citronella, lemongrass and vetiver in human lymphocytes. Food Chem. Toxicol. 68, 71–77.

Sornprasert, R., Aroonsrimorakot, S., 2014. Utilization of *Vetiveria zizaniodes* (L.) Nash leaves in *Ganoderma lucidum* cultivated. APCBEE Procedia 8, 47–52.

Sripen, S., 1996. Growth potential of vetiver grass in relation to nutrients in waste water of Changwat Phetchaburi. In: Proc. First International Conference on Vetiver, Chiang Rai, Thailand, 4–8 February 1996, pp. 45–52.

Stevenson, F.L., Cole, M.A., 1999. Cycles of Soil: Carbon, Nitrogen, Phosphorus, Sulfur, Micronutrients. John Wiley & Sons, Inc., New York, USA.

Sujaritjun, W., Uawongsuwan, P., Pivsa-Art, W., Hamada, H., 2013. Mechanical property of surface modified natural fiber reinforced PLA biocomposites. Energy Procedia 34, 664–672.

Sukwat, P.S., Thiengkamol, N., Navanugraha, C., Thiengkamol, C., 2012. Development of prototype of young Buddhist environmental education. Social Sciences 7 (1), 56–60.

Sutapun, W., Raksakulpiwat, Y., Suppakarn, N., 2012. Effect of heat treatment on chemical structure of a bio-filler from vetiver grass. Adv. Mater. Res. 410, 71–74.

Tayfur, G., Baba, A., 2011. Groundwater contamination and its effect on health in turkey. Environ. Monit. Assess. 183 (1–4), 77–94.

Thammathavorn, S., Khanema, P., 2010. Sequestration potential of vetiver. Report to the Office of the Royal Development Projects Board by the Biology Branch, Science Office, Suranaree Technical University, Nakhon Ratchasima, Thailand (in Thai).

Tordoff, G.M., Baker, A.J.M., Willis, A.J., 2000. Current approaches to the revegetation and reclamation of metalliferous mine wastes. Chemosphere 41 (1–2), 219–228.

Truong, P., 1999. Vetiver Grass Technology for Mine Rehabilitation. Office of the Royal Development Projects Board, Bangkok. Technical Bulletin No. 1999/2.

Truong, P., 2000. Vetiver grass technology for environmental protection. In: Proc. second International Vetiver Conferences: Vetiver and the Environment, Cha Am, Thailand, January 2000.

Truong, P.N., 1996. Vetiver grass for land rehabilization. In: Proc. First International Vetiver Conferences, Thailand, pp. 49–56.

Truong, P.N., Baker, D., 1996. Vetiver grass for the stabilization and rehabitation of acid sulfate soils. In: Proc. Second National Conference on Acid Sulfate soils. Coffs Harbour, Australia, pp. 196–198.

Truong, P.N., Baker, D., 1998. Vetiver grass for environment protection, Pacific Rim Vetiver Network, Techical Pollution No.1998/1. Pacific Rim. Office of the Royal Development Project Board, Bangkok, Thailand.

Truong, P.N., McDowell, M., Christiansen, I., 1995. Stiff grass barrier with vetiver grass. A new Approach to erosion and sediment control. In: Proc. Downstream Effects of Land Use Conference, Rockhamton, Australia, 24–27 November 1995, pp. 301–304.

Truong, P., Van, T.T., Pinners, E., 2008. Vetiver System Application: A Technical Reference Manual. The Vetiver Network International, San Antonio, TX.

Tulachan, B., Meena, S.K., Rai, R.K., Mallick, C., Kusurkar, T.S., Teotia, A.K., Sethy, N.K., Bhargava, K., Bhattacharya, S., Kumar, A., Sharma, R.K., Sinha, N., Singh, S.K., Das, M., 2014. Electricity from the silk cocoon membrane. Sci. Rep. 4, 5434. http://dx.doi.org/10.1038/srep05434.

Vandermeer, J., 1989. The Ecology of Intercropping. Cambridge University Press, Cambridge. p. 237.

Verma, S.K., Singh, K., Gupta, A.K., Pandey, V.C., Trivedi, P., Verma, R.K., Patra, D.D., 2014. Aromatic grasses for phytomanagement of coal fly ash hazards. Ecol. Eng. 73, 425–428.

Vinayagamoorthy, R., Rajeswari, N., 2014. Mechanical performance studies on *Vetiveria zizanioides*/jute/glass fiber-reinforced hybrid polymeric composites. J. Reinf. Plast. Comp. 33 (1), 81–92.

Vollú, R.E., Blank, A.F., Seldin, L., Coelho, M.R.R., 2012. Molecular diversity of nitrogen-fixing bacteria associated with *Chrysopogon zizanioides* (L.) Roberty (Vetiver), an essential oil producer plant. Plant Soil 356 (1–2), 101–111.

Walker, D.J., Clemente, R., Roig, A., Bernal, M.P., 2003. The effects of soil amendments on heavy metal bioavailability in two contaminated Mediterranean soils. Environ. Pollut. 122, 303–312.

Wilde, E.W., Brigmon, R.L., Dunn, D.L., Heitkamp, M.A., Dagnan, D.C., 2005. Phytoextraction of lead from firing range soil by vetiver grass. Chemosphere 61, 1451–1457.

Wong, J.W.C., Ip, C.M., Wong, M.H., 1998. Acid-forming capacity of lead-zinc mine tailings and its implications for mine rehabilitation. Environ. Geochem. Health 20 (3), 149–155.

Wong, M.H., 2003. Ecological restoration of mine degraded soils, with emphasis on metal contaminated soils. Chemosphere 50 (6), 775–780.

Wu, F.Y., Chung, A.K.C., Tam, N.F.Y., Wong, M.H., 2012. Root exudates of wetland plants influenced by nutrient status and types of plant cultivation. Int. J. Phytorem. 14 (6), 543–553.

Xia, H.P., 2004. Ecological rehabilitation and phytoremediation with four grasses in oil shale mined land. Chemosphere 54, 345–353.

Yang, B., Shu, W.S., Ye, Z.H., Lan, C.Y., Wong, H.M., 2003. Growth and metal accumulation in vetiver and two *Sesbania* species on lead/zinc mine tailings. Chemosphere 52, 1593–1600.

Yaseen, M., Singh, M., Ram, D., 2014. Growth, yield and economics of vetiver (*Vetiveria zizanioides* (L.) Nash) under intercropping system. Ind. Crop. Prod. 61, 417–421.

Ye, M., Sun, M., Liu, Z., Ni, N., Chen, Y., Gu, C., Kengara, F.O., Li, H., Jiang, X., 2014a. Evaluation of enhanced soil washing process and phytoremediation with maize oil, carboxymethyl-β-cyclodextrin, and vetiver grass for the recovery of organochlorine pesticides and heavy metals from a pesticide factory site. J. Environ. Manage. 141, 161–168.

Ye, M., Sun, M., Wan, J., Fang, G., Li, H., Hu, F., Jiang, X., Kengara, F.O., 2014b. Evaluation of enhanced soil washing process with tea saponin in a peanut oil-water solvent system for the extraction of PBDEs/PCBs/PAHs and heavy metals from an electronic waste site followed by vetiver grass phytoremediation. J. Chem. Technol. Biotechnol. http://dx.doi.org/10.1002/jctb.4512.

Ye, Z.H., Wong, J.W.C., Wong, M.H., 2000. Vegetation response to lime and manure compost amendments on acid lead/zinc mine tailings: a green house study. Restor. Ecol. 8, 289–295.

Yong, R.N., Mohamed, A.M.O., Warkentin, B.P., 1992. Principles of Contaminant Transport in Soils. Elsevier, Amsterdam.

Zhang, K., Chen, Q., Luo, H.B., Li, X.T., 2014a. Accumulation of chromium in *Vetiveria zizanioides* assisted by earthworm (*Eisenia foetida*) in contaminated soil. Adv. Mater. Res. 989–994, 1313–1318.

Zhang, K., Chen, Q., Luo, H.B., Li, X.T., 2014b. Application of *Vetiveria zizanioides* assisted by different species of earthworm in chromium-contaminated soil remediation. Adv. Mater. Res. 1010–1012, 564–569.

Zhang, X., Gao, B., Xia, H., 2014c. Effect of cadmium on growth, photosynthesis, mineral nutrition and metal accumulation of bana grass and vetiver grass. Ecotoxicol. Environ. Safe. 106, 102–108.

Zheljazkov, V.D., Craker, L.E., Baoshan, X., 2006. Effects of Cd, Pb and Cu on growth and essential oil contents in dill pepper mint and basil. Environ. Exp. Bot. 58, 9–16.

Zhu, B.C.R., Henderson, G., Chen, F., Fei, H.X., Laine, R.A., 2001. Evaluation of vetiver oil and seven insect-active essential oils against the Formosan subterranean termite. J. Chem. Ecol. 27, 1617–1625.

Zhuang, Z., Hazelton, P., 2014. An investigation of the efficiency of fulvic acid and straw water amendments for arsenic uptake from groundwater by *Vetiveria zizanioides*. Adv. Mater. Res. 864–867, 1233–1239.

UTILIZATION OF CONTAMINATED LANDS FOR CULTIVATION OF DYE PRODUCING PLANTS

K. Schmidt-Przewoźna, A. Brandys

Institute of Natutral Fibres and Medicinal Plants, Poznan, Poland

1 INTRODUCTION

Natural plants that serve as a source of colors are of great economic importance. In recent years we have observed a revival of interest in natural raw materials and in extraction techniques that had often been completely forgotten. The sudden revival of interest in natural dyes not only concerns textile dyeing, but also dyes used in cosmetics, such as hair dye, food coloring, and other products. We can find dyes in fruits, flowers, roots, bark, leaves, seeds of trees, plants, insects, and lichens. Different parts of the same plants may give different colors and shades. Moreover, there is a big potential for the application of the compounds found in dyeing trees in various fields of science and economy. Natural dyes obtained from plants are seen as an ecological solution, and as eco-friendly dyes with natural antibacterial and antimicrobial properties. Plant extracts contain tannins, flavonoids, saponins, essential oils, mucilage, vitamins, and many other valuable nutritional substances. They have various properties; for example, medicinal, soothing, caring, and disinfecting. From this point of view they are well known for their curative properties. Many of them are extensively used in Ayurvedic and homeopathic medicine in Asia and Africa (Cardon, 2010).

Many plants with dyeing properties do not require special conditions for cultivation. They can grow on sands and wetlands as well as contaminated lands. The environment adapts to climate and soil changes. Natural dyes derived from plant materials have a wide range of applications. They can be used to dye textiles, food products, cosmetics, composites, leather, and wood. Of course, in the food and cosmetics industries, numerous requirements must be met for the cultivation, acquisition, and extraction of the dye. In contrast, dyes obtained for dyeing fabrics, wood, and composites can be obtained from contaminated lands. Given the huge potential of plant materials, there is a niche demand for products dyed with plant materials. These dyes are intended for use in the development of organic clothing collections with health properties. They can be and are used in the cosmetic and food industries. The research field concerns the issues of the rational use of natural resources as well as environmental management and protection. This research requires the participation and cooperation of scientists from several scientific fields: the natural sciences, technology, and social science.

Bioremediation and Bioeconomy. http://dx.doi.org/10.1016/B978-0-12-802830-8.00014-9
Copyright © 2016 Elsevier Inc. All rights reserved.

2 HISTORICAL BACKGROUND

The search for color is present in all cultures in the world and dates back to the dawn of human civilization. Archeological findings indicate that dyes from natural sources have been used to color textiles for the last 6000 years. Since the beginning of their existence, people have tried to mark their ethnic and individual character by means of, for example, colors. Color determined their affiliation to the group— the identity that gave them a sense of security and safety; it also expressed their emotional states. Color became a means of expression, an essential element of the evolutionary change from the world of nature to the world of culture. Observing colorful nature, primitive people wished to use nature's colors also for themselves. They found it in roots, berries, branches, in the bark of trees, leaves, insects, shellfish, and minerals. The art of using vegetable dyes, and particularly the skill of using auxiliary chemicals, were a secret art for ages. The art of dyeing developed on different continents independently, and as a result of migrations of peoples and the import of dyes. The cradle of dyeing knowledge and art is central and south Asia and the great civilizations of antiquity (Schmidt-Przewozna and Skrobak, 2013). In ancient Egypt dyeing was at an advanced level. The earliest methods of dyeing using mordant and mineral dyes were well known. This knowledge was adopted from the Egyptians by the Israelites and Phoenicians and later by the Greeks and Romans (Schmidt-Przewozna, 2009).

The discovery of America and its exploration revealed that independent dyeing techniques had been developed in what is today Peru and Mexico. Thanks to the existing sea routes to East India, the following dyes became known there: cochineal, indigo, dyeing trees, red sandalwood *Pterocarpus santalinus*, cutch *Acacia catechu*, brazil wood *Caesalpinia brasiliensis* (red), and logwood *Haemotoxylon campechianum* (blue and violet). The ground root turmeric *Curcuma longa* was imported from China and India. This dyestuff allowed for obtaining yellow and orange colors. Annatto *Bixa orellana* and indigo *Indigofera tinctoria* were also known and used in small amounts. Color can be obtained from virtually every plant; however, its durability and saturation differs. The keys to dyeing are knowledge and experience, which have been passed on from generation to generation over the course of human culture (Schmidt-Przewozna, 2009; Cardon, 2007; Schmidt-Przewozna, 2001).

Until 1856, all dyes were obtained from natural raw materials. The first synthetic dye, mauveine, was invented by William Henry Perkin. Some sources say that it was magenta, which was invented by the Polish chemist Jakub Natanson in 1855. The outstanding success of Perkin's mauve gave a great impetus to the discovery of new dyes in a period receptive to new scientific advances. The development of chemistry slowly led to the departure away from complex dyeing technologies based on natural raw materials. It is widely believed that synthetic dyes are easier to use, that the colors are more intense, and that their resistance to washing and light is better.

Taking part in this discussion, one should pay attention to historic artifacts. Many of them, for hundreds and even thousands of years, kept their beautiful and clean colors. Looking at these historic fabrics, we may notice the difference in the colors' durability. Textiles found in the graves of the pharaohs in Egypt are more than 4000 years old. Some colors were partially degraded, but many others are of fascinating intensity (Schmidt-Przewozna, 2002). Natural dyeing techniques require extensive knowledge related not only to the dyeing of fabrics, but also to the fabrics' preparation for dyeing. The knowledge of plants and the coloring capabilities that can be achieved through the use of various mortars were important. The dyeing process lasted longer and was carried out in stages. In modern times, characterized by rapid consumption and seasonally changing fashion, color does not have to last for years. In the twentieth century, historical techniques of dyeing with natural plant extracts, developed over generations, slowly disappeared. However, at the turn of the twenty-first century, the positive impact of natural-fiber fabrics dyed with plant extracts on the human body began to be noticed. These interests are connected to the general processes of an increased use of ecological natural resources, which are tested for their health benefits, as well as with their wide application in fashion and design. Accordingly prepared fibers of modern clothing, among others, through the use of natural dyeing, the individual compounds contained therein act on the human skin.

Such research is conducted in India, in other Asian countries, and in Europe, as well as Poland, in the Institute of Natural Fibres and Medicinal Plants in Poznań. There are regions in India, East Asia, Central and South America, where natural dyeing has been preserved on a very limited scale. In the 1990s, there was a debate in India (one of the main centers of cultivation of dyeing plants) as to whether using natural dyes is useful. As a result, Indian research institutes launched a series of studies on the improvement of the fastness of dyes. Such studies are carried out at the India Institute of Technology in New Delhi. As a result of these works, collections of clothes dyes naturally produced by local society of manufacturing engineers (SME) have been introduced to the market. One of the companies that works on launching fabrics dyed with natural dyestuffs is Aura Herbal Textiles Ltd.

Plant extracts are essences from plants: their flowers, leaves, bark, fruit, or roots. They contain the following: tannins, flavonoids, saponins, essential oils, vitamins, and other nourishing substances. Depending on the content of active substances and their precious ingredients they can have the following properties: medicinal, soothing, moistening, anti-inflammatory, regenerative, antiviral properties, antifungal, antioxidant, and protection against UV radiation (UVR) (Schmidt-Przewoźna and Zimniewska, 2005). Many of them are extensively used in Ayurvedic and homeopathic medicine in Asia and Africa. Oak galls, *Terminalia chebula*, pomegranate peel *Punica granatum*, and oak, willow, and buckthorn bark are rich sources of tannins. When used in premordanting they affect the hue and the durability of colors. Some of them have also been used to intensify the color of other natural dyes and to improve its fastness.

3 POTENTIAL AND FUTURE PROSPECTS OF DYEING PLANTS

All around us, new factories, highways, and concrete housing estates are being created. People do not think about how they change the environment until they begin to suffer from civilizational diseases, such as allergies, atherosclerosis, or cancer. Pollution is predominant in industrial and urban areas. In areas where there are a lot of cars, the air is filled with gaseous pollutants and suspended particulate matter. Car exhaust gases include heavy metals, toxic nitrogen oxides, and—most recently, recognized as one of the most dangerous—suspended microscopic particulates. These particles penetrate the lung alveoli, and the body is unable to defend against them. In the surroundings of communication routes, just a few dozen meters from the lanes, there is not only contaminated air, but also water and soil contamination. It turns out that the only solution in such a changed environment is plants. Plants, as organisms leading a sedentary lifestyle, in the process of evolution had to develop defense mechanisms allowing survival even in extreme conditions. They not only produce oxygen and water vapor, but they are also capable of taking up and accumulating contaminants from soil, water, and air in their leaves, stems, and roots. They retain them thanks to the wax and hairs on the leaf surface or, when the poison penetrates beneath the surface of the skin of the leaf, they immobilize and neutralize them. In this way, plants also catch the dangerous microdusts emitted by car engines. The cleaning activity of plants also involves cooperation with the soil microflora. Microorganisms working on the breakdown of complex chemical compounds appear in the vicinity of the roots. Scholars claim that the number of microorganisms there are 10 times higher than in plantless areas. The treatment of the environment with the use of plants is called phytoremediation. This method has only advantages; it is effective and relatively inexpensive. In areas degraded by industry, perennial plantations of so-called energy plants are created, such as miscanthus, willow, and poplar. Herbaceous plants are also useful (Figure 1).

In order to breathe clean air, it is worth having an oak, ash, maple, or common pear tree near the house: Thanks to their leaf hairs or wax coating, these trees will remove the biggest amount of suspended particulate matter. Common weeds also play an important role in areas where municipal waste is stored. They form a vegetation cover over the polluted area. We can find the following perennials in landfills: tansy *Tanacetum vulgare*, broadleaf dock *Rumex obtusifolius*, stinging nettle *Urtica dioica*, greater celandine *Chelidonium majus*, annual bedstraw *Galium aparine*, common yarrow *Achillea millefolium* (Figure 2).

On former landfill sites, especially on the top of slag heaps and their gentle slopes, perennial plants have made their home. Many of them have dyeing properties. Researchers from the University of Ecology and Management in Warsaw have defined sets of the most expansive plants in areas degraded by municipal landfills in Poland: late goldenrod *Solidago gigantea*, lady's thumb *Polygonum persicaria*, knotgrass *Polygonum aviculare*, tansy *Tanacetum vulgare*, bitter dock *Rumex obtusifolius*, greater burdock *Arctium lappa*, stinging nettle *Urtica dioica*, dandelion *Taraxacum officinale*, common yarrow *Achillea millefolium*, and greater celandine *Chelidonium majus* (Dygus et al., 2012). These common herbs are also dyeing plants. After a few years, postmining sites were populated with shrubs, such as willows *Salix* sp. and trees. Among the trees, the most common ones were silver birch *Betula pendula*, aspen *Populus tremula*, Scots pine *Pinus sylvestris*, and the English oak *Quercus robur.* The fastest-growing species used in the recultivation of degraded areas are the "Royal Purple" smoke tree *Cotinus coggygria* and staghorn sumac *Rhus typhina* (Figure 3).

FIGURE 1

Common yarrow *Achillea millefolium* on contaminated lands.

Photo by K. Schmidt-Przewoźna.

FIGURE 2

Goldenrod *Solidago gigantea*.

Photo by K. Schmidt-Przewoźna.

FIGURE 3

Elderberry *Sambucus nigra* in areas degraded by municipal landfills in Poland.

Photo by K. Schmidt-Przewoźna.

4 SOURCES OF NATURAL DYEING PLANTS

4.1 DYEING TREES

Many trees, thanks to the compounds contained in them, are extensively used in traditional medicine and homeopathy. The ingredients derived from trees, such as bark, seeds, pith, and leaves, constitute a valuable resource for traditional and unconventional medicine thanks to the compounds they contain. Moreover, as they contain biologically active substances, they have a wide spectrum of pharmacological effects. They have also been used in the furniture, cosmetics, dyeing, and tanning industries. Dyeing trees are not only a rich source of dyes, but due to the high content of tannins, they are also used in premordating fabrics. Oak galls, myrabalan, oak bark, pomegranate peel, and other plants have been used in the process of mordating fabrics in different cultures around the world. This process is important in achieving lasting, more intense colors and their resistance to external factors, such as water and light (Przewoźna et al., 2011).

4.1.1 Sandalwood

Common Names: red sanders, red sandalwood, sanders wood (Figure 4).

Red sandalwood *Pterocarpus santalinus* is a medium-sized tree of about 10-15 m in height. It grows on dry, rocky soils, as well as the dry slopes of deciduous forests. It has been a valuable resource in centime memorial (Dean, 1999).

It can be found in India, Indonesia, Ceylon, Australia, and the islands of the Pacific. Documented sources confirm that it has been used 4000 years ago in India, Egypt, Greece, and Rome. Many temples and buildings have been built using this material. In Egypt, sandalwood was also used to cremate dead bodies.

The plant contains alkaloids, phenolics, saponins, glycosides, flavonoids, and tannins. Valuable essential oils, dyes, and wood are obtained from red sandalwood (Cardon, 2007).

Valuable components are present in the leaves, fruit, and bark of the tree.

Medical and pharmaceutical uses

Essential oils support the treatment of urethra and urinary tract infections. Moreover, sandalwood oil helps to remove bloating and colic, as it acts as a relaxant on the intestine. It also acts as an expectorant.

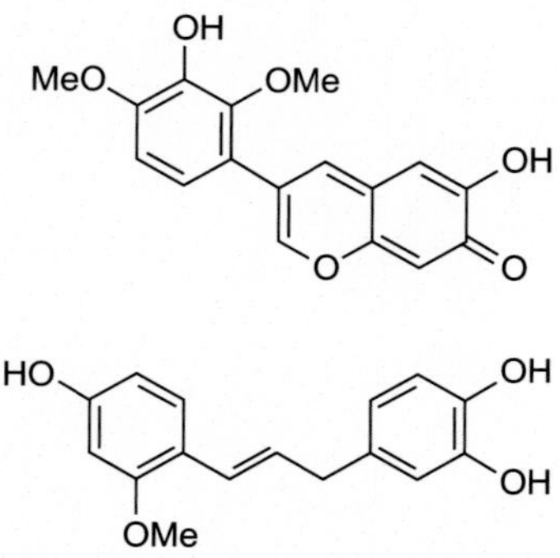

FIGURE 4

Santalin (dye from red sanders) chemical structure.

http://onlinelibrary.wiley.com/doi/10.1002/anie.201302317/abstract, accessed 13.04.2015, 13:52 a.m.

Red sandalwood tree oil has both excellent medicinal and cosmetic properties. It helps in the treatment of urethra and urinary tract infections. Moreover, sandalwood oil helps to remove bloating and colic, as it acts as a relaxant on the intestine. It also acts as an expectorant. It has astringent, antiseptic, antibacterial, disinfectant, and anti-inflammatory properties. It has a positive effect on dry skin and is efficient in the treatment of acne and lichens. It is widely used in aromatherapy for its calming effect, stimulating imagination and relieving anxiety.

The antibacterial activity of the leaves and stem bark of *Pterocarpu santalinus* was investigated by Manjunatha from the National Institute of Technology, Karnataka, India (Manjunatha, 2006). The stem bark extract showed maximum activity against *Enterobacter aerogenes*, *Alcaligenes faecalis*, *Escherichia coli*, *Pseudomonas aeruginosa*, *Proteus vulgaris*, *Bacillus cereus*, *Bacillus subtilis*, and *Staphylococcus aureus*. The leaf extract showed maximum activity against *Escherichia coli*, *Alcaligenes faecalis*, *Enterobacter aerogenes*, and *Pseudomonas aeruginosa* (Manjunatha, 2006; Bojase et al., 2002; Vanita et al., 2013).

Other uses

The oil is used in the production of cosmetics-shampoos, shower gels, body lotions, and aromatic oils.

The dyeing extracts are used to dye pharmaceuticals, food (coloring sodas reddish-orange and intensifying the color of seafood, meat, bread, and alcoholic drinks), paper, textiles, as well as in the tanning and cosmetic industries. It is a component of perfumes as well as an aphrodisiac, valued by many peoples of Asia and Africa.

Sandalwood is also a valuable building and sculpting material.

Red sandalwood: natural dye

The red wood yields a natural dye, santalin, which dyes objects red, brown, and russet/ginger. The dye has been widely used to color wool in different parts of the world. It was imported to Europe in mass amounts from India, as a substitute for other red dyes. In the eighteenth century, Jan Hellot, a French chemist and tanner, recommended the following recipe to obtain red color:

Tannining process: 12 pounds of alder bark, 10 pounds of sumac, and ½ pound of ground gallnuts.
Dyeing process: 4 pounds of red sandalwood for each piece of cloth (Cardon, 2007).

4.1.2 *Indian mulberry* Morinda citrifolia

Common Names: great morinda, Indian mulberry, mengkudu-malai, noni, beach mulberry, cheese fruit (Figure 5).

Indian mulberry *Morinda citrifolia* is a tree that grows in a warm climate in Asia, Australia, and Africa. It grows in shady forests as well as on open rocky and sandy shores. It grows well on sands, saline soils, and degraded areas. On the volcanic island of Krakatoa, degraded by lava flows, the tree grows without any problems. Indian mulberry does not require good soil. The tree reaches a height of approximately 5-9 m.

Its leaves, fruit, bark, and roots are used for different purposes.

Medical and pharmaceutical uses

The leaves, fruit, and roots of *Morinda citrifolia* have been used in folk medicine for hundreds of years.

Noni fruit juice contains many vitamins, minerals, natural antioxidants, as well as proteins, enzymes that work synergistically and support the work of the whole body in many ways (Morton, 1992). The juice

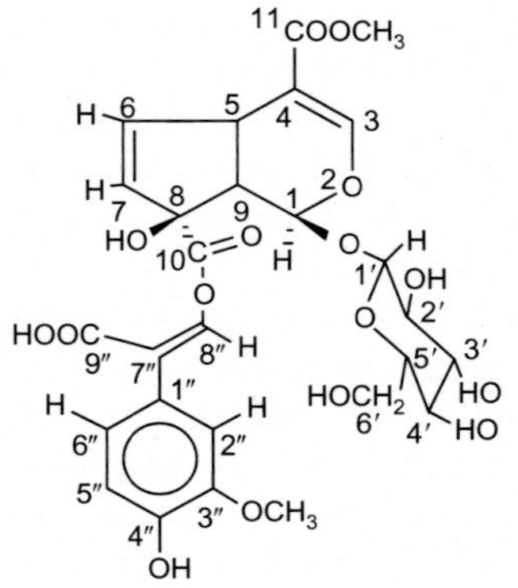

FIGURE 5

Morinda citrifolia dye (alizarin) chemical structure.

http://www.prodifact.com/1morinda-citrifolia.html, accessed 13.04.2015, 13:56 a.m.

supports the treatment of many diseases, including diabetes, arthritis, gout, and cancer. It contains compounds with analgesic, antibacterial, and antiviral properties, which support the body's resistance to infections (Saleem et al., 2005). The characteristic polysaccharides discovered in it enhance the immune system. It is recommended for people with chronic diseases and convalescents. Thanks to the wealth of natural antioxidants contained in it—polyphenols, phenolic acids, and coumarin, which support the fight against free radicals and delay the aging process of the body—noni fruit juice constitutes a great component of slimming and detox diets. It helps to remove toxins from the body. The fruit of the Indian mulberry is consumed for remedying lumbago, asthma, and dysentery in India and China (West et al., 2008). Noni fruit juice is considered to be an elixir of vitality: It regulates metabolism, helps in removing toxins, improves digestion, boosts energy levels, and restores the natural homeostasis of the organism. It also accelerates the regeneration of damaged skin cells, promoting faster healing of wounds, bruises, and burns. The populations of Africa, Asia, and Australia often use the fruits in their diets and medicine (Mohd Zin et al., 2007).

In Hawaii, the fruit juice is used to treat lice and make shampoo.

In Malaysia, mengkudu leaf tea is drunk to alleviate the symptoms of cough and to treat nausea and colic. Most of the folk uses for diabetes involve chewing the leaves or a combination of the plant and leaves.

Other uses

Noni fruits are not tasty and have an unpleasant odor. However, they are consumed in some places in the world. On the islands of Samoa and Fiji they are food for the poor. Morinda extract is also used to stain food and alcohol.

The young leaves can also be eaten as a vegetable and they contain protein (4-6%). The seeds can be eaten after being roasted.

Morinda is also used as a dietary supplement in tablets.

Indian mulberry: natural dye

Drury (1873) wrote that a dye was derived from the "heart of the wood in older trees."

Chemists at Delhi University, India, isolated morindone, damnacanthal, and nordamnacanthal from the shavings of the heartwood, the latter not previously found in plants (Murti et al., 1959).

Later, Balakrishna et al. (1960) discovered a new glycoside of morindone, which they named morindonin, in the root bark. The main dyeing substance in *Morinda* is alizarin. Other important substances are rubiadin, lucidin, damnacanthal, normancanthal, soranjidiol, antharagallol, pseudopurpurin, purpurin, and morindone (Dominique P. 676, 777). Morindone is a mordant dye that gives a yellowish-red color when used with an aluminum mordant, brown with a chromium mordant, and dull purple to black with an iron mordant. The dye from the roots of the tree is very popular in India and in the Pacific islands. In Borneo they call it *Iban mengkudu* and use it to dye fabrics called *ikat*, and in Java batik painting. The red color is obtained in the complex process of dyeing, where an important step is the premordating of the cotton used for *ikat*. In Rumah Garie's longhouses, in Borneo, the cotton yarn is mordated for 2 weeks. Ginger, palm salt, and palm oil are used in the process. Red-maroon and russet colors are obtained during dyeing. To get a deeper red color, sappan wood and alum are often added (Figure 6).

The bark of the Indian mulberry plant produces a reddish-purple and brown-colored dye, which is used in making *batik*. The tree is extensively grown for the purpose of obtaining dye in Java.

4.1.3 *Kamala* Mallotus philippensis

Common Names: kamala tree, dyer's rottlera, monkey face tree, orange kamala, red kamala, scarlet croton, kamala, raini, rohan, rohini, jia ma la.

Kamala *Mallotus philippensis* is a tree that grows up to 25 m in height. It can be found in the Himalayas, India, Sri Lanka, China, Taiwan, Australia, and Africa. The tree's leaves, fruits, and boughs are used. The fruit is covered with red granules from which powder is obtained. The powder constitutes only 1.5-4% of their weight, which makes this valuable resource expensive.

FIGURE 6

Indian mulberry *Morinda citrifolia*.

Photo by K. Schmidt-Przewoźna.

Medical and pharmaceutical uses

Kamala medicinal uses include treatment for afflictions of the skin, inflammations, eye diseases, bronchitis, abdominal diseases, and spleen enlargement, but there is no scientific evidence to prove its efficiency (Sharma and Varma, 2011).

The roots are used for dissolving coagulated blood. The leaves and powdered fruit are used in treating skin diseases and hard-to-heal wounds. Kamala extracts help in the treatment of anorexia and increase the feeling of hunger. *Mallotus philippensis* has antibacterial, antifungal, antitumor, and laxative properties (Thakur et al., 2015; Rao and Seshadri, 1947).

Other uses

The fruit of *Mallotus philippensis* is covered with a red powder called *kamala*, and it is used locally to make textile dye, syrup, and as an old remedy for tapeworm, because it has a laxative effect. The wood is used for rafters, tool handles, and matchboxes. The oil of the seeds is used as a substitute for tung oil (*Vernicia* Lour., Euphorbiaceae) in the formulation of rapid-drying paints, varnishes, hair conditioners, and ointments.

The dye composition of kamala makes it useful in food and drinks. The leaves are used as fodder and the wood is used to make paper.

Kamala: natural dye

The powder obtained from ripe fruits contains rottlerin (Figures 7), iso-rotten, citric acid, tannin, and wax. Kamala is a mortar dye and, therefore, to obtain a durable color, cotton fabrics should be mordated before dyeing. It is used primarily for dyeing silk, but it can also be used to dye wool, cotton, and linen. In Japan, on the island of Mijako, *Mallotus philippensis* leaves are used in the traditional indigo dyeing's finishing processes. As a result, the color blue acquires a beautiful copper shade. The colors that we obtain from kamala using different dyes (Figure 8) and methods are yellow, cream, olive, and red.

Experimental findings

See Table 1.

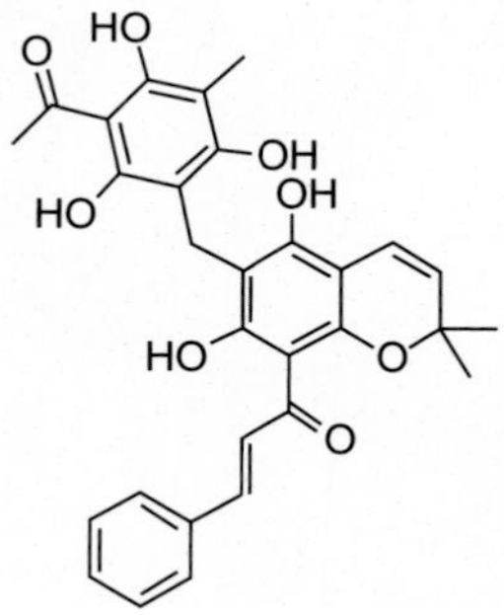

FIGURE 7

Rottlerin chemical structure.

http://en.wikipedia.org/wiki/Rottlerin accessed, 13.04.2015, 13:51 a.m.

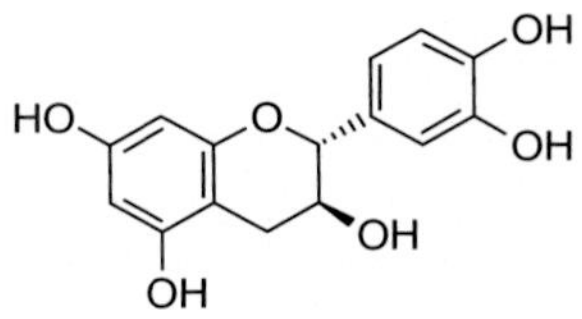

FIGURE 8

Catechin chemical compound.

http://en.wikipedia.org/wiki/Catechin accessed, 13.04.2015, 13:50 a.m.

4.1.4 *Logwood* Haematoxylum campechianum *L.*

Common Names: bois campeche, bloodwood tree, campeche, campeche wood, logwood, palo de campeche, palo de tinta, palo de tinte, palo negro, tinto, ek (Figures 9 and 10).

Logwood *Haematoxylum campechianum* L. is a small, spiny tree native to the tropical regions of America and Mexico, extending south through the tropical dry forests of Oaxaca, Guatemala, Costa Rica, and Colombia (Cardon, 2007).

The tree grows up to 12 m in height. Its wood is hard and dark brown.

The generic name of logwood *Haematoxylum* (often spelled *Haematoxylon*) means *bloodwood*, referring to the dark red heartwood that is a source of dye. The name *campechianum* refers to the city of Campeche on the Yucatan Peninsula. It is possible to find this tree, containing valuable heartwood, in the vicinity of this town.

The parts of the tree that are used are the wood, heartwood, and leaves. It was brought to Europe after Christopher Columbus's discovery of America. The English brought huge amounts of it in large pieces from their colonies. During the Industrial Revolution in the UK, the tree's dye was needed for dyeing cotton and wool.

As it was not a durable dye, its use was banned at the turn of the seventeenth and eighteenth centuries in many European countries.

Medical and pharmaceutical uses

Logwood has also been used traditionally in folk medicine. The tree's bough and its other parts were used to treat many diseases. In a publication from 1981 describing medical plants from Mexico, one of the 184 plants is *Haematoxylum campechianum* L. It was used in the pre-Columbian era and was one of the most important medical plants.

It is used as a mild astringent and tonic. It is administered in the form of a decoction and liquid extract. It is also useful against diarrhea, dysentery, and dyspepsia as well as for menstrual disorders.

Other uses

The heartwood may be used to dye not only linen and cotton, but also leather as well as fur, silk, feathers, paper, bones, and in the manufacturing of inks (Schmidt-Przewozna and Brandys, 2014).

Logwood, which is called palo de tinto in Mayan cultures, was used a 1000 years ago to make wooden constructions in temples and palaces. Even today, the peoples living in that area build wooden house constructions from this hardwood. The Mayans use the natural shapes of the branches and tree trunks in their constructions.

Table 1 Kamala Dyeing Results on Linen and Cotton Samples

No.	Fabric	Mordant	Spectrophotometric Result					Fastness	
			L^*	a^*	b^*	C^*	$h°$	Wash	Light
1	Linen—5 g	No mordant	78.7	4.5	33.8	34.1	82.4	5	4
2	Cotton—5 g	No mordant	82.1	3.2	31.8	32.0	84.2	3	3-4
3	Linen—5 g	Alum—2 g	84.8	3.8	31.5	31.7	83.1	5	4
4	Cotton—5 g	Alum—2 g	84.4	2.0	29.2	29.2	86.0	4	4
5	Linen—5 g	Soda—2 g	87.1	2.3	15.8	16.0	81.9	3	3
6	Cotton—5 g	Soda—2 g	89.4	1.1	12.9	13.0	85.2	3	3
7	Linen—5 g	Citric acid—2 g	87.3	-0.7	24.2	24.3	91.6	3	3
8	Cotton—5 g	Citric acid—2 g	88.2	-1.6	21.2	21.3	94.3	3-4	3
9	Linen—5 g	Copper—2 g	82.7	-2.5	19.0	19.2	97.5	4	4
10	Cotton—5 g	Copper—2 g	85.4	-2.1	18.2	18.3	96.6	4	4
11	Linen—5 g	Iron—2 g	66.6	4.6	18.0	18.6	75.6	4	4
12	Cotton—5 g	Iron—2 g	70.5	4.5	20.5	21.0	77.7	4	4

Fabrics: L, linen; C, cotton.
Mordant: Alum, $Al_2K_2(SO_4)_4 \cdot 24H_2O$; Copper, $CuSO_4 \cdot 5H_2O$; Iron, $FeSO_4 \cdot 7H_2O$; Citric acid, $C_6H_8O_7 \cdot H_2O$; Washing soda, Na_2CO_3.

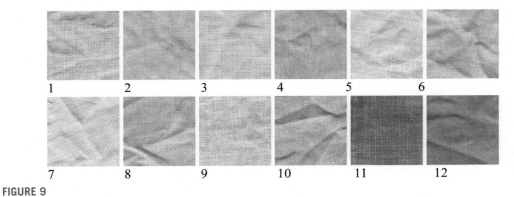

FIGURE 9

Kamala *Mallotus philippensis* colors results with different mordants.

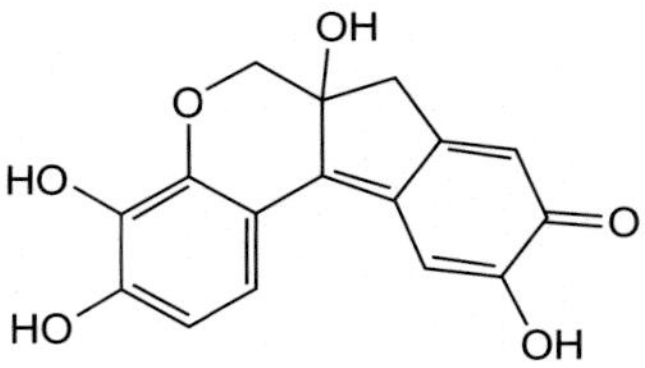

FIGURE 10

Hematoxylin chemical structure.

http://pl.wikipedia.org/wiki/Hematoksylina accessed 13.04.2015, 13:50 a.m.

Logwood: natural dye

The dye has been known since pre-Columbian times. In the Mayan language, the name of the tree, "ek," also means the color black.

The dye obtained from logwood comes from its heartwood. The main dyeing substance in logwood is hematoxylin, which was isolated for the first time by Chevreul in 1810. Other dyes in this plant are small amounts of brazilein, hematein, and large amounts of tannins. Hematoxylin is extracted by boiling chips or shavings of logwood in water. Logwood powder has a red color and in the dyeing process it can be a source of violet, gray-violet, lavender, brown, black, light brown, purplish-red, and indigo blue colors.

The whole range of colors that can be obtained from logwood has to do with the use of dyeing mortars. The pH of the water is also important. Violet and lavender are obtained with the use of alum and copper sulfate, blacks and navy blues are obtained with the use of ferrous sulfate, and the colors red and purple are obtained without the use of mortar (Schmidt-Przewozna and Brandys, 2014).

Experimental findings

See Table 2 and Figure 11.

4.1.5 *Cutch tree* Acacia catechu

Common Names: cutch, cashoo, khoyer, terra Japonica, acacia à cachou, caciu, kath, kachu, cacho, Japan earth, black cutch, cachou, sa-che, seesit, sha (Figure 12).

Table 2 Logwood Dyeing Results on Linen and Cotton Samples

No.	Fabric	Mordant	Spectrophotometric Result						Fastness	
			L^*	a^*	b^*	C^*	$h°$		Wash	Light
1	Linen—5 g	No mordant	57.8	3.7	8.5	9.3	66.5		4	3
2	Linen/G—5 g	No mordant	43.5	4.5	8.8	9.9	63.1		4-5	3
3	Cotton—5 g	No mordant	33.6	3.9	3.8	5.5	43.8		4	2
4	Linen—5 g	Alum—0.35 g	51.2	8.7	−137	16.3	302.4		4	3
5	Linen/G—5 g	Alum—0.35 g	36.4	9.0	−14.5	17.1	301.9		4-5	3-4
6	Cotton—5 g	Alum—0.35 g	54.2	6.9	−12.3	14.1	299.3		4	2
7	Linen—5 g	Soda—0.35 g	79.3	2.0	8.9	9.1	77.5		4	3
8	Linen/G—5 g	Soda—0.35 g	66.5	3.6	17.6	17.9	78.4		4	3-4
9	Cotton—5 g	Soda—0.35 g	72.1	3.3	10.8	11.3	73.0		4	2
10	Linen—5 g	Citric acid—0.35 g	60.9	7.0	5.0	8.6	35.1		4	3
11	Linen/G—5 g	Citric acid—0.35 g	47.3	7.3	4.5	8.6	31.5		4-5	3-4
12	Cotton—5 g	Citric acid—0.35 g	53.2	9.5	8.1	12.5	40.4		4	2
13	Linen—5 g	Copper—0.35 g	43.3	−3.9	−3.9	5.5	225.1		4	3
14	Linen/G—5 g	Copper—0.35 g	33.2	−1.8	−2.4	3.0	232.5		4-5	3-4
15	Cotton—5 g	Copper—0.35 g	41.3	−4.5	−4.9	6.6	227.6		4	3
16	Linen—5 g	Iron—0.35 g	53.6	1.2	−1.8	2.2	304.0		4	4
17	Linen/G—5 g	Iron—0.35 g	29.7	2.2	−4.2	4.7	297.8		4-5	4-5
18	Cotton—5 g	Iron—0.35 g	41.6	1.6	−3.3	3.7	296.4		4	3

Fabrics: L, linen; C, cotton; Linen/G, linen premordating in oak gall.
Mordant: Alum, $Al_2K_2(SO_4)_4 \cdot 24H_2O$; Copper, $CuSO_4 \cdot 5H_2O$; Iron, $FeSO_4 \cdot 7H_2O$; Citric acid, $C_6H_8O_7 \cdot H_2O$; Washing soda, Na_2CO_3.

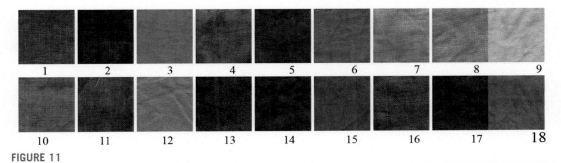

FIGURE 11

Logwood color results with different mordants.

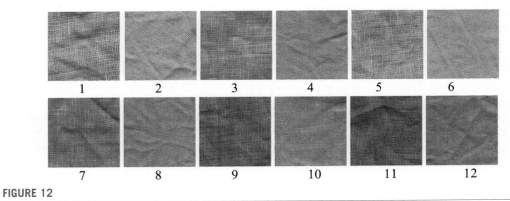

FIGURE 12

Cuth color results with different mordants.

The source of color from the cutch tree is the heartwood. The cutch tree is small and usually typical in arid regions with sandy soils (e.g., in southern Pakistan, India, Burma, Thailand, and China). The heart of cutch wood, when extracted, evaporated, cooled, and filtrated, is a source of red-brown color (Cardon, 2007).

Medical and pharmaceutical uses

The bark and heartwood of the cutch tree are used in traditional medicine. Catechu wood extract has been commonly used in medicine, health products, and in food and chemical industry. It has the following properties: antioxidative, antiaging, reducing blood lipids, reducing blood sugar, anticancer, antiradiation, antiultraviolet, and removing free radicals.

Other uses

Cutch heart extract is used in tanning leather, dyeing, and as a preservative for fishing nets. Blavod black vodka is very popular in the UK. Its unique black color is derived from the black catechu tree. The most important use for the cutch tree is its high usage for furniture industry.

Cutch tree: natural dye

Dried aqueous extract of wood and bark of cutch tree is a well-known natural dye. This is obtained by leaching of wood. Extract is the best for cotton and silk dyeing, and pretanning leathers. The extract of the cutch tree also has astringent properties that facilitate penetration and brightening of the skin.

Experimental findings

See Table 3 and Figure 12.

4.1.6 *Annatto* Bixa orellana

Common Names: lipstick tree, achiote, bija, roucou, onoto, atsuete, colorau (Figure 13).

The annatto plant grows in tropical areas. Annatto is a profusely fruiting shrub or small tree that grows 5–10 m in height. Approximately 50 seeds grow inside prickly, reddish-orange, heart-shaped pods. One small annatto tree can produce up to 270 kg of seeds.

Annatto is found in Ecuador, Bolivia, Brazil, and Mexico. Annatto's Latin designation (*Bixa orellana* L.) was chosen for the Spanish conquistador Francisco de Orellana during his exploration of the Amazon River (Cardon, 2007).

Medical and pharmaceutical uses

Annatto seeds contain a characteristic pleasant-smelling oil. The composition of the seed oil is similar to that of soybean oil. Fats in plant can be used for production of butter (Lauro and Jack, 2000; Morton, 1960). A decoction of annatto root is taken orally to control asthma. A macerated seed decoction is taken orally for relief of fever, and the pulp surrounding the seed is made into an astringent drink used to treat dysentery and kidney infection. Infusions of root in water and rum are used to treat venereal diseases. The dye is used as an antidote for prussic acid poisoning. Extracts from different parts of the tree kill bacteria, fight free radicals, kill parasites, increase urination, stimulate digestion, lower blood pressure, act as mildly laxative, protect the liver, reduce fever, arterial hypertension, high cholesterol, cystitis, obesity, renal insufficiency, and eliminate uric acid.

Annatto is a noncarcinogenic and nontoxic native plant, used as haemostatic, antioxidant, astringent and antibacterial, antidysenteric, diuretic, aphrodisiac, and an effective febrifuge, digestive, and gentle purge. It is prescribed also for epilepsy, erysipelas and sundry skin diseases, and throat infections.

Other uses

Annatto has the common name "lipstick tree." In the past it was in use for coloring women's lips and for body painting. From seeds we can obtain dye and oil extract for cosmetics, essential oils for shampoos, soaps, etc. From color cosmetics list code of annatto seeds is the same like extract from annatto. Code is: E160b; CI number is 75120.

Annatto is very popular as a food colorant. It is used for cheese, sausages, meats, and candies and also beverages, cosmetics, pharmaceutical products, and traditional spices for food.

Annatto: natural dye

Pigment from seeds of this tree contains bixin (Figure 14) and norbixine. Because of this, annatto is an excellent natural colorant used for food industry and as natural dyes for textiles and also for coloring other things like feathers, sheepskin, ivory, bones, bamboo mats, and rattan. The color may be fixed with tamarind leaves or with barks containing tannin. These pigments are a source of tones between yellow and red. Specialist literature reports that green shades can be obtained from the leaves.

The powder from seeds is soaked in hot water overnight, boiled 1 h, and filtered.

Table 3 Cutch Dyeing Results on Linen and Cotton Samples

No.	Fabric	Mordant	Spectrophotometric Result					Fastness	
			L*	a*	b*	C*	h°	Wash	Light
1	Linen	No mordant	52.9	17.7	16.7	24.3	43.4	5	5
2	Cotton	No mordant	53.4	16.4	18.0	24.3	47.8	4	4
3	Linen	Alum—2 g	64.0	13.5	21.2	25.1	57.4	5	5-6
4	Cotton	Alum—2 g	68.1	13.7	21.1	25.1	57.0	4	5
5	Linen	Soda—2 g	76.2	11.5	9.9	15.2	40.6	5	5
6	Cotton	Soda—2 g	69.7	11.2	9.6	14.7	40.6	4	4-5
7	Linen	Citric acid—2 g	61.8	12.8	17.4	21.5	53.7	4-5	4
8	Cotton	Citric acid—2 g	63.7	11.1	14.8	18.5	53.0	3	3
9	Linen	Copper—2 g	57.9	10.2	15.2	18.3	56.1	4-5	5
10	Cotton	Copper—2 g	63.1	8.8	14.0	16.5	57.9	4	4-5
11	Linen	Iron—0.5 g	54.2	4.7	7.4	8.8	57.8	5	5
12	Cotton	Iron—0.5 g	51.2	4.5	7.05	8.3	57.5	4-5	5

Fabrics: L, linen; C, cotton.
Mordant: Alum, $Al_2K_2(SO_4)_4 \cdot 24H_2O$; Copper, $CuSO_4 \cdot 5H_2O$; Iron, $FeSO_4 \cdot 7H_2O$; Citric acid, $C_6H_8O_7 \cdot H_2O$; Washing soda, Na_2CO_3.

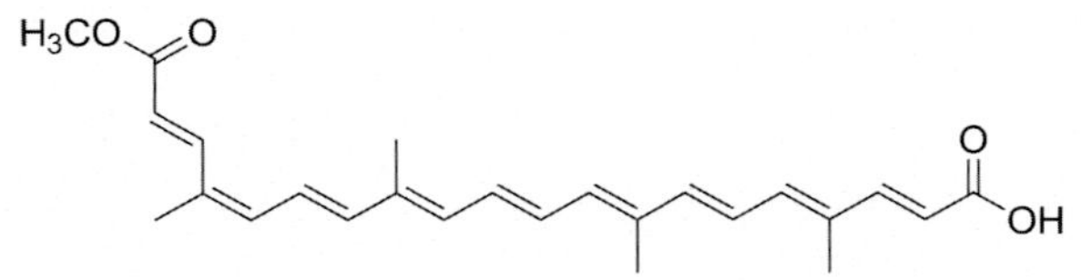

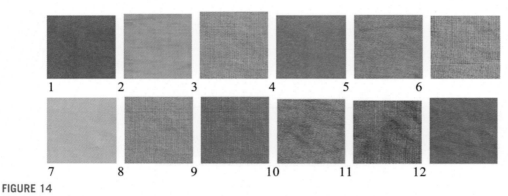

FIGURE 13

Bixin chemical structure. The color of annatto comes from various carotenoid pigments.

http://pl.wikipedia.org/wiki/Biksyna, accessed 13.04.2015, 13:49 a.m.

FIGURE 14

Annatto colors results with different mordants.

In addition to bixin and norbixin, annatto contains bixaghanene, bixein, bixol, crocetin, ellagic acid, ishwarane, isobixin, phenylalanine, salicylic acid, threonine, tomentosic acid, and tryptophan (Silva et al., 2008).

Experimental findings
See Table 4 and Figure 14.

4.2 TANNIN SOURCES: TANNIN COMPOUNDS

Natural sources rich in tannin include oak gall *Quercus infectoria* Oliv., Chebulic Myrobalan *Terminalia chebula*, oak bark *Quercus robur*, and pomegranate *Punica granatum*. Descriptions and details of this plants follow.

4.2.1 Oak gall Quercus infectoria
Quercus infectoria Oliv is a small tree and shrub mainly found in Greece, Asia Minor, Syria, and Iran. Galls are formed on the leaves, roots, and bark of trees. We can usually find them on oaks and Chinese sumac *Rhus* species (Figure 15).

Oak galls contain tannic and gallic acid. Gallnuts are used to obtain tannin, which allows for better dyeing of fabrics and is used in tanning. In some galls the compounds (tannins) constitute half or even more of the substance's dry matter. The tannin compound is gallotannin, which is an ester of glucose, ellagic, and gallic acid. Gallnuts, which are widely used in medicine, usually come from oaks. They

Table 4 Annatto Dyeing Results on Linen and Silk Samples

No.	Fabric	Mordant	Spectrophotometric Result					Fastness	
			L^*	a^*	b^*	C^*	$h°$	Wash	Light
1	Silk—4.5 g	No mordant	59.0	33.3	43.6	54.9	52.6	4	4
2	Linen—4.5 g	No mordant	68.1	28.2	32.1	42.8	48.7	4	3
3	Linen—4.5 g	Alum—2 g	70.1	24.5	28.1	44.7	50.6	5	5
4	Silk/G—4.5 g	Alum—2 g	77.2	26.4	32.7	46.3	57.3	5	5
5	Linen/G—4.5 g	Soda—2 g	75.5	16.1	38.4	41.6	67.0	5	5
6	Linen—4.5 g	Soda—2 g	70.5	22.5	22.4	31.8	44.9	5	4
7	S/B—4.5 g	Citric acid—2 g	54.3	31.8	33.8	44.2	52.0	4	4
8	Linen/G—4.5 g	$CuSO_4$—2 g	28.2	11.8	7.4	13.8	33.5	5	5
9	Linen—4.5 g	$CuSO_4$—2 g	24.0	10.2	6.6	12.2	32.6	4	5
10	L/B—4.5 g	Iron—2 g	51.8	30.9	41.7	56.2	48.9	5	5
11	Linen/G—4.5 g	Iron—2 g	53.2	32.1	44.3	58.1	50.2	5	5
12	S/B—4.5 g	Iron—2 g	58.1	35.0	49.0	60.2	54.5	5	5

Fabrics: Linen—Linen/G, linen premordating in galas; Linen/O, linen premordating in bark of oak; Silk—Silk/G, silk premordating in galas; Silk/O, silk premordating in bark of oak.
Mordant: Alum, $Al_2K_2(SO_4)_4 \cdot 24H_2O$; Copper, $CuSO_4 \cdot 5H_2O$; Iron, $FeSO_4 \cdot 7H_2O$; Citric acid, $C_6H_8O_7 \cdot H_2O$; Washing soda, Na_2CO_3.

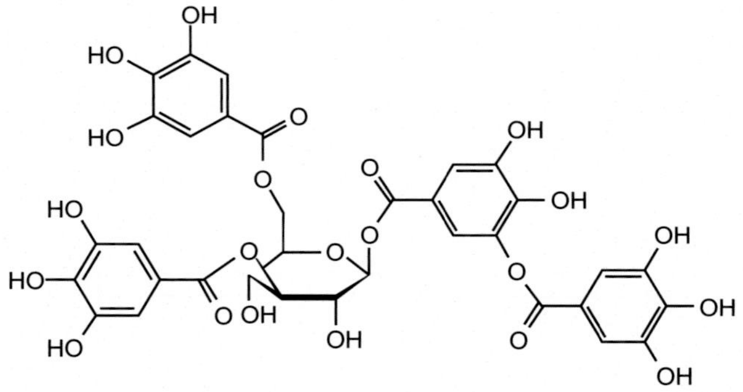

FIGURE 15

Tannin acid chemical structure.

http://pl.wikipedia.org/wiki/Kwas_taninowy, accessed 13.04.2015, 13:48 a.m.

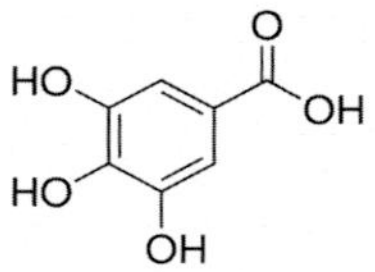

FIGURE 16

Gallic acid chemical structure.

https://en.wikipedia.org/wiki/Gallic_acid.

have an astringent, sedative, antipyretic and antidiabetic effect, and they are widely used in the medical and pharmaceutical industries. They are also widely used as an additive to food and animal feed, as well as in paint and ink production and metallurgy. Galls give black, navy blue, and brown colors (Kannan et al., 2009; Suchalata and Devi, 2005) (Figure 16).

Experimental findings
See Table 5 and Figure 17.

4.2.2 *Chebulic Myrobalan* Terminalia chebula
Myrobalan is a tree that grows up to approximately 30 m. It can be found in the natural environment in Asia, in the Himalayas, northern India, Sri Lanka, Burma, Thailand, Indochina, and southern China. It is an undemanding tree because it can grow on sands and in clay soil. Thanks to the compounds contained in it, such as chebulinic acid, ellagic acid, gallic acid, and flavonoids (routine and quercetin), it is often used in homeopathic medicine in Asia and Africa. *Terminalia chebula* has antibacterial, antiviral, antifungal properties (Kannan et al., 2009; Suchalata and Devi, 2005). Thanks to the wealth of its ingredients, it is used in making therapeutic infusions for rinsing the mouth, throat, and eyes. Drugs containing Myrobalan lower cholesterol levels, regulate metabolism and digestive disorders, are used in the treatment of dysentery, and auxiliarily, of HIV infections. Infusions prepared from its fruit have proven helpful for people with chronic fever, and they are also used as a mild sedative. *Terminalia*

Table 5 Gall Oak Dyeing Results on Linen and Cotton Samples

| No. | Fabric | Mordant | Spectrophotometric Result | | | | | | Fastness | |
			L^*	a^*	b^*	C^*	$h°$		Wash	Light
1	Linen—5 g	No mordant	70.2	3.9	19.4	19.8	78.7		4	5
2	Linen/G—5 g	No mordant	69.2	2.4	23.5	23.6	84.1		5	6
3	Cotton—5 g	No mordant	74.6	3.3	18.9	19.2	80.0		4	5
4	Linen—5 g	Alum—0.35 g	68.2	4.9	21.7	22.3	77.2		4-5	5
5	Linen/G—5 g	Alum—0.35 g	66.4	5.7	23.6	24.3	76.5		5	6
6	Cotton—5 g	Alum—0.35 g	72.5	5.1	20.4	21.0	76.0		4	5
7	Linen—5 g	Soda—0.35 g	62.4	4.2	21.4	21.8	78.8		4-5	5
8	Linen/G—4.5 g	Soda—0.35 g	58.3	4.6	22.8	23.2	78.5		4-5	6
9	Cotton—5 g	Soda—0.35 g	69.1	3.4	20.2	20.5	80.3		4	5
10	Linen—g	Citric acid—0.2 g	69.5	4.9	19.8	20.4	76.1		4-5	5
11	Linen/G—5 g	Citric acid—0.2 g	70.5	4.5	21.2	21.6	78.1		5	6
12	Cotton—5 g	Citric acid—0.2 g	73.4	3.9	19.2	19.6	78.5		4	5
13	Linen—5 g	$CuSO_4$—0.35 g	50.2	5.6	24.3	24.9	77.1		5	5
14	Linen/G—4.5 g	$CuSO_4$—0.35 g	51.2	5.1	23.5	24.1	77.7		5	5-6
15	Cotton—5 g	$CuSO_4$—0.35 g	56.9	5.2	25.1	25.7	78.3		4	5
16	Linen—5 g	Fe—0.35 g	39.5	3.7	3.0	4.7	38.8		4	5
17	Linen/G—5 g	Fe—0.35 g	35.7	3.2	2.8	4.3	41.3		4-5	5-6
18	Cotton—5 g	Fe—0.35 g	45.7	3.2	4.3	5.4	53.5		4	5

Fabrics: Linen—Linen/G, linen premordating in galas; Cotton.
Mordant: Alum, $Al_2K_2(SO_4)_4 \cdot 24H_2O$; Copper, $CuSO_4 \cdot 5H_2O$; Iron, $FeSO_4 \cdot 7H_2O$; Citric acid, $C_6H_8O_7 \cdot H_2O$; Washing soda, Na_2CO_3.

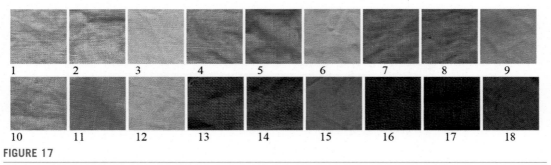

FIGURE 17

Oak galls colors results with different mordants.

chebula contains 58-60% of tannin compounds, which makes it a valuable dye and mordant. It is the source of yellow color.

4.2.3 Oak bark Quercus robur

The most important parts of the oak bark are tannins, which have an astringent effect. Oak bark is used externally in the treatment of inflammatory conditions of the skin, mucous membranes, and mild burns. Oak bark decoction has an astringent effect on damaged skin and mucous membranes. Apart from tannins, oak bark also contains free phenolic acids (ellagic and gallic acid), flavonoids (quercetin), resin compounds, and mineral salts. Oak bark is the source of brown colors which, due to the presence of tannin, are used in the premordating of fabrics (Hwang et al., 2000; Nagesh et al., 2012) (Figures 18 and 19).

4.2.4 Pomegranate Punica granatum

Pomegranate is a shrub or small tree of up to 5 m in height. We can find valuable resources in the peel of the fruit and the bark of the branches and roots. Pomegranate can be found in most warm countries. The rind of the pomegranate contains a considerable amount of tannin, about 19% with pelletierine (Adeel et al., 2009; Tiwari et al., 2010). The main coloring agent is granatoine. Pomegranates also have antioxidant, antiviral, and antitumor properties. The rind of the fruit is a valuable source of vitamins A and E and folic acid. In folk medicine, unripe fruit and peel infusions are administered to stop diarrhea and in the treatment of dysentery. In Africa, it was used to treat leprosy.

Its fruits are used to make refreshing juices, sorbets, and even wine. Pomegranate was used as a dye in antiquity, in Mediterranean civilizations and the cultures of the East. In ancient civilizations, it was seen as a symbol of love. Its dried peels are used to dye fabrics yellow, brown, and olive. Often, in order to revive the color, in India and South Asia turmeric root extract is added. In conjunction with indigo, green or dark green colors have been obtained. Pomegranate extract is used in tanning, particularly of saffian.

4.3 WILD PLANTS

Experiments have been conducted with wild plants, trying to achieve three primary colors—red, yellow, and blue. The next step was a combination between all shades and tints of this color to get a wide palette of hues for cellulose and animal fibers. During our research a group of herbal plants showing good dyeing properties has been selected. Those plants have been classified into three color groups: blue, yellow, and red (Tables 6–8 and Figure 20).

FIGURE 18

Quercus robur.

Photo by K. Schmidt-Przewoźna.

FIGURE 19

Quercus robur bark.

Photo by K. Schmidt-Przewoźna.

Table 6 Wild Plants with Yellow Color Properties

Family	Scientific Name	Common Name	Parts Containing Dyestuff
Papaveraceae	*Fumaria officinalis* L.	Drug fumitory	Plant
Equisetaceae	*Equisetum avense* L.	Common horsetail	Plant
Polygonaceae	*Polygonum persicaria* L.	Heart's-ease	Plant
Polygonaceae	*Polygonum aviculare* L.	Common knotweed	Plant
Polygonaceae	*Rumex acetosella* L.	Sorrel	Plant
	Rumex acetosa L.		
Iridaceae	*Iris pseudoscorus* L.	Yellow flag	Flowers
Compositae	*Bidens tripartitus* L.	Trifid bur-marigold	Plant
Compositae	*Achillea millefolium* L.	Yarrow	Plant
Compositae	*Arctium lappa* L.	Łopian większy	Plant
Compositae	*Taraxacum officinale* L.	Dandelion	Flowers
Compositae	*Calendula officinalis* L.	Pot marigold	Flowers
Lycopodiaceae	*Lycopodium complanatum* L.	Ground cedar	Plant
Papilionaceae	*Genista tinctoria*	Dyer's greenweed	Plant tops
Resedaceae	*Reseda luteola* L.	Weld	Plant tops
Compositae	*Solidagp serotina* Ait.	November goldenrod	Plant tops
Compositae	*Tanaceum vulgare* L.	Tansy	Plant tops
Compositae	*Anthemis tinctoria* L.	Chamomile	Plant tops
Compositae	*Anthemis arvensis* L.	Chamomile	Plant tops
Compositae	*Matricaria chamomilla* L.	Wild chamomile	Plant tops
Geraniamceae	*Geranium robertianum* L.	Herb-Robert	
	Serratula tinctoria L.	Saw-wort	
Rdestowate	*Polygonium persarica* L.	Spotted ladysthumb	Plant
	Urtica dioica L.	Nettle	Yellow, green
Ericaceae	*Calluna vulgaris* L.	Heather	Yellow, green

Table 7 Wild Plants with Red Color Properties

Family	Scientific Name	Common Name	Parts Containing Dyestuff
Rubiceae	*Galium aparine* L.	Annual bedstraw	Roots
Rubiceae	*Galium mollugo* L.	White bedstraw, wild madder	Roots
Labiatae	*Origanum vulgare* L.	Wild marjoran	Plant
Rubiceae	*Asperula tinctoria* L.	Dyer's woodruff	Roots
Guttiferae	*Hypericum perforatum* L.	St. John's wort	Plant
Rubiceae	*Galium verum* L.	Yellow bedstraw	Roots
Rubiceae	*Galium palustre* L.	Marsh bedstraw	Roots
Boraginaceae	*Lithospermum officinale* L.	Common gromwell	Plant
Boraginaceae	*Lithospermum arvense*	Corn gromwell	Plant

Table 8 Wild Plants with Blue Color Properties

Family	Scientific Name	Common Name	Parts Containing Dyestuff
Cruciferae	*Isatis tinctoria* L.	Woad	Leaves
Polygonaceae	*Polygonum aviculare* L.	Common knotweed	Plant
Compositae	*Cirsium arvense* L.	Canada thistle	Flowers
Polygonaceae	*Polygonum tinctorium* L.	Rdest barwierski	Plant
Papilionacaeae	*Indygofera tinctoria* L.	Indigo	Leaves
Ericaceae	*Vaccinum myrtillus* L.	Bilberry	Fruits, leaves

FIGURE 20

Golden chamomile *Anthemis tinctoria* in Wielkopolska region in Poland—reclaimed land.

Photo by K. Schmidt-Przewoźna.

5 EXPERIMENTAL FINDINGS

5.1 METHOD OF DYEING

5.1.1 Extraction of dye

The plants are crushed to small pieces and soaked in hot water overnight, boiled 1 h, and filtrated.

5.1.2 Mordants

In our methods we used oak galls, sodium carbonate anhydrous, copper sulfate, citric acid, iron (ferrous sulfate), and alum (potassium aluminum sulfate).

5.1.3 Development of color on linen, cotton, and silk fabrics

5.1.4 Equipment

Laboratory dyeing machine: the Easykrome (Ugolini).

5.2 COLOR SPECTROPHOTOMETRIC MEASUREMENTS AND COLOR FASTNESS

The main research goal was the application of vegetable dyestuffs in natural fabrics and the development of a fashion collection dyed with those dyestuffs. After the preliminary selection of plants showing dyeing properties, a few species were selected for cultivation and laboratory trials. A dyeing method was developed (temperature, kind, and quantity of mordant, pH) for plant dyestuffs testing to obtain a suitable and diverse palette of colors for one dye. When the dyeing plant database was established, the following species were selected: annatto, wild madder, juniper, and turmeric. The color of samples prepared during trials has been expressed by color parameters obtained in spectrophotometric measurements, according to the CIELab system of color spacing. Changes in color parameters of samples (L, a, b) were tested using three textile raw materials dyed with five natural dyestuffs with added different metallic mordants. The resulting color palette covers many shades of yellow, orange, red, green, and brown. Testing of spectrum characteristics in visible light wavelength range was performed using a Macbeth 2020+ spectrophotometer. The measurements were used to determine the chromaticity coordinates CIE: X, Y, Z, which were used to calculate the coordinates of color in the CIELab system. The system uses three coordinates—L, a, and b—that can be calculated from the trichromatic components X, Y, Z of color, thus facilitating understandable description of color. The L, a, b changes of color parameters of all tested samples were tested against the effect of different mordants and raw materials. Different mordants yield differences in color parameter values. Similarly, different raw materials dyed with the same dye and with the same mordant added, produce differences in the color (ΔE—a difference of color). Obtained L, a, b color parameters are presented in Tables 1–5 and illustrated graphically. The samples dyed with addition of mordants display good resistance of the color to washing, sweat, friction, and light. Here at least is one example where a return to an ancient technology enhanced with modern science can result in a more efficient commercial process and one that is more environmentally friendly. For the results of color measurement, see Table 1 for kamala, Table 2 for logwood, Table 3 for cutch, Table 4 for annatto, and Table 5 for gall oak.

5.2.1 Light fastness

5.2.2 Color fastness to sunlight

To observe the effect of sunlight on color fastness, linen and silk samples were tested on Laboratory Machine Xenotest 150. The test was carried out according to the standard PN-ISO 105-B02:1997.

Thirty naturally dyed samples were exposed to sunlight for 200 h, then graded for color fastness.

5.2.3 Measurement of color fastness to wash

The changes of color on linen and silk samples were assessed in gray scale (1–5).

Testing Washing Fastness with the Laboratory Dyer Ugolini according to the standard PN-ISO 105-C06:1996:

Preparation of washing bath: 4 g of washing agent per 1 l of water
Preparation of the samples of naturally dyed and reference fabrics
For Tests A and B: Reference fabrics for linen were linen and wool:
    Linen: reference fabrics—linen and wool
    Silk: reference fabrics—silk and cotton
Test conditions A1M: temp 40 °C, time: 45 min
For natural silk crepe and silk shantung, temperature of 30 °C and duration of 45 min were applied.

5.2.4 UV Protection Factor of Naturally Dyed Linen and Silk (Schmidt-Przewozna and Kowalinski, 2006)

The Laboratory of Physiological Influence of Textiles on Human Body has done research to compare the results for ultraviolet protection factor (UPF) on linen and silk samples dyed with natural dyes. The transmission, absorption, and reflection of UVR are in turn dependent on the fiber, fabric construction (thickness and porosity), and finishing. Many dyes used in the finishing process absorb UVR. Darker colors of the same fabric type (black, navy, and dark red) will usually absorb UVR more than light pastel shades and consequently will have a higher UPF rating. See Table 9 for the UPF classification system.

In this study all shades of red and purple applied on silk and linen were analyzed. Determination of the UVR transmission of a dry textile was done in accordance with Australian/New Zealand Standard and British Standard for sun protection clothing with the use of Cary 50 Solascreen apparatus.

The study also determined the UPF of linen, hemp, and silk fabrics dyed by natural dyestuffs. It also examined the influence of fabrics' structure, color, and methods of dyeing on level UV protection. Results are shown in Table 10 and detailed below.

Fabrics

Fabric A: 43002 thin linen Silk A—silk knitwear 100%
Fabric B: 30187 thick linen Silk B—silk shantung 100%

Results

1. The value of UPF linen and silk fabrics depends on product structure, density of thread, thickness, type of dyestuffs used, color and kind of fabrics.
2. The result of the comparison of UPF on linen and silk fabrics:
   Excellent UVR protection was obtained on samples:
   - India madder (linen B)—no mordant, 50+
   - India madder (linen B)—no mordant, 50+
   - India madder (linen B)—soda, 50+
   - Dyer's Coreopsis (linen)—soda, 50+
   - Dyer's Coreopsis (linen)—copper, 50+
   - Dyer's Coreopsis (linen)—iron, 50+
   - Gall oak (linen A)—iron, 50+
   - Black Myrobalan (linen Fabric A) iron, 50+
   - Laka (linen B)—copper sulfate, 45
   - Indian mulberry (linen B) copper, 45
   - Dyer's Coreopsis (linen)—soda, 40

Table 9 UPF Classification System (According to the Australian Standard)		
UPF Range	**UVR Protection Category**	**UPF Ratings**
15-24	Very good but insufficient protection	15, 20
25-39	Very good protection	25, 30, 35
40-50, 50+	Excellent protection	40, 45, 50, 50+

Table 10 The Results of UPF on Linen and Silk Samples Dyed by Natural Dyestuffs

No.	Samples	Natural Dyestuff	Mordant	Color	UVA	UVB	UPF
	Fabric A—linen		No mordant	White	13.828	19.205	5
	Fabric B—linen		No mordant	White	3.966	2.878	20
	Silk A		No mordant	White	21.911	27.249	0
1.	Fabric A—linen	Wild madder	No mordant	Light violet	7.948	7.821	10
2.	Fabric A—linen	Wild madder	Copper sulfate	Dark violet	3.270	3.334	20
3.	Linen A	Gallas	Ferrous sulfate	Dark beige	1.180	1.230	50+
4.	Fabric A—linen	Logwood	Washing soda	Dark violet	2.951	2.771	30
5.	Fabric A—linen	Wild madder	Citric acid	Coral	6.043	6.297	10
6.	Fabric B—linen	Indian mulberry	Copper sulfate	Light violet	1.326	1.299	45
7.	Fabric B—linen	Logwood	Washing soda	Dark violet	1.823	1.714	35
8.	Fabric B—linen	Indian mulberry	Citric acid	Dark pink coral	1.966	2.133	35
9.	Fabric B—linen	India madder	No mordant	Pink bisque	0.922	0.992	50+
10.	Fabric B—linen	India madder	Washing soda	Light salmon	1.058	1.089	50+
11.	Fabric A—linen	Wild madder	Premordant, alum	Red	2.221	2.275	25
13.	Silk B	Bilberry	Alum + iron	Violet	3.645	3.885	20
14.	Silk B	Sandalwood	Citric acid	Madder red	1.857	1.781	35
15.	Linen	Turmeric	Premordant	Yellow	4.456	3.036	25
16.	Linen	Turmeric	No mordant	Light yellow	3.543	3.305	20
17.	Linen	Turmeric	Premordant + copper	Olive yellow	2.907	2.011	30
18.	Linen	Turmeric	Premordant + citric acid	Sun yellow	4.022	3.364	30
19.	Linen	Dyer's Coreopsis	No mordant	Old gold	1.646	1.979	30
20.	Linen	Turmeric	Premordant + soda	Sahara yellow	3.302	2.53	30
21.	Linen	Turmeric	Premordant + iron	Olive brown	2.772	2.532	30
22.	Linen	Dyer's Coreopsis	No mordant	Old gold	1.646	1.979	30
23.	Linen	Dyer's Coreopsis	Soda	Old gold	1.854	2.137	40
24.	Linen	Dyer's Coreopsis	Citric acid	Gold	2.237	2.939	35
25.	Linen	Dyer's Coreopsis	Copper	Old gold	0.967	1.255	50
26.	Linen	Dyer's Coreopsis	Iron	Dark brown	0.881	1.117	50
27.	Linen	Common knotweed	No mordant	Brown	1.816	2.178	35
28.	Linen	Indigo	No mordant	Dark blue	0.840	0.794	50+
29.	Linen	Henna	No mordant	Rust	1.739	2.438	35
30.	Linen A	Fengurek	Copper	Yellow	1.83	1.94	40
31.	Linen A	Dyer's chamomile	Alum	Light yellow	1.383	1.249	50+
32.	Linen A	Black Myrobalan	Iron	Beige	1.156	1.276	50+

Very good protection was obtained on samples:
- Cochineal (silk B)—citric acid, 35
- Cochineal (silk B)—alum + ferrous sulfate, 35
- Henna (linen)—no mordant, 35
- Common knotweed (linen)—no mordant, 35
- Dyer's Coreopsis (linen)—citric acid, 35
- Dyer's Coreopsis (linen)—no mordant, 30
- Laka (linen Fabric A)—copper sulfate, 20
- Laka (linen Fabric A)—washing soda, 30
- Laka (linen Fabric B)—washing soda, 35
- Wild madder (linen A)—premordant, alum, 25

6 CONCLUSION

Plants play a huge role in the remediation of sites contaminated by external factors. They restore the site by cleaning, loosening, and extraction of heavy metals from the soil. They fertilize the areas affected by drought and regulate soil water management.

Dyeing trees have valuable properties that can be used for multifarious purposes. Dyes derived from them have a number of compounds with health benefits. Primitive peoples used them to paint their bodies, dye fabrics, and during magic and ritual ceremonies. Alternative medicine is based on a deep knowledge of the compounds found in plants. Furthermore, tannin compounds, flavonoids, and anthocyanins, extracted from plants, have great medical potential. This special group of plants shows antibacterial, antiviral, antifungal, and antioxidant properties. Anthocyanins have an antioxidant and anti-inflammatory effect. They play a significant role in the strengthening of blood vessels, and they lower cholesterol levels in the blood. They play an important role in the prevention of cancer and atherosclerosis, as well as in the treatment of irregularities in the functioning of the eye. Flavonoids have antioxidant and anti-inflammatory properties, and they reduce the harmful effects of UVR.

A wide range of colors can be obtained from dyeing plants. The use of a mortar has a significant impact on the color as it modulates it. In view of increasing pollution and emerging civilizational diseases, including allergies, scientific research has been conducted on the construction of fabrics that have a positive, even healing effect on human skin. Such research has been pursued at the Institute of Natural Fibres and Medicinal Plants in Poznań under Project BIOAKOD (bioactive curing of clothing based on natural fibers). Fabrics made from natural fibers, linen and organic cotton, were dyed using plant dyes: Dyer's broom *Genista tintoria*, Coreopsis *Coreopsis tintoria*, logwood *Haematoxylum campechianum*, madder *Rubia tinctorium* L., oak gall *Quercus infectoria* Oliv., Chebulic Myrobalan *Terminalia cebula*. Microbiological and medical tests have confirmed the positive effect of naturally dyed fabrics on human skin.

Moreover, some natural dyes increase skin protection against UVR. Linen fabrics dyed with Dyer's chamomile, Black Myrobalan, Dyer's Coreopsis, and India madder received a +50 UV factor; that is, very good protection. The finishing of textiles plays a very important role in ecoproduction. Natural fibers show good sun protection due to components of natural pigments like lignin, waxes, and pectins that act as UVR absorbents. The UPF barrier effect can be also achieved with the use of special UV blockers, which are generally used in medicinal products and cosmetics. Herbal ingredients contained in dyeing trees and herbs have many important medicinal properties.

REFERENCES

Adeel, S., Ali, S., Bhatti, I.A., Zsila, F., 2009. Dyeing Of Cotton Fabric Using Pomegranate (Punica granatum) Aqueous Extract. Asian J. Chem 21 (5), 3493–3499.

Balakrishna, S., Seshadri, T.R., Venkataramani, B., 1960. Special chemical components of commercial woods and related plant materials. IX. Morindonin, a new glycoside of morindone. J. Sci. Ind. Res. 19, 433–436.

Bojase, G., Wanjala, C.C.W., Gashe, B.A., Majinda, R.R.T., 2002. Antimicrobial flavonoids from, Bolusanthus speciosus. Planta Med. 68, 615–620.

Cardon, D., 2007. Natural Dyes: Sources, Tradition, Technology and Science. Archetype Publications, London, UK pp. 66,115,146, 143–148, 163–164,193, 693–694.

Cardon, D., 2010. Natural dyes, our global heritage of colors. In: Textile Society of America Symposium Proceedings.

Dean, J., 1999. Wild Colour. Octopus Publishing Group, London, UK 114 pp.

Drury, H., 1873. The Useful Plants of India, second ed. William H. Allen & Co., London.

Dygus, K., Wasiak, J.S.G., Madej, M., 2012. Roslinosc odpadow komunalnych i przemysłowych. Wydawnictwo Naukowe Gabriel Borowski, Warszawa pp. 100–104.

Hwang, J.K., Kong, T.W., Baek, N.I., Pyun, Y.R., 2000. Alpha-glycosidase inhibitory activity of hexagalloylglucose from the galls of Quercus infectoria. Planta Med. 66, 273–274.

Kannan, P., Ramadevi, S.R., Hopper, W., 2009. Antibacterial activity of Terminalia chebula fruit extract. Afr. J. Microbiol. Res. 3 (4), 180–184.

Lauro, G.J., Jack, F.F., 2000. Natural Food Colorants: Science and Technology. IFT Basic Symposium Series. Marcel Dekker, New York.

Manjunatha, B.K., 2006. Antibacterial activity of Pterocarpus santalinus. Indian J. Pharm. Sci. [serial online] 115–116.

Mohd Zin, Z., Abdul Hamid, A., Osman, A., Saari, N., Misran, A., 2007. Isolation and identification of antioxidative compound from fruit of Mengkudu (Morinda citrifolia L.). Int. J. Food Prop. 10 (2), 363–373.

Morton, J.F., 1960. Can Annatto (Bixa orellana L), An Old Source of Food Color, Meet New Needs for Safe Dye. Florida State Horticultural Society, USA.

Morton, Julia F., 1992. The ocean-going noni, or Indian mulberry (Morinda citrifolia, Rubiaceae) and some of its "colorful" relatives. Econ. Bot. 46 (3), 241–256.

Murti, V.V.S., Neelakantan, S., Seshadri, T.R., Venkataramani, B., 1959. Special chemical components of commercial woods and related plant materials. VIII. Heartwood of Morinda tinctoria. J. Sci. Ind. Res. 18B, 367–370.

Nagesh, L., Sivasamy, S., Muralikrishna, K.S., Bhat, K.G., 2012. Antibacterial potential of gall extract of Quercus infectoria against Enterococcus faecalis—an in vitro. Pharmacogn. J. 4 (30), 47–50.

Przewoźna, K., Kowaliński, J., Tomassini, M., 2011. Potential and Future Prospects of Dyeing Plants: Madder Rubia tinctorium L., Annatto Bixa orellana L., Dyer's Broom Genista tinctoria L., Woad Isatis tinctoria L. NOVA Science, New York pp. 49–57.

Rao, V.S., Seshadri, T.R., 1947. Kamala dye as an anthelmintic. Proc. Indian Acad. Sci. A 26 (3), 178–181.

Saleem, M., Kim, H.J., Ali, M.S., Lee, Y.S., 2005. An update on bioactive plant lignans. Nat. Prod. Rep. 22 (6), 696–716.

Schmidt- Przewoźna, K., Zimniewska, M., 2005. The Effect of Natural Dyes Used for Linen on UV Blocking Renewable Resources and Plant Biotechnology. NOVA Science, New York pp. 110–117.

Schmidt-Przewozna, K., 2001. The sources and properties of natural dyestuff applied in east and west tradition. J. Dyes Hist. & Archaeol. 20, Archetype Publication, UK.

Schmidt-Przewozna, K., 2002. The return to natural dyeing methods – properties and threats. In: Procedings of The Textile Institute 82nd World Conference, Cairo, Egypt. pp. 220–226.

Schmidt-Przewozna, K., 2009. Barwienie Metodami Naturalnymi. In: ECO-Press, Białystok, ISBN: 978-83-929329-0-1 pp. 1–20.

Schmidt-Przewozna, K., Brandys, A., 2014. Dyeing trees as a source of dyestuffs, cosmetic and herbal ingredients. Proc.Narossa. cd.

Schmidt-Przewozna, K., Kowalinski, J., 2006. Light fastness properties & UV protection factor on naturally dyed linen and hemp. In: Proceedings Saskatoon Conference.

Schmidt-Przewozna, K., Skrobak, E., 2013. A Thread of Colour Yarn. Regional Invest-Druk, Warszawa. pp. 1–10.

Sharma, J., Varma, R., 2011. A review on endangered plant of *Mallotus philippensis* (Lam.). Pharmacologyonline 3, 1256–1265.

Silva, G.F., Gamarra, F.M.C., Oliveira, A.L., Cabral, F.A., 2008. Extraction of bixin from annatto seeds using supercritical carbon dioxide. Braz. J. Chem. Eng. 25 (2) 419–426.

Suchalata, S., Devi, C.S., 2005. Antioxidant activity of ethanolic extract of *Terminalia chebula* fruit against isoproterenol – induced oxidative stress in rats. Indian J. Biochem. Biophys. 42, 242–249.

Thakur, S., Thakur, S., Chaube, S., Singh, S., 2015. An etheral extract of Kamala (*Mallotus philippinensis* (*Moll.Arg*) *Lam.*) seed induce adverse effects on reproductive parameters of female rats. In: Reproductive Toxicology, 20. Elsevier, USA, pp. 149–156.

Tiwari, H.C., Singh, P., Mishra, P.K., Srivastava, P., 2010. Evaluation of various techniques for extraction of natural colorants from pomrgranate rind - Ultrasound and enzyme assisted extraction. Indian J Fibre & Textile Res. 25, 272–276.

Vanita, P., Nirali, A., Khyati, P., 2013. Effect of phytochemical constituents of *Ricinus communis*, *Pterocarpus santalinus*, *Terminalia belerica* on antibacterial, antifungal and cytotixic activity. Int. J. Toxicol. Pharmacol. Res. 5 (2), 47–54.

West, B.J., Claude, J.J., Westendorf, J., 2008. A new vegetable oil from noni (*Morinda citrifolia*) seeds. Int. J. Food Sci. Technol. 43 (11), 1988–1992.

BROWNFIELD RESTORATION AS A SMART ECONOMIC GROWTH OPTION FOR PROMOTING ECOTOURISM, BIODIVERSITY, AND LEISURE: TWO CASE STUDIES IN NORD-PAS DE CALAIS

15

G. Lemoine

Établissement Public Foncier Nord-Pas de Calais, Euralille, France

1 INTRODUCTION

Remediation provides a solution for reduction of pressure on land resources. Land resources from a global perspective are under immense pressure. The pressure on available land resources is increasing because of land degradation, the growing world population, and urbanization. Mining is one of the major factors in land degradation in addition to natural processes. Therefore biological remedial planning is essential. The primary objectives of biological remediation are to (a) make the site fit for use; (b) remove all contact risks; (c) prevent further spreading of pollutants in air, water, and soil; and (d) restore contaminated sites for real estate development as well as recreation sites. Green-rehabilitated and ecorestored industrial sites have been transformed into new vibrant centers of biodiversity conservation, ecotourism, and leisure with a significantly increased value.

Nord-Pas de Calais is one of the old industrial regions of Western Europe. Since coal was discovered in 1720, heavy industrialization has been promoted in this area, leaving about 700 pits and 340 mining waste deposits. Near mining industries, the abundance and the moderate cost of the energy produced by coal extraction favored the installation and the development of steel and metallurgic factories in the regional coalfield.

Economic recession since the 1970s closed a lot of firms and created significant unemployment. Many firms and factories that closed left behind numerous fallow lands, brownfields, and polluted areas. Nord-Pas de Calais holds the dubious distinction of having the worst number of industrial derelict sites (10,000 ha, 50% of the French total). Since 1990 the regional reconversion used mining slag

Bioremediation and Bioeconomy. http://dx.doi.org/10.1016/B978-0-12-802830-8.00015-0
Copyright © 2016 Elsevier Inc. All rights reserved.

heaps for new remediation projects and new local land appropriations for inhabitants, for the creation of greenbelt and nature areas for human recreation, leisure, and biodiversity protection. The presence of land and soils polluted by heavy metal elements remains problematic. Among the various types of projects of restoration or reuse (phytorestoration, phytoextraction, miscanthus crops for biomass or energy, etc.), the most unusual project is probably the protection of the most interesting and polluted calaminarian grassland, which allows the presence of rich vegetation and rare species of metallophytes. This chapter presents experiences both of nature protection of landscapes and of biotopes of industrial heritage. The first takes place on a large scale for slag heaps (coal mining wastes), the second on a small but complex local scale (Figure 1a,b).

Case Study 1: Ecological Interests, Restoration and Valuation of Coal Mining Slag Heaps

Several hundred black hills are evident here, bearing witness to the region's industrial history and to the extraction of coal, which lasted for more than two centuries. Composed of sandstone and coal shale, the majority of the slag heaps are being, or have already been, redeveloped. Now it is a question of the fate of the others. Advocates for the protection of natural areas have been interested in this question for more than 25 years; the well-drained and mineral nature of the materials of which they are composed, their black color, and their landform create very different geotopes and biotopes and surprising landscapes (Figure 2). These slag heaps are formed from rapidly drying, more or less acidic materials that are relatively unstable and are easily reheated by the sun's rays or by combustion phenomena; therefore, they provide elements of biodiversity in an area of the north where the land is typically limestone and clay with a damp climate and moderate temperatures. Thus they present large ruptures or ecological contrasts (in landform, climate, soil, usage), like several quarries in the region that encourage the presence of numerous nonregional thermophilic and acidocline species. In this way the slag heaps create unique habitats and uncommon and rare animal and plant species, often unknown in the region or existing with difficulty in their natural and original environments. The different new habitats created by mining exploitation have thus encouraged exotic species or more specialized elements of our own flora and characteristic of natural environments with edaphic and climatic conditions existing in limestone or areas of dune. Faced with this surprising geological, ecological, and threatened heritage, the local authorities of the region have acted together to ensure its preservation and have urged the Etablissement Public Foncier Nord-Pas de Calais (EPF) to buy back the slag heaps and the brownfield sites and return them to the local authorities. A considerable amount of this heritage has thus been integrated into the protection of the Départements du Nord et du Pas-de-Calais (county councils) and some local authorities. The Départements du Nord et du Pas-de-Calais have thus become owners of nearly 20 mining sites, making up more than a thousand hectares. Before their transfer the EPF, in connection and in concert with the local authorities, undertook redevelopment and security of the brownfield sites where it was deemed to be necessary. Some of this restorative work has allowed certain slag heaps to become strong links in blue and green belts and wildlife corridors in the mining basin (Figure 3). At the same time, they provide biodiversity in a very industrial region with a large population and an efficient farming system.

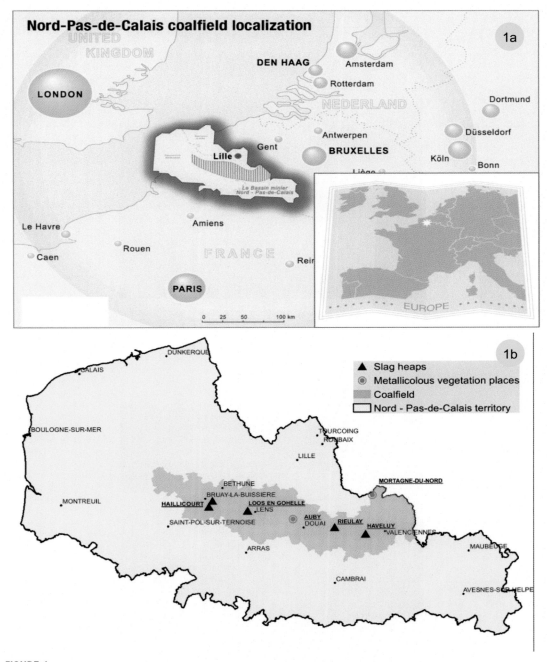

FIGURE 1

Figure 1a: Location of the Nord - Pas de Calais territory, Figure 1b: Location of mining and calaminarian sites mentioned in the chapter.

FIGURE 2

View of the Nord-Pas de Calais coalfield basin (Hallicourt site).

FIGURE 3

General view of ecologic restoration of the Rieulay mining site.

2 GENERAL DESCRIPTION OF STUDY AREA

The Nord-Pas de Calais region is located in the north of France near the Belgium border. Spread over more than 120 km, the Nord-Pas de Calais coalfield offers an outstanding diversity and impressive density of coal mining (340 slag heaps) and inhabitants (400 inhabitants per square kilometer) (O'Miel, 2008; Conservatoire des Sites Naturels du Nord and du Pas-de-Calais, 2005). Up to 1950, the Nord-Pas de Calais coalfield produced more than 50% of the total French coal production and played a strategic role in the reconstruction of the whole country, severely damaged by the Second World War. A total of 29 million tons were produced in the early 1950s (Conservatoire des Sites Naturels du Nord and du Pas-de-Calais, 2005). Spread over 3500 ha, slag heaps, results of the coal extraction, have a combined weight of 700 million tons of waste deposits for 2.3 thousand millions of tons of coal extracted (Conservatoire des Sites Naturels du Nord and du Pas-de-Calais, 2005; Kaszynski, 2008).

The coal exploitation in northern France, which began in 1720 in Fresnes-sur-Escaut (near the Belgium-Wallonia border), produced injured landscapes and derelict lands with a very rich biodiversity, but has also been accused of creating ugly landscapes. Many, therefore, would not object if they were to disappear, with the loss of their historical and biological significance, in the creation of roads or other infrastructures.

In reaction to this threat, some public partners and private organizations, associations for nature protection who appreciate the importance of this exceptional and unusual patrimony, asked the EPF Nord-Pas de Calais (the public land body) to intervene. The EPF was created in December 1990 to redevelop old industrialized and disaster areas in the Nord-Pas de Calais region. EPF Nord-Pas de Calais is a public land agency dedicated to public communities. The agency operates through one of its current missions; namely, the management and the renaturation of land located in hazardous areas (Lemoine, 2005). Biodiversity has also been one of its missions since the end of the 1990s. Many brownfield sites have been rehabilitated since then with the help of EPF Nord-Pas de Calais. Examples of types of brownfield that have been rehabilitated for biodiversity preservation include wastelands, quarries, sand and gravel pits, and lots of slag heaps in the coalfield basin. Slag heaps have been, also, compulsorily acquired by the EPF in 2002, which was also project manager for the redevelopment of numerous mining sites (natural or architectural places) for local communities (Lemoine, 2005).

3 INITIATIVE BY ÉTABLISSEMENT PUBLIC FONCIER

In a north French coalfield, the communities Départements du Nord and Département du Pas-de-Calais (councils of the North and the Pas-de-Calais districts/counties), along with municipalities and associations of local authorities, are closely associated with the EPF Nord-Pas de Calais for the preservation and the requalification of the postmining landscape heritage. The postmining heritage site managed by EPF consists of 2160 ha that include more than 50 slag heaps. The heritage site is the object of a politically ambitious acquisition on the part of the northern departments and the Pas de Calais districts thanks to the creation of preemption zones under the title of the policy "Espaces Naturels Sensibles" (ENS) for the purchase of certain of them (Lemoine, 2007). The land purchases were completed with the help of the EPF, which buys all slag heaps for public purposes with compulsory purchase orders.

Properties were acquired from the Terrils SA company, a subsidiary of the Charbonnages de France (the national coal mining company). Mining areas were purchased to be sold to local communities. The acquisitions of these different bodies (local municipalities and EPF) are dedicated to preserving the basic heritage of mining activities. This ambitious requalification/recultivation was funded by the European Union and with regional and state financial aid. After restoration work the slag heaps with biological interest were then purchased by the regional authorities, Département du Nord and Département du Pas-de-Calais Councils (Lemoine, 2007), or by a grouping of local municipalities. The aim is to plan new infrastructures, both for public access and for the protection and development of biodiversity in these particular areas.

4 SLAG HEAPS: A VERY SPECIAL ENVIRONMENT

Slag heaps have created a unique biological community. The extraction of the coal shale (schist), a black, porous, and easily heated material, has caused topographical features that are not naturally present on the region's surface. New habitats are thus created, which attract and select only a certain part of the local fauna and flora, usually coming from environments with similar edaphoclimatic conditions, like limestone slopes and coastal dunes. Particular soils (stones depots/deposits, stones accumulations) and local microclimates help new species, normally found in the south of France, in the Mediterranean, or more exotic regions, to become established. These new elements have colonized an environment complementary to their original biotopes. For them, mining sites are particularly welcoming environments that allow for an extension of their original area of distribution (Lemoine, 2005, Lemoine, 2010).

Six major ecological effects (Voeltzel and Février, 2010) of the mining ecological perturbations have been identified, which are Edaphic break, climatic break, topographic break, combo effects, agricultural break, and wetlands.

5 THE EDAPHIC BREAK EFFECT

In the Nord-Pas de Calais region, with its heavy, calcareous, and fresh soils, slag heaps create xerothermophilic steppe, with dry and warm conditions that support such species as dog figwort (*Scorphularia canina*) that come from Italian regions. The acid character of certain types of coal shale further accentuates the unusual nature of these artificial soils vis-à-vis the regular regional features (neutral or alkaline soil).

In these mining areas some plants typical of very poor soils and acidic conditions have been identified. Among the typical mosses and lichen, the more interesting species were *Filago minima*, *Filago vulgaris*, *Spergularia rubra*, *Trifolium arvense*, *Ornithopus perpusillus*, *Minuartia hybrida*, *Teesdalia nudicaulis*, *Nardurus maritimus*, and annual grasses like *Aira praecox* and *Aira caryophyllea* (Petit, 1972; Toussaint et al., 2008; Lemoine, 2010). In slag heaps in Belgium you can also find prickly salt-wort (*Salsola kali*). *Digitalis purpurea* is also a typical acidic weed found in areas of mining waste.

Because of its oligotrophic nature and a high filtering capacity, the soil of slag heaps is home to an amazing combination of mosses and lichens and also to species favoring dry grasslands. These exceptional environments are threatened by the development of fallow land and scrub even though rabbits slow down the succession and afforestation process. As regards afforestation, slag heaps are characterized by the spontaneous presence of the birch (*Betula verrucosa*), the fittest pioneer. The rapid decomposition of its leaves helps soil development. Generally speaking, all local trees and shrubs are found on the slag heaps as well as exotic trees like black locust (*Robinia pseudoacacia*) and butterfly bush (*Buddlelja davidii*), which are masters of barren soils. There is also the occasional chestnut tree (*Castanea sativa*), normally almost nonexistent in this limestone region. We also found another exotic tree, the wild black cherry (*Prunus serrotina*), which became invasive like other introductions. The glades of certain wooded areas have also attracted a remarkable bird for this region: the European nightjar (*Caprimulgus europaeus*).

6 THE CLIMATIC BREAK EFFECT

With the dry and mineral conditions, the black stone fields created by mining waste deposits give a particular climate because of the accumulation of the heat from the sun in the black material. Scientists estimate the gain of temperature of the black mining soil to be 5 °C on account of the local conditions. Generally speaking, all the common thermophilic plants are found on the slag heaps, including blue-weed (*Echium vulgare*), mullein (*Verbascum* sp.), mouse-ear hawkweed (*Hieracium pilosella*), and *Potentilla argentea* and *Potentilla intermedia*. In addition, *Hieracium bauhinii*, *Astragalus glyciphyllos*, *Lathyrus sylvestris*, and *Oenothera subterminalis* are some of the most interesting species.

Similar conditions found in dunes and limestone grassland areas help the colonization of *Carlina vulgaris*, *Veronica officinalis*, *Carex arenaria*, *Plantago arenaria*, and *Plantago coronopus*. Even the St Lucie cherry (*Prunus mahaleb*) absent in northern climes is found here.

The most original plants found on the slag heaps are certainly the Australian goosefoot (*Chenopodium pumilio*) and the South African groundsel (*Senecio inaequidens*), which has now become an alien (invasive weed!) like the Guernsey fleabane (*Conyza sumatrensis*), which came from Sumatra carried by the winds or, more likely, carried via international trade (wool) and the settlement of Commonwealth troops during the two world wars (Petit, 1972; Toussaint et al., 2008; Lemoine, 2010).

Other species have come from Mediterranean regions, such as *Micropyrum tenellum*, *Chenopodium botrys*, *Dittrichia graveolens*, *Galeopsis angustifolia*, *Digitaria sanguinalis*, *Astragalus cicer*, etc.

Iberis umbellata, *Centranthus ruber*, *Saponaria ocymoide*, *Lychnis coronaria*, *Iris Germanica*, and the common fig (*Ficus carica*), which had probably escaped from nearby gardens, were also found on the sites.

Some insects have also appreciated the thermophilic mining soil conditions; for example, the blue-winged grasshopper (*Oedipoda caerulescens*), the Italian tree cricket (*Oecanthus pellucens*) (*Orthopterous*), and two kinds of tiger beetles (*Cicindela campestris* and *Cicindela hybrida*) (*Coleopterous*) (Lemoine, 2005, Lemoine, 2010).

The coallike nature of slag heaps allows them, in particular circumstances, to enter a process of combustion. Exceptionally the spontaneous oxydation of one deposited material of the mining waste extracted with mining schist and sandstone creates strong warm conditions and situations. Spontaneous combustion generates dangerous situations (fires, explosion, gas emission), for that reason, public access is forbidden in many sites. The exothermic oxidation of the iron pyrite helps the combustion of coal dusts (Figure 4). The slag heaps also register extremes in temperature (700 °C in depth, 50 °C in surface!). Therefore the slag heaps are protected from the frosts in winter, which of course favors southern plants and explains why a moss normally found growing on Vesuvius was discovered on a slap heap's hillside! At this kind of

FIGURE 4

Spontaneous combustion in Haveluy slag heap.

site we also found green purslane (*Portulaca oleracea*), common in France's southern vineyards, and the house cricket (*Acheta domesticus*), common in houses and bakeries and in the Paris underground. After this first localization, certain south-oriented parts are even home to *Portulaca oleracea*.

7 THE TOPOGRAPHIC BREAK EFFECT

Coal exploitation made a chain of hills in a northern area locally called "flat country." The climatic effect is complemented by the landform with slag heaps and by the natural and artificial dynamic, moving aspect of these sites. The black shale (schist) becomes oxidized and vitrified with the combustion. It turns red and acquires good mechanical properties. Slag heaps are sometimes reexploited for coal residue (because of inefficient initial sorting) or for red shale. These human interventions create yet more disturbances (Figure 5). The repercussions of industrial activities (excavations, material deposits, roads, perpetual escarpment creations) have helped and continue to help pioneer species to become established from pioneer environments (amphibians, plants), which can regularly find new suitable biotopes (Lemoine, 1999; Godin, 2002). These new habitats can replace more unstable zones such as banks that were home to the sand martin (*Riparia riparia*), which is now found on the "cliffs" created by reexploited slag heaps. Among the most spectacular species is the buckler sorrel (*Rumex scutatus*) (Figure 6), never seen before in the region, which has found a home of its own on the slag heaps' unstable sections (mass of fallen rocks). Probably coming from mountainous regions (the Alps), it was carried to the region Nord-Pas de Calais on the trains transporting conifer trees for the mines' galleries. In other cases, the beach poppy (*Glaucium flavum*) has colonized unstable slopes on many slag heaps.

FIGURE 5

Ecological restoration in progress at Rieulay.

FIGURE 6

Buckler sorrel (*Rumex scutatus*) is a nonnative plant of the Nord-Pas de Calais region. It comes from the Alps and grows on unstable slopes of slag heaps.

8 A COMBINATION OF EFFECTS

The combination of different edaphic, climatic, and topographic conditions favors very interesting fauna in the herpetological group. Common wall lizards (*Podarcis muralis*) found here, are at their geographic distribution northmost limit. Numerous amphibians are xerothermophilic pioneer species adapted to treeless and steppe environments. The natterjack toad (*Bufo calamita*) joins the midwife toad (*Alytes obstetricans*) and the parsley frog (*Pelodytes punctatus*) (Godin, 2002; Parent, 1970; Figure 7). The latter is a Franco-Iberian species found at this latitude in these warm conditions at the most northern unit of its distribution. In slag heaps in Lorraine (northeast of France, near German border), the green toad (*Bufo viridis*) found some very interesting habitats for its life cycle.

As already seen, the permeable nature of the shale makes for very dry conditions, but in a few cases the water accumulations create pools with oligotrophic levels, and the richness of ecological conditions (temporary presence) favor typical amphibians and vegetations like brookweed (*Samolus valerandi*), strapwort (*Corrigiola litoralis*), *Potamogeton coloratus*, and others (Lemoine, 2010).

9 THE AGRICULTURAL BREAK EFFECT

The agronomically "difficult" nature of these environments further accentuates their pioneer component, even if, as time goes by, they lose some of their pedological and climatic originality. They also become exceptional refuges for the once common flora and fauna of the countryside because they escape the intensive pressure brought on the countryside by intensive farming. The mining fallow land habitats, free from pesticides and gradually recolonized by both regional and new vegetation,

FIGURE 7

Parsley frog (*Pelodytes punctatus*) is a Franco-Iberian species found in Nord-Pas de Calais coalfield in the warm conditions.

become beautiful and efficient refuges for numerous insects (wild bees, butterflies, locusts) threatened elsewhere in the region; for example, the old world swallowtail (*Papilio machaon*). In order to understand the importance of the slag heap in the region Nord-Pas de Calais, one has to understand the context of our industrial regions. Nord-Pas de Calais is characterized by a high population density (4 million inhabitants) and great conurbations, where the natural world has been sacrificed to overintensive farming (trees & wood account for only 7% for example) (Kaszynski, 2008; Lemoine, 2007).

10 AN IN SITU LABORATORY

As time goes by, the slag heaps begin a process of stabilization and start to be covered in vegetation, and the steps to colonization are as diverse as they are interesting. After a phase of complete bareness and very little vegetation, they then become covered with grass (scattered thermophilics or high fallow vegetation with blueweed (*Echium vulgare*) and wild carrots (*Daucus carota*)) and later they reach a copselike stage before forest takes over. Another curious fact worth mentioning is the presence of species of fallow land at the top of the slag heaps. Once at a new destination, they do not share their adaptation faculty. Scientists have observed the oats species (*Arrhenatherum elatius*), which adds a genetic isolation to their geographical isolation by bringing forward their flowering periods (Petit, 1980).

Aware of this rich and original heritage, several groups (e.g., municipality, county council) are working closely with EPF, the project manager. For the fauna and flora whose environments are slowly vanishing (e.g., dunes), the slag heaps offer a new alternative. Because of their geological composition and their steep sides, they offer a new kind of environment to the region; hence, allowing new habitats and species to develop. In fact, in the Nord-Pas de Calais region, industrial mining activities tend to complete, or even to act as a substitute, for the natural rate of disturbance that expresses itself with more and more difficulties. The occurring disruptions lead to the arrival and preservation of pioneer animal and vegetal species that come and find a substitute or additional habitats to their primary original environment. Slag heaps (like sandpits and quarries), which provide open environments and thermophilous conditions, contain a patchwork of habitats generally without chemical agricultural treatments (biocides) and allow the development of a rich fauna, particularly insects, reptiles, and amphibians, and rich flora, particularly dunes and Mediterranean species (Lemoine, 2005).

Slag heaps, despite making colonization difficult for local flora, are thus excellent disseminators of flora on a world scale. The movement of species or their seeds and, in certain cases, their amazing capacity for adaptation, reveals the role of wind, transportation, and commercial trades.

11 THE CREATION OF WETLANDS

Mining development is, in certain cases, accompanied by the creation of humid zones. In effect the subsoil, made fragile by the creation of galleries, and the significant weight of slag heaps have led to slight modifications in the topography of the ground. The ground, sinking by 2 or 3 m, has changed the direction of the flow of surface water in a region with little noticeable relief. Some vast, humid zones have been created and are still being created around some of the largest slag heaps. To limit this phenomenon in urban or inhabited areas (isolated homes), a great part of the mining basin in the Nord-Pas de Calais is artificially pumped in order to avoid flooding. These humid zones (and their ecological wealth,

FIGURE 8

Works in Hallicourt slag heaps: Mining decant pools are restored on wetland for bird conservation.

especially ornithological, if they are not hunted) owe their existence, therefore, to mining development (Figures 3 and 8). They participate, in an important way, in the conservation of regional biodiversity. Wetlands are, in effect, fragile and threatened areas in the northwest of Europe and their disappearance in the region of the Nord-Pas de Calais was very significant in the past. (Wetlands have receded from 30% of the surface of the region in the Middle Ages to 0.8% today.)

12 AN AMBITIOUS PROGRAM FOR THE REGENERATION OF NATURE AND ITS ACCESSIBILITY TO THE PUBLIC

Today in the northern slag heaps nature protection and public access extend to 2000 ha over 40 sites. A part of these sites (30% of their surface) used to belong to the Département councils' preemption zones before it was bought by the public land body (EPF). Then the Nord Département Council (Conseil général du Nord) bought it, in accordance with its policy to protect sensitive natural areas (ENS) At the end of 2002, a partnership was established between the partners to redevelop (for requalification) numerous mining sites; to plan a new infrastructure (Briand et al., 2007; Kaszynski, 2008), both for public and family access (including, in some cases, access for the disabled); to create bike paths (Figure 9) and birdwatching facilities and in certain cases access for mountain bikes and horseback riding. They planned for the protection, development, and ecological management of the area. Département teams took part in construction site meetings so that planning will take into consideration the criteria for the conservation of nature both in the conception and implementation phases (Lemoine, 2007; Briand et al., 2007).

FIGURE 9

Pedestrian access for the twin slag heaps in Loos-en-Gohelle (Europe's biggest slag heaps, 175 m high).

13 NATURE PROTECTION

The goal has been to identify the most unusual and rare species and to help accelerate their development. The targets were to restore habitats and ecological conditions through restoration works and management, to enhance the local biodiversity by using the biological potential of mining soils, creating receptive habitats for aquatic, wetland, meadow, natural sward, and glade-forest species, resulting in a living mosaic of biotopes.

For example, biodiversity-related actions at mining sites mainly aimed at maintaining unstable zones and protecting and restoring the dry and acidic grounds and lawns or grasses of the mining shale with horse and sheep grazing to stop the afforestation and maintain dry and open grasses. This action was completed by the conservation of pebble steppes for characteristic birds of open areas like the small plover (*Charadrius dubius*), in complement defining the appropriate and adapted mix of trees and shrubs for moderate tree-planting actions. In the local forest that grows on mining waste deposits EPF created glades of birch trees for the European nightjar (*Caprimulgus europaeus*) and the tree pipit (*Anthus trivialis*).

The management of the water aspects encouraged EPF to create temporary humid depressions and small ponds for the natterjack toad (*Bufo calamita*) and other patrimonial amphibians (Lemoine, 1999). Large reed beds and muddy zones provided for wetland birds in the pools localized in mining sites (Figure 3). EPF and their partners regularly work on the protection of local small cliffs for particular birds like the sand martin (*Riparia riparia*) and on the creation of bat refuges for the winter period in old mining infrastructures.

14 ACTIONS FOR PUBLIC ACCESS AND LEISURE AREAS

As regards the politics of natural spaces, Départements are obliged to act on two accounts. There is the preservation of the natural heritage and its access to the public. The public cannot, in effect, be excluded from natural spaces unless the latter are deemed to be too fragile and unless any disturbance would threaten the species and their environment (Figure 10). These measures of exclusion are very rare as access to green spaces and to nature is considered a real social need (the extent of public green spaces in the mining basin is close to 1 m² per inhabitant) (Lemoine, 2007; Briand et al., 2007).

The works carried out by EPF at the request of the Département du Nord, the Département du Pas-de-Calais, and municipalities groupings have therefore integrated this essential approach. The whole of the renovated area presents extensive paths for pedestrians and for biking (e.g., 12 km of paths at the Chabaud Latour site). They are connected to urban access and to various car parks. In some cases the sites are equipped with trails for mountain bikes and horses (e.g., the slag heaps of Argales in Rieulay). The latter site was renovated with help from Association des Paralysés de France (APF) (association for the disabled) in order to welcome members of the public with disabilities. This project is in the process of being categorized "tourism and handicap" by the French ministry of tourism. The majority of local mining sites are also connected to the regional and national paths and benefit from a system of markers. The presence of disused railway tracks, now also the property of the two Départements or local municipalities and renovated into footpaths, completes the possibilities for walking and for making human and ecological connections between the sites (Briand et al., 2007).

FIGURE 10

Leisure area for boating and swimming at mining pool in Rieulay site.

Départements and local municipalities and groups have also become involved in education programs on nature and the environment. Regular nature walks, free for the public and for schoolchildren, have been set up to inform and to encourage appreciation of this particular heritage. Some sites are also equipped, not only with classic information and welcome boards, but also with boards pertaining to nature. The boards on the site at Argales (Rieulay) present the geological history of the mining basin and of the different rocks extracted from the subsoil (Figure 11). Information plaques (of which one is written in the local dialect) exist on certain sites. Permanent departmental rangers and, in the summer, additional backups on horseback, complete the contingent of information and supervision.

The economic projects of reusing mining sites are, in themselves, numerous. From the emblematic ski run at Noeux-les-Mines can now be added the architectural value of coal mining housing estates and villages, and the buildings around the pits, which are today increasingly recognized. These major components of coal mining heritage and the four remaining big pits are being renovated, which completes the cultural and touristic coal mining offer, symbolized by the Regional Coalmining Historic Centre in Lewarde (one of the first regional museums with 150,000 visitors per year) and by the Louvre-Lens museum located in a old mining area. The shape and height of the slag heaps and the former coal

FIGURE 11

Geologic interpretation and mining stones exhibit at Rieulay site.

mining railway tracks break the monotony of the "flat country" and become "hotspots" for important sport events (slag heap races, "coalfield raids," mountain biking). The natural mining sites (slag heaps and ponds from mining subsidence) are also the object of projects for sustainable tourism. The very good quality of the water at the Argales in Rieulay site and the space available is encouraging the commune and its partners to develop a mining museum, restaurants, hotels, and recreation based on open-air activities (walks, swimming, and nature watching). The development of the mining basin has become a wider process in which it has been proposed and accepted that the mining area of the Nord-Pas de Calais coalfield be registered on the UNESCO World Heritage List.

15 CONCLUSION

In Nord-Pas de Calais, the production chapter of the "coal story," started in 1720, came to an end in December 1990 when the last pit closed. But the "coal story" still continues with the recognition of the Nord-Pas de Calais coalfield in the World Heritage List! Launched in 2002, the candidacy of the Nord-Pas de Calais coalfield to the UNESCO World Heritage List comes into the category of "Evolved Cultural Landscape." Landscape enables the coalfield to be presented, not as a collection of sites and monuments, but as an organic and coherent entity, combining history and memory, nature and buildings. Today, one can see how, in many fields, local people and institutions have taken ownership of their industrial heritage. Both public and private initiatives have set up footpaths for recreational use and for the discovery of natural history, the geological heritage, and the industrial past. For years, the Nord-Pas de Calais ex-coalfield suffered from a poor image (hard work, unemployment, social struggles, and so on), inherited from its industrial past, but this negative perception has changed now. Slag heaps and former pits have turned into natural spaces, protected for their specific qualities and high biodiversity. Indeed these sites have been analyzed by numerous teams of experts (e.g., Mission Bassin Minier, CPIE-Chaîne des Terrils, and the Bailleul National Botanic Conservatory) and the work carried out by EPF with the help of national, regional, and local authorities is seen as exemplary in the environmental regeneration of these sites, to facilitate public access and to help people to appreciate its scientific and landscape importance. The slag heaps, as with the whole mining heritage, are now the object of renewed interest. Now slag heaps are an image of revival and dynamism in the region. The international recognition for this tangible and intangible heritage of mining supported by local and regional elected officials for 10 years, for enlistment by UNESCO was accepted in June 2012. In the same way, a demand for legal and regulatory protection is in progress by the French state for certain of its elements in respect of historic heritage, beauty, landscape, or nature.

Case Study 2: Environmental Protection and Management of Two Metalliferous Sites

1 INTRODUCTION

For a century and a half, agricultural soils and natural areas located in Nord-Pas de Calais (France) have been greatly polluted by three major lead and zinc metallurgical plants. These soils have been impacted by the storage of by-products (slag heaps exposed to weathering) and by the effects of factory smoke. Then, industrial activities indirectly helped the formation of calamine substrates. The presence of these contaminated soils raises numerous questions such as:

– Do the metallophyte species growing on these anthropogenic soils present an ecological interest?
– Do calaminarian soils need protection?
– Are xenophyte species natural patrimony?
– Where and how do metallophyte species come from?
– Should we perform soil remediation in order to prevent health risks and ecosystem deterioration? On the other hand, should we protect this "natural" heritage? If so, how to proceed and how to convince people about the preservation of polluted soils and the maintenance of metallicolous vegetation?
– What plans can be developed for protection, restoration, and management of the calaminarian grasslands in the Nord-Pas de Calais region?

We cannot forget, however, that polluted areas have no money value, present strong ecological and health risks, and are derelict lands or brownfields. Paradoxically, if we want to preserve them, these areas are also threatened by urbanization and tree plantations.

This part of the chapter describes the actions taken to preserve the two last calamine substrates in Nord-Pas de Calais.

2 HISTORICAL SETTINGS

For nearly a century and a half, the industrial history of Nord-Pas de Calais has been characterized by intense mining and metallurgical activities. These industries have greatly impacted the landscape. Concerning metallurgical activities, there were three major industrial sites:

– A lead and zinc metallurgical plant at Noyelles-Godault, property of Peñarroya-Metaleurop. It was built in 1893, closed in 2003, and deconstructed in 2005. This plant was one of the biggest in all of Europe.
– A zinc-producing plant at Auby since 1869, also called "zinc capital city." It was the property of the *Compagnie royale asturienne des mines*, then called Nyrstar. Nowadays it is the property of Umicore. Its annual production capacity reaches 220,000 ton.
– A plant producing lead and zinc at Mortagne-du-Nord. This plant worked from 1901 to 1963. It was the property of the *Compagnie royale asturienne des mines*.

One can easily imagine the environmental impacts on the suroundings. Some can be easily identified by looking at the landscape (e.g., the presence of slag heaps). Others are very hard to identify, such as the contamination of large areas including urban zones and agronomical soils caused by the effects of factory smoke (Figure 12). Some of these soils are characterized by heavy metal concentrations several orders of magnitude greater than those of natural soil: a hundred and a thousand times greater for lead and zinc, respectively; a dozen times greater for cadmium. Nowadays, polluted areas are regularly investigated by state services with environmental and population health monitoring. The aim is to establish restrictions concerning town planning and agricultural and gardening practices.

FIGURE 12

Armeria halleri: An effective indicator of the pollution of districts (railway near the zinc factory).

3 FAUNA, FLORA AND NATURAL HABITATS IN POLLUTED AREAS

On such type of polluted substrates, one can easily take stock of four species belonging to the dicotyledonous angiosperms: Halleri cress *Arabidopsis halleri* (Brassicaceae) (Figure 13), Halleri marsh daisy, or thrift *Armeria maritima* subsp. *halleri* (Plumbaginaceae) (Figure 14), yellow zinc violet *Viola calaminaria* (Violaceae) (Figure 15), and bladder campion *Silene vulgaris* var. *humilis* (Caryophyllaceae) (Figure 16), and one bryophyte: *Scopelophila cataractae* (LECRON comm pers.). In the past (1980), *Microbryum starkeanum* (*Pottia starkaena*) was found (Van Haluwyn et al., 1987). Calaminarian soils are usually characterized by the combination of these species (Figures 17 and 18). Under quite natural conditions, combinations of *Arabidopsis halleri*, *Armeria maritima* subsp. *halleri*, and *Silene vulgaris* var. *humilis* can be found only on soils formed from rocks similar to metal deposits. In middle Europe, these soils are located in Wallonia (Belgium) and Germany. Human activities have progressively created conditions similar to these natural soils.

One must keep in mind that the specific flora living on what can be called "industrial soils" are very far from their original location. Calamine vegetation in Nord-Pas de Calais comes from anthropogenic conditions. At Auby, the metalliferous vegetation has an exceptional composition. The calaminarian flora is very abundant (Figures 18 and 19). The four metallophyte species previously cited constitute the majority of the "vegetal carpet," in association with three other species tolerant to heavy metal. These are *Arrhenatherum elatius*, *Agrostis capillaris*, and *Cerastium fontanum* subsp. *vulgare*. At Auby, *Viola calaminaria* is very active and colonized an area of nearly 2 ha in only two decades. This is surprising when we consider that this species has difficulty in maintaining itself in some of its

FIGURE 13

Halleri cress (*Arabidopsis halleri*), a hyperaccumulator of heavy metal elements.

FIGURE 14

Halleri marsh daisy or thrift (*Armeria halleri*) in Mortagne-du-Nord site.

FIGURE 15

The yellow zinc violet (*Viola calaminaria*), a new accidental introduction from East Belgium.

FIGURE 16

The fourth metallophyte dicotyledonous: Bladder campion (*Silene vulgaris* var. *humilis*).

FIGURE 17

Typical vegetation in Auby (*Viola calaminaria* and *Arabidopsis halleri*).

FIGURE 18

Metalliferous vegetation with three characteristic European species: *Viola calaminaria*, *Armeria halleri*, and *Arabidopsis halleri*.

FIGURE 19

General view of Péru park in Auby (gramineous colonization against typical metallophyte species).

original sites, notably in Germany (BOTHE, com pers.). Two invertebrate species can also be found in such type of artificial environment: the Mazarine Blue butterfly *Cyaniris semiargus* and the 24-spot ladybird beetle *Subcoccinella 24-punctata*. The butterfly *Cyaniris semiargus*, unknown in the rest of the industrial territory and commonly associated with Fabaceae, may be using *Armeria maritima* subsp. *halleri*, as a host plant (Lemoine, 2013; Figure 20) just like the 24-spot ladybird uses the *Silene* genus plant as its host plant for food.

4 WHERE DOES THIS PARTICULAR FLORA COME FROM?

No calaminarian soils were naturally formed in Nord-Pas de Calais, so the cited species were not present prior to the industrial settlement. This vegetation has been introduced recently into polluted soils and therefore corresponds to xenophytes. Recent approaches imply elimination of nonindigenous plants and use of pest controls. For this reason, the four cited metallophyte plants are not considered as patrimony species by the botanical national conservatory of Bailleul. According to this body, only metallophyte vegetations (e.g., *Violetalia calaminariae*, which has a phytosociologic association) have interest.

So, some questions come to mind: Where do these species come from and how?

The most studied scenario is that of the *Arabidopsis halleri*. It is a mountain species. Its geographical distribution spreads mainly from Central Europe to the southern Carpathian mountains. So, this species is observed hundreds of kilometers from its natural spreading areas. In Nord-Pas de Calais, *Arabidopsis halleri* was first observed in 1944 by the botanist André Berton (1892-1982). Berton (1946) collected a testimonial from a former employee who worked for the Compagnie

FIGURE 20

Blue Mazarine (*Cyaniris semiargus*) on its host plant, *Armeria maritima* subsp. *halleri.*

royale asturienne des mines. According to him, *Arabidopsis halleri* was introduced near 1920 as a honey plant by one of the directors. Recent studies of phytogeography indicate the geographical origin of the *Arabidopsis halleri* at Auby. Genetics and molecular markers suggest that it comes from Germany and, more precisely, from the Harz. The areas where the plant comes from are also polluted by heavy metals (Zn, Cd, and Pb) due to mining activities (Pauwels, 2006; Pauwels et al., 2005).

Concerning *Armeria maritima* subsp. *halleri*, many hypotheses were proposed. This species is known as a mid-European one. It is assumed that its introduction is due to industrial activities (Centre régional de phytosociologie agréé (CRP), 2005). *Armeria maritima* subsp. *maritima* is also found along the coastline of the Nord-Pas de Calais region. More precisely, it grows in areas subjected to the spray of the sea, on the cliffs just near the coast. Mériaux (1984) indicates that *Armeria* was introduced at Mortagne-du-Nord. This was done by engineers of the *Compagnie royale asturienne des mines*. In Spain, these engineers had noticed the ability of the plant to grow on contaminated soils. At Auby, *Armeria maritima* subsp. *halleri* was found by Berton in the middle of the twentieth century. A former employee told him this species had already been present for five or six decades. *Armeria* was seeded at Mortagne-du-Nord with seeds taken at Auby (Berton, 1946), but some inhabitants of Mortagne-du-Nord told us that engineers may also have taken this species from the coast of French Brittany in order to seed areas without vegetation around production sites. Further details on the origin of *Armeria maritima* subsp. *maritima* and *Armeria maritima* subsp. *halleri* should be obtained by genetic investigations.

The yellow zinc violet *Viola calaminaria* is an endemic species of calaminarian soils in Belgium and Germany. This species was first observed in a wood in the neighborhood of the industrial site of

Auby in 1995 (Petit, 2002). The way it was introduced is unknown. Petit supposes that seeds could have been unintentionally scattered by researchers working on calaminarian grasslands. In Nord-Pas de Calais calaminarian soils, this species is only located at Auby. Genetic studies indicate that *Viola calaminaria* from the north of France are very similar to those present in Belgium (Bizoux, 2006).

Silene vulgaris var. *humilis* is the calaminarian subspecies of *Silene vulgaris*. *Silene vulgaris* is found throughout Nord-Pas de Calais. One can suppose that plant colonization of calamine sites was made from the neighboring populations of *Silene vulgaris*.

5 NATURA 2000 PRESERVATION AND PRESERVATION STATE

Plain calaminarian grasslands are rare in the region and unknown in the rest of France. Auby, Roost-Warendin, Flers-en-Escrebieux, Douai, Noyelles-Godault, Courcelles-les-Lens, Château-l'Abbaye, and Mortagne-du-Nord are the only French municipalities where calaminarian grasslands are plainly observed. The French state has defined two special conservation zones within the Natura 2000 network: Auby (and Roost-Warendin) for a surface of 17 ha and Mortagne-du-Nord (and Château-l'Abbaye) for 17 ha. Considering the residual calaminarian grasslands (within brownfields and on roadsides), the total area of calaminarian grasses and grasslands is estimated at 40 ha for the Nord-Pas de Calais. *Armeria maritima* has been listed as a protected species in Nord-Pas de Calais since 1991. The two Natura 2000 sites are mainly planted with poplar trees (*Populus x canadensis*), are exposed to partial shade, are marginally threatened by urbanization, and are colonized by grasses. In some places calaminarian grasslands are abundantly invaded by *Arrhenatherum elatius* and thick organic matter accumulations without degradation that progressively replace *Arabidopsis halleri* and *Armeria maritima* subsp. *halleri*. In Mortagne-du-Nord, within the frame of the Natura 2000 plan, experimental work is being done to restore the calaminarian vegetation (Parc Naturel Régional Scarpe-Escaut, 2011).

6 PROTECTING A POLLUTED AREA!

Near the historically known calaminarian sites in Auby (bois des Asturies), new calaminarian grasslands have been discovered in the Péru park, following the diminution of the number of grass cuttings associated with ecological gardening practices (Lemoine, 2011, 2012). This 2-ha site has a high ecological diversity and a high patrimonial interest with calaminarian grasslands in a very good state of conservation and development. This zone was the object of a depollution plan by state injunction. The soil, heavily polluted with Zn, Cd, and Pb, was about to be removed.

Considering the richness of the site, a new project has been submitted to the state services. The most interesting area was to be protected while the rest of the park was to be converted to a leisure park after evacuation of the polluted earth. Numerous representatives (elected officials) and neighbors questioned the relevance of maintaining a polluted soil in order to preserve its flora. A negotiation with the state allowed the preservation of a calaminarian grassland on a surface of 1.2 ha. Important information campaigns and consultation procedures were conducted with inhabitants living in the neighborhoods. A visit to Liège (Belgium) was organized in order to discover actions taken to protect and give value to a calaminarian grassland. Numerous conferences and guided visits in Auby allowed the audience to

FIGURE 21

Permanent notice board for flora presentation in Auby site.

understand why the most polluted soil of the public park was not depolluted while private gardens or schools were. Workshops for neighbors, young adults, and children were organized to imagine a new project for Péru park. Actions taken in concert allowed the establishment of local rules for uses of the Péru park. Emphasis was placed on the patrimonial aspect of this flora, in order to develop and maintain similar ecological restoration in the future. The aim of the project was to transform perception of calaminarian grasslands. From similar actions, polluted areas will be transformed into industrial heritage and biological patrimony (Delhaye et al., 2011; Figure 21).

7 SCIENTIFIC MONITORING

The Péru park and the Mortagne-du-Nord grassland are monitored within the scope of a scientific project financed by the Nord-Pas de Calais regional council and the Foundation for Biodiversity Research. This project aims at identifying the geographic origin of metallophyte plants and the dates of their introduction. Several ecological restoration plans are also applied in the Péru park and neighboring areas in order to avoid colonization by common grasses like *Agrostis capillaris* and *Arrhenatherum elatius* (Figure 22). Management experiments were also undertaken to restore the Mortagne-du-Nord calaminarian grassland (Parc Naturel Régional Scarpe-Escaut, 2011; Figure 23) and to conserve the best vegetations developed in Péru park (Mériaux, 1984).

8 CONCLUSION

Local calaminarian grasslands are poorly known in France and in Europe, and are a specificity of the Nord-Pas de Calais region, inherited from its industrial history. Their preservation raises important questions as they develop on heavily polluted soils, inherited from human activities, and correspond to species introduced on the territory. It is therefore relevant to question to which extent these species can

FIGURE 22

Management of Péru parc for calaminarian grassland restoration with gardening methods.

FIGURE 23

The spectacular calaminarian grassland restored, located in Mortagne-du-nord.

be considered to belong to the natural patrimony of the region. Metallophyte species are important pollution markers and often indicate the presence of invisible pollutants. In addition to their interest as testimonies to the industrial history of the region, and their role for phytoremediation (phytostabilization) of polluted soils, calaminarian and metallophyte species have an important scientific and educational interest. Within the context of biological crisis, metallophyte vegetation shows the ability of certain species to adapt to extreme and toxic ecological conditions. Calaminarian grasslands are natural laboratories allowing study of the mechanisms involved in the maintainance and evolution of biodiversity.

ACKNOWLEDGMENTS

The author wishes to express sincere thanks to Lynn Seddon and Nicolas Seignez for the translation of the French text into English, and to Laurence Deruet (EPF Nord-Pas de Calais), Jord Maitte, and Sandrine Belland (Mission Bassin Minier) for maps of the study area. Thanks are also due to Professor Majeti Narasimha Vara Prasad for continued support and several suggestions that resulted in this chapter.

REFERENCES

Berton, A., 1946. Présentation des plantes *Arabis Halleri*, *Armeria elongata*, *Oenanthe fluviatilis*, *Galinsoga parviflora discoidea*. Bull. Soc. Bot. Fr. 93, 139–145.

Bizoux, J.-P., 2006. Biologie de la conservation d'une métallophyte endémique: *Viola calaminaria*, Communauté française de Belgique, Académie universitaire Wallonie-Europe – Faculté universitaire des sciences agronomiques de Gembloux, 132 pp.

Briand, G., Lemoine, G., Belland, S., Mensah, J., Huttner, B., 2007. Guide Pour L'Ouverture au Public d'un Terril, Quelles Démarches Comment Aménager et Gérer? Mission Bassin Minier éditeur, Oignies 22 pp.

Centre régional de phytosociologie agréé (CRP), 2005. Plantes Protégées et Menacées de la Région Nord/Pas-de-Calais. Conservatoire botanique national de Bailleul, Bailleul 434 pp.

Conservatoire des sites naturels du Nord, du Pas-de-Calais, 2005. Les terrils, livret nature. Wambrechies, 26 pp.

Delhaye, E., Lemoine, G., Rivet, F., Grandjacques, C., Top, A., 2011. Le parc Péru: d'un espace pollué voué à la destruction, à la reconnaissance partagée d'un patrimoine naturel d'exception. Journées techniques nationales «Reconversion des friches urbaines polluées», Paris, 11 et 12 octobre 2011, ADEME, 10 pp.

Godin, J., 2002. Degré de rareté, évolution de la distribution et particularités de l'herpétofaune de la région nord-Pas de Calais. Bull. Soc. Herp. Fr. 104, 16–35.

Kaszynski, M., 2008. Les enjeux d'une gestion de transition du patrimoine industriel. In: Les Paysages de la Mine, un Patrimoine Contesté? Centre historique minier édit., Lewarde, pp. 186–190.

Lemoine, G., 1999. Prise en compte des crapauds calamites dans la requalification de friches industrielles dans le nord-Pas de Calais. Supplément du bulletin de la Société Herpétologique de France 91, 6–7.

Lemoine, G., 2005. Nature et Espaces Industriels; Terrils Miniers, Carrières et Sablières. Conseil général du Nord éditeur, Lille 38 pp.

Lemoine, G., 2007. Les terrils miniers intègrent les espaces naturels sensibles. Revue Espaces naturels 19.

Lemoine, G., 2010. Die Abraumhalden in den Nordfranzösischen Kohlerevieren. Bergbeau Folge Landschaft. In: Internationale Bauausstellung (IBA) Fürst-Pückler-Land 2000–2010. Jovis éditeur, Berlin, pp. 42–51.

Lemoine, G., 2011. Le Conseil général protège ses pelouses calaminaires. Espaces naturels 33, 51–52.

Lemoine, G., 2012. De l'importance des pelouses calaminaires d'Auby et notamment du parc Péru. Bull. Soc. Bot. N. Fr. 65 (1–4), 51–58.

Lemoine, G., 2013. Le demi-argus *Cyaniris semiargus* (Rottemburg, 1775) apprécie les pelouses calaminaires de la région nord – Pas-de-Calais!. Le Héron 2012-45 (1), 59–70.

Mériaux, J.-L., 1984. Les biotopes particuliers du nord de la France: la pelouse métallicole de mortagne in Le patrimoine naturel régional nord – Pas-de-Calais. In: Actes du colloque organisé par l'Association multidisciplinaire des biologistes de l'environnement (A.M.B.E.) des 23–25 novembre 1983, Lille. pp. 227–230.

O'Miel, C., 2008. La Procédure d'inscription du Bassin Minier du Nord-Pas de Calais sur la Liste du Patrimoine Mondial de l'Unesco. In Les Paysages de la Mine, un Patrimoine Contesté? Centre historique minier éditeur, Lewarde. pp. 192–201.

Parc Naturel Régional Scarpe-Escaut, 2011. Document d'objectifs du site FR3100505, Pelouses métallicoles de Mortagne du Nord et de Château l'Abbaye. Chambre d'agriculture du Nord, CRPF Nord – Pas de Calais – Picardie, PNR Scarpe-Escaut pour le Ministère de l'écolgie et du développement durable. Non paginé.

Parent, G.-H., 1970. Le pélodyte ponctué, *Pelodytes punctatus* (daudin), existe-t-il en Belgique et au grand-duché du Luxembourg? note préliminaire. Bulletin Les Naturalistes Belges t.51-7, 333–337.

Pauwels, M., 2006. Origine et Évolution de la tolérance au zinc chez Arabidopsis halleri (Brassicaceae). Approches phénotypique et génétique. Thèse de doctorat, Université de Lille 1.

Pauwels, M., Saumitout-Laprade, P., Holl, A.-C., Petit, D., Bonnin, I., 2005. Multiple origin of metallicolous populations of the pseudometallophyte *Arabidopsis halleri* (Brassicaceae) in central Europe: the cDNA testimony. Mol. Ecol. 14, 4403–4414.

Petit, D., 1972. Les végétaux thermophiles peu communs de la région minière du nord et du Pas-de-Calais. Le Monde des Plantes 375, 5.

Petit, D., 1980. La végétation des terrils du Nord de la France, écologie, phytosociologie, dynamisme. Thèse, USTL, Lille, 250 pp.

Petit, D., 2002. Viola calaminaria dans le bois des Asturies (Auby, 62). Bull. Soc. Bot. N. Fr. 55 (1–2), 48.

Toussaint, B., Mercier, D., Bedouet, F., Hendoux, F., Duhamel, F., 2008. Flore de la Flandre Française. Centre régional de phytosociologie agréé Conservatoire botanique national de Bailleul, Bailleul pp. 556.

Van Haluwyn, C., Petit, D., Mériaux, J.-L., 1987. Végétations métallicoles dans la région nord – Pas-de-Calais. Bull. Soc. Bot. N. Fr. 40 (1–2), 7–15.

Voeltzel, D., Février, Y., 2010. Gestion et Aménagement Écologique des Carrières de Roches Massives. Guide Pratique à l'usage des Exploitants de Carrières. ENCEM et CNC – UNPG, SFIC et UPC, Paris pp. 230.

BIOLOGICAL RECULTIVATION OF MINE INDUSTRY DESERTS: FACILITATING THE FORMATION OF PHYTOCOENOSIS IN THE MIDDLE URAL REGION, RUSSIA

T.S. Chibrik[1], N.V. Lukina[1], E.I. Filimonova[1], M.A. Glazyrina[1], E.A. Rakov[1], M.G. Maleva[1], M.N.V. Prasad[1,2]

Ural Federal University named after First President of Russia B.N. Yeltsin, Ekaterinburg, Russia[1]

University of Hyderabad, Hyderabad, Telangana, India[2]

1 INTRODUCTION

Industrial deserts or lunarscapes are large areas overloaded with technogenic waste (Peterson, 1995). Cleanup of such waste is cost prohibitive. Therefore, one emerging approach to this problem is biological recultivation (Bolshakov and Chibrik, 2007). It has been satisfactorily implemented in various countries by selecting and growing perennial grasses, trees, and bushy plant species (Bell, 2001; Bengson, 1995; Bradshaw, 1997, 2000; Frontasyeva et al., 2004).

Economic and ecological aspects of land development options are important in the field of bioeconomy (Doetsch et al., 1999; Eydenzon et al., 2013). On a site-specific basis, there are two types of principal restoration options:

- *ameliorative*: improvement of the physical and chemical nature of the site
- *adaptive*: ecological restoration through establishing ecosystem structure and function and thus biodiversity

Ecological restoration usually depends on careful selection of suitable substrates for plant growth. Species that would provide wildlife habitat (and forage for domestic animals) and improve esthetics are generally preferred. However, native species that are available as propagules often do not satisfy the above criteria. In this situation, a rapid solution to problems can be addressed by selecting species that enable colonization, facilitating succession and restoration of the native ecosystem. *Agrostis capillaris* L.

Bioremediation and Bioeconomy. http://dx.doi.org/10.1016/B978-0-12-802830-8.00016-2
Copyright © 2016 Elsevier Inc. All rights reserved.

and *Festuca rubra* L. are examples. These grasses have a proven reclamation function on a variety of industrial waste-contaminated soils (Dulya et al., 2013). The revegetated contaminated area must meet two basic objectives: forage and habitat for livestock and wildlife.

There is not a single best method in all circumstances for any reclamation operation. The procedures and techniques described in this chapter and in Chapter 15 by G. Lemoine (Brownfield restoration as a smart economic growth option for promoting ecotourism and leisure: Two case studies in Nord-Pas De Calais) are successful examples. Further examples in the literature indicate that a huge amount of knowledge dealing with various types of abandoned mine waste rehabilitation is available (Table 1).

Table 1 Contribution to the Knowledge of Abandoned Mine Waste Rehabilitation (In Reverse Chronology—This List is Not Exhaustive)

Reference	Observation
Ors et al. (2015)	Reclamation of saline sodic soils with the use of mixed fly ash and sewage sludge
Bing-Yuan, and Li-Xun (2014)	Mine land reclamation and eco-reconstruction in Shanxi province
Channabasava et al. (2015)	Mycorrhizoremediation of fly ash using *Paspalum scrobiculatum* L., inoculated with *Rhizophagus fasciculatus*
Li et al. (2014)	Synthesis of merlinoite from Chinese coal fly ashes and its potential utilization as slow release K-fertilizer
Shin et al. (2014)	Trial construction of ground reclamation using dredged soil mixed with coal combustion products
Srivastava et al. (2014)	Reclamation of overburden and lowland in coal mining area with fly ash and selective plantation
Das et al. (2013)	Ash ponds reclamation through green cover development
Guo et al. (2013)	Geopolymer based reclamation of used sand and fly ash
Hu et al. (2013)	Reclamation of soil in coal-mining subsidence areas
Kovshov (2013)	Biological ground recultivation and increase of soil fertility
Liu and Lal (2013)	A laboratory study on amending mine soil quality
Liu et al. (2013)	Reclaimation of land by fly ash
Wang et al. (2013)	Modification of fly ash and its reclamation applications
Yang et al. (2013)	Using kenaf (*Hibiscus cannabinus*) to reclaim multi-metal contaminated acidic soil
Yu et al. (2013)	Metal elements utilization by mycorrhizal fungi in fly ash reclamation
Free (2012)	Surface mine reclamation in the appalachian coal basin
Lal et al. (2012)	Release and uptake of potassium and sodium with fly ash application in rice on reclaimed alkali soil
Martin et al. (2012)	Characterization of coal combustion byproducts
Phommachanh et al. (2012)	Expression of self-hardening property of coal fly-ash
Qian et al. (2012)	Effects of AMF on soil enzyme activity and carbon sequestration capacity in reclaimed mine soil
Skousen et al. (2012)	Use of coal combustion by-products in mine reclamation
Vacenovska and Drochytka (2012)	Development of a new reclamation material by hazardous waste solidification/stabilization
Visa (2012)	Tailoring fly ash activated with bentonite as adsorbent for complex wastewater treatment
Visa et al. (2012)	Fly ash adsorbents for multi-cation wastewater treatment

Table 1 Contribution to the Knowledge of Abandoned Mine Waste Rehabilitation (In Reverse Chronology—This List is Not Exhaustive)—Cont'd

Reference	Observation
Wang et al. (2012)	Co-detoxification of transformer oil-contained PCBs and heavy metals in medical waste incinerator fly ash under sub- and supercritical water
Yan et al. (2012)	Feasibility of fly ash-based composite coagulant for coal washing wastewater treatment
Zhao et al. (2012)	Integrated coagulation-trickling filter-ultrafiltration processes for domestic wastewater treatment and reclamation
Zhou and Xu (2012)	Reclamation technology in complex matrix of Vesicular Arbuscular [VA] mycorrhiza on coal mine abandoned wasteland
Abdelhadi et al. (2011)	Utilization of fly ash materials as an adsorbent of hazardous chemical compounds
Awang et al. (2011)	Coal ash mixtures as backfill materials
Babu and Reddy (2011)	Inoculation of arbuscular mycorrhizae with phosphate solubilizing fungi contributes revevetation of fly ash ponds
Butalia and Kirch (2012)	Reclamation of asphalt pavements using coal combustion byproducts
Babu and Reddy (2011)	Diversity of arbuscular mycorrhizal fungi associated with plants growing in fly ash pond and their potential role in ecological restoration
Park et al. (2011)	Artificial soil mixture with coal combustion byproduct
Qian et al. (2011a,b)	Variation of microbial activity in reclaimed soil in mining area
Qian et al. (2011a,b)	An environmentally sound usage of both coal mining residue and sludge
Qian et al. (2011a,b)	Environmental impact of APC residues from municipal solid waste incineration: Reuse assessment based on soil and surface water protection criteria
Rai and Paul (2011)	Physical characterisation of fly ash from coal fired thermal power plants, Jharia Coalfield, Jharkhand
Shin et al. (2011)	Effect of coal ash contents on the acceleration of settling and self-weight consolidation of clayey ground
Shou et al. (2011)	Study on the contents of heavy metal in soil and vegetation in filling reclaimed land in Xuzhou Jiuli mining area
Sun et al. (2011)	Improving the mechanical characteristics and restraining heavy metal evaporation from sintered municipal solid waste incinerator fly ash by wet milling
Tripathi et al. (2011)	Reclamation of wasteland for cultivation of cotton crop through application of pond ash and its leachate
Viestová et al. (2011)	Conditions of waste and waste mixture utilization in technical land reclamation.
Zhang et al. (2011a,b)	Research on modified fly ash for high iron and high manganese acid mine drainage treatment
Zhang et al. (2011a,b)	Research progress of ecological restoration for wetlands in coal mine areas
Chaudhary et al. (2011)	Growth and metal accumulation potential of Vigna radiata L. grown under fly-ash amendments
Haibin and Zhenling (2010)	Recycling utilization patterns of coal mining waste in China Resources
Ren et al. (2010)	Kinetic and equilibrium studies of Cr(VI) from wastewater with modified fly ashes
Srivastava and Ram (2010)	Reclamation of coal mine spoil dump through fly ash and biological amendments
Zhao et al. (2010)	Effects of ameliorants addition on Cd contents and yield of crop in Cd-rich reclaiming substrates
Ouyang et al. (2010)	Municipal sewage sludge used as reclaiming material for abandoned mine land

(Continued)

Table 1 Contribution to the Knowledge of Abandoned Mine Waste Rehabilitation (In Reverse Chronology—This List is Not Exhaustive)—Cont'd

Reference	Observation
Tanhan et al. (2007)	Uptake and accumulation of cadmium, lead and zinc by Siam weed [*Chromolaena odorata* (L.) King & Robinson]
Lai et al. (2006)	Subcellular distribution of rare earth elements and characterization of their binding species in a newly discovered hyperaccumulator *Pronephrium simplex*
Pratas et al. (2005)	Plants growing in abandoned mines of Portugal are useful for biogeochemical exploration of arsenic, antimony, tungsten and mine reclamation
Paschke et al. (2005)	Manganese toxicity thresholds for restoration grass species
Neagoe et al. (2005)	The effect of soil amendments on plant performance in an area affected by acid mine drainage
Kothe et al. (2005)	Molecular mechanisms in bio-geo-interactions: From a case study to general mechanisms
Johnson and Hallberg (2005)	Acid mine drainage remediation options: a review
Hallberg and Johnson (2005)	Biological manganese removal from acid mine drainage in constructed wetlands and prototype bioreactors
Gilbert et al. (2005)	Municipal compost-based mixture for acid mine drainage bioremediation
García et al. (2004)	Performance and use of *Piptatherum miliaceum* (Smilo grass) in Pb and Zn phytoremediation
Dell'Amico et al. (2005)	Analysis of rhizobacterial communities in perennial Graminaceae from polluted water meadow soil, and screening of metal-resistant, potentially plant growth-promoting bacteria
Caille et al. (2005)	Metal transfer to plants grown on a dredged sediment: use of radioactive isotope ^{203}Hg and titanium
Bethwell and Mutz (2005)	Effect of acid mine drainage on the chemical composition and fall velocity of fine organic particles
Bednar et al. (2005)	Effects of iron on arsenic speciation and redox chemistry in acid mine water
Banuelos et al. (2005)	Selenium volatilization in vegetated agricultural drainage sediment from the San Luis Drain
Kramer (2005)	Phytoremediation: novel approaches to cleaning up polluted soils
Prasad (2004)	Phytoremediation of metals and radionuclides in the environment: The case for natural hyperaccumulators, metal transporters, soil amending chelators and transgenic plants
Pratas et al. (2004)	*Pinus pinaster* Aiton (maritime pine): a reliable indicator for delineating areas of anomalous soil composition for biogeochemical prospecting of As (Arsenic), Sb (Antimony) and W (Tungsten)
Rai et al. (2004)	Revegetating fly ash landfills with *Prosopis juliflora* L.: impact of different amendments and *Rhizobium* inoculation
Lai and Chen (2004)	Role of EDTA on solubility of cadmium, zinc, and lead and their uptake by rainbow pink and vetiver grass
Gaur and Adholeya (2004)	Prospects of arbuscular mycorrhizal fungi in phytoremediation of heavy metal contaminated soils
Freitas et al. (2004)	Analysis of serpentinophytes from north-east of Portugal for trace metal accumulation—relevance to the management of mine sites

Table 1 Contribution to the Knowledge of Abandoned Mine Waste Rehabilitation (In Reverse Chronology—This List is Not Exhaustive)—Cont'd

Reference	Observation
Chen et al. (2004)	The use of vetiver grass (*Vetiveria zizanioides*) in the phytoremediation of soils contaminated with heavy metals
Wong (2003)	Ecological restoration of mine degraded soils, with emphasis on metal contaminated soils
Pang et al. (2003)	Physiological aspects of vetiver grass for rehabilitation in abandoned metalliferous mine wastes
Malcová et al. (2003)	Effects of inoculation with *Glomus intraradices* on lead uptake by *Zea mays* L. and *Agrostis capillaris* L
Madyiwa et al. (2003)	Greenhouse studies on the phyto-extraction capacity of *Cynodon nlemfuensis* for lead and cadmium under irrigation with treated wastewater
Caille et al. (2004)	Influence of liming and phosphate fertilisation on revegetation arsenic contaminated soils using *Pteris vittata*
Chen et al. (2004)	The use of vetiver grass (*Vetiveria zizanioides*) in the phytoremediation of soils contaminated with heavy metals
Prasad and Freitas (2003)	Metal hyperaccumulation in plants—biodiversity prospecting for phytoremediation technology
Freitas et al. (2004)	Plant community tolerant to trace elements growing on the degraded soils of São Domingos mine in the south east of Portugal: environmental implications
Shu et al. (2002)	Lead, zinc and copper accumulation and tolerance in populations of *Paspalum distichum* and *Cynodon dactylon*
Schwitzguébel et al. (2002)	Phytoremediation: European and American trends
Bleeker et al. (2002)	Revegetation of the acidic, As contaminated Jales mine spoil tips using a combination of spoil amendments and tolerant grasses
Bashmakov et al. (2002)	Zinc hyperaccumulating weeds from temperate Russia
Prasad (2001a)	Bioremediation potential of Amaranthaceae
Kidd et al. (2001)	The role of root exudates in aluminium resistance and silicon-induced amelioration of aluminium toxicity in three varieties of maize (*Zea mays* L.)
Dinelli et al. (2001)	Metal distribution and environmental problems related to sulfide oxidation in the Libiola copper mine area (Ligurian Apennines, Italy)
Cherry et al. (2001)	An integrative assessment of a watershed impacted by abandoned mined land discharges
Black and Craw (2001)	Arsenic, copper and zinc release from the Wangaloa coal mine, southeast Otago, New Zealand
Prasad (2001b)	Metals in the environment: analysis by biodiversity
Schutzendubel and Polle (2002)	Plant responses to abiotic stresses: heavy metal-induced oxidative stress and protection by mycorrhization
Tordoff et al. (2000)	Current approaches to the revegetation and reclamation of metalliferous mine wastes
McLaughlin et al. (2000)	A bioavailability-based rationale for controlling metal and metalloid contamination of agricultural land in Australia and New Zealand
Simon et al. (1999)	Pollution of soils by the toxic spill of a pyrite mine (Aznalcóllar, Spain)
Prasad and Freitas (1999)	Feasible biotechnological and bioremediation strategies for serpentine soils and mine spoils

(Continued)

Table 1 Contribution to the Knowledge of Abandoned Mine Waste Rehabilitation (In Reverse Chronology—This List is Not Exhaustive)—Cont'd

Reference	Observation
Glass (1999)	U.S. and international markets for phytoremediation
Vangronsveld and Cunningham (1998)	Metal-contaminated soils: *in-situ* inactivation and phytorestoration
Panin (1998)	Influence of antropogenic activity and human argochemical activity on migration of heavy metals in system "soil-plant"
Lan et al. (1998)	Reclamation of Pb/Zn mine tailings at Shaoguan, Guangdong Province, People's Republic of China: the role of river sediment and domestic refuse
Brooks (1998)	Plants that hyperaccumulate heavy metals
Rao and Tarafdar (1998)	Selection of plant species for rehabilitation of gypsum mine spoil in arid zon
Kraemer et al. (1997)	Nickel localization in leaves of the hyperaccumulator plant *Alyssum lesbiacum by* micro-PIXE technique
Leblanc et al. (1996)	Accumulation of arsenic from acidic mine waters by ferruginous bacterial accretions (stromatolites)
Singh (1996)	*Prosopis juliflora* in an alkali soil
Gonsalves et al. (1997)	Mycorrhizae in a portuguese serpentine community
Mesjasz-Przybylowicz et al. (1994)	Proton microprobe and X-ray fluorescence investigation of nickel distribution in serpentine flora from South Africa
Baker et al. (1991)	*In situ* heavy metal decontamination of polluted soils using crops of metal-accumulating crops
Brooks (1987)	Serpentine vegetation
Smith and Bradshaw (1979)	The use of metal tolerant plant populations for the reclamation of metalliferous waste

The Middle Ural is one of the oldest mining regions of Russia (Figure 1). It is also one of the largest territories of mineral resources in Russia. This has led to intensive development of metallurgy, construction, chemical industry, and mining operations, including gold mining (Khokhryakov, 2003; Koptsik, 2014). These lands have lost their economic value due to mining and mineral exploration. The mining industry areas have had a harmful impact on the environment because of soil damage and changes in hydrological conditions (Belskaya and Vorobeichik, 2013; Belskii and Belskaya, 2009; Brooks et al., 2005). The environmental impacts of mining-related activities in and around middle Ural region including Karabash are extremely severe (Williamson et al., 2004a,b, 2008). The area has been affected by gaseous and particulate emissions from a copper smelter, acid drainage from abandoned mine workings, and leachates and dusts from waste dumps (Chukanov et al., 1993; Udachin et al., 2003; Spiro et al., 2004; Williamson et al., 2008; Spiro et al., 2012). The extent of environmental pollution in Russia is shown in Figure 2. In this chapter, some results of biological recultivation of various mine industry damaged sites, including fly ash dumps and dumps from coal mining and iron ore mining, are presented. The generalized procedures for establishing and revegetating mine industry waste-contaminated sites are shown in Figure 3. The emerging practices of reclamation of mine waste are depicted in Figure 4.

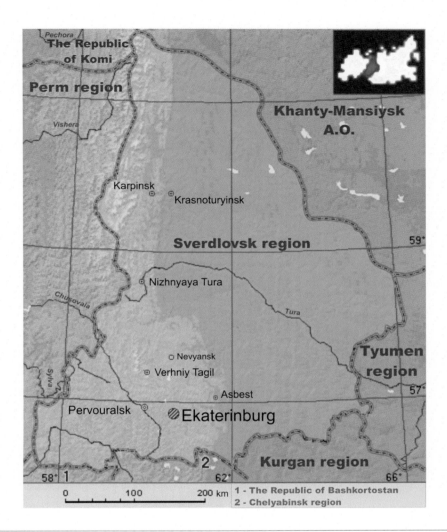

FIGURE 1

Sverdlovsk region is in the middle of Euro Asiatic continent. There is central part of the Ural Mountains from the north to the south of Sverdlovsk region. The border of Europe-Asia lays at the watershed of the Urals. The area of Sverdlovsk region is about 194 km². Sverdlovsk region is one of the oldest mining regions in Russia—it has great amount of mineral resources; there are about 200 deposits of ferrous metals, nonferrous metals, rare metals, noble metals, nonmetal ores and others. Some power plants work on Ural coal and lignite; some deposits are at the closing stage. Sverdlovsk region has one of the biggest deposits of chrysotile-asbestos in the world (Asbest town). Sverdlovsk region territory is 1% of Russian total territory. At the same time area of destructed territories is 5.3% of Russian total technogenic loads in Sverdlovsk region are higher than mean Russian.

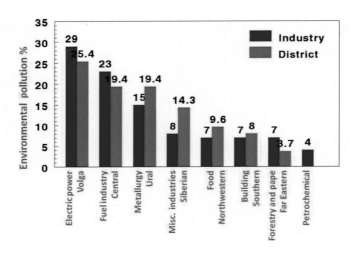

FIGURE 2

Extent of from environmental pollution in Russia (%) by industry and federal district wise.

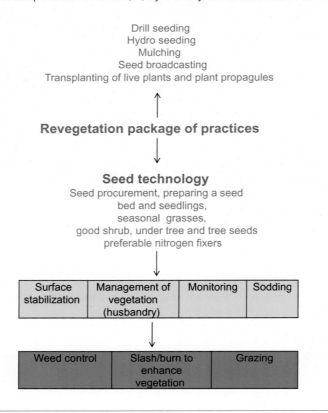

FIGURE 3

Approaches for restoration of mine industry ravaged sites.

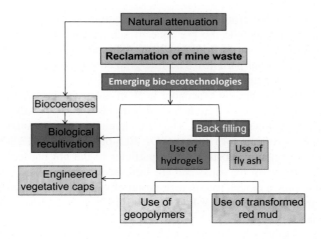

FIGURE 4

The emerging practices of reclamation of mine waste.

Key components of biological recultivation of an area include:

- structure and functioning of technogenic ecosystems
- monitoring of the ecological situation of the mine industry-ravaged lands
- development of biological recultivation
- floristic composition, which indicates conditions of the environment
- structure of the plant populations (finding the dominant species)
- dynamics and structure of populations (mycorrhiza show the community preparedness)
- productivity, chemical compounds, and the quality of phytoproducts
- Transformation of ash dumps to phytocoenose that are sterile substrata

2 THE PURPOSE AND METHODS OF RESEARCH

The aim of this work is to assess the success of biological recultivation and transformation of plant communities on the Verkhnetagilskaya power plant ash dump, located in the Sverdlovsk region (eastern slope of the Middle Urals, taiga zone, subzone of the southern taiga). The total area is 125 ha.

Biological recultivation on the ash dump started in 1968-1970 (i.e., 3 years after the end of ash feeding) and continued in subsequent years. Biological recultivation quickly restored vegetation on the ash dump and controlled wind erosion of the substrate. The end result was the creation of vegetation with economic importance. Clay strips, ranging in thickness from 10 cm to 15 cm were laid to regulate erosion. Most of the bands were sown with perennial grasses (*Agropyron cristatum* (L.) Beauv., *Bromopsis inermis* (Leyss.) Holub., *Festuca rubra* L., *Medicago media* Pers., *Onobrychis arenaria* (Kit.) DC, etc.). Portion of area was left for the self-overgrowing. As a result of this, a diverse group of ecotopes were formed. The study of the formation of vegetation on the ash dump was carried out according to conventional techniques for 30 years. Researchers began studying 10-year-old plant communities in 1980, and followed them to the age of 40 years.

Ash dumps are also formed from the heat power stations working on high-ash content coals. Structurally these ash dumps include the presence of slag and are characterized by different sized particles. The thickness of ashes varies from 2 to 20 m. Ashes are light or dark gray in color with black inclusions of the not burned coal particles. The fine-grained ash factions contain coarse particles of physical sand (1-0.05 mm) and dust (0.05-0.001 mm).

Mechanical ash substratum structure (account in % of air to dry probe) is shown in Table 2, and the chemical structure of the ash substratum is shown in Table 3. The analytical chemistry of water extract of ash substratum is shown in Table 4. The total chemical composition of the ash dump substratum corresponds to aluminosilicate formations (SiO_2, 40.5-60.3%; Al_2O_3, 12.9-32.4%). Ashes contain practically no nitrogen or organic substance. The particles of the not burned coal (potential humus) are connected with the silicate of the ashes and are subjected to slow physicochemical and biochemical transformations that play a crucial role in restoration of substratum fertility. The content of mobile phosphorus changes from 1.25 up to 32.5 P_2O_5 mg/100 g of ashes; exchangeable potassium changes from 1.6 up to 25.0 mg/100 g of ashes.

Coal ashes have a small absorption capacity, similar to the structure of soils, because of the low content of highly dispersed organic substance and silt particles.

The absorption ability of the external layers of ash dumps can be enhanced by the covering with peat and soil, on the surface of ash dumps, as well as by introducing mineral fertilizers and the neutralized wastewater.

The (pH) of ashes changes from low acid (pH = 5.9) up to alkaline (pH = 8.5), but in the ranges suitable for growth and development of plants.

The ash dump's substratum contains a large macro- and microelements spectrum.

On a temperature mode, the coal ashes belong to low thermally conductive substratum with sharp oscillation frequencies of temperature from surface to depth.

The water-physical properties of the substratum of ashes are rather peculiar. The coal ashes have a good porosity and good air and water penetrability, from deep to middle at level of sandy loam and sandy soils. The unfavorable conditions arise during the germination of seeds and in the first period of plant life. At this time, the roots are located in a stratum of 0-10 cm, which is subjected to rapid drying.

Thus, on water-physical properties and chemical structure, the coal ashes can be related to substratum suitable for the existence of plants. However, for the creation of a long, productive culturphytocoenosis, the agrotechnical measures for improving the properties of coal ashes as a substratum for the cultivation of plants is necessary.

3 PHYTOCOENOSIS FORMATION ON ASH DUMPS

The ashes were released by the hydraulic method via pipelines. The recultivation using the strips of a ground was conducted on part of ash dump after the termination of release of ashes. The ground was seeded by the perennial grass.

Twenty years after the recultivation work on the ash dumps, a rather diverse ecotope spectrum was observed. This caused formation of unique biotope and vegetative communities. The recultivation actions influenced this process.

Table 2 Mechanical Ash Substratum Structure

Ash Dump Zone	Hygroscopical Moisture	Amount of Particles								The Sum of Fractions		Substratum Mechanical Structure
		Sand (Diameter, mm)		Dust (Diameter, mm)			Silt			Physical—Sand	Physical—Clay	
		Average	Small-Sized	Large	Average	Small-Sized						
		1-0.25	0.25-0.05	0.05-0.01	0.01-0.005	0.005-0.001	<0.001			(<0.01)	(>0.01)	
Ash dump in the forest zone	–	0.53	4.23	56.08	5.45	9.33	4.91			60.84	19.69	Soup

Table 3 Ash Substratum Chemical Structure

| The Name of Ash Dump | Loss at Calcinations (%) | The Total Contents of Basic Elements (% in Calcinated Probe) | | | | | | | | | | The Contents of Mobile Elements | | | |
		SiO_2	Al_2O_3	Fe_2O_3	CaO	MgO	MnO	P_2O_5	SO_3	K_2O	Na_2O	Nitrogen (%)	P_2O_5 mg/100 g ashes	K_2O mg/100 g ashes	pH (KCl)
Ash dump forest zone	2.4	48.4	23.4	14.2	4.9	2.9	–	–	3.8	–	–	Trace	23.5	7.0	8.5

Table 4 The Analysis of the Water Extract of Ash Substratum

The Name of Ash Dump	Dense Residual (%)	The Contents (% from Absolute Dry Probe)					
		CO_3^{-2}	HCO_3^{-1}	Cl^{-1}	SO_4^{-2}	Ca^{+2}	Mg^{+2}
Forest zone	0.200	Not present	0.023	0.004	0.118	0.034	Not present

It is possible to describe the initial ecotopes with the following scheme:

I. Nonrecultivated area
   Ia. initial ecotope: dry ash dump, "pure ashes"
   Ib. moderate moistering, "pure ashes", favorable conditions for drift seeds;
   Ic. residual depressions periodically flooded by water (thawed waters, filtration from ash dumps, etc.)
II. The area with the strips of a soil deposited during the primary recultivation:
   IIa. ashes with deposition of a ground and crop of perennial grasses;
   IIb. ashes with placement of ground and without any seeding;
   IIc. space between bands of substratum
III. Second phase of recultivation:
   – bushes sawing, full covering by the stratum of peat.

The scheme of phytocoenosis formation on ash dump depending on ecotope is shown in Table 5 and Figure 5.

Concerning phytocenosis formation, the following is understood: development of a vegetative grouping takes place from a stage of a settlement of separate to its grouping with the certain density irrespective of the phytocoenosis dynamic status.

In the formation of communities on industrially disturbed lands with rather mentioned process of self-overgrowing, the following stages are selected: ecotopic groups (projective cover 0.1%); simple groups (0.1-5%); complicated groups (6-50%); and phytocoenosis (projective cover more than 50%).

The scheme of phytocoenosis formation on ash dumps, depending on ecotope, was constructed on the basis of actual dated of geobogtanical descriptions, which were done on selected ecotopes. Through the generalization of given and similar material from other ash dumps, the creation of a generalized model of phytocenosis formation on ash dumps became possible. This model was used for the phyto-coenosises located in different zones and climatic conditions. Phytocoenosis is considered to be a main component of industrial ecosystems that were formed from conditions of thermal power station ash dumps (Figure 6).

Ten years after biological reclamation in dry areas, the "clean" ash *Chenopodium album* L. (Ia) dominates. Areas with sufficient moisture ash substrate (Ib) are formed by Puccinella groups (*Puccinella distans* (Jacq.) Parl., *P. hauptiana* Krecz.). On recultivated areas (IIa), cultural phyto-cenosis is formed with the domination of seeded grasses, such as *Agropyron cristatum* (L.) Beauv., *Bromopsis inermis* (Leyss.) Holub., *F. rubra* L., *Medicago media* Pers., and others. At overgrown reclaimed territory on strips of soil and grass, forb plant communities (IIb) formed with a predomi-nance of *Elytrigia repens* (L.) Nevski, *Poa pratensis* L., *Deschampsia cespitosa* (L.) Beauv., and *Artemisia vulgaris* L. At the space of "pure" fly ash, there is depleted species composition, sparse

Table 5 The Scheme of Phytocoenosis Formation on the Ahs Dump

Ecotope	Age (Years)				
	5	10	15	.....	25
Ia	Ecotopical different-grass vegetative grouping (EG)		Simple diverse-herbacious-cereal or diverse-herbacious-cereal vegetative grouping (payload)	↑↑	Over with sparse of willows and birches (P) Dense were over of willows with impurity of birches and aspens (P)
Ib	EG: – water-plant-moss; – diverse-herbacious-cereal	Payload: – diverse-herbacious-cereal	Complicated vegetative grouping (CG): – cereal; – diverce-grass-cereal; – bean -cereal	↑↑↑↑↑	Meadow with growth of birches Fluffy and willows (P) Dense bushes atussock grass meadow (P) White clover-rough meadow grass (P)
Ic			Single islands Vegetative groupings of a hydro-hydrophit type	↑↑	Various variants of coastal vegetation
IIa	Culture phytocoenosis + woods-apofites	CG: – highgrass-cereal; – diverse-herbacious-cereal from a large share (long) of cultural kinds	Diverse-herbacious-cereal phytocoenosis with growing woods	↑↑	Wood phytocoenosis with the rather produced circles: wood—sparse of a pine, birches, aspens; grassy—different grass cereal (P)
IIb	Payload: – diverse-herbacious-cereal with wood	CG: – diverse-herbacious-cereal also grow up—volume of trees and bushes	Woodphytocoenosis: were over wood with poorly produced grass-bush with a circle	↑↑	Wood phytocoenosis with clearly expressed wood, bush, grassy, and moss by circles (P)
IIc	EG: – diverse-herbaceous	Payload: – diverse-herbaceous; – diverse-herbacious-cereal; – young growth wood	CG: – diverse-herbaceous with wood young growth – diverse-herbaceous-cereal – power young growth and wood young growth	↑↑	Were over of deciduous breeds with poorly produced grassy-bush with a circle, strong moss (P)

EG, ecotopical vegetative grouping—0.1%; payload, simple vegetative grouping—0.1-5%; CG, complicated vegetative grouping—6-50%; P, phytocoenosis—more than 50%. Percent means projective cover of ash dump surface by plants.

plant communities (IIc) with a high abundance of *Melilotus officinalis* (L.) Pall. and *M. albus* Medik. and significant participation of *D. cespitosa* (L.) Beauv. and *F. rubra* L. The appearance of regrowth of trees and shrubs was noted.

On "pure ashes" (Ia), the overgrowth of *Calamagrostis epigeios* (L.) Roth was generated in 20 years, with minor participation of other species. *Salix* and *Betula* were sparse and there was very dense overgrowth of Salix (6 species) with impurities of *Betula pendula* Roth., *Betula pubescens* Ehrh., and *Populus tremula* L. At favorable humidity (Ib) that makes a substratum stable, the formation of communities is accelerated. Within 10 years, a tussock grass meadow (= dominant grass *Deschampsia cespitosa* (L) Beauv), dense bushes, an atussock grass meadow, and white clover (rough meadow grass) (*Amoria repens* L. C. Presl *Poa trivialis* L.) were generated over many hectares. Development of moss cover (20-40%) was observed everywhere. Nearby the lowerings, filled with water (Ic), has been colonized by *Equisetum palustre* L.

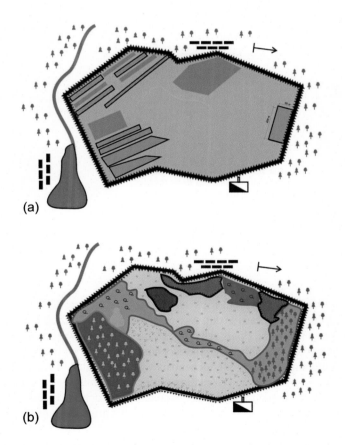

(a)

(b)

FIGURE 5

Scheme-map of ash dump of Verkhnetagilskaya power station: (a) 1971 and (b) 2012.

Explanation to sybmols in Figure 5

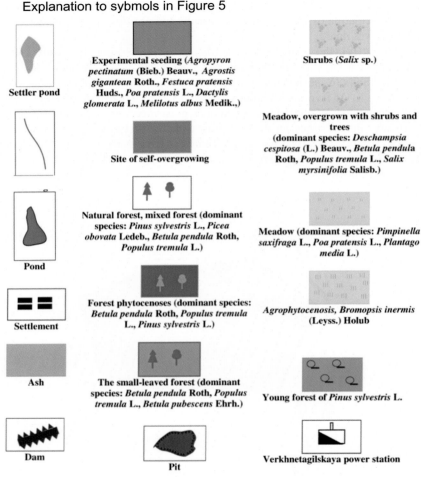

FIGURE 5—CONT'D

On a primary recultivated territory, on strips with a deposited organic debris and perennial grass (IIa), at the first year, partial cutting was carried out. As a result, the diverse herbaceous-cereal and herbaceous-vegetative communities with sparse *Pinus sylvestris* L. and *Betula sp.* were generated in 20 years. These trees strengthened their role in establishing phytocoenose. On strips with organic debris without crops of grass (IIb), the formation of woody phytocoenosis was accelerated. These reasons exclude cutting, retarding of the sodding of a surface and formation of grassy communities of a meadow type.

It is necessary to take into account that the delivered organic debris contained certain propagules of wood phytocoenosis species. As a result, a mixed forest with a prevalence of *Pinus*, and less often with *Betula pendula* was formed.

On the "pure ashes" between the sod strips, the forest communities composed of *Betula,* sparse *Salix,* and *Populus tremula* formed with a delay of 5-10 years (IIc). A layer of grassy bushes was

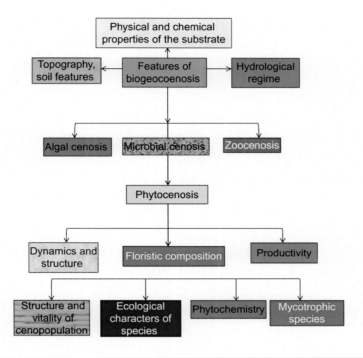

FIGURE 6

Scheme of phytocenosis investigations at mine industry ravaged sites.

poorly produced. It is possible that the formation of this phytocoenosis ash dump was connected with the receipt of seeds from an adjacent, earlier overgrowth. It is natural that improving ash properties by proper management controlled wind erosion. The opposite process, drift of ashes by wind on bands of a ground, also takes place.

The bushes sowing, continuous plotting of a stratum of peat, crop of long-term grass was conducted on the second time recultivated area. With the use of a complex organic and mineral fertilizer, productive moving and grazing fields were created.

By 2013, on the ash dump of the Verkhnetagilskaya power plant (40 years after biological recultivation), at site of the "clean" ash *Calamagrostis epigeios,* (L.) Roth thickets of vegetation had been formed (Figures 7–10) (Ia), with codominants of *Cirsium setosum* (Willd.), Bess, and *Deschampsia cespitosa* (L.) Beauv, with tree groups of *Betula pendula* Roth and *Betula pubescens* Ehrh. *Deschampsia* meadows were formed (IIb) dominated by *Deschampsia cespitosa* (L.) Beauv., with codominants: *Calamagrostis epigeios* (L.) Roth., *Poa pratensis* L., *Hieracium umbellatum* L., and *Chamaenerion angustifolium* (L.) Holub. Tree species are classified as undergrowth (up to 1.5 m) *Populus tremula* L., *Betula pendula* Roth, *Pinus sylvestris* L., and *Salix myrsinifolia* Salisb. On the ashes along the periphery, a small-leaved forest was formed, characterized by a relatively high canopy cover of trees and a complex vertical structure. The upper tree layer is dominated by small-leaved species, such as *Populus tremula, Betula pendula* Roth and *Betula pubescens* Ehrh., *Salix caprea* L., coniferous species—Pinus sylvestris L., and Picea

FIGURE 7

(a) Bazhenovsky asbestos mine over 100 years (mining from 1889) (Asbest). (b) Ash dump of Bogoslovskaya power station 37 years (Krasnoturyinsk, small "islands" of vegetation at located organic debris. *Betula pendula* Roth, *Pinus sylvestris* L., *Salix* sp., *Festuca rubra* L., *Melilotus albus* Medik.). (c) Ash dump of Bogoslovskaya power station 37 years (Krasnoturyinsk, overwatered territory, self over growing with *Betula pendula* Roth, *Pinus sylvestris* L., *Salix* sp., *Festuca rubra* L., *Melilotus albus* Medik. at located organic debris). (d) Agrophytocenosis of *Bromopsis inermis* (Leyss.) Holub at the Verkhnetagilskaya power station ash dump 20 years (Verhniy Tagil). (e) Agrophytocenosis of *Bromopsis inermis* (Leyss.) Holub at the Verkhnetagilskaya power station ash dump 20 years (Verhniy Tagil). (f) Red mud deposit 30 years (Krasnoturyinsk, self overgrowing with *Betula pendula* Roth, *Pinus sylvestris* L., *Salix caprea* L.).

FIGURE 8

(a) Ash dump of Verkhnetagilskaya power station 20 years (Verhniy Tagil, agrophytocenosis of *Bromopsis inermis* (Leyss.) Holub at recultivated site: ash + soil + peat). (b) Ash dump of Verkhnetagilskaya power station 20 years (Verhniy Tagil, agrophytocenosis of *Bromopsis inermis* (Leyss.) Holub at recultivated site: ash + soil + peat). (c) Ash dump of Reftinskaya power plant station 5 years (Ekaterinburg, recultivated with seedings of *Pinus sylvestris* L.). (d) Ash dump of Reftinskaya power station 10 years (Ekaterinburg, recultivated site seedings of *Pinus sylvestris* L. + self-overgrowing with *Hippophae* sp.). (e) Ash dump of Nizneturinskaya power plant 45 years (Nizhnaya Tura, dominated by *Calamagrostis epigeios* (L.) Roth, *Salix* sp.). (f) South-Veselovskiy coal-mine dump 40 years (Karpinsk, seedings of *Pinus sylvestris* L., upper part of dump). (g) South-Veselovskiy coal-mine dump 40 years (Karpinsk, seedings of *Pinus sylvestris* L., middle part of dump).

obovata Ledeb. were included in the lower group. The shrub layer is composed of *Chamaecytisus ruthenicus* (Fisch. Ex Woloszcz.) Klásková, *Rosa acicularis* Lindl., *Salix myrsinifolia* Salisb., and *Salix pentandra* L. The undergrowth is composed of *Sorbus aucuparia* L., *Viburnum opulus* L., and *Padus avium* Mill., the height of which varies from 0.7-0.8 to 3.5 m (projective cover is 15-20%, sometimes up to 30%). The total projective cover of herbaceous species is 30-35%. The greatest values of herbaceous plants have *Amoria repens* (L.), *C. Presl*, and *Trifolium pratense* L., *Festuca rubra* L., *Poa pratensis* L., *Calamagrostis epigeios* (L.) Roth, and *Vicia cracca* L. There are also typical forest species of the boreal zone: *Pyrola chlorantha* Sw., *Chimaphila umbellata* (L.) W. Barton, as well as a young population of Orchidáceae (*Platanthéra bifólia* (L.) Rich.).

In the primary recultivated territories on strips coated with organic debris (IIa), forb grass and forb plant communities were formed. Total projective cover on strips of soil reaches 90-100%, and 60-80% on the ash. The species on the ashes and the strips of soil include *Pimpinella saxifraga* L., *Euphorbia virgata*, Waldst.et Kit., *Achillea millefolium*, L., *Picris hieracioides*, and L., *F. rubra* in high abundance. On the organic debris, *Poa pratensis*, *Centaurea scabiosa* L., *Lathyrus pratensis* L., and *V. cracca* L. prevail, in addition. On the organic debris, *Stellaria graminea* L. and *Melandrium album* (Mill.) Garcke prevail, in addition.

At a substantial part of the ash dump, as a result of ash and soil overgrowth (IIb), forest communities close to the zonal type with substantial interests, and sometimes with the dominance of *Pinus sylvestris* L., *Betula pendula* Roth, *Betula pubescens* Ehrh., *Populus tremula* L., are formed. In the form of undergrowth, *Picea obovata* Ledeb. and *Pinus sibirica* (Rupr.) Mayr were found, together with *Larix sibirica* Ledeb. and *Abies sibirica* Ledeb.

The shrub layer is formed by *Sorbus aucuparia* and *Padus avium*. With increasing age and degree of development of the ash dump, forest communities strengthen their impact on the environment. The transformation of herbaceous vegetation in these communities toward increasing the diversity of forest types has been seen. This has been accompanied by a decrease in the abundance and fallout from the emerging plant communities of some weed-ruderal species. Bush cover species have appeared, such as *Vaccinium vitis-idaea* L., *Orthilia secunda* (L.) House, *Pyrola rotundifolia* L., *P. media,* and *Monese suniflora* (L.) A. Gray. In a grassy layer, *Fragaria vesca* L., *Aegopodium podagraria* L., and *Rubus saxatilis* L prevail.

FIGURE 9

(a) Agrophytocenosis of *Bromopsis inermis* at the Turinskiy coal-mine dump 40 years (Karpinsk). (b) Dumps of the Shuralino-Yagodnoe gold mining 16 years (Nevyansk, self-overgrowing peat-covered site: *Betula pendula* Roth, *Chamaenerion angustifolium* (L.) Scop., *Calamagrostis epigeios* (L.) Roth, *Achillea millefolium* L.). (c) Experimental seeding of *Festuca pratensis* Huds. on the Shuralino-Yagodnoe dumps after gold mining 3 years (Nevyansk). (d) Self-overgrowing of the Shuralino-Yagodnoe dumps after gold mining 18 years (*Betula pendula* Roth, *Pinus sylvestris* L., *Salix caprea* L., *S. myrsinifolia* Salisb., *S.triandra* L.), Nevyansk. (e) Self-overgrowing of recultivated sites at ash dump of Verkhnetagilskaya power station 40 years (VerhniyTagil, formation of forest phytocenosis: *Picea obovata* Ledeb., *Betula pendula* Roth). (f) Limestone (marble) deposit 45 years (Pervouralsk, *Betula pendula* Roth., *Betula pubescens* Ehrh., *Populus tremula* L., *Pinus silvestris* L.). (g) *Pyrola rotundifolia* L. in forest communities at ash dump of Verkhnetagilskaya power station 40 years (VerniyTagil). (h) Ash dump of Yuzhnouralskaya power station 30 years (Yuzhnouralsk, dominated: *Elaeagnus angustifolia* L., *Artemisia dracunculus* L.).

FIGURE 10

(a) Iron deposit 10 years (Iron tailings damps) (Nizhniy Tagil, undergrowth of *Betula pendula* Roth).
(b) Iron deposit 30 years (Nizhniy Tagil, *Pinus sylvestris* L., *Betula pendula* Roth.). (c) Limestone ($CaCO_3$)
(marble) deposit 20 years (Pervouralsk, undergrowth of *Betula pendula* Roth.). (d) Dolomite deposit 15 years
(Pervouralsk, *Betula pendula* Roth., *Betula pubescens* Ehrh., *Populus tremula* L., *Pinus sylvestris* L.). (e)
Dolomite deposit 15 years [$CaMg(CO_3)^2$] (Pervouralsk, undergrowth of *Betula pendula* Roth.). (f) Red mud
deposits 30 years (Krasnoturyinsk, undergrowth of *Pinus sylvestris* L., *Betula pendula* Roth., *B. pubescens*
Ehrh, *Salix caprea* L., *Salix myrsinifolia* Salisb., *Populus tremula* L.). (g) Asbestos deposit 30 years (Asbest,
Betula pendula Roth., *Betula pubescens* Ehrh., *Pinus sylvestris* L.). (h) Limestone (marble) deposit 45 years
(Pervouralsk, *Betula pendula* Roth., *Betula pubescens* Ehrh., *Populus tremula* L., *Pinus sylvestris* L.).

On the secondary recultivated area, application of a layer of peat in the early 1990s, promoted the growth of perennial grasses. When using complex organic and mineral fertilizers, productive pasture grasslands composed of forb grass plant communities were created and maintained for 15 years (long-term care and mowing). These communities were dominated by *Bromopsis inermis* and *Festuca rubra* and covered the entire area (90-100%).

After termination of hay mowing there is a decrease in the density of *Bromopsis inermis* hay shoots from 408 (2004), 250 (2010) to 173 (2011) pcs./m^2. Portions of vegetative shoots of *Bromopsis inermis* in the cultural phytocenosis increased from 75% to 91%, and then decreased to 64%.

When the economic assessment of plant communities is concerned, one of the most important indicators is their productivity. At 10-year-old plant grouping, the average weight of the air dried shoots was an 4.56 cwt/ha (ranged from 0.22 cwt/ha to 7.32 cwt/ha) on only ash; however, with organic matter the air-dry weight of shoots was 17.4 t/ha (ranged from 4.04 to 24.6 cwt/ha). Thus, the average productivity of the communities on ashes with organic matter was four times greater compared to productivity on ash alone.

4 CONCLUSION

The conducted observations enable an evaluation of the biological recultivation of ash dump sites with strips, by covering the sites with a ground and perennial grass.

The stabilization of the substratum of the ashes is achieved at the expense of bands. This results in the diminution or even the termination of dust storms.

It is necessary to recognize the importance of the improvement of the water-physical and agrochemical properties of ashes, via the washing of a ground from bands to space between the strips with "pure ashes." The crop of perennial grass and its consequent cutting accelerates the gardening of bands with a ground, but retards the establishment of trees, bushes, and the formation of a wood circle of the formed forest phytocoenosis.

Biological recultivation of ash dumps with the strips of organic material produces fodder but the quality of fodder produced need to be evaluated.

Experience gained in creating phytocenosis on the ash dump of the Verkhnetagilskaya power plant showed that *Bromopsis inermis* (Leyss.) Holub. is a key species for restoration. One of the most important indicators of the economic assessment of phytocenosis is community productivity. At 10 years, the shoots of plant groups in ash dump conditions had the average dry weight phytomass of 4.56 c/ha (min, 0.22 c/ha; max, 7.32 c/ha) on the ash, and 17.4 c/ha on the ash with organic matter (min, 4.04 c/ha; max, 24/6 c/ha). On average, the productivity of forming communities on the ash-organic matter mixture is 4 times higher than the productivity on empty ash.

After 40 years of biological recultivation, the productivity of plant communities at soil strips is 1.3 times higher than the productivity at ash strips and 1.5 times higher than productivity at clean fly ash (Figure 11).

ACKNOWLEDGMENTS

The authors are thankful for the support of the scientific researches of higher educational institutions within the state task force of the Russian Federation no 2014/236, project no 2485. M.N.V. Prasad is thankful to the Ural Federal University (UrFU), Ekaterinburg, for the invitation to be a visiting professor in November 2014.

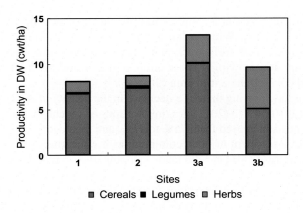

FIGURE 11

Productivity of herbaceous communities at different sites of ash dump.

REFERENCES

Abdelhadi, M.A., Yoshihiko, T., Abdelhadi, N.A., Matouq, M., 2011. Utilization of fly ash linear materials as an adsorbent of hazardous chemical compounds. Sci. Res. Essays 6 (15), 3330–3334.

Awang, A.R., Marto, A., Makhtar, A.M., 2011. Geotechnical properties of Tanjung Bin coal ash mixtures for backfill materials in embankment construction. Electron. J. Geotech. Eng. 16L, 1515–1531.

Babu, A.G., Reddy, M.S., 2011. Dual inoculation of arbuscular mycorrhizal and phosphate solubilizing fungi contributes in sustainable maintenance of plant health in fly ash ponds. Water Air Soil Pollut. 219 (1–4), 3–10.

Baker, A.J.M., Reeves, R.D., McGrath, S.P., 1991. In situ decontamination of heavy metal polluted soils using crops of metal-accumulating plants - a feasibility study. In: Hinchee, R.E., Olfenbuttel, R.F. (Eds.), In situ Bioreclamation: Application and Investigation for Hydrocarbon and Contaminates Site Remediation. Butterworth-Heinemann, Boston & London, pp. 600–605.

Banuelos, G.S., Lin, Z.Q., Arroyo, I., Terry, N., 2005. Selenium volatilization in vegetated agricultural drainage sediment from the San Luis Drain, Central California. Chemosphere 60, 1–203.

Bashmakov, D.I., Lukatkin, A.S., Prasad, M.N.V., 2002. Zinc hyperaccumulating weeds from temperate Russia. Zeszyty Naukowe PAN 33, 309–313.

Bednar, A.J., Garbarino, J.R., Ranville, J.F., Wildeman, T.R., 2005. Effects of iron and on arsenic speciation and redox chemistry in acid mine water. J. Geochem. Explor. 85, 55–61.

Bell, L.C., 2001. Establishment of native ecosystems after mining—Australian experience across diverse biogeographic zones. Ecol. Eng. 17, 179–186.

Bethwell, C., Mutz, M., 2005. Effect of acid mine drainage on the chemical composition and fall velocity of fine organic particles. Ecol. Eng. 24, 75–87.

Belskaya, E.A., Vorobeichik, E.L., 2013. Responses of leaf-eating insects feeding on aspen to emissions from the Middle Ural copper smelter. Russ. J. Ecol. 44 (2), 108–117.

Belskii, E.A., Belskaya, E.A., 2009. Composition of pied flycatcher (*Ficedula hypoleuca* pall.) nestling diet in industrially polluted area. Russ. J. Ecol. 40 (5), 368–371.

Bengson, S.A., 1995. Stabilization of copper mine tailings: two decades of management in the arid Southwest. Min. Environ. Manag. 3, 14–17.

Bing-Yuan, H., Li-Xun, K., 2014. Mine land reclamation and eco-reconstruction in Shanxi province: mine land reclamation model. Sci. World J. 2014, 1–9. Art. no. 483862.

Black, A., Craw, D., 2001. Arsenic, copper and zinc occurrence at the Wangaloa coal mine, southeast Otago, New Zealand. Int. J. Coal Geol. 45, 181–193.

Bleeker, P.M., Assunção, A.G., Teiga, P.M., de Koe, T., Verkleij, J,.A., 2002. Revegetation of the acidic, as contaminated Jales mine spoil tips using a combination of spoil amendments and tolerant grasses. Sci. Total Environ. 300, 1–13.

Bolshakov, V., Chibrik, T., 2007. Biological recultivation. Sci. Russ. 3, 106–112. Russian Academy of Sciences.

Bradshaw, A., 1997. Restoration of mined lands—using natural processes. Ecol. Eng. 8, 255–269.

Bradshaw, A., 2000. The use of natural processes in reclamation—advantages and difficulties. Landsc. Urban Plan. 51, 89–100.

Brooks, R.R., 1987. Serpentine and its Vegetation: A Multiclisciplinary Approach. Croom Helm. Dioscorides Press, Kent, England. viii + 454 p.

Brooks, R. (Ed.), 1998. Plants that Hyper Accumulate Heavy Metals. CAB International, Wallingford. pp. 384.

Brooks, S.J., Udachin, V., Williamson, B.J., 2005. Impact of copper smelting on lakes in the southern Ural Mountains, Russia, inferred from chironomids. J. Paleolimnol. 33 (2), 229–241.

Butalia, T.S., Kirch, J., 2012. Full depth reclamation of asphalt pavements using coal combustion by products: ASTM Special Technical Publication, 1540, 100 West Conshohocken, PA, USA, pp. 376–411.

Caille, N., Swanwick, S., Zhao, F.J., Mcgrath, S.P., 2004. Arsenic hyperaccumulation by *Pteris vittata* from arsenic contaminated soils and the effect of liming and phosphate fertilisation. Environ. Pollut. 132, 113–120.

Caille, N., Vauleon, C., Leyva, l C., Morel, J.-L., 2005. Metal transfer to plants grown on dredged sediment: use of radioactive isotope 203Hg and titanium. Sci. Total Environ. 341 (1–3), 227–239.

Channabasava, A., Lakshman, H.C., Muthukumar, T., 2015. Fly ash mycorrhizoremediation through *Paspalum scrobiculatum* L., inoculated with *Rhizophagus fasciculatus*. C.R. Biol. 338 (1), 29–39.

Chaudhary, S.K., Rai, U.N., Mishra, K., Huang, H.G., Yang, X.E., Inouhe, M., Gupta, D.K., 2011. Growth and metal accumulation potential of *Vigna radiata* L. grown under fly-ash amendments. Ecol. Eng. 37, 1583–1588.

Chen, Y., Shen, Z., Li, X., 2004. The use of vetiver grass (*Vetiveria zizanioides*) in the phytoremediation of soils contaminated with heavy metals. Appl. Geochem. 19, 1553–1565.

Cherry, D.S., Currie, R.J., Soucek, D.J., Latimer, H.A., Trent, G.C., 2001. An integrative assessment of a watershed impacted by abandoned mined land discharges. Environ. Pollut. 111, 377–388.

Chukanov, V.N., Volobuev, P.V., Poddubnyj, V.A., Trapeznikov, A.V., 1993. Ecological problem of Ural. Defektoskopiya 7, 38–46.

Das, M., Agarwal, P., Singh, R., Adholeya, A., 2013. A study of abandoned ash ponds reclaimed through green cover development. Int. J. Phytoremediation 15 (4), 320–329.

Dell' Amico, Cavalca, L., Andreoni, V., 2005. Analysis of rhizobacterial communities in perennial Graminaceae from polluted water meadow soil, and screening of metal-resistant, potentially plant growth-promoting bacteria. FEMS Microbiol. Ecol. 52, 153–162.

Dinelli, E., Lucchini, F., Fabbri, M., Cortecci, G., 2001. Metal distribution and environmental problems related to sulfide oxidation in the Libiola copper mine area, Ligurian Apennines, Italy. J. Geochem. Explor. 74, 141–152.

Doetsch, P., Rüpke, A., Burmeier, H., 1999. Brownfields versus Green-Fields—Economic and Ecological Aspects of Land Development Options. Federal Environmental Agency, Berlin. Contaminated Land Section, Berlin, Germany.

Dulya, O.V., Mikryukov, V.S., Vorobeichik, E.L., 2013. Strategies of adaptation to heavy metal pollution in *Deschampsia caespitosa* and *Lychnis flos-cuculi*: analysis based on dose-response relationship. Russ. J. Ecol. 44, 4271–4281.

Eydenzon, D., Ganieva, I., Shpak, N., 2013. Socio-economic and environmental aspects of the industry imbalances in the regional economy. Eco. Region 4, 115–122.

Free, B.M., 2012. Flash forward: improving vegetation establishment practices for surface mine reclamation in the appalachian coal basin. In: 43rd International Erosion Control Association Annual Conference. Los Vegas, USA, pp. 161–167.

Freitas, H., Prasad, M.N.V., Pratas, J., 2004. Plant community tolerant to trace elements growing on the degraded soils of São Domingos mine in the south east of Portugal environmental implications. Environ. Int. 30, 65–72.

Frontasyeva, M.V., Smirnov, L.I., Steinnes, E., Lyapunov, S.M., Cherchintsev, V.D., 2004. Heavy metal atmospheric deposition study in the South Ural Mountains. J. Radioanal. Nucl. Chem. 259, 19–26.

García, G., Faz, A., Cunha, M., 2004. Performance of *Piptatherum miliaceum* (Smilo grass) in edaphic Pb and Zn phytoremediation over a short growth period. Int. Biodeterior. Biodegrad. 54, 245–250.

Gaur, A., Adholeya, A., 2004. Prospects of arbuscular mycorrhizal fungi in phytoremediation of heavy metal contaminated soils. Curr. Sci. 86 (4), 528–534.

Gilbert, O., de Pablo, J., Cortina, J.L., Ayora, C., 2005. Municipal compost-based mixture for acid mine drainage bioremediation: metal retention mechanisms. Appl. Geochem. 20, 1648–1657.

Glass, D.J., 1999. U.S. and International Markets for Phytoremediation (1999-2000). D.J. Glass Associates, Needham, MA. 266 pp.

Gonsalves, S.C., Gonalves, M.T., Freitas, H., Martins-Loucao, M.A., 1997. Mycorrhizae in a portuguese serpentine community. In: Jaffr, T., Reeves, R.D., Becquer, T. (Eds.), The Ecology of Ultramafic and Metalliferous Areas. ORSTOM, New Caledonia, pp. 87–89.

Guo, L.-J., Wu, Y.-S., Li, R.-Y., Wang, M., 2013. Properties of geopolymer based on dust in hot-reclamation of used sand and fly ash. Foundry (Zhuzao)62 (9), 905–909.

Haibin, L., Zhenling, L., 2010. Recycling utilization patterns of coal mining waste in China. Resour. Conserv. Recycl. 54 (12), 1331–1340.

Hallberg, K.B., Johnson, D.B., 2005. Biological manganese removal from acid mine drainage in constructed wetlands and prototype bioreactors. Sci. Total Environ. 338, 115–124.

Hu, Z., Li, L., Zhao, Y., Feng, X., 2013. Morphology development evaluation of reclaimed soil in coal-mining subsidence areas with high groundwater levels. Trans. Chin. Soc. Agri. Eng. (Nongye Gongcheng Xuebao)29 (5), 95–101.

Johnson, D.B., Hallberg, K.B., 2005. Acid mine drainage remediation options: a review. Sci. Total Environ. 338, 3–14.

Khokhryakov, A.V., 2003. Environmental problems of mining in the Ural. Erzmetall: J. Exploration, Min. Metall. 56 (10), 595–600.

Kidd, P.S., Llugany, M., Poschenrieder, C., Gunsé, B., Barceló, J., 2001. The role of root exudates in aluminium resistance and silicon-induced amelioration of aluminium toxicity in three varieties of maize, Zea mays L. J. Exp. Bot. 52 (359), 1339–1352.

Koptsik, G.N., 2014. Modern approaches to remediation of heavy metal polluted soils: a review. Eurasian Soil Sci. 1064-2293. 47 (7), 707–722. © Pleiades, 2014. Original Russian Text published in Pochvovedenie, 2014, No. 7, pp. 851–868.

Kothe, E., Bergmann, H., Büchel, G., 2005. Molecular mechanism in bio-geo-interactions: from a case study to general mechanisms. Chem. Erde 65 (S1), 7–27.

Kovshov, S., 2013. Biological ground recultivation and increase of soil fertility. IJED 25, 105–113.

Kraemer, U., Grime, G.W., Smith, J.A.C., Hawes, C., Baker, A.J.M., 1997. Micro-PIXE as a technique for studying nickel localization in leaves of the hyper accumulator plant alyssum lesbiacum. Nucl. Instrum. Meth. Phys. Res. 130 (1), 346–350.

Kramer, U., 2005. Phytoremediation: novel approaches to cleaning up polluted soils. Curr. Opin. Biotechnol. 16, 133.

Lai, H.Y., Chen, Z.S., 2004. Effects of EDTA on solubility of cadmium, zinc, and lead and their uptake by rainbow pink and vetiver grass. Chemosphere 55 (3), 421–430.

Lai, Y., Wang, Q., Yang, L., Huang, B., 2006. Subcellular distribution of rare earth elements and characterization of their binding species in a newly discovered hyperaccumulator *Pronephrium simplex*. Talanta 70, 26.

Lal, K., Chhabra, R., Mongia, A.D., Meena, R.L., Yadav, R.K., 2012. Release and uptake of potassium and sodium with fly ash application in rice on reclaimed alkali soil. J. Indian Soc. Soil Sci. 60 (3), 181–186.

Lan, C.Y., Shu, W.S., Wong, M.H., 1998. Reclamation of Pb/Zn mine tailings at Shaoguan, Guangdong Province, People's Republic of China: the role of river sediment and domestic refuse. Bioresour. Technol. 65, 117–124.

Leblanc, M., Achard, B., Ben Othman, D., Luck, J.M., Bertrand-Sarfati, Personné, J.Ch., 1996. Accumulation of arsenic from acidic mine waters by ferruginous bacterial accretions, stromatolites. Appl. Geochem. 11 (4), 541–554.

Li, J., Zhuang, X., Font, O., Moreno, N., Vallejo, V.R., Querol, X., Tobias, A., 2014. Synthesis of merlinoite from Chinese coal fly ashes and its potential utilization as slow release K-fertilizer. J. Hazard. Mater. 265, 242–252.

Liu, R., Lal, R., 2013. A laboratory study on amending mine soil quality. Water Air Soil Pollut. 224 (9), 1679.

Liu, F., Yan, J., Marx, B., Weiss, E., 2013. The related research about the thickness of covered soil in reclaimed land filled by fly ash. Adv. Mater. Res. 664, 251–255.

Madyiwa, S., Chimbari, M.J., Schutte, C.F., Nyamangara, J., 2003. reenhouse studies on the phyto-extraction capacity of cynodon nlemfuensis for lead and cadmium under irrigation with treated wastewater. Phys. Chem. Earth 28, 859–867.

Malcová, R., Vosatka, M., Gryndler, W., 2003. Effects of inoculation with glomus intraradices on lead uptake by Zea mays L. and agrostis capillaris L. Appl. Soil Ecol. 23, 53–67.

Martin, L.C., Branam, T.D., Naylor, S., Olyphant, G.A., 2012. Characterization of coal combustion byproducts fifteen years after emplacement in an abandoned mine land site. In: 29th Annual National Conference of the American Society of Mining and Reclamation 2012, pp. 346–358.

McLaughlin, M.J., Hamon, R.E., McLaren, R.G., Speir, T.W., Rogers, S.L., 2000. Review: a bioavailability-based rationale for controlling metal and metalloid contamination of agricultural land in Australia and New Zealand. Aust. J. Soil Res. 38 (6), 1037–1086.

Mesjasz-Przybylowicz, J., Balkwill, K., Przybyłowicz, W.J., Annegarn, H.J., 1994. Proton microprobe and X-ray fluorescence investigations of nickel distribution in serpentine flora from South Africa. Nucl. Instrum. Meth. Phys. Res. 89 (1), 208–212.

Neagoe, A., Eben, G., Carlsson, E.., 2005. The effect of soil amendments on plant performance in an area affected by acid mine drainage. Chemder. Erde 65 (S1), 115–129.

Ors, S., Sahin, U., Khadra, R., 2015. Reclamation of saline sodic soils with the use of mixed fly ash and sewage sludge. Arid Land Res. Manag. 29 (1), 41–54.

Ouyang, J., Chen, K., He, S., 2010. Municipal sewage sludge used as reclaiming material for abandoned mine land. In: International Conference on Multimedia Technology (ICMT), Ningbo, China. pp. 1–4. http://dx.doi.org/10.1109/ICMULT.2010.5629698.

Pang, J., Chan, G.S., Zhang, J., Liang, J., Wong, M.H., 2003. Physiological aspects of vetiver grass for rehabilitation in abandoned metalliferous mine wastes. Chemosphere 52, 1559–1570.

Panin, M.S., 1998. Influence of antropogenic activity and human argochemical activity on migration of heavy metals in system "soil-plant. In. Conf. The state and rational use of soils in Kazkhstan" Almaty. pp. 76–79.

Park, S.-W., Chae, D., Kim, K.-O., Kwon, O., Cho, W., 2011. Consolidation characteristics of artificial soil mixture with coal combustion byproduct. In: Proceedings of the International Offshore and Polar Engineering Conference, pp. 531–535.

Paschke, M.W., Valdecantos, A., Redente, E.F., 2005. Manganese toxicity thresholds for restoration grass species. Environ. Pollut. 135 (2), 313–322.

Peterson, D.J., 1995. Russia's environment and natural resources in light of economic regionalization. Post Sov. Geogr. 36, 291–309.

Phommachanh, V., Maegawa, F., Kawai, K., Iizuka, A., 2012. Expression of self-hardening property of coal fly-ash with the constitutive model. In: Volume II: Soil Improvement. 5th Asia-Pacific Conference on Unsaturated Soils 2012. pp. 424–429.

Prasad, M.N.V., Freitas, H., 1999. Feasible biotechnological and bioremediation strategies for serpentine soils and mine spoils. Electron. J. Biotechnol. 2 (1), 36–50.

Prasad, M., 2001a. Bioremediation potential of amaranthaceae. In: Leeson, A., Foote, E.A., Banks, M.K., Magar, V.S. (Eds.), . Phytoremediation, Wetlands, and Sediments, 6.,5.:165–172. Proc, 6th Int In Situ & On-Site Bioremediation Symposium, Battelle Press, Columbus, OH.

Prasad, M.N.V. (Ed.), 2001b. Metals in the Environment: Analysis by biodiversity. Marcel Dekker Inc, New York, p. 504.

Prasad, M.N.V., Freitas, H., 2003. Metal hyperaccumulation in plants – biodiversity prospecting for phytoremediation technology. Electron. J. Biotechnol. 6 (3), 275–321.

Prasad, M.N.V., 2004. Phytoremediation of metals and radionuclides in the environment: the case for natural hyperaccumulators, metal transporters, soil amending chelators and transgenic plants. In: Prasad, M.N.V. (Ed.), Heavy Metal Stress in Plants: From Biomolecules to Ecosystems, second ed. Springer-Verlag, Heidelberg. Narosa New Delhi, pp. 345–392.

Pratas, J., Prasad, M.N.V., Freitas, H., Conde, L., 2004. Pinus pinaster aiton (maritime pine).: a reliable indicator for delineating areas of anomalous soil composition for biogeochemical prospecting of arsenic, antimony. and tungsten. Eur. J. Miner. Process. Environ. Protect. 4, 136–143.

Pratas, J., Prasad, M.N.V., Freitas, H., Conde, L., 2005. Plants growing in abandoned mines of Portugal are useful for biogeochemical exploration of arsenic, antimony, tungsten and mine reclamation. J. Geochem. Explor. 85, 99–107.

Qian, K.-M., Wang, L.-P., Li, J., 2011a. Variation of microbial activity in reclaimed soil in mining area. J. Ecology Rural Environ. 27 (6), 59–63.

Qian, K.-M., Zhang, L., Wang, L.-P., 2011b. An environmentally sound usage of both coal mining residue and sludge. Adv. Mater. Res. 183–185, 595–599.

Qian, K., Wang, L., Yin, N., 2012. Effects of AMF on soil enzyme activity and carbon sequestration capacity in reclaimed mine soil. Int. J. Mining Sci. Tech. 22 (4), 553–557.

Rao, A.V., Tarafdar, J.C., 1998. Title selection of plant species for rehabilitation of gypsum mine spoil in arid zone. J. Arid Environ. 39, 559–567.

Rai, U.N., Pandey, K., Sinha, S., Singh, A., Saxena, R., Gupta, D.K., 2004. Revegetating fly ash landfills with Prosopis juliflora L.: impact of different amendments and Rhizobium inoculation. Environ. Int. 30, 293–300.

Rai, A.K., Paul, B., 2011. Physical characterisation of fly ash from coal fired thermal power plants, Jharia Coalfield, Jharkhand. Ecol. Environ. Conserve. 17 (3), 553–556.

Ren, R.-S., Chen, Y.-N., Shi, F.-E., Jiang, D.-H., 2010. Kinetic and equilibrium studies of Cr (VI) from wastewater with modified fly ashes. In: 4th International Conference on Bioinformatics and Biomedical Engineering, iCBBE 2010. Art. no. 5517082.

Schutzendubel, A., Polle, A., 2002. Plant responses to abiotic stresses: heavy metal-induced oxidative stress and protection by mycorrhization. J. Exp. Bot. 53 (372), 1351–1365.

Schwitzguébel, J.P., van der Lelie, D., Baker, A., Glass, D.J., Vangronsveld, J., 2002. Phytoremediation: European and American trends successes, obstacles and needs. J. Soils Sediments 2, 91–99.

Shin, H.Y., Kim, K.O., Kim, Y.S., Kim, T.H., 2011. Effect of coal ash contents on the acceleration of settling and self-weight consolidation of clayey ground. In: Proceedings of the International Offshore and Polar Engineering Conference, pp. 546–549.

Shin, H.-Y., Kim, K.O., Kim, Y.-J., Kang, B.-Y., Cho, Y.-K., 2014. Trial construction of ground reclamation using dredged soil mixed with coal combustion products. In: Proceedings of the International Offshore and Polar Engineering Conference, pp. 760–763.

Shu, W.S., Ye, Z.H., Lan, C.Y., Zhang, Z.Q., Wong, M.H., 2002. Lead, zinc and copper accumulation and tolerance in populations of paspalum distichum and cynodon dactylon. Environ. Pollut. 120 (2), 445–453.

Simon, M., Ortiz, I., Garcia, I., Fernandez, E., Fernandez, J., Dorronsoro, C., Aguilar, J., 1999. Pollution soils by the toxic spill of a pyrite mine, Aznalcollar, Spain. Sci. Total Environ. 242 (1–3), 105–115.

Singh, G., 1996. Effect of site preparation techniques on prosopis juliflora in an alkali soil. Forest Ecol. Manag. 80, 267–278.

Shou, H., Chang, X., Pan, Q., Huang, M., Zhang, L., Dong, J., 2011. Study on the contents of heavy metal in soil and vegetation in filling reclaimed land in Xuzhou Jiuli mining area. In: International Conference on Remote Sensing, Environment and Transportation Engineering, RSETE 2011—Proceedings, pp. 1691–1694. Art. no. 5964617.

Skousen, J., Ziemkiewicz, P., Yang, J.E., 2012. Use of coal combustion by-products in mine reclamation: review of case studies in the USA. Geosystem Eng. 15 (1), 71–83.

Smith, R.A.H., Bradshaw, A.D., 1979. The use of metal tolerant plant populations for the reclamation of metalliferous wastes. J. Appl. Ecol. 16, 595–612.

Spiro, B., Weiss, D., Purvis, O., Mikhailova, I., Williamson, B., Coles, B., Udachin, V., 2004. Lead isotopes in lichen transplants around a Cu smelter in Russia determined by MC-ICP-MS reveal transient records of multiple sources. Environ. Sci. Technol. 38, 6522–6528.

Spiro, B., Udachin, V., Williamson, B., Purvis, O., Tessalina, S., Weiss, D., 2012. Lacustrine sediments and lichen transplants: two contrasting and complimentary environmental archives of natural and anthropogenic lead in the South Urals, Russia. Aquat. Sci. 75 (2), 185–198. http://dx.doi.org/10.1007/s00027-012-0266-3.

Srivastava, N.K., Ram, L.C., 2010. Reclamation of coal mine spoil dump through fly ash and biological amendments. Int. J. Ecol. Dev. 17 (F10), 17–33.

Srivastava, N.K., Ram, L.C., Masto, R.E., 2014. Reclamation of overburden and lowland in coal mining area with fly ash and selective plantation: a sustainable ecological approach. Ecol. Eng. 71, 479–489.

Sun, C.-J., Li, M.-G., Gau, S.-H., Wang, Y.-H., Jan, Y.-L., 2011. Improving the mechanical characteristics and restraining heavy metal evaporation from sintered municipal solid waste incinerator fly ash by wet milling. J. Hazard. Mater. 195, 281–290.

Tanhan, P., Kruatrachue, M., Pokethitiyook, P., Chaiyarat, R., 2007. Uptake and accumulation of cadmium, lead and zinc by Siam weed [*Chromolaena odorata* (L.) King & Robinson]. Chemosphere 68 (2), 323–329.

Tordoff, G.M., Baker, A.J.M., Willis, A.J., 2000. Current approaches to the revegetation and reclamation of metalliferous mine wastes. Chemosphere 41, 219–228.

Tripathi, R.C., Jha, S.K., Ram, L.C., Singh, G., 2011. Reclamation of wasteland for cultivation of cotton crop through application of pond ash and its leachate. In: Jenkins, P.T. (Ed.), The Sugar Industry and Cotton Crops. Nova Science, New York, pp. 35–64.

Udachin, V., Williamson, B.J., Purvis, O.W., Spiro, B., Dubbin, W., Brooks, S., Coste, B., Herrington, R.J., Mikhailova, I., 2003. Assessment of environmental impacts of active smelter operations and abandoned mines in karabash, Ural mountains of Russia. Sustain. Dev. 11, 1–10.

Vacenovska, B., Drochytka, R., 2012. Development of a new reclamation material by hazardous waste solidification/stabilization. Adv. Mater. Res. 446–449, 2793–2799.

Vangronsveld, J., Cunningham, S.D., 1998. Metal-contaminated soils: in-situ in activation and phytorestoration. Springer, Berlin, Heidelberg. 265 pp.

Viestová, Z., Hlavatá, M., Wilkosz, A., 2011. Conditions of waste and waste mixture utilization in technical land reclamation (Uwarunkowania wykorzystania odpadów i mieszaniny odpadów do technicznej rekultywacji terenu). Inzynieria Mineralna 12 (1), 51–60.

Visa, M., 2012. Tailoring fly ash activated with bentonite as adsorbent for complex wastewater treatment. Appl. Surf. Sci. 263, 753–762.

Visa, M., Isac, L., Duta, A., 2012. Fly ash adsorbents for multication wastewater treatment. Appl. Surf. Sci. 258 (17), 6345–6352.

Wang, C., Zhu, N., Wang, Y., Zhang, F., 2012. Co-detoxification of transformer oil-contained PCBs and heavy metals in medical waste incinerator fly ash under sub- and supercritical water. Environ. Sci. Technol. 46 (2), 1003–1009.

Wang, Z.H., Zhou, B., Sun, X.J., Peng, J.W., 2013. Modification of fly ash and its application state research in wastewater treatment. Adv. Mater. Res. 726–731, 2455–2460.

Wong, M.H., 2003. Ecological restoration of mine degraded soils, with emphasis on metal contaminated soils. Chemosphere 50, 775–780.

Williamson, B.J., Mikhailova, I., Purvis, O.W., Udachin, V., 2004a. SEM-EDX analysis in the source apportionment of particulate matter on *Hypogymnia Physodes* lichen transplants around the Cu smelter and former mining town of Karabash, South Urals, Russia. Sci. Total Environ. 322 (1–3), 139–154.

Williamson, B.J., Udachin, V., Purvis, O.W., Spiro, B., Cressey, G., Jones, G.C., 2004b. Characterisation of airborne particulate pollution in the Cu smelter and former mining town of Karabash, South Ural Mountains of Russia. Environ. Monit. Assess. 98 (1–3), 235–259.

Williamson, B.J., Purvis, O.W., Mikhailova, I.N., Spiro, B., Udachin, V., 2008. The lichen transplant methodology in the source apportionment of metal deposition around a copper smelter in the former mining town of Karabash, Russia. Environ. Monit. Assess. 141 (1–3), 227–236.

Yan, L., Wang, Y., Ma, H., Han, Z., Zhang, Q., Chen, Y., 2012. Feasibility of fly ash-based composite coagulant for coal washing wastewater treatment. J. Hazard. Mater. 203–204, 221–228.

Yang, Y.-X., Lu, H.-L., Zhan, S.-S., Deng, T.-H.-B., Lin, Q.-Q., Wang, S.-Z., Yang, X.-H., Qiu, R.-L., 2013. Using kenaf (*Hibiscus cannabinus*) to reclaim multi-metal contaminated acidic soil. J. Appl. Ecol. 24 (3), 832–838 (Chinese).

Yu, M., Bi, Y.-L., Zhang, C.-Q., Yin, N., 2013. Metal elements utilization by mycorrhizal fungi in fly ash reclamation. J. China Coal Soc. (Meitan Xuebao)38 (9), 1675–1680.

Zhang, L., Xie, M., Pang, J., Shi, Y., Yang, M., Pan, W., 2011a. Research on modified fly ash for high iron and high manganese acid mine drainage treatment. In: International Conference on Remote Sensing, Environment and Transportation Engineering, RSETE 2011—Proceedings, pp. 7053–7056. Art. no. 5965989.

Zhang, Q., Wang, F., Wang, R., 2011b. Research progress of ecological restoration for wetlands in coal mine areas. Procedia Environ. Sci. 10 (PART C), 1933–1938.

Zhao, J., Feng, Y., Hao, G., Fan, R., 2010. Effects of ameliorants addition on Cd contents and yield of crop in Cd-rich reclaiming substrates. Trans. Chin. Soc. Agric. Eng. (Nongye Gongcheng Xuebao)26 (6), 292–295.

Zhao, Q.-L., Zhong, H.-Y., Liu, J.-L., Liu, Y., 2012. Integrated coagulation-trickling filter-ultrafiltration processes for domestic wastewater treatment and reclamation. Water Sci. Technol. 65 (9), 1599–1605.

Zhou, H., Xu, G., 2012. The research and application of reclamation technology in complex matrix of VA mycorrhiza on coal mine abandoned wasteland. Adv. Mater. Res. 550–553, 2224–2227.

FIGURE 1

Phycoremediation plant in operation at SNAP industries, India from 2006.

4 WORLD'S FIRST PHYCOREMEDIATION PLANT AT RANIPET, INDIA

Large-scale phycoremediation projects lab/pilot and filed scale have been considered for their dependable service to treat waste water and industrial effluents (Levoie and de la Noue, 1985; Mezzari et al., 2014; Murugesan et al., 2007; Olguín and Sánchez-Galván, 2012; Padmapriya et al., 2012; Parameswari et al., 2010; Prajapati et al., 2013a,b; Rajamani et al., 2007; Ranjith Kumar et al., 2011; Ratha et al., 2011; Rawat et al., 2011; Renuka et al., 2015; Rorrer, 2005; Sankaran et al., 2014; Sharara et al., 2014; Singh et al., 2012; Sivasubramanian et al., 2009; Solovchenko et al., 2014; Suresh Kumar et al., 2007, 2015; Tripathi et al., 2001; Zainal et al., 2012).

India's largest phytoremediation plant is in operation at SNAP Natural and Alginate Products, Ranipet, India from September 2006 (Figure 1). The industry generates 30-40 kL of highly acidic effluent every day which is being pH corrected and evaporated using an algae based treatment

Table 1 A Comparison of Physic-Chemical Parameters of Raw Effluent with Effluent Taken from the Bottom of the Tank after 2 Years of Phycoremediation (Evaporation at 30 kL/day/2 years) (values ± SD) (Sivasubramanian et al., 2009)

Parameters	Raw Effluent	Effluent Taken from the Bottom of the Tank after 2 Years of Phycoremediation
Turbidity NTU	106 ± 7.20	5.9 ± 0.15
TDS (mg/L)	27,600 ± 600	49,220 ± 230
Electrical conductivity (μmho/cm)	36,430 ± 120	69,896 ± 354
pH	1.66 ± 0.15	7.02 ± 0.17
Alkalinity pH (mg/L)	0	0
Alkalinity (mg/L)	Nil	2916 ± 160.8
Total hardness (mg/L)	2100 ± 123	5125 ± 110
Ca (mg/L)	520 ± 15.7	1120 ± 49.3
Mg (mg/L)	192 ± 4.2	558 ± 7.7
Na (mg/L)	6800 ± 143.7	7750 ± 123.5
K (mg/L)	700 ± 15.3	8125 ± 150.7
Fe (mg/L)	17.99 ± 0.34	4.13 ± 0.19
Mn (mg/L)	Nil	Nil
Free ammonia (mg/L)	56 ± 7.2	13.44 ± 0.28
NO_2 (mg/L)	0.43 ± 0.04	Nil
NO_3 (mg/L)	22 ± 1.5	16 ± 1.2
Chloride (mg/L)	3216 ± 14.6	12,189 ± 110
Fluoride (mg/L)	0.62 ± 0.14	0
Sulfate (mg/L)	5221 ± 110.8	1195 ± 89.3
Phosphate (mg/L)	28.62 ± 1.22	169 ± 5.67
SiO_2 (mg/L)	5.48 ± 0.23	79.49 ± 3.45
BOD (mg/L)	44 ± 5.6	960 ± 23.5
COD (mg/L)	148 ± 12.5	3266 ± 24.6

technology developed by Sivasubramanian and his team from Phycospectrum Environmental Research Centre (PERC), Chennai, India (Sivasubramanian et al., 2009). The physico-chemical characteristics of effluent are provided in Table 1. There is 100% reduction in sludge by phycoremediation.

5 THE TECHNOLOGY AND BIOMASS BASED COMMERCIALLY VALUABLE PRODUCTS

Phycoremediation plant is used to treat the acidic effluent from this alginate industry. The liquid effluent is highly acidic. Conventionally, sodium hydroxide has been used for the neutralization of the acidic effluent which results in an increase in total dissolved solids (TDS) and the generation of solid waste. The study was conducted in three stages. In the first stage, the solar ponds used for evaporating the effluent were converted into high rate algal ponds with *Chroococcus turgidus*, a blue green alga. Based on the

FIGURE 2

Biofertilizer and aquaculture product produced by SNAP industry.

results of pilot plant studies, a full scaling up of the slope tank was made. With the addition of around 30kL of acidic effluent every day, the pH of the effluent remained constant around 7.02 and total dissolved salts stabilized at 49g/L. There was no sludge formation even after 2 years of operation. With just one circulation of the effluent at a pumping rate of 80kl/h on the slopes, the desired evaporation of 30kL was achieved. Algal cell density is maintained at $2300 \times 10^4 \, \text{mL}^{-1}$ (0.75g/L on dry weight basis). Excess algal biomass is regularly harvested by the industry to produce two important products viz., biocompost, biofertilizer, and aquaculture feed (Figure 2). Both these products (06 EMMA and PLANK-10) are produced at 2 tons per month ($2 million per year). PLANK-10 is widely used aquaculture farms all over India to reduce harmful algae, improve plankton levels, reduce BOD and COD and to improve dissolved oxygen in fisheries ponds. Algal mats growing on slope surface of ETP plant is also harvested on a regular basis. Algal biomass also settles down at the bottom of ETP plant which is used as manure.

Phycoremediation is a sophisticated and multifaceted effluent treatment process. And each case of application will address a different set of operational imperatives. The summary given in Table 2 depicts the business case for this technology in the case of this particular application.

6 LEATHER PROCESSING CHEMICAL INDUSTRY

Leather processing industry in Tamil Nadu, India was selected for investigation. The industry manufactures dyes, binders and pigments. The effluent and sludge generated has heavy metals and residual chemicals used in production. Algal technology is employed in treating the effluent and sludge generated by the industry (60-70 kL/day). Table 3 gives information on the physic-chemical parameters of the effluent. Analysis shows that effluent has almost all the essential nutrients needed for algal growth.

Treated water is recycled by the industry. The sludge produced by the industry also supports very good growth of algae. *Chlorella vulgaris* grows luxuriantly in leather processing chemical industry (reaching 1 g/L dry weight) (Figures 3 and 4). Regular harvesting of micro algal biomass is done and the treated water is reused.

Table 2 SNAP Phycoremediation Plant—Operational and Financial Comparison

Cost Parameter	Conventional Effluent Treatment	Phycoremediation	Annual Cost Benefit
Acidity—high levels of dissolved carbon dioxide	Neutralization with caustic soda	Algal treatment to absorb the acidic contents and neutralize the effluent	Rs. 50 lakhs spent for caustic soda annually is saved (100%) The total cost for the utilities (labor/electricity, etc.) used in the operation is almost identical. At around Rs. 2 lakhs p.a.*
Sludge formation	Evaporation of effluents deposits sludge. That needs to be buried in a land fill About 290 tons of sludge produced annually from this treatment	Algal remediation produces a nutrient rich, commercially valuable fertilizer that is highly demanded in the market There is no residual sludge	The sludge disposal used to cost an estimated Rs. 3 lakhs annually. This cost is saved* Additionally revenues from the sale of algal biomass fertilizer
Structures and space	11,000 m² of masonry tank for evaporating the effluent	3000 m² of tank for containing and evaporating the effluent	About 75% of the effluent treatment facility space is released. This very valuable real estate structural space is now being used for other productive uses

*3000 USD

Table 3 Physico-Chemical Characteristics of Effluent from a Leather Processing Industry in Tamil Nadu, India

Parameter	Raw	Treated	Parameter	Raw	Treated
Turbidity NTU	87.3	21.3	Total Kjeldahl Nitrogen (TKN)	72.8	63.84
Total solids, mg/L	3401	3275	Free ammonia (as NH_3), mg/L	70.56	50.4
Total suspended solids, mg/L	187	503	Nitrite (as NO_2), mg/L	0.85	0.10
Total dissolved solids, mg/L	3214	2772	Nitrate (as NO_3), mg/L	19	3
Electrical conductivity, μmho/cm	4561	3876	BOD, mg/L	210	35
pH	8.08	8.55	COD, mg/L	602	103
Alkalinity total (as $CaCO_3$), mg/L	2048	1512	Iron (as Fe), mg/L	2.43	0.13
Total hardness (as $CaCO_3$), mg/L	500	323	Chloride (as Cl), mg/L	458	208
Calcium (as Ca), mg/L	110	81	Sulfate (as SO_4), mg/L	184	52
Magnesium (as Mg), mg/L	54	29	Phosphate (as PO_4), mg/L	83.66	0.16
Sodium (as Na), mg/L	800	880	Silica (as SiO_2), mg/L	54.12	45.46
Potassium (as K), mg/L	150	70	Tidy's test (as O), mg/L	5.5	16

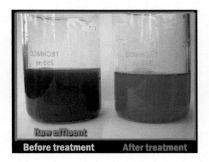

FIGURE 3

Shows algal growth (450 cells × 10^4 mL^{-1} of *Chlorella vulgaris*) in effluent from leather processing chemical industry in Tamil Nadu, India.

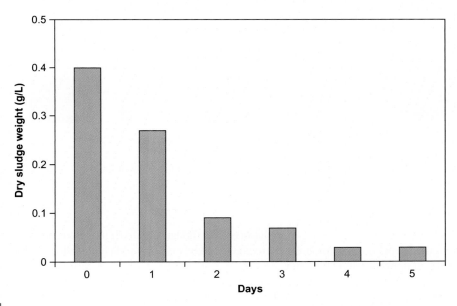

FIGURE 4

Growth of *Chlorella vulgaris* and sludge reduction in effluent from a leather processing industry.

7 EFFLUENT FROM CONFECTIONERY INDUSTRY

A Confectionery industry in Tamil Nadu, India was selected for the study. The total effluent generated per day amounts to an average of 50-70 kL. The plant effluent generated is divided into two streams viz., industrial effluent stream from the production process and sewage effluent stream from the human activities. These are mixed prior to sending to the equalization tank. The effluent for phycoremediation

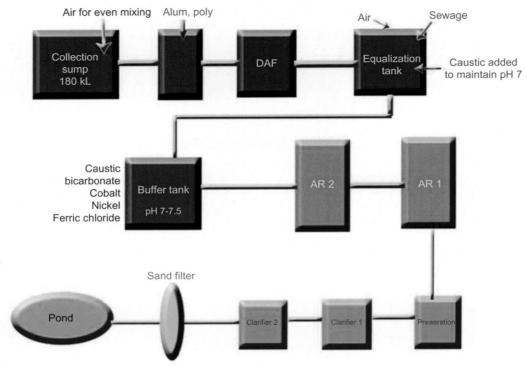

FIGURE 5

ETP facility at the confectionery industry (algal remediation was introduced at the fourth stage, equalization tank, to correct pH, reduce sugar levels, reduce BOD and COD).

treatment is taken after it goes through the dissolved air floatation (DAF) in the conventional treatment method (Figure 5). pH is conventionally corrected by adding caustic soda at the equalization tank stage which results in doubling of TDS. After pH correction the effluent is sent to buffer tank and anaerobic reactors (AR) to digest rest of the organic compounds. After digestion in AR effluent goes through a series of clarifiers and sand filter and finally taken to R/O for recycling.

8 EFFLUENT CHARACTERISTICS

The characteristic raw effluent produced by this confectionery industry is characterized by its organic content, which is composed of easily biodegradable compounds such as sugars, sweeteners, casein, vegetable oils, acacia gums, condensed milk, food coloring and flavoring agents, etc. This confectionery industry is using major ingredients such as sugar and sweeteners, natural colors, acacia gum, sugar substitutes, gum base, and flavors for all type of confectionery products. They use number a chemicals during the process and they all become a part of the effluent. This liquid effluent is acidic in pH, dominated by yeast cell population. Table 4 shows the physic-chemical parameters of effluent from confectionery industry.

Table 4 Physico-Chemical Characteristics of Effluent from a Confectionery Industry Near Chennai, India

S. No.	Parameters	Raw Effluent
1	pH	6.5
2	Turbidity NTU	116.3
3	Total dissolved solids, mg/L	3528
4	Sodium (as Na), mg/L	925
5	Potassium (as K), mg/L	75
6	Electrical conductivity, μmho/cm	5606
7	Alkalinity total (as $CaCO_3$), mg/L	1859
8	Nitrite (as NO_2), mg/L	Nil
9	Nitrate (as NO_3), mg/L	29
10	Phosphate (as PO_4), mg/L	5.78
11	Sulfate (as SO_4), mg/L	135
12	Tidy's test (as O), mg/L	781
13	Oil and grease, mg/L	0.0094

9 ALGAL TREATMENT

Chlorella vulgaris grows very well in the raw effluent. It utilizes sugar present in the effluent. The effluent becomes less turbid and this reduces the load for the AR (Table 5). Table 6 shows growth of *Chlorella vulgaris* in pilot tank. Phycoremediation helps to correct pH, reduce sugar levels, reduces BOD and COD. Effluent is added and removed from the algal treatment tank at 3500 L/h which is the flow rate requirement to anaerobic digesters. Algal biomass reaches to 1.5 g dry weight/L.

9.1 BIOCHEMICAL ANALYSIS OF *CHLORELLA VULGARIS* GROWN IN CONFECTIONERY INDUSTRY EFFLUENT

Figure 2 shows the results of biochemical analysis of *Chlorella vulgaris* grown in confectionery industry effluent compared to control. There is considerable increase in proteins and total lipids when grown in effluent. A preliminary trial conducted using effluent from a confectionery industry in Trivandrum, India showed rich growth of algae in the raw effluent (Figures 6 and 7). *Chlorococcum humicola* grew luxuriantly reaching a cell number of $3160 \times 10^4 \, mL^{-1}$ (4.8 g dry weight/L).

Table 5 Uptake of Sugar from Confectionery Effluent by *Chlorella vulgaris*

Organism Name and Treatment Method	Total Chlorophyll	Total Carbohydrate	Total Protein	Total Lipid
Before treatment (raw effluent), mg/L	–	13,918	15,918	18,972
After treatment (supernatant) (treated with *Chlorella vulgaris* VIAT027), mg/L	–	9896	10,961	10,987
After treatment (micro algal pellet), μg/10^6 cells	0.0068	0.4323	0.6001	0.6923

Table 6 Growth of *Chlorella vulgaris* (Circulation Flow Rate 900 L/h)

Day	Initial Dip (cm)	Addition of Raw Effluent in cm (Whole Day)	Evaporation (cm)	Final Dip (cm)	Cell Count×10⁴ cells/mL *Chlorella vulgaris* VIAT027		Cell Count×10⁴ cells/mL Yeast Cells	
					Initial	Final	Initial	Final
1	20.0	5.0	1	24	230	262	35	5
2	17.5	7.5	0.5	24.5	262	273	56	21
3	12.5	12.5	1	24	273	285	120	15
4	7.5	17.5	1	24	285	305	131	19
5	5.0	20.0	1	24	305	320	152	10
6	1.0	24.0	0.7	24.3	320	338	160	2

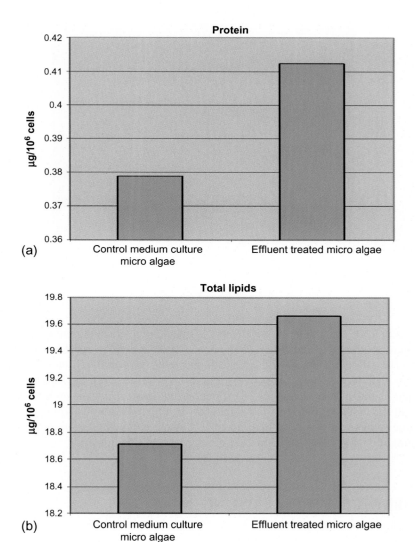

FIGURE 6

Biochemical characteristics of *Chlorella vulgaris* grown in effluent from a confectionery industry near Chennai, India.

10 EFFLUENT FROM SOFT DRINK MANUFACTURING INDUSTRY

Soft drink manufacturing units in India use ground water for the production. The ground water is filtered and softened using chemicals and sent to R/O and nano filtration (N/F) for further TDS reduction. The effluent generated by the industry includes R/O reject, reject from N/F, utilities, cleaning, softener regeneration, bottle wash, and cleaning in process (CIP). In some units the treated

FIGURE 7

Growth of *Chlorococcum humicola* in confectionery effluent from Trivandrum, India (productivity = 3160×10^4 cell/mL; 68 g/L wet weight; 4.8 g/L dry weight).

effluent is sent to R/O for recycling. The effluent is treated with conventional chemical and physical methods (Figure 8). Removal of nutrients esp. nitrates and phosphates is the major problem faced by this industry. The effluent contains high inorganic nutrients (especially nitrates and phosphates) and low pH. Micro algal technology is effectively employed to remove nitrates and phosphates. Table 7 shows physico-chemical characteristics of effluent from a soft drink manufacturing industry. The effluent has all the essential nutrients for algal growth. Figure 9a and b shows nutrient removal efficiency of micro algae employed in this industry. Algae employed could remove nutrients (nitrate and phosphate) at a rapid rate well within the requirement of the industry.

11 EFFLUENT FROM TEXTILE DYEING INDUSTRIES

Phycoremediation technology is employed in a few textile dyeing industries by the author and his team. Two textile dyeing industries were selected for the present discussion (one near Chennai and the second one at Ahmedabad; Figures 10 and 11). In Chennai industry the effluent is generated from various sources like dye bath, mercerizer, wash water, desizing water, and printing. The industry at Tamil Nadu generates around 200 kL of effluent per day. The effluent is treated with conventional physical and chemical methods resulting in huge amount of sludge and water effluent. The dyeing industrial effluents is rich in various dyes and high pH and TDS, because of various chemicals being used, like sodium bicarbonate, sodium chloride, etc. The second industry generates around 84,000 m³ effluents every day. The effluent generated is highly alkaline and treated with conventional chemical methods. Dye removal, reduction of BOD and COD are the major problems associated with effluent treatment. This effluents contain various dyes and high pH, TDS, BOD, and COD because of various chemicals being used, like sodium bicarbonate, sodium chloride, etc.

1. Oil and grease gravity settling tank
2. Equalization tank
3. pH correction after secondary treatment
4. Primary Clarifier
5. Aeration 1
6. Aeration 2 – Shut down   (Algal sludge)
Final treated effluent in the pond

FIGURE 8

ETP plant at a soft drink industry (raw effluent as well as treated effluent are rich in nutrients suitable for algal biomass production).

Table 7 Physico-Chemical Characteristics of Effluent from a Soft Drink Manufacturing Industry in Ahmedabad, India—Treated with Micro Algae

Parameters	Soft Drink Industry Effluent		
	Initial	*Final*	*% Reduction*
Physical examination			
Turbidity NTU	17.8	40	–
Total solids, mg/L	3960	3621	8.6
Total dissolved solids (TDS), mg/L	3864	3518	8.9
Total suspended solids (TSS), mg/L	96	103	–
Electrical conductivity, μmho/cm	5496	4872	10.9
Chemical examination			
pH	7.21	8.78	–
Alkalinity pH (as $CaCO_3$), mg/L	0	36	–
Alkalinity total (as $CaCO_3$), mg/L	1296	1325	–
Total hardness (as $CaCO_3$), mg/L	1220	360	70.5
Calcium (as Ca), mg/L	320	81	74.7
Magnesium (as Mg), mg/L	101	36	64.4
Sodium (as Na), mg/L	690	690	0
Potassium (as K), mg/L	20	40	–
Iron (as Fe), mg/L	2.91	3	–
Manganese (as Mn), mg/L	Nil	Nil	–
Free ammonia (as NH_3), mg/L	21.28	32.8	–
Nitrite (as NO_2), mg/L	Nil	Nil	–
Nitrate (as NO_3), mg/L	21	8	61.9
Chloride (as Cl), mg/L	792	760	4.0
Flouride (as F), mg/L	1.08	1	–
Sulfate (as SO_4), mg/L	288	121	58
Phosphate (as PO_4), mg/L	8.05	3	62
Tidy's test (as O), mg/L	106	13.2	–
Silica (as SiO_2), mg/L	44.18	24	45.7
BOD, mg/L	360	50	86.1
COD, mg/L	998	141	85.9
Total Kjeldhal nitrogen, mg/L	22.4	50	–
Copper (as Cu), mg/L	0.00321	0.0003	–
Zinc (as Zn), mg/L	0.148	0.1	–
Chromium (as Cr), mg/L	0.00236	0.0001	–

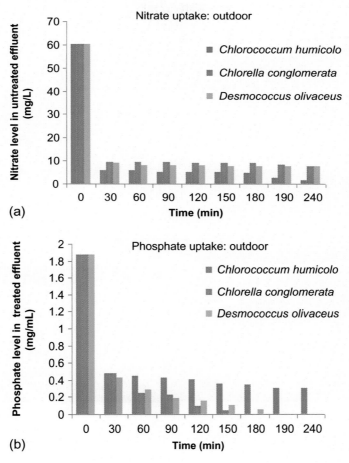

FIGURE 9

Nutrient removal from soft drink industry effluent using micro algae.

12 PHYCOREMEDIATION OF TEXTILE DYEING EFFLUENTS

Table 8 gives details of the characteristics of textile dyeing effluent from a factory near Chennai, India treated with micro algae in the lab as well as field. Effective color removal, pH correction, BOD and COD reduction and sludge reduction achieved by algal treatment. Effluent provides all the necessary nutrients for the algal growth. Cell number is maintained around $250 \times 10^4 \, mL^{-1}$ (0.75 g dry weight/L). Table 9 provides details of physic-chemical characteristics of textile dyeing effluent from Ahmedabad treated with micro algae. In the Chennai unit the algal biomass is harvested, dried and used as fuel. The dried algal pellet has a high calorific value.

FIGURE 10

Phycoremediation of textile dyeing effluent (near Chennai) and algal biomass production.

FIGURE 11

Algal growth in the raw effluent of textile dyeing industry at Ahmedabad, India. *Chlorococcum humicola*, *Chroococcus turgidus*, and *Desmococcus olivaceous* grow very well in the raw effluent. Color removal, pH correction, reduction of sludge, BOD and COD achieved without using chemicals.

guidelines for disposal in to lined pit. The oil drilling industrial effluents has higher inorganic chemicals from oil separation process which utilizes HCl, Triton X, sodium bicarbonate, etc. This effluent is alkaline and high in TDS. Phycoremediation technology is now being employed to treat waste water and mud with mineral oil (Figure 12). Excellent sludge reduction and reduction of mineral oil (<2%) have been achieved. The waste water and sludge support excellent growth of selected micro algae. Table 12 gives a comparison of analytical report on the physic-chemical characteristics of oil drilling effluent treated with micro algae in lab, pilot tank, and scaled up tank. There is a significant reduction in all major parameters and there was 70% reduction in sludge. The mud supports very

FIGURE 12

ETP and phycoremediation plant at oil drilling industrial treatment plant—Kakinada, Andhra Pradesh (effluent storage tanks, algal growth on sludge, phycoremediation plant, algal growth in effluent are shown).

Table 12 Phycoremediation of Oil Drilling Industrial Effluent: A Comparison of Lab, Pilot Plant, and Scaled Up Plant Trials

Parameters	Raw Effluent	Effluent Treated with Alga (Lab)	Effluent from Pilot Slope Tank Treated with Alga	Effluent from Scaled up Tank Treated with Alga	% Reduction (Lab)	% Reduction (Pilot Tank)	% Reduction (Scaled Up Tank)
Physical examination							
Turbidity NTU	56.3	23.2	42.5	7.3	32.55	24.51	87.03
Total solids, mg/L	16,564	15,316	14,544	3175	7.53	12.19	80.83
Total dissolved solids (TDS), mg/L	16,492	15,262	14,410	3168	7.45	12.62	80.79
Total suspended solids (TSS), mg/L	72	54	134	7	25	–	90.27
Electrical conductivity, μmho/cm	23,488	21,770	20,561	4501	7.31	12.46	80.83
Chemical examination							
pH	6.45	7.20	7.48	7.78	–	–	–
Alkalinity pH (as $CaCO_3$), mg/L	0	0	0	0	0	0	0
Alkalinity total (as $CaCO_3$), mg/L	390	198	115	310	49.23	70.51	20.51
Total hardness (as $CaCO_3$), mg/L	375	975	5850	875	–	–	–
Calcium (as Ca), mg/L	90	210	1300	200	–	–	–
Magnesium (as Mg), mg/L	36	108	624	90	–	–	–
Sodium (as Na), mg/L	525	1725	6900	810	–	–	–
Potassium (as K), mg/L	88	175	850	35	–	–	60.22
Iron (as Fe), mg/L	1.38	1.56	22.64	20	–	–	–
Manganese (as Mn), mg/L	Nil	Nil	Nil	Nil	Nil	Nil	Nil

Parameter							
Free ammonia (as NH₃), mg/L	2.24	3.24	10.08	22	–	–	–
Nitrite (as NO₂), mg/L	Nil	0.22	Nil	Nil	–	Nil	–
Nitrate (as NO₃), mg/L	41	9	48	38	78.04	–	7.31
Chloride (as Cl), mg/L	351	1224	6393	896	–	–	–
Sulfate (as SO₄), mg/L	215	200	9	235	6.97	95.81	–
Phosphate (as PO₄), mg/L	3.42	2.82	19.01	6.48	17.54	–	–
Tidy's test (as O), mg/L	445.2	43	328	90	90.34	26.32	79.78
Silica (as SiO₂), mg/L	13.53	32.47	49.38	23	–	–	–
BOD, mg/L	4710	180	1400	320	96.17	70.27	93.20
COD, mg/L	1050	502	3423	892	52.19	–	15.04
Total Kjeldhal nitrogen, mg/L	3.36	31.36	10.08	114	–	–	–
Copper (as Cu), mg/L	0.00812	0.00096	0.0080	0.00628	88.17	1.47	22.66
Zinc (as Zn), mg/L	0.147	0.090	0.143	0.114	38.77	2.72	22.44
Chromium (as Cr), mg/L	0.00485	0.00036	0.00485	0.00410	92.57	–	15.46

good growth of algae. Micro algal remediation technology is now employed to reduce mineral oil in the mud. Phycoremediation has improved the texture, organic content and water holding capacity of mud and it is now used to grow plant.

15 EFFLUENT FROM DETERGENT MANUFACTURING INDUSTRY

The detergent company at Ranipet, India, generates nearly 60-70 kL of effluent per day during production. Blue wash and Sulphonation plant are two different plants in the industry. Effluent is generated from both the plants. Blue wash plant works in a stepwise process. Each step leads to generate effluent as first wash, second wash, third wash, and fourth wash. The TDS will be very high in the first wash as compared to the fourth wash. The effluent generated from different washes is accumulated. The first blue wash has a TDS of 150,000 mg/L and the last wash will have the lowest TDS of 30,000 mg/L. The blue wash had very high TDS due to its high inorganic salt content. The pH of the blue wash is acidic in nature. Blue wash effluent was not taken for the study due to its very minimum output. The effluent generated from the sulphonation plant has maximum output. The effluent of sulphonation plant was taken for the phycoremediation study. The TDS of the effluent of sulphonation plant was in the range of 75,000-150,000. The pH of the sulphonation effluent was also acidic in nature. The high TDS content and acidic pH of the effluent has made it difficult for the industry in disposing it in the environment.

The company was dealing this problem in a conventional method. They treated the effluent with lime to increase its pH. The treated effluent was kept in solar evaporation ponds. Over a period of time the water of the effluent evaporates in to the atmosphere. The sludge was remaining in the ponds after evaporation. The sludge accumulated in the solar evaporating ponds were removed and stored. The sludge generated due to this method amassed to an extent of 500 tons per annum.

The TDS in the effluent were due to high organic and inorganic content. The chemical constituents like calcium, chloride, magnesium, nitrate, nitrite, potassium, sodium, and sulfate were responsible for the high level of TDS in the ultramarine effluent.

Phycoremediation technology is now employed in the detergent manufacturing industry to correct pH and to reduce sludge. Water is completely evaporated. There is more than 70% reduction in sludge achieved in the treatment process. The industry now avoids chemicals and treats the effluent with algae. Detergent industry effluent seems to support luxuriant growth of selected species of micro algae. Micro algae degrade sulfates and sulphites present in the effluent. Table 13 shows physic-chemical characteristics of effluent from the detergent manufacturing industry treated with algae. Figure 13 shows growth of *Chlorococcum humicola* in detergent effluent. The cell density reaches up to $450 \times 10^4 mL^{-1}$ (1.5 g dry weight/L) showing thereby detergent effluent can be a very good medium for growing algae.

16 EFFLUENT FROM ELECTROPLATING INDUSTRY

The electroplating industrial effluents are both acidic and alkaline. The acidic effluent contains heavy metals like chromium, nickel, cadmium, etc., and high TDS. Alkaline effluent is generated from coating of paint to wheels, this effluent contains calcium chloride and heavy metals.

Table 13 Physico-Chemical Characteristics of Effluent from a Detergent Industry Treated with Micro Algae (Pilot Plant Trials)

Parameters	Physico-Chemical Parameters of Raw Effluent	Physico-Chemical Parameters of Treated Effluent
Physical examination		
Appearance	Whitish	Whitish
Odor	Offensive smell	Offensive smell
Turbidity NTU	14.1	74.6
Total solids, mg/L	63,650	56,662
Total suspended solids, mg/L	140	2280
Total dissolved solids, mg/L	63,510	54,382
Electrical conductivity, μmho/cm	90,846	77,707
Chemical examination		
pH	3.8	6.7
Alkalinity pH (as $CaCO_3$), mg/L	0	0
Alkalinity total (as $CaCO_3$), mg/L	Nil	Nil
Total hardness (as $CaCO_3$), mg/L	5800	2100
Calcium (as Ca), mg/L	1360	560
Magnesium (as Mg), mg/L	576	168
Sodium (as Na), mg/L	18,500	16,100
Potassium (as K), mg/L	500	600
Iron (as Fe), mg/L	0.82	0.99
Manganese (as Mn), mg/L	Nil	Nil
Free ammonia (as NH_3), mg/L	0.37	16.28
Nitrite (as NO_2), mg/L	2.67	2.58
Nitrate (as NO_3), mg/L	11	12
Chloride (as Cl), mg/L	27,400	24,200
Flouride (as F), mg/L	0.97	0.84
Sulfate (as SO_4), mg/L	2007	2920
Phosphate (as PO_4), mg/L	3.38	73.84
Copper (as Cu), mg/L	0.00128	0.00634
Chromium (as Cr), mg/L	0.079	0.057
Zinc (as Zn), mg/L	0.304	1.843
Tidy's test (as O), mg/L	651	516
Silica (as SiO_2), mg/L	31.66	71.9
Total Kjeldahl nitrogen (as N), mg/L	11.2	34.72
BOD, mg/L	860	740
COD, mg/L	2894	2512
Oil and grease, mg/L	0.0064	0.0152

FIGURE 13

Showing growth of *Chlorococcum humicola* in detergent effluent treated using pilot tanks.

The sludge is rich in heavy metals especially chromium (Balaji et al., 2015; Gomes and Asaeda, 2009). Table 14 shows physic-chemical characteristics of effluent from electroplating industry treated with algae. Figure 14 shows algal growth in electroplating sludge. Chrome sludge from the electroplating industry supported very good algal growth (1.5 g dry weight/L). Using open raceway pond chrome sludge was treated with micro alga, *Desmococcus olivaceus*. There was a considerable amount of sludge reduction and biomass production in open raceway pond amended with chrome sludge. A remarkable reduction was found in TDS, sodium, potassium, and phosphate.

17 SEWAGE AND ALGAL BIOMASS PRODUCTION

Sewage is the most favorable medium for algae production (Kalaivani et al., 2009; Singh et al., 2012). It provides all the necessary nutrients needed for algal growth. New Zealand's Aquaflow Bionomic Corp. has become the World's first producer of biofuel from sewage-pond-grown algae (Ahmad et al., 2013; Azarpira et al., 2014). One particular advantage of the human-sewage approach is that algae from sewage tends to have a lot of oil according to Cary Bullock, CEO of Greenfuel Technologies, a company cultivating algae to convert emissions into biofuel.

Bacteria digestion of organics is the known method to reduce BOD, COD, TSS, TDS, etc. in Sewage. Anaerobic bacteria proliferate in untreated sewage giving rise to H_2S gas that produces the obnoxious smell. Aerobic bacteria require plenty of dissolved oxygen to do the organic digestion. Untreated sewage is let out into the nearest water bodies where anaerobic bacteria slowly consume

Table 14 Comparison of Parameters of Raw and Algae Treated Electroplating Industrial Chrome Sludge in Open Pond

Parameters	Raw Sludge	Algal Treated Sludge	% of Reduction
Physical examination			
Turbidity NTU	69.0	18.9	41.6
Total solids, mg/L	3022	1955	35.30
Total dissolved solids (TDS), mg/L	66	17	74.24
Total suspended solids (TSS), mg/L	2956	1938	34.43
Electrical conductivity, μmho/cm	4702	2745	41.42
Chemical examination			
pH	7.95	8.91	
Alkalinity pH (as $CaCO_3$), mg/L	8	16	
Alkalinity total (as $CaCO_3$), mg/L	1393	1027	26.27
Total hardness (as $CaCO_3$), mg/L	680	520	23.52
Calcium (as Ca), mg/L	176	198	
Magnesium (as Mg), mg/L	58	48	17.24
Sodium (as Na), mg/L	680	340	50
Potassium (as K), mg/L	50	20	60
Iron (as Fe), mg/L	6.48	4.70	27.46
Manganese (as Mn), mg/L	Nil	Nil	Nil
Free ammonia (as NH_3), mg/L	38.08	23.52	38.23
Nitrite (as NO_2), mg/L	0.84	0.36	41.14
Nitrate (as NO_3), mg/L	45	23	48.88
Chloride (as Cl), mg/L	351	176	49.85
Flouride (as F), mg/L	1.12	0.99	11.60
Sulfate (as SO_4), mg/L	186	129	30.64
Phosphate (as PO_4), mg/L	43.51	6.02	86.16
Tidy's test (as O), mg/L	67.2	58.4	13.09
Silica (as SiO_2), mg/L	40.12	47.68	15.85
BOD, mg/L	643	497	22.70
COD, mg/L	210	160	23.80
Total Kjeldhal nitrogen, mg/L	51.52	59.36	
Copper (as Cu), mg/L	0.01371	0.01256	08.38
Zinc (as Zn), mg/L	0.285	0.238	16.49
Chromium (as Cr), mg/L	0.024	0.016	33.33

the organics and produce H_2S that gives rise to bad smell. Further the waters become infested with water weeds and plants like hyacinth. These then become the breeding place of mosquitoes and disease producing organisms. Utilization of domestic sewage for algal biomass production has been studied extensively by employing a variety of micro algal species (Levoie and de la Noue, 1985; Oswald, 2003; Pushparaj et al., 1997; Kong et al., 2010).

FIGURE 14

Growth of *Desmococcus olivaceous* in chrome sludge from an electroplating industry near Chennai, India.

Sewage samples collected from the city of Chennai, India were screened and treated with micro algae. *Chlorococcum humicola*, *Chlorella conglomerata*, *Chroococcus turgidus,* and *Desmococcus olivaceus* were inoculated into sewage samples. Growth and biochemical composition of micro algae and physico- chemical parameters of sewage samples were monitored. Biodiesel potentials of algal biomass were also determined. *Chlorella* sp. grew very well followed by *Desmococcus* sp. Table 15 shows the physico-chemical characteristics of sewage treated with algae. Phycoremediation could reduce important parameters like BOD, COD, and other major nutrients and minerals to a great extent. Figure 15 shows the results of FAME analysis of algal biomass grown in sewage compared to control. FAME % increases by 50% when alga was grown in sewage.

Table 15 Physico-Chemical Characteristics of Sewage Treated with Micro Algae

Parameters	Raw	*Chlorococcum* sp.	*Chroococcus* sp.	*Chlorella* sp.	*Desmococcus* sp.
Physical examination					
Appearance	Blackish	Greenish	Greenish	Greenish	Greenish
Odor	Offensive odor	Algal smell	Algal smell	Algal smell	Algal smell
Turbidity NTU	14.5	22.9	38.1	53.7	35.1
Total solids, mg/L	3224	2174 (32.57)	3240	2914 (9.61)	2990 (7.26)
Total dissolved solids (TDS), mg/L	2784	1736 (37.64)	2572 (7.61)	2548 (8.48)	2630 (5.53)
Total suspended solids (TSS), mg/L	440	438 (0.45)	668	366 (16.81)	360 (18.18)
Electrical conductivity, µmho/cm	3974	2486 (37.44)	3673 (7.57)	3643 (8.33)	3982
Chemical examination					
pH	7.05	9.1	9.54	9.37	9.44
Alkalinity pH (as $CaCO_3$), mg/L	0	20	20	40	40
Alkalinity total (as $CaCO_3$), mg/L	255	159 (37.65)	195 (23.53)	160 (37.25)	147 (42.35)
Total hardness (as $CaCO_3$), mg/L	520	184 (64.61)	164 (68.46)	156 (70)	112 (78.46)
Calcium (as Ca), mg/L	120	42 (65)	37 (69.17)	34 (71.67)	26 (78.33)
Magnesium (as Mg), mg/L	53	19 (64.15)	17 (67.92)	17 (67.92)	11 (79.24)
Sodium (as Na), mg/L	580	420 (27.59)	550 (5.17)	540 (6.9)	550 (5.17)
Potassium (as K), mg/L	50	40 (20)	50 (0)	40 (20)	40 (20)
Iron (as Fe), mg/L	0.78	0.65 (16.67)	0.64 (17.95)	0.70 (10.26)	0.76 (2.56)
Manganese (as Mn), mg/L	0	0	0	0	0
Free ammonia (as NH_3), mg/L	1.69	1.05 (37.87)	1.24 (26.63)	0.83 (50.89)	1.29 (23.67)
Nitrite (as NO_2), mg/L	1.41	0.43 (69.5)	0.39 (72.34)	0.57 (59.57)	0.66 (53.19)
Nitrate (as NO_3), mg/L	14	11 (21.43)	12 (14.28)	12 (14.28)	13 (7.14)
Chloride (as Cl), mg/L	1061	680 (35.9)	1002 (5.56)	1020 (3.86)	1050 (1.04)
Sulfate (as SO_4), mg/L	90	72 (20)	86 (4.44)	88 (2.22)	89 (1.11)
Phosphate (as PO_4), mg/L	0.63	0.50 (20.63)	0.58 (7.94)	0.34 (46.03)	0.60 (4.76)
Tidy's test (as O), mg/L	236	59 (75)	38 (83.9)	36 (84.74)	37 (84.32)
Silica (as SiO_2), mg/L	28	27 (3.57)	25 (10.71)	23 (17.86)	28
BOD, mg/L	2612	190 (92.72)	260 (90.04)	140 (94.64)	120 (95.4)
COD, mg/L	5200	657 (87.36)	1680 (67.7)	555 (89.33)	384 (92.61)
Total Kjeldhal nitrogen, mg/L	37	13 (64.86)	9 (75.67)	8 (78.38)	7 (81.08)
	0.00092	0.00044 (52.17)	0.00018 (80.43)	0.00023 (75)	0.00011 (88.04)
Zinc (as Zn), mg/L	0.9	0.4 (55.55)	0.6 (33.33)	0.09 (90)	0.45 (50)
Chromium (as Cr), mg/L	0.00567	0.00448 (20.99)	0 (100)	0.00224 (60.49)	0.0034 (40.03)
Oil and grease, mg/L	0.0148	0.0044 (70.27)	0.0018 (87.84)	0.00122 (91.76)	0.00244 (83.51)

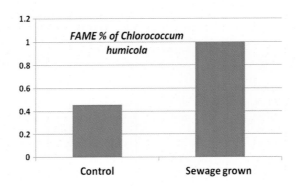

FAME % of micro alga grown in sewage compared to control.

18 EFFLUENT FROM A CHEMICAL INDUSTRY WHICH PRODUCES ORGANIC ACIDS

The main products of the chemical industry are petrochemicals and a range of food acids. Phthalic anhydride, maleic anhydride, succinic anhydride derivatives, and various downstream products, including food ingredients, etc., generates different streams of effluents, which are treated in various conventional ways to finally bring down the COD to the required level. The effluent produced is highly acidic (pH 0.6-1.0) and with very high TDS. For an initial screening acidic effluent (malic acid—1%; citraconic acid—2.3%; fumaric acid—1.5%; phthalic acid—1.7%; benzoic acid—0.8%; pH—less than 2) was added to the cultures daily (amount of addition was determined based on pH change). Feasibility study conducted in the laboratory gave encouraging results. Selected micro algae were tried in the pilot tanks. *Desmococcus olivaceous* and *Chlorococcum humicola* grew very well and selected for scaling up (Figure 16). Micro

FIGURE 16

Growth of micro algae in the acidic effluent from a chemical industry producing organic acids.

algae could correct the pH and completely degrade all the organic acids present in the effluent. There was a huge reduction in BOD, COD, and sludge. Algal biomass could be developed to a maximum of 0.75 g dry weight/L/day). Biochemical analysis done on the algal biomass grown in effluent revealed very high increase in proteins, carbohydrates, and lipids when compared to controls (Figures 17–20).

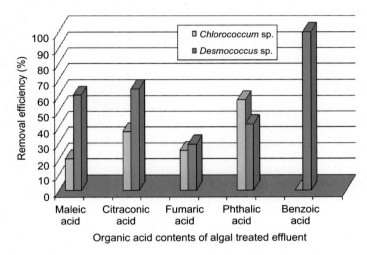

FIGURE 17

Organic acid removal efficiency of micro algae from the chemical industry effluent at the lab trial.

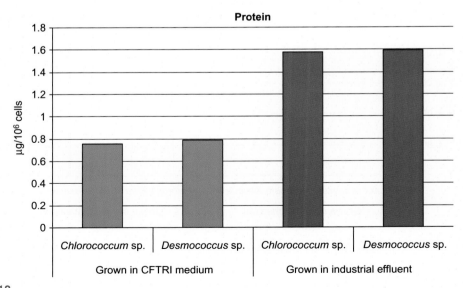

FIGURE 18

Protein levels in the micro algae grown in the chemical industry effluent.

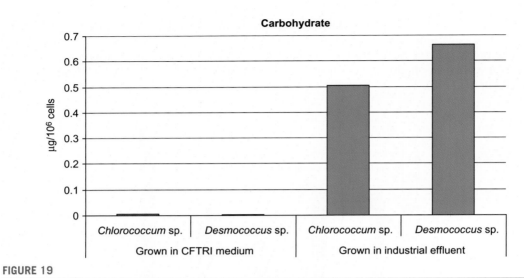

FIGURE 19

Carbohydrate levels in the micro algae grown in chemical industry effluent.

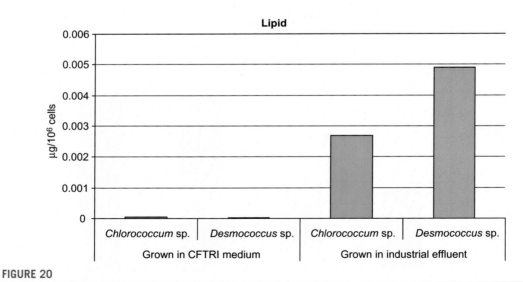

FIGURE 20

Lipid levels in the micro algae grown in chemical industry effluent.

Table 16 Advantages of Integration of Algal Biomass Production with Phycoremediation

Industry in India	Type of Effluent/ Sludge/Waste	Benefits of Phycoremediation				Algal Biomass Production Potential (kg dry biomass/kL/day)
		pH Correction	Reduction of BOD, COD	Sludge Reduction	Reduction in Operation Cost	
SNAP Natural and alginate products, Ranipet	Acidic effluent with high TDS	H	H	H	H	0.75
Leather processing chemical industry, Ranipet	Effluent and sludge	L	H	H	H	1.0
Confectionery industry 1—Chennai	Acidic effluent	H	H	H	H	1.5
Confectionery industry 2—Trivandrum	Acidic effluent	H	H	M	H	4.8
Soft drink industry, Ahmedabad	Neutral effluent	L	L	L	H	0.5
Textile dyeing industry, Chennai	Effluent and sludge	L	H	H	H	0.75
Oil drilling industry, Kakinada	Effluent and sludge	L	H	H	H	0.5
Detergent industry, Ranipet	Acidic effluent with high TDS	H	H	H	H	1.5
Electroplating industry, Chennai	Effluent and sludge	L	H	H	H	1.5
Chemical industry, Ranipet	Acidic effluent with high TDS	L	H	H	H	0.75

L: low; M: medium; H: high.

19 CONCLUSIONS

From the discussion so far regarding phycoremediation it is evident that most of the industrial effluents and sludge support very good growth of micro algae (Table 16). This is very significant information to be accommodated when the economics of mass cultivation of algae is taken up for discussion. The economic viability is based on the twin benefits of phycoremediation. On the one hand it handles the removal and degradation of most of the toxic chemicals and on the other it provides the same chemicals as nutrients for the growing algae. Thus we not only save on the nutrients but also see that the effluent is treated and that most of the toxic chemicals degraded and converted into valuable algal biomass. It is equally pertinent here to note that the algal strains capable of delivering this twin advantage have to be developed under careful lab and field studies and be kept in scrupulously maintained culture banks. Any presence of the residue of the degraded toxic chemicals from the effluent reflecting in the resulting algal biomass is negated in all our experiments through standard studies. However this should become a standard quality control procedure during large-scale productions.

Availability of the feed stock consistently throughout the year and the ensuing economic viability are the major hurdles in the biomass based technologies. Algal biomass production generally suffers due to expensive nutrient inputs, water scarcity, and non-availability of land. Most of the industries dealing with mass cultivation of micro algae depend on chemical fertilizers for cheaper nutrient inputs. Integrating algal biomass production with Phycoremediation seems to address most of these problems associated with mass cultivation. Apart from cleaning the environment from the onslaught of ever increasing pollution, valuable, and cheaper biomass is generated from wastewater, effluents, and sludge if appropriate algal species are identified and grown in industrial wastes.

REFERENCES

Ahmad, F., Khan, A.U., Yasar, A., 2013. Comparative phycoremediation of sewage water by various species of algae. Proc. Pakistan Acad. Sci. 50 (2), 131–139.

Azarpira, H., Dhumal, K., Pondhe, G., 2014. Application of phycoremediation technology in the treatment of sewage water to reduce pollution load. Adv. Environ. Biol. 8 (7), 2419–2423.

Balaji, S., Kalaivani, T., Rajasekaran, C., Siva, R., Shalini, M., Das, R., Madnokar, V., Dhamorikar, P., 2015. Bioremediation potential of *Arthrospira platensis* (*Spirulina*) against chromium (VI). Clean: Soil, Air, Water 43 (7), 1018–1024.

Chinnasamy, S., Bhatnagar, A., Hunt, R.W., Das, K.C., 2010. Micro algae cultivation in a wastewater dominated by carpet mill effluents for bio-fuel applications. Bioresour. Technol. 101, 3097–3105.

Dayananda, C., Sarada, R., Bhattacharya, S., Ravishankar, G.A., 2005. Effect of media and culture conditions on growth and hydrocarbon production by *Botryococcus braunii*. Process Biochem. 40, 3125–3131.

Dayananda, C., Sarada, R., Srinivas, P., Shamala, T.R., Ravishankar, G.A., 2006. Presence of methyl branched fatty acids and saturated hydrocarbons in botryococcene producing strain of *Botryococcus braunii*. Acta Physiol. Plant. 28, 251–256.

Dayananda, C., Sarada, R., Kumar, V., Ravishankar, G.A., 2007. Isolation and characterization of hydrocarbon producing green alga *Botryococcus braunii* from Indian freshwater bodies. Electron. J. Biotechnol. 10 (1).

de-Bashan, L.E., Hernandez, J.-P., Bashan, Y., 2012. The potential contribution of plant growth-promoting bacteria to reduce environmental degradation—a comprehensive evaluation. Appl. Soil Ecol. 61, 171–189.

Table 1 Comparison of Phycoremediation and Conventional Waste Treatment

Phycoremediation	Conventional Waste Treatment
Rely on microalgae and macroalgae	Rely on physical and chemical treatments
Utilize natural oxygen generated from photosynthesis for oxidation of pollutants	Utilize artificial oxidation process such as stirred tank reactors and activated sludge process
Wide range of waste substrate specificity (municipal/domestic wastewaters, industrial effluents, agricultural/food processing wastes, etc.), including waste gases	Substrate specific requiring different methods of treatment for different effluent types
Generally *in situ* process	Either *in situ* or *ex situ* process
Compatible with conventional processes	Highly specific and noncompatible
Single-step processes for pH correction, TDS reduction, BOD/COD removal, color or odor removal	Multistage process for each parameter
Negligible generation of sludge and treatment rejects	High sludge generation
Waste remediation can be coupled with biomass valorization	Sludge used as fertilizers or as landfills
Process can be linked with algae-based bioenergy production (biodiesel, bioethanol, and biogas)	Highly energy consuming
Lower capital and operational costs	High capital and maintenance costs
Slower process with long retention times	Faster process

utilized for production of specialty lipids like polyunsaturated fatty acids (PUFAs) (Becker, 2007). The hydrocarbon fractions can be used as biofuel additives, and the polysaccharides can be used for bioethanol production (Olguin, 2012). In addition to the extractable metabolites, the defatted or depigmented algal biomass (DAB) can be used as raw material for animal feed, biogas generation, or as low-cost adsorbent for removal of synthetic dyes (Sarat Chandra et al., 2014; Vidyashankar et al., 2014). Thus the algal biomass offer a complete range of valuable coproducts making the whole phycoremediation process economically viable. The potential bioproducts that can be obtained from microalgae are listed in Table 2.

Several reports are available on the use of microalgae in waste remediation focusing specifically on single type of wastes. Though plenty of information is available from past literature, a summary of essential processes and possible benefits in terms of sustainable algae biomass production and value addition to the traditional processes is lacking. Thus this chapter focuses on different ways of phycoremediation and biomass valorization. The chapter deals with utilization of microalgae for treating industrial and domestic effluents, heavy metal sequestration, xenobiotics degradation, flue gas treatment, and CO_2 mitigation with various methods of integrated algal cultivation for VAP thereof.

2 MICROALGAE FOR WWT AND XENOBIOTICS BREAKDOWN

Wastewater can be defined as polluted water that is nonpotable and generated from anthropogenic activities. There are two main types of wastewater: municipal or domestic generated from households and industrial generated from various industrial processes. The untreated domestic/sewage wastewaters have high concentrations of nutrients, mainly nitrogen and phosphorus, causing eutrophication

Table 2 Microalgal Biomass Applications

Value-Added Products	Algal Source	References
Whole biomass: food supplements, SCPs, protein source, animal feed	*Spirulina* sp., *Chlorella* sp., *Scenedesmus* sp. Seaweeds: *Palmaria palmata*, *Porphyra tenera*, *Ulva* sp., *Euchema*, etc.	Becker, (2004), Ravishankar et al. (2008), and Vidyashankar et al. (2014)
Lipids/fatty acids, PUFAs	*Spirulina* (gamma linolenic acid), *Porphyridium creuntum* (arachidonic acid), *Isochrysis* sp. *Pavlova* sp., *Nannochloropsis* sp. (eicosapentenoic acid), *Schizochytrium* sp., *Crypthecodinium* sp. (docosahexenoic acid)	Spolaore et al. (2006), Guschina and Harwood (2009), Doughman et al. (2007), and Kathiresan et al. (2007)
Natural colorants (carotenoids)	*Dunaliella* (beta carotene), *Haematococcus pluvialis* (astaxanthin), green microalgae (Lutein), *Spirulina* (phycobilin-C-phycocyanin)	Ben-Amotz et al. (1989), Sarada et al. (2006), and Rangarao et al. (2014)
Source of dietary fibers	*Chlorella* sp., *Undaria pinnatifida* (wakame), *Ulva sp.* (sea lettuce), *Porphyra tenera* (nori)	Burtin (2003) and Lahaye (2006)
Minerals	*Spirulina* (iron), brown seaweeds: *Fucus* sp. (iodine)	Becker (2004)
Vitamins	*Spirulina* (vitamin B12), *Ascophyllum*, *Fucus* (Vitamin E)	Kumudha et al. (2010)
Hydrocolloids: carragenan, agar, alginates, etc.	*Kappaphycus alvarezii*, *Chondrus crispus*, *Euchema* sp., *Gelidium* sp., *Gracilaria* sp., *Laminaria* sp., *Ascophyllum* sp., *Macrocystis* sp.	Santos (1989) and McHugh (2003)

of receiving water bodies. Eutrophication causes uncontrolled growth of certain aquatic plants and undesirable algal blooms that cause a shift in aquatic environmental equilibrium by negatively affecting the natural nutrient recycling process. The wastewater nitrogen occurs in different forms such as nitrates (NO_3^-), nitrites (NO_2^-), and ammonia nitrogen (NH_4-N) while phosphorus is generally present as inorganic orthophosphates (PO_4^{3-}). The ammonia N present in influent waters is toxic to most aquatic life, especially young fish, and causes depletion of dissolved oxygen in water as a result of subsequent nitrification reactions while nitrates pollutes groundwater and make it unsuitable for human consumption (Garcia et al., 2000). Nitrites act as precursors of N-nitroso compounds such as nitrosamines, which have teratogenic and carcinogenic properties. Excessive presence of nitrogen in waters interferes during chlorine disinfection process leading to generation of chlorine residues and causes methemoglobinemia (Abdel-Raouf et al., 2012). Similarly, excessive phosphate levels lead to excessive nutrient accumulation and are contributed mainly from human wastes in domestic sewage. The conventional processes of N and P removal are called nutrient stripping (Horan, 1990), in which various physical and chemical methods are employed for nutrient removal. For example, phosphorus removal is performed by chemical precipitation, such as addition of inorganic compounds of calcium, aluminum, and iron leading to precipitation of metal phosphates. These precipitation processes are performed prior to primary settling, during secondary treatment, or as part of a tertiary treatment process. Addition of inorganic salts efficiently removes inorganic P, however, leading to formation of additional sludge posing disposal problems (Hoffman, 1998).

Microalgae could be a potential agent for removal for N and P since these nutrients are recycled and converted into algae biomass in a cost-effective manner utilizing solar energy. The secondary treated wastewaters generally contain ammonia nitrogen in the concentration of 20-40 mg L^{-1} and orthophosphates in concentrations between 1 and 10 mg L^{-1}, adequate enough to support the microalgal growth (McGinn et al., 2011). Domestic wastewater generally harbors many planktonic algal species mainly belonging to cyanobacterial and chlorophyceaen members. Cyanobacteria dominate the planktonic community with an abundance of >90% (Furtado et al., 2009). Among the cyanobacterial community, strains belonging to *Phormidium* sp., *Plankothrix* sp., *Limnotrix* sp., and *Synechocystis* sp., are generally dominant (Martin et al., 2010; Gupta et al., 2013). However, *Oscillatoria* sp. and *Lyngbya* sp. were predominant in aerobic treatment tanks of the agrofood industry (Vasconcelos and Pereira, 2001). Among the chlorphycean members, *Chlorella* sp., *Scenedesmsus* sp., *Monoraphidium* sp., and *Chlorococcum* sp. are commonly occurring. Among these species, *Chlorella* (*C. vulgaris* and *C. minutissima*) and *Scenedesmus* (*S. obliquus*) have been studied widely in treating sewage effluents. These two species show a very high nutrient removal efficiency, >80% for both N and P and almost 90% reduction in COD from secondary treated wastewater (Cho et al., 2011). In addition to chlorophycean members, several euglenophytes (*Euglena gracilis*, *Euglena viridis*) have been identified in wastewaters with higher organic loading. These euglenophytes derive nutrients from the environments using mixotropic mode of nutrition (Chanakya et al., 2012). Apart from domestic sewage waters, algal forms were identified to remove nutrients from agricultural wastewater and livestock waste slurries that are high in N and P. For example, *Botryococcus braunii* was reported to grow at higher growth rates in piggery wastewater comprising 800 mg L^{-1} nitrates and showed ~80% removal of nitrate-N (An et al., 2003).

In the case of industrial effluents, water-intensive industries like pulp and paper, agroindustries, tanneries, distilleries, and textile industries are the major contributors. These untreated effluents contain potentially harmful substances such as high levels of organic material (cellulose, hemicelluloses, starch, and carbohydrates), pathogenic microorganisms, and toxic heavy metals, etc. (Rawat et al., 2011). Industrial effluents are generally characterized by high BOD and COD (Table 3). The chemical composition, BOD, and COD of wastewaters vary from different industries. Agricultural and food processing wastewaters contain high BOD and suspended solids with variable pH depending on the type of products being processed. For example, confectionary wastewaters contain simple sugars while waters generated from meat processing contain body fluids, antibiotics, hormones, and other organic wastes. In the case of paper and pulp industries, waters are contaminated with high levels of suspended

Table 3 BOD and COD of Various Industrial Effluents

Effluent Source	COD (mg L^{-1})	BOD (mg L^{-1})
Wool-scouring effluent	45,000	17,500
Distillery	60,000	30,000
Dairy	1800	900
Tannery	13,000	1270
Textile	1360	660
Kraft mill	620	226
Oil-drilling effluent	1050	4710

Adapted from Doble and Kumar (2005).

solids and are abundant in salts of inorganic acids and alkali. In steel or iron industries the waters are contaminated with gasification products such as benzene, naphthalene, anthracene, cyanide, ammonia, phenols, cresols together with a range of more complex organic compounds known collectively as polycyclic aromatic hydrocarbons (PAH) (Doble and Kumar, 2005).

These complex organic compounds that are present in effluents are unnatural to the aquatic ecosystems and are broadly termed as xenobiotics. Xenobiotics include pesticides and industrial-process residues like pharmaceutically active compounds, organochlorides, dye effluents, plasticizers (phthalates, bisphenol), dioxins, PAH, etc. Xenobiotics are often persistent in the environment; they are recalcitrant to most degradation processes and are termed as persistent organic pollutants (POPs) (Rieger et al., 2002). These POPs disrupt the aquatic ecosystem and have negative physiological and acute toxicity effects such as endocrine disruption, neural toxicity, and carcinogenicity at very low concentration ($1\,\mu g\,mL^{-1}$). Conventional treatment of sewage and industrial effluents focus on removal of simple nutrients, mainly nitrogen and phosphorus, thus leaving out these organic compounds. The POPs are commonly treated using a variety of physicochemical methods such as ozonation, photochemical degradation, activated carbon absorption, membrane filtration, ion exchange, and electrokinetic coagulation (Robinson et al., 2001). These methods are often costly and are highly specific in their action requiring different processes for each type of pollutant. Biodegradation offers a cost-effective treatment option utilizing the photosynthetic and heterotrophic properties of microorganisms to degrade these complex organic molecules.

Microbial degradation is often a complex process involving different consortia of bacteria, fungi, and photosynthetic microalgae (both cyanobacteria and eukaryotic algae). These consortia exist as microbial mats constituted by cyanobacteria, diatoms, anoxygenic phototrophic bacteria, sulfate-reducing bacteria, etc. The microbial members are arranged in layers depending on the sequence of metabolic reactions determined by gradients of light and redox potentials (Subashchandrabose et al., 2011). The interactions between microalgae and bacteria can be either cooperative or competitive. For example, microalgae require bacterial symbionts for meeting the nutritional requirements such as vitamins like biotin, thiamine, cobalamine, and bacterial siderophores during iron-deficient conditions (Butler, 1998; Croft et al., 2006), while microalgal or cyanobacterial, extracellular, polymeric substances and their exudates such as mannitol, arabinose, and glycolate under hyperoxic and alkaline conditions and acetate, propionate, lactate, and ethanol as fermentation products serve as bacterial growth substrates (Abed et al., 2007). Various microalgae and bacterial consortia have been reported to utilize or degrade complex organic compounds such as phenols, nonchlorinated aliphatic, aromatic hydrocarbons, and petroleum hydrocarbons, etc., as sources of carbon (Table 4).

Apart from the algal bacterial consortia, several studies on microalgae individually involved in removal of xenobiotics have been reported. However, the toxicity and transformation of pollutants by cyanobacteria or microalgae may change depending on the species. In general, chlorophycean microalgae such as *Chlorella* sp., *Scenedesmus* sp., and *Selenastrum capricornutum* have been widely reported in breakdown of these POPs. These microalgae have been reported to remove the xenobiotics from the aqueous environment by either the process of bioaccumulation or biotransformation (Table 5).

The bioaccumulation involves active uptake of these organic compounds and their storage intracellular. For example, in a study with *Chlorella* sp. VT-1 strain, Scragg et al. (2003) reported removal of 2,4-dichlorophenol, a precursor used in the synthesis of the herbicide 2,4-dichlorophenoxyacetic acid with 6 days of incubation. A 2% decrease in aqueous concentration of 2,4-dichlorophenol was observed compared to initial levels, of which 1.5% was extracted from the cells; however, no breakdown

Table 4 Microalgae-Bacteria Consortia in Wastewater Remediation

Cyanobacterium/ Microalga/Microalgal Consortia	Bacterium/ Consortia	Pollutant and Initial Concentration	Removal Efficiency (%)/Rate of Removal/Residual Concentration	References
Synechocystis sp. PCC6803	*Pseudomonas* related strain GM41	Phenanthrene; 0.15 mmol L^{-1}	0.8 µg day^{-1}	Abed (2010)
Pseudoanabaena PP16	*Pseudomonas* sp. P1	Phenol; 1 mmol L^{-1}	95%	Kirkwood et al. (2006)
Consortia of *Phormidium* sp., *Oscillatoria* sp., *Chroococcus* sp.	*Burkholderia cepacia*	Diesel; 0.6 v/v	Hydrocarbon removal up to 99%	Chavan and Mukherjee (2010)
Chlorella sorokiniana 211/8k	*Ralstonia basilensis*	Sodium salicylate; 5 mmol L^{-1}	1 mmol L^{-1}day^{-1}	Guieysse et al. (2002)
Chlorella sorokiniana	*Comamonas* sp.	Acetonitrile; 1 gL^{-1}	0.44 g L^{-1}day^{-1}	Muñoz et al. (2005)
Chlorella sorokiniana IAM C-212	*Microbacterium* sp. CSSB-3	Propionate; 125 mg L^{-1}	100% removal	Imase et al. (2008)
Chlorella sorokiniana	Pseudomonas migulae	Phenanthrene; 200-500 mg L^{-1}	24.2 g m^{-3}h^{-1}	Muñoz et al. (2003a,b)
Scenedesmus obliquus GH2	Consortia of *Sphingomonas* sp., GY2B *B. cepacia* GS3C *Pseudomonas* GP3A *Pandoraea pnomenusa* GP3B	Straight-chain alkanes, alkylcycloalkanes, alkylbenzenes, naphthalene, fuorene and phenanthrene; Crude oil at 0.3% v/v	100% removal	Tang et al. (2010)
Consortia of *Chlorella* sp., *Scenedesmus obliquus, Stichococcus* sp., *Phormidium* sp.	Consortia of *Rhodococcus, Kibdelosporangium aridum*	Phenols; 0.48 mg L^{-1} Oil; 40 mg L^{-1}	85% removal 96% removal	Safonova et al. (2004)

Adapted from Subashchandrabose et al. (2011).

products were observed. It was hypothesized that *Chlorella* VT-1 act like higher plants and transform and store the chlorophenol within cells, rather than degrade it, while the biotransformation process involves active uptake of xenobiotics inside the cell and their degradation to simple molecules using the complex intracellular enzymatic systems.

Friesen-Pankartz et al. (2003) evaluated the removal of the herbicide atrazine and the insecticide lindane using the green algae *Selenastrum capricornutum*. Cultivation of the alga with both the pesticides for 11 days resulted in decreased aqueous presence and hypothesized that algal cells biosorb the pesticides to facilitate their degradation. Similarly Cáceres et al. (2008) reported breakdown of organophosphorus insecticide fenamiphos (ethyl 4-methylthio-m-tolyl isopropyl phosphoramidate) by *Chlorella* sp., during 4 days of treatment with a 99% biotransformation efficiency. The organophosphorous pesticide was

Table 5 Microalgae in Bioaccumulation and Biotransformation of Xenobiotics

Microalgae	Bioaccumulation	Biotransformation
Chlamydomonas sp.	Mirex	Lindane, naphthalene, phenol
Chlorella sp.	Toxaphene, methoxychlor	Lindane, chlordimeform
Chlorococcum sp.	Mirex	
Cylindrotheca sp.	DDT	
Dunaliella sp.	Mirex	DDT, naphthalene
Euglena gracilis	DDT, parathion	Phenol
Scenedesmus obliquus	DDT, parathion	Naphthalene sulfonic acid
Selenastrum capricornutum	Benzene, toluene, chlorobenzene, 1,2-dichlorobenzene, nitrobenzene naphthalene, 2,6-dinitrotoluene, phenanthrene, di-nbutylphthalate, pyrene	Benzo[a]pyrene

Adapted from Priyadarshini et al. (2011).

mainly converted to its primary oxidation product, fenamiphos sulfoxide, during the biotransformation process. Similarly, several cyanobacteria have been implicated in the metabolism of organophosphate insecticides such as monocrotophos, quinalphos, and methyl parathion (Megharaj et al., 1987, 1994). In the case of aromatic compounds such as naphthalene, the compound was transformed to its hydroxylated intermediate such as cis-naphthalene dihydrodiol, 1-naphthol, naphthalene 1,2-oxide by cyanobacterium *Oscillatoria* sp., and microalga *Agmenellum quadruplicatum* (Cerniglia et al., 1979, 1980).

Most of the algal-mediated degradation process involves biotransformation of these pollutants to their hydroxylated/oxidized intermediates. In most of these degradation processes, the critical step is the opening of the aromatic ring. The ring opening could be either by ortho- or metacleavage to form 1,2-dihydroxybenzoidal moiety. Dioxygenase enzymes were reported to be involved in opening of the aromatic ring. In the case of green alga *Selenastrum capricornutum*, a dioxygenase enzyme system was found to be involved in oxidation of benzo[α]pyrene to *cis*-dihydrodiols. The dioxygenase enzyme induces oxidation by hydroxylating the aromatic rings. The mechanism was found to be similar to that of bacterial degradation of PAH where enzymes such as cytochrome P-450 monooxygensaes and epoxide hydrolases were involved in biotransformation of PAH to *trans*-dihydrodiols (Schoeny et al., 1988; Warshawsky et al., 1995). Similarly, in *Ochromonas danica*, the breakdown of phenol was reported to be aided by a series of oxido-reductase enzymes such as catechol dioxygenase (1,2 dioxygenase or 2,3 dioxygenase), NAD^+-dependent dehydrogenase to an intermediate called 4-oxalocrotonate, which was finally metabolized to pyruvate (Semple and Cain, 1996). These initial biotransformations are extremely important in the overall degradation of pollutants in the environment as they change the physical and chemical nature of the pollutants. The biotransformation of pollutants leads to reduced K_{ow} (octanol water partition coefficient) value, which makes the organic molecule more soluble in aqueous phase and susceptible to attack by other microorganisms (Semple et al., 1999).

In addition to breakdown of xenobiotics, microalgae-bacteria consortia are involved in the uptake and reduction in nutrients such as ammonia N, phosphates, sulfates, organic carbon, COD, etc., and heavy metal absorption. For example, in a tannery effluent, *Chlorella* sp. (*C. sorokiniana* and *C. vulgaris*) and cyanobacteria such as *Spirulina* sp. were identified with bacterial species

Azospirillum brasilense in reduction of sulfate levels by up to 80% efficiency (Rose et al., 1998). Similarly, in piggery effluent microalgal strains such as *Chlorella sorokiniana* and *Euglena viridis* were identified along with activated sludge bacteria in reducing the total organic carbons, ammonia N, and phosphorus with efficiencies of 47-50%, 21%, and 50-55%, respectively (de Godos et al., 2010). The heavy metal absorption in oil-contaminated wastewaters by microalgal consortia consisting of *Chlorella* sp., *Scenedesmus obliquus*, *Stichococcus* sp., and *Phormidium* sp., with bacteria such as *Rhodococcus* sp. and *Kibdelosporangium aridum* was observed (Safonova et al., 2004). The consortium could remove metal ions such as copper, nickel, zinc, iron, and manganese with a removal efficiency of 62%, 62%, 90%, 64%, and 70%, respectively (Safonova et al., 2004). Likewise a consortium of microalga *C. sorokiniana* and *R. basilensis* was identified in removal of copper with an efficiency of 57.5% from pollutant water diluted with Bristol medium (Muñoz and Guieysse, 2006). Based on the literature survey it is evident that microalgae can be a potential biocontrol agent for degradation of POPs. However, precaution is necessary in application of microalgae forms during biodegradation process with respect to the toxicity of breakdown products. Detailed reports pertaining to toxicity and life cycle of the partially metabolized POPs are not available. Hence a concerted research is required in bioprospection of microalgal diversity for biodegradation applications with high POP removal efficiency, toxicity of breakdown products to aqueous environments, downstream application of microalgae biomass posttreatment, etc.

3 HEAVY METAL REMOVAL BY MICROALGAE

Heavy metals are another major pollutant in wastewaters that affect the aquatic ecosystem. Heavy metals' mainly transition elements (d-block group) (elements with atomic density $>5\,\mathrm{g\,cm^{-3}}$) are required at low concentrations for various physiological functions in living organisms, mainly as enzyme cofactors, and are commonly referred as micronutrients or trace elements. However, elevated levels of these heavy metals induce toxic effects on the metabolism (Kaplan, 2013). These heavy metals exist as different forms (speciation) such as free ions, vapors, salts, minerals or bound to organic compounds. The chemical forms of these heavy metals constantly change due to dynamic biotic and abiotic interactions and are quite stable and persistent in environment. The toxicity of the metals depends on the chemical form of speciation in which they exist. The equilibrium between these chemical forms in aquatic system are interchangeable and depend on both biotic and abiotic factors such as microorganism type, temperature, pH, and alkalinity (Hafeburg and Kohe, 2007). Although a metal can exist in multiple forms, the free ionic species are most toxic to aquatic biota compared to bound form. Further, these heavy metals are persistent and cannot be degraded.

The free ionic species of metals possess unpaired electrons and show multiple oxidation state. The variable valency and ability to undergo changes in oxidation state involving one electron cause generation of free radicals. These metallic free radicals react with intracellular oxygen, H_2O_2 and hydroxyl groups in macromolecules and initiate a series of redox reactions unnatural to the system (Mallick, 2004). Further, this hampers regular electron transport in chloroplast, blockage of functional groups in macromolecules, lipid peroxidation, and enzyme inactivation, etc. (Foyer, 1997; Mallick and Rai, 2002).

Microalgae respond to the slightest change in the concentrations of various pollutants including heavy metals and show significant changes in their metabolism. Hence, microalgae can be used as a biological indicator for determining these pollutants. However, algae have developed a variety of mechanisms to counter the metal toxicity. These include extracellular detoxification, reduced metal uptake, efflux, metal sequestration by peptides such as phyhelatins (Cobbett and Goldsbrough, 2002; Tripathi et al., 2006). In addition to these tolerance mechanisms, algae have generated various enzymatic and nonenzymatic antioxidant defense mechanisms to counter the oxidative stress induced by heavy metals. Enzymatic defense involves elevated expression of antioxidant enzymes such as superoxide dismutase, catalase, and ascorbate peroxidase, whereas nonenzymatic defense involves production and accumulation of intracellular antioxidants such as glutathione, tocopherols, ascorbate, carotenoids, and proline in response to heavy metal stress (Pinto et al., 2003; Tripathi and Gaur, 2004).

The heavy metal tolerance and sequestration by microalgae vary from different species and are mainly affected by the efficiency of the intracellular metal homeostasis mechanisms and environmental conditions such as ionic strength and pH of the water (Rangsayatron et al., 2002). Higher ionic strength of the media reduce heavy metal uptake due to the competition for the functional groups by nonmetallic ionic species (Chojnacka et al., 2005). On the contrary, increasing the pH improves heavy metal uptake since the negative sites on algal cells are available for metal binding (Crist et al., 1994). Several microalgae have been reported to tolerate multiple heavy metal species; for example, *Phormidium* sp. could successfully hyperaccumulate cadmium, zinc, lead, nickel, and copper (Shanaab et al., 2012; Rana et al., 2013) while *Scenedesmus* sp. could tolerate the metals like copper, nickel, cadmium, and zinc at lower concentrations (2-5 mg L^{-1}) and lead at relatively higher concentrations up to 30 mg L^{-1} (Shehata and Badr, 1980). *Spirulina platensis*, a blue-green alga, showed tolerance to cadmium up to 100 mg L^{-1} and exhibited removal of about 98.04 mg Cd g^{-1} biomass (Rangsayatron et al., 2002). The marine diatom *Phaeodactylum tricornutum* was found to possess high cadmium tolerance up to 22 mg L^{-1} (Torres et al., 1997). *Chlorella* sp. was reported to remove cadmium and nickel from growth medium under laboratory conditions with a removal efficiency ranging from 70% to 95% (Rehman and Shakoori, 2004). While *Dunaliella salina*, a marine green alga, exhibited highest removal efficiency toward zinc followed by cobalt and copper and least removal tendency toward cadmium (Shafik, 2008).

3.1 MECHANISM OF METAL SEQUESTRATION IN MICROALGAE

The metal homeostasis in microalgae is generally maintained by two phenomena: metal avoidance and metal tolerance. The phenomenon of metal avoidance occurs by the process of adsorption of metals onto cell surface. The extracellular components (mucilage) and cell wall polysaccharides, together known as extracellular polysaccharides (EPS), bind to these metals and prevent their entry into the cells. The EPS contain negatively charged functional groups like sulfydryl (SH$^-$), phosphate (PO$_4^{3-}$), carboxyl (COO$^-$), hydroxyl (OH$^-$), etc., that bind randomly (nonspecifically) to metal cations present in the aquatic environment (Pereira et al., 2011). This process of extracellular binding occurs rapidly and passively in both living and dead cells. Another mechanism generally observed in algae is by exudation of organic compounds to the surrounding environment, which binds to these metals.

In the case of tolerance to metals, the microalgal cells employ the phenomenon of active intracellular absorption of metals. This is a biphasic process, where the cells first bind randomly to the metal ions using the extracellular polymers followed by an active transport of metals inside by specific transmembrane transporter proteins. Once transported inside the cells, these metals are detoxified by binding to specific metal-binding proteins and transport of these metal protein complexes inside vacuoles for sequestration (Figure 1) (Kaplan, 2013).

Metal absorption and sequestration in microalgae begin with active uptake of metals via metal transporters. These metal transporters are an important component in the interaction of algae with their

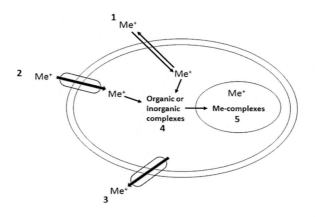

FIGURE 1

Schematic representation of various mechanisms of heavy metal tolerance in microalgae: (1) diffusion (inactive process), (2) active uptake via ion transporters, (3) active efflux of metal ions, (4) formation of complexes with free metal ions and organic/inorganic molecules, and (5) transport to vacuoles and sequestration.

Adapted from Kaplan 2013

environment and represent the first line of defense to cellular perturbations in metal concentrations (Blaaby-Haas and Merchant, 2012). A number of membrane transport protein families have been identified with metal homeostasis in eukaryotes and plants. These transporters can be broadly divided into two groups based on their function. Group A transporters are involved in the increase of intracellular content (cytoplasmic) of metals such as ZIP (zrt-, irt-like proteins), FTR (Fe transporter), NRAMP (natural resistance-associated macrophage proteins), and CTR (Cu transporter) families. These are assimilative transporters found in the plasma membrane and vacuole membranes. These transporters increase the intracellular concentration of metal when the equilibrium between chelating sites and metal ions is perturbed due to deficiency. Group B transporters are distributive in nature and decrease the cytoplasmic concentration of metal by providing metals to the organelle-localized metal-dependent proteins. Further, these transporters mediate exocytosis of excess metals when present in the membranes of the secretory pathway. Transporter protein families such as CDF (cation diffusion facilitator), P-type ATPases, FPN (FerroPortiN), and Ccc1 (Ca(II)-sensitive cross-complementer 1)/VIT1 (vacuolar iron transporter 1) belong to this group (Williams et al., 2000; Mäser et al., 2001; Cobbett and Goldsbrough, 2002; Hall and Williams, 2003). These transporter proteins are encoded by multigene families and were characterized using standard eukaryotic yeast model *Saccharomyces cerevisiae* and plant model *Arabidopsis* sp. Based on system biology approach and high throughput data mining from the genome sequences of *Chlamydomonas reinhardtii* (green alga) and *Cyanidioschyzon merole* (red alga), a total of 41 and 25 putative metal transporters in *C. reinhardtii* and *C. merolae*, respectively, were identified (Hanikenne et al., 2005; Castruita et al., 2011). The identification was performed by comparing algal genomes with human, yeast, and Arabidopsis CDF, ZIP, CAX, COPT, HMA, yellow-stripe 1-like (YSL), FTR, NRAMP, and iron-regulated 1 (IREG1)-like (sub)families, as well as multidrug resistance-associated protein (MRP) and ABC transporter of the mitochondria (ATM)/heavy metal tolerance (HMT) subfamilies of ABC transporters (Hanikenne et al., 2005). The list of metal transporters and their structural variations and function in *Chlamydomonas reinhardtii* and *Cyanidioschyzon merolae* is presented in Table 6.

Table 6 Metal Transporters Identified in *Chlamydomonas reinhardtii* and *Chlamydomonas merolae*

Metal Transporter Family	Function	Structure	Subcellular Location
Cation diffusion facilitators (CDF)	Metal homeostasis and tolerance Catalyze the efflux of transition metal cations, like Zn^{2+}, Cd^{2+}, Co^{2+}, Ni^{2+}, or Mn^{2+} from the cytoplasm to the outside of the cell or into subcellular compartments	Six transmembrane domains N and C termini predicted to be cytoplasmic His-rich loop region between transmembrane domains IV and V Signature sequence between transmembrane domains I and II Cation efflux domain comprising transmembrane domains I-VI	Vacuole membrane and secretory pathway
Zrt-, irt-like proteins (ZIP)	Metal homeostasis. Generally mediate the influx of metal cations, like Zn^{2+}, Fe^{2+}, Cd^{2+}, Mn^{2+} from outside the cell or from a subcellular compartment into the cytoplasm	Eight transmembrane domains N and C termini are extracytoplasmic ZIPs possess a long cytoplasmic loop (variable region) between transmembrane domains III and IV The variable region contains a probable metal-binding His-rich domain Transmembrane domains IV and V are amphipathic and believed to form a polar cavity required for the cation metal transport The loop between transmembrane domains II and III could be the site of initial binding of the substrate	Vacuole and plasma membrane
Cation exchangers (CAX)	Divalent cation/H1 antiporters Plays a role in transition metal homeostasis in algae	Contain 10-14 transmembrane domains	Vacuole membrane
Copper transporters (COPT)	Copper transport and homeostasis	Three transmembrane domains with an extracellular N terminus and a cytoplasmic C terminus Presence of two Met-rich regions in the N terminus that may act as a copper scavenger in the extracellular regions Additional Met-rich region in transmembrane domain II, probably involved in copper coordination during transmembrane transport	Plasma membrane and vacuole membrane

Table 6 Metal Transporters Identified in *Chlamydomonas reinhardtii* and *Chlamydomonas merolae*—cont'd

Metal Transporter Family	Function	Structure	Subcellular Location
Heavy metal P-type ATPases (HMA) HMAs CPx ATPases subfamily	P-type ATPases transport a broad range of small cations, and possibly phospholipids Phosphorylation-mediated metal transport Heavy metal transport Classified based on substrate specificity Monovalent—Cu^+/Ag^+ Divalent—$Zn^{2+}/Co^{2+}/Cd^{2+}/Pb^{2+}$ cations	Typically contain 8-12 transmembrane domains and a large cytoplasmic loop Cytoplasmic loop contain ATP binding and phosphorylation sites Eight transmembrane domains The sixth domain contains a conserved Cys-Pro-Cys/His/Ser motif (CPx motif) believed to be involved in metal cation translocation across the membrane	Chloroplast organelle membrane
ABC transporters MRP subfamily (Glutathione-S conjugate pumps) ATM/HMT subfamily	Involved in various physiological process Transport of bis(glutathionato) cadmium complexes from the cytoplasm into the vacuole Heavy metal transport, cadmium detoxification Transport of cadmium-phytochelatin complexes into vacuole	Two conserved nucleotide-binding folds responsible for ATP hydrolysis Alternating two highly hydrophobic domains (containing 4-6 transmembrane spans) that specify the substrates to be transported	Organelle membrane Mitochondria and vacuoles Mitochondrial membrane, export of cadmium from mitochondria
Natural-resistance-associated macrophage protein 1 (NRAMP)	Use the transmembrane proton gradient to facilitate transport of divalent cations particularly iron	NRAMPs possess 12 transmembrane domains Transmembrane domain 6 contains two conserved His residues A transport motif in the intracellular loop between transmembrane domains 8 and 9	Vacuole
Iron (ferrous) transporter family (FTR) Multi copper ferroxidase (MCO) subfamily	Copper-dependent ferrous transporters Soil acidification by H^+ ATPases to solubilize iron Reduction of ferric iron [Fe(III)] by plasma membrane ferric chelate reductases Uptake of ferrous iron [Fe(II)] by AtIRT1, a member of the ZIP family	Contain seven potential TMs REXXE motifs in TMs I and IV are proposed to be involved in iron translocation The extracellular loop between TMs 6 and 7 involved in trafficking of iron between the oxidase and the permease	Plasma membrane

Source: Hanikenne et al. (2005) and Castruita et al. (2011).

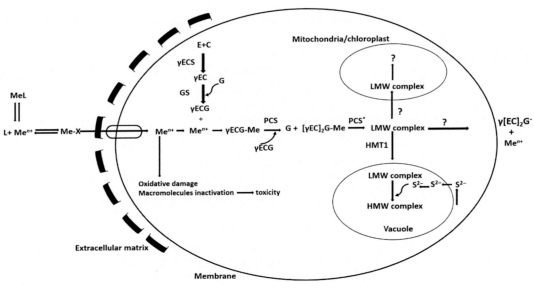

FIGURE 2

General scheme of heavy metal detoxification mechanism mediated by MtIII in microalgae.

Adapted from Perales-Vela et al. (2006).

3.2 METAL-BINDING PROTEINS AND SEQUESTRATION OF METALS

The metal ions that get transported into the cytoplasm are generally unchelated and are highly toxic, affecting the regular physiological process. However, microalgae have developed a variety of mechanisms to overcome the metal toxicity. Among the various detoxifying mechanisms, chelation of metals by macromolecules is primary. These metal-chelating macromolecules are proteins called phytochelatins or metallothioneins, which are either enzymatically synthesized (class III metallothioneins, MtIII) or gene encoded (class II metallothioneins). Phytochelatins (PCs, class III) are heat-stable, metal-binding polypeptides with a general structure $(\gamma\text{-Glu-Cys})_n$-Gly, where n varies between 2 and 11. The unique feature of these polypeptides are that the glutamic acid and cysteine residues are bonded by γ-carboxyl group and not the regular α-carboxyl group. Their molecular weight ranges from 2000 to 10,000 Da. Based on the number of Glu-Cys units, PCs have been classified as PC2, PC3, PC4, etc. PCs are synthesized on elevated metal concentrations in a wide variety of plants and algae. PCs are synthesized enzymatically by the addition of glutathione (GSH) to $(\gamma\text{-Glu-Cys})$n by the phytochelatin synthase to produce a $n+1$ oligomer. The activation of phytochelatin synthase occurs by the binding of metal ions and metal-glutathione (Perales-Vela et al., 2006) (Figure 2).

4 MICROALGAE-MEDIATED CO$_2$ MITIGATION AND FLUE GAS TREATMENT

Carbon is one of the key resources that influence the microalgae production as it constitutes about 35-60% of the dry biomass (Sydney et al., 2010). The major sources of carbon are from atmospheric CO_2, soluble carbonate salts (Wang et al., 2008; McGinn et al., 2011), and industrial flue gases. CO_2 sequestration is a natural phenomenon occurring in microalgae where the cells concentrate relatively low concentrations of atmospheric CO$_2$ up to three folds for efficient functioning of Rubisco (Ribulose 1,5, bisphosphate carboxylase/oxygenase) enzyme which catalyses the first step in carbon fixation cycle and also for maintenance of photosynthesis rates (Raven, 2010). The carbon concentrating mechanism (CCM) involves series of transmembrane and intracellular carbonic anhydrase enzymes that convert CO$_2$ into bicarbonate (HCO$_3^-$) followed by dehydration of HCO$_3^-$ ions to CO$_2$ near Rubisco sites. This evolutionary mechanism is developed to counter the low affinity of Rubisco to CO$_2$ compared to O$_2$ (Reinfelder, 2011). The ability of microalgae to sequester CO$_2$ varies from different algal groups and is mainly affected by the efficiency of CCM function. In general, coccolithophores have low-efficiency CCMs, and diatoms and the haptophyte genera have high-efficiency CCMs with a C4 mechanism of carbon fixation while dinoflagellates possess moderately efficient CCMs (Reinfelder, 2011).

Although microalgae have been reported to efficiently absorb CO$_2$ over terrestrial plants (Rittmann, 2008), their tolerance levels to CO$_2$ vary from different species. Most of the species show a negative response to CO$_2$ levels as low as 2-5% with hampered photosynthesis function (Solovchenko and Khozin-Goldberg, 2013). For example, the growth of certain species like *N. occulata* and *Chlorella* sp. were completely inhibited at CO$_2$ concentrations above 5% (Chiu et al., 2008, 2009). De Morais and Vieira Costa (2007) reported the CO$_2$ fixation rate of *Dunaliella tertiolecta*, *Botryococcus braunii*, *Spirulina platensis*, and *Chlorella vulgaris* to be high at 6% CO$_2$ while Tang et al. (2011) reported maximum biofixation efficiency at 10% CO$_2$ in *Chlorella pyrenoidosa* and *Scenedesmus obliquus*. Based on their tolerance, microalgae can be divided into CO$_2$-sensitive (tolerate <2-5%) and CO$_2$-tolerant groups (tolerate >10% and above). High CO$_2$-tolerating microalgae identified from various reports are listed in Table 7. High CO$_2$ tolerance is a complex process involving a variety of physiological adjustments such as state transition of photosystem to increased cyclic electron transport, higher ATP synthesis, higher H$^+$ ATPase activity to maintain cellular pH homeostasis and shutting down of CCM, etc. (Solovchenko and Khozin-Goldberg, 2013).

For industrial production of microalgal biomass, CO$_2$ from the atmosphere would not be sufficient as its concentration is very low (about 380 ppm). Compression of CO$_2$ or supply of soluble carbonates are highly unfeasible for bioremediation applications and increase the operation and biomass production costs (Molina Grima et al., 2003; Becker, 2008). To achieve an enhanced biomass productivity in a cost-effective manner, direct use of flue gas generated from industries seems to be a viable option (Hende et al., 2012). Flue gases are combustion products containing hundreds of different compounds such as oxides of nitrogen and sulfur, halogenated compounds, unburned PAHs, etc. However, nitrogen (72-77% v/v) and CO$_2$ (10-27% v/v) contribute maximum to the flue gas composition (Hende et al., 2012).

Table 7 CO$_2$ Tolerance of Microalgae

Microalgae	Maximum Tolerated CO$_2$ Levels (% v/v)	References
Cyanidium caldarium	100	Seckbach et al. (1970)
Scenedesmus sp.	80	Hanagata et al. (1992)
Chlorococcum littorale	60	Kodama et al. (1993)
Synechococcus elongatus	60	Miyairi (1995))
Euglena gracilis	45	Nakano et al. (1996)
Chlorella sp.	40	Hanagata et al. (1992)
Eudorina sp.	20	Hanagata et al. (1992)

Source: Raven et al. (2008) and Solovchenko and Khozin-Goldberg (2013).

The conventional process of treating flue gases involves use of chemical and physical processes such as catalytic reduction, adsorption, wet scrubbing, electrostatic precipitation, photochemical oxidation, and fungi-based bioreactors. However, most of these processes are energy intensive and involve high maintenance costs. The main advantage of microalgal cultivation combined with flue gas treatment option is that microalgae do not need a pure supply of CO_2 for their growth, and flue gas contains sufficient levels of CO_2 and other carbon compounds that may promote growth. Although a conceivable option, utilization of flue gas is not easier as flue gas carries higher temperatures and needs a certain degree of pretreatment prior to feeding algal cultures. The presence of toxic compounds such as sulfur and nitrogen oxides may affect the algal growth and induce deleterious effects such as pH shock or oxidative stress that hamper the cell growth (Hende et al., 2012). Hence thorough understanding of the flue gas composition and tolerance levels of algae is required. Table 8 lists some of the compounds present in flue gas generally and their effects on microalgae.

Most of the research on flue gas/CO_2 mitigation has been focused on species selection, optimization of reactors, fundamental research on CCM, and tracking of biochemical changes in biomass under flue gas/CO_2 supplementation (Jansson and Northen, 2010). CO_2 bioremediation by microalgae depends on bioreactor geometry and mass flow, input CO_2 concentrations, cell density, light intensity, and temperature (Raeesossadati et al., 2014). However, for direct mitigation of flue gas two properties are

Table 8 Effect of Various Flue Gas Components on Microalgae

Flue Gas Components	Positive Effects	Negative Effects
N_2	Nitrogen source for cyanobacteria	–
CO_2	Carbon source for microalgal growth	Excess concentration affects growth medium pH leading to CO_2 injury
	Increases CO_2:O_2 ratio in environment and consequently increased affinity of Rubisco to CO_2	Inhibits photosystem II at higher concentrations
	Counteracts pH increase due to microalgal growth	
	Changes the biochemical composition of microalgal biomass—positive influence on lipid and polysaccharide composition	
O_2	Electron acceptor during photorespiration	Excess concentration may induce oxidative stress
		Competes for Rubisco
NO_x	Source of nitrogen	Toxic at higher concentration
SO_x	Source of sulfur when converted to sulfate	Induces pH reduction
		Induces bisulfite toxicity
CO	Source of carbon	–
C_xH_y	Source of carbon	–
Halogen acids	–	Reduces culture pH
Heavy metals	Source of trace elements	Induces oxidative stress at high concentration
	Co-factors for enzymes; e.g., Zn for carbonic anhydrase	

Adapted from Hende et al. (2012).

essential—thermotolerance and high CO_2 tolerance—since raw flue gases are hotter as they are combustion products and carry high CO_2 concentrations (10-27%) as mentioned earlier. Most of the microalgal species are mesophilic in nature, having a temperature optimum between 20 and 30 °C that restricts their ability to utilize flue gas. This necessitates bioprospection of thermotolerant strains, which are common in hot water springs and water bodies surrounding geothermal power stations. The ability of microalgae to withstand high temperatures is a promising aspect to increase efficiency of biomass cultivation, particularly by reducing contamination risk and lowering operation costs of bioreactors used for outdoor cultivation at warm climates by eliminating the need for heat exchangers, water cooling systems, or enclosed greenhouses (Onay et al., 2014). Further, adaptation of these thermophilic strains to higher CO_2 concentration would be beneficial in treating raw flue gas and simultaneous production of thermostable proteins and enzymes for wider biotechnological applications.

Screening studies have revealed the presence of certain thermophilic cyanobacteria such as *Thermosynechococcus* sp. (*T. elongates* and *T. vulcanus*) and *Chlorogleopsis* sp., which have a temperature optima of around 55-60 °C (Iwai et al., 2004). Pan et al. (2011) reported a high-lipid-accumulating, thermotolerant, green microalga, *Desmodesmus* sp., which survived a temperature of 45 °C continuously over a time of 24 h. Similarly, Hu et al. (2013) identified a thermotolerant green alga *Coelastrella* sp. (F50) that sustained temperatures up to 50 °C and could accumulate pigments such as astaxanthin, lutein, canthaxanthin, and β-carotene under stress conditions. The CO_2 fixation potential and thermal tolerance of a few microalgal strains are listed in Table 9. Thus, employing a thermal and CO_2-tolerant microalgae would be beneficial in flue gas mitigation and for industrial production of algal metabolites in tropical areas.

5 INTEGRATED WWT AND ALGAL CULTIVATION

The most commonly used method for WWT using microalgae is by their cultivation in high rate algal ponds (HRAP). HRAPs are shallow, raceway-type ponds and have depths of 0.2-1 m. The water is mixed by a paddlewheel to give a mean horizontal water velocity of ~0.15-0.3 m s^{-1} (Craggs, 2005). The raceway configuration may be as a single loop or multiple loops around central dividing walls called baffles (Figure 3). The pond bottom may be either lined or unlined depending on soil conditions. HRAPs fall under the category of advanced pond system of WWT, which incorporates anaerobic digestion pits, HRAPs, algal settling ponds, and maturation ponds in series (Craggs, 2005). The capital and operational costs of advanced pond systems are significantly less (one-half to one-fifth) compared to activated sludge systems. HRAPs offer a cheaper option for production of algal biomass. Commercial algal biomass production requires freshwater, nutrients, and carbon dioxide, which contribute to 10-30% of total production costs (Borowitzka, 2005; Benemann, 2008; Tampier, 2009; Clarens et al., 2010). In HRAP-associated process, the costs of algal biomass production and harvest are essentially covered by the WWT plant capital and operation costs. Further, HRAPs have significantly less environmental impact in terms of water footprint, energy, and fertilizer use and the resulting spent growth medium can be discharged to water streams (Park et al., 2011a,b).

The removal of nutrients in HRAPs is achieved by growth of algae in the mixed liquor of primarily treated waters after sludge removal. As mentioned in the earlier section, microalgae exist in symbiotic relation with bacteria and fungi during the nutrient removal process in HRAPs. The role of algae can be direct or indirect in nutrient removal. Direct nutrient removal involves absorption of heavy metals,

Table 9 CO$_2$ Fixation and Temperature Tolerance of Microalgae

Microalgae	Inlet CO$_2$ Concentration (% v/v)	Temperature (°C)	Biomass Production System	CO$_2$ Fixation Rate (mg L^{-1} day^{-1})	References
Chlorococcum littorale	40	30	Fermenter (PBR)	1.0	Iwasaki et al. (1998)
Chlorella kessleri	18	30	Conical flask-based PBR	0.163	De Morais and Vieira Costa (2007)
Chlorella vulgaris LEB 104	5	25	Fermenter (PBR)	0.251	Sydney et al. (2010)
Chlorella vulgaris CCAP 211/11B	12	26	Bubble column PBR	0.86	Hulatt et al. (2012)
Scenedesmmus obliquus	18	30	Bubble column PBR	0.26	De Morais and Vieira Costa (2007)
Spirulina sp.	12	30	Bubble column PBR	0.413	De Morais and Vieira Costa (2007)
Spirulina platensis LEB 52	5	25	Fermenter (PBR)	0.318	Sydney et al. (2010)
Haematococcus pluvialis	16-34	20	PBR	0.143	Huntley and Redalje (2007)
Dunaliella sp.	3	27	PBR	0.313	Kishimoto et al (1994)
Dunaliella tertiolecta SAG 13.86	5	25	Fermenter (PBR)	0.272	Sydney et al. (2010)
Dunaliella tertiolecta SAG 13.86	12	26	Bubble column PBR	1.2	Hulatt et al. (2012)
Scenedesmus dimorphus CFR-01-1/FW	15	25	Low-density polyethylene-based PBR		Vidyashankar et al (2013)
Botryococcus braunii SAG 30.81	5	25	Fermenter (PBR)	0.496	Sydney et al. (2010)
Thermosynechococcus sp. CL-1	10	50	Bubble column PBR	0.141	Hsueh et al. (2009)
Chlorogleopsis sp. (SC2)	5	50	PBR	0.204	Ono and Cuello (2007)

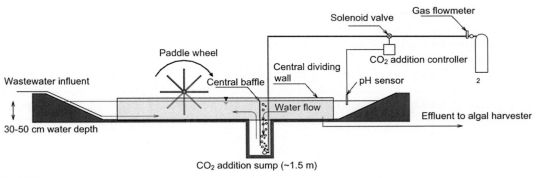

FIGURE 3

Schematic representation of HRAP.

Adapted from Park et al. (2011a,b).

nitrate uptake, CO_2 uptake, etc. Indirect removal occurs during algal photosynthesis where the rise in liquor pH (due to photosynthesis) results in ammonia-N stripping and orthophosphate precipitation, causing the nutrient removal from wastewater. Thus, nutrient removal efficiency in an HRAP is controlled by the parameters that determine algae growth and activity, such as cellular retention time (also known as hydraulic retention time, or HRT), solar radiation, and temperature (Garcia et al., 2000).

HRT is the amount of time spent by the influent wastewater in HRAPs during the nutrient removal process. HRT plays an important role in determining the efficiency of the HRAP-based nutrient removal process as it determines certain essential parameters like microalgae distribution (species domination) and algal/bacterial ratio during the treatment process (Mehrabadi et al., 2014). A shorter HRT is ideal for increased algal biomass proportion since the algae proliferate under high nutrient concentration. Once the nutrient levels decrease, the algal growth is slowed as the cells reach stationary phase while bacterial population becomes dominant, resulting in decreased pond efficiency. Park and Craggs (2011) reported significant increase in algal population with over 80% distribution when cultivated at shorter HRT (2 days) compared to longer HRT periods of 8 days with only 56% algal distribution. Further to change in the algal/bacterial ratio, HRT affects the algal population dynamics. The microalgal diversity in HRAPs is essential in nutrient removal process. Microscopic analysis indicated that the chlorophyceae class, including *Chlorella* sp., *Scenedesmus* sp., and *Stigeoclonium* sp., were the dominant microalgae (Garcia et al., 2000; Kim et al., 2014). Microalgae for WWT by HRAP systems must possess attributes such as high growth rate (productivity), high tolerance to nitrogen and phosphorus, tolerance to seasonal variations and diurnal cycles when cultivated in open ponds, easily harvestable (colonial morphology), and accumulation of high-value metabolites for downstream applications such as lipids, pigments, or carbohydrates. Strains that can reduce or absorb maximum nutrients from wastewater in a short time are ideal for WWT (Park et al., 2011a,b).

The other important parameter that affects algal growth in HRAPs is the rate of mixing. Mechanical mixing of HRAPs results in increased cell light exposure promoting algal photosynthesis and decreased HRT. As mentioned earlier, increased algal photosynthesis leads to nutrient stripping resulting in reduction in influent wastewater N and P levels. Further, mixing results in reduction of algal cell boundary layer and reduction in the thermal stratification in the pond (Mehrabadi et al., 2014).

Unmixed ponds result in generation of anaerobic conditions at the bottom layers causing decreased nutrient availability to algal cells, thereby reducing the nutrient removal efficiency. In addition, higher photosynthetic rates during mixing can result in excessive dissolved oxygen accumulation suppressing algal growth.

The third and important nutrient factor affecting HRAP performance is CO_2 availability. It was observed that the nutrient removal efficiency of HRAPs improves with CO_2 addition. The addition of CO_2 increases the C:N ratio of the pond from 2.5 to 4:1 to 6:1, significantly improving the algal biomass productivity by 30% (Park and Craggs, 2011; Park et al., 2013). Further, they reported improvement of N availability to algal cells during CO_2 supplementation. The shift in the culture pH to acidic environment during CO_2 supplementation resulted in conversion of free ammonia to ammonium ions, making N more bioavailable to algal cells. Apart from the benefits offered to the nutrient removal process, CO_2 supplementation affects the biomass composition significantly. CO_2 supplementation resulted in 30% increase in algal cellular fatty acid content and overall energy content of the algal biomass (Muradyan et al., 2004). Further, several authors have reported an increase in cellular pigment composition such as chlorophylls and carotenoids that have potential industrial applications with CO_2 supplementation (Dayananda et al., 2006; Vidyashankar et al., 2013). The CO_2 supply to HRAPs can be obtained from industrial flue gases as described in the earlier sections. However, care must be taken in terms of the gas temperatures and its composition, which are known to significantly affect algal growth and HRAP and algal performance (Hende et al., 2012). Hence identification of thermotolerant strains becomes necessary during such integrated cultivation processes.

The major disadvantage with HRAPs is the presence of zooplanktons (herbivorous grazers) in wastewaters generally belonging to the classes of cladocerans and rotifers (Park et al., 2011a,b). The commonly contaminating grazers are *Paraphysomonas imperferata*, a nonspecific heterotrophic flagellate, and *Euplotes* sp., a ciliate that grazes on algal cells and causes aggregation of cells (Zmora and Richmond, 2004). These zooplanktons have the potential to reduce the algal concentrations by 90% within 2 days. Cauchie et al. (1995) reported a 99% reduction in microalgal chlorophyll content due to the presence of *Daphnia* grazing. In addition to the grazer attacks, the HRAPs can be contaminated by fungal or viral infection. The reduction in algal population can be controlled by maintaining selective population of algae in the HRAP. Park et al. (2011a,b) reported that selective recycling of colonial microalgae such as *Pediastrum* sp., *Dictyosphaerium* sp., and *Coelastrum* sp. could control grazer population since the colonial cells are larger for grazer uptake. Further, the presence of colonial microalgae helps in easy harvesting of algae during the downstream processing.

6 VALUE ADDITION TO WWT: ALGAL BIOMASS APPLICATIONS

The main advantage of microalgae-mediated WWT is their photosynthetic efficiency and higher biomass productivity ranging between 12 and $40\,g\,m^{-2}day^{-1}$ (Park et al., 2011a,b). In addition to offering a cheaper option of WWT, microalgal biomass derived from the WWT process could be utilized for a range of applications such as biofertilizers, source of lipids and biodiesel, feedstock for biogas production, etc. The various biomass application options from microalgae are represented in Figure 4. The algal biomass can have a range of important industrial applications such as source of nutraceuticals that can be extracted from the biomass, bioenergy precursors such as lipids and hydrocarbons, nutrient-rich biofertilizers, use as animal feed, etc.

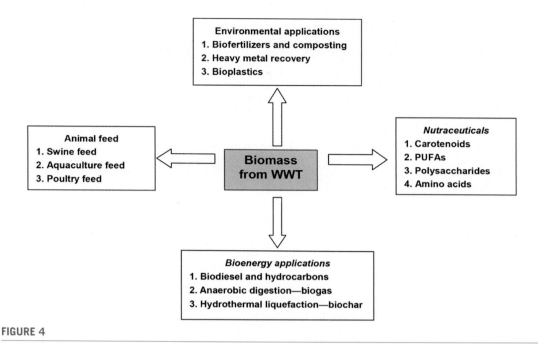

FIGURE 4

Various applications of microalgal biomass obtained from wastewater treatment process.

6.1 ALGAL BIOMASS AS SOURCE OF NUTRACEUTICALS AND ANIMAL FEED

Microalgae possess an immense chemical diversity that can be exploited for nutritional and therapeutic applications. Some of the essential metabolites from microalgae such as carotenoids, phycobiliproteins, and PUFAs have high nutraceutical value fetching tremendous commercial benefits. The safety of algal-derived metabolites has been evaluated and commercially approved in several countries. Many reviews detailing the potential uses of microalgae as food sources and nutraceuticals are summarized in Table 10.

Another application for microalgae is as an alternative source of protein in livestock feed industry. Conventionally soybean meal, groundnut cake, rice bran meal, etc., are used as protein source. However, cultivation of these crops for livestock feed competes with food production both in acreage and water footprint (Vidyashankar et al., 2014). With global increase in consumption of animal products, identification of sustainable feed sources that can support growing demand for meat and the animal industry becomes important. Microalgal biomass obtained from WWT can be targeted as a source of protein in animal feed, reducing the burden on food crops. The protein content of dry microalgal biomass ranges between 20% and 30%; however, *Spirulina platensis* contain proteins up to 60%. The algal proteins are rich in essential amino acids and score equally in terms of essential amino acid index compared to soy protein and casein (Vidyashankar et al., 2014). Pioneering studies on use of wastewater-grown alga as animal feed were performed by Hintz and Heitmann (1967). The study involved use of sewage-grown algae *Chlorella* and *Scenedesmus* as a replacement for soybean and cotton seed meal. They reported that incorporation of algal biomass at 2.5%, 5%, and 10% levels substitution during a 7-week feeding

Table 10 List of High-Value Nutracueticals Produced from Algal Biomass

Nutraceutical Product	Organism	Uses	Company	Country
Phycobiliproteins	Cyanobacteria, red algae	Anti-oxidant, natural food colorant, diagnostic fluorescence dye	Cyanotech Inc.	United States
C-phycocyanin	*Spirulina platensis*		Parry Nutraceuticals	India
R-phycoerythrin	*Porphyridium cruentum*		Dainippon Ink and Chemicals	Japan
Beta carotene	*Dunaliella* sp.	Antioxidant and natural food colorant	Cognis Nutrition & Health	Australia
			Parry Nutraceuticals	India
Astaxanthin	*Haematococcus pluvialis*	Antioxidant, natural food colorant, feed supplement	Valensa International	United States
			Algaetech International	Malaysia
Polyunsaturated fatty acids	*Spirulina platensis*	Nutritional and therapeutic applications	(*Spirulina* biomass only)	India
Omega-6 fatty acid				
Gamma linolenic acid (GLA)			Parry Nutraceuticals	
Arachidonic acid (AA)	*Parietochloris incise*		Avantha Holdings Pvt Ltd	India
Docosahexaenoic acid (DHA)	*Crypthecodinium, Thraustochytrium, Isochrysis, Pavlova,* and *Schizochytrium*	Nutritional and therapeutic applications	Martek Biosciences	United States
			Source Omega LLC	United States
Omega-3 fatty acid			Aurora Algae	United States
			Lonza	Switzerland
Eicosapentaenoic acid (EPA)	*Phaeodactylum tricornutum, Chlorella* sp., *Monodus subterraneus,* and *Nannochloropsis* sp.	Nutritional and therapeutic applications	NA	NA
Omega-3 fatty acid				
Protein	*Spirulina* sp., *Chlorella vulgaris, Scenedesmus* sp.	Single-cell proteins, animal feed	Parry Nutraceuticals	India
			Avantha Holdings	India
			Sun Chlorella Corporation	Japan
			Earthrise Farms	United States

Source: Becker (2004), Pulz and Gross (2004), Spolaore et al. (2006), Ravishankar et al. (2008), Plaza et al. (2009), and Milledge (2011).

period resulted in similar performance of animals compared to control group in terms of feed efficiency ratio, body weight gain, and meat quality i.e., pig mash and the quality of pork. Several reports on use of microalgae such as *Scenedesmus dimorphus*, diatoms like *Staurosira* sp., *Nannochloropsis oculata*, *Spirulina platensis, Dunaliella salina*, etc., as source of feed replacing conventional feedstocks are available (Ravishankar et al., 2012; Lum et al., 2013; Vidyashankar et al., 2014). These experimental studies suggested that microalgae could be supplemented up to 20-30% replacement to food crops sometimes even completely.

Based on the ability of microalgae to absorb nutrients, Neori et al (2000) discussed sustainable integrated mariculture systems involving the cultivation of fish, shrimps, and oysters along with algae. The effluents from ponds with fish and shrimp fed with high nutrients have high BOD and excessive nutrients. The discharge of effluent water causes nitrification of coastal waters causing pollution (Neori and Shpigel, 2002). These waters can be reused and treated by passing through ponds with algae as biofilters. The algae consume the nutrients in the effluents, thereby reducing the treatment costs and also the production costs. The algae are later consumed by invertebrate bivalves such as *Crassostrea* sp., *Artemia*, etc., which are used as primary feed in cultivation of many aquaculture forms.

6.2 BIOENERGY OPTIONS FROM MICROALGAL BIOMASS

Microalgae have been explored as a potential source of renewable fuel due to their photosynthetic nature, simple growth requirements, adaptability to a variety of environmental conditions, rapid growth, and simple triggering mechanisms to accumulate oil, all of which have made microalgae a popular choice as a source for renewable fuels (Brennan and Owende, 2010). The microalgae accumulate important fuel precursors such as triacylglycerol and hydrocarbons that can be converted to biodiesel or directly blended with petroleum fuels. The harvestable algal biomass from HRAP-based WWT have been reported to contain a heating value of about $18\text{-}22 \times 106\,\text{MJ ton}^{-1}$, thus having a huge potential for conversion of the biomass to various biofuels (Park et al., 2013; Mehrabadi et al., 2014). The main advantage of integration of WWT and algal biofuel production is that the costs of algal cultivation (nutrient costs) and harvesting are covered by tertiary water treatment process. The harvesting of algal biomass becomes easier if colonial species are selectively cultivated (Craggs et al., 2013). Algal biomass offers four modes of energy production such as (a) biogas by conversion of biomass to methane by anaerobic digestion; (b) bio-oil by conversion of biomass by thermochemical liquefaction; (c) biodiesel production by extraction of lipids from biomass and transesterification of lipids to biodiesel; and (d) fermentation of carbohydrate fraction and distillation of bioethanol (Ravishankar et al., 2012; Park et al., 2013; Mehrabadi et al., 2014). The various modes of energy production from microalgal biomass are presented in Figure 5.

The most commonly followed route of bioenergy production is extraction of lipids followed by transesterification and conversion to biofuels. This mode of biofuel production demands relatively higher lipid content in the alga biomass. Hyperlipid-producing algae are listed in Table 11.

Generally increased lipid accumulation in algal cells are achieved under stressful environments such as nutrient deprivation (mainly N limitation), high light intensity, and salinity stress (Rodolfi et al., 2009; Guschina and Harwood, 2009; Vidyashankar et al., 2013). Contrary to the requirement, the wastewaters have high nutrients (N and P), thus leading to lower lipid content of the biomass. Further, cyanobacterial populations dominate wastewaters, which are naturally low-lipid accumulators. Therefore the WWT ponds must have relatively larger populations of microalgae that accumulate lipids such as green algae or diatoms (Griffiths and Harrison, 2009; Ravishankar et al., 2012). *Botryococcus braunii*, colonial freshwater algae, suits some of the requirements for integrated WWT and biofuel process such as having colonial morphology, hence easily harvestable, high oil content, and fuel grade oil profile. *Botryococcus braunii* accumulates a rare group of hydrocarbons that are not observed in other species. These hydrocarbons are secreted in the extracellular layers, which are easily extracted by nonpolar solvents. Based on the hydrocarbon composition seen in them, *Botryococcus braunii* is categorized into three races. Race A produces odd-numbered, fatty acid-derived, n-alkadiene type

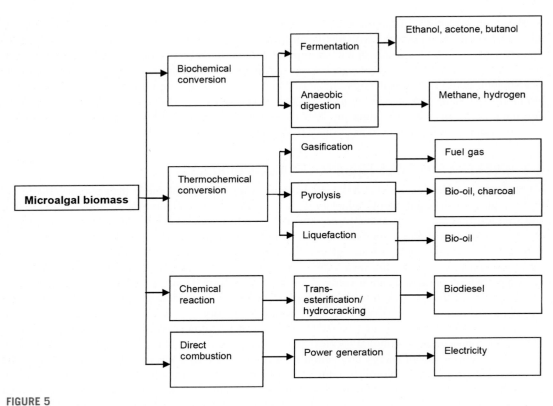

FIGURE 5

Different forms of energy production from microalgal biomass.

Adapted from Wang et al. (2008).

hydrocarbons ranging from C23 to C33. Race B produces unsaturated hydrocarbons called botryococcenes and methyl-branched squalenes. Race L produces tetra-terpenoid hydrocarbons known as lycopadiene (Metzger and Largeau, 2005). Hydrocarbon obtained from *Botryococcus braunii* when hydrocracked produced a distillate with good fuel properties, comprising 67% gasoline fraction, 15% aviation fuel, 15% diesel fraction, and remaining residual oil (Banerjee et al., 2002; Dayananda et al., 2006). However, the major disadvantage in utilization of *Botryococcus braunii* for commercial biofuel production is that they have a slow growth rate with over 36.5h as doubling time. Hence, the necessity of supplementation of nutrients for slow-growing organisms affects the economics of biofuel processes (Krzeminska et al., 2014).

However, the primary aim in the integrated WWT process is nutrient removal and since there is no lack of nutrient supply, *Botryococcus braunii* could find potential use in biofuel production. Based on these assumptions, An et al. (2003) reported *Botryococcus braunii* cultivation and hydrocarbon production from tertiary-treated piggery wastewater. A biomass yield of 8.5 g dry cell weight L^{-1} was achieved with a hydrocarbon content of 0.95 g hydrocarbon L^{-1} after 12 days of incubation. A total of $620\,mg\,N\,L^{-1}$

Table 11 Hyper-Lipid-Producing Microalgae

Microalgae	Class	Lipid Content (% Dry Weight)	Habitat
Botryococcus braunii	Chlorophyceae	25-75	Freshwater
Chlorella vulgaris	Chlorophyceae	30-35	Freshwater
Chlorococcum	Chlorophyceae	19.3	Freshwater
Scenedesmus spp.	Chlorophyceae	21.1	Freshwater
Neochloris oleoabundans	Chlorophyceae	35-54	Freshwater
Crypthecodinium cohnii	Dinoflagellate	20	Marine
Tetraselmis suecia	Chlorophyceae	36.4	Marine
Dunaliella primolecta	Chlorophyceae	23	Marine
Isochrysis spp.	Haptophyceae	25-33	Marine
Pavlova sp.	Haptophyceae	25-30	Marine
Skeletonema costatum	Bacillariophyceae	17.4	Marine
Chaetoceros muelleri	Bacillariophyceae	21.8	Marine
Chaetoceros calcitrans	Bacillariophyceae	17.6	Marine
Nitzschia spp.	Bacillariophyceae	45-47	Marine
Nannochloropsis spp.	Eumastigophyceae	31-68	Marine
Schizochytrium spp.	Thraustochytriidae	50-77	Marine
Monodus subterraneus	Eumastigophyceae	30.4	Marine
Thalsssiosira pseudonana	Bacillariophyceae	17.4	Marine

Sources: Chisti (2007), Rodolfi et al. (2009), Illman et al. (2000), Mutanda et al. (2011), and Vidyashankar et al. (2013).

was removed by the alga with 80% removal efficiency. These results suggested that pretreated piggery wastewater provides a good culture medium for the growth and hydrocarbon production by *Botryococcus braunii*.

Several reports on production biodiesel from algal lipids are available (Francisco et al., 2010; Prommuak et al., 2012; Velasquez-Orta et al., 2013; Cheng et al., 2014). This route of bioenergy production has high conversion rate >90%, in terms of conversion of triacylglycerol to fatty acid alkyl esters (biodiesel). However, the quality of biodiesel is significantly affected by the fatty acid composition (Shekh et al., 2013). Microalgae that accumulate PUFAs become unsuitable since they have poor oxidative stability, while microalgae that accumulate high amounts of saturated fatty acids have poor cold flow properties leading to solidification of fuels at colder climates (Giakoumis, 2013). Hence a blend of monounsaturated fatty acids and saturated fatty acids with PUFA content less than or equal to 12% weight of total fatty acids has been suggested as ideal for biodiesel applications (Knothe, 2011). Obtaining such fatty acid composition in microalgal cells is difficult. Further, generally only one-third or one-fourth of the algal lipids are converted to biodiesel, giving rise to only a 10% yield of biodiesel with respect to dry biomass (Zhu et al., 2013). In addition, transesterification is a costlier process with a number of parameters such as solvent requirement, catalyst ratio, and high process temperatures leading to negative energy balance; that is, higher input energy over output energy, nearly 2 (Mehrabadi et al., 2014).

The sustainability and economic viability of the biodiesel production from microalgal biomass can be significantly improved by following a biorefinery approach. In this process, the residues or coproducts generated during biodiesel production are effectively utilized for further applications. According to Sander and Murthy (2010), for every 24 kg algal biodiesel produced, 34 kg of coproducts are generated such as defatted biomass, glycerol, and unsaponifiable lipids. Glycerol and unsaponifiable lipids can be used as precursors for production of industrial chemicals such as 1,3-propanediol used in polymer and paint industries (Yazdani and Gonzalez, 2007). The DAB can find application in biogas production through anaerobic digestion (Sarat Chandra et al., 2014) or as a source of carbon-neutral animal feed supplement (Vidyashankar et al., 2014).

As mentioned earlier, microalgae offer other routes of bioenergy production of which the least energy-intensive route is biogas production by anaerobic digestion (Mehrabadi et al., 2014).

Hydrothermal liquefaction (HTL) is another method of conversion of wet algae biomass to energy-dense heavy oil or tarry material. This process can convert wet algal biomass with a moisture content above 75% by heating at subcritical temperatures (200-350 °C) and high pressure (5-20 MPa). The biomass is broken down by hot compressed water to energy-dense hydrocarbons of chain length C17-C18 and PAHs. The resultant biocrude oil has heating value of about 30-40 kJ g^{-1} with an average yield ranging between 30% and 50% of initial wet biomass. The HTL process generates various by-products that can be recycled for energy production. They include: mixture of gases such as CO_2, H_2, CH_4, C_2H_4, C_2H_6, N_2 with 20% yield, residual solids of about 10% yield, aqueous phase of about 20-30% yield rich in nutrients that can be recycled back to WWT (López Barreiro et al., 2013). The residual solids generally have C content >20%, which can be further used for energy conversion either by anaerobic digestion or by biomass gasification.

The biocrude oil yield is significantly affected by the biomass composition, where higher lipid content results in higher yield of biocrude. For example, when *Chlorella vulgaris*, *Nannochloropsis oculata*, and *Porphyridium cruentum* were liquefied at 350 °C for 1 h in the presence of Na_2CO_3, HCOOH, and water, *Nannochloropsis oculata* and *Chlorella vulgaris* resulted in higher biocrude oil compared to *Porphyridium cruentum* due to their higher lipid content. Brown et al. (2010) reported 43% yield of biocrude oil from *Nannochloropsis* sp. at 350 °C, almost 34% higher than *Nannochloropsis* lipid yield (28%). Yu et al. (2011) reported a yield of 34.4% biocrude oil for *Chlorella pyrenoidosa* when liquefied at temperatures ranging from 240 to 300 °C for 30-120 min. He further observed that liquefaction at lower temperatures resulted in asphaltlike crude and when operated at higher temperatures yielded a pourable biocrude. However, biocrude obtained by HTL contained significant amounts of nitrogen (generated from decomposition of proteins) up to 5% by weight, which when combusted generated NO_X (López Barreiro et al., 2013).

Pyrolysis offers another method for thermal decomposition where the dry biomass is cracked in the absence of oxygen at very high temperatures around 500 °C for a time of 30-120 min (López Barreiro et al., 2013). The process contains three steps: dehydration (vaporization of intracellular components at 80-190 °C), volatilization and condensation (volatile components are produced in the temperature ranges of 190-500 °C and condensed to form stable liquid or gases), and decomposition at temperatures equal to or above 500 °C to produce solid biochar (Porphy and Farid, 2012; López Barreiro et al., 2013). Pyrolysis biocrude generally contains aliphatic hydrocarbons, aromatic hydrocarbons, phenols, and long-chain fatty acids; however, the composition varies from biomass composition, biomass pretreatment, conversion temperature, heating rate, residence type, catalyst type, and condensation process (Babich et al., 2011; Porphy and Farid, 2012; López Barreiro et al., 2013). Pyrolysis results in yield 20-45% of bio-oil with an average energy content of 35 kJ g^{-1}. However, the main problem with pyrolysis biocrude is that it is acidic in nature with pH ranging between 2.5 and 3.7, leading to storage issues. Babich et al. (2011)) reported that use of catalyst like Na_2CO_3 resulted in higher pH above 3.5, which may reduce the corrosiveness of the biocrude. However, use of catalyst resulted in increased gas product above 15% resulting in reduced yield of biocrude.

In addition to HTL and pyrolysis, direct biomass gasification offers a third route for bioenergy production based on thermal decomposition strategy from dry algal biomass. Gasification is a partial oxidation of biomass at high temperatures ranging between 800 and 1000 °C. Gasification produces a mixture of combustible gases such as H_2, CH_4, CO_2, C_2H_4 (López Barreiro et al., 2013). Gasification is a highly energy-intensive process with low calorific value of the product (only 4-6 MJ kg^{-1}). The mixture of gases can be directly used for burning, hence used in engines, or can be used as syngas, feedstock for chemical production like methanol (Amin, 2009). However, this process is highly energy consuming with low economic profitability.

6.3 ECONOMIC VIABILITY OF THE WASTEWATER-INTEGRATED ALGAL BIOMASS PRODUCTION

Integration of WWT and algal cultivation could be a viable technology for production of microalgal biomass for commercial applications such as biofuels, industrial chemicals, animal feed application, etc. The integration of the two processes could reduce the capital costs (land use and pond construction) significantly for algal biomass production by approximately reducing US $0.1 to $0.25 million toward construction of raceway ponds (Park et al., 2011a,b). Further, the operation and maintenance costs are covered under WWT. However, the most important parameter that determines the success of algal cultivation and biomass production is nutrient and water footprint. Freshwater consumption costs are at maximum during the biomass production. For understanding the economics of biomass production, the process needs thorough technoeconomic modeling and resource utilization analysis, together known as life-cycle analysis (LCA) (Fishman et al., 2010). LCA is an approach to assessment of the resource use and environmental impacts of industrial processes, mainly the greenhouse gas emissions, carbon footprint, and water footprint. Yang et al. (2011) examined the LCA of biofuel production from microalgae and reiterated the necessity of recycling water or of using marine or wastewater for making microalgae-based biofuel production an economically competitive technology. According to them, to generate 1 kg of biodiesel, about 3726 kg water, 0.33 kg nitrogen, and 0.71 kg phosphate are required if freshwater is used without recycling. Recycling of water after the harvest of biomass, or use of sea or wastewater, decreases water requirement by 90% and eliminates the need for addition of nutrients except for phosphates. In another study Norsker et al. (2011) calculated the microalgal biomass production costs for three different production systems operating at commercial scale: open ponds, horizontal tubular photobioreactors, and flat-panel photobioreactors. The resulting biomass production costs for these systems were €4.95, €4.15, and €5.96 kg^{-1}, respectively. The parameters included for the costing were irradiation conditions, mixing, photosynthetic efficiency of systems, medium, carbon dioxide costs, and dewatering. Lundquist et al. (2010) thoroughly assessed the technical and engineering systems involved in biofuel production from microalgae. According to their report, the overall production costs for biofuel production as a coproduct of WWT was of high economic feasibility, at US $28/barrel of oil or US $0.17 kWh^{-1} electricity produced through biogas generation. However, in the case of biofuel or biogas production as the major objective, the system was very costly, at US $332/barrel of oil and US $0.72 kWh^{-1} electricity. Some of the technical assumptions of the study were 25% recoverable lipid content, dry biomass yield of 22 g m^{-2} day^{-1} (80 ton ha^{-1} year^{-1}), and a resulting oil yield of $20,000 \, L \, ha^{-1} \, year^{-1}$. CO_2 was supplied from a flue gas source. Further, they recommended use of bioflocculation for harvesting biomass, wet extraction, and emulsification of oil, which together could reduce operating costs by 20-25% avoiding the energy-intensive steps of dewatering and extraction. Ravishankar et al, 2012 discussed extensively the biorefinery approach to utilization of biomass for economic benefit. Herein, all the cell inclusions of utility value can be separated for various uses and the spent biomass can be used as fertilizer or for energy needs. Such an orientation will add value to the process as various components of the biomass will realize its potential for economic viability. Thus, a biorefinery approach supports the enormous potential of algae for remediation, energy, and production of novel utility chemicals for various applications.

7 CONCLUSION

Phycoremediation is realizing its potential as a powerful agent of removing toxicants from the environment and often detoxifying the hazardous substances. The algae are able to thrive in varied habitats and have the powerful machinery to carry out the process in an effective manner. The research on the ability of algal forms to effectively uptake heavy metals and to degrade xenobiotics and toxicants offer innumerable opportunities to find remediation measures specific for the needs. Though the algal processes are not fully exploited to deal with the treatment processes, their potential is being realized and is being explored for the remedial measures toward mitigation of pollution problems. However, there is need for research in exploring the potentials of the regional or site-specific algal consortia that can be recruited to do the job of mitigation by selecting the right organism(s) that can be effectively used in the process. The process of degradation of toxicants needs to be coupled to the utilization of biomass for the purposes of economic value. Though utilization for food and feed uses demands the highest degree of safety requirements, the alternate uses for nonfood uses could be of value too. In this context utilization of biomass for bioenergy is an interesting area. Biomass cultivation for use in mitigating climate change is an issue attracting attention. If properly adopted it would be one of the solutions to near-neutral CO_2 balance, which is being planned as a remedial measure of excessive CO_2 buildup as a result of industrial activities. However, the need of the hour is to find the right algal form(s) that can carry out the specific tasks and to develop a process of mitigation of environmental pollution in an eco-friendly and economical manner. This necessitates undertaking of R&D with basic and applied research components where we can find the right algae with the ability to detoxify the target molecules in an effective manner with least energy inputs. Use of closed and open cultivation systems as applicable to the environmental conditions needs evaluation and adoption. Genetically modified algal forms that can do the specific tasks need to be developed, and such organisms can be grown in closed photobioreactor systems without being released to the environment. This technology holds promise in view of its potential for eco-friendly solutions to bioremediation business.

ACKNOWLEDGMENT

GAR is grateful to Dr. Premachandra Sagar, Vice Chairman, Dayanada Sagar Institutions, for his encouragement and facilities.

REFERENCES

Abdel-Raouf, N., Al-Homaidan, A.A., Ibraheem, I.B.M., 2012. Microalgae and wastewater treatment. Saudi J. Biol. Sci. 19 (3), 257–275.

Abed, R.M.M., 2010. Interaction between cyanobacteria and aerobic heterotrophic bacteria in the degradation of hydrocarbons. Int. Biodeterior. Biodegrad. 64, 58–64.

Abed, R.M.M., Zein, B., Al-Thukair, A., de Beer, D., 2007. Phylogenetic diversity and activity of aerobic heterotrophic bacteria from a hypersaline oil-polluted microbial mat. Syst. Appl. Microbiol. 30, 319–330.

Aitken, D., Ladislao, B.A., 2015. Achieving a green solution: limitations and focus points for sustainable algal fuels. Energies 5, 1613–1647.

Amin, S., 2009. Review on biofuel oil and gas production processes from microalgae. Energ. Convers. Manage. 50 (7), 1834–1840.

An, J.Y., Sim, S.J., Lee, J.S., Kim, B.W., 2003. Hydrocarbon production from secondarily treated piggery wastewater by the green alga *Botryococcus braunii*. J. Appl. Phycol. 15, 185–191.

Babich, I.V., Van der Hulst, M., Lefferts, L., Moulijn, J.A., O'Connor, P., Seshan, K., 2011. Catalytic pyrolysis of microalgae to high-quality liquid bio-fuels. Biomass Bioenergy 35 (7), 3199–3207.

Banerjee, A., Sharma, R., Chisti, Y., 2002. *Botryococcus braunii*: a renewable source of hydrocarbons and other chemicals. Crit. Rev. Biotechnol. 22 (3), 245–279.

Becker, W., 2004. Microalgae in human and animal nutrition. In: Richmond, A. (Ed.), Handbook of Microalgal Culture: Biotechnology and Applied Phycology. Blackwell Science, Ames, IA, pp. 312–352.

Becker, E.W., 2007. Micro-algae as a source of protein. Biotechnol. Adv. 25, 207–210.

Becker, E.W., 2008. Microalgae: biotechnology and microbiology. Cambridge Studies in Microbiology, vol. 10.Cambridge University Press, Cambridge. 293 pp.

Ben-Amotz, A., Avron, M., 1990. The biotechnology of cultivating the halotolerant alga *Dunaliella*. Trends Biotechnol. 8, 121–126.

Ben-Amotz, A., Shaish, A., Avron, M., 1989. Mode of action of the massively accumulated beta carotene of *Dunaliella bardawil* in protecting the alga against damage by excess irradiation. Plant Physiol. 91 (3), 1040–1043.

Benemann, J.R., 2008. Open ponds and closed photobioreactors—comparative economics. In: 5th Annual World Congress on Industrial Biotechnology and Bioprocessing, Chicago, April 30.

Blaaby-Haas, C.E., Merchant, S.S., 2012. The ins and outs of algal metal transport. Biochim. Biophys. Acta 1823 (9), 1531–1552.

Borowitzka, M.A., 2005. Culturing microalgae in outdoor ponds. In: Andersen, I.R.A. (Ed.), Algal Culturing Techniques. Elsevier, Academic Press, New York, pp. 205–218.

Brennan, L., Owende, P., 2010. Biofuels from micro-algae—a review of technologies for production, processing, and extractions of biofuels and coproducts. Renewable Sustainable Energy Rev. 14 (2), 557–577.

Brown, T.M., Duan, P., Savage, P.E., 2010. Hydrothermal liquefaction and gasification of *Nannochloropsis* sp.. Energy Fuels 24 (6), 3639–3646.

Burtin, P., 2003. Nutritional value of seaweeds. Electron. J. Environ. Agric. Food Chem. 2 (4), 498–503.

Butler, A., 1998. Acquisition and utilization of transition metal ions by marine organisms. Science 281, 207–210.

Cauchie, H.M., Hoffman, L., Jaspar-Versali, M.F., Salvia, M., Thome, J.P., 1995. Daphnia magna Straus living in an aerated sewage lagoon as source of chitin: ecological aspects. J. Zool. 125, 67–78.

Cáceres, T., Megharaj, M., Naidu, R., 2008. Biodegradation of the pesticide fenamiphos by ten different species of green algae and cyanobacteria. Curr. Microbiol. 57, 643–646.

Castruita, M., Casero, D., Karpowicz, S.J., Kropat, J., Vieler, A., Hsieh, S.I., Yan, W., Cokus, S., Loo, J.A., Benning, C., Pellegrini, M., Merchant, S.S., 2011. Systems biology approach in *Chlamydomonas* reveals connections between copper nutrition and multiple metabolic steps. Plant Cell 23, 1273–1292.

Cerniglia, C.E., Gibson, D.T., van Baalen, C., 1979. Algal oxidation of aromatic hydrocarbons: formation of 1-naphthol from naphthalene by *Agmenellum quadruplicatum*, strain PR-6. Biochem. Biophys. Res. Commun. 88, 50–58.

Cerniglia, C.E., Gibson, D.T., van Baalen, C., 1980. Oxidation of naphthalene by cyanobacteria and microalgae. J. Gen. Microbiol. 116, 495–500.

Chanakya, H.N., Mahapatra, D.M., Ravi, S., Chauhan, V.S., Abitha, R., 2012. Sustainability of large-scale algal biofuel production in India. J. Indian Inst. Sci. 92 (1), 63–98.

Chavan, A., Mukherji, S., 2010. Effect of co-contaminant phenol on performance of a laboratory scale RBC with algal–bacterial biofilm treating petroleum hydrocarbon-rich wastewater. J Chem. Technol. Biotechnol. 85, 851–859.

Cheng, J., Huang, R., Yu, T., Li, T., Zhou, J., Cen, K., 2014. Biodiesel production from lipids in wet microalgae with microwave irradiation and bio-crude production from algal residue through hydrothermal liquefaction. Bioresour. Technol. 151, 415–418.

Chinnasamy, S., Bhatnagar, A., Hunt, R.W., Das, K.C., 2010. Microalgae cultivation in a wastewater dominated by carpet mill effluents for biofuel applications. Bioresour. Technol. 101 (9), 3097–3105.

Chisti, Y., 2007. Biodiesel from micro-algae. Biotechnol. Adv. 25, 294–306.

Chisti, Y., 2013. Constraints to commercialization of algal fuels. J. Biotechnol. 167, 201–214.

Chiu, S.Y., Kao, C.Y., Chen, C.H., Kuan, T.C., Ong, S.C., Lin, C.S., 2008. Reduction of CO_2 by a high-density culture of *Chlorella* sp. in a semi-continuous photobioreactor. Bioresour. Technol. 99, 3389–3396.

Chiu, S.Y., Kao, C.Y., Tsai, M.T., Kuan, T.C., Ong, S.C., Chen, C.H., Lin, C.S., 2009. Lipid accumulation and CO_2 utilization of *Nannochloropsis oculata* in response to CO_2 aeration. Bioresour. Technol. 100, 833–838.

Cho, S., Luong, T.T., Lee, D., Oh, Y.-K., Lee, T., 2011. Reuse of effluent water from a municipal wastewater treatment plant in microalgae cultivation for biofuel production. Bioresour. Technol. 102, 8639–8645.

Chojnacka, K., Chojnacki, A., Gorecka, H., 2005. Biosorption of Cr^{3+}, Cd^{2+} and Cu^{2+} ions by blue-green algae *Spirulina* sp.: kinetics, equilibrium and the mechanism of the process. Chemosphere 59, 75–84.

Clarens, A.F., Resurreccion, E.P., White, M.A., Colosi, L.M., 2010. Environmental life cycle comparison of algae to other bioenergy feedstocks. Environ. Sci. Tech. 44, 1813–1819.

Cobbett, C., Goldsbrough, P., 2002. Phytochelatin and metallothioneins: roles in heavy metal detoxification and homeostasis. Annu. Rev. Plant Biol. 53, 159–182.

Craggs, R.J., 2005. Advanced integrated wastewater ponds. In: Shilton, A. (Ed.), Pond Treatment Technology. IWA Scientific and Technical Report Series, IWA, London, pp. 282–310.

Craggs, R.J., Lundquist, T.J., Benemann, J.R. 2013. Wastewater treatment and algal biofuel production. In: Borowitzka, M.A., Moheimani, N.R. (Eds.), Algae Biofuels Energy, Developments in Applied Phycology, vol. 5, pp. 153–163.

Crist, R.H., Martin, J.R., Carr, D., Watson, J.R., Clarke, H.J., 1994. Interaction of metals and protons with algae. 4. Ion exchange vs adsorption models and a reassessment of Scatehard plots; ion-exchange rates and equilibria compared with calcium alginate. Environ. Sci. Technol. 28 (11), 1859–1866.

Croft, M., Warren, M., Smith, A., 2006. Algae need their vitamins. Eukaryot. Cell 5, 1175–1183.

Dalrymple, O.K., Halfhide, T., Udom, I., Gilles, B., Wolan, J., Zhang, Q., Ergas, S., 2013. Wastewater use in algae production for generation of renewable resources: a review and preliminary results. Aquat. Biosyst. 9 (2), 1–11.

Dayananda, C., Sarada, R., Srinivas, P., Shamala, T.R., Ravishankar, G.A., 2006. Presence of methyl branched fatty acids and saturated hydrocarbons in botryococcene-producing strain of *Botryococcus braunii*. Acta Physiol. Plant. 28, 251–256.

de Godos, I., Vargas, V.A., Blanco, S., Garcia Gonzalez, M.C., Soto, R., Garcia-Encina, P.A., Becares, E., Munoz, R., 2010. A comparative evaluation of microalgae for the degradation of piggery wastewater under photosynthetic oxygenation. Bioresour. Technol. 101 (14), 5150–5158.

De Morais, M.G., Vieira Costa, J.A., 2007. Biofixation of carbon dioxide by *Spirulina* sp. and *Scenedesmus obliquus* cultivated in a three-stage serial tubular photobioreactor. J. Biotechnol. 129, 439–445.

de-Bashan, L.E., Bashan, Y., 2010. Immobilized microalgae for removing pollutants: review of practical aspects. Bioresour. Technol. 101, 1611–1627.

Doble, M., Kumar, A., 2005. Biotreatment of Industrial Effluents. Elsevier Butterworth-Heinemann, Burlington, MA, ISBN: 978-0-7506-7838-4. pp. 1–11.

Doughman, S.D., Krupanidhi, S., Sanjeevi, C.B., 2007. Omega-3 fatty acids for nutrition and medicine: considering microalgae oil as a vegetarian source of EPA and DHA. Curr. Diabetes Rev. 3, 198–203.

Field, C.B., Behrenfeld, M.J., Randerson, J.T., Falkowski, P., 1998. Primary production of biosphere: integrating terrestrial and oceanic components. Science 281, 237–240.

Fishman, D., Majumdar, R., Morello, J., Pate, R., Yang, J., 2010. National algal biofuels technology roadmap in workshop and road map by U.S. DOE. Available at www.Biomass.energy.gov. (accessed 07.12.14.).

Fogliano, V., Andreoli, C., Martello, A., Caiazzo, M., Lobosco, O., Formisano, F., Carlino, P.A., Meca, G., Graziani, G., Rigano, V.M., Vona, V., Carfagna, S., Rigano, C., 2010. Functional ingredients produced by culture of *Koliella antarctica*. Aquaculture 299, 115–120.

Foyer, C.H., 1997. Oxygen metabolism and electron transport in photosynthesis. In: Scandalios, J. (Ed.), Molecular Biology of Free Radical Scavenging Systems. Cold Spring Harbor Laboratory Press, New York, pp. 587–621.

Francisco, E.C., Neves, D.B., Jacob-Lopes, E., Franco, T.T., 2010. Microalgae as feedstock for biodiesel production: carbon dioxide sequestration, lipid production and biofuel quality. J. Chem. Technol. Biotechnol. 85, 395–403.

Friesen-Pankartz, B., Doebel, C., Farenhorst, A., Goldsborough, L.G., 2003. Interactions between algae (*Selenastrum capricornutum*) and pesticides: implications for managing constructed wetlands for pesticide removal. J. Environ. Sci. Health B 38 (2), 147–155.

Furtado, A.L.F.F., Calijuri, M.C., Lorenzi, A.S., Honda, R.Y., Genuario, D.B., Fiore, M.F., 2009. Morphological and molecular characterization of cyanobacteria from Brazilian facultative wastewater stabilization pond and evaluation of microcystis production. Hydrobiologia 627, 195–209.

Garcia, J., Mujeriego, M., Marine, M.H., 2000. High rate algal pond operating strategies for urban wastewater nitrogen removal. J. Appl. Phycol. 12, 331–339.

Giakoumis, E.G., 2013. A statistical investigation of biodiesel physical and chemical properties, and their correlation with the degree of unsaturation. Renew. Energy 50, 858–878.

Griffiths, M.J., Harrison, S.T.L., 2009. Lipid productivity as a key characteristic for choosing algal species for biodiesel production. J. Appl. Phycol. 21 (5), 493–507.

Guieysse, B., Borde, X., Munoz, R., Hatti-Kaul, R., Nugier-Chauvin, C., Patin, H., 2002. Influence of the initial composition of algal bacterial microcosms on the degradation of salicylate in a fed-batch culture. Biotechnol. Lett. 24, 531–538.

Gupta, V., Ratha, S.K., Sood, A., Chaudhary, V., Prasanna, R., 2013. New insights into the biodiversity and applications of cyanobacteria (blue-green algae)—prospects and challenges. Algal Res. 2, 79–97.

Guschina, I.A., Harwood, J.L., 2009. Algal lipids and effect of the environment on their biochemistry. In: Arts, M.T., Brett, M.T., Kainz, M. (Eds.), Lipids in Aquatic Ecosystems. Springer, New York, pp. 1–24.

Hafeburg, G., Kohe, E., 2007. Microbes and metals: interactions in the environment. J. Basic Microbiol. 47, 453–467.

Hall, J.L., Williams, L.E., 2003. Transition metal transporters in plants. J. Exp. Bot. 54, 2601–2613.

Hanagata, N., Takeuchi, T., Fukuju, Y., Barnes, D.J., Karube, I., 1992. Tolerance of microalgae to high CO_2 and high temperature. Phytochemistry 31, 3345–3348.

Hanikenne, M., Krämer, U., Demoulin, V., Baurain, D., 2005. A comparative inventory of metal transporters in the green alga *Chlamydomonas reinhardtii* and the red alga *Cyanidioschizon merolae*. Plant Physiol. 137, 428–446.

Hende, S.V.D., Vervaeren, H., Boon, N., 2012. Flue gas compounds and microalgae: (bio-) chemical interactions leading to biotechnological opportunities. Biotechnol. Adv. 30 (6), 1405–1424.

Hintz, H.F., Heitmann, H., 1967. Sewage grown algae as a protein supplement for swine. Anim. Prod. 9, 135–141.

Hoffman, J.P., 1998. Wastewater treatment with suspended and non-suspended algae. J. Phycol. 34, 757–763.

Horan, N.J., 1990. Biological Wastewater Treatment Systems. Theory and operation. John Wiley and Sons, Chickester.

Hsueh, H.T., Li, W.J., Chen, H.H., Chu, H., 2009. Carbon bio-fixation by photosynthesis of *Thermosynechococcus* sp. CL-1 and *Nannochloropsis oculta*. J. Photochem. Photobiol. B Biol. 95 (1), 33–39.

Hu, C., Chuang, L., Yu, P., Chen, C.N., 2013. Pigment production by a new thermotolerant microalga *Coelestrella* sp. F50. Food Chem. 138, 2071–2078.

Hulatt, C.J., Lakaniemi, A.M., Puhakka, J.A., Thomas, D.N., 2012. Energy demands of nitrogen supply in mass cultivation of two commercially important microalgal species, *Chlorella vulgaris* and *Dunaliella tertiolecta*. Bioenergy Res. 5 (3), 669–684.

Huntley, M., Redalje, D., 2007. CO_2 mitigation and renewable oil from photosynthetic microbes: a new appraisal. Mitigat. Adapt. Strategies Glob. Change 12, 573–608.

Illman, A.M., Scragg, A.H., Shales, S.W., 2000. Increase in *Chlorella* strains calorific values when grown in low-nitrogen medium. Enzyme Microb. Technol. 27 (8), 631–635.

Imase, M., Watanabe, K., Aoyagi, H., Tanaka, H., 2008. Construction of an artificial symbiotic community using a Chlorella-symbiont association as a model. FEMS Microbiol. Ecol. 63, 273–282.

Iwai, M., Katoh, H., Katayama, M., Ikeuchi, M., 2004. Improved genetic transformation of the thermophilic cyanobacterium, *Thermosynechococcus elongatus* BP-1. Plant Cell Physiol. 45 (2), 171–175.

Iwasaki, I., Hu, Q., Kurano, N., Miyachi, S., 1998. Effect of extremely high-CO_2 stress on energy distribution between photosystem I and photosystem II in a high-CO_2 tolerant green alga, *Chlorococcum littorale* and the intolerant green alga *Stichococcus bacillaris*. J. Photochem. Photobiol. B Biol. 44, 184–190.

Jansson, C., Northen, T., 2010. Calcifying cyanobacteria—the potential of biomineralization for carbon capture and storage. Curr. Opin. Biotechnol. 21, 1–7.

Kaplan, D., 2013. Absorption and adsorption of heavy metals by microalgae. In: Richmond, A., Hu, Q. (Eds.), Handbook of Microalgal Culture: Applied Phycology and Biotechnology, second ed. John Wiley and Sons, Oxford.

Kathiresan, S., Sarada, R., Bhattacharya, S., Ravishankar, G.A., 2007. Culture media optimization for growth and phycoerythrin production from *Porphyridium purpureum*. Biotechnol. Bioeng. 96 (3), 456–463.

Kim, B.H., Kang, Z., Ramanan, R., Choi, J.E., Cho, D.H., Oh, H.M., Kim, H.S., 2014. Nutrient removal and biofuel production in high rate algal pond using real municipal wastewater. J. Microbiol. Biotechnol. 24 (8), 1123–1132.

Kirkwood, A., Nalewajko, C., Fulthorpe, R., 2006. The effects of cyanobacteria exudates on bacterial growth and biodegradation of organic contaminants. Microb. Ecol. 51, 4–12.

Kishimoto, M., Okakura, T., Nagashima, H., Minowa, T., Yokoyama, S.-Y., Yamaberi, K., 1994. CO_2 fixation and oil production using micro-algae. J. Ferment. Bioeng. 78, 479–482.

Knothe, G., 2011. A technical evaluation of biodiesel from vegetable oils vs. algae. Will algae-derived biodiesel perform? Green Chem. 13 (11), 3048–3065.

Kodama, M., Ikemoto, H., Miyachi, S., 1993. A new species of highly CO_2-tolerant fast-growing marine microalga suitable for high-density culture. J. Mar. Biotechnol. 1, 21–25.

Krishna, A.R., Dev, L., Thankamani, V., 2012. An integrated process for industrial effluent treatment and biodiesel production using microalgae. Res. Biotechnol. 3 (1), 47–60.

Krzeminska, I., Pawlik-Skowronska, B., Trzcinska, M., Tys, J., 2014. Influence of photoperiods on the growth rate and biomass productivity of green microalgae. Bioprocess Biosyst. Eng. 37 (4), 735–741.

Kumudha, A., Kumar, S.S., Thakur, M.S., Ravishankar, G.A., Sarada, R., 2010. Purification, identification, and characterization of methylcobalamin from *Spirulina platensis*. J. Agric. Food Chem. 58, 9925–9930.

Larkum, A.W.D., 2010. Limitation and prospects of natural photosynthesis for bio-energy production. Curr. Opin. Biotechnol. 21, 271–276.

López Barreiro, D., Prins, W., Ronsse, F., Brilman, W., 2013. Hydrothermal liquefaction (HTL) of microalgae for biofuel production: state of the art review and future prospects. Biomass Bioenergy 53, 113–127.

Lum, K.K., Kim, J., Lei, X.G., 2013. Dual potential of microalgae as a sustainable biofuel feedstock and animal feed. J. Anim. Sci. Biotechnol. 4, 53.

Lundquist, T.J., Woertz, I.C., Quinn, N.W.T., Benemann, J.R., 2010. A Realistic Technology and Engineering Assessment of Algae Biofuel Production. Energy Biosciences Institute, Berkeley, CA. 178 pp. http://www.ascension-publishing.com/BIZ/Algae-EBI.pdf (accessed 29.11.11).

Malla, F.A., Khan, S.A., Rashmi, Sharma, G.K., Gupta, N., 2014. Phycoremediation potential of *Chlorella minutissima* on primary and tertiary treated wastewater for nutrient removal and biodiesel production. Ecol. Eng. 75, 343–349.

Mallick, N., 2002. Biotechnological potential of immobilized algae for wastewater N, P and metal removal: a review. BioMetals 15, 377–390.

Mallick, N., and Rai, L.C., (2002). Physiological responses of non-vascular plants to heavy metals. In: Prasad, M.N.V., Strzalka, K. (Eds.), Physiology and Biochemistry of Metal Toxicity and Tolerance in Plants. Kluwer Academic Publishers, Dordrecht, The Netherlands, pp. 111–147.

Mallick, N., 2004. Copper-induced oxidative stress in the chlorophycean microalga. J. Plant Physiol. 161, 591–597.

Mallick, N., Mohn, F.H., 2000. Reactive oxygen species: response of algal cells. J. Plant Physiol. 157, 183–193.

Markou, G., Angelidaki, I., Georgakakis, D., 2012. Microalgal carbohydrates: an overview of the factors influencing carbohydrates production, and of main bioconversion technologies for production of biofuels. Appl. Microbiol. Biotechnol. 96, 631–645.

Martin, J., Peixe, L., Vasconcelos, V., 2010. Cyanobacteria and bacteria co-occurrence in a wastewater treatment plant: absence of allelopathic effects. Water Sci. Technol. 62, 1954–1962.

Mäser, P., Thomine, S., Schroeder, J.I., Ward, J.M., Hirschi, K., Sze, H., Talke, I.N., Amtmann, A., Maathuis, F.J., Sanders, D., Harper, J.F., Tchieu, J., Gribskov, M., Persans, M.W., Salt, D.E., Kim, S.A., Guerinot, M.L., 2001. Phylogenetic relationships within cation transporter families of Arabidopsis. Plant Physiol. 126, 1646–1667.

McGinn, P.J., Dickinson, K.E., Bhatti, S., Frigon, J.C., Guiot, S.R., O'Leary, S.J.B., 2011. Integration of microalgae cultivation with industrial waste remediation for biofuel and bioenergy production: opportunities and limitations. Photosynth. Res. 109, 231–247.

McHugh, D.J., 2003. A Guide to the Seaweed Industry. http://www.fao.org/docrep/006/y4765e/y4765e00.htm#contents. (accessed 17.02.2015.).

Megharaj, M., Venkateswarulu, K., Rao, A.S., 1987. Metabolism of monocrotophos and quinalphos by algae isolated from soil. Bull. Environ. Contam. Toixcol. 39 (2), 251–256.

Megharaj, M., Madhavi, D.R., Sreenivasulu, C., Umamaheswari, A., Venkateswarulu, K., 1994. Biodegradation of methyl parathion by soil isolates of microalgae and cyanobacteria. Bull. Environ. Contam. Toixcol. 53 (2), 292–297.

Mehrabadi, A., Craggs, R., Farid, M.M., 2014. Wastewater treatment high rate algal ponds (WWT HRAP) for low-cost biofuel production. Bioresour. Technol. 184, 202–214.

Metzger, P., Largeau, C., 2005. *Botryococcus braunii*: a rich source for hydrocarbons and related other lipids. Appl. Microbiol. Biotechnol. 66 (5), 486–496.

Milledge, J.J., 2011. Commercial application of micro-algae other than as biofuels: a brief review. Rev. Environ. Sci. Biotechnol. 10 (1), 31–41.

Miyairi, S., 1995. CO_2 assimilation in a thermophilic cyanobacterium. Energy Convers. Manage. 36, 763–766.

Molina Grima, G.E., Belarbi, E.H., Acién Fernández, F.G., Medina, R.A., Chisti, Y., 2003. Recovery of microalgal biomass and metabolites: process options and economics. Biotechnol. Adv. 20, 491–515.

Muñoz, R., Guieysse, B., 2006. Algal-bacterial processes for the treatment of hazardous contaminants: a review. Water Res. 40, 2799–2815.

Muñoz, R., Guieysse, B., Mattiasson, B., 2003a. Phenanthrene biodegradation by an algal-bacterial consortium in two-phase partitioning bioreactors. Appl. Microbiol. Biotechnol. 61, 261–267.

Muñoz, R., Köllner, C., Guieysse, B., Mattiasson, B., 2003b. Salicylate biodegradation by various algal-bacterial consortia under photosynthetic oxygenation. Biotechnol. Lett. 25, 1905–1911.

Muñoz, R., Rolvering, C., Guieysse, B., Mattiasson, B., 2005. Aerobic phenanthrene biodegradation in a two-phase partitioning bioreactor. Water Sci. Technol. 52, 265–271.

Muradyan, E.A., Klyachko-Gurvich, G.L., Tsoglin, L.N., Sergeyenko, T.V., Pronina, N.A., 2004. Changes in lipid metabolism during adaptation of the *Dunaliella salina* photosynthetic apparatus to high CO_2 concentration. Russ. J. Plant Physiol. 51 (1), 53–62.

Mutanda, T., Ramesh, D., Karthikeyan, S., Kumari, S., Anandraj, A., Bux, F., 2011. Bioprospecting for hyper-lipid producing micro-algal strains for sustainable biofuel production. Bioresour. Technol. 102, 57–70.

Norsker, N.-H., Barbosa, M.J., Vermue, M.H., Wijffels, R.H., 2011. Microalgal production — A close look at the economics. Biotechnol. Adv. 29, 24–27.

Nakano, Y., Miyatake, K., Okuno, H., Hamazaki, K., Takenaka, S., Honami, N., Kiyota, M., Aiga, I., Kondo, J., 1996. Growth of photosynthetic algae Euglena in high CO_2 conditions and its photosynthetic characteristics. In: International Symposium on Plant Production in Closed Ecosystems 440, pp. 49–54.

Neori, A., Shpigel, M., 2002. Mariculture sustainability by integration of algae/algivores to fish/shrimp ponds. Abstract the first congress of the International Society for Applied Phycology, Aquadulce, Roquetas de Mar, Almeria (Spain). Universidad de Almeria Servicio de Publicaciones.

Neori, A., Shpigel, M., Ben-Ezra, D., 2000. Sustainable integrated system for culture of fish, seaweed and abalone. Aquaculture 186, 279–291.

Olguin, E.J., 2012. Dual purpose microalgae-bacteria-based systems that treat wastewater and produce biodiesel and chemical products within a biorefinery. Biotechnol. Adv. 30, 1031–1046.

Onay, M., Sonmez, C., Oktem, H.A., Yucel, A.M., 2014. Thermo-resistant green microalgae for effective biodiesel production: isolation and characterization of unialgal species from geothermal flora of Central Anatolia. Bioresour. Technol. 169, 62–71.

Ono, E., Cuello, J.L., 2007. Carbon dioxide mitigation using thermophilic cyanobacteria. Biosyst. Eng. 96 (1), 129–134.

Osundeko, O., Davies, H., Pittman, J.K., 2013. Oxidative stress-tolerant microalgae strains are highly efficient for biofuel feedstock production on wastewater. Biomass Bioenergy 56, 284–294.

Pan, Y.Y., Wang, S., Chuang, L., Chang, Y., Chen, C.N., 2011. Isolation of thermo-tolerant and high lipid content green microalgae: oil accumulation is predominantly controlled by photosystem efficiency during stress treatments in *Desmodesmus*. Bioresour. Technol. 102, 10510–10517.

Park, J.B.K., Craggs, R.J., 2011. Algal production in wastewater treatment high rate algal ponds for potential biofuel use. Water Sci. Technol. 63 (10), 2403–2410.

Park, J.B.K., Craggs, R.J., Shilton, A.N., 2011a. Recycling algae to improve species control and harvest efficiency from a high rate algal pond. Water Res. 45 (20), 6637–6649.

Park, J.B.K., Craggs, R.J., Shilton, A.N., 2011b. Wastewater treatment high rate algal ponds for biofuel production. Bioresour. Technol. 102, 35–42.

Park, J.B.K., Craggs, R.J., Shilton, A.N., 2013. Enhancing biomass energy yield from pilot-scale high rate algal ponds with recycling. Water Res. 47 (13), 4422–4432.

Perales-Vela, H.V., Pena-Castro, J.M., Canizares-Villanueva, R.O., 2006. Heavy metal detoxification in eukaryotic microalgae. Chemosphere 64, 1–10.

Pereira, S., Micheletti, E., Zille, A., Santos, A., Ferreira, P.M., Tamagnini, P., Philippis, R.D., 2011. Using extracellular polymeric substances (EPS)-producing cyanobacteria for the bioremediation of heavy metals: do cations compete for the EPS functional groups and also accumulate inside the cell. Microbiology 157, 451–458.

Pinto, E., Sigaud-Kutner, T.C.S., Leitao, M.A.S., Okamoto, O.K., Morse, D., Colepicolo, P., 2003. Heavy metal induced oxidative stress in algae. J. Phycol. 39, 1008–1018.

Pittman, J.K., Dean, A.P., Osundeko, O., 2011. The potential of sustainable algal biofuel production using wastewater resources. Bioresour. Technol. 102, 17–25.

Plaza, M., Herrero, M., Cifuentes, A., Ibanez, E., 2009. Innovative natural functional ingredients from microalgae. J. Agric. Food Chem. 57 (16), 7159–7170.

Porphy, S.J., Farid, M.M., 2012. Feasibility study for production of biofuel and chemicals from marine microalgae *Nannochloropsis* sp. Based on basic mass and energy analysis. ISRN Renew. Energy 2012. http://dx.doi.org/10.5402/2012/156824. 11 pp.

Priyadarshini, I., Sahu, D., Rath, B., 2011. Microalgal bioremediation: current practices and perspectives. J. Biochem. Tech. 3 (3), 299–304.

Prommuak, C., Pavasant, P., Quitain, A.T., Goto, M., Shotipruk, A., 2012. Microalgal lipid extraction and evaluation of single-step biodiesel production. Eng. J. 16 (5), 157–166.

Pulz, O., Gross, W., 2004. Valuable products from biotechnology of microalgae. Appl. Microbiol. Biotechnol. 65, 635–648.

Raeesossadati, M.J., Ahmadzadeh, H., McHenry, M.P., Moheimani, N.R., 2014. CO_2 bioremediation by microalgae in photobioreactors: impacts of biomass and CO_2 concentrations, light, and temperature. Algal Res. 6, 78–85.

Rana, L., Chhikara, S., Dhankar, R., 2013. Assessment of growth rate of indigenous cyanobacteria in metal enriched culture medium. Asian J. Exp. Biol. 4 (3), 465–471.

Rangarao, A., Moi, P.S., Sarada, R., Ravishankar, G.A., 2014. Astaxanthin: sources, extraction, stability, biological activities and its commercial applications—a review. Mar. Drugs 12, 128–152.

Rangsayatron, N., Upatham, E.S., Kruatrachue, M., Pokethitiyook, P., Langa, G.R., 2002. Phytoremediation potential of *Spirulina* (*Arthrospira*) *platensis*: biosorption and toxicity studies of cadmium. Environ. Poll. 119, 45–53.

Raven, J.A., 2010. Inorganic carbon acquisition by eukaryotic algae: four current questions. Photosynth. Res. 106, 123–134.

Raven, J., Cockell, C., De La Rocha, C., 2008. The evolution of inorganic carbon concentrating mechanisms in photosynthesis. Philos. Trans. B 363, 2641–2650.

Ravishankar, G.A., Sarada, R., Kamath, B.S., Namitha, K.K., 2008. Food applications of algae. In: Shetty, K., Paliyath, G., Pometto, A., Levin, R.E. (Eds.), Food Biotechnology. CRC Press, Boca Raton, FL, pp. 491–524.

Ravishankar, G.A., Sarada, R., Vidyashankar, S., VenuGopal, K.S., Kumudha, A., 2012. Cultivation of micro-algae for lipids and hydrocarbons, and utilization of spent biomass for livestock feed and for bio-active constituents. In: Makkar, H.P.S. (Ed.), Biofuel Co-Products as Livestock Feed—Opportunities and Challenges. Food and Agriculture Organization, Rome, ISBN: 978-92-5-107299-8, pp. 423–446.

Rawat, I., Ranjith Kumar, R., Mutanda, T., Bux, F., 2011. Dual role of microalgae: phycoremediation of domestic wastewater and biomass production for sustainable biofuels production. Appl. Energy 88 (10), 3411–3424.

Rehman, A., Shakoori, A.R., 2004. Tolerance uptake of cadmium and nickel by *Chlorella* sp. isolated from tannery effluents. Pakistan J. Zool. 36 (4), 327–331.

Reinfelder, J.R., 2011. Carbon concentrating mechanisms in eukaryotic marine phytoplankton. Ann. Rev. Mar. Sci. 3, 291–315.

Rieger, P.G., Meier, H.M., Gerle, M., Vogt, U., Groth, T., Knackmuss, H.J., 2002. Xenobiotics in the environment: present and future strategies to obviate the problem of biological persistence. J. Biotechnol. 94 (1), 101–123.

Rittmann, E.B., 2008. Opportunities for renewable bioenergy using microorganisms. Biotechnol. Bioeng. 100, 203–212.

Robinson, T., McMullan, G., Marchant, R., Nigam, P., 2001. Remediation of dyes in textile effluents: a critical review on current treatment technologies with a proposed alternative. Bioresour. Technol. 77, 247–255.

Rodolfi, L., Zittelli, C.G., Bassi, N., Padovani, G., Biondi, N., Bonini, G., Tredici, M.R., 2009. Micro-algae for oil: strain selection, induction of lipid synthesis and outdoor mass cultivation in a low cost photobioreactor. Biotechnol. Bioeng. 102 (1), 100–112.

Rose, P.D., Boshoff, G.A., van Hille, R.P., Wallace, L.C.M., Dunn, K.M., Duncan, J.R., 1998. An integrated algal sulphate reducing high rate ponding process for the treatment of acid mine drainage wastewaters. Biodegradation 9, 247–257.

Safonova, E., Kvitko, K.V., Iankevitch, M.I., Surgko, L.F., Afti, I.A., Reisser, W., 2004. Biotreatment of industrial wastewater by selected algal-bacterial consortia. Eng. Life Sci. 4, 347–353.

Sander, K., Murthy, G.S., 2010. Life cycle analysis of algae biodiesel. Int. J. Life Cycle Assess. 15, 704–714.

Santos, G.A., 1989. Carrageenans of species of *Eucheuma* J. Agardh and *Kappaphycus* Doty (Solieriaceae, Rhodophyta). Aquat. Bot. 36 (1), 55–67.

Sarada, R., Vidhyavathi, R., Usha, D., Ravishankar, G.A., 2006. An efficient method for extraction of astaxanthin from green alga *Haematococcus pluvialis*. J. Agric. Food Chem. 54, 7585–7588.

Sarat Chandra, T., Suvidha, G., Mukherji, S., Chauhan, V.S., Vidyashankar, S., Krishnamurthi, K., Sarada, R., Mudliar, S.N., 2014. Statistical optimization of thermal pretreatment conditions for enhanced biomethane production from defatted algal biomass. Bioresour. Technol. 162, 157–165.

Schoeny, R., Cody, T., Warshawsky, D., Radike, M., 1988. Metabolism of mutagenic polycyclic aromatic hydrocarbons by photosynthetic algal species. Mutat. Res. 197, 289–302.

Scragg, A.H., Spiller, L., Morrison, J., 2003. The effect of 2,4-dichlorophenol on the microalga Chlorella VT-1. Enzyme Microb. Technol. 32 (5), 616–622.

Seckbach, J., Baker, F.A., Shugarman, P.M., 1970. Algae thrive under pure CO_2. Nature 227, 744–745.

Semple, K.T., Cain, R.B., 1996. Biodegradation of phenolics by *Ochromonas danica*. Appl. Environ. Microbiol. 62, 1265–1273.

Semple, K.T., Cain, R.B., Schmidt, S., 1999. Biodegradation of aromatic compounds by microalgae. FEMS 170, 291–300.

Shafik, M.A., 2008. Phytoremediation of some heavy metals by *Dunaliella salina*. Global J. Environ. Res. 2 (1), 1–11.

Shanaab, S., Essa, A., Shalaby, E., 2012. Bioremoval capacity of three heavy metals by some microalgae species (Egyptian isolates). Plant Signal. Behav. 7 (3), 392–399.

Shehata, S.A., Badr, S.A., 1980. Growth response of *Scenedesmus* to different concentrations of copper, cadmium, nickel, zinc and lead. Environ. Int. 4, 431–434.

Shekh, A.Y., Shrivastava, P., Krishnamurthi, K., Mudliar, S.N., Devi, S.S., Kanade, G.S., et al., 2013. Stress-induced lipids are unsuitable as a direct biodiesel feedstock: a case study with *Chlorella pyrenoidosa*. Bioresour. Technol. 138, 382–386.

Sivakumar, G., Xu, J., Thompson, R.W., Yang, Y., Smith, P.R., Weathers, P.G., 2012. Integrated green algal technology for bioremediation and biofuel. Bioresour. Technol. 107, 1–9.

Sivasubramanian, V., Muthukumaran, M., 2012. Large scale phycoremediation of oil drilling effluent. J. Algal Biomass Util. 3 (4), 5–17.

Sivasubramanian, V., Subramanian, V.V., Murali, R., Vijayalakshmi, T.M., 2012. Algal biomass production integrated with sewage treatment and utilization as feedstock for bio-fuel. J. Algal Biomass Util. 3 (2), 65–70.

Solovchenko, A., Khozin-Goldberg, I., 2013. High-CO_2 tolerance in microalgae: possible mechanisms and implications for biotechnology and bioremediation. Biotechnol. Lett. 35, 1745–1752.

Spolaore, P., Joannis-Cassan, C., Duran, E., Isambert, A., 2006. Commercial applications of microalgae. J. Biosci. Bioeng. 101 (2), 87–96.

Subashchandrabose, S.R., Ramakrishnan, B., Megharaj, M., Kadiyala, V.K., Naidu, R., 2011. Consortia of cyanobacteria/microalgae and bacteria: biotechnological potential. Biotechnol. Adv. 29, 896–907.

Sydney, E.B., Sturm, W.S., de Carvalho, J.C., Thomaz-Soccol, V., Larroche, C., Pandey, A., et al., 2010. Potential carbon dioxide fixation by industrially important microalgae. Bioresour. Technol. 101, 5892–5896.

Tampier, M., 2009. Microalgae technologies and processes for biofuels/bioenergy production in British Columbia: current technology, suitability and barriers to implementation. In: The British Columbia Innovation Council, 14 January 2009.

Tang, X., He, L.Y., Tao, X.Q., Dang, Z., Guo, C.L., Lu, G.N., et al., 2010. Construction of an artificial microalgal-bacterial consortium that efficiently degrades crude oil. J. Hazard. Mater. 181, 1158–1162.

Tang, D., Han, W., Li, P., Miao, X., Zhong, J., 2011. CO_2 biofixation and fatty acid composition of *Scenedesmus obliquus* and *Chlorella pyrenoidosa* in response to different CO_2 levels. Bioresour. Technol. 102, 3071–3076.

Torres, E., Cid, A., Fidalgo, P., Herrero, C., Abalde, J., 1997. Long chain class III metallothioneins as a mechanism of cadmium tolerance in the marine diatom *Phaeodactylum tricornutum* Bohlin. Aquat. Toxicol. 39, 231–246.

Tripathi, B.N., Gaur, J.P., 2004. Relationship between copper and zinc induced oxidative stress and proline accumulation in *Scenedesmus* sp. Planta 219, 397–404.

Tripathi, B.N., Mehta, S.K., Amar, A., Gaur, J.P., 2006. Oxidative stress in *Scenedesmus* sp. during short- and long term exposure to Cu^{2+} and Zn^{2+}. Chemosphere 62, 538–544.

Vasconcelos, V.M., Pereira, E., 2001. Cyanobacteria diversity and toxicity in a wastewater plant (Portugal). Water Res. 35, 1354–1357.

Velasquez-Orta, S.B., Lee, J.G.M., Harvey, A.P., 2013. Evaluation of FAME production from wet marine and freshwater microalgae by in situ transesterification. Biochem. Eng. J. 76, 83–89.

Vidyashankar, S., Deviprasad, K., Chauhan, V.S., Ravishankar, G.A., Sarada, R., 2013. Selection and evaluation of CO_2 tolerant indigenous microalga *Scenedesmus dimorphus* for unsaturated fatty acid rich lipid productionunder different culture conditions. Bioresour. Technol. 144, 28–37.

Vidyashankar, S., VenuGopal, K.S., Chauhan, V.S., Muthukumar, S.P., Sarada, R., 2014. Characterisation of defatted *Scenedesmus dimorphus* algal biomass as animal feed. J. Appl. Phycol, 1–9. http://dx.doi.org/10.1007/s10811-014-0498-9.

Wang, B., Li, Y., Wu, N., Lan, C.Q., 2008. CO_2 bio-mitigation using microalgae. Appl. Microbiol. Biotechnol. 79, 707–718.

Warshawsky, D., Cody, T., Radike, M., Reilman, R., Schumann, B., LaDow, K., Schneider, J., 1995. Biotransformation of benzo[a]pyrene and other polycyclic aromatic hydrocarbons and heterocyclic analogs by several green algae and other algal species under gold and white light. Chem. Biol. Interact. 97, 131–148.

Wijfells, R.H., Barbosa, M.J., 2010. An outlook on microalgal biofuels. Science 329 (5993), 796–799.

Wilkie, A.C., Mulbry, W.W., 2002. Recovery of dairy nutrients by benthic freshwater algae. Bioresour. Technol. 84, 81–91.

Williams, L.E., Pittman, J.K., Hall, J.L., 2000. Emerging mechanisms for heavy metal transport in plants. Biochim. Biophys. Acta 1465, 104–126.

Yang, J., Xu, M., Zhang, X., Hu, Q., Sommerfeld, M., Chen, Y., 2011. Life-cycle analysis on biodiesel production from micro-algae: water footprint and nutrients balance. Biotechnology 102, 159–165.

Yazdani, S.S., Gonzalez, R., 2007. Anaerobic fermentation of glycerol: a path to economic viability for the biofuels industry. Curr. Opin. Biotechnol. 18, 213–219.

Yu, G., Zhang, Y., Schideman, L., Funk, T.L., Wang, Z., 2011. Hydrothermal liquefaction of low lipid content microalgae into bio-crude oil. Trans. ASABE 54 (1), 239–246.

Zhu, L., Wang, Z., Shu, Q., Takala, J., Hiltunen, E., Feng, P., Yuan, Z., 2013. Nutrient removal and biodiesel production by integration of freshwater algae cultivation with piggery wastewater treatment. Water Res. 47 (13), 4294–4302.

Zmora, O., Richmond, A., 2004. Microalgae for aquaculture: microalgae production for aquaculture. In: Richmond, A. (Ed.), Handbook of Microalgal Culture: Biotechnology and Applied Phycology. Blackwell Publishing, Oxford.

BIOPROCESSES, BIOENGINEERING FOR BOOSTING BIO-BASED ECONOMY

BUILDING A BIO-BASED ECONOMY THROUGH WASTE REMEDIATION: INNOVATION TOWARDS SUSTAINABLE FUTURE

19

K. Amulya, Shikha Dahiya, S. Venkata Mohan

Bioengineering & Environmental Sciences; CSIR-Indian Institute of Chemical Technology (CSIR-IICT), Hyderabad, India

1 INTRODUCTION

Albert Einstein in his school essay themed "My Future Plans" wrote the less familiar quotation: "A happy man is too satisfied with the present to dwell too much on the future." This aptly fits the attitude of the current society (Brehmer, 2008). In the early days, the abundance of natural resources made the society affluently use everything available, without dwelling too much on the future. Fossil fuels have been indiscriminately used to obtain various forms of cheaper energy and fuels, and we are now enjoying the benefits of it. However, the question arises, on how long can we continue this way, because everything that we use has an expiration date including fossil fuels. Over the coming years we must work toward building a future that is not based on resource scarcity. In order to achieve this, we need to address all the environmental, social, and economic challenges. Although these issues have made headlines over the past few years, not much action has been taken, until recently, when governments have given environment protection its due importance while drafting national policies. Therefore, looking out for alternate renewable sources of energy is the need of the hour (Ragauskas et al., 2006).

Apart from diminishing fossil fuels, another problem the world is facing is the enormous quantities of waste being generated. Millions of tons of waste is generated worldwide every year, making its management and treatment a challenging ordeal (Arneil et al., 2013). Waste takes many forms, from household and industrial wastes to gasses emitted from industries. Although society is aware of this problem, considerable effort has not been put into managing the wastes. However, the research fraternity worldwide has gauged the magnitude of the problem and has been working relentlessly toward waste remediation. Apart from treatment, the focus has also been on developing several sustainable technologies that could maximize the value of waste by producing industrially important chemicals, materials, and fuels that not only reduce our dependency on fossil-based fuels and products but also help in mitigating environmental pollution.

Many scientists working with waste believe that it contains enough energy to replace a major fraction of the world's energy crisis, if it can be economically converted to useful energy forms. Although it would be presumptuous to believe that waste will be able to completely meet the world's energy needs, it is still achievable if a coalesced approach to recycle and reuse waste is developed, wherein the waste

Bioremediation and Bioeconomy. http://dx.doi.org/10.1016/B978-0-12-802830-8.00019-8

Copyright © 2016 Elsevier Inc. All rights reserved.

coming from one industry or sector becomes the raw material for another. In line with the above mentioned facts, this chapter attempts to elucidate the various forms of bioenergy and biofuels that can be generated from waste in a biorefinery approach. The potential of various existing technologies for waste remediation and energy generation along with limitations have been discussed. Future perspectives toward developing a bio-based economy have been delineated keeping sustainable society in the forefront.

2 WASTE AND ENERGY: A PARTNERSHIP FOR A SUSTAINABLE FUTURE

The proposition that energy can be obtained from wastewater at a scale that is sufficiently large to substantially meet both the energy demand and sustainability is quite apparent from several recent studies (Venkata Mohan, 2014). This idea is much more widely accepted now than only a few years ago due to the fact that it carries along with it environmental benefits like reducing environmental pollution, providing renewable source of energy, greenhouse gas (GHG) mitigation, and improved quality of life (Lynd et al., 2008). The arguments in favor of obtaining products from waste are persuasive and will only become more intense with the growing urgency of tackling the twin problems of climate change and energy demand (Marshall et al., 2013). Realizing even a tiny proportion of the benefits anticipated will motivate us to move forward in this direction and transform waste treatment toward sustainability.

3 VALUE ADDITION FROM WASTE
3.1 BIOLOGICAL HYDROGEN PRODUCTION

Hydrogen is considered to be an important alternative energy carrier and a bridge to achieve sustainable energy in the future (Venkata Mohan et al., 2009a). Hydrogen neither exists freely in nature nor is an energy source. It is a secondary form of energy that has to be generated. To be precise, it is an energy carrier. Hydrogen has a high energy yield of 122 kJ/g, which is 2.75 times greater than hydrocarbon fuels (Tawfik et al., 2011). Hydrogen production is carried out using different technologies, of which half of the production is carried out either by thermocatalytic or gasification processes that make the use of fossil fuels. Biologically, hydrogen can be produced through two main pathways: dark fermentation and photosynthesis. Hydrogen generated by biological machinery can be termed as *biohydrogen* (bio-H_2).

All the biological production methods are controlled by hydrogen-producing enzymes, such as hydrogenases and nitrogenases (Skizim et al., 2012). Based on light dependency, the bio-H_2 production processes can be classified into light-independent (dark) fermentation and light-dependent photosynthetic processes (Chandra and Venkata Mohan, 2014). Depending on the carbon source and the microorganisms involved, the photosynthetic process can again be classified into photosynthetic or fermentation process. Dark fermentative hydrogen production is carried out by obligate and facultative anaerobes under strictly anaerobic conditions. The dark fermentation process is confined to anaerobic metabolism where anaerobic microorganisms (mostly acidogenic bacteria) generate H_2 metabolically through an acetogenic process along with the generation of volatile fatty acids (VFA) and CO_2 (Eqs. 1–5).

$$C_6H_{12}O_6 + 2H_2O \rightarrow 2CH_3COOH + 2CO_2 + 4H_2 \qquad (1)$$

$$C_6H_{12}O_6 \rightarrow CH_3CH_2CH_2COOH + 2CO_2 + 2H_2 \qquad (2)$$

$$C_6H_{12}O_6 + 2H_2 \rightarrow 2CH_3CH_2COOH + 2H_2O \qquad (3)$$

$$C_6H_{12}O_6 + 2H_2 \rightarrow COOHCH_2CH_2OCOOH + CO_2 \qquad (4)$$

$$C_6H_{12}O_6 \rightarrow CH_3CH_2OH + CO_2 \qquad (5)$$

Light-dependent processes can be through direct and indirect biophotolysis of water using green algae and cyanobacteria or via photofermentation mediated by photosynthetic bacteria. In direct biophotolysis the solar energy directly converts water into hydrogen via the photosynthetic reaction (Eq. 6).

$$H_2O + hv \ (\text{solar energy}) \rightarrow 2H_2 + O_2 \qquad (6)$$

Cyanobacteria that participate in the indirect biophotolysis have unique characteristics of using carbon dioxide (CO_2) from the air as a carbon source and solar energy as an energy source to produce biomass (Eqs. 7 and 8) that is subsequently used for hydrogen production.

$$6H_2O + 6CO_2 + hv \ (\text{solar energy}) \rightarrow C_6H_{12}O_6 + 6O_2 \qquad (7)$$

$$C_6H_{12}O_6 + 6H_2O + hv \ (\text{solar energy}) \rightarrow 12H_2 + 6CO_2 \qquad (8)$$

Photosynthetic bacteria use simple organic acids (acetic and butyric acids) as electron donors and liberate molecular hydrogen, using nitrogenase enzyme under nitrogen-deficient, anaerobic, and in the presence of light energy.

3.1.1 Waste to Bio-H₂

Organic material in wastes, agricultural residues, wood and wood waste, aquatic plants, and algae can be considered as a good substrate for bio-H$_2$ production. (Saratale et al., 2008). Biomass derived from plant crops, agricultural residues, and woody biomass are also being used for generating bio-H$_2$ via both thermochemical and biological routes (Venkata Mohan et al., 2009b; Subhash and Venkata Mohan, 2014; Venkata Mohan, 2009). In order to make the organic fraction available for the bacteria, the cellulosic material requires an initial pretreatment step, which adds to the cost of the entire process. Unlike cellulosic materials, waste contains readily available carbon, thus making it a highly preferred substrate for bio-H$_2$ production. Dark fermentation process is widely used for bio-H$_2$ production as it is relatively less energy intensive and more environmentally sustainable due to its operational feasibility (Figure 1; Mohanakrishna and Venkata Mohan, 2013; Venkata Mohan et al., 2008a). Despite the advantages, low substrate conversion efficiency, accumulation of carbon-rich acid intermediates, and drop in system pH impede upscaling of bio-H$_2$ production (Pasupuleti et al., 2014). If these factors are optimized, then they can be the game changers, and H$_2$ production at a practical scale in a cost-effective way and with high efficiency might become a reality.

With petroleum-based fuels reaching the verge of depletion and the concern about continued emissions of additional CO_2 to the atmosphere intensifying, bio-H$_2$ from waste materials will be considered as a key technology for future sustainable energy supply. Renewable shares of 36% (2025) and 69% (2050) on the total energy demand will lead to hydrogen shares of 11% in 2025 and 34% in 2050. With

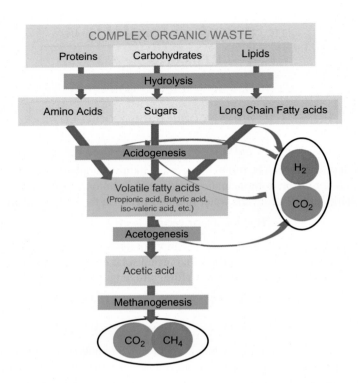

FIGURE 1

Schematic representation of the acidogenic fermentation process, generating volatile fatty acids and biohydrogen.

increasing reports on bio-H_2 production and several funded programs from many national government agencies all over the world, it is quite evident that hydrogen is being propagated as the fuel for the future (Balat and Kırtay, 2010).

3.2 VOLATILE FATTY ACIDS

Considering the various possibilities of resource recovery, production of volatile fatty acids (VFAs) as the intermediates of anaerobic digestion form different organic wastes is deemed to be one of the prominent areas with good commercial interest (see Figure 1) (Singhania et al., 2013). VFAs are short chain fatty acids (acetic, propionic, butyric, etc.) consisting of six or fewer carbon atoms (C2-C6), with pKa values between 4.7-4.9, and can be distilled at atmospheric pressure. VFAs can be produced chemically or synthetically from fossil resources, or as metabolic intermediates during acidogenic or dark fermentation (acidogenesis, acetogenesis, and methanogenesis). These VFAs are a part of the bacterial metabolism and can be produced either using mixed cultures or by pure strains under anaerobic conditions. Although commercial production of VFAs is mostly accomplished by chemical routes (Huang et al., 2002), its production using biological methods and organic-rich wastes is steadily increasing.

Biologically produced VFA when separated properly can be used as one of the most important and valuable chemical compounds with diverse applications in the market.

In biological VFA production, substrates from pure sugars such as glucose, sucrose to organic-rich wastes such as sludge generated from wastewater treatment plant, food waste, municipal solid waste, and industrial wastewaters have been commonly employed as the main carbon source (Dahiya et al., 2015; Amulya et al., 2014; Reddy et al., 2011). Production of VFAs is the function of individual bacteria in the pure culture or the syntrophic relation between the groups of organisms present in the open microbiome. In open mixed microbiome, acetic acid production generally occurs by the action of acetogens. These acetogens can degrade long chain fatty acid to acetic acid and H_2. These acetogens are mainly of two types: obligate H_2-producing acetogenic bacteria that are capable of producing acetate and H_2 from higher fatty acids, and the hydrogen-consuming acetogenic bacteria such as *Clostridium aceticum*, which are also considered as homoacetogens. These bacteria utilize H_2 and CO_2 to form acetic acid. The overall equation for the production of acetic acid from glucose is given in Eq. (9):

$$C_6H_{12}O_6 + 2NAD^+ + 2ADP + 2Pi \rightarrow 2CH_3CO_2COO^- + 2NADH + 2ATP + 2H_2O + 2H^+ \qquad (9)$$

In broader terms, acetogens are obligately anaerobic bacteria that use the reductive acetyl-CoA or Wood-Ljungdahl pathway as their main mechanism for energy conservation and for synthesis of acetyl-CoA and cell carbon from CO_2. An acetogen is sometimes called a "homoacetogen" (meaning that it produces only acetate as its fermentation product) or a "CO_2-reducing acetogen." Wood-Ljungdahl pathway is the hallmark of acetogens, but is also used by other bacteria, including methanogens and sulfate-reducing bacteria, for both catabolic and anabolic purposes. The key aspects of the acetogenic pathway are several reactions that include the reduction of CO_2 to carbon monoxide (CO) and the attachment of the carbon monoxide to a methyl group. The first process is catalyzed by enzymes called carbon monoxide dehydrogenase. The coupling of methyl group (provided by methylcobalamin) and the CO is catalyzed by acetyl CoA synthetase.

$$2CO_2 + 4H_2 \rightarrow CH_3COOH + 2H_2O \qquad (10)$$

Syntrophic bacteria produce acetic acid from higher fatty acids (propionic acid and butyric acid). Two pathways for propionate metabolism are known, the methylmalonyl-CoA pathway and a dismutation pathway, where two propionate molecules are converted to acetate and butyrate, the butyrate being degraded to acetate and hydrogen. In the methylmalonyl-CoA pathway, propionate is first activated to propionyl-CoA and then carboxylated to methylmalonyl-CoA. Methylmalonyl-CoA is rearranged to form succinyl-CoA, which is converted to succinate. Succinate is oxidized to fumarate, which is then hydrated to malate and oxidized to oxaloacetate. Pyruvate is formed by decarboxylation, and is further oxidized in a HS-CoA-dependent decarboxylation to acetyl-CoA and finally to acetate. Butyrate is degraded via β-oxidation pathway (McInerney et al., 2008). In a series of reactions acetyl groups are cleaved off yielding acetate and hydrogen. To metabolize fatty acids, first activation to a HS-CoA derivative takes place. The HS-CoA derivative is then dehydrogenated to form an enoyl-CoA. After water addition, a second dehydrogenation takes place to form a ketoacylacetyl-CoA. After hydrolysis acetyl-CoA and acyl-CoA are formed, which enters another cycle of dehydrogenation and the cleaving off of acetyl-CoA leading to acetic acid generation.

Propionic acid is produced by the propionic acid bacteria, which include *Corynebacteria*, *Propionibacterium*, and *Bifidobacterium*. The formation of propionate is a complex and indirect process involving five or six reactions. Overall, 3 mol of lactate are converted to 2 mol of propionate + 1 mol of acetate + 1 mol of CO_2, and 1 mol of ATP is squeezed out in the process. Most of the butyric acid-producing bacteria produce acetic acid in addition to butyric acid as their major fermentation products. In general, glucose (hexose) is catabolized via EMP (Embden-Meyerhof-Parnas) pathway and xylose (pentose) is catabolized by HMP (hexose monophosphae) pathway to pyruvate, which is then oxidized to acetyl-CoA and CO_2 with concomitant reduction of ferredoxin (Fd) to FdH_2. FdH_2 is then oxidized to Fd to produce hydrogen, which is catalyzed by hydrogenase, with excess electrons released to convert NAD^+ to NADH. Acetyl-CoA is the key metabolic intermediate at the node dividing the acetate formation branch from butyrate-formation branch. Phosphotransacetylase (PTA) and acetate kinase (AK) are two enzymes that convert acetyl-CoA to acetic acid, whereas phosphotransbutyrylase (PTB) and butyrate kinase (BK) catalyze the production of butyric acid from butyryl-CoA.

On a broader context, any organic waste (1 kg of chemical oxygen demand (COD)) on its 100% bioconversion can produce 0.62 kg VFA (acetate pathway), 0.41 kg (butyrate pathway), and 0.97 kg of VFA (propionate pathway). The VFA produced from acidogenic fermentation of waste are the building blocks for many chemicals synthesis and serve as a valuable substrate to a variety of applications such as the production of biodegradable plastics, generation of bioenergy, and biological nutrient removal. It is often feasible, and certainly highly desirable, to utilize the VFA-rich fermented waste directly in these applications. VFAs serve as potential substrate for the production of hydrogen via photofermentation by mixed microbial cultures (Srikanth et al., 2009). Apart from bio-H_2 and electricity, they can also serve as a carbon source for lipid storage in microalgae (Venkata Mohan and Devi, 2012), synthesis of single cell oils (SCOs) by oleaginous yeast (Christophe et al., 2012), and production of bioplastics (polyhydraoxyalkanates). VFAs can also serve as substrate for microbial fuel cell (Chen et al., 2012) as only soluble organic compounds can be directly fed into anode of microbial fuel cells (MFCs) to generate electricity, and single chain fatty acids are the most preferred one (Venkata Mohan et al., 2008b). In addition, VFA also assist the biological removal of nitrogen and phosphorus from wastewater through aerobic nitrification followed by anoxic denitrification (Lee et al., 2014). All these applications show the potential of VFA to bring about a revolution in the field of bioenergy. VFA production can be considered an effective strategy to bridge the gap between waste remediation and product recovery. Although today's major challenge is VFA separation, the zero cost of substrate and use of technologies like membrane separation and extractive fermentation can make the process inexpensively feasible.

3.3 MICROBIAL ELECTROCHEMICAL SYSTEMS (MECS)

3.3.1 Microbial Fuel Cell (MFC) for Bioelectricity Generation

MFC are a type of MECS that are used for the generation of bioelectricity. The use of MFC for power generation was first documented in 1915 by Potter and his coworkers (Potter, 1910). Many years later in 1980, studies on mediators that improve the performance of MFCs were studied (Du et al., 2007; Venkata Mohan et al., 2014a). After nearly 2 decades, studies on electrochemically active bacteria that transfer electrons from the cytoplasm to the electrode surface (Bond and Lovley, 2003) started to emerge, resulting in the development of mediator-less MFCs (Kim et al., 2002). Since then intense research has been carried out to improve the power densities in MFCs. For obvious reasons, the power densities obtained in MFCs cannot be compared with that obtained in conventional fuel cells, because the latter uses pure and expensive chemicals and fuels rather than wastes (Suresh Babu et al., 2014). Nevertheless, with major breakthroughs in reactor configurations, electrode materials and enrichment of electroactive of microbes can help improve the performance of MFCs to a greater extent (Zhou et al., 2012).

3.3.1.1 Bioelectrogenesis in MFC

Normally, bacteria utilize the carbon (substrate) available in the wastewaters to generate reducing equivalents (protons [H$^+$] and electrons [e$^-$]), in order to carry out their metabolic activities. Redox components/carriers (NAD, FAD, FMN, etc.) carry these e$^-$ and H$^+$ toward an available terminal electron acceptor, thus generating proton motive force, which in turn facilitates generation of energy-rich phosphate bonds that are useful for the microbial growth and metabolism (Venkata Mohan et al., 2014b; Zhao et al., 2009). In MFCs the electrodes introduced into the cell act as intermediary electron acceptors. Oxidation occurs at the anode, generating (H$^+$) and electrons (e$^-$), while reduction takes place at the cathode. The electrons remain at the anode generating negative anodic potential, while H$^+$ move toward the cathode via the proton exchange membrane, generating positive cathode potential. The reactions occurring at anode and cathode can be represented as follows:

$$C_6H_{12}O_6 + 6H_2O \rightarrow 6CO_2 + 24H^+ + 24e^- \ (anode) \tag{11}$$

$$4e^- + 4H^+ + O_2 \rightarrow 2H_2O \ (cathode) \tag{12}$$

$$C_6H_{12}O_6 + 6H_2O + 6O_2 \rightarrow 6CO_2 + 12H_2O \ (overall) \tag{13}$$

The transfer of electrons from bacterial cytoplasm to the electrode surface is an important factor that governs the performance of MFC. There are two possible mechanisms by which the electrons are transferred onto the electrode surface: direct electron transfer (DET) and mediated electron transfer (MET). DET occurs via the physical contact of the microbial cell wall and anode without the involvement of any redox species or mediators (Schroder, 2007). The membrane-bound electron transport proteins could permit the transfer of electrons from the outer membrane of the bacteria to anode (Venkata Mohan et al., 2014b). Formation of biofilm through thecoic acids that help in the adherence of the bacteria to electrode surface by many of the Gram-positive bacteria also helps in DET (Liu et al., 2012; Annie Modestra et al., 2014). C-type cytochromes and multiheme proteins, are identified as one of the possible routes for DET. One of the major disadvantages of this process is that only a monolayer of cells participate in the transfer of electrons to the anode (Gupta et al., 2013). Studies have identified that a few electrochemically active bacteria have shown efficient DET mechanism; for example, *Geobacter*, *Rhodoferax*, and *Shewanella* (Lovley and Nevin, 2013; Chang et al., 2006). Some bacteria like *Geobacter* and *Shewanella* carry out DET through conductive pilli (electrically conductive nanowires) formed on the bacterial cell surface (Malvankar et al., 2012). MET is carried out through the redox shuttles (that are either added as artificial mediators or secreted as soluble mediators such as primary and secondary metabolites from bacterial metabolism) which mediate the electron flow from bacterial metabolism toward electrode. The oxidized mediators penetrate the membrane of the organism and get reduced by the electrons. These reduced mediators again pass through the bacterial membrane and reach the anode, where they become reoxidized by losing electrons (Neto et al., 2010). In this manner the electrons are transferred from inside the bacterial cell to the electrode, thus accelerating electron transfer and enhancing power output. The examples of artificial mediators include inorganic (potassium ferricyanide) or organic (benzoquinone) group dyes and metallorganics such as neutral red (NR), methyleneblue (MB), thionine, Meldola's blue (MelB), 2-hydroxy-1,4-naphthoquinone (HNQ), and Fe (III) EDTA (Choi et al., 2003; McKinlay and Zeikus, 2004). Since the synthetic mediators are unstable and toxic, their use in MFCs are limited. However, various naturally secreted secondary metabolites by the bacteria act as good mediators. These include phenazines, phenoxazines, quinines, pyocyanin, etc. (Bennetto et al., 1983; Hernandez and Newman, 2001).

MFC can use a diverse range of substrates (anolyte fuel) as electron donors for anodic oxidation to generate reducing equivalents that in turn generate power. Utilization of wastewater as well as solid waste as anodic fuel has been well established since it facilitates simultaneous treatment (Venkata Mohan et al., 2008b,c, 2013a,b). The performance of MFC depends on several factors such as configuration of fuel cell, nature of anolyte, electrode materials, spacing between the electrodes, proton exchange membrane properties, nature of microbes, electron transfer mechanism, redox condition (pH), anolyte microenvironment, etc. (Venkata Mohan and Chandrasekhar, 2011; Raghavulu et al., 2009). MFCs primarily make use of anaerobic bacteria, but other types of organisms like photosynthetic bacteria have also been used in MFCs for tapping solar energy, thus widening the scope of MFC applications (Srikanth et al., 2009b; Raghavulu et al., 2013; Chandra et al., 2012). These fuel cells can also be designed with different configurations; for example, benthic MFCs, plant-based MFC, and stacked MFC. In benthic MFCs bioelectricity is harnessed from an aquatic ecosystem using natural habitants (Chiranjeevi et al., 2013a) while in plant-based MFCs, indirect utilization of solar radiation takes place for generation of green electricity by integrating the roots of living plant (rhizosphere) with electrodes. Rhizodeposits secreted from roots are organic in nature and provide favorable microenvironment for proliferation of bacteria (Chiranjeevi et al., 2013b). In stacked MFC, individual MFCs are stacked in series and parallel connections to increase the power output (Kim et al., 2012).

3.3.2 Bioelectrochemical treatment system (BET)

If MCES are used exclusively for the treatment of wastewater, then it is termed as a BET. Multiple reactions, namely biochemical, physical, physicochemical, electrochemical, oxidation, etc., can be triggered in the anodic chamber of BET operation for the treatment of complex pollutants present in wastewaters (Venkata Mohan et al., 2009c; Velvizhi and Venkata Mohan, 2015). Although the anode chamber of BET resembles the conventional anaerobic bioreactor and electrochemical cell, the simultaneous occurrence of both oxidation and reduction reactions gives it an edge over the conventional treatment processes where reduction reactions are predominant (Chandrasekhar and Venkata Mohan, 2012). BET have been used for multi-pollutant removal like perchlorate (Thrash et al., 2007), aromatic organic compounds (Venkata Mohan and Chandrasekhar, 2011; Kiran Kumar et al., 2012), total dissolved solids (TDS) (Velvizhi et al., 2014) and sulfides/nitrates (Velvizhi et al., 2014). Designing an appropriate configuration of BET can facilitate its integration with existing wastewater treatment processes, which is a critical aspect for further development of this process (Kelly and He, 2014; Vamshi Krishna et al., 2014).

3.3.3 Microbial Electrolysis Cell (MEC)

Like BET, the configuration of MFC can be modified to generate bio-H_2. Applying a small amount of external potential to achieve value addition in the form of bio-H_2 is the main principle underlying the functioning of MEC. It facilitates the conversion of electron equivalents in organic compounds to H_2 gas by combining microbial metabolism with bioelectrochemical reactions. Low energy consumption compared to the conventional water electrolysis, high product (H_2) recovery, and substrate degradation are some of the potential benefits that favor MEC as an alternate process (Cheng and Logan, 2007) for value addition.

3.3.4 Bioelectrochemical Systems (BES)

In addition, certain microorganisms in a defined fuel cell system are capable of generating biofuels, hydrogen gas, methane, acetate (Annie Modestra et al., 2015), ethanol, hydrogen peroxide, butanol, and other valuable inorganic and organic chemicals and this electrically driven reduction of electron acceptors in cathode chamber is termed as BES (Rabaey and Rozendal, 2010; Gong et al., 2013).

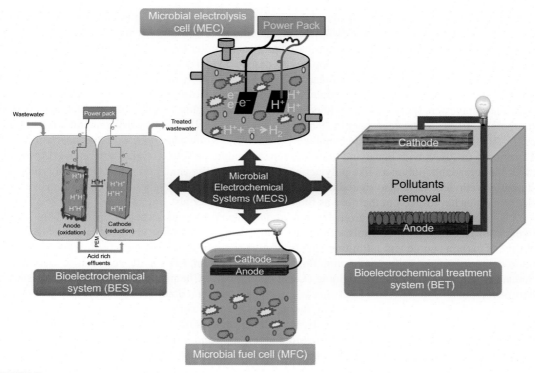

FIGURE 2

Overview of the multifaceted applications of microbial electrochemical systems (MECS).

The essence of these MCES is the feasibility of achieving value addition with simultaneous waste remediation (Figure 2). Despite the advantages and the advances made in MCES research, the output achieved either in terms of power or value-added products in these systems is a major limitation. The performance of these systems is dependent on the efficiency of the electron transfer machinery in the organisms. From the extensive research being carried out in microbial dynamics, it is evident that microbial limitations will have a profound influence on the performance of these systems rather than limitations in configurations. Therefore, attention is required to enhance the electron transfer capabilities of the organisms. Genetic manipulation of specific proteins involved in electron transfer can help increase the electron transfer by the organisms. A new era for MFC research might evolve where efforts will be directed toward gaining thorough understanding on the biocatalyst's capabilities, electron transfer mechanisms, and kinetics.

3.4 BIOPLASTICS

The flexibility and durability of synthetic plastics has made them one of the most used materials in a variety of industrial and day-to-day applications (Keshavarz and Roy, 2010). Ever since their discovery, they have been used indiscriminately, but now people have realized the harmful impact of these plastics on the environment. Environmental issues like global warming and the greenhouse effect are caused when these materials are used for a short time span and then are often dumped in landfills or are incinerated. In addition to the environmental issues, the dwindling price of crude oil is another factor of immense uncertainty especially for the production of petroleum-based plastics. In order to avert the problems associated with synthetic plastics, the exigency for developing bio-based polymeric materials is increasing.

The solution to this problem was already given in the 1920s by the French microbiologist Maurice Lemoigne, who discovered the intracellular granules in Gram-positive bacterium *Bacillus megaterium*, which are polyesters (poly[3-hydroxybutyrate],P[3HB]) belonging to polyhydroxyalkanoates (PHAs) (Lemoigne, 1926). Many years later in 1974, PHA containing 3HB and 3-hydroxyvalerate (HV) (Arneil et al., 2013) were found in activated sludge by Wallen and Rohwedder (1974). After nearly a decade, 11 varieties of PHAs with linear and branched repeating units of four to eight were reported. Although other biodegradable polymers like polylactic acid, polyglycollic acid, and starch-based polymers like starch-polyethylene have also shared the limelight, lack of variability in structure and extensive material properties have ruled out their use as future polymeric materials.

PHAs are polyesters synthesized by numerous prokaryotes and archea from renewable sources like carbohydrates, lipids, alcohols, or organic acids, under unfavorable growth conditions like availability of carbon source and a limiting supply of nitrogen, phosphate, or dissolved oxygen or certain microcomponents like sulfur, potassium, tin, iron, or magnesium. PHA in the cells can be identified by Sudan black staining or Nile red staining (Legat et al., 2010). Chemically PHAs are polyoxyesters of hydroxyalkanoic acids (HAs). PHAs appear as granules having a diameter of 0.2-0.7 μm. The granule is made of a hydrophobic core containing the polymer (97.7%) surrounded by a membrane coat made up of proteins (1.8%) and lipids (0.5%). Every granule consists of at least a thousand polymer chains forming right-handed 21-helices with a twofold screw axis and a fiber repeat of 0.596 nm, stabilized by hydrogen bonds and van der Waals intramolecular forces (Braunegg and Lefebvre, 1998; Figure 3). Based on the number of carbon atoms in the monomers, PHAs are divided into three different groups:

- short-chain-length (scl) PHAs containing 3-5 carbon atoms
- medium-chain-length (mcl) PHAs containing 6-14 carbon atoms
- long-chain-length (lcl) PHAs containing more than 15 carbon atoms (these have not been detected in naturally occurring PHAs and only in vitro production has been carried out).

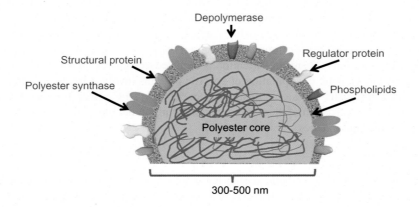

FIGURE 3

Diagrammatic representation of PHA granule formed in bacterial cells.

The production of either scl-PHAs or mcl-PHAs is confined only to specific organisms. *Ralstonia eutropha* and *Azohydromonas lata* synthesize scl-PHAs while *Pseudomonas* sp. produce mcl-PHAs (Koller et al., 2010). Under balanced growth conditions, the amount of CoA is high, and the first enzyme of PHA synthesis beta ketothiolase is inhibited and acetyl-coA is metabolized in the tricarboxylic acid (TCA) cycle (Anderson and Dawes, 1990). Under stress conditions, proteins cannot be synthesized and NADH is accumulated inside the cell. NADH inhibits citrate synthase, acetyl CoA accumulates in the cell, and PHB synthesis is initiated. The synthesis of scl-PHA begins with the conversion of carbon source to acetate. To this acetate molecule an enzyme co-factor is attached via a thioester bond forming acetyl-coA. Two molecules of acetyl-coA are condensed to form acetoacetyl coA, and this is subsequently reduced to monomer unit (R)-3-hydroxybutyryl-CoA. The enzymes involved in these three steps are the beta ketothiolase (phaA), NADH- dependent acetoacetyl-coA reductase (phaB), and PHA synthase (phaC). Mcl PHAs are mainly synthesized either by the fatty acid oxidation pathway or the fatty acid de novo synthesis pathway (Klinke et al., 1999; Figure 4).

Despite the striking advantages of PHA, competing with the well established synthetic plastic manufacturing industry is a daunting task. Not only the synthetic plastics but also to stand out among the other bioplstics like PLA, various process modifications right from production to downstream processing should be carried out. One of the major constraints of PHA production is the cost of the substrate which accounts to 50-60% of the total production cost. Considerable efforts have been made to use a

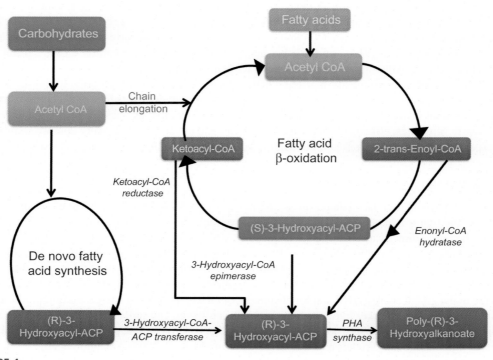

FIGURE 4

Schematic pathway depicting the production of short-chain- and medium-chain-length PHAs.

variety of affordable and renewable wastes for PHA production. Extensive studies were carried out using wastes from different sources like olive oil mill effluents (Beccari et al., 2009), palm oil mill (Bhubalan et al., 2007), spent wash effluents (Amulya et al., 2014; Reddy et al., 2014), corn steep liquor (Nikel et al., 2006), food waste (Reddy and Venkata Mohan, 2012), etc. Detailed studies have been carried out to develop efficient culture conditions for enhancing PHA production (Chen et al., 2013; Albuquerque et al., 2011; Venkateswar Reddy and Venkata Mohan, 2012; Venkata Mohan and Venkateswar Reddy, 2013). Different fermentation strategies like batch, particularly fed-batch and continuous fermentations, have been investigated (Wang and Lee, 1997). Studies have also been carried out using mixed cultures with bacterial species and strains often obtained from the same waste streams. The most preferred strategy for PHA production in mixed cultures is the aerobic dynamic feeding strategy, wherein the organisms are subjected to alternate "feast" and "famine" conditions (Villano et al., 2010). During the feast condition, excess carbon is available for growth of the organism and storage of polymer. During the famine condition, there is a deficit of carbon source, which is compensated by the use of the stored polymer. Using this strategy, most of the PHA fermentations are carried out in two stages. The objective is to produce a high cell density culture in the first stage (growth) and then to increase the concentration of PHAs during the second stage (Serafim et al., 2004).

The applications of PHA are not limited to just packaging, but can be used for a wide variety of purposes like medical applications, pharmaceuticals, agroindustrial products, etc. (Saharan et al., 2011; Akaraonye et al., 2010). PHAs are completely degradable and are degraded to water and CO_2 as the final products of their oxidative breakdown. The final products obtained are the basic materials that can be used by green plants to generate carbohydrates. It has to be emphasized at this point that petroleum-derived plastics on burning release CO_2 that is not a part of the natural carbon cycle and gets accumulated in the environment. PHAs, on the other hand, are embedded in the natural closed carbon cycle (Koller et al., 2010). Although extensive studies have been carried out for PHA production, commercial production becomes economically feasible when high cell densities and high polymer production are obtained at low cost. In order to achieve this, the two most important factors that need attention are substrate and microorganisms. Developing highly productive strains that are capable of using a wide variety of waste substrates, effective fermentation strategies and lowering the recovery cost are some of the key aspects that need to be worked on to achieve economically viable, industrial production of PHAs.

3.5 BIODIESEL

In 1900, Rudolph Diesel used peanut oil as fuel in the World Exhibition in Paris, marking that date in history as the first use of vegetable oil as an alternative fuel. However, the major problem with vegetable oil when used directly as fuel is its high viscosity, which interferes with engine functioning leading to engine failure (Knothe, 2005; Singh and Rastogi, 2009). Therefore, vegetable oils were modified to bring their combustion-related properties closer to those of diesel fuel. In the present day these fuels are known as biodiesel. Currently, biodiesel is one of the renewable biofuels that is being explored as a solution to the global energy crisis and for abatement of GHG emissions. First-generation biodiesel was produced using edible sources like soybean, rapeseed, sunflower, and safflower. The second-generation biodiesel was produced by using nonedible oil sources like used frying oil, grease, tallow, lard, karanja, jatropha, and mahua oils (Francis and Becker, 2002; Dorado et al., 2002; Alcantara et al., 2000). First- and second-generation sources are costly and seem to initiate land clearing and potentially compete with net food production (Chisti, 2008; Marsh, 2009). Therefore, the interest of scientists is now deviated toward the use of cheap, reliable, and renewable sources, which are more promising and economically viable. In this category of renewable sources, algae are the forerunners (Venkata Mohan et al., 2011, 2014c). Using algae for biodiesel production can solve the problem of reducing the large land requirements, more CO_2 sequestration, and rapid growth when compared with the crop-based terrestrial sources of biomass. In addition, the fatty acid profile of microalgal oil is suitable for the synthesis of biodiesel (Gouveia and Oliveira, 2009). Algae are photosynthetic microorganisms that live in water and grow hydroponically, capable of assimilating carbon heterotrophically and mixotrophically (Subhash et al., 2014; Subhash and Venkata Mohan, 2015; Chandra et al., 2014; Devi et al., 2012). This property of the microalgae is an advantage and hence algal biodiesel production can be integrated with wastewater treatment (Venkata Mohan et al., 2015; Figure 5). Several wastewaters are potential sources of carbon, which can be taken up by microalgae and converted to biodiesel. Under stress conditions microalgae synthesize triacylglycerols (TAG) in the form of storage lipids. Several species of microalgae such as *Prymnesium paryum*, 22-38%; *Chlorella vulgaris*, 22%; *Chlamydomonas rheinhardii*, 21%; *Spirogyra* sp., 11-21%; *Scenedesmus dimorphous*, 16.40%; *Scenedesmus obliquus*, 12-12%; *Synchoccus* sp., 11%; *Porphyridium cruentum*, 4-14%; *Dunaliella bioculata*, 8%; and *Tetraselmis maculate*, 3%; based on dry biomass (Becker, 1994, 2004), are rich in lipids, which can be converted to biodiesel using extraction strategies.

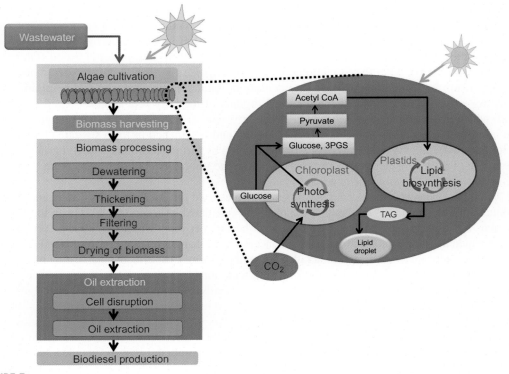

FIGURE 5

Schematic representation of biodiesel and lipid production pathway.

Algae fix CO_2 during the day via photophosphorylation (thylakoid) and produce carbohydrate during the Calvin cycle (stroma), which converts into various products, including TAGs, depending on the species of algae or specific conditions (Liu and Benning, 2012). The biosynthetic pathway of lipid in algae starts with the carbohydrates' accumulation inside the cell. This carbon accumulation varies with both autotrophic and heterotrophic organisms. Heterotrophic organisms assimilate it from outside the cell, whereas autotrophs synthesize their own carbon through photosynthesis. In photoautotrophs, photosynthates provide an endogenous source of acetyl-CoA, which is a key link in further lipid biosynthetic pathways. Heterotrophic nutrition is further light dependent and light independent, where the carbon uptake will be through an inducible active hexose symport system from outside the cell (Perez-Garcia et al., 2011; Tanner, 1969; Komor, 1973; Komor and Tanner, 1974). This process in the cell invests energy in the form of adenosine triphosphate (ATP). In light-independent processes (dark heterotrophic), light inhibits the expression of the hexose/Hþ symport system (Perez-Garcia et al., 2011; Kamiya and Kowallik, 1987), which decreases glucose transport inside the cell. Once carbon enters the cytosol, it follows cytosolic conversion of glucose to pyruvate through glycolysis and leads to the generation of acetyl-CoA. In the second step the produced acetyl coA is transported from the cytosol to the plastid, where it is converted to the fatty acid and subsequently to TAG, which again is transported to the cytosol and forms the lipid bodies. The acetyl-CoA pool will be maintained through the Calvin cycle, glycolysis, and pyruvate kinase (PK) mediated synthesis of pyruvate from phosphoenolpyruvate (PEP). The acetyl CoA then moves toward the formation of malonyl CoA. This malonyl Co-A undergoes synthesis of long carbon-chain fatty acids through repeating multistep sequences. A saturated acyl group produced by this set of reactions becomes the substrate for subsequent condensation with an activated malonyl group (Ohlroggeav and Browseb, 1995). After the formation of seven malonyl-CoA molecules, a four-step repeating cycle (extension by two carbons/cycle)—condensation, reduction, dehydration, and reduction—takes place for seven cycles and forms the principal product of the fatty acid synthase systems, palmitic acid, which is the precursor of other long chain fatty acids (Fan et al., 2011; Alban et al., 1994). In the last step, the palmitic acid gets modified further and lengthened to form stearate (18:0) or even to longer saturated fatty acids (oleiceate, linealate, etc.) by further additions of acetyl groups through the action of fatty acid elongation systems present in the smooth endoplasmic reticulum (ER) and in mitochondria (Thelen and Ohlrogge, 2002).

There have been several recent leaps in technologies used for microalgal cultivation and biodiesel production. The theoretical maximum possible yield calculated by Joseph C. Weissman for algal productions is $100\,g\,day^{-1}$ or 365 tons dry biomass per hectare per year (Weissman, 2008). Current limitations include that CO_2 is not free at the high concentrations that are required for peak algal growth and that algal grazers are a significant but a rather ignored problem. Cost-effective harvesting has been and still is a major limiting factor. As a step further, integration of algae-based biodiesel production with wastewater treatment is now considered as an economically viable platform for bioenergy generation (Stephens et al., 2010). In a nutshell, wastewater has the requisite potential to support algal biomass growth and lipid accumulation, and if various parameters influencing biodiesel production are optimized, it can become a viable technology. Conventionally, wastewater is treated in large volumes in bioreactors where penetration of light into a dense culture is limited. Light-independent heterotrophic cultivation can come to the rescue and facilitate integration of wastewater treatment and biodiesel production. The aim of an algal biorefinery is to achieve optimum utilization of resources and maximum remediation of environment during the course of biodiesel production. Autotrophic algal cultivation can be supplemented with CO_2 originating from industrial waste gases. The effluents from these dark fermentations can again be used for heterotrophic algal cultivation, making the whole process highly sustainable.

Algal biofuels are apparently the only current renewable source that can effectively meet the global demand for transport fuels. Moreover, microalgal biofuels have much lower impacts on the environment and the world's food supply than conventional biofuel-producing crops. The main reasons for this are high yields, a near-continuous harvest stream, and the potential to site the algal bioreactors on nonarable land. The biggest challenge over the next few years in the biodiesel field will be to reduce costs for cultivation and to further improve the biology of oil production. There is promise in the algal biorefinery concept where novel designs and materials will ensure highly efficient closed bioreactors for algal biomass growth, and the process will be integrated to several other processes for optimum utilization of algal biomass.

4 BIO-BASED ECONOMY

The bio-based economy is a term that encapsulates our vision of a future that is not based on resource scarcity, one no longer entirely dependent on fossil fuels for energy and industrial raw materials (Venkata Mohan, 2014). Bio-based economy is fueled by research and innovation in the multidisciplinary areas of science, management, and engineering converging toward biological sciences and is one of the rapidly growing segments of the world's economy due to the multitude of societal benefits it offers (White House, 2012). The concept of bioeconomy is not just limited to one sector but spans all the areas that revolve around development, processing, production, handling, or utilization of any form of biological resources, such as plants, animals, and microorganisms. Even in the energy sector, the role of bio-based economy cannot be denied. Carbon-based fossil fuels have been the backbone for our energy and chemical sectors, as we primarily rely on coal, oil, and natural gas from running our cars to producing a wide range of plastic products (Federal Ministry of Education and Research, 2011). However, decades of research has led to advancements in technologies which provide a picture of the previously unimaginable future offering ready-to-use fuels, biodegradable plastics, bio-based materials and chemicals from wastes/CO_2 (White House, 2012). All these products can be broadly classified as "bio-based products"; which mainly include biofuels, bioenergy, and bio-based chemicals. Production of these bio-based products from waste materials is increasing tremendously, after knowing their importance and role in environment. These biofuels can save nature from destruction and can also help in solving the problem of fossil fuels depletion. The researchers are working on the fields where agricultural waste, forestry residue, and solid municipal wastes can be converted into advanced biofuels, chemicals, and other valuable coproducts (Canadian Renewable Fuels association, n.d.).

The production of biofuels from waste gives a new dimension to the world as it can resolve the major problem of pollution and energy in a unified approach. This strategy manifests a clear waste-to-value addition. Biofuels produced from waste also are the cleanest and most sustainable source of fuel. Biofuels production is now becoming a priority in developing a bio-based economy (OECD, 2013). Biofuels are gaining perceptible importance within the transportation sector to reduce GHG as much as 99% when compared to petroleum-based fuels, depending on feedstock used for their production. Biofuels generate significantly lower GHG emissions than petroleum and diesel, especially CO_2.

Expanding the use of biofuels would further lower GHG emissions from carbon-intensive industries such as mining, transportation, etc. Thus, a bio-based economy can offer a unique opportunity to address interconnected challenges of resource scarcity and climate change, with simultaneous achievement of green economic growth.

In India, about 70% of energy generation capacity is from fossil fuels, and it is estimated that by 2030 India's dependence on energy imports is expected to exceed 53% of the country's total energy consumption. India has increased its share of renewable energy (electricity) from 2% (1628 MW) in 2002 to 11% (18,155 MW) in 2010 (Strategic Research Agenda, 2014). India's twelfth 5-year (2012-2017) plan emphasizes the projected derivatives from a biomass or bio-based economy route for utilization of agrowastes as energy sources. In India, the plan to develop a bio-based economy is solely based on growing nonfood feedstocks on degraded lands or wastelands that are not suited to agriculture, thus averting the conflict of fuel versus food security (National Biofuel Policy of India).

The bio-based economy in India is expected to grow by 2017, and there is a plan to achieve 55-GW power generation capacities from renewable resources. The energy policy in India extends its support to the renewable energy sector by providing incentives at the federal and state government levels. India is rich in biodiversity as well as in biomass resources. Recognizing the potential of such resources is imperative, not only for meeting energy demands, but also for monetizing these resources to enhance rural livelihoods and to substitute conventional feed-stocks for bio-based industries. Currently the main focus in India is on the supply of energy and electricity to rural areas rather than on developing bio-based commodities and products (Strategic Research Agenda, 2014).

4.1 TRANSITION TOWARD A BIO-BASED ECONOMY

Four main societal challenges that can be seen as the drivers for the bio-based economy are (Strategic Research Agenda, 2014)

- Food security (production, quality, fair consumption)
- Climate change (mitigation and adaptation)
- Resource security (energy, scarce materials)
- Ecosystem services (soil, water, biodiversity)

Addressing these multidimensional issues requires a strategic and comprehensive approach involving different policies. Although the concept of developing biorefineries and bio-based products, including biofuels, has been extensively supported in different regions of the world, the lack of comprehensive policies limits the development of a bio-based economy. Europe and the United States are the only two countries where these policies seem to emerge, due to the well-established chemical and biotech industries that drive the use of renewable sources of energy. Thus, the transition toward a bio-based economy can be achieved by developing policies that affect multiple sectors such as agriculture, research, industry, and trade. Gaging the impacts of the policies is also important, given the variety of policy instruments available (taxes, subsidies, price support, etc.) and the way they are applied. The policies implemented should be able to stimulate efficient use of resources to achieve a sustainable bio-based economy.

Not only the government, but also the consumers and producers have the responsibility of acting in a responsive way to support the realization of the sustainable bio-based-economy. Fortunately, there is a growing population that realizes this grave situation, but more awareness and education is required

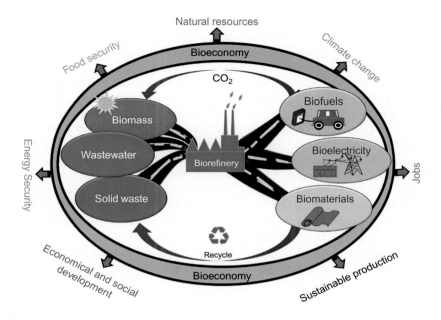

FIGURE 6

Overview of the bio-based economy.

to realize the implementation of the sustainable bio-based-economy. Although the bio-based economy proposes to offer various prospects, many efforts are still needed in order to fully exploit its advantages. In order to achieve this, renewable sources of energy should be available at affordable prices, market value of the products should be stimulated, and research must be more focused to develop innovative products. Regulatory moves will also play a role in ensuring that new technologies come to market (EuropaBio, 2010). Shifting society's dependence away from petroleum to renewable resources and realizing the full potential of waste in an integrated and holistic approach is essential and will help in developing a sustainable and competitive bio-based economy (Ragauskas et al., 2006). We can succeed in transitioning toward a bio-based economy, if food security, environmental, climate, and biodiversity protection are coalesced with it (Figure 6).

5 FUTURE PERSPECTIVES

The current industrial system that generates products required by the continuously developing world is not sustainable. Earth is being depleted of its resources at an alarming rate that needs to be halted. Recent data show that 25% of the population from industrialized countries consumes about 75% of the world's natural resources and controls about 88% of the total production, 80% of trade, and 94% of industrial products of the entire world (Hens and Quynh, 2008). Additionally, 90% of the resources used during the production process approach obsolescence and end up as waste, creating environmental problems. All these factors generate premises to evaluate and develop ways of producing products in a sustainable and environmentally friendly manner. These processes should be based on the use of renewable sources like biomass or biowastes, which are abundantly available, are the most promising carbon- neutral source of energy that can mitigate greenhouse emissions, and are perceived to be an essential part of the bio-based economy (Kamm et al., 2006). Developing an environmental biorefinery that can be defined as "a facility that converts waste materials into fuels, energy, chemicals, and materials" should be our agenda (Figure 7). Similar to a petrochemical refinery, an environmental biorefinery should be able to produce one or a few principal products initially and then slowly evolve with time by technology development and various integration strategies (Venkata Mohan, 2014). However, the major difference between petroleum-based biorefinery and environmental biorefinery is the variation and complexity of the waste and wastewaters. These variations and complexity can be viewed as an advantage in obtaining a diverse range of end products. Research efforts in these areas of renewable energy should be expedited and technologies for solving the gargantuan problems of waste management and languishing fossil fuels should be developed at the earliest. Efforts are underway and more and more researchers are turning toward waste as a potential feedstock for generation of different forms of bioenergy and for resource recovery. The perspective is changing; waste is no longer a menace; it is being perceived as treasure not trash. It is only a matter of a few decades before the energy generation potential of waste will be realized by novel technological innovations.

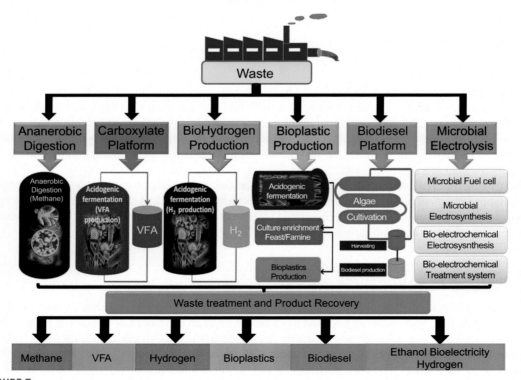

FIGURE 7

Schematic diagram representing the production of various bio-based products in an environmental biorefinery.

ACKNOWLEDGMENTS

The authors wish to thank the director of CSIR-IICT for support and encouragement in carrying out this work. The research was supported by CSIR-XII plan project, (SETCA; CSC-0113) and Department of Biotechnology (DBT), Government of India through National Bioscience Award 2012 (BT/HRD/NBA/34/01/2012(VI)) and New INDIGO project (DST/IMRCO/New INDIGO/Bio-e-MAT/2014/(G/ii)).

REFERENCES

Akaraonye, E., Keshavarz, T., Roy, I., 2010. Production of polyhydroxyalkanoates: the future green materials of choice. J. Chem. Technol. Biotechnol. 85, 732–743.

Alban, C., Baldet, P., Douce, R., 1994. Localization and characterization of two structurally different forms of acetylCoA carboxylase in young pea leaves, of which one is sensitive to aryloxyphenoxypropionate herbicides. Biochem. J. 300, 557–565.

Albuquerque, M.G.E., Martino, V., Pollet, E., Avérous, L., Reis, M.A.M., 2011. Mixed culture polyhydroxyalkanoate (PHA) production from volatile fatty acid (VFA)-rich streams: effect of substrate composition and feeding regime on PHA productivity, composition and properties. J. Biotechnol. 151, 66–76.

Alcantara, R., Amores, J., Canoira, L., Fidalgo, E., Franco, M.J., Navarro, A., 2000. Catalytic production of biodiesel from soybean oil, used frying oil and tallow. Biomass Bioenergy 18, 515–527.

Amulya, K., Venkateswar Reddy, M., Venkata Mohan, S., 2014. Acidogenic spent wash valorization through polyhydroxyalkanoate (PHA) synthesis coupled with fermentative biohydrogen production. Bioresour. Technol. 158, 336–342.

Anderson, A.J., Dawes, E.A., 1990. Occurrence, metabolism, metabolic role, and industrial uses of bacterial polyhydroxyalkanoates. Microbiol. Rev. 54, 450–472.

Annie Modestra, J., Venkata Mohan, S., 2014. Bio-electrocatalyzed electron efflux in Gram positive and Gram negative bacteria: an insight into disparity in electron transfer kinetics. RSC. Adv. 4, 34045–34055.

Annie Modestra, J., Navaneeth, B., Venkata Mohan, S., 2015. Bio-electrocatalytic reduction of CO_2: Enrichment of homoacetogens and pH optimization towards enhancement of carboxylic acids biosynthesis. Journal of CO_2 Utilization, 10, 78–87.

Arneil, D.R., Carol, A., Lin, S.K., Chan, K.M., Him Kwan, T.H., Luque, R., 2013. Advances on waste valorization: new horizons for a more sustainable society. Energy Sci. Eng. 1, 53–71.

Balat, H., Kırtay, E., 2010. Hydrogen from biomass—present scenario and future prospects. Int. J. Hydrog. Energy 35, 7416–7426.

Beccari, M., Bertin, L., Dionisi, D., Fava, F., Lampis, S., Majone, M., Valentino, F., Vallini, G., Villano, M., 2009. Exploiting olive oil mill effluents as a renewable resource for production of biodegradable polymers through a combined anaerobic–aerobic process. J. Chem. Technol. Biotechnol. 84, 901–908.

Becker, E.W., 1994. Microalgae: Biotechnology and Microbiology. Cambridge University Press, Cambridge, UK.

Becker, E.W., 2004. Microalgae in human and animal nutrition. In: Richmond, A. (Ed.), Handbook of Microalgal Culture: Biotechnology and Applied Phycology. Blackwell Science, Oxford, pp. 312–351.

Bennetto, H.P., Stirling, J.L., Tanaka, K., Vega, C.A., 1983. Anodic reactions in microbial fuel cells. Biotechnol. Bioeng. 25, 559–568.

Bhubalan, K., Lee, W.H., Loo, C.Y., Yamamoto, T., Doi, Y., Sudesh, K., 2007. Controlled biosynthesis and characterization of poly (3-hydroxybutyrate-co-3-hydroxyvalerate-co-3- hydroxyhexanoate) from mixtures of palm kernel oil and 3HV-precursors. Polym. Degrad. Stab. 93, 17–23.

Bond, D.R., Lovley, D.R., 2003. Electricity production by Geobacter sulfurreducens attached to electrodes. Appl. Environ. Microbiol. 69, 1548–1555.

Braunegg, G., Lefebvre, K.F.G., 1998. Polyhydroxyalkanoates, biopolyesters from renewable resources: physiological and engineering aspects. J. Biotechnol. 65, 127–161.

Brehmer, B., 2008. Chemical Biorefinery Perspectives: The Valorisation of Functionalised Chemicals from Biomass Resources Compared to the Conventional Fossil Fuel Production Route. Wageningen Universiteit (Wageningen University), The Netherlands.

CRFA Action plan on biofuels to bioeconomy, 2014. Canadian Renewable Fuels Association.

Chandra, R., Venkata Mohan, S., 2014. Enhanced bio-hydrogenesis by co-culturing photosynthetic bacteria with acidogenic process: augmented dark-photo fermentative hybrid system to regulate volatile fatty acid inhibition. Int. J. Hydrog. Energy 39, 7604–7615.

Chandra, R., Subhash, G.V., Venkata Mohan, S., 2012. Mixotrophic operation of photo-bioelectrocatalytic fuel cell under anoxygenic microenvironment enhances the light dependent bioelectrogenic activity. Bioresour. Technol. 109, 46–56.

Chandra, R., Rohit, M.V., Swamy, Y.V., Venkata Mohan, S., 2014. Regulatory function of organic carbon supplementation on biodiesel production during growth and nutrient stress phases of mixotrophic microalgae cultivation. Bioresour. Technol. 165, 279–287.

Chandrasekhar, K., Venkata Mohan, S., 2012. Bio-electrochemical remediation of real field petroleum sludge as an electron donor with simultaneous power generation facilitates biotransformation of PAH: effect of substrate concentration. Bioresour. Technol. 110, 517–525.

Chang, I.S., Moon, H., Bretschger, O., Jang, J.K., Park, H.I., Nealson, K.H., Kim, B.H., 2006. Electrochemically active bacteria (EAB) and mediator-less microbial fuel cells. J. Microbiol. Biotechnol. 16, 163–177.

Chen, Z., Huang, Y.C., Liang, J.H., Zhao, F., Zhu, Y.G., 2012. A novel sediment microbial fuel cell with a biocathode in the rice rhizosphere. Bioresour. Technol. 108, 55–59.

Chen, H., Meng, H., Nie, Z., Zhang, M., 2013. Polyhydroxyalkanoate production from fermented volatile fatty acids: effect of pH and feeding regimes. Bioresour. Technol. 128, 533–538.

Cheng, S., Logan, B.E., 2007. Ammonia treatment of carbon cloth anodes to enhance power generation of microbial fuel cells. J. Electrochem. Commun. 9, 492–496.

Chiranjeevi, P., Chandra, R., Venkata Mohan, S., 2013a. Ecologically engineered submerged and emergent macrophyte based system: an integrated ecoelectrogenic design for harnessing power with simultaneous wastewater treatment. Ecol. Eng. 51, 181–190.

Chiranjeevi, P., Chandra, R., Venkata Mohan, S., 2013b. Rhizosphere mediated electrogenesis with the function of anode placement for harnessing bioenergy through CO_2 sequestration. Ecol. Eng. 51, 181–190.

Chisti, Y., 2008. Biodiesel from microalgae beats bioethanol. Trends Biotechnol. 26, 126–131.

Choi, Y., Kim, N., Kim, S., Jung, S., 2003. Dynamic behaviors of redox mediators within the hydrophobic layers as an important factor for effective microbial fuel cell operation. Bull. Korean Chem. Soc. 24, 437–440.

Christophe, G., Lara Deo, J., Kumar, V., Nouaille, R., Fontanille, P., Larroche, C., 2012. Production of oils from acetic acid by the oleaginous yeast *Cryptococcus curvatus*. Appl. Biochem. Biotechnol. 167, 1270–1279.

Dahiya, S., Sarkar, O., Swamy, Y.V., Venkata Mohan, S., 2015. Acidogenic fermentation of food waste for volatile fatty acid production along with co-generation of biohydrogen. Bioresour. Technol. 182, 103–113. http://dx.doi.org/10.1016/j.biortech.2015.01.007.

Devi, M.P., Subhash, G.V., Venkata Mohan, S., 2012. Heterotrophic cultivation of mixed microalgae for lipid accumulation during sequential growth and starvation phase operation: effect of nutrient supplementation. Renew. Energy 43, 276–283.

Dorado, M.P., Ballesteros, E., Almeida, J.A., Schellert, C., Lohrlein, H.P., Krause, R., 2002. An alkali-catalyzed transesterificationprocess for high free fatty acid waste oils. Trans. ASAE 45, 525–529.

Du, Z., Li, H., Gu, T., 2007. A state of the art review on microbial fuel cells: a promising technology for wastewater treatment and bioenergy. Biotechnol. Adv. 25, 464–482.

Europa Bio Policy Guide on Building a Bio-based Economy for Europe in, 2020. 2010. European Commission, Belgium.

Fan, J., Andre, C., Xu, C., 2011. A chloroplast pathway for the de novo biosynthesis of triacylglycerol in Chlamydomonas reinhardtii. FEBS Lett. 585, 1985–1991.

Federal Ministry of Education and Research, 2011. Report on National Research Strategy BioEconomy, 2030, Our Route towards a biobased economy. Federal Ministry of Education and Research, Germany.

Francis, G., Becker, K., 2002. Biodiesel from Jatropha Plantations on Degraded Land. University of Hohenheim, Stuttgart, Germany. p. 9. http://www.youmanitas.nl/pdf/Bio-diesel.pdf.

Gong, Y., Ebrahim, A., Feist, A.M., Embree, M., Zhang, T., Lovely, D., Zengler, K., 2013. Sulfide-driven microbial electrosynthesis. Environ. Sci. Technol. 47, 568–573.

Gouveia, L., Oliveira, A.C., 2009. Microalgae as a raw material forbiofuels production. J. Ind. Microbiol. Biotechnol. 36, 269–274.

Gupta, V. K., Tuohy, M., Kubicek, C. P., Saddler, J., & Xu, F. (Eds.). 2013. Bioenergy Research: Advances and Applications. Newnes. Elsevier B.V. ISBN: 978-0-444-59561-4. http://dx.doi.org/10.1016/B978-0-444-59561-4.01001-9

Hens, L., Quynh, L.X., 2008. Environmental space. In: Erik Jorgensen, S., Fath, B. (Eds.), Encyclopedia of Ecology. Academic Press, Oxford, pp. 1356–1363.

Hernandez, M.E., Newman, D.K., 2001. Extra cellular electron transfer cell. Cell. Mol. Life Sci. 58, 1562–1571.

Huang, Y.L., Wu, Z., Zhang, L., Cheung, C.M., Yang, S., 2002. Production of carboxylic acids from hydrolyzed corn meal by immobilized cell fermentation in a fibrous-bed bioreactor. Bioresour. Technol. 82, 51–59.

Kamiya, A., Kowallik, W., 1987. The inhibitory effect of light on proton-coupled hexose uptake in *Chlorella*. Plant Cell Physiol. 28, 621–625.

Kamm, B., Kamm, M., Schmidt, M., Hirth, T., Schulze, M., 2006. Lignocellulose-based chemical products and product family trees. In: Kamm, B., et al. (Eds.), In: Biorefineries – Industrial Processes and Products: Status Quo and Future Directions., vol. 2. Wiley-VCH, Weinheim, pp. 97–149.

Kelly, P.T., He, Z., 2014. Review nutrients removal and recovery in bioelectrochemical systems: a review. Bioresour. Technol. 153, 351–360.

Keshavarz, T., Roy, I., 2010. Polyhydroxyalkanoates: bioplastics with a green agenda. Curr. Opin. Microbiol. 13, 321–326.

Kim, H.J., Park, H.S., Hyun, M.S., Chang, I.S., Kim, M., Kima, B.H., 2002. A mediator-less microbial fuel cell using a metal reducing bacterium, *Shewanella putrefaciens*. Enzym. Microb. Technol. 30, 145–152.

Kim, D., An, J., Kim, B., Jang, J.K., Kim, B.H., Chang, I.S., 2012. Scaling-up microbial fuel cells: configuration and potential drop phenomenon at series connection of unit cells in shared anolyte. ChemSusChem 5, 1086–1091.

Kiran Kumar, A., Reddy, M.V., Chandrasekhar, K., Srikanth, S., Venkata Mohan, S., 2012. Endocrine disruptive estrogens role in electron transfer: bioelectrochemical remediation with microbial mediated electrogenesis. Bioresour. Technol. 104, 547–556.

Klinke, S., Ren, Q., Witholt, B., Kessler, B., 1999. Production of medium-chain-length poly (3-hydroxyalkanoates) from gluconate by recombinant *Escherichia coli*. Appl. Environ. Microbiol. 65 (2), 540–548.

Knothe, G., 2005. Dependence of biodiesel fuel properties on the structure of fatty acid alkyl esters. Fuel Process. Technol. 86, 1059–1070.

Koller, M., Salerno, A., Miguel Dias, M., Reiterer, A., Braunegg, G., 2010. Modern biotechnological polymer synthesis: a review biotechnological polymer synthesis. Food Technol. Biotechnol. 48, 255–269.

Komor, E., 1973. Proton-coupled hexose transport in *Chlorella vulgaris*. FEBS Lett. 38, 16–18.

Komor, E., Tanner, W., 1974. The hexose-proton symport system of *Chlorella vulgaris*: specificity, stoichiometry and energetic of sugar-induced proton uptake. Eur. J. Biochem. 44, 219–223.

Lee, W.S., Chua, A.S.M., Yeoh, H.K., Ngoh, G.C., 2014. A review of the production and applications of waste-derived volatile fatty acids. Chem. Eng. J. 235, 83–99.

Legat, A., Gruber, C., Zangger, K., Wanner, G., Stan-Lotter, H., 2010. Identification of polyhydroxyalkanoates in *Halococcus* and other haloarchaeal species. Appl. Microbiol. Biotechnol. 87 (3), 1119–1127.

Lemoigne, M., 1926. Products of dehydration and of polymerization of β-hydroxybutyric acid. Chem. Biol. 8, 770–782.

Liu, B., Benning, C., 2012. Lipid metabolism in microalgae distinguishes itself. Curr. Opin. Biotechnol. 24, 300–309.

Liu, L., Bryan, S.J., Huang, F., Yu, J., Nixon, P.J., Rich, P.R., Mullineaux, C.W., 2012. Control of electron transport routes through redox regulated redistribution of respiratory complexes. Proc. Natl. Acad. Sci. 109, 6431–6436.

Lovley, D.R., Nevin, K.P., 2013. Electro biocommodities: powering microbial production of fuels and commodity chemicals from carbon dioxide with electricity. Curr. Opin. Biotechnol. 24, 385–390.

Lynd, L.R., Laser, M.S., Bransby, D., Dale, B.E., Davison, B., Hamilton, R., Himmel, M., Keller, M., McMillan, J.D., Sheehan, J., Wyman, C.E., 2008. How biotech can transform biofuels. Nat. Biotechnol. 26, 169–172.

Malvankar, N.S., Tuominen, M.T., Lovley, D.R., 2012. Lack of cytochrome involvement in long-range electron transport through conductive biofilms and nanowires of *Geobacter sulfurreducens*. Energy Environ. Sci. 5, 8651–8659.

Marsh, G., 2009. Small wonders: biomass from algae. Renew. Energy Focus 9, 74–78.

Marshall, C.W., LaBelle, E.V., May, H.D., 2013. Production of fuels and chemicals from waste by microbiomes. Curr. Opin. Biotechnol. 24, 391–397.

McInerney, M.J., Struchtemeyer, C.G., Sieber, J., Mouttaki, H., Stams, A.J., Schink, B., Rohlin, L., Gunsalus, R.P., 2008. Physiology, ecology, phylogeny, and genomics of microorganisms capable of syntrophic metabolism. Ann. N. Y. Acad. Sci. 1125, 58–72.

McKinlay, J.B., Zeikus, J.G., 2004. Extracellular iron reduction is mediated in part by neutral red and hydrogenase in *Escherichia coli*. Appl. Environ. Microbiol. 70, 3467–3474.

Mohanakrishna, G., Venkata Mohan, S., 2013. Multiple process integrations for broad perspective analysis of fermentative H_2 production from wastewater treatment: technical and environmental considerations. Appl. Energy 107, 244–254.

Neto, S.A., Forti, J.C., Andrade, A.R., 2010. An overview of enzymatic biofuel cells. Electrocatalysis 1, 87–94.

Nikel, P.I., Almeida, A.D., Melillo, E.C., Galvagno, M.A., Pettinari, M.J., 2006. New recombinant *Escherichia coli* strain tailored for the production of poly (3-hydroxybutyrate) from agro-industrial by-products. Appl. Environ. Microbiol. 72, 3949–3954.

OECD, 2013. Biotechnology for the Environment in the Future: Science, Technology and Policy. OECD Publishing, Paris. OECD Science, Technology and Industry Policy Papers, No. 3. http://dx.doi.org/10.1787/5k4840hqhp7j-en.

Ohlroggeav, J., Browseb, J., 1995. Lipid biosynthesis. Plant Cell 7, 957–970.

Pasupuleti, S.B., Sarkar, O., Venkata Mohan, S., 2014. Upscaling of biohydrogen production process in semi-pilot scale biofilm reactor: evaluation with food waste at variable organic loads. Int. J. Hydrog. Energy 39 (14), 7587–7596.

Perez-Garcia, R.O., Bashan, Y., Puente, M.E., 2011. Organic carbon supplementation of municipal wastewater is essential for heterotrophic growth and ammonium removing by the microalgae *Chlorella vulgaris*. J. Phycol. 47, 190–199.

Potter, M.C., 1910. On the difference of potential due to the vital activity of microorganisms. Proc. Univ. Durham Philos. Soc. 3, 245–249.

Rabaey, K., Rozendal, R.A., 2010. Microbial electrosynthesis—revisiting the electrical route for microbial production. Nat. Rev. Microbiol. 8, 706–716.

Ragauskas, A.J., Williams, C.K., Davison, B.H., Britovsek, G., Cairney, J., Eckert, C.A., Frederick Jr., W.J., Hallett, J.P., Leak, D.J., Liotta, C.L., Mielenz, J.R., Murphy, R., Templer, R., Tschaplinski, T., 2006. The path forward for biofuels and biomaterials. Science 311, 484.

Raghavulu, S.V., Venkata Mohan, S., Venkateswar Reddy, M., Mohanakrishna, G., Sarma, P.N., 2009. Behavior of single chambered mediatorless microbial fuel cell (MFC) at acidophilic, neutral and alkaline microenvironments during chemical wastewater treatment. Int. J. Hydrog. Energy 34, 7547–7554.

Raghavulu, S.V., Modestra, A.J., Amulya, K., Reddy, C.N., Venkata Mohan, S., 2013. Relative effect of bioaugmentation with electrochemically active and non-active bacteria on bioelectrogenesis in microbial fuel cell. Bioresour. Technol. 146, 696–703.

Reddy, M.V., Venkata Mohan, S., 2012. Influence of aerobic and anoxic microenvironments on polyhydroxyalkanoates (PHA) production from food waste and acidogenic effluents using aerobic consortia. Bioresour. Technol. 103 (1), 313–321.

Reddy, M.V., Chandrasekhar, K., Venkata Mohan, S., 2011. Influence of carbohydrates and proteins concentration on fermentative hydrogen production using canteen based waste under acidophilic microenvironment. J. Biotechnol. 155, 387–395.

Reddy, M.V., Amulya, K., Rohit, M.V., Sarma, P.N., Venkata Mohan, S., 2014. Valorization of fatty acid waste for bioplastics production using *Bacillus tequilensis*: integration with dark-fermentative hydrogen production process. Int. J. Hydrog. Energy 39, 7616–7626.

Saharan, B.S., Sahu, R.K., Sharma, D., 2011. A review on biosurfactants: fermentation, current developments and perspectives. Genet. Eng. Biotechnol. J. 2011, 1.

Strategic Research Agenda supporting the Roadmap - 2014, SAHYOG Project.

Saratale, G.D., Chen, S., Lo, Y., Saratale, J.L.G., Chang, J., 2008. Outlook of biohydrogen production from lignocellulosic feedstock using dark fermentation: a review. J. Sci. Ind. Res. 67, 962–979.

Schroder, U., 2007. Anodic electron transfer mechanisms in microbial fuel cells and their energy efficiency. Phys. Chem. Chem. Phys. 9, 2619–2629.

Serafim, L.S., Lemos, P.C., Oliveira, R., Reis, M.A.M., 2004. Optimization of polyhydroxybutyrate production by mixed cultures submitted to aerobic dynamic feeding conditions. Biotechnol. Bioeng. 87, 145–160.

Singh, I., Rastogi, V., 2009. Performance analysis of a modified 4-stroke engine using biodiesel fuel for irrigation purpose. Int. J. Environ. Sci. 4, 229–242.

Singhania, R.R., Patel, A.K., Christophe, G., Fontanille, P., Larroche, C., 2013. Biological upgrading of volatile fatty acids, key intermediates for the valorization of biowaste through dark anaerobic fermentation. Bioresour. Technol. 145, 166–174.

Skizim, N.J., Ananyev, G.M., Krishnan, A., Dismukes, G.C., 2012. Metabolic pathways for photobiological hydrogen production by nitrogenase and hydrogenase-containing unicellular cyanobacteria cyanothece. J. Biol. Chem. 287, 2777–2786.

Srikanth, S., Venkata Mohan, S., Devi, M.P., Peri, D., Sarma, P.N., 2009. Acetate and butyrate as substrates for hydrogen production through photo-fermentation: process optimization and combined performance evaluation. Int. J. Hydrog. Energy 34 (17), 7513–7522.

Stephens, E., Ross, I.L., King, Z., Mussgnug, J.H., Kruse, O., Posten, C., Borowitzka, M.A., Hankamer, B., 2010. An economic and technical evaluation of microalgal biofuels. Nat. Biotechnol. 28, 126–128.

Subhash, G.V., Venkata Mohan, S., 2014. Deoiled algal cake as feedstock for dark fermentative biohydrogen production: an integrated biorefinery approach. Int. J. Hydrog. Energy 39 (18), 9573–9579.

Subhash, G.V., Venkata Mohan, S., 2015. Sustainable biodiesel production through bioconversion of lignocellulosic wastewater by oleaginous fungi. Biomass Convers. Biorefin. 5 (2), 215–226. http://dx.doi.org/10.1007/s13399-014-0128-4.

Subhash, G.V., Rohit, M.V., Devi, M.P., Swamy, Y.V., Venkata Mohan, S., 2014. Temperature induced stress influence on biodiesel productivity during mixotrophic microalgae cultivation with wastewater. Bioresour. Technol. 169, 789–793.

Suresh Babu, P., Srikanth, S., Dominguez-Benetton, X., Venkata Mohan, S., Pant, D., 2014. Dual gas diffusion cathode design for microbial fuel cell (MFC): optimizing the suitable mode of operation in terms of bioelectrochemical and bioelectro-kinetic evaluation. J. Chem. Technol. Biotechnol. http://dx.doi.org/10.1002/jctb.4613.

Tanner, W., 1969. Light-driven active uptake of 3-*O*-methylglucose via an inducible hexose uptake system of *Chlorella*. Biochem. Biophys. Res. Commun. 36, 278–283.

Tawfik, A., Salem, A., El-Qelish, M., Abdullah, A.M., Abou Taleb, E., 2011. Feasibility of biological hydrogen production from kitchen waste via anaerobic baffled reactor (ABR). Int. J. Sustain. Water Environ. Syst. 2, 117–122.

Thelen, J.J., Ohlrogge, J.B., 2002. Metabolic engineering of fatty acid. Biosynth. Plants Metab. Eng. 4, 12–21.

Thrash, J.C., Trump, V.J.I., Weber, K.A., Miller, E., Achenbach, L.A., Coates, J.D., 2007. Electrochemical stimulation of microbial perchlorate reduction. Environ. Sci. Technol. 41, 1740–1746.

Vamshi Krishna, K., Sarkar, O., Venkata Mohan, S., 2014. Bioelectrochemical treatment of paper and pulp wastewater in comparison with anaerobic process: integrating chemical coagulation with simultaneous power production. Bioresour. Technol. 174, 142–151.

Velvizhi, G., Goud, R.K., Venkata Mohan, S., 2014. Anoxic bio-electrochemical system for treatment of complex chemical wastewater with simultaneous bioelectricity generation. Bioresour. Technol. 151, 214–220.

Velvizhi, G., Venkata Mohan, S., 2015. Bioelectrogenic role of anoxic microbial anode in the treatment of chemical wastewater: microbial dynamics with bioelectro-characterization. Water Res. 70 (1), 52–63.

Venkata Mohan, S., 2009. Harnessing of biohydrogen from wastewater treatment using mixed fermentative consortia: process evaluation towards optimization. Int. J. Hydrog. Energy 34, 7460–7474.

Venkata Mohan, S., 2014. Sustainable waste remediation: a paradigm shift towards environmental biorefinery. Chem. Eng. World 49 (12), 29–35.

Venkata Mohan, S., Chandrasekhar, K., 2011. Solid phase microbial fuel cell (SMFC) for harnessing bioelectricity from composite food waste fermentation: influence of electrode assembly and buffering capacity. Bioresour. Technol. 102 (14), 7077–7085.

Venkata Mohan, S., Devi, M.P., 2012. Fatty acid rich effluents from acidogenic biohydrogen reactor as substrates for lipid accumulation in heterotrophic microalgae with simultaneous treatment. Bioresour. Technol. 123, 627–635.

Venkata Mohan, S., Venkateswar Reddy, M., 2013. Optimization of critical factors to enhance polyhydroxyalkanoates (PHA) synthesis by mixed culture using Taguchi design of experimental methodology. Bioresour. Technol. 128, 409–416.

Venkata Mohan, S., Mohankrishna, G., Sarma, P.N., 2008a. Integration of acidogenic and methanogenic processes for simultaneous production of biohydrogen and methane from wastewater treatment. Int. J. Hydrog. Energy 33, 2156–2166.

Venkata Mohan, S., Mohanakrishna, G., Reddy, B.P., Sarvanan, R., Sarma, P.N., 2008b. Bioelectricity generation from chemical wastewater treatment in mediator less (anode) microbial fuel cell (MFC) using selectively enriched hydrogen producing mixed culture under acidophilic microenvironment. Biochem. Eng. J. 39, 121–130.

Venkata Mohan, S., Mohana Krishna, G., Sarma, P.N., 2008c. Effect of anodic metabolic function on bioelectricity generation and substrate degradation in single chambered microbial fuel cell. Environ. Sci. Technol. 42, 8088–8094.

Venkata Mohan, S., Mohanakrishna, G., Goud, R.K., Sarma, P.N., 2009a. Acidogenic fermentation of vegetable based market waste to harness biohydrogen with simultaneous stabilization. Bioresour. Technol. 100, 3061–3068.

Venkata Mohan, S., Babu, M.L., Mohanakrishna, G., Sarma, P., 2009b. Harnessing of biohydrogen by acidogenic fermentation of *Citrus limetta* peelings: effect of extraction procedure and pretreatment of biocatalyst. Int. J. Hydrog. Energy 34 (15), 6149–6156.

Venkata Mohan, S., Veer Raghuvulu, S.V., Dinakar, P., Sarma, P.N., 2009c. Integrated function of microbial fuel cell (MFC) as bio-electrochemical treatment system associated with bioelectricity generation under higher substrate load. Biosens. Bioelectron. 24, 2021–2027.

Venkata Mohan, S., Prathima Devi, M., Mohanakrishna, G., Amarnath, N., Lenin Babu, M., Sarma, P.N., 2011. Potential of mixed microalgae to harness biodiesel from ecological water-bodies with simultaneous treatment. Bioresour. Technol. 102, 1109–1117.

Venkata Mohan, S., Srikanth, S., Velvizhi, G., Lenin Babu, M., 2013a. Microbial fuel cells for sustainable bioenergy generation: principles and perspective applications. In: Gupta, V.K., Tuohy, M.G. (Eds.), Biofuel Technologies: Recent Developments. Spinger, Berlin, ISBN: 978-3-642-34518-0. Chapter 11.

Venkata Mohan, S., Venkateswar Reddy, M., Chandra, R., Venkata Subhash, G., Prathima Devi, M., Srikanth, S., 2013b. Bacteria for bioenergy: a sustainable approach towards renewability. In: Gaspard, S., Ncib, M.C. (Eds.), Biomass for Sustainable Applications: Pollution, Remediation and Energy. RSC Publishers, Cambridge, ISBN: 9781849736008, pp. 251–289.

Venkata Mohan, S., Velvizhi, G., Vamshi Krishna, K., Babu, M.L., 2014a. Microbial catalyzed electrochemical systems: a bio-factory with multi-facet applications. Bioresour. Technol. 165, 355–364.

Venkata Mohan, S., Velvizhi, G., Modestra, A.J., Srikanth, S., 2014b. Microbial fuel cell: critical factors regulating bio-catalyzed electrochemical process and recent advancements. Renew. Sust. Energ. Rev. 40, 779–797.

Venkata Mohan, S., Devi, M.P., Subhash, G.V., Chandra, R., 2014c. In: Pandey, A., Lee, D.J., Chisti, Y. (Eds.), Biofuels from ALGAE. Elsevier, CJL, School, ISBN: 9780444595584, pp. 155–187 (Chapter 8).

Venkata Mohan, S., Rohit, M.V., Chiranjeevi, P., Chandra, R., Navaneeth, B., 2015. Heterotrophic microalgae cultivation to synergize biodiesel production with waste remediation: progress and perspectives. Bioresour. Technol. 184, 169–178. http://dx.doi.org/10.1016/j.biortech.2014.10.056.

Venkateswar Reddy, M., Venkata Mohan, S., 2012. Effect of substrate load and nutrients concentration on the polyhydroxyalkanoates (PHA) production using mixed consortia through wastewater treatment. Bioresour. Technol. 114, 573–582.

Villano, M., Beccari, M., Dionisi, D., Lampis, S., Miccheli, A., Vallini, G., Majone, M., 2010. Effect of pH on the production of bacterial polyhydroxyalkanoates by mixed cultures enriched under periodic feeding. Process Biochem. 45, 714–723.

Wallen, L.L., Rohwedder, W.K., 1974. Poly-betahydroxyalkanoate from activated sludge. Environ. Sci. Technol. 8, 576–579.

Wang, F., Lee, S.Y., 1997. Poly (3-hydroxybutyrate) production with high productivity and high polymer content by a fed-batch culture of *Alcaligenes latusunder* nitrogen imitation. Appl. Environ. Microbiol. 63, 3703–3706.

Weissman, J.C., 2008. Factors limiting photosynthetic efficiency in outdoor mass culture of microalgae. In: 11th International Conference on Applied Phycology, Galway, Ireland, pp. 52.

White House, 2012. National Bioeconomy Blueprint. White House, USA. April.

Zhao, F., Slade, R.C., Varcoe, J.R., 2009. Techniques for the study and development of microbial fuel cells: an electrochemical perspective. Chem. Soc. Rev. 38, 1926–1939.

Zhou, M., Jin, T., Wu, Z., Chi, M., Gu, T., 2012. Microbial fuel cells for bioenergy and bioproducts. In: Gopalakrishnan, K., Leeuwen, J.V., Brown, R. (Eds.), Sustainable Bioenergy and Bioproducts. Springer-Verlag, Berlin, New York, pp. 131–172.

ENERGY FROM WASTEWATER TREATMENT

S.Z. Ahammad, T.R. Sreekrishnan

Indian Institute of Technology, Delhi, India

1 INTRODUCTION

Increasing energy demands along with the progress of human civilization is a major global concern as the dominant energy-yielding resources are heavily dependent on natural reserves of fuels available in nature in solid, liquid, and gaseous forms. The limited natural sources are bound to run out, and there is a dire need to develop alternate solutions in order to meet human needs (Rittmann and McCarty, 2001). Due to this tremendous pressure, industries are adopting more energy-efficient modern technologies to replace energy-intensive manufacturing processes. While the search for energy-efficient technologies is being done to improve the fuel economy, focus has also been on use of eco-friendly, green technologies to reduce the burden on the environment. The gross domestic product (GDP) of any country is highly dependent on environmental pollution and its consequent effect on public health (The World Bank Report, 2003). Use of green technology and alternate energy resources, commonly known as renewable energy resources, would reduce the pollution load and dependency on fossil fuels.

Waste generation is an inevitable consequence of human activities, and it has undergone major changes in its quantity as well as quality due to scientific and technological advancements. Technological advancements have resulted in the generation of more goods and services aimed at improving the quality of human life. Unfortunately, many of the so-called advanced technologies are not sustainable and have resulted in larger quantities of wastes with more complex constituents being generated. Waste is generated directly by human activities as well as by indirect means through different industrial manufacturing processes. Treatment of waste is a must to protect human beings and the entire ecosystem. Different treatment processes are used in various sectors to treat waste and reduce its burden on the environment. Technological developments have also made us rethink the possible use and reuse of waste materials. Wastes are no longer considered unusable; rather, they are often used as a resource or raw material to generate energy and/or by-products. Anaerobic digestion of organic matter to produce biogas, use of microbial electrolysis cells (MECs) and microbial fuel cells (MFCs) to generate electricity while treating wastes, and use of algal systems to sequester atmospheric carbon dioxide to produce biofuels are a few prominent examples of such technologies where energy in various forms is being produced while treating the waste materials. Considering that treatment and stabilization of wastes is a necessary activity, energy minimization in the process is also equally important where such energy-positive systems (systems that produce more energy than they consume) are not suitable to be used. Based on the net energy requirement, processes are classified into three major groups: energy-negative processes, where energy is supplied to accomplish the process, energy-neutral processes where net energy use is zero, and energy-positive processes where the process results in net energy production.

Waste treatment processes are most often not well practiced in instances where strict implementation of regulations are not followed and also due to possible expenses in treating the waste. Use of energy-neutral and/or energy-positive treatment process can definitely change the scenario. A World Bank report (2003) indicates that after incorporation of strict norms for environment protection and use of energy-efficient technologies the GDP and pollution problem could be decoupled and substantial reduction of environmental pollution (at very minimal level) is achievable without compromising the growth rate of GDP.

Biological waste treatment processes are always preferred over chemical and/or physicochemical processes because of energy and cost benefits. A number of biological treatment options are available to treat various kinds of waste streams. Sometimes, due to the persistent nature of the pollutants, a combination of biological and chemical processes is also used. Biological processes are broadly classified into two categories based on the microbes employed in the treatment process: aerobic microbe-dominated systems are known as aerobic systems and anaerobic systems, which depend on the activity of anaerobic microorganisms. Reactor operations are also classified into two major groups depending on the biomass density required in the reactor system. Attached cell systems or high cell density reactors and suspended cell systems are the two classes of reactor systems where the biomass concentration required is the major guiding parameter to select the reactor configuration.

2 WASTEWATER TREATMENT

Waste generated by domestic and industrial activities are of different scales. They differ not only in their quantity but also in their nature. In developed countries the domestic waste generated by a person in a day is about 80 g BOD5 (biochemical oxygen demand) (Mara, 2004) and average waste generation by the world population is 60-120 g chemical oxygen demand (COD) per person per day (Kiely, 1997). If the recoverable energy available in the waste is 14.7 kJ/g COD (Shizas and Bagley, 2004), and considering world population as 6.8 billion people, one can expect $2.2\text{-}4.4 \times 10^6$ TJ of energy per year as available in the waste. This can produce a continuous supply of 70-140 GW of energy. We need to burn 52-104 million tonnes of oil in a modern power station, or use 12,000-24,000 of the world's largest wind turbines working continuously, to produce such a huge amount of energy. This estimation considers only the energy potential of domestic wastes. The figures will be much higher if we can utilize energy potential of industrial wastes as well.

The best overall treatment method depends on the source and nature of the waste, such as waste production rates, waste constituents, and concentrations of the different constituents. Optimal process technologies and designs used for treating wastewater should be as simple as possible in design and operation, while being efficient in removing key pollutants, minimizing energy usage, and reducing production of negative by-products. More complex operations are used only when it is absolutely necessary.

Within a typical wastewater treatment plant, each type of treatment has a different purpose. For example, the main objective of a biological treatment is to degrade soluble organic matter in the wastewater, which often requires physical pretreatment to remove solids before applying the biological treatment. For domestic wastewater, the main objective is to reduce the organic content and, in growing numbers of cases, secondary nutrients (nitrogen and phosphorus). For industrial wastewaters, the objective is usually to remove or reduce the concentration of organic compounds, especially specific toxicants that can be present in some wastewaters. This is why chemical processes are also included in many industrial wastewater treatment systems. However, biological processes are almost always used when possible.

Biological degradation of organics is accomplished through the combined activity of microorganisms, including bacteria, fungi, algae, protozoa, and rotifers. To maintain the ecological balance in the receiving water, regulatory authorities have set standards for the maximum amount of the undesirable compounds present in the discharge water. In a typical wastewater treatment plant, the necessary steps and operations are carried out to achieve the desired quality of the effluent before it can be safely discharged into the receiving water.

3 BIOLOGICAL TREATMENT OPTIONS

Biological processes are classified according to the primary metabolic pathways present in the dominant microorganisms active in the treatment system. As per the availability and utilization of oxygen, the biological processes are classified as aerobic, anoxic, and anaerobic (Shizas and Bagley, 2004).

3.1 AEROBIC PROCESSES

Treatment processes that occur in the presence of molecular oxygen (O_2) and use aerobic respiration to generate cellular energy are called aerobic processes. They are most metabolically active, but also generate more residual solids as cell mass. In addition to routine energy-demanding operations such as process fluid flow, they require a major quantum of energy to ensure that the process is not limited by the availability of dissolved oxygen (Metcalf and Eddy, 2003).

3.2 ANOXIC PROCESSES

Anoxic processes occur in the absence of free molecular oxygen (O_2) and generate energy through anaerobic respiration. The microorganisms employed are different from the typical "obligate anaerobe" in the sense that they can (will) use dissolved oxygen when it is available. In the absence of dissolved oxygen, they use combined oxygen from inorganic material in the waste (e.g., nitrate) as their terminal electron acceptor. Anoxic processes are common in biological nitrogen removal systems through the coupled nitrification-denitrification route. It was observed that the combined oxic-anoxic process is more economical compared to only aerobic process. The net energy requirement of such process to treat wastewater generated by 700 person equivalent (PE) is <200 Wh per PE per day (Battistoni et al., 2003).

3.3 ANAEROBIC PROCESSES

Anaerobic processes occur in the absence of free or combined oxygen, and result in sulfate reduction and methanogenesis. They usually produce biogas, a mixture of mostly methane and carbon dioxide, as a useful by-product and tend to generate lower amounts of biosolids (sludge) as by-product. Theoretically, removal of 1 kg of COD from the wastewater will result in the production of $0.35 \, m^3$ of methane (at STP, standard temperature and pressure). This computation of methane production is "theoretical" because it assumes that all the COD removed appears as methane (since carbon dioxide, the other component of biogas, has zero COD). In actual practice, the process generates 0.27-0.30 m³ methane per kg COD removed. Pure methane has a calorific value of 39,800 kJ m⁻³. Biogas will have 55-70% methane (the rest is carbon dioxide) depending on the type of reactor and reactor operating conditions (Ahammad et al., 2010). Even then, this works out to a tidy bit of net energy generation, after deducting all the energy-consuming steps such as pumping, mixing, and heating (wherever applicable). In fact, according to the current state of the art, the maximum energy production through wastewater treatment occurs through anaerobic treatment of the wastewater.

Apart from a classification based on microbial metabolism and/or oxygen utilization, biological wastewater treatment processes can also be classified based on the growth conditions in the reactor (Figure 1). In this case, the two main categories are suspended growth and attached growth processes.

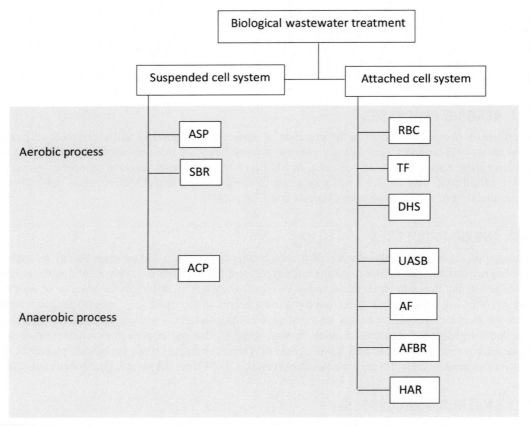

FIGURE 1

Different biological treatment systems: ASP, activated sludge process; SBR, sequential batch reactor; RBC, rotating biological contactor; TF, trickling filter; DHS, downflow hanging sponge reactor; ACP, anaerobic contact process; UASB, upflow anaerobic sludge blanket; AF, anaerobic filter; AFBR, anaerobic fluidized bed reactor; HAR, hybrid anaerobic reactor.

4 SUSPENDED GROWTH PROCESSES

In suspended growth processes, the microorganisms that are responsible for the conversion of organic matter present in the waste to simpler compounds and biomass are maintained in suspension within the liquid phase. However, there are different types of aerobic and anaerobic suspended growth processes. Aerobic processes include activated sludge, aerated lagoons, and sequencing batch reactors, whereas anaerobic processes include bag digesters, plug-flow digesters, stirred-tank reactors, and baffled reactors with organisms primarily in the liquid phase.

5 ATTACHED GROWTH PROCESS

In attached growth processes, the microorganisms responsible for degrading the waste are attached to surfaces (e.g., stones, inert packing materials), or are self-immobilized (microorganisms being attached to one another) as flocs or granules in the system (Ahammad et al., 2013a). Attached growth processes can be aerobic or anaerobic. Aerobic attached growth processes include trickling filters, roughing filters, rotating biological contactors, and packed-bed reactors. Anaerobic systems include upflow packed-bed reactors, downflow packed bed reactors, anaerobic rotating biological contactors, anaerobic fluidized bed reactors (AFBRs), upflow anaerobic sludge blanket (UASB) reactors, and various hybrid anaerobic reactors (HARs). UASBs are widely used reactors for the anaerobic treatment of industrial and domestic wastewater.

6 ANAEROBIC WASTEWATER TREATMENT PROCESSES

Anaerobic treatment technologies are widely practiced in different industries on the basis of their requirement and suitability. The processes have some advantages and disadvantages in treating different wastes. Under anaerobic conditions, organic matter is degraded through the sequential and syntrophic metabolic interactions of various trophic groups of prokaryotes, including fermenters, acetogens, methanogens, and sulfate-reducing bacteria (SRB) (Ahammad et al., 2008). Metabolic interactions between these microbial groups lead to the transformation of complex organic compounds to simple compounds such as methane, carbon dioxide, hydrogen-sulfide, and ammonia. The digestion process is essentially accomplished in four major reaction stages involving different microorganisms in each stage (Gunaseelan, 1997).

Stage 1: Hydrolysis—The organic waste material mainly consists of carbohydrates, proteins, and lipids. Complex and large substances are broken down into simpler compounds by the activity of the microbes and the extracellular enzymes released by these microbes. The hydrolysis or solubilization is mainly done by hydrolytic microbes such as *Bacteroides*, *Bifidobacterium*, *Clostridium*, and *Lactobacillus*. These organisms hydrolyze complex organic molecules (cellulose, lignin, proteins, lipids) into soluble monomers such as amino acids, sugars, fatty acids, and glycerol. These hydrolysis products are used by the fermentative acidogenic bacteria in the next stage.

Stage 2: Acidogenesis—Fermentative acidogenic bacteria convert simple organic materials such as sugars, amino acids, and long chain fatty acids into short chain organic acids such as formic, acetic, propionic, butyric, valeric, isobutyric, isovaleric, lactic, and succinic acids; alcohols and ketones (ethanol, methanol, glycerol, and acetone); carbon dioxide; and hydrogen. Generally, acidogenic bacteria have high growth rates and are the most abundant bacteria in any anaerobic digester. The high activity of these organisms implies that acidogenesis is seldom the rate-limiting step in the anaerobic digestion process. The volatile acids produced in this stage are further processed by microorganisms characteristic of the acetogenesis stage.

Stage 3: Acetogenesis—In this stage, acetogenic bacteria, also known as obligate hydrogen-producing acetogens, convert organic acids and alcohols into acetate, hydrogen, and carbon dioxide, which are subsequently used by methanogens and SRB. There is a strong symbiotic relationship between acetogenic bacteria and methanogens. Methanogens and SRB use hydrogen, which helps achieve the low hydrogen concentration conditions required for acetogenic conversions. A different group of bacteria called homoacetogens, which convert hydrogen (and carbon dioxide) to acetate, are also seen in anaerobic reactors.

Stage 4: Methanogenesis—It is the final stage of anaerobic digestion where methanogenic archaea convert the acetate, methanol, methylamines, formate, and hydrogen produced in the earlier stages into methane. The growth rate of methanogens is very low, and therefore, in most cases, this step is considered as the rate-limiting step of the anaerobic process, although there are also examples where hydrolysis is rate limiting.

7 UPFLOW ANAEROBIC SLUDGE BLANKET REACTORS

Presently, the most common and widely used anaerobic wastewater treatment reactor is the UASB reactor. It is an attached, self-immobilized cell system, which consists of a bottom layer of settled sludge bed (sludge blanket) and an upper liquid layer, as shown in Figure 2. Wastewater flows upward through the sludge bed consisting of bacterial aggregates, and the microbes present in the sludge bed convert the complex organic materials to methane, carbon dioxide, and hydrogen. The granular sludge (1-5 mm in diameter) has high biomass (mixed liquor volatile suspended solids, MLVSS) content and specific activity, and good settling properties. The upward flow of the liquid inside the reactor is obtained by means of effluent recirculation. Because of the high density of biomass present in the self-immobilized granular sludge, the reactor is able to support a high solids retention time, which is different from the hydraulic retention time (HRT) and requires no support material (Seghezzo et al., 1998). Typically, the reactor operates as a once-through system (no liquid recirculation) and therefore requires less energy input. But maintaining the necessary HRT without liquid recirculation calls for lower liquid velocities (0.5-1.0 m/h) inside the reactor. This can lead to buildup of concentration gradients within the bulk liquid phase, lowering the reaction rate in the biomass granules. These problems are overcome by using a HAR where the positive aspects of an AFBR are coupled with those of a UASB by maintaining a high upflow velocity (4-8 m h^{-1}) inside the reactor. With higher upflow velocity, better mass transfer is obtained in the reactor, which reflects on the higher degradation with less HRT required for the operation. The main purpose of these reactors is to achieve better degradation of waste and increase the production of biogas (methane) in a substantially reduced size anaerobic reactor.

8 ANAEROBIC FLUIDIZED BED REACTORS

In an AFBR, the wastewater is pumped up a bed of carrier material, typically sand or pumice stone of less than 5 mm size. The bed is fluidized by the incoming wastewater. A biofilm, comprised of all the different groups of anaerobic microorganisms, is made to develop on the surface of the carrier material. Depending on the wastewater characteristics and reactor operating conditions, this film has a thickness

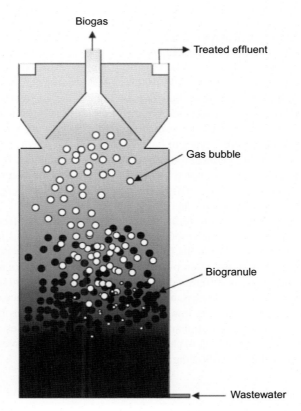

FIGURE 2

Upflow anaerobic sludge blanket reactor.

of 100-300 μm. The dissolved organic compounds in the wastewater diffuse into the biofilm where they become degraded into fatty acids and finally, methane and carbon dioxide. These reactors are designed with a gas-liquid-solid separator at the top so that the biofilm-covered carrier particles are retained in the system, the treated wastewater goes out, and the gas produced in the process comes out at the top. In these reactors, since the upflow superficial liquid velocity has to be greater than the minimum fluidization velocity, the normally encountered liquid velocities are in the range of 12-30 m h^{-1} (Heijnen et al., 1989). This is made possible by recycling a substantial part of the liquid. This type of operation results in higher energy requirements for pumping. This does have the advantage of very good mixing and therefore very high conversion rates.

9 HYBRID ANAEROBIC REACTOR

Very slow growth rate of anaerobic consortia is a bottleneck to enhance the treatment rate. In most of the cases, during anaerobic treatment of wastewaters, the methanogenesis step is considered as the rate-limiting one. To enhance the degradation rate, higher methanogen concentration in the reactor is desired. In HARs, the biomass concentration is enhanced through the use of biogranules, which consist of all the four different groups of anaerobic microorganisms. Higher superficial liquid velocity is maintained in the reactor to fluidize the biogranules (Ahammad et al., 2010; Saravanan and Sreekrishnan, 2005). Though maintenance of higher upflow velocity requires extra energy, it enhances the mass transfer rate to a substantial level. In a typical fluidized bed reactor, the biogranules are developed by growing biofilm onto inert support material. As inert support has its own weight, it demands extra energy to fluidize. In an HAR, use of self-immobilized granules can eliminate extra energy demand while fluidizing the granules. In an HAR this is made possible by growing self-immobilized anaerobic granules, which consist of all different types of bacteria and archaea required for anaerobic treatment, and they are present at a very high concentration in the compact granules. It results in lower energy usage (than in AFBR) in operating such reactor as well as improving the anaerobic treatment and subsequently increasing production of biogas (Ahammad et al., 2010). Net energy obtained from such a system makes it an energy-positive wastewater treatment option.

The energy audit is always done considering the existing practices in different treatment facilities. There is no doubt that the activated sludge process (ASP), an aerobic process, is the most common and widely used treatment technology despite its high demand on energy for aeration. Its robustness and simplicity of design and operation are often overshadowed by its drawbacks associated with the high energy input. Status of energy use in different processes is changing due to increase in the awareness of carbon footprints associated with the process, and efforts are in place to reduce carbon footprints. More and more processes are adapting greener, low carbon footprint technologies by working to eliminate the problems associated with energy-neutral and energy-positive technologies.

Understanding of the ASP would help to choose the appropriate alternatives to replace it. The main features of ASP are discussed in the following section.

10 ACTIVATED SLUDGE PROCESS

Classic ASPs are aerobic suspended cell systems. Mineralization of waste organic compounds is accompanied by the formation of new microbial biomass and sometimes the removal of inorganic compounds, such as ammonia and phosphorus, depending on the particular process design. ASPs were first conceived in the early 1900s with the word "activated" referring to solids that catalyze the degradation of the waste. It was subsequently discovered that the "activation" part of the sludge was a complex mixture of microorganisms. The liquid in activated sludge systems is called the "mixed liquor," which includes both wastewater and the resident organisms. There have been several incarnations of the ASP. The most common designs use conventional, step aeration, and continuous flow stirred tank reactors. A conventional ASP consists of standard pretreatment steps, an aeration tank, and a secondary clarifier, an example of which is shown in Figure 3. The aeration tank can be aerated by surface aerators or submerged, diffused air aerators connected to a compressor or blower designed to supply adequate dissolved oxygen to the water for the microorganisms to thrive. The wastewater flows through the tank and resident microorganisms consume organic matter in the wastewater. The aeration tank effluent flows to the clarifier where the microorganisms are removed. The clarifier supernatant is then transferred to disinfection or treatment units, and ultimately discharged to the receiving water. Part of the settled sludge (biosolids) from the clarifier bottom are recycled back to the head of the treatment system (aeration tank) and the rest sent to sludge stabilization, dewatering, and final disposal.

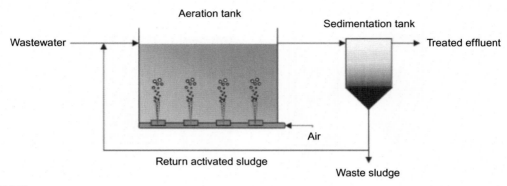

FIGURE 3

Activated sludge process.

10.1 AERATION TANKS

Aeration tanks are usually designed uncovered, open to the atmosphere. Air is supplied to the microorganisms by two primary methods: mechanical aerators or diffusers. Mechanical aerators, such as surface aerators and brush aerators, aerate the surface of the water mechanically and promote diffusion of oxygen to water from the atmosphere. The concentration of dissolved oxygen in the liquid can be controlled by adjusting the speed of the rotors. Both mechanical aerators and diffusers are the largest energy consumers in aerobic biological wastewater treatment processes. Diffusers bubble air directly into the tank at a depth and are usually preferred because of higher oxygen transfer efficiencies. As previously indicated, aeration provides O_2 to the microorganisms and also serves to mix the liquor in the tank. Although complete mixing is desired, there are usually "dead zones" in the tank where anaerobic/anoxic conditions develop in poorly mixed areas. It is desirable to keep these zones to a minimum to minimize undesired odors and also problems with sludge bulking, which can reduce settling efficiency in (secondary) clarifiers.

10.2 (SECONDARY) CLARIFIERS

Clarifiers are used to separate the biomass and other solids coming out of the aeration tank by means of gravity settling. The flow rate of the liquid is maintained in such a way that the upflow velocity of the liquid is less than the settling velocity of the biosolids present in the liquid. Some of the settled biosolids are returned back to the aeration tank to increase the solids' contact time with the wastes and also maintain the desired biomass (MLVSS) concentration levels in the aeration tank.

10.2.1 Combination of Anaerobic and Aerobic Treatment Processes

Often, combinations of anaerobic and aerobic treatment processes are used to achieve complete treatment of wastewater (Figure 4). The anaerobic system alone is not enough to remove nitrogen and phosphorus present in the wastewater. Effluent obtained from an anaerobic system requires aerobic treatment to remove ammonia present in the effluent. Anaerobic followed by aerobic system is quite common in a typical waste treatment plant. Apart from ASP, other aerobic treatment systems are also used as posttreatment systems. ASP requires active aeration whereas other aerobic processes such as modified trickling filter, downflow hanging sponge (DHS) reactor, aerated lagoon, etc., do not require

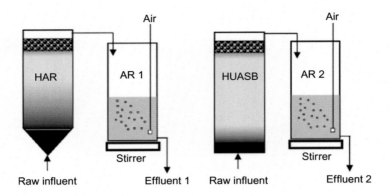

FIGURE 4

Combination of anaerobic and aerobic reactor systems: HAR, hybrid anaerobic reactor; AR, aerobic reactor; HUASB, hybrid upflow anaerobic sludge blanket.

any active aeration. Passive aeration used in the system not only reduces the operating cost, but it also reduces the energy usage and contributes to lower carbon footprints.

A typical activated sludge system having sewage treatment capacity of 1 million liters per day consumes approximately 390 MWh electrical energy per year (Electric Power Research Institute (EPRI), 2002). The enormous energy requirement is mostly contributed by the active aeration required to operate the system. Systems operated under passive aeration or no active air supply would reduce the energy demand and the carbon footprint of the technology to a substantial level. This could make the overall process energy neutral. Passive aeration is commonly used in the trickling filter system where wastewater is treated by the immobilized biofilm with the supply of very low aeration or zero active air supply.

11 TRICKLING FILTER

Trickling filter is a widely used aerobic biological treatment system. Also called a biofilter, it is a downflow packed bed type of reactor. It consists of a fixed bed made up of different inert materials. Biofilm grows on the surface of the inert bed. Different cheap and porous materials such as rocks, lava, coke, gravel, slag, pumice stone, polyurethane foam, peat moss, ceramic, or plastic media can be used for making the porous bed. Wastewater enters from the top of the fixed bed making use of a rotating arm distributor or static nozzles fed with a variable head feed source. Microbial biofilm grown on the surface of the inert support helps to degrade the waste. Aerobic condition is achieved by active or passive aeration by using either a blower or fan (forced aeration) or natural convection of air due to the temperature difference between the water and ambient air. Low strength wastewaters ($COD < 1000$ mg l^{-1}) such as sewage (domestic wastewater) can easily be treated using the system and desired effluent quality can be achieved by maintaining a typical HRT of 1 day. Clogging and channeling are two very common problems associated with its operation. Periodic cleaning of the bed (biomass removal) is required to get desired performance. A typical trickling filter used for treating sewage uses 0.22 kWh energy to remove 1 kg-COD (Evans et al., 2004).

12 DOWNFLOW HANGING SPONGE REACTOR

A DHS reactor is a type of trickling filter where sponge media is used to prepare the bed. Microbes grow on the sponge bed, which is used as the inert support for the biofilm. Due to its very highly porous structure, sponge provides very high surface-to-volume ratio. Higher available surface area in the support material (sponge) helps to get good biofilm formation and results in very high concentration of biomass in a given volume.

DHS is used as the aerobic posttreatment of the anaerobic reactor effluent and treated water quality reaches the level adequate to satisfy discharge standards. The effluent from anaerobic treatment system is discharged to the upper side of the DHS reactor, and it flows (trickles) downward due to gravity.

12.1 LOW ENERGY WASTEWATER TREATMENT PROCESSES

Various reactors are used in treating industrial and domestic wastewaters. The data on energy usage in the treatment processes to remove COD from the effluent are given in Table 1. The data (see Table 1) show that use of an anaerobic system is always beneficial compared to conventional ASP (Ahammad et al., 2013b; Christgen et al., 2015).

The sun is the most abundant and sustainable energy source for Earth. Every year, more than 3800 ZJ ($1 ZJ = 1 \times 10^{21}$ J) of solar energy are absorbed by the atmosphere and surface of the Earth but only about 0.05% of this energy is captured in biomass through the process of photosynthesis (Mohammed et al., 2014).

Processes such as algal pond– or microalgae-based treatment are gaining popularity due its energy-neutral operation. While treating the wastewater these treatment processes also help in sequestrating CO_2 and effectively reduce the carbon footprint of the entire process. Biomass generation and biodiesel production are two other major advantages one can derive from such algal systems. The requirements of appropriate light source and land area to cultivate the algae are the major bottlenecks to use this technology on a larger scale. The process is still well accepted and practiced in many countries where appropriate weather conditions and required land are available.

In the anaerobic process, biogas is produced and can be used as an energy source. It is a proven technology to convert waste to energy. A few other possibilities have been explored to get alternate energy sources such as hydrogen, ethanol, butanol, and other high-value products while treating waste materials. Most often the processes are not capable of making any impact in the larger scale because of inherent limitations, issues with upscaling, and various associated problems, which are mostly neglected or not observed in laboratory-scale settings. These technologies are not in a position now to be put into practice in industries. But in future, process improvement and change in relative acceptance criteria can make them appealing to use. Two such processes, biohydrogen production through dark fermentation to produce hydrogen, and use of MFC to produce electricity, are discussed in the following sections.

Table 1 Energy Usage in Treating Industrial Effluent Using Different Reactor Configuration

Reactor	Mean COD Removal Efficiency (%)	Energy Used per kg COD Removed (kWh kg^{-1} COD)
UASB + ASP	87.9 ± 0.4	0.307
HAR + ASP	86.8 ± 0.5	0.375
Anaerobic CSTR (continuous stirred tank reactor) + ASP	76.2 ± 0.6	0.612
ASP	68.7 ± 1.4	2.356

13 HYDROGEN PRODUCTION IN DARK FERMENTATION

Hydrogen production in dark fermentation is a fermentative process to degrade bigger organic molecules to smaller fractions and produce hydrogen. It is a complex syntrophic process obtained by the activity of diverse groups of bacteria under anaerobic condition. The heterotrophic microorganisms hydrolyze complex organic substrates to their respective monomers and subsequently to small chain fatty acids and alcohols. A specific group of bacteria known as acidogens plays a key role in this process (Ntaikou et al., 2010).

Industrial wastewater as a fermentative substrate for H_2 production addresses most of the criteria required for substrate selection: availability, cost, and biodegradability. The efficiency of the dark fermentative H_2 production process was found to depend on pretreatment of the mixed consortia used as a biocatalyst, operating pH, and organic loading rate apart from wastewater characteristics.

In spite of its advantages, the main challenge observed with fermentative H_2 production process is the relatively low energy conversion efficiency from the organic source. Typical H_2 yields range from 1 to 2 mol of H_2/mol of glucose, which results in 80-90% of the initial COD remaining in the wastewater in the form of various volatile fatty acids (VFAs) and solvents, such as acetic acid, propionic acid, butyric acid, and ethanol. Even under optimal conditions about 60-70% of the original organic matter remains in solution. Bioaugmentation with selectively enriched acidogenic consortia could enhance the hydrogen production and incorporation of methanogenesis immediately after dark fermentation could improve the process economy.

14 MICROBIAL FUEL CELLS

Waste degradation by the microbial community is the result of a series of complex metabolic reactions. Combination of reduction and oxidation reactions is the key to all metabolic activities. Microbial consortia present in the MFC help to generate electrical energy by converting the chemical energy by the catalytic reaction of microorganisms (Logan et al., 2006).

A typical MFC consists of anode and cathode compartments separated by a cation-permeable barrier or membrane. In the anode compartment, wastewater is oxidized by microorganisms, generating CO_2, electrons and protons. Electrons generated in the anode compartment are transferred to the cathode compartment through an external electric circuit, while protons are transferred to the cathode compartment through the proton-permeable barrier or membrane. In the cathode compartment, electrons and protons are utilized to produce water with the help of oxygen.

Two distinct classes of MFCs are available. One requires a mediator to facilitate electron transfer and another does not require any addition of mediator. Mediators are used to enhance the electrochemical activity in the MFC. Electron transfer from microbial cells to the electrode surface is facilitated by these polar compounds. Some commonly used mediators are methyl viologen, methyl blue, and neutral red. Most of the mediators are toxic in nature. Introduction of mediator is always a debatable issue when process economics is considered (Benetton et al., 2012).

Mediator-free MFC uses electrochemically active bacteria for electron transfer to the electrode. Different *Geobacter* are well-known electrochemically active bacteria and commonly available in mediator-free MFC. *Shewanella putrefaciens*, *Aeromonas hydrophila*, and *Geobacter sulfurreducens* are such bacteria commonly found in MFCs.

Though the concept of MFC shows a promising option to obtain green energy from waste, the inherent problems of low power output and other upscaling issues are posing restriction to use them on a field scale. Laboratory-scale systems are also not very promising in terms of productivity and waste degradation. In addition, some of the best available proton exchange membranes are extremely expensive, being proprietary (monopoly) products. Use of air-cathode MFCs arranged in banks consisting of several MFCs are currently under exploration in order to improve the power output and power sustainability issues.

14.1 MICROBIAL ELECTROLYSIS CELL

MEC is a variation of the mediator-less MFC. Energy is generated by means of electricity in MFCs by the metabolic activity of the bacteria during decomposition of organic compounds present in wastewater. In MECs, the metabolic process is partially reversed to generate hydrogen or methane by applying a potential difference across the electrodes. The externally supplied electrical energy to the system is used to supplement the voltage generated by the microbial decomposition of organics by the bacteria. The resultant electrical energy should be sufficient to electrolyze the water to produce hydrogen or to produce methane by reducing carbon dioxide. A complete reversal of the MFC principle is found in microbial electrosynthesis, in which carbon dioxide is reduced by bacteria using an external electric current to form multicarbon organic compounds (Call and Logan, 2008).

REFERENCES

Ahammad, S.Z., Gomes, J., Sreekrishnan, T.R., 2008. Wastewater treatment for production of H_2S-free biogas. J. Chem. Technol. Biotechnol. 83 (8), 1163–1169.

Ahammad, S.Z., Gomes, J., Sreekrishnan, T.R., 2010. A comparative study of two high cell density methanogenic bioreactors. Asia Pac. J. Chem. Eng. 6, 95–100.

Ahammad, S.Z., Davenport, R.J., Read, L.F., Gomes, J., Sreekrishnan, T.R., Dolfing, J., 2013a. Rational immobilization of methanogens in high cell density bioreactors. RSC Adv. 3, 774–781.

Ahammad, S.Z., Bereslawski, J.L., Dolfing, J., Mota, C., Graham, D.W., 2013b. Anaerobic–aerobic sequencing bioreactors improve energy efficiency for treatment of personal care product industry wastes. Bioresour. Technol. 139, 73–79.

Battistoni, P., Angelis, A.D., Boccadoro, R., Bolzonella, D., 2003. An automatically controlled alternate oxic–anoxic process for small municipal wastewater treatment plants. Ind. Eng. Chem. Res. 42 (3), 509–515.

Benetton, X.D., Sevda, S., Dalak, E., Sreekrishnan, T.R., Vanbroekhoven, K., Pant, D., 2012. Microbial electrochemical cells (MXCs): novel approaches for sustainable energy production and fuel recovery from wastes and wastewaters. In: Ginneken, L.V., Pelkmans, L. (Eds.), Biofuels in Practice: Technological, Socio-economical and Sustainability Perspectives. ILM Publications, Hertfordshire, UK. ISBN: 9781906799113, pp. 270.

Call, D., Logan, B.E., 2008. Hydrogen production in a single chamber microbial electrolysis cell lacking a membrane. Environ. Sci. Technol. 42 (9), 3401–3406.

Christgen, B., Yang, Y., Ahammad, S.Z., Li, B., Rodriguez, C.D., Zhang, T., Graham, D.W., 2015. Metagenomics shows low energy anaerobic-aerobic treatment reactors reduce antibiotic resistance gene dissemination from domestic wastewater. Environ. Sci. Technol. 49 (4), 2577–2584.

Electric Power Research Institute (EPRI), 2002. Water & Sustainability: U.S. Electricity Consumption for Water Supply & Treatment–The Next Half Century. vol. 4 EPRI, Palo Alto, CA. pp. 3–5.

Evans, E.A., Ellis, T.G., Gullicks, H., Ringelestein, J., 2004. Trickling filter nitrification performance characteristics and potential of a full-scale municipal wastewater treatment facility. J. Environ. Eng. 130 (11), 1280–1288.

Gunaseelan, V.N., 1997. Anaerobic digestion of biomass for methane production: a review. Biomass Bioenergy 13 (1–2), 83–114.

Heijnen, J.J., Mulder, A., Enger, W., Hoeks, F., 1989. Review on the application of anaerobic fluidized bed reactors in waste-water treatment. Chem. Eng. J. 41 (3), B37–B50.

Kiely, G., 1997. Environmental Engineering. McGraw-Hill, New York.

Logan, B.E., Hamelers, B., Rozendal, R., Schröder, U., Keller, J., Freguia, S., Aelterman, P., Verstraete, W., Rabaey, K., 2006. Microbial fuel cells: methodology and technology. Environ. Sci. Technol. 40 (17), 5181–5192.

Mara, D., 2004. Domestic Wastewater Treatment in Developing Countries. Earthscan, London.

Metcalf, Eddy, 2003. Wastewater Engineering: Treatment Disposal Reuse. McGraw Hill, New York.

Mohammed, K., Ahammad, S.Z., Sallis, P.J., Mota, C.R., 2014. Energy-efficient stirred-tank photobioreactors for simultaneous carbon capture and municipal wastewater treatment. Water Sci. Technol. 69 (10), 2106–2112.

Ntaikou, I., Antonopoulou, G., Lyberatos, G., 2010. Biohydrogen production from biomass and wastes via dark fermentation: a review. Waste Biomass Valorization 1 (1), 21–39.

Rittmann, B.E., McCarty, P.L., 2001. Environmental Biotechnology: Principles and Applications. McGraw Hill, New York.

Saravanan, V., Sreekrishnan, T.R., 2005. Hydrodynamic study of biogranules obtained from an anaerobic hybrid reactor. Biotechnol. Bioeng. 91 (6), 715–721.

Seghezzo, L., Zeeman, G., van Lier, J.B., Hamelers, H.V.M., Lettinga, G., 1998. A review: the anaerobic treatment of sewage in UASB and EGSB reactors. Bioresour. Technol. 65 (3), 175–190.

Shizas, I., Bagley, D.M., 2004. Experimental determination of energy content of unknown organics in municipal wastewater streams. J. Energy Eng. 130 (2), 45–53.

World Bank, 2003. Environmental health. Public Health at a Glance; HNP notes. World Bank, Washington, DC. http://documents.worldbank.org/curated/en/2003/06/12005673/environmental-health.

BIOPROCESSES FOR WASTE AND WASTEWATER REMEDIATION FOR SUSTAINABLE ENERGY

G. Mohanakrishna, S. Srikanth, D. Pant

VITO—Flemish Institute for Technological Research, Mol, Belgium

1 INTRODUCTION TO ENERGY THROUGH BIOREMEDIATION

Bioremediation is a natural process that decomposes or detoxifies or removes pollutants by the action of biological catalyst, that occurs at different levels of the ecosystem. Bioremediation encompasses water, soil, and air environments. Pollution rates in these areas are increasing rapidly with the industrialization and evolution of human needs. In the early stages of microbiology development, various biological processes for treating pollutants were understood. As the importance of remediation or treatment is intensifying, different processes have been developed for the treatment, and it has also been observed that biological treatment processes are superior to the other physicochemical processes. Bioremediation disintegrates pollutants and debris, which are stacking up in the environment and causing deterioration. The benefits of bioremediation also extend from the environment to health, life, and the world economy. Since bioremediation proceeds with biological components, limited energy input and mild operating conditions are sufficient for the mineralization or detoxification of pollutants. Traditional wastewater treatment processes, such as activated sludge process (ASP), treat soluble organic materials and suspended solids present in low-strength wastewater efficiently, but the process demands a high amount of energy (Abbassi et al., 2000; Judd, 2010). Later, anaerobic digestion (AD) was discovered as an efficient treatment process that also generates methane as the value-added energy product. Treating a wide variety of wastewaters and solid waste are added advantages of AD. The diverse microbial metabolisms present in AD led to the process modification for generation of energy and other chemical products such as hydrogen, volatile organic products, alcohols, etc. This aspect became very encouraging in the search for nontraditional and renewable energy (Lettinga et al., 1980; Nishio and Nakashimada, 2007; Mohan et al., 2008a). Both ASP and AD are prokaryotic processes, demonstrated by bacteria. Besides bacteria, eukaryotic microorganisms such as algae and fungi are also involved in waste remediation. Photoautotrophic and heterotrophic metabolisms of the microalgae help in the treatment process, which is called phycoremediation. Mycoremediation is the process that proceeds by fungi, which are one of the best-known catalysts to treat complex organic materials (Pant and Adholeya, 2009, 2010). The global research fraternity is greatly focused on bioenergy generation from waste though fermentative process. Use of these processes has been increasing exponentially in the last decade, underscoring the potential of waste treatment for bioenergy generation (Figure 1).

Bioremediation and Bioeconomy. http://dx.doi.org/10.1016/B978-0-12-802830-8.00021-6
Copyright © 2016 Elsevier Inc. All rights reserved.

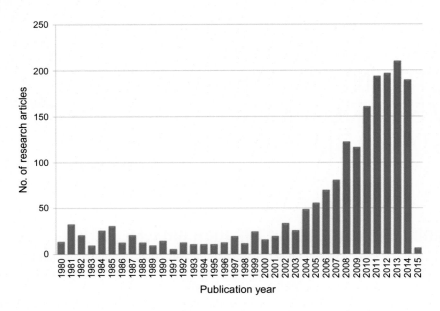

FIGURE 1

Scopus research results number of articles published from 1980 to till early 2015 on the energy production from waste fermentation (keywords: energy AND fermentation AND waste).

Any kind of energy generated from waste is due to the oxidation of organic pollutants present in waste, which means that the energy generation efficiency of any process is always proportional to the waste/contaminant removal. Even though, the side reactions are possible with all the biological processes, they should be eliminated to transfer the energy generated from the oxidation of waste to the desired product (Elmekawy et al., 2013a). This is also the basic principle of sustainable development, where both treatment/remediation and energy generation will be merged and materialized. This chapter presents comprehensive information about various biological treatment or remediation processes that also generate energy in different forms.

2 VERSATILE COMPONENTS OF BIOENERGY GENERATION THROUGH REMEDIATION

2.1 C-SOURCES

Biological energy generation is mostly possible during carbon metabolism by microorganisms. The organic matter that is present in the wastewater, solid waste, and agricultural-based biomass oxidized to simpler organic molecules, generates the energy (catabolic process) (Mohanakrishna and Mohan, 2013; Niessen et al., 2004; Madsen, 2011). Carbon dioxide (CO_2) reduction and fixing to biomass or biomolecules also leads to energy equivalents such as saccharides, lipids, and chemicals (anabolism). Algae that grow in wastewaters and produce lipids and carbohydrate biomass exhibit mixotrophic metabolism (Mohan et al., 2011a; Gentili, 2014). Irrespective of biocatalyst, all the mechanisms are linked to the organic/carbon pollutants emphasizing their importance. Two different types of substrates are available, solid and liquid wastes, for bioenergy-generating processes, that are mainly differ based on total solids, total carbon content, and moisture/water content (Pant et al., 2012; Mohan et al., 2011b). Diverse energy-producing trajectories from the biological treatment of waste and wastewater and their respective products are depicted in Figure 2 and Table 1.

2.1.1 Solid waste

Solid waste contains low water content and high solids/organic matter that is not amenable for treatment. Based on solids concentration, it needs dilution with water or domestic sewage before being subjected to biological treatment. In recent days, solid state fermentation is also gaining importance to treat directly or with a limited amount of water supplementation to reduce the bioreactor volume, to avoid secondary clarifier, and to improve the process economics. In general, food-based, food industry-based, and agro-based wastes were extensively studied for energy generation (Elmekawy et al., 2014a). Although petroleum sludge treatment is also being evaluated for different bioprocesses, its energy generation potential is limited. Pretreatment of wastes is a process that converts organic matter to a biologically available form. Agro-based wastewaters contain high content of cellulosic materials and can be pretreated with biological (enzymatic solubilization), physical (grinding, ultrasound, thermal, and freeze-thawing), chemical (acid and alkaline), and physical-chemical (thermo-acid) processes (Guo et al., 2010; Nah, 2000; Marañón et al., 2012). Poultry and swine waste contains ammonia that inhibits the biological process, which requires the pretreatment for ammonia oxidation (Lei et al., 2007).

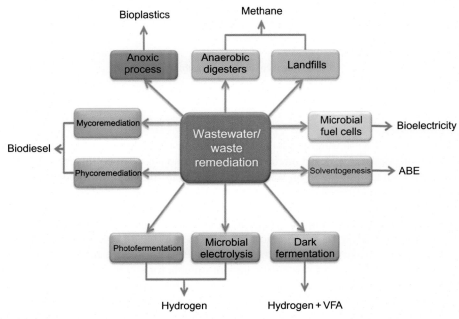

FIGURE 2

Energy-generating processes in wastewater/waste remediation. [VFA, volatile fatty acid; ABE, Acetone–butanol–ethanol].

Table 1 Various Biofuels Generated as the Value Added Products from Different Bioremediation Processes with Their Target Compounds/Pollutants/Substance

Bioremediation Process	Treatment Output	Energy Output
Anaerobic digestion (AD) and landfills	Organic matter removal from wastewater	Methane
Acidogenic fermentation Photofermentation	Organic matter removal from wastewater	Hydrogen
AD integration with acidogenic fermentation	Organic matter removal from wastewater	Hythane
Heterotrophic microalgae cultivation	Nutrients and organic matter removal from wastewater; CO_2 reduction	Lipids for biodiesel and carbohydrates for biofuels production
Microbial fuel cells	Waste/wastewater treatment Desalination Nutrient removal	Bioelectricity
Microbial electrolysis cells	Waste/wastewater treatment Fermentation effluents	Hydrogen
Microbial electrosynthesis	Waste/wastewater treatment Carbon dioxide reduction Specific pollutant removal	Acetate Ethanol Butyrate Butanol Methane
Autotrophic metabolic functions	Carbon dioxide and carbon monoxide, industrial emissions	Biodiesel
Anoxic or microaerophilic process	Wastewater Effluents of dark fermentation	Bioplastics

2.1.2 Liquid waste

The rate and volume of wastewater production is increasing continuously with increase in industrialization and human population. Based on the concentration and nature of pollutants, wastewater can be rated as high strength and low strength wastewaters as well as high and low biodegradable wastewaters. Wastewater from food-based industries, domestic origin, and agroindustries that contains a higher amount of biodegradable organic material can be treated efficiently (Zhang et al., 2014; Hamawand, 2015). The wastewater from chemical, pharmaceutical, metal- and rubber-based industries, which has a low fraction of biodegradable organic material, tends to have lower treatment efficiency. The carbon content of the wastewater also determines the potential of the wastewater toward application for energy generation. All types of wastewater are not suitable for all bioenergy-generating processes. The selection of bioenergy-generating process depends on the nature and composition of the wastewater. For example, distillery wastewater and dairy wastewater are suitable for methane generation by AD, biohydrogen production through dark fermentation, and bioelectricity generation through microbial fuel cells (MFCs), whereas domestic sewage is suitable for biohydrogen production by photofermentation and biodiesel production through microalgae cultivation (Mohan et al., 2011b; Zhou et al., 2012; Tyagi, 2013).

2.2 BIOCATALYST

2.2.1 Whole cell

Microorganisms have diverse metabolic functions that help in the degradation of a wide variety of substrates. Microorganisms ranging from eukaryotic algae and fungi to prokaryotic bacteria and archaea. These microorganisms are not only involved in the metabolic functions of degradation but also in the conversion of higher organic molecules to lower organic molecules and lower to higher organic molecules. Present-day technology has engineered such bioprocesses for various health, food, and environmental applications. Microbial environmental applications have been evaluated for wastewater/waste remediation; such a process is known as bioremediation or biodegradation. In the milieu of energy crisis and intense need for the world, various biodegradation and bioremediation technologies are being redirected toward simultaneous energy generation another advantage of biocatalyst is, reusability. Among all known bacterial processes for energy generation, the biocatalyst used for polyhydroxyalkanoates (PHAs) production cannot be reused as the product can be separated only through breaking the cell wall (Venkateswar Reddy et al., 2014; Jia et al., 2013). Similarly, in the case of lipid production from algae, the biocatalyst cannot be reused.

2.2.2 Mixed cultures

Wastewater treatment process proceeds more comfortably with mixed bacterial cultures as the biocatalyst than any other pure bacteria strains (Pearce, 2003; Fang and Liu, 2002). The heterogeneity of the substrate and unsterilized conditions during the process are the main factors for the effective performance of mixed cultures (Mohanakrishna et al., 2010a). The importance of mixed cultures has become prominent as the waste remediation process is engineered to generate biofuels or/and chemicals and has led to the distinct field of mixed culture biotechnology (MCB). The diversity or composition of mixed culture is not similar for every waste-remediating process. Based on the process, the substrate and operating conditions, the microbial population corresponding to the specific process can be effectively enriched from the natural environment. Compared to pure culture-based energy or chemicals production, the advantages of MCB include: (1) no sterilization required, (2) adaptive capacity owing to microbial diversity, (3) generates a narrow product spectrum from the degradation of mixed substrate, (4) biologically robust system, and (5) continuous process with less complexity (Kleerebezem and van Loosdrecht, 2007; Mohan et al., 2007; Han et al., 2011).

2.3 BIOPROCESSES

2.3.1 Anaerobic process/AD

AD is the classical bioprocess that works competently for removal of organic compounds from waste with simultaneous generation of a valuable energy product, methane. Even though the process was known for more than a century, it has been studied in depth only in the last 3 decades for different aspects of commercialization. AD is a process in which microorganisms oxidize the organic matter under anaerobic conditions into biogas (a mixture of methane (CH_4) and CO_2) (Angenent et al., 2004; Mohan et al., 2007). This basic process comprises four different phases in which the substrate undergoes biochemical changes: hydrolysis, fermentation/acidogenesis, acetogenesis, and methanogenesis (Figure 3). Since each phase of AD has individual importance with respect to metabolites and the AD process can be controlled at each phase for different forms of energy and chemicals production, it has become a versatile process in environmental biotechnology (Li and Yu, 2011).

As shown in Figure 3, complex organic molecules such as polysaccharides, proteins, and lipids are hydrolyzed to their respective monomers (proteins to amino acids; polysaccharides to sugar; lipids to

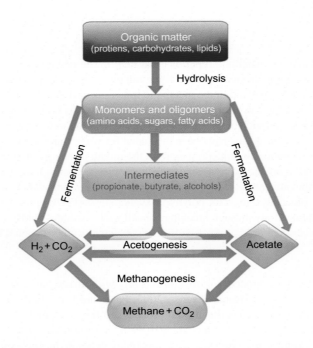

FIGURE 3

A network of microbial processes of anaerobic digestion, treating complex wastewaters into CH_4 through four different phases. Inhibiting the final step of methanogenesis, known as acidogenic fermentation, results in H_2 and VFAs.

fatty acids) by hydrolyzing bacteria. This process is referred as hydrolysis. Through fermentation, fermentative bacteria convert the monomers to a mixture of low molecular weight organic acids (VFAs) and alcohols. Further, these fermentation products can be oxidized to acetic acid and hydrogen by hydrogen-producing acetogenic bacteria. This process, known as acetogenesis (Mohanakrishna et al., 2010b), also includes acetate production from the reduction of hydrogen and carbon dioxide. The bacterial group that reduces CO_2 and H_2 is known as homoacetogens. Finally, the acetate is converted to methane by methanogenesis; the bacterial group involved is acetoclastic methanogens. The pathways involved in methanogenesis constitute a complex metabolic network. On complete degradation of organic matter, methanogenesis ultimately leads to the production of CH_4 and CO_2. When the metabolism of methanogenesis is visualized in depth, this process utilizes three types of substrates. They are C1–C6 short chain fatty acids (also known as VFAs), n- or i-alcohols, and gases such as CO, CO_2, and H_2. Methane production can lead through two groups of methanogens. One group mainly uses acetate for methane production, whereas the second group uses H_2 and CO_2 or H_2/CO_2. During methane formation process, the coenzymes M play a significant role. If the formate and CO are the substrates for methane production, first CO and formate will be transformed to CO_2 by coenzyme F420, and then CO_2 will be reduced to methane by the action of coenzyme M. As this process alone has various pathways and the intermediary products have economic viability, understanding this metabolic network helps to engineer the process for the production of biohydrogen, VFAs, and alcohols. Integration with other biological processes for which the intermediates of AD act as substrates will lead to the production of alcohols, solvents, and PHAs, as visualized in Figure 3.

2.3.2 Algal metabolism

Algae are one of the oldest life forms on earth. Microalgae are unicellular and simple multicellular microorganisms that include both prokaryotic microalgae (cyanobacteria) and eukaryotic microalgae [green algae (Chlorophyta), red algae (Rhodophyta), and diatoms (Bacillariophyta)]. Eukaryotic algae have been found to be the most important classes of microalgae for biofuels generation. Algae can either be autotrophic (photosynthetic, requires only inorganic compounds such as CO_2 and salts) or heterotrophic (nonphotosynthetic, requires organic source). Some photosynthetic algae are mixotrophic (they have ability to perform the processes) (Brennan and Owende, 2010). Several advantages of microalgae-derived biofuels can be foreseen, based on these attributes of microalgae: (1) capable of year round production; (2) require less water than terrestrial oil-producing crops, therefore reducing the load on freshwater sources; (3) can be cultivated in polluted or brackish water, therefore not compromising the production of food and arable land; (4) rapid growth rate (exponential growth rates is 3.5 h) and oil content will be in the range of 20-50% of dry cell weight; (5) higher potential for CO_2 fixation (1.83 kg of CO_2 can generate 1 kg of dry biomass); (6) wastewater acts as source for nutrients (nitrogen, phosphorus, and other micronutrients) required for microalgae cultivation; (7) microalgae can produce valuable coproducts like proteins and de-oiled biomass (after lipid extraction), which can be used as carbohydrate source for methane and ethanol production and as fertilizer. With these advantages, microalgae involves different types of biofuels generation such as biodiesel, biomethane, biohydrogen production with simultaneous accomplishment of wastewater treatment. Apart from these, cyanobacteria produce biohydrogen through photofermentation through treatment of effluents rich in organic acids. Most of the microalgae accumulate lipids under environmental stress conditions like nitrogen or phosphate limitation. Therefore such environmental conditions are controlled to improve microalgal lipid accumulation (Bellou et al., 2014; Mohan et al., 2011a; Amaro et al., 2011; Courchesne et al., 2009). The metabolic response of microalgal strains to the different wastewaters is not similar and has been found to depend on various factors such as type of energy source, pH, temperature, etc (Table 2). Figure 4 elucidates the detailed mechanism of lipid production in microalgae. Apart from lipid production, algae is also cultivated to capture the sunlight and atmospheric CO_2 into biomass. Compared to simple structure of the biomolecules of the algae, it is being used as substrate for several bioenergy-generating processes such as biohydrogen, biomethane alcohols production, etc. (Turon et al., 2014). The residue of the algal biomass from biodiesel production process is rich in carbohydrate and was also found to have less protein content.

Table 2 Lipid Content for Biodiesel Production Through Microalgal Cultivation in Different Wastewaters

Type of Wastewater or Waste	Lipid Content	Biocatalysts	References
Industrial wastewater	18.4%	*Chlamydomonas* sp. TAI-2	Wu et al. (2012)
Biogas plant effluent	35%	*Monoraphidium* sp. KMN5	Tale et al. (2014)
Cheese whey permeate	37.8 mg/L/day	*Scenedesmus obliquus*	Girard et al. (2014)
Secondary treated sewage	17%	*Botryococcus braunii*	Órpez et al. (2009)
Dairy effluent	42%	*Chlorococcum* sp. RAP13	Ummalyma and Sukumaran (2014)
Anaerobic digestate of food wastewater	35.06%	*Scenedesmus bijuga*	Shin et al. (2014)

Table 2 Lipid Content for Biodiesel Production Through Microalgal Cultivation in Different Wastewaters—Cont'd

Type of Wastewater or Waste	Lipid Content	Biocatalysts	References
Synthetic feed (fructose, glucose, and acetate)	52.6%	*Scenedesmus* sp.	Ren et al. (2013)
Agro-industrial co-products	43%	*Chlorella vulgaris*	Mitra et al. (2012)
Piggery wastewater	6.3 mg/L/day	*Chlorella pyrenoidosa*	Wang et al. (2012)
Dairy and municipal wastewater	29%	Mixed microalgae	Woertz et al. (2009)
Domestic wastewater	26%	Mixed microalgae	Mohan et al. (2011a)
Domestic wastewater	28.2%	Mixed microalgae	Prathima Devi et al. (2012)
Tertiary wastewater of CETP	0.171 g lipid/g cell/day	*Chlorella minutissima*	Malla et al. (2015)
Primary treated wastewater	0.132 g lipid/g cell/day	*Chlorella minutissima*	Malla et al. (2015)

CETP, common effluent treatment plant.

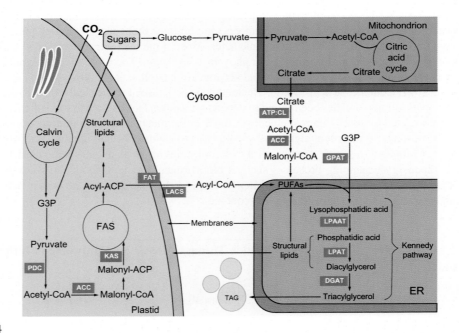

FIGURE 4

A simplified scheme showing lipid synthesis in microalgae (Bellou et al., 2014). ACC, acetyl-CoA carboxylase; ACP, acyl-carrier protein; LACS, long-chain acyl-CoA synthetase; ATP:CL, ATP-dependent citrate lyase; CoA, coenzyme A; DGAT, diacylglycerol acyltransferase; ER, endoplasmic reticulum; FAS, fatty acid synthase; FAT, fatty acyl-ACP thioesterase; G3P, glycerate-3-phosphate; GPAT, glycerol-3-phosphate acyltransferase; KAS, 3-ketoacyl-ACP synthase; LPAAT, lyso-phosphatidic acid acyltransferase; LPAT, lyso-phosphatidylcholine acyltransferase; PDC, pyruvate dehydrogenase complex; TAG, triacylglycerol.

3 BIOREMEDIATION PROCESSES THAT GENERATE ENERGY
3.1 ANAEROBIC DIGESTION

AD is a well-studied biological process that treats waste and wastewater to produce a methane-rich biogas. AD provides a spectrum of striking advantages such as pollution reduction and waste stabilization, energy recovery in the form of methane, cost-effective treatment process, suitable for a wide range of wastewaters and limited environmental impact. In addition, the solid remnants of the process can be used as fertilizer. With extensive research on various aspects of the AD process, it is now an established and commercially proven approach for treatment and recycling of biomass, waste, and wastewater (Mohan et al., 2008a; Kothari et al., 2014).

Various categories of wastewaters and solid wastes from industries, agriculture, and domestic activities were used for the generation of methane through AD. Critical factors influencing the process were identified. In general terms, the CH_4 production efficiency is dependent on the biodegradable nature of substrate. For example, the effluents of chemical, pharmaceutical wastewaters contain very low biodegradability and exhibit low efficiency for CH_4 production. As the biodegradability increases, CH_4 production potential increases. Biological/biochemical methane potential (BMP) is a standard test protocol that provides maximum CH_4 of any substrate. The substrate needs to be evaluated for physical characteristics such as carbon oxygen demand (COD), biological oxygen demand (BOD), total solids (TS), volatile solids (VS), total carbon (TC), and total organic carbon (TOC) and then the BMP will be evaluated in batch tests. Operational parameters such as time, temperature, pH, etc., can be optimized from this BMP test (Mata-Alvarez et al., 2014).

During the operation of anaerobic reactor, the different categories of complex waste/substrate (carbohydrate, lipids, and protein) should simultaneously digest all organic substrates in a single bioreactor. The composition of all the wastes/wastewaters is not identical. In general to AD, the key parameters that govern the process are temperature, VFAs, pH, ammonia, nutrients, trace elements. Additional parameters might be included, based on the composition and nature of the waste (Appels et al., 2011).

Temperature is the most significant factor that influences microbial activity, which in turn influences the CH_4 and quality of the effluent/digestate. Among the three thermo-dependent bacterial groups, mesophilic bacteria that perform at 30-40°C condition are observed in many digesters. However, methane yield improves with increase in temperature. Thermophilic bacteria have accelerated metabolic rates and higher growth rates, which improves the biogas formation (El-Mashad et al., 2004; Sánchez et al., 2001). Biogas production under thermophilic condition (55 °C) is found to be double that of psychrophilic condition (15 °C), thus highlighting the key role of temperature (Sánchez et al., 2001). Higher temperature also increases degradation of substrates by benefiting endergonic reactions. Organic nitrogen degradation and phosphorus assimilation were also found to increase with increase in temperature. VFAs are intermediary products of AD, which include acetic acid, propionic acid, butyric acid, and valeric acid. By the action of syntrophic acetogenic bacteria, these VFAs can be converted to CH_4. Among the four VFAs, propionic acid to acetic acid ratio is considered as the indicator for the performance of the AD. The higher value of this ratio infers digestion imbalance (Marchaim and Krause, 1993; Buyukkamaci and Filibeli, 2004). VFA determines the system pH, and it is observed that both VFA concentration and pH are dependent factors. pH in the range of 6.5-7.2 was observed to be optimum for methanogenesis. Higher VFA concentration leads to decrease in pH and acts as vice versa phenomenon. The rate of shift in pH with respect to VFA concentration change is also dependent on the buffering nature of the bulk liquid. High buffering systems provide more stability with change in VFA concentration. C/N ratio is another key parameter needed for the optimum nutrient balance required for the growth of bacteria. Generally, the C/N ratio in the range of 20-30, is considered as optimum (Li et al., 2011; Puyuelo et al., 2011). If the C/N ratio of any waste is not suitable for methanogenesis, a co-substrate addition strategy can maintain optimum ratio in the digester. Waste or wastewater from animal origin such as meat-processing wastewater or manure of swine or poultry contains high protein content. With treatment in anaerobic digester, it generates ammonia. The ammonia generation process consumes essential nutrients required for bacterial growth. Also, ammonia in higher concentration is toxic to the microbial metabolism.

3.2 DARK FERMENTATION

3.2.1 Biohydrogen

Biohydrogen production through dark fermentation is referred as the truncated version of methanogenesis. During methanogenesis, if the final/fourth stage is inhibited or stopped, the end products of the acetogenesis (one of the intermediary steps of AD) are H_2 and acetate. Based on the microbiological and operational behavior of the biohydrogen production process, several strategies were developed. Compared to pure cultures, mixed cultures perform better for hydrogen production from wastewater or waste remediation. This finding led to the enrichment of hydrogen-producing bacterial consortia from the methane-producing (anaerobic) consortia. Treating anaerobic sludge with acid, alkaline, heat, and chemical (BESA, bromoethane sulphonic acid) were proved as efficient methods (Wang and Wan, 2009; Mohan et al., 2007, 2008c; 2012a,b; Mohanakrishna et al., 2011). The common mechanism behind these methods is elimination of methanogens from the total bacterial consortia of AD for methane production. The methanogens are spore-forming bacteria; they are sensitive to extreme environments such as heat, acid, and alkaline conditions, whereas BESA inhibits enzyme cofactor, which plays a key role in methanogenesis (Mohanakrishna and Mohan, 2013; Wang and Wan, 2009). Optimum operational conditions of the process such as pH, time of operation, VFA concentration, and organic loading rate were used to control the hydrogen-producing system for maximum production efficiencies. Mild acidic conditions (pH 5.5-6.5), lower retention time (than methanogenesis), and mild VFA concentrations are optimum operation conditions. Compared to a continuous mode, batch mode operation is very much suitable for hydrogen production. Several categories of industrial wastewaters, viz., food processing, dairy-based, alcohol-based, plant and agricultural-based wastewater, were evaluated and optimized for the hydrogen production. However, H_2 production and treatment efficiencies vary with each type of waste. Similar to AD, biohydrogen production process was also influenced by various factors like organic loading rate, temperature, pH, etc., (Zhu and Beland, 2006).

Compared to AD, dark fermentation is found to have less substrate degradation efficiency. The final product, hydrogen, generates along with the VFAs and will not be further degraded to carbon dioxide. Since the carbon persist in the VFAs, the effluent of the hydrogen process contains at least 70% of the carbon residue. Practically, during bioreactor operation, various other microbial metabolisms also act on the organic matter, and because of it the maximum residual organic matter is found around 30-35% (maximum substrate degradation efficiency, 65-70%). Higher concentrations of residual VFA in dark fermentation against the targeted research focus of complete waste degradation, has directed towards the bioprocess integration approach, which, will be discussed in later sections.

3.3 PHOTOFERMENTATION

Apart from the dark fermentation, photofermentation is one of the possible alternatives for energy generation from waste streams. Photofermentation is the process of fermentative conversion of organic carbon, in the presence of light unlike dark fermentation, manifested mostly by photosynthetic bacteria (PSB) similar to anaerobic fermentation. Photofermentation is most studied for H_2 production with respect to bioenergy generation, especially as an integration to the acidogenic fermentation to valorize the residual organic fraction rich in volatile acid intermediates (Srikanth et al., 2009a; Srikanth et al., 2009b; Mohan et al., 2013a). Processing of unutilized carbon sources in wastewater for additional H_2 production will sustain practical applicability of the process. Combination of dark and photofermentation processes is considered to be an ideal route leading to nearly the highest theoretically possible yield (Srikanth et al., 2009a; Srikanth et al., 2009b).

Photobiological H_2 production is feasible through two diverse photosynthetic mechanisms, oxygenic and anoxygenic. Oxygenic photosynthesis is catalyzed by microalgae and cyanobacteria, while anoxygenic photosynthesis is catalyzed by PSB. Anoxygenic process for H_2 production is advantageous over oxygenic process due to the absence of oxygen, which is considered to be a scavenging molecule. During oxygenic photosynthesis, O_2 gets released as a by-product during photolysis of water, where H_2 gets generated through either direct or indirect photolysis by hydrogenase, with a marginal variation in the mechanism (Kruse et al., 2005; Beer et al., 2009). Generally, hydrogenase function is based on the O_2 availability in the system and supply of e^- and H^+, either directly by photosynthetic water splitting (driven by photosystem (PSII)) or indirectly from the degradation of organic molecules (starch and nitrogen fixation). Hydrogenase acts as H^+/e^- release valve by recombining H^+ (from the medium) and e^- (from reduced ferredoxin) to produce H_2. Nitrogenase can also catalyze the H_2 production during nitrogen fixation but the process is extremely O_2 sensitive. Localization of nitrogenase in the heterocysts of filamentous cyanobacteria protects it from O_2 (Vyas, 1995). Further investigations on this process also revealed that the H_2 production in cyanobacteria is stimulated by nitrogen starvation and is catalyzed by "reversible hydrogenase" or "bidirectional hydrogenases" (Horch et al., 2012). Bidirectional hydrogenases are mainly involved in the disposal of excess reducing power derived from fermentation and photosynthesis resulting in H_2 evolution.

PSB are potent H_2 producers compared to cyanobacteria in the presence of sunlight as they neither use water as an e^- source nor produce O_2 photosynthetically. PSB can use a variety of carbon sources (carbohydrates and VFA), and the process is strictly anoxygenic in nature (Wilhelm, 1996; Mohan et al., 2013a). The enzymes responsible for H_2 production in PSB are also hydrogenase and nitrogenase, but their activities are not inhibited due to the nonproduction of O_2. The efficiency of light energy conversion to H_2 by PSB is much higher than cyanobacteria due to less quantum of light energy requirement than the water photolysis (Batyrova et al., 2012; Vyas, 1995). Among the photosynthetic processes of H_2 production, the photofermentation route seems to be favorable due to relatively higher substrate to H_2 yields, its ability to trap energy over a wide range of the light spectrum, and versatility in sources of metabolic substrates (Srikanth et al., 2009a).

3.4 BIOELECTROCHEMICAL SYSTEMS

Research on bioenergy generation from renewable feedstocks has discovered the most transdisciplinary system with high treatment efficiencies along with multifaceted applications; namely, bioelectrochemical systems (BES) (Pant et al., 2012). BES works at the interface of fermentation and electrochemistry as basic research areas, although various other sciences also contribute to make the system complete. The applications of BES include current generation, bioremediation of a wide range of wastes, specific pollutants/xenobiotics removal, desalination, color removal, toxicity reduction, gaseous pollutants treatment, synthesis of commercially viable chemicals and solvents, CO_2 reduction, etc. (Venkata Mohan et al., 2014a). In this chapter, however, the focus is more on considering waste as bioenergy source based on three well-established forms of BES: MFCs, microbial electrolysis cell (MEC), and microbial desalination cell (MDC).

3.4.1 Microbial fuel cells
Harnessing of bioelectricity from waste streams using MFC has garnered significant interest in both basic and applied research, in the recent past, due to its sustainable and renewable nature. Creating an electrical circuit between e^- source (substrates) and the e^- sink (molecular oxygen) by placing noncatalyzed electrodes facilitates the development of a potential gradient against which the e^- flow in the circuit and can be harnessed as bioelectricity (Venkata Mohan et al., 2014a; Pant et al., 2012; ElMekawy et al., 2013b). MFC facilitates direct transformation of chemical energy stored in the waste to electrical energy via microbial catalyzed redox reactions under ambient temperature and pressure (Figure 5). Anodic oxidation and cathodic reduction reactions are the basis for MFC function. The reducing equivalents H^+ and e^- generated through a series of bioelectrochemical redox reactions at anode (Eq. 1) will reach the cathode and become reduced in the presence of an electron acceptor (Eq. 2).

$$C_6H_{12}O_6 + 6H_2O \rightarrow 6CO_2 + 24H^+ + 24e^- \, (Anode) \tag{1}$$

$$4e^- + 4H^+ + O_2 \rightarrow 2H_2O \, (Cathode) \tag{2}$$

$$C_6H_{12}O_6 + 6H_2O + 6O_2 \rightarrow 6CO_2 + 12H_2O \, (Overall) \tag{3}$$

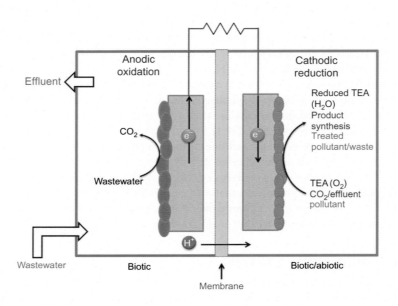

FIGURE 5

Schematic of a typical Bioelectrochemical system with multiple possible application. [TEA, terminal electron acceptor].

In general, the anode chamber of MFC is operated under anaerobic condition, but few reports are available in the literature pertaining to the application of aerobic or microaerophilic metabolic function at anode (Ringeisen et al., 2007; Rodrigo et al., 2010; Mohan et al., 2008b). Apart from power generation, MFC application has been extended toward multiple functions including waste remediation, specific pollutants removal, and recovery of value-added products based on the electron donor and acceptor conditions, bringing these functions together under one heading, BES. In fact, MFC operation was mostly reported with respect to waste remediation and removal of specific pollutants, rather than power generation. Combined function of all the biological, physical, and chemical components of MFC provides an opportunity to trigger multiple reactions at once as a result of microbial metabolism and subsequent secondary reactions (Mohan and Srikanth, 2011). The anode chamber of MFC mimics the functions of both conventional anaerobic bioreactor and electrochemical unit, where the combination of these redox reactions helps in enhancing the degradation of organic matter and toxic or xenobiotic pollutants. Similarly, the cathode chamber also depicts effective waste treatment with *in situ* generation of strong oxidants. The potential difference between anodic oxidation and cathodic reduction reactions helps in the enhancement of the degradation of different pollutants at both the electrodes. Synergistic interaction between the MFC components and the biocatalyst needs to be understood and optimized to fully exploit the capacities of these systems in order to maximize the treatment efficiency and energy generation (Srikanth et al., 2011).

Effective utilization of waste and specific pollutants as substrates in MFC has the dual advantage of bioenergy generation as well as waste treatment. Origin and composition of waste can influence the treatment efficiency of MFC as well as energy recovery. A diverse range of wastewaters has been studied as substrates in MFC from domestic to industrial origin (Pant et al., 2010; Mohan et al., 2013b). Table 3 presents a detailed list of waste used in MFC as substrate and the degree of treatment efficiencies reported. However, the efficiency of electron recovery depends on the oxidation state of electron donor and the ratio of the electron donor to the microbe that can oxidize it. Waste having higher biodegradability, from dairy, food, vegetable, kitchen, etc., origin, will have higher power generation capacity, over the waste originated from complex industrial environments with low biodegradability

Table 3 Various Food Industry Based Wastewater Used as Anodic Fuels in MFC and Their Respective Performances

Wastewater	Power Density	ξCOD (%)	References
Chemical wastewater	339.87-862.85 mA/m^2	55.76-62.9	Mohan et al. (2008c,d) and Raghavulu et al. (2009)
Slaughter house wastewater	578 mW/m^2	93 ± 1	Katuri et al. (2012)
Animal carcass wastewater	2.19 W/m^3	50.66	Li et al. (2013b)
Swine wastewater	228 mW/m^2	84	Kim et al. (2008)
Beer brewery wastewater	483 mW/m^2	87	Wang et al. (2008)
Beer brewery wastewater	264 mW/m^2	43	Wen et al. (2009)
Dairy wastewater	1.1 W/m^3 ($\sim$36 mW/m^2)	95.49	Venkata Mohan et al. (2010a)
Dairy wastewater	161 mW/m^2	90	Elakkiya and Matheswaran (2013)
Cheese whey	1.3 ± 0.5 W/m^3	59.0 ± 9.3	Kelly and He (2014)
Chocolate industry wastewater	1500 mW/m^2	74.77	Patil et al. (2009)

Table 3 Various Food Industry-Based Wastewater Used as Anodic Fuels in MFC and Their Respective Performances—Cont'd

Wastewater	Power Density	ξCOD (%)	References
Molasses wastewater	1410 mW/m^2	53.2	Zhang et al. (2009)
Distillery wastewater	124.35-245.34 mW/m^2	49.32-72.84	Mohanakrishna et al. (2010a, 2012)
Fermented distillery wastewater	224.93 mA/m^2	86.67	Mohan et al. (2011c)
Molasses wastewater mixed with sewage	382 mW/m^2	59	Sevda et al. (2013)
Palm oil mill effluent	44.6 mW/m^2	~90	Cheng et al. (2010)
Vegetable waste	57.38 mW/m^2	62.86	Venkata Mohan et al. (2010b)
Fermented vegetable waste	111.76 mW/m^2	80	Mohanakrishna et al. (2010b)
Cereal-processing wastewater	81 ± 7 mW/m^2	95	Oh and Logan (2005)
Canteen based food waste	~556 mW/m^2	86.4	Jia et al. (2013b)
Food waste leachate	432 mW/m^3	87-92	Li et al. (2013c)
Mustard tuber wastewater	246 mW/m^2	57.1	Guo et al. (2013)
Protein food industry wastewater	45 mW/m^2	86	Mansoorian et al. (2013)
Cattle dung	220 mW/m^3	73.9 ± 1.8	Zhao et al. (2012)
Cattle manure slurry	765 mW/m^2	41.9-56.7	Inoue et al. (2013)
Cow manure	349 ± 39 mW/m^2	~50 (carbon)	Wang et al. (2014a)
Diluted wheat straw hydrolysate	148 mW/m^2	95 (xylan and glucan)	Thygesen et al. (2011)
Steam exploded corn stover hydrolysate	367 ± 13 mW/m^2	94 ± 1	Zuo et al. (2006)
Powdered raw corn stover	331 mW/m^2	42 ± 8 (cellulose)	Wang et al. (2009)
Steam exploded corn stover residue	406 mW/m^2	60 ± 4 (cellulose)	Wang et al. (2009)
Rice straw hydrolysate	137.6 ± 15.5 mW/m^2	49.4-72	Wang et al. (2014b)

(ElMekawy et al., 2014a). Still, the wastewater could be a potential carbon source for MFC due to the possibility of converting negative-valued waste into bioenergy. However, optimization is still required for upscaling the process with concerted efforts (Pasupuleti et al., 2015). MFC was also reported as a secondary integration unit to the fermentation and preliminary treatment processes, to treat the residual organics present in the effluent. Few studies have been reported in the literature based on this concept but only a very few of them used real field effluents, and the coulombic efficiencies are between 12% and 45%. The biocatalyst enriched in presence of acid metabolites was reported to depict higher treatment efficiencies and power output (ElMekawy et al., 2014b) due to the adaptation of biocatalyst. Treatment gained in this type of integrated system is an addition to the first process, increasing the total valorization capacity of the waste.

Apart from wastewater and effluents, water with specific pollutants can also contribute to energy generation in MFC with their electron donor and acceptor functions. In general, the chemotrophic (autotrophic/heterotrophic) microbes can utilize various inorganic components as electron donors for their metabolism, where some of the pollutants of wastewater can take the donor job and oxidize them at anode. Similarly, a few biocatalysts can also utilize these pollutants as electron acceptors in respiration (in absence of O_2) at cathode, facilitating their remediation. Apart from this, some pollutants such as sulfur, metals, estrogens, etc., can act as redox mediators for the electron transfer at anode (Kumar et al., 2012). Removal of sulfides (Rabaey et al., 2006), nitrates (Clauwaert et al., 2007; Virdis et al., 2008), perchlorate (Butler et al., 2010), chlorinated organic compounds (Venkata Mohan et al., 2014b), and estrogens (Kumar et al., 2012) was reported in MFC along with power generation. Some of the known metal pollutants such as iron, manganese, selenium, uranium, chromium, arsenic, vanadium, and cobalt can also enhance the power generation by acting as electron acceptors. Also, the colored dye compounds can act as alternate electron acceptors resulting in power generation and their removal. Nitrobenzenes, polyalcohols, and phenols were also studied for their treatment in MFC.

Similar to fermentation, MFC can also be operated using both dark and photosynthetic modes. However, very few and specific reports are available on the usage of photosynthetic mechanism in MFC operation (ElMekawy et al., 2014c). Most of these studies are related to single strains of green algae, cyanobacteria, and PSB acting as anodic biocatalyst under photoautotrophic or photoheterotrophic mode. Energy from the sun will be used to activate the photosystem and an organic or inorganic electron source will be used to generate energy. Light energy captured by pigment molecules channels to the bacteriochlorophyll (Bchl) and triggers a series of photochemical reactions separating the positive and negative charges across the membrane (Chandra et al., 2012). This charge separation initiates electron transfer coupled to the translocation of protons across the membrane, generating a potential gradient and resulting in electrogenesis (Mohan et al., 2013a). In algae, the mechanism of electrogenesis is little different and is based on membrane-bound protein complexes (photosystem (PSI), PSII, and cytochrome bf complex) and O_2 generating (oxygenic photosynthesis). Henceforth, the power generation will be on the lower side but the treatment efficiency is on the higher side compared to the anoxygenic photosynthesis (bacterial). *Chlamydomonas reinhardtii, Phormidium, Nostoc, Spirulina, Anabaena, Synechocystis PCC-6803*, etc., are the most studied species with oxygenic photosynthesis in MFC (Mohan et al., 2013a; Rosenbaum et al., 2010), while anoxygenic photosynthesis was evaluated using PSB like *Rhodopseudomonas palustris*, *Rhodobacter sphaeroides*, *Rhodobacter*, *Rhodopseudomonas*, etc., as biocatalyst in MFC (Rosenbaum et al., 2010; Mohan et al., 2013a). The application of oxygenic photosynthesis was also studied at cathode of MFC (algal biocathode), where the *in situ*-generated oxygen replaces the mechanical aeration that reduces the overall input energy for the system (Mohan et al., 2009; Venkata Mohan et al., 2014c).

3.4.2 Microbial electrolysis cells

Hydrogen is one of the alternative fuels having high impact in the present bioenergy research due to its high energy value (122 KJ/kg) and non-emission of greenhouse gases at combustion. The biological route of H_2 production involves more sustainable chemical, physical, and electrochemical methods due to less energy input needed and the possibility of avoiding pollution during its production (Mohan et al., 2011b). Furthermore, the carbon source present in wastewater can also be considered as substrate for the biological H_2 production due to its dual benefits, H_2 production and waste remediation. An enormous amount of literature is available on the biological route of H_2 production, and the technology is well established and understood with respect to the operational and regulating factors (Mohan et al., 2011b). In the course of understanding the process, various constraints such as dominance of methanogenic activity, low substrate conversion efficiency, accumulation of acid metabolites, sudden pH drop, etc., were observed (Srikanth and Venkata Mohan, 2014). Different strategies were also developed to overcome these constraints such as selective enrichment of inoculum (Srikanth et al., 2010), integration of secondary systems such as photofermentation (Srikanth et al., 2009a), MFC (ElMekawy et al., 2014b; Mohanakrishna et al., 2010b), methanogenesis (Mohanakrishna and Mohan, 2013), PHAs production (Venkateswar Reddy and Venkata Mohan, 2012), etc. In summary, the fermentative H_2 production alone cannot compete with the existing wastewater treatment methodologies because of the limited treatment ability.

In this context, researchers tried to combine the acidogenic fermentation with the electrolysis cell, to reduce the energy required for the water hydrolysis through microbial metabolism, naming this process MECs (Jeremiasse et al., 2011; Jeremiasse, 2011; Logan et al., 2008; Cheng and Logan, 2007). MEC is based on the principle that the reducing equivalents generated at anode from the oxidation of organic matter will get reduced to hydrogen at cathode (Liu et al., 2005). However, the supply of a small amount of external energy makes the process cross the endothermic barrier to form H_2. The standard redox potential for H_2 formation is −0.414 V, which is not possible to generate from microbial metabolism, and hence an additional energy supply is required to favor the formation of H_2. Low energy consumption compared to water electrolysis and high product (H_2) formation and substrate degradation rather than fermentative process make MEC a potential replacement for the existing methods and escalating energy demands. The performance of MEC has significantly improved within just a few years of its discovery (Logan et al., 2008; Liu et al., 2005). A few research groups also started working toward integrating the wastewater treatment with MEC to define an economically feasible hydrogen production methodology. A few pilot-scale operations also have been demonstrated, in spite of some drawbacks in process optimization (Cusick et al., 2011). MEC was also reported to utilize the acid-rich effluents generated from fermentation bioreactor (Lalaurette et al., 2009). Further to this research, a new process has been developed to convert the inherent energy of the wastewater into commercially viable commodities. Using the protons and electrons (energy) generated in the system, the microbial reduction process is being carried out in similar kinds of systems instead of generating H_2, naming the process microbial electrosynthesis (Pant et al., 2012; Venkata Mohan et al., 2014a; Rabaey and Rozendal, 2010; Sharma et al., 2013). CO_2 from the atmosphere and industrial off-gases is being considered as a major carbon source for this process. However, substantial input is needed to understand this process and make it a potential tool for mitigating environmental pollution and complementing global energy demands.

3.4.3 Microbial desalination cells

MDCs are one of the forms of BES and are being considered a promising technology for clean water and energy production (ElMekawy et al., 2014d). MDC technology is derived from MFC, where a third chamber is introduced between anode and cathode chambers to drive the energy towards water desalination. The ion exchange membranes, anion exchange membrane (AEM) and cation exchange membrane (CEM), used on both sides of middle chamber (filled with saline water) separating the anode and cathode, respectively, will do the job of desalination by separating the ions based on charge toward respective electrodes (anions to anode and cations to cathode) as depicted in Figure 6. Anodic biofilm triggers the electron flow in the circuit by oxidizing the organics and deploys the migration of anions and cations from the middle chamber with simultaneous power generation (Jacobson et al., 2011; Cao et al., 2009; Mehanna et al., 2010).

Migration of ions in the MDC is powered by the potential difference between the electrodes, and this process can reach a desalination level up to 99% along with energy generation (Jacobson et al., 2011; Cao et al., 2009). The first report on MDC was published by Cao and his coworkers in 2009 with a volume of 3 mL, which later increased to 1 L by Jacobson and his coworkers (9-11).

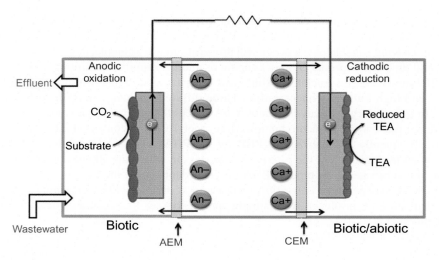

FIGURE 6

Microbial desalination cell (MDC) depicting removal of salts through three-chambered design.

Since then, the MDC technology has evolved on several fronts including reactor design, process optimization, and productivity. Most of the MDC studies are based on synthetic/designed salt waters, and a very few studies were based on brackish water or real seawater. However, this technology is proven to treat high concentration of salt water (35 g/L) with a desalination efficiency of 93%. Experimental output from several studies confirmed the high desalination efficiency of MDC and its suitability for being utilized as either a stand-alone technology or pretreatment process for the conventional desalination technologies. Various attempts were also made to improve the charge transfer, balance the pH in anode and cathode, increase the desalination efficiency and continuous operation of the cell (ElMekawy et al., 2014d). Irrespective of its advantages, MDC technology has various limitations, including the inability to concentrate huge amounts of acid or salt solutions by ion exchange process, prolonged time needed due to the decelerated anodic metabolism (Brastad and He, 2013), competitive ions other than those representing hardness, gradual increment of salt concentration in the anode and cathode compartments, etc. Overall, the MDC technology can be more effective in desalinating the water along with power generation, with optimal operating conditions.

3.5 ANOXIC PROCESS AND BIOPLASTICS

In wastewater treatment, the bacteria is subjected to feast and famine regimes. It is very often observed during sequencing batch processing of wastewater. When the substrate is present in excess or under stress conditions, the bacteria accumulates the storage polymers in the cytoplasm (Amulya et al., 2014; Bengtsson et al., 2008). Known as PHAs, these are considered as bioplastics from bacterial origin. Under external substrate limited conditions, these storage polymers depleted within the bacterial cell provide required energy for the cell metabolism. The accumulation of biopolymers further engineered for the large-scale production of bioplastics. PHAs are the polymers of hydroxy fatty acids that are naturally produced by specific groups of bacteria (Moralejo-Gárate et al., 2014). PHAs are group of polymers in which the polyhydroxybutyrate (PHB) and polyhydroxyvalerate (PHV) were found as dominant and having suitable physicochemical properties to use as bioplastics.

In general the PHA is produced under anoxic or microaerophilic conditions (low dissolved oxygen in bulk liquid) and in association with a few other factors: high substrate availability, nutrient limitation, and high biomass concentration. Three different pathways were reported for the PHA synthesis. In the primary pathway the acetyl-CoA acts as the precursor on which the PHA synthase plays a crucial role for the PHA production. Here, acetyl-CoA is converted into 3-hydroxybutyryl-CoA and subsequently polymerized to PHB by the PHA synthase. In another pathway, PHA is produced as an intermediate of the β-oxidation process in which the alkanes, alkanols, and alkanoates act as the substrates. The type of PHA polymer depends on the type of substrate. Fatty acid *de novo* biosynthesis pathway is the third pathway that produces PHA, in which various types of hydroxyalkanoates are produced from simple carbon sources like acetate, fructose, glycerol, and lactate (Urtuvia et al., 2014).

PHA production was well studied in wastewater treatment plants that are processing a wide variety of wastewaters (Table 4). Established industrial processes consume expensive sugars like glucose, sucrose, or other agri-products such as corn for PHA production (Venkateswar Reddy and Venkata, 2012). Utilizing inexpensive substrates such as waste, wastewater, and microbial consortia as catalyst for PHA production is making the process economical (Lemos et al., 2006). In recent years, this is one of the reasons for decreasing the polymer production cost. In general, during the bacterial energy generation processes through fermentation, the product is exocellular, and the biocatalyst can be reused or regenerated. In this case, the PHA production happens at the expense of biomass (biocatalyst). In terms

Table 4　Bioplastics Production from Wastewater and Waste (PHB or Any Other)

Type of Wastewater or Waste	PHA Production Efficiency	Biocatalysts	References
Candy bar factory	0.70 gPHA/gVSS	*Plasticicumulans acidivorans*	Tamis et al. (2014)
Municipal wastewater	25% of dry biomass	Mixed consortia	Morgan-Sagastume et al. (2010)
Dairy industry wastewater	0.5 gPHA/gVSS	Mixed consortia	Chakravarty et al. (2010)
Food industry wastewater	0.60 gPHA/gVSS	Mixed consortia	Anterrieu et al. (2014)
Excess sludge fermentation liquid	59.5% of DCW	Mixed consortia	Jia et al. (2014)
Wood mill effluents	37% of storage yield	Mixed consortia	Mato et al. (2010)
Spent wash effluents	40% of DCW	*Bacillus tequilensis*	Amulya et al. (2014)
Acidogenic fermented food waste	59% of DCW	*Bacillus tequilensis*	Venkateswar Reddy et al. (2014)
Unfermented food waste	35.6% of DCW	Aerobic mixed culture	Reddy and Mohan (2012)
Feedstock from pulp industry	67.6% of DCW	Mixed microbial culture	Queirós et al. (2014)

DCW, dry cell weight.

of energy, the regeneration of new biomass is consuming more energy. Composition of the PHA polymer is determined by the different metabolisms that prevail in the bacteria/microbial consortia. PHB, PHV, and copolymer of PHB-PHV are dominant polymers in PHA group. PHV is the main product from propionate, whereas PHB is the only product from acetate. Yield of PHB-PHV mixtures and its ratio is directly related to the acetate and propionate uptake rate (Jiang et al., 2011). Pilot-scale operations have also been conducted for the synthesis of PHA from different types of wastewaters. A successful demonstration at a Mars candy factory in the Netherlands using *Plasticicumulans acidivorans* as biocatalyst reported higher PHA production of 0.70 gPHA/gVSS (volatile suspended solids) within 4h (Tamis et al., 2014; Figure 7).

PHA was produced from the different types of wastewater that are originated from the food and agro-product-based industries, which have different types of carbon sources ranging from acetate to glucose and sucrose. Compared to hexoses and pentoses, simple substrates such as VFAs are more suitable for the PHA production. In this direction the effluents of biohydrogen production process are used as substrate. It resulted in at least 10% higher accumulation of PHA in bacterial cell (Reddy and Mohan, 2012).

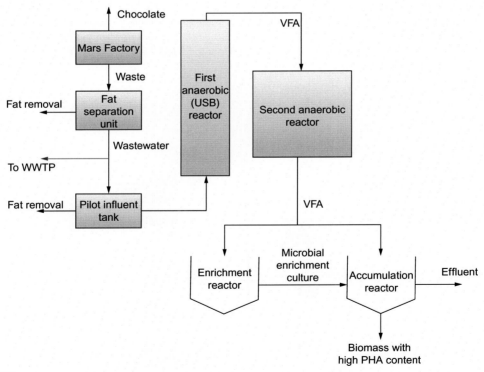

FIGURE 7

Overview of the pilot-scale system for PHA production from wastewater at the Mars candy bar factory, Veghel, The Netherlands (Tamis et al., 2014). [WWTP, wastewater treatment plant; USB, upflow sludge blanket].

4 ENERGY GENERATION THROUGH BIOPROCESS INTEGRATIONS

The yield exhibited by any process is always different from the theoretical yield due to several operational, biological, and environmental factors. Complete oxidation of the waste or residual organics by secondary processes yields additional value-added products (Guwy et al., 2011; Laurinavichene et al., 2010; Mohanakrishna and Mohan, 2013). Sequential integration of one bioenergy-generating process to another bioenergy-generating process has led to the biorefinery approach, which improves the degradation efficiency, process economics. The integration of one process to another is selected based on the substrate nature and the possible high energy yield in individual process. The best examples of such an integration approaches were shown in Figure 8, where dark fermentation is integrated with five other processes such as solventogenesis (for alcohols), MFC (for bioelectricity), MEC (for hydrogen), photofermentation (for hydrogen), and anoxic process (for bioplastics). The maximum reported substrate degradation using dark fermentation alone is 70%. An effluent of effective dark fermentation is rich in VFAs and the integrated five processes were proved to work efficiently with VFA-rich substrates. The residual pollutants of the primary effluent were further treated in secondary process, and the secondary effluents were found to have lesser organic matter. Based on the residual organic matter and nature of composition, tertiary treatment is also possible for the additional energy generation (Mohanakrishna and Mohan, 2013). Dark fermentation, photofermentation, and methanogenesis were integrated in 11 different combinations and evaluated for substrate degradation efficiency and energy generation from each combination. Results showed 87.5% of substrate degradation and 85.3% of carbon conversion efficiency toward H_2 and CH_4 (Mohanakrishna and Mohan, 2013).

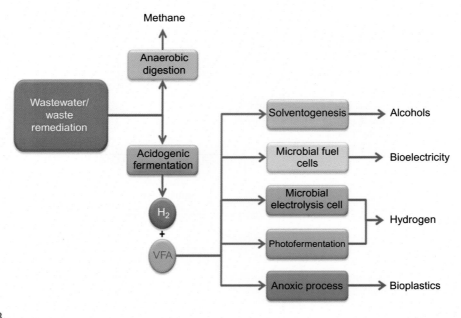

FIGURE 8

Dark fermentation-centered bioprocess integration diagram explaining various other energy-generating processes that can be integrated with biohydrogen production process.

5 FUTURE PERSPECTIVES

Removing gaseous pollutants through biological processes is less studied. CO_2 fixation for energy generation is going to be a pioneer technology in the near future. Microalgae processes for biomass and biodiesel production is one of the major available routes. In addition, the very recent concept of microbial electrosynthesis (MES) has foreseen several advantages such as CO_2 fixation with simultaneous bio-commodities synthesis with low energy demand. This process can be integrated with the other renewable energy-generating technologies. Biorefinery concepts that integrate different bioprocesses for different energy are at infant stage. A comprehensive program for the technoeconomic evaluation for such processes needs to be developed for commercial scale. Metabolic engineering for biorefineries needs to be simulated at the laboratory scale to achieve broad and rapid understanding on energy efficiencies. On the other hand, molecular, genetic tools accelerate designed microorganisms for remediation and simultaneous energy generation. The knowledge of the microbiological pathways and enzymes that involves energy generation from waste and their metabolic pathways can be advantageous in the quest for alternative energy (Singh et al., 2008).

REFERENCES

Abbassi, B., Dullstein, S., Räbiger, N., 2000. Minimization of excess sludge production by increase of oxygen concentration in activated sludge flocs; experimental and theoretical approach. Water Res. 34, 139–146.

Amaro, H.M., Guedes, A.C., Malcata, F.X., 2011. Advances and perspectives in using microalgae to produce biodiesel. Appl. Energy 88, 3402–3410.

Amulya, K., Reddy, M.V., Mohan, S.V., 2014. Acidogenic spent wash valorization through polyhydroxyalkanoate (PHA) synthesis coupled with fermentative biohydrogen production. Bioresour. Technol. 158, 336–342.

Angenent, L.T., Karim, K., Al-Dahhan, M.H., Wrenn, B.A., Domíguez-Espinosa, R., 2004. Production of bioenergy and biochemicals from industrial and agricultural wastewater. Trends Biotechnol. 22, 477–485. http://dx.doi.org/10.1016/j.tibtech.2004.07.001.

Anterrieu, S., Quadri, L., Geurkink, B., Dinkla, I., Bengtsson, S., Arcos-Hernandez, M., Alexandersson, T., Morgan-Sagastume, F., Karlsson, A., Hjort, M., Karabegovic, L., Magnusson, P., Johansson, P., Christensson, M., Werker, A., 2014. Integration of biopolymer production with process water treatment at a sugar factory. N. Biotechnol. 31, 308–323.

Appels, L., Lauwers, J., Degrève, J., Helsen, L., Lievens, B., Willems, K., et al., 2011. Anaerobic digestion in global bio-energy production: potential and research challenges. Renew. Sustain. Energy Rev. 15, 4295–4301.

Batyrova, K.A., Tsygankov, A.A., Kosourov, S.N., 2012. Sustained hydrogen photoproduction by phosphorus-deprived *Chlamydomonas reinhardtii* cultures. Int. J. Hydrogen Energ. 37, 8834–8839.

Beer, L.L., Boyd, E.S., Peters, J.W., Posewitz, M.C., 2009. Engineering algae for biohydrogen and biofuel production. Curr. Opin. Biotechnol. 20, 264–271.

Bellou, S., Baeshen, M.N., Elazzazy, A.M., Aggeli, D., Sayegh, F., Aggelis, G., 2014. Microalgal lipids biochemistry and biotechnological perspectives. Biotechnol. Adv. 32, 1476–1493. http://dx.doi.org/10.1016/j.biotechadv.2014.10.003.

Bengtsson, S., Werker, A., Welander, T., 2008. Production of polyhydroxyalkanoates by glycogen accumulating organisms treating a paper mill wastewater. Water Sci. Technol. 58, 323–330. http://dx.doi.org/10.2166/wst.2008.381.

Brastad, K.S., He, Z., 2013. Water softening using microbial desalination cell technology. Desalination 309, 32–37.

Brennan, L., Owende, P., 2010. Biofuels from microalgae – a review of technologies for production, processing, and extractions of biofuels and co-products. Renew. Sustain. Energy Rev. 14, 557–577. http://dx.doi.org/10.1016/j.rser.2009.10.009.

Butler, C.S., Clauwaert, P., Green, S.J., Verstraete, W., Nerenberg, R., 2010. Bioelectrochemical perchlorate reduction in a microbial fuel cell. Environ. Sci. Technol. 44, 4685–4691.

Buyukkamaci, N., Filibeli, A., 2004. Volatile fatty acid formation in an anaerobic hybrid reactor. Process Biochem. 39, 1491–1494.

Cao, X., Huang, X., Liang, P., Xiao, K., Zhou, Y., Zhang, X., et al., 2009. A new method for water desalination using microbial desalination cells. Environ. Sci. Technol. 43, 7148–7152.

Chakravarty, P., Mhaisalkar, V., Chakrabarti, T., 2010. Study on poly-hydroxyalkanoate (PHA) production in pilot scale continuous mode wastewater treatment system. Bioresour. Technol. 101, 2896–2899.

Chandra, R., Venkata Subhash, G., Venkata, M.S., 2012. Mixotrophic operation of photo-bioelectrocatalytic fuel cell under anoxygenic microenvironment enhances the light dependent bioelectrogenic activity. Bioresour. Technol. 109, 46–56.

Cheng, S., Logan, B.E., 2007. Sustainable and efficient biohydrogen production via electrohydrogenesis. Proc. Natl. Acad. Sci. U. S. A. 104, 18871–18873.

Cheng, J., Zhu, X., Ni, J., Borthwick, A., 2010. Palm oil mill effluent treatment using a two-stage microbial fuel cells system integrated with immobilized biological aerated filters. Bioresour. Technol. 101, 2729–2734. http://dx.doi.org/10.1016/j.biortech.2009.12.017.

Clauwaert, P., Rabaey, K., Aelterman, P., de Schamphelaire, L., Pham, T.H., Boeckx, P., et al., 2007. Biological denitrification in microbial fuel cells. Environ. Sci. Technol. 41, 3354–3360.

Courchesne, N.M.D., Parisien, A., Wang, B., Lan, C.Q., 2009. Enhancement of lipid production using biochemical, genetic and transcription factor engineering approaches. J. Biotechnol. 141, 31–41.

Cusick, R.D., Bryan, B., Parker, D.S., Merrill, M.D., Mehanna, M., Kiely, P.D., et al., 2011. Performance of a pilot-scale continuous flow microbial electrolysis cell fed winery wastewater. Appl. Microbiol. Biotechnol. 89, 2053–2063.

Elakkiya, E., Matheswaran, M., 2013. Comparison of anodic metabolisms in bioelectricity production during treatment of dairy wastewater in Microbial Fuel Cell. Bioresour. Technol. 136, 407–412.

El-Mashad, H.M., Zeeman, G., Van Loon, W.K.P., Bot, G.P.A., Lettinga, G., 2004. Effect of temperature and temperature fluctuation on thermophilic anaerobic digestion of cattle manure. Bioresour. Technol. 95, 191–201.

Elmekawy, A., Diels, L., De Wever, H., Pant, D., 2013a. Valorization of cereal based biorefinery byproducts: reality and expectations. Environ. Sci. Technol. 47, 9014–9027.

Elmekawy, A., Hegab, H.M., Dominguez-Benetton, X., Pant, D., 2013b. Internal resistance of microfluidic microbial fuel cell: challenges and potential opportunities. Bioresour. Technol. 142, 672–682. http://dx.doi.org/10.1016/j.biortech.2013.05.061.

Elmekawy, A., Srikanth, S., Bajracharya, S., Hegab, H.M., Nigam, P.S., Singh, A., et al., 2014a. Food and agricultural wastes as substrates for bioelectrochemical system (BES): the synchronized recovery of sustainable energy and waste treatment. Food Res. Int. 73, 213–225.

Elmekawy, A., Srikanth, S., Vanbroekhoven, K., De Wever, H., Pant, D., 2014b. Bioelectro-catalytic valorization of dark fermentation effluents by acetate oxidizing bacteria in bioelectrochemical system (BES). J. Power Sources 262, 183–191. http://dx.doi.org/10.1016/j.jpowsour.2014.03.111.

Elmekawy, A., Hegab, H.M., Vanbroekhoven, K., Pant, D., 2014c. Techno-productive potential of photosynthetic microbial fuel cells through different configurations. Renew. Sustain. Energy Rev. 39, 617–627.

Elmekawy, A., Hegab, H.M., Pant, D., 2014d. The near-future integration of microbial desalination cells with reverse osmosis technology. Energy Environ. Sci. 7, 3921–3933.

Fang, H.H.P., Liu, H., 2002. Effect of pH on hydrogen production from glucose by a mixed culture. Bioresour. Technol. 82, 87–93.

Gentili, F.G., 2014. Microalgal biomass and lipid production in mixed municipal, dairy, pulp and paper wastewater together with added flue gases. Bioresour. Technol. 169, 27–32.

Girard, J.-M., Roy, M.-L., Hafsa, M. Ben, Gagnon, J., Faucheux, N., Heitz, M., Tremblay, R., Deschênes, J.-S., 2014. Mixotrophic cultivation of green microalgae *Scenedesmus obliquus* on cheese whey permeate for biodiesel production. Algal Res. 5, 241–248.

Goud, R. K., Raghavulu, S. V., Mohanakrishna, G., Naresh, K., Mohan, S. V., 2012b. Predominance of Bacilli and Clostridia in microbial community of biohydrogen producing biofilm sustained under diverse acidogenic operating conditions. Int. J. Hydrogen Energ. 37(5), 4068–4076.

Guo, X.M., Trably, E., Latrille, E., Carrère, H., Steyer, J.-P., 2010. Hydrogen production from agricultural waste by dark fermentation: a review. Int. J. Hydrogen Energ. 35, 10660–10673. http://dx.doi.org/10.1016/j.ijhydene.2010.03.008.

Guo, F., Fu, G., Zhang, Z., Zhang, C., 2013. Mustard tuber wastewater treatment and simultaneous electricity generation using microbial fuel cells. Bioresour. Technol. 136, 425–430.

Guwy, A.J., Dinsdale, R.M., Kim, J.R., Massanet-Nicolau, J., Premier, G., 2011. Fermentative biohydrogen production systems integration. Bioresour. Technol. 102, 8534–8542.

Hamawand, I., 2015. Anaerobic digestion process and bio-energy in meat industry: a review and a potential. Renew. Sustain. Energy Rev. 44, 37–51. http://dx.doi.org/10.1016/j.rser.2014.12.009.

Han, J.-L., Liu, Y., Chang, C.-T., Chen, B.-Y., Chen, W.-M., Xu, H.-Z., 2011. Exploring characteristics of bioelectricity generation and dye decolorization of mixed and pure bacterial cultures from wine-bearing wastewater treatment. Biodegradation 22, 321–333. http://dx.doi.org/10.1007/s10532-010-9401-9.

Horch, M., Lauterbach, L., Lenz, O., Hildebrandt, P., Zebger, I., 2012. NAD(H)-coupled hydrogen cycling – structure-function relationships of bidirectional [NiFe] hydrogenases. FEBS Lett. 586, 545–556.

Inoue, K., Ito, T., Kawano, Y., Iguchi, A., Miyahara, M., Suzuki, Y., Watanabe, K., 2013. Electricity generation from cattle manure slurry by cassette-electrode microbial fuel cells. J. Biosci. Bioeng. 116, 610–615. http://dx.doi.org/10.1016/j.jbiosc.2013.05.011.

Jacobson, K.S., Drew, D.M., He, Z., 2011. Efficient salt removal in a continuously operated upflow microbial desalination cell with an air cathode. Bioresour. Technol. 102, 376–380.

Jeremiasse, A.W., 2011. Cathode Innovations for Enhanced H_2 Production Through Microbial Electrolysis. Wagenigen University, Wageningen.

Jeremiasse, A.W., Bergsma, J., Kleijn, J.M., Saakes, M., Buisman, C.J.N., Cohen Stuart, M., et al., 2011. Performance of metal alloys as hydrogen evolution reaction catalysts in a microbial electrolysis cell. Int. J. Hydrogen Energ. 36, 10482–10489. http://dx.doi.org/10.1016/j.ijhydene.2011.06.013.

Jia, Q., Wang, H., Wang, X., 2013. Dynamic synthesis of polyhydroxyalkanoates by bacterial consortium from simulated excess sludge fermentation liquid. Bioresour. Technol. 140, 328–336. http://dx.doi.org/10.1016/j.biortech.2013.04.105.

Jia, J., Tang, Y., Liu, B., Wu, D., Ren, N., Xing, D., 2013b. Electricity generation from food wastes and microbial community structure in microbial fuel cells. Bioresour. Technol. 144, 94–99. http://dx.doi.org/10.1016/j.biortech.2013.06.072.

Jiang, Y., Hebly, M., Kleerebezem, R., Muyzer, G., van Loosdrecht, M.C.M., 2011. Metabolic modeling of mixed substrate uptake for polyhydroxyalkanoate (PHA) production. Water Res. 45, 1309–1321.

Jia, Q., Xiong, H., Wang, H., Shi, H., Sheng, X., Sun, R., Chen, G., 2014. Production of polyhydroxyalkanoates (PHA) by bacterial consortium from excess sludge fermentation liquid at laboratory and pilot scales. Bioresour. Technol. 171, 159–167. http://dx.doi.org/10.1016/j.biortech.2014.08.059.

Judd, S., 2010. The MBR Book: Principles and Applications of Membrane Bioreactors for Water and Wastewater Treatment, vol. 26. Elsevier, Amsterdam.

Katuri, K.P., Enright, A.-M., O'Flaherty, V., Leech, D., 2012. Microbial analysis of anodic biofilm in a microbial fuel cell using slaughterhouse wastewater. Bioelectrochemistry 87, 164–171.

Kelly, P.T., He, Z., 2014. Understanding the application niche of microbial fuel cells in a cheese wastewater treatment process. Bioresour. Technol. 157, 154–160. http://dx.doi.org/10.1016/j.biortech.2014.01.085.

Kim, J.R., Dec, J., Bruns, M.A., Logan, B.E., 2008. Removal of odors from Swine wastewater by using microbial fuel cells. Appl. Environ. Microbiol. 74, 2540–2543. http://dx.doi.org/10.1128/AEM.02268-07.

Kleerebezem, R., Van Loosdrecht, M.C., 2007. Mixed culture biotechnology for bioenergy production. Curr. Opin. Biotechnol. 18, 207–212. http://dx.doi.org/10.1016/j.copbio.2007.05.001.

Kothari, R., Pandey, A.K., Kumar, S., Tyagi, V.V., Tyagi, S.K., 2014. Different aspects of dry anaerobic digestion for bio-energy: an overview. Renew. Sustain. Energy Rev. 39, 174–195. http://dx.doi.org/10.1016/j.rser.2014.07.011.

Kruse, O., Rupprecht, J., Bader, K.-P., Thomas-Hall, S., Schenk, P.M., Finazzi, G., et al., 2005. Improved photobiological H_2 production in engineered green algal cells. J. Biol. Chem. 280, 34170–34177.

Kumar, A.K., Reddy, M.V., Chandrasekhar, K., Srikanth, S., Mohan, S.V., 2012. Endocrine disruptive estrogens role in electron transfer: bio-electrochemical remediation with microbial mediated electrogenesis. Bioresour. Technol. 104, 547–556. http://dx.doi.org/10.1016/j.biortech.2011.10.037.

Lalaurette, E., Thammannagowda, S., Mohagheghi, A., Maness, P.-C., Logan, B.E., 2009. Hydrogen production from cellulose in a two-stage process combining fermentation and electrohydrogenesis. Int. J. Hydrogen Energ. 34, 6201–6210.

Laurinavichene, T.V., Belokopytov, B.F., Laurinavichius, K.S., Tekucheva, D.N., Seibert, M., Tsygankov, A.A., 2010. Towards the integration of dark- and photo-fermentative waste treatment. 3. Potato as substrate for sequential dark fermentation and light-driven H2 production. Int. J. Hydrogen Energ. 35, 8536–8543. http://dx.doi.org/10.1016/j.ijhydene.2010.02.063.

Lei, X., Sugiura, N., Feng, C., Maekawa, T., 2007. Pretreatment of anaerobic digestion effluent with ammonia stripping and biogas purification. J. Hazard. Mater. 145, 391–397.

Lemos, P.C., Serafim, L.S., Reis, M.A.M., 2006. Synthesis of polyhydroxyalkanoates from different short-chain fatty acids by mixed cultures submitted to aerobic dynamic feeding. J. Biotechnol. 122, 226–238.

Lettinga, G., van Velsen, A.F.M., Hobma, S.W., De Zeeuw, W., Klapwijk, A., 1980. Use of the upflow sludge blanket (USB) reactor concept for biological wastewater treatment, especially for anaerobic treatment. Biotechnol. Bioeng. 22, 699–734.

Li, W.W., Yu, H.Q., 2011. From wastewater to bioenergy and biochemicals via two-stage bioconversion processes: a future paradigm. Biotechnol. Adv. 29, 972–982. http://dx.doi.org/10.1016/j.biotechadv.2011.08.012.

Li, Y., Park, S.Y., Zhu, J., 2011. Solid-state anaerobic digestion for methane production from organic waste. Renew. Sustain. Energy Rev. 15, 821–826.

Li, X., Zhu, N., Wang, Y., Li, P., Wu, P., Wu, J., 2013b. Animal carcass wastewater treatment and bioelectricity generation in up-flow tubular microbial fuel cells: effects of HRT and non-precious metallic catalyst. Bioresour. Technol. 128, 454–460. http://dx.doi.org/10.1016/j.biortech.2012.10.053.

Li, X.M., Cheng, K.Y., Selvam, A., Wong, J.W.C., 2013c. Bioelectricity production from acidic food waste leachate using microbial fuel cells: effect of microbial inocula. Process Biochem. 48, 283–288.

Liu, H., Grot, S., Logan, B.E., 2005. Electrochemically assisted microbial production of hydrogen from acetate. Environ. Sci. Technol. 39, 4317–4320.

Logan, B.E., Call, D., Cheng, S., Hamelers, H.V.M., Sleutels, T.H.J.A., Jeremiasse, A.W., et al., 2008. Microbial electrolysis cells for high yield hydrogen gas production from organic matter. Environ. Sci. Technol. 42, 8630–8640.

Madsen, E.L., 2011. Environmental microbiology: from genomes to biogeochemistry. John Wiley & Sons.

Malla, F.A., Khan, S.A., Sharma, G.K., Gupta, N., Abraham, G., 2015. Phycoremediation potential of *Chlorella minutissima* on primary and tertiary treated wastewater for nutrient removal and biodiesel production. Ecol. Eng. 75, 343–349.

Marañón, E., Castrillón, L., Quiroga, G., Fernández-Nava, Y., Gómez, L., García, M.M., 2012. Co-digestion of cattle manure with food waste and sludge to increase biogas production. Waste Manag. 32, 1821–1825.

Mansoorian, H.J., Mahvi, A.H., Jafari, A.J., Amin, M.M., Rajabizadeh, A., Khanjani, N., 2013. Bioelectricity generation using two chamber microbial fuel cell treating wastewater from food processing. Enzyme Microb. Technol. 52, 352–357. http://dx.doi.org/10.1016/j.enzmictec.2013.03.004.

Marchaim, U., Krause, C., 1993. Propionic to acetic acid ratios in overloaded anaerobic digestion. Bioresour. Technol. 43, 195–203.

Mata-Alvarez, J., Dosta, J., Romero-Güiza, M.S., Fonoll, X., Peces, M., Astals, S., 2014. A critical review on anaerobic co-digestion achievements between 2010 and 2013. Renew. Sustain. Energy Rev. 36, 412–427. http://dx.doi.org/10.1016/j.rser.2014.04.039.

Mato, T., Ben, M., Kennes, C., Veiga, M.C., 2010. Valuable product production from wood mill effluents. Water Sci. Technol. 62, 2294–2300.

Mehanna, M., Saito, T., Yan, J., Hickner, M., Cao, X., Huang, X., et al., 2010. Using microbial desalination cells to reduce water salinity prior to reverse osmosis. Energy Environ. Sci. 3, 1114.

Mitra, D., van Leeuwen, J. (Hans), Lamsal, B., 2012. Heterotrophic/mixotrophic cultivation of oleaginous *Chlorella vulgaris* on industrial co-products. Algal Res. 1, 40–48.

Mohan, S.V., Srikanth, S., 2011. Enhanced wastewater treatment efficiency through microbially catalyzed oxidation and reduction: synergistic effect of biocathode microenvironment. Bioresour. Technol. 102, 10210–10220. http://dx.doi.org/10.1016/j.biortech.2011.08.034.

Mohan, S.V., Mohanakrishna, G., Raghavulu, S.V., Sarma, P.N., 2007. Enhancing biohydrogen production from chemical wastewater treatment in anaerobic sequencing batch biofilm reactor (AnSBBR) by bioaugmenting with selectively enriched kanamycin resistant anaerobic mixed consortia. Int. J. Hydrogen Energ. 32, 3284–3292. http://dx.doi.org/10.1016/j.ijhydene.2007.04.043.

Mohan, S.V., Mohanakrishna, G., Sarma, P.N., Venkatamohan, S., 2008a. Integration of acidogenic and methanogenic processes for simultaneous production of biohydrogen and methane from wastewater treatment. Int. J. Hydrogen Energ. 33, 2156–2166. http://dx.doi.org/10.1016/j.ijhydene.2008.01.055.

Mohan, S.V., Mohanakrishna, G., Sarma, P.N., 2008b. Effect of anodic metabolic function on bioelectricity generation and substrate degradation in single chambered microbial fuel cell. Environ. Sci. Technol. 42, 8088–8094.

Mohan, S.V., Babu, V.L., Sarma, P.N., 2008c. Effect of various pretreatment methods on anaerobic mixed microflora to enhance biohydrogen production utilizing dairy wastewater as substrate. Bioresour. Technol. 99, 59–67.

Mohan, S.V., Mohanakrishna, G., Reddy, B.P., Saravanan, R., Sarma, P.N., 2008c. Bioelectricity generation from chemical wastewater treatment in mediatorless (anode) microbial fuel cell (MFC) using selectively enriched hydrogen producing mixed culture under acidophilic microenvironment. Biochemical Engineering Journal 39 (1), 121–130.

Mohan, S.V., Mohanakrishna, G., Srikanth, S., Sarma, P.N., 2008d. Harnessing of bioelectricity in microbial fuel cell (MFC) employing aerated cathode through anaerobic treatment of chemical wastewater using selectively enriched hydrogen producing mixed consortia. Fuel 87 (12), 2667–2676.

Mohan, S.V., Srikanth, S., Raghuvulu, S.V., Mohanakrishna, G., Kumar, A.K., Sarma, P.N., 2009. Evaluation of the potential of various aquatic eco-systems in harnessing bioelectricity through benthic fuel cell: effect of electrode assembly and water characteristics. Bioresour. Technol. 100, 2240–2246. http://dx.doi.org/10.1016/j.biortech.2008.10.020.

Mohan, S.V., Devi, M.P., Mohanakrishna, G., Amarnath, N., Babu, M.L., Sarma, P.N.N., 2011a. Potential of mixed microalgae to harness biodiesel from ecological water-bodies with simultaneous treatment. Bioresour. Technol. 102, 1109–1117.

Mohan, S.V., Mohanakrishna, G., Chiranjeevi, P., 2011c. Sustainable power generation from floating macrophytes based ecological microenvironment through embedded fuel cells along with simultaneous wastewater treatment. Bioresour. Technol. 102 (14), 7036–7042.

Mohan, S. V., Chiranjeevi, P., Mohanakrishna, G. 2012a. A rapid and simple protocol for evaluating biohydrogen production potential (BHP) of wastewater with simultaneous process optimization. Int. J. Hydrogen Energ. 37(4), 3130–3141.

Mohan, S.V., Reddy, M.V., Chandra, R., Subhash, G.V., Devi, M.P., Srikanth, S., 2013a. Biomass for Sustainable Applications. Royal Society of Chemistry, Cambridge.

Mohan, S.V., Srikanth, S., Velvizhi, G., Babu, M.L., 2013b. Biofuel Technologies. Springer, Berlin/Heidelberg. http://dx.doi.org/10.1007/978-3-642-34519-7.

Mohanakrishna, G., Mohan, S.V., 2013. Multiple process integrations for broad perspective analysis of fermentative H_2 production from wastewater treatment: technical and environmental considerations. Appl. Energy 107, 244–254.

Mohanakrishna, G., Mohan, S.V., Sarma, P.N., 2010a. Bio-electrochemical treatment of distillery wastewater in microbial fuel cell facilitating decolorization and desalination along with power generation. J. Hazard. Mater. 177, 487–494. http://dx.doi.org/10.1016/j.jhazmat.2009.12.059

Mohanakrishna, G., Mohan, S.V., Sarma, P.N., 2010b. Utilizing acid-rich effluents of fermentative hydrogen production process as substrate for harnessing bioelectricity: an integrative approach. Int. J. Hydrogen Energ. 35, 3440–3449.

Mohanakrishna, G., Subhash, G.V., Mohan, S.V., Venkata Subhash, G., Venkata, M.S., 2011. Adaptation of biohydrogen producing reactor to higher substrate load: redox controlled process integration strategy to overcome limitations. Int. J. Hydrogen Energ. 36, 8943–8952. http://dx.doi.org/10.1016/j.ijhydene.2011.04.138.

Mohanakrishna, G., Mohan, S.K., Mohan, S.V., 2012. Carbon based nanotubes and nanopowder as impregnated electrode structures for enhanced power generation: evaluation with real field wastewater. Applied Energy 95, 31–37.

Moralejo-Gárate, H., Kleerebezem, R., Mosquera-Corral, A., Campos, J.L., Palmeiro-Sánchez, T., van Loosdrecht, M.C.M., 2014. Substrate versatility of polyhydroxyalkanoate producing glycerol grown bacterial enrichment culture. Water Res. 6, 2–10. http://dx.doi.org/10.1016/j.watres.2014.07.044.

Morgan-Sagastume, F., Karlsson, A., Johansson, P., Pratt, S., Boon, N., Lant, P., Werker, A., 2010. Production of polyhydroxyalkanoates in open, mixed cultures from a waste sludge stream containing high levels of soluble organics, nitrogen and phosphorus. Water Res. 44, 5196–5211. http://dx.doi.org/10.1016/j.watres.2010.06.043.

Nah, I., 2000. Mechanical pretreatment of waste activated sludge for anaerobic digestion process. Water Res. 34, 2362–2368.

Niessen, J., Schroder, U., Scholz, F., 2004. Exploiting complex carbohydrates for microbial electricity generation – a bacterial fuel cell operating on starch. Electrochem. Commun. 6, 955–958. http://dx.doi.org/10.1016/j.elecom.2004.07.010.

Nishio, N., Nakashimada, Y., 2007. Recent development of anaerobic digestion processes for energy recovery from wastes. J. Biosci. Bioeng. 103, 105–112.

Oh, S.E., Logan, B.E., 2005. Hydrogen and electricity production from a food processing wastewater using fermentation and microbial fuel cell technologies. Water Res. 39, 4673–4682.

Órpez, R., Martínez, M.E., Hodaifa, G., El Yousfi, F., Jbari, N., Sánchez, S., 2009. Growth of the microalga *Botryococcus braunii* in secondarily treated sewage. Desalination 246, 625–630.

Pant, D., Adholeya, A., 2009. Concentration of fungal ligninolytic enzymes by ultrafiltration and their use in distillery effluent decolorization. World J. Microbiol. Biotechnol. 25, 1793–1800.

Pant, D., Adholeya, A., 2010. Development of a novel fungal consortium for the treatment of molasses distillery wastewater. Environmentalist 30, 178–182.

Pant, D., Van Bogaert, G., Diels, L., Vanbroekhoven, K., 2010. A review of the substrates used in microbial fuel cells (MFCs) for sustainable energy production. Bioresour. Technol. 101, 1533–1543. http://dx.doi.org/10.1016/j.biortech.2009.10.017.

Pant, D., Singh, A., Van Bogaert, G., Irving Olsen, S., Singh Nigam, P., Diels, L., et al., 2012. Bioelectrochemical systems (BES) for sustainable energy production and product recovery from organic wastes and industrial wastewaters. RSC Adv. 2, 1248. http://dx.doi.org/10.1039/c1ra00839k.

Pasupuleti, S.B., Srikanth, S., Dominguez-Benetton, X., Mohan, S.V., Pant, D., 2015. Dual gas diffusion cathode design for microbial fuel cell (MFC): optimizing the suitable mode of operation in terms of bioelectrochemical and bioelectro-kinetic evaluation. J. Chem. Technol. Biotechnol. doi: 10.1002/jctb.4613.

Patil, S.A., Surakasi, V.P., Koul, S., Ijmulwar, S., Vivek, A., Shouche, Y.S., Kapadnis, B.P., 2009. Electricity generation using chocolate industry wastewater and its treatment in activated sludge based microbial fuel cell and analysis of developed microbial community in the anode chamber. Bioresour. Technol. 100, 5132–5139. http://dx.doi.org/10.1016/j.biortech.2009.05.041.

Pearce, C., 2003. The removal of colour from textile wastewater using whole bacterial cells: a review. Dyes Pigments 58, 179–196.

Prathima Devi, M., Venkata Subhash, G., Venkata Mohan, S., 2012. Heterotrophic cultivation of mixed microalgae for lipid accumulation and wastewater treatment during sequential growth and starvation phases: effect of nutrient supplementation. Renew. Energy 43, 276–283.

Puyuelo, B., Ponsá, S., Gea, T., Sánchez, A., 2011. Determining C/N ratios for typical organic wastes using biodegradable fractions. Chemosphere 85, 653–659.

Queirós, D., Rossetti, S., Serafim, L.S., 2014. PHA production by mixed cultures: a way to valorize wastes from pulp industry. Bioresour. Technol. 157, 197–205.

Rabaey, K., Rozendal, R.A., 2010. Microbial electrosynthesis – revisiting the electrical route for microbial production. Nat. Rev. Microbiol. 8, 706–716. http://dx.doi.org/10.1038/nrmicro2422.

Rabaey, K., Van De Sompel, K., Maignien, L., Boon, N., Aelterman, P., Clauwaert, P., et al., 2006. Microbial fuel cells for sulfide removal. Environ. Sci. Technol. 40, 5218–5224.

Raghavulu, S.V., Mohan, S.V., Reddy, M.V., Mohanakrishna, G., Sarma, P.N., 2009. Behavior of single chambered mediatorless microbial fuel cell (MFC) at acidophilic, neutral and alkaline microenvironments during chemical wastewater treatment. International journal of hydrogen energy 34 (17), 7547–7554.

Reddy, M.V., Mohan, S.V., 2012. Influence of aerobic and anoxic microenvironments on polyhydroxyalkanoates (PHA) production from food waste and acidogenic effluents using aerobic consortia. Bioresour. Technol. 103, 313–321.

Ren, H.-Y., Liu, B.-F., Ma, C., Zhao, L., Ren, N.-Q., 2013. A new lipid-rich microalga *Scenedesmus* sp. strain R-16 isolated using nile red staining: effects of carbon and nitrogen sources and initial pH on the biomass and lipid production. Biotechnol. Biofuels 6, 143.

Ringeisen, B.R., Ray, R., Little, B., 2007. A miniature microbial fuel cell operating with an aerobic anode chamber. J. Power Sources 165, 591–597.

Rodrigo, M.A., Cañizares, P., Lobato, J., 2010. Effect of the electron-acceptors on the performance of a MFC. Bioresour. Technol. 101, 7025–7029.

Rosenbaum, M., He, Z., Angenent, L.T., 2010. Light energy to bioelectricity: photosynthetic microbial fuel cells. Curr. Opin. Biotechnol. 21, 259–264.

Sánchez, E., Borja, R., Weiland, P., Travieso, L., Martín, A., 2001. Effect of substrate concentration and temperature on the anaerobic digestion of piggery waste in a tropical climate. Process Biochem. 37, 483–489.

Sevda, S., Dominguez-Benetton, X., Vanbroekhoven, K., De Wever, H., Sreekrishnan, T.R., Pant, D., 2013. High strength wastewater treatment accompanied by power generation using air cathode microbial fuel cell. Appl Energy 105, 194–206.

Sharma, M., Aryal, N., Sarma, P.M., Vanbroekhoven, K., Lal, B., Benetton, X.D., et al., 2013. Bioelectrocatalyzed reduction of acetic and butyric acids via direct electron transfer using a mixed culture of sulfate-reducers drives electrosynthesis of alcohols and acetone. Chem. Commun. (Camb.) 49, 6495–6497. http://dx.doi.org/10.1039/c3cc42570c.

Shin, D.Y., Cho, H.U., Utomo, J.C., Choi, Y.-N., Xu, X., Park, J.M., 2014. Biodiesel production from *Scenedesmus bijuga* grown in anaerobically digested food wastewater effluent. Bioresour, Technol.

Singh, S., Kang, S.H., Mulchandani, A., Chen, W., 2008. Bioremediation: environmental clean-up through pathway engineering. Curr. Opin. Biotechnol. 19, 437–444. http://dx.doi.org/10.1016/j.copbio.2008.07.012.

Srikanth, S., Venkata, M.S., 2014. Regulating feedback inhibition caused by the accumulated acid intermediates during acidogenic hydrogen production through feed replacement. Int. J. Hydrogen Energ. 39, 10028–10040. http://dx.doi.org/10.1016/j.ijhydene.2014.04.152.

Srikanth, S., Venkata Mohan, S., Prathima Devi, M., Lenin Babu, M., Sarma, P.N., 2009a. Effluents with soluble metabolites generated from acidogenic and methanogenic processes as substrate for additional hydrogen production through photo-biological process. Int. J. Hydrogen Energ. 34, 1771–1779. http://dx.doi.org/10.1016/j.ijhydene.2008.11.060.

Srikanth, S., Mohan, S.V., Prathima Devi, M., Peri, D., Sarma, P.N., 2009b. Acetate and butyrate as substrates for hydrogen production through photo-fermentation: process optimization and combined performance evaluation. Int. J. Hydrogen Energ. 34, 7513–7522.

Srikanth, S., Mohan, S.V., Lalit Babu, V., Sarma, P.N., 2010. Metabolic shift and electron discharge pattern of anaerobic consortia as a function of pretreatment method applied during fermentative hydrogen production. Int. J. Hydrogen Energ. 35, 10693–10700. http://dx.doi.org/10.1016/j.ijhydene.2010.02.055.

Srikanth, S., Pavani, T., Sarma, P.N., Venkata, M.S., 2011. Synergistic interaction of biocatalyst with bio-anode as a function of electrode materials. Int. J. Hydrogen Energ. 36, 2271–2280. http://dx.doi.org/10.1016/j.ijhydene.2010.11.031.

Tale, M., Ghosh, S., Kapadnis, B., Kale, S., 2014. Isolation and characterization of microalgae for biodiesel production from Nisargruna biogas plant effluent. Bioresour. Technol. 169, 328–335.

Tamis, J., Lužkov, K., Jiang, Y., van Loosdrecht, M.C., Kleerebezem, R., 2014. Enrichment of *Plasticicumulans acidivorans* at pilot-scale for PHA production on industrial wastewater. J. Biotechnol. 192, 161–169. http://dx.doi.org/10.1016/j.jbiotec.2014.10.022.

Turon, V., Baroukh, C., Trably, E., Latrille, E., Fouilland, E., Steyer, J.-P., 2014. Use of fermentative metabolites for heterotrophic microalgae growth: yields and kinetics. Bioresour. Technol. 175, 342–349. http://dx.doi.org/10.1016/j.biortech.2014.10.114.

Thygesen, A., Poulsen, F.W., Angelidaki, I., Min, B., Bjerre, A.-B., 2011. Electricity generation by microbial fuel cells fuelled with wheat straw hydrolysate. Biomass and Bioenergy 35, 4732–4739. http://dx.doi.org/10.1016/j.biombioe.2011.09.026.

Tyagi, S.K. (Ed.), 2013. Recent Advances in Bioenergy Research, vol. II. SSS-NIRE, Kapurthala.

Ummalyma, S.B., Sukumaran, R.K., 2014. Cultivation of microalgae in dairy effluent for oil production and removal of organic pollution load. Bioresour. Technol. 165, 295–301.

Urtuvia, V., Villegas, P., González, M., Seeger, M., 2014. Bacterial production of the biodegradable plastics polyhydroxyalkanoates. Int. J. Biol. Macromol. 70, 208–213. http://dx.doi.org/10.1016/j.ijbiomac.2014.06.001.

Venkata Mohan, S., Mohanakrishna, G., Sarma, P.N., 2010a. Composite vegetable waste as renewable resource for bioelectricity generation through non-catalyzed open-air cathode microbial fuel cell. Bioresour. Technol. 101, 970–976. http://dx.doi.org/10.1016/j.biortech.2009.09.005.

Venkata Mohan, S., Mohanakrishna, G., Velvizhi, G., Babu, V.L., Sarma, P.N., 2010b. Bio-catalyzed electrochemical treatment of real field dairy wastewater with simultaneous power generation. Biochem. Eng. J. 51, 32–39. http://dx.doi.org/10.1016/j.bej.2010.04.012.

Venkata Mohan, S., Mohanakrishna, G., Srikanth, S., 2011. Biohydrogen production from industrial effluents. In: Pandey, A., Ricke, S.C., Larroche, C., Dussap, C.-G., Gnansounou, E. (Eds.), Biofuels: Alternative Feed stocks and Conversion Processes. Elsevier Science and Technology, Imprint: Academic Press, pp. 499–524 (Chapter 23). ISBN 13: 978-0-12-385099-7.

Venkata Mohan, S., Velvizhi, G., Annie Modestra, J., Srikanth, S., 2014a. Microbial fuel cell: critical factors regulating bio-catalyzed electrochemical process and recent advancements. Renew. Sustain. Energy Rev. 40, 779–797.

Venkata Mohan, S., Velvizhi, G., Vamshi Krishna, K., Lenin, B.M., 2014b. Microbial catalyzed electrochemical systems: a bio-factory with multi-facet applications. Bioresour. Technol. 165, 355–364.

Venkata Mohan, S., Srikanth, S., Chiranjeevi, P., Arora, S., Chandra, R., 2014c. Algal biocathode for in situ terminal electron acceptor (TEA) production: synergetic association of bacteria-microalgae metabolism

BIOREMEDIATION IN BRAZIL: SCOPE AND CHALLENGES TO BOOST UP THE BIOECONOMY

G. Labuto[1], E.N.V.M. Carrilho[2]

Universidade Federal de São Paulo, São Paulo, SP, Brazil[1]

Universidade Federal de São Carlos, Araras, SP, Brazil[2]

1 THE BIOECONOMY IN BRAZIL

According to the document "The European Bioeconomy in 2030," which was drawn up by The European Plant Science Organisation (2011), there are six major challenges to be faced and overcome by humanity so as to make sure of a healthy and peaceful future: (1) the sustainable management of natural resources, (2) sustainable production, (3) improvement of public health services, (4) the mitigation of climate changes, (5) integration and equilibrium of social evolution among different countries, and (6) sustainable global development.

In the next few decades, the exploitation of bioresources will need a process of innovation that will make the world economy more dynamic, with the utilization of biomass for the sustainable and economical generation of products needing investment in research and development, for the competitive insertion of the Brazilian industry in a new market, bioeconomy (Bomtempo, 2014).

In this way, bioeconomics is closely linked to the issue of sustainability, as it seeks the production and conversion of biomass from renewable sources coming from agroindustrial handling and management in products that supply the chemical and pharmaceutical industries, as also products for the food industry and for energy production (The European Plant Science Organisation, 2011; Enriquez-Cabot, 1998).

Associated with the added impulse of the economy and of sustainable development, bioeconomy shows advantages such as the production of safe energy, improvements to public health, sustainable industrial production, expansion of social development, production of safe foodstuffs, reduction of impact on the environment, and promotion of safe preservation of the environment, as this is based on a production chain that depends on natural resources to exist (The European Plant Science Organisation, 2011).

The estimated world demand for raw materials exceeded the figure of 10.5 tons per capita in 2008, when we consider the term biomass as including all products coming from live matter, whether animal, plant, or derived products (Wiedmann et al., 2015). The study says that Brazil is the country with the

Bioremediation and Bioeconomy. http://dx.doi.org/10.1016/B978-0-12-802830-8.00022-8
Copyright © 2016 Elsevier Inc. All rights reserved.

highest per capita consumption of biomass, within a universe of 186 countries that were assessed between 1998 and 2008, with a level exceeding 10 tons per capita. However, this enormous quantity of biomass also reflects the production of raw materials of animal and plant origin that are produced and then exported by Brazil to many countries throughout the world.

The estimated production of biomass by agribusiness alone in Brazil, in tons/year for 2013, based on the 2009-10 crop, was 334 million for sugarcane bagasse, 822 thousand for dry yeast, 4.3 million from rice chaff, 48.9 million from corn chaff, and 171.8 million from soybean chaff (Embrapa Informação Tecnológica, 2006; Ferreira-Leitão et al., 2010). Even though part of this biomass is consumed in the country for energy production and for biofuels, with Brazil being the most important producer of energy from renewable sources in the whole world (Agência Nacional de Energia Elétrica, 2005; Cardoso, 2012; Brand and Neves, 2014; Júnior et al., 2009), biomass is normally used in its natural state, without any addition of value or study of new applications for it.

The enormous production of biomass associated with the holding of a share of some 15% of the world's biodiversity and abundant natural resources, reflecting in a vast genetic pool, puts Brazil in a key position to become a significant country on the international scenario.

2 THE USE OF BIOMASS FOR ENVIRONMENTAL REMEDIATION AND THE BIOECONOMY

The use of biomass as a biosorbent has kindled interest from the world scientific community due to the wide range of sorption sites available to retain metals and metallic ions, as well as other organic compounds; to the high selectivity for different chemical forms; to the possibility of using the biosorbent on live, deactivated, or dead materials; sorption capacity similar to that of synthetic materials; its use being less prone to interference by alkali and alkaline earth metals when compared with resins of ionic exchange; and ease of use in continuous form or in flow systems, and also its relative abundance (Madrid and Cámara, 1997).

Biomass can be used either live or inactivated, depending on the contaminants or chemicals of interest and also the physical compartment where these are found, considering the need, or not, for metabolic action by live organisms for the retainment, bioaccumulation, or biotransformation of chemicals. Thus, it is possible to list the mechanisms for bioaccumulation, biodegradation, and biotransformation that need the intervention of metabolism in order to occur (main organisms used: algae, fungi, yeasts, bacteria, and plants, with this last process being known as phytoremediation) and also biosorption, the physical and chemical retainment of chemicals of interest through the use of live or inactivated biomass for the promotion of the bioremediation of contaminated systems (Volesky, 2004; Vegliò and Beolchini, 1997; Carrilho et al., 2002, 2003; Carrilho and Gilbert, 2000; Trama et al., 2014; Araújo et al., 2007, 2009; Gonzalez et al., 2008).

The main advantages of the use of biosorbents, whether inactivated or killed, are the following: It is not necessary to maintain the biotic conditions; easy to keep; the sorption of the analytes does not depend on metabolism; the toxicity of the analyte is not a problem for the maintenance of the biosorbent; the more precise control of the quantity of the analyte retained; the possibility of improving the sorption properties through the method of inactivation; ease in ready desorption of the metals; the potential health hazard is reduced even when pathogenic organisms are used (Sağ and Kutsal, 1996; Brady et al., 1994; Machado et al., 2009).

In 2000, the world expenditure on depollution of the environment was estimated at between US $25 and $30 billion, with the United States alone having invested between US $55 and $103 million on phytoremediation alone (Glass, 1998), which shows a promising market for the prospection, investigation, strengthening, and innovation in relation to biomass and its usability as a remediator for the environment.

The main advantage of the use of biomass for phytoremediation is the possibility that the contaminant can be effectively removed from the soil or the body of water by means of the removal of the plant used in the process, normally a bioaccumulator (Koelmel et al., 2015). Here we must mention that, apart from finding plants that can operate as phytoremediators, these may also be genetically modified to serve specific purposes. However, it is very important, first and foremost, to establish an appropriate end destination for the contaminated residue before promoting its use, which could even involve the recovery of metals of commercial interest, thereby turning a liability into an environmental asset.

The term phytoremediation can be subdivided into others, according to the phenomenon involved in the remediation thus made (Pletsch et al., 1999; Prasad, 2011):

- Phytoextraction—Absorption of the soil and storage in the roots or in other tissues, without any change: the disposal of contaminated matter is thus made easier;
- Phytotransformation—Absorption and bioconversion of the contaminant, turning it into less toxic forms, in the roots or other plant tissues (catabolism or anabolism);
- Phytovolatilization—Absorption and conversion of the contaminant into a volatile form, which is released into the atmosphere;
- Phytostimulation—Stimulation of microbial biodegradation through the exudates of the roots;
- Rhizofiltration—Absorption and concentration of the contaminant in plant tissues, and possible disposal of plant materials, appropriate for aqueous mediums; immobilization, lignification, or humification of the contaminant in the soil.
- Phytostabilization—Immobilization, lignification, or humification of the contaminant in the soil.

Among the advantages of phytoremediation we could mention the following: lower cost; applicable on site; the soil can be reused after the treatment; applicable to a wide spectrum of pollutants, including recalcitrants; applicable to large areas that are prohibitive for other types of technology; could be a permanent solution as organic pollutants could be mineralized; transfer is faster than a natural process of attenuation; high public acceptance; fewer air and water emissions; generation of a lower volume of secondary waste (Pletsch et al., 1999; Prasad, 2011).

Its disadvantages are the following: the processes tend to be slower than those of other technologies; highly dependent on seasonality and also on the soil, according to the plant used, the climate, and also the local soil; potential toxicity to the plant; not 100% effective; can only be applied to the surface of the soil or shallow streams and groundwater; mass transfer limitations associated with other biotreatments; slower than mechanical treatments; the contaminants may be mobilized into the groundwater; only effective for moderately hydrophobic contaminants; needs appropriate disposal of the plant residue when contaminated; the toxicity and bioavailability of degradation products are not known (Pletsch et al., 1999; Prasad, 2011).

Biosorption is a passive phenomenon, meaning that biomass is applied, in its inactive form, for the removal of organic and inorganic contaminants (Aksu, 2005; Das et al., 2008), normally from aqueous systems, with applications to treatment of water and effluents. Biosorption is characterized by the interaction of functional groups present in the composition of materials of biological origin and also contaminants or chemicals of interest, through phenomena such as adsorption, chelation, ionic exchange, microprecipitation, or complexation (Volesky, 2004). Different types of biomass have been studied and used for biosorption, such as algae, fungi, yeasts, bacteria, and also residue from the food and the agriculture industries (Volesky and Holan, 1995; Vieira and Volesky, 2000).

Among the possible uses given to postsorption biomass, we have incineration, encapsulation in concrete, or storage in landfills (Wase and Forster, 1997). This means that the application of environmentally friendly methodologies for treatment and disposal of biomass containing the contaminant or chemical of interest could be the subject of research, and lead to more appropriate ways to dispose of residue and/or those that allow the recovery of species of economic interest, in a cleaner and more effective manner.

Considering potentiality, diversity, and also the advantages of using biomass for bioremediation, the technological, industrial, and commercial application of established products could be an option to boost bioeconomy throughout the world.

3 RESEARCH ON THE USE OF BIOMASS FOR ENVIRONMENTAL REMEDIATION IN BRAZIL

The use of live or inactivated biological materials for the removal and/or degradation of organic and inorganic contaminants in aqueous systems and soils has been discussed at length in the scientific disclosure media throughout the world, there having been, over the last 20 years, a total of 24,501 and 32,744 listed in the Web of Science and Scopus sites, respectively, searched by the keywords bioremediation, biosorption, and phytoremediation.

Despite Brazil's significant potential in production of biomass, biota, and microbiota to be studied, the participation of Brazilian researchers in publications during this period of 20 years has been very limited, accounting for around 4%, 875 and 866 articles recorded in the Web of Science and Scopus database, respectively. The number of citations of these articles is around 12,000, and they have been published in 235 scientific journals. Out of all, 80% of the journals hold an impact factor, in a total of 762 articles, which reflects some 90% of Brazilian scientific production in bioremediation, biosorption, and phytoremediation. Out of all these articles, 46% show an impact factor higher than 1.5 (385 articles) and 18.7% hold an impact factor higher than 3 (185 articles). Although the majority of these articles have been published in lower impact factor journals (below 1.5), the number of citations reported (around 14/article) is quite significant, which indicates a fair quality of Brazilian production in this field. In addition, these figures account for only the articles and citations of papers found with a limited number of keywords.

Out of all the articles published by Brazilian researchers (Figure 1), the main points of remediation using live or inactivated biomass are heavy metals, petroleum and petroleum products (hydrocarbons in general), and textile dyes. For the term phytoremediation (205 articles in all), 63.9% are related to metal retainment, and 36.1% refer to removal or degradation of organic compounds (such as herbicides, petroleum, and petroleum by-products that contaminate soil and water). For the term biosorption (total 383 articles), 77.3% of the work is about retainment of metals, while 22.2% of the work focuses on the sorption of chemical products, mostly textile dyes. Among the articles about bioremediation (389 in all), there is an inversion of the focus of species of interest for removal: 79.7% focused on the removal or degradation of organic compounds, 4.4% concern the prospection of study of metabolites of

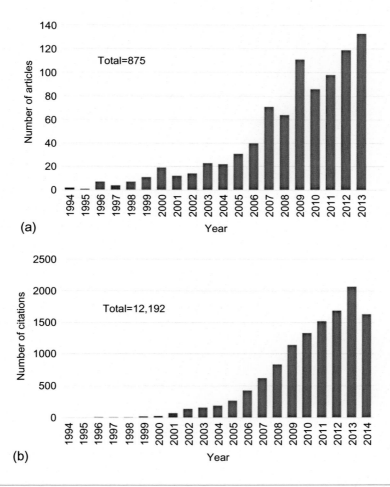

FIGURE 1

(a) Number of articles published by Brazilian scientists and (b) number of articles published by Brazilians over the last 20 years, searched by the keywords: Brazil, biosorption, phytoremediation, and bioremediation.

Source: Web of Science and Scopus websites in 11/30/2014.

microorganisms exposed to contaminants, and 15.2% focus on the removal of metals, which reflects the interest for the phenotypic potential of live organisms and also their genetic changes to make them able to take up behavior patterns of interest to humans. Figures 2–4 show the distribution by type of biomass, the analytes or contaminants of interest that have been investigated, and also the distribution of the types of analytes and/or contaminants in relation to the biomasses used by Brazilians and published in scientific periodicals between 1994 and 2014.

Most published studies about phytoremediation are geographically from regions with a temperate climate, which brings weather limitations, different from what happens in Brazil, which has a tropical climate in most of its territory (Wase and Forster, 1997). If, on the one hand, the scarcity of studies

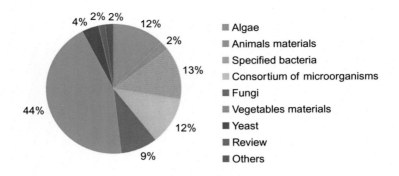

FIGURE 2

Types of biomass used as biosorbents in articles published by Brazilians between 1994 and 2014, searched by the keywords: phytoremediation, bioremediation, and biosorption.

Source: Web of Science and Scopus websites in 11/30/2014.

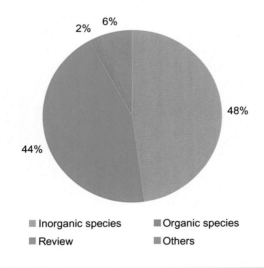

FIGURE 3

Types of contaminants or analytes of interest in biosorption studies published by Brazilians between 1994 and 2014, searched by the keywords: phytoremediation, bioremediation, and biosorption.

Source: Web of Science and Scopus websites in 11/30/2014.

about phytoremediation and prospection of microbial communities in soils of tropical regions together with the lack of instruments to check results and decision making has harmed the use and recommendation of phytoremediation as an alternative to remediation and also soil and bodies of water, on the other hand it becomes an attractive niche to be exploited, both for research and for the development of products with good commercial potential.

The precariousness of regulatory benchmarks, as also the lack of investment in research, have been given as limiting factors hindering the boosting of phytoremediation as an active front of bioeconomy

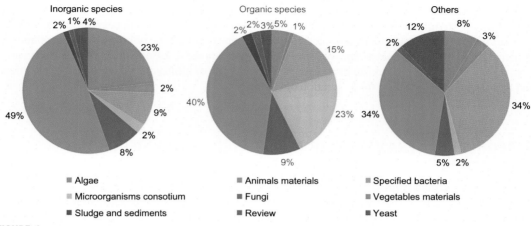

FIGURE 4

Distribution of the types of biomass in articles published by Brazilians between 1994 and 2014 searched by the keywords phytoremediation, bioremediation, and biosorption for metals, organic compounds, and other compounds (studies seeking prospection of biosorbents, production of substrates by biosorption, description of the metabolic route, etc.).

in Brazil. In 2004, only ten research groups found in the keyword phytoremediation were registered on the Lattes Platform website (a website of active researchers in Brazil) and on the Directory of Research Groups of the Brazilian National Council for Scientific and Technological Development (CNPq), involving 225 PhDs (Marques et al., 2011). After 10 years, in 2014, the number of research groups and PhDs associated with this keyword are 59 and 587, respectively. This indicates growing interest in this area.

For the terms biosorption and bioremediation, there are no previous reports that could allow the comparison to check for progression of interest in these areas. However, considering the rise in the number of publications and also the growth profile as observed for the term phytoremediation, it is believed that the number of researchers associated to the terms mentioned earlier is also growing. For biosorption and bioremediation, there are, respectively, 19 and 144 research groups registered on the platform of Directories of Research Groups of CNPq, and the number of doctors that show up in association with these terms comes to 357 and 1158, respectively, according to the Lattes Platform.

It is important to stress that only eight groups have been registered in association with the keyword bioeconomy on the platform of Directories of Research Groups of CNPq, compared with 190 groups registered in relation to the keywords biosorption, bioremediation, or phytoremediation. Out of all the groups related to bioeconomy, four are also connected with areas of Social Sciences and Humanities (Economics and Business Administration), one with Forest Resources and Forest Engineering, one with Fishing Resources and Fishing Engineering, one with Agronomy, and one with Ecology. In addition, according to the Lattes Platform, only 53 researchers have been listed in connection with the term bioeconomy.

The bank of theses and dissertations at the Coordination for the Improvement of Higher Education Personnel (CAPES) (available at http://capesdw.capes.gov.br/) has records of theses and dissertations defended since 2010, and the respective figures for these are 150 and 48 for the term bioremediation, 29 and 17 for phytoremediation, 7 and 2 for biosorption, and 2 and 2 for bioeconomy. These figures confirm the need for investments and research in relation to the use of biomass as a remediation alternative for contaminated areas and systems. These also represent the nonrecognition, on the part of researchers, of the possible impact of research concerning use of biomass for the removal of chemicals, and their possible application for generation of business.

4 REGULATORY FRAMEWORKS ASSOCIATED WITH BIOECONOMY IN BRAZIL

In relation to Brazilian regulatory benchmarks, the most recent legislation concerning the investigation and the use and trade of biomass comes from 2005 (Table 1), the Biosafety Law, which regulates the production and commercialization of genetically modified organisms (GMOs) and research with stem cells, classifying what is research or commercial activity. Brazil also upholds international treaties in relation to the protection of intellectual property. Table 1 depicts the main regulatory frameworks related to biomass exploration in Brazil.

Here it is important to highlight that there is a Bill of Law currently going through the Brazilian Federal Senate, seeking the establishment of the government agency Management of Research

Table 1 Main Regulatory Frameworks Related to Biomass Exploration in Brazil

Regulatory Framework	Date	Determination of Subject of Regulatory Framework
Law No. 11105	03/24/2005	Biosafety Law
Law No. 10603	12/17/2002	Resolves about the protection of information not disclosed submitted for approval to commercialization of products and settles other measures
Law No. 10196	02/14/2001	Changes and adds to the terms of Law No. 9279 of May 14, 1996, regulating rights and obligations regarding industrial property and presents other terms
Law No. 9782	01/26/1999	Defines the National Sanitary Surveillance System, creates the National Health Surveillance Agency, and settles other measures
Law No. 9610	02/19/1998	Changes, updates, and consolidates copyright law and settles other measures
Law No. 9456	04/25/1997	Establishes the Law of Cultivated Plant Protection and settles other measures
Law No. 9279	05/14/1996	Regulates rights and obligations relating to industrial property
Law No. 8078	09/11/1990	Explain consumers' protection and settles other measures
Law No. 6938	08/31/1981	Defines the National Environmental Policy, its purposes and mechanisms of formulation and implementation, and settles other measures
Law No. 13123	05/20/2015	Resolves about the access to genetic resources, protection and access to associated traditional knowledge, and benefits sharing for the conservation and sustainable use of biodiversity. Revoke the provisional measure No 2186-16 of August, 2001, and settles other measures
Decree No. 6041	02/08/2007	Establishes the development policy of biotechnology, creates the National Biotechnology Committee, and settles other measures

Table 1 Main Regulatory Frameworks Related to Biomass Exploration in Brazil—Cont'd

Regulatory Framework	Date	Determination of Subject of Regulatory Framework
Decree No. 5950	10/31/2006	Regulates Article 57-A of Law 9985, of July 18, 2000, to establish the limits for planting genetically modified organisms (GMOs) in surrounding protected areas
Decree No. 5591	11/22/2005	Regulates provisions of Law 11105, of March 24, 2005, which regulates items II, IV, and V of Paragraph 1 of Article 225 of the Brazilian Federal Constitution and settles other measures
Decree No. 4680	04/24/2003	Regulates the right to information, as provided by Law 8078 of September 11, 1990, as foods and food ingredients intended for human or animal consumption containing or produced from GMOs, subject to compliance with other rules applicable
Decree No. 4074	01/04/2002	Regulates Law No. 7802, of July 11, 1989, which provides for research, experimentation, production, packaging and labeling, transport, storage, marketing, commercial advertising, use, import, export, waste disposal and packaging, registration, classification, control, inspection and surveillance of pesticides, their components and the like, and settles other measures
Decree No. 4339	08/22/2002	Establishes principles and guidelines for the implementation of the National Biodiversity Policy
Decree No. 2519	03/16/1998	Enacts the convention on biological diversity signed in Rio de Janeiro on June 5, 1992
Decree No. 2366	11/05/1997	Regulates Law No. 9456, of April 25, 1997, establishing the cultivated plant protection on the National Service for Plant Variety Protection (SNPC) and other measures
Legislative Decree No. 2	02/03/1994	Approves the text of the convention on biological diversity, signed at the United Nations Conference on Environment and Development held in the city of Rio de Janeiro, from June 5 to June 14, 1992
Decree No. 635	21/08/1992	Promulgates the Paris Convention for the Protection of Industrial Property, revised in Stockholm on July 14, 1967
Decree No. 75572	04/08/1975	Promulgates the Paris Convention for the Protection of Industrial Property Review, Stockholm, 1967
CNS Resolution No. 292	07/08/1999	Complementary rule to CNS Resolution 196/96, on the specific area of research on human beings, outside of coordinates or with foreign participation and research involving the sending of biological material abroad
CNS Resolution No. 251	07/08/1997	Approves guidelines for research involving human beings in the thematic area of research with new drugs, medicines, vaccines, and diagnostic tests
CNS[a] Resolution No. 196	10/10/1996	Approves the regulatory guidelines and standards for research involving human subjects
CONAMA[b] Resolution No. 314	10/29/2002	Resolves about the product registration for remediation and settles other measures
IBAMA[c] Normative Instruction No. 5	05/17/2010	Establishes the procedures and the requirements to be adopted for purposes of registration, registration renewal, and prior consent to carry out research and experimentation with remediation products

(Continued)

Table 1 Main Regulatory Frameworks Related to Biomass Exploration in Brazil—Cont'd

Regulatory Framework	Date	Determination of Subject of Regulatory Framework
Normative Act No. 127	05/03/1997	Provides for the application of the Industrial Property Law regarding patents and certificates of addition to invention
CTNBio[d] Resolution No. 1	30/10/1996	Approving the Internal Regulations of the National Biosafety Technical Commission (CTNBio)
Treaty of Budapest	04/28/1977	Provides for the International Recognition of the Deposit of Microorganisms for the purposes of proceedings relating to patents
Treaty of Cooperation in Patents (PCT)	06/19/1970	Establishes the principle of a single application for patent valid for all countries adhering to the PCT
Paris Convention of Union for the Protection of Intellectual Property	03/20/1883	Provides for international criteria for the granting and validity of industrial patents

[a] *Conselho Nacional de Segurança (National Health Council).*
[b] *Conselho Nacional do Meio Ambiente (National Council for the Environment).*
[c] *Instituto Brasileiro do Meio Ambiente e dos Recursos Renováveis (Brazilian Institute for the Environment and Renewable Resources).*
[d] *Coordenação-Geral da Comissão Técnica Nacional de Biossegurança (Board of the National Technical Committee for Biosecurity).*

Licensing in Brazilian Biomes and for Monopoly of Patents (EMGEBIO). This Bill of Law sets out that this agency would have the main purpose of managing the licensing of research at Brazilian biomes and would hence hold the monopoly of all patents arising from these studies, for a period of 10 years. The creation of this agency is enshrouded in controversy, as this monopoly could repel private investments in the area of patent production, which would be unsatisfactory for new developments in bioeconomy.

Those in favor of its establishment defend the idea that Brazil, being the holder of the fifth largest genetic pool in the world, should create conditions for the sustainable exploitation of these resources, as the access to these has recently been regulated by the law 13123 of May 20, 2015. Those in opposition to the need to modernize and debureaucratize the national regulatory benchmark, in a way that would make it easier to research and to gain access to the Brazilian genetic pool within national territory, say that most research about the Amazon region, for example, is now carried out abroad, while control over knowledge would lead to cognitive sovereignty, which would be more powerful than laws.

There are also those who defend the approval of a new regulatory benchmark for the sector and the taking up of strategies to encourage private investment in development of research, and who support sharing the resources arising from the exploitation of Brazilian biomes with traditional and native Brazilian populations that have guarded this biodiversity. The Brazilian Society for Science Progress (SBPC) was opposed to part of the text of the Law No. 13123. According to the SBPC, not attending the requirement, which states that any foreign legal company interested in access to any Brazilian components of the genetic heritage or traditional knowledge will be associated with Brazilian science and technology institutions and must sign an Agreement of Benefit Sharing as a condition to obtain a permit, is a serious risk to Brazilian sovereignty over their genetic heritage. On the other hand, the SBPC pointed out the importance of full involvement of the civil society on the Board of Management of Genetic Heritage (CGEN) (sbpcnet, 2015).

In relation to the existence and the study of products that could be used as environmental remediators in Brazil, the current legislation requires that these be registered at the Brazilian Institute for the Environment and Natural Renewable Resources (IBAMA), be they for production, importation, commercialization, or use (Article 1, CONAMA No. 314/2002). In a sole paragraph within this article, the legislation says that registration is not required in the case of remediators to be used for research and experimentation, with the prior acquiescence of IBAMA being required for these activities.

Here we stress that there are plans to implement penalties and other sanctions (Law No. 9605, of 12 February 1998 and Decree No. 3179, of 21 September 1999) for anyone who fails to comply with the legislation in force, that stresses that "in spite of the benefits that could arise from the appropriate use of remediators, these products could lead to an imbalance of the ecosystem and damage to the environment, due to its peculiarities or inappropriate use" (Normative Instruction No. 5/2002). It is also pointed out that the control over the existence and study of such remediators in Brazil has its importance based on the possibility of there being impurities in the chemical products; GMOs, pathogenic microorganisms, or exotic microorganisms in the products of biological origin; the production of some remediators could occur through the use of the Brazilian genetic pool.

The relevant legislation also considers a remediator to be "a product, whether composed of microorganisms or not, intended for the recovery of environments and ecosystems that have been contaminated, treatment of effluents and other residue, unclogging and cleaning of ducts and equipment also as an agent of a physical, chemical, or biological process, or a mixture of these" (CONAMA 314/2002). There is also the classification of three types of environmental remediators, which are shown in Table 2.

For the registration of products that have the role of bioremediators and biostimulants, which are considered biomass and are therefore possible boosters of the bioeconomy, there is a series of requirements that must be fulfilled.

For bioremediators, it is necessary to present the main data of the manufacturer, the producer, the importer, and also the party handling the product; declaration of quali-quantitative composition with a statement from a laboratory; the taxonomic classification; biological cycle; declaration and proof of absence of the most important pathogenic microorganisms (*Escherchia coli*, *Pseudomonas aeruginosa*, *Shigella*, and *Salmonella*); declaration of presence or absence of GMOs— a technical statement provided copy of the technical statement issued by CTNBio (National Biosafety Technical Commission); information about the process of production and quality control; physical and chemical properties (physical state, color, odor, form of presentation, density, pH, and miscibility in water); packaging (type, material, and capacity); conditions of keep and storage; indication of recommended use (locations, uses, contaminants to be biodegraded); a copy of the registration certificate granted either by the Brazilian Health Surveillance Agency (ANVISA) or by the Ministry of Agriculture, Livestock and Food Supply (MAPA); instructions for use: means of dilution, dose, frequency, and method of application of the product, together with any restrictions on use; method of action; proof of efficiency; undesirable environmental impacts; withdrawal of the product; final destination of empty packaging; personal protection equipment to be used for handling; accidental spillage and first aid; registration or authorization in other countries (for imported products).

Table 2 Classification of Remediators According to Brazilian Legislation (Normative Instruction No. 5 of May 17, 2010, Article 2)

Type of Remediator	Definition
Bioremediator	The active ingredient is microorganisms, which can reproduce and also biochemically degrade contaminant compounds and substances
Chemical or physicochemical remediator	The active ingredient is a chemical compound or substance, being either an oxidant, a surfactant, or dispersant, or also polymers or enzymes, among others, that can degrade, adsorb, or absorb contaminants
Biostimulator	Contains nutrients in its composition, favoring the growth of microorganisms that are naturally present in the environment where the substance will be applied, thereby speeding up the bioremediation process

For biostimulators the following information shall be provided: data of manufacturer, preparer, importer, and handler of the product, declaration of the quali-quantitative composition, with a statement issued by a laboratory, physical, and chemical properties (physical state, color, odor, presentation, density, pH, and miscibility in water), description of the production process, packaging (type, material, and capacity), indications of use (locations, destination, contaminants to be biodegraded), copy of the registration certificate granted by ANVISA or by MAPA.

The requests for acquiescence for research and experimentation should have data from the requesting party, legal representative, the technical party responsible for the execution of the project, the manufacturer and supplier of the product, remediator, commercial brand, type, composition, research project (location, purpose, methodology, quantity of product, schedule, etc.); information about the presence or absence of pathogenic microorganisms; information about the presence of GMOs.

Table 3 shows the registration situation of remediators in Brazil in 2013, with their respective recommended uses, and shows the small number of available products.

It is also important to stress that there is no legislation currently in force that regulates the use and destination of dead or inactivated biomass for remediation, or the use of superior organisms such as bioremediators.

Table 3 Registration Status of Remediators in Brazil, as of 2013

Type of Remediator	Registered	Under Analysis	To Be Analyzed	Indication for Water Environments
Bioremediator	33	13	4	1
Biostimulant	4	1	0	1
Physicochemical	2	8	6	2

Source: IBAMA, available at: http://www.mma.gov.br/port/conama/processos/9A16ED0C/RegistroProdutosRemedia.dores_CintiaAraujo.pdf.

5 VIEW OF THE INDUSTRIAL SECTOR ABOUT BIOECONOMY IN BRAZIL

In Brazil, the discussions about bioeconomy at industrial firms expanded in 2012, when the Brazilian National Confederation of Industry (CNI), in partnership with the Harvard Business School (HBS), held the first forum on bioeconomy in Brazil, an event that has since been repeated every year; in 2013, there was the launch of a specific schedule for Brazil, showing some strategic areas that need specific action and mapping some niches for investment in research and the perception of business people, researchers, and those responsible for devising government policies about the issue, seeking political, social, and economic actions to boost bioeconomy in the country.

The main goal of the research run by the CNI was to map the general bioeconomy scene in Brazil, and also identify obstacles and opportunities to consolidate this area in the country (Marques et al., 2011). Among those interviewed, there were representatives from the Brazilian federal government (40), from academia (20), and also from companies associated with the bioeconomy chain (100), with 50% of those interviewed hailing from the state of São Paulo, which is the Brazilian state with the highest number of industrial firms and which has over 50% of the Brazilian production of scientific articles published in international journals (Pesquisa sobre Bioeconomia no Brasil, 2004).

The data of the interview show that the term bioeconomy is known by 81.2% of the interviewees, with 92.2% having a positive perception of the issue. In a question asking for the first words that came to mind when the theme of bioeconomy was mentioned, the order of the top five spontaneous responses was that of economy associated with: biological processes, biotechnology, conscientious use of natural resources, sustainable economy, and biodiversity.

For those interviewed, with the possibility of choosing more than one answer to the query, 81.5% believe that Brazil has advantages and 92.3% that it also has disadvantages compared with the other countries, regarding the bioeconomy. Among the five main advantages spontaneously mentioned by these interviewees, in decreasing order of appearance, the result was as follow: (1) biodiversity, (2) natural resources, (3) qualified scientific body, (4) territorial size, and (5) professional qualification and biofuels (both with similar frequency). On the other hand, the five main disadvantages spontaneously reported were: (1) inappropriate regulatory benchmark, (2) bureaucracy, (3) lack of professional qualification, lack of investment, lack of government incentive (with similar frequency), (4) lack of preparation of Brazilian industry and poor degree of technological development, and (5) lack of investment in development research.

Over 70% of the respondents say that Brazil is not making full use of its potential in the area of bioeconomy, and 65.6% do not consider the Brazilian regulatory benchmark as being suitable for the development of bioeconomy in Brazil, and the lack of suitable labor was also mentioned by 55.6% of those interviewed.

About 77% of those interviewed consider that investments made in research in this area are not sufficient for the progress of bioeconomy studies in Brazil, and 92% agreed that there should be tax incentives for companies that invest in bioeconomy. Almost 79% say there is no legal security for the development of bioeconomy in Brazil.

Even though over 55.1% of the representatives of industrial firms as interviewed have said that investments in sectors related to the biotechnology of their companies would be expanded over the next 3 years, and 53% said that investments made by their companies in areas related to bioeconomy increased between 2011 and 2013, according to these companies, the total investment volume involved has not exceeded 40% of the total invested for 20% of these companies, with another 24% who did not reply or declined to provide the information.

On the other hand, 95% of the researchers and 70% of the representatives of the government segment that were interviewed did not consider Brazilian industry as being innovative, compared with only 45% among those interviewed from the industrial sector.

Considering the data presented in the research study (Marques et al., 2011), we can conclude that, even though all categories covered in the interview do indeed consider bioeconomy important for Brazil, there is also evidence that there is a need for evolution in Brazilian regulatory benchmarks, in order to promote the elimination of bureaucracy, and also that research investments must be stepped up, and also that there is a need to qualify human resources specifically to work in this area. On the other hand, the data show that there is disagreement about the innovative characteristics of Brazilian industry.

The document "Bioeconomy: An Agenda for Brazil," prepared by Harvard Business Review Analytic Services, in partnership with the CNI (Harvard Business Review Analytic Services and Confederação Nacional da Indústria, 2013), defends a National Bioeconomy Policy as a way to spread the culture of innovation throughout industrial and academic environments; promote the coordination between government policy and legislation that is in need of updating, also allowing the flexibilization of government institutions in relation to hirings and acquisitions of goods and services; intensification of the inborn Brazilian potential through the exploitation of the country's biodiversity, capacity to produce biomass cheaply, and to use technological and scientific knowledge of high quality and already applied to Brazilian agroindustry, allowing the country to enter the age of biotechnological language, which could have a long-term and medium-term impact on development in different areas of knowledge, thereby bringing new funds into Brazil while also boosting the population's standard of living.

The document says that bioeconomy needs polyvalent professionals, where the researcher acts as an innovator and entrepreneur, for multidisciplinary research groups that are skillfully related to the business sector. In addition, the document highlights the need to break barriers when transferring scientific and technological knowledge to the business sector, also expanding knowledge of issues concerning the protection, commercialization, and management of intellectual property assets, particularly patents.

The following modulators are mentioned for the construction of an active sector concerning bioeconomy: restrictions in the generation of knowledge; the challenge of modernizing the regulatory benchmark to favor the nurturing of science, innovation, and production; and the construction of favorable conditions that allow the development of a base of scientists and technologists that are entrepreneurs and also innovators.

The following are also mentioned as convergent and critical actions to achieve the development of the Brazilian bioeconomy:

a. Modernize the regulatory benchmark;
b. Expand investments in research, development, and innovation;
c. Achieve densification of the contingent dedicated to science and technology;
d. Expand and modernize research infrastructure;
e. Encourage entrepreneurship; and
f. Spread the culture of innovation.

This document also brings courses of action to be carried out so that each of the proposed goals can be achieved.

In spite of the quality of the document and also the many reports highlighted by the CNI about the importance of this issue and how it should be treated, the participation of the public sector is pointed out as the boosting agent that proposes actions. This has been indicated in the research carried out with 362 people of the industrial segment interested in the development of bioeconomy in Brazil: 62% of those interviewed have said that their companies do not have funding to invest in bioeconomy over the forthcoming 5 years, and 68% expect Brazil to become a potential nation in this area within a time frame longer than 9 years. The proper role of the industrial segment as a nurturing factor is yet to be better defined.

6 REMOVAL OF BARRIERS TO BOOST UP THE USE OF BIOMASS FOR BIOREMEDIATION IN BRAZIL

The boosting of bioeconomy in Brazil permeates, in the first instance, the actions of government policies based on its strategic vision so that there is the creation of a new chain of lucrative business, based on new strategies and models that can expand the funds held by the country.

In the document "Constructing Bioeconomy Analysing the National Strategies for Development of the Biotechnology Industry," produced by Pugatch Consilium and commissioned by the Biotechnology Industry Organization (BIO) (Harvard Business Review Analytic Services and Confederação Nacional da Indústria, 2013), particular mention is given to the need for a country to make moves to promote innovation at different levels, with coherent plans that promote synergy of actions and initiatives as defined, to promote innovation.

In this interim, the creation of generic and specific policies to favor the structuring of the bases for the development of a certain field of knowledge and that also leads to the expansion of funds held by a country will be established through dialog between the public and private sectors, academia as well as society. It is almost impossible to have an efficient policy for encouraging innovation in any area if there is not the basic infrastructure that allows for the qualification, training, and development of scientists and researchers.

Thus, Brazil needs, first and foremost, to launch an official state policy, rather than a government strategy, to plan and establish courses of public action to favor the qualification of human capital: the creation, expansion, and development of infrastructure for research and development; assigning security to the protection of intellectual property; modernization of regulatory benchmarks; establishment of requirements and standards for transfer of technology; define and create commercial and market incentives; and, also, ensure the legal security that is necessary to promote innovation and development in the area of bioeconomy.

In relation to the qualification of human capital able to work in the bioeconomy area, Brazil needs effective action to improve the quality of education, allowing ample and constitutional access to state-run education, free of charge and of good quality, at all levels of academic qualification. It has been proved that investment in quality education is directly related to the scientific, technological, economic, and social development of a country (Pugatch Consilium, 2014).

Once the challenge of universalization of basic education has been overcome, Brazil then needs to make it have sufficient quality so that the country may make the leap it has been longing for, and scientific education is closely attached to this need for development. According to Raupp (Psacharopoulos, 1994), in a world that is molded by science, quality education, especially in science, is essential for citizen development, so that these are qualified to solve problems with creativity and to actively and conscientiously participate in the collective demands of society (Raupp, 2009; Colclough, 1982).

In this way, investment in qualification, training, and valuation of teachers in areas akin to science, together with the schools' structure to allow scientific experience, are fundamental elements for the qualification of future researchers and workers who can carry out scientific development, in the different roles as requested by society, and also cater to new possibilities of economic and sustainable development, as defended by bioeconomy.

As far as bioeconomy is concerned, with it being an interdisciplinary area, research in areas such as agriculture, engineering, chemistry, bioscience, information technology, robotics, materials, and economics should also be the focus of improvements in basic quality formation. The expansion in the interest for such areas by means of incentives to propose interdisciplinary graduate programs related to bioeconomy by CAPES and also to invest in the qualification of researchers associated with the nurturing of public-private partnerships, could boost the area of bioeconomy in Brazil.

Regarding the creation, expansion, and development of Infrastructure for Research and Development and also for investment in research, related to financing, the Brazilian Government investment, considering only the data of the federal government, is higher than 55%, compared with 45% for private investments (source: Rodrigo Teixeira, seminar of the Committee on Science, Technology, Innovation, Communications and Information Technology—CCT, of the Brazilian Senate). Brazil is in the opposite direction in comparison with scientifically developed countries (the United States, Germany, and South Korea) and also countries undergoing rapid development (China), where the proportion of private investment in research and innovation is over 70%.

It is also important to note that the Brazilian government investment is around 1.7% of the Gross Domestic Product (some US $24.2 billion per year), which is about 17 times less than that invested by the United States. However, efforts have been made, including the establishment of the National Program for Knowledge Platforms (Decree No. 8269 of 06/25/2014) by the federal government, in a move to raise the standard and impact of science, technology, and innovation in Brazil, seeking to implement actions for the expansion and enhancement of basic education, new systems for access and expansion of college education, the supply of 101,000 grants in a time range of 3 years, for international student exchange, improvement of graduate studies, increasing of the research productivity grants, and calls on the public for financing of research, creation of special science and technology committees to establish a policy for nurturing and financing and the creation of the Brazilian Company for Corporate Research and Innovation (EMBRAPII), in an attempt to establish dialog between the economic sector, the academic environment, and the government.

In relation to private efforts, considering the prospects of investments in research and innovation in the bioeconomy sector as suggested by the survey run in 2014 by the CNI, it is extremely important that Brazilian entrepreneurs have a change of culture and are informed, confident, and also willing to step up investment. EMBRAPII suggested, back in 2012, a model for investment in innovation where the institutes of government, science, and technology and also private companies could have a tripartite contribution through injection of resources.

The establishment of strategic areas for nurturing research for generating knowledge and innovation should be made together with different segments of society, to make government resources less and less dependent on individual positions and on the interests of groups dominating the national scientific scene, thereby bringing about the appearance of new exponents and new opportunities for Brazil.

One must stress also that, in Brazil, bioeconomy has one aspect that is almost exclusively related to the generation of biofuels, boosted by the ethanol industry, and government investments in the area are restricted to windows of opportunities rather than continuous financing programs, which is the case in other countries (Bomtempo, 2014; Enriquez-Cabot, 1998). Brazilian investments could be inappropriate for the kind of demands due to the technological requirements of bioeconomy, and assistance from American and European programs is needed to boost development of bioeconomy, particularly in relation to biofuels, are guided by a strategic plan defined by government policies, while in Brazil no such strategic plan on a national scale has been devised for the medium term (Bomtempo, 2014).

Data related to specific national investments for research into bioeconomy, biosorption, bioremediation, and phytoremediation could not be found. However, the Foundation in Support of Research of the State of São Paulo (FAPESP), Brazil's largest state agency for nurturing research, which applies 1% of the GDP of the State of São Paulo, one of the states with greatest production of biomass and the most industrialized in Brazil, has registered the financing of only 235 projects involving these terms (including grants and financing of research projects), compared with a total of 198,722 assistances already granted, which works out at 0.12% of this total, which is very small considering that there are 27 research groups based in São Paulo related to the above-mentioned subjects duly registered on the platform of Directories of Research Groups, of the CNPq.

In order to establish bioeconomy as a strategic area for the development of Brazil, programs and investments aimed at the segment would significantly expand the number of research projects and grants for financed students in the area and, hence, the innovation necessary for bioeconomic growth.

Concerning the Brazilian regulatory benchmark, it is necessary to review the overlap of competences of government organizations, correction of contradictions, and reduction in excessive use of provisional measures, making the process of environmental licensing less political, with reduced cost, less ideological and bureaucratic, while more active and transparent, based on technical and scientific information, while maintaining sustainability and environmental preservation, without ever ceasing to be an inductor of innovation.

The time frame for concession of patent rights stands at 10.8 years, and, according to the CNI, this long time frame demotivates development of innovation within the industrial segment. In addition, one must increase the incentives set out in Law No. 11196 of 11/21/2005, which, among other issues, suggests tax exemption for technological innovation for small and micro businesses, which would also help the increase of the universe of the companies now carrying out research and development in the country, currently 6000, according to the CNI. This entity also suggests there should be a preference in government purchases for companies that invest in research, the expansion of scientific cooperation between companies and scientific institutions, through nurturing laws and the creation of a public model of reimbursable government financing, where companies would provide resources for the government to invest in research through the National Fund for Scientific and Technological Development (FNDCT), which could be reimbursed with the profits arising from the commercialization of the innovation.

Considering that the use of biomass in Brazil is focused mainly on energy production, there is establishment of the possibilities of expansion to the commercial uses of biomass, to strengthen bioeconomy with different applications. However, the legal conditions for this to occur should be established.

In the case of bioremediation, the legislation dealing with regulation of products does not address the use of phytoremediators or inactive biosorbents, be they biomass of microbiological origin or other materials arising from agribusiness or environmental exploitation. The review of the legislation is therefore urgent, considering the number of researchers active in the fields involving the use of these materials, and the expansion possibilities of licensed products to be commercialized.

The debureaucratization to acquire the permits for studies using Brazilian genetic materials becomes essential so that Brazilian researchers may carry out their works. Currently, depending on the issue and the origin of the biomass, researchers must hold authorizations from the National Indian Foundation (FUNAI), from IBAMA, the Committee for Management of Genetic Assets (CGEN) of the Brazilian Ministry for the Environment, among others.

7 CONCLUSIONS

Brazil needs to recognize the strategic character of bioeconomy for the country and then articulate a national agenda with the following aims: (1) to modernize the regulatory benchmarks applicable to access Brazilian genetic materials, to strengthen the environment of legal security; (2) to invest in basic education, especially the teaching of science, to supply the need for persons trained and qualified to participate in the research; (3) to set strategic guidelines to promote nurturing of research into bioeconomy; and (4) to attract the industry sector to invest in research to develop the area of bioeconomy.

Sovereignty over genetic assets is not only geographic but also a matter of knowledge, and political tools need to be applied so that Brazilian researchers and industrial firms may access such assets legally, without forsaking sustainability, for the creation of new services and products that benefit society and promote the economic growth of Brazil.

REFERENCES

Agência Nacional de Energia Elétrica, 2005. Atlas de Energia Elétrica, second ed., 77–92. Available in: http://www.aneel.gov.br/aplicacoes/atlas/pdf/05-Biomassa(2).pdf.

Aksu, Z., 2005. Application of biosorption for the removal of organic pollutants: a review. Process Biochem. 40, 997–1026.

Araújo, G.C.L., Lemos, S.G., Ferreira, A.G., Freitas, H., Nogueira, A.R.A., 2007. Effect of pre-treatment and supporting media on Ni(II), Cu(II), Al(III) and Fe(III) sorption by plant root material. Chemosphere 68, 537–545.

Araújo, G.C.L., Lemos, S., Nabais, C., 2009. Nickel sorption capacity of ground xylem of *Quercus ilex* trees and effects of selected ligands present in the xylem sap. J. Plant Physiol. 166, 270–277.

Bomtempo, J.V., 2014. Bioeconomia em construção II – Os grants e subvenções às empresas: comparando o Biomass Program do DOE e o PAISS do BNDES/FINEP. Boletim Infopetro: Petróleo e Gás Brasil vol. 2, 50–54. Available in: https://infopetro.files.wordpress.com/2014/07/infopetro05062014.pdf.

Brady, D., Stoll, A., Duncan, J.R., 1994. Biosorption of heavy metal cations by non-viable yeast biomass. Environ. Technol. 15, 429–438.

Brand, M.A., Neves, M.D., 2014. Situação atual e perspectivas da utilização da biomassa para a geração de energia, 1–7. Available in: http://www.tractebelenergia.com.br/wps/wcm/connect/3133ef5d-0680-4c2f-a617-3b4794b2266b/0403-001-2005_art.pdf?MOD=AJPERES&CACHEID=3133ef5d-0680-4c2f-a617-3b4794b2266b.

Cardoso, B.M., 2012. Uso da Biomassa como Alternativa Energética, Master, Escola Politécnica da Universidade Federal do Rio e Janeiro. Available in: http://monografias.poli.ufrj.br/monografias/monopoli10005044.pdf.

Carrilho, E.N.V.M., Gilbert, T.R., 2000. Assessing metal sorption on the marine alga *Pilayella littoralis*. J. Environ. Monit. 2, 410–415.

Carrilho, E.N.V.M., Ferreira, A.G., Gilbert, T.R., 2002. Characterization of sorption sites on *Pilayella littoralis* and metal binding assessment using 113Cd and 27Al nuclear magnetic resonance. Environ. Sci. Technol. 36, 2003–2007.

Carrilho, E.N.V.M., Gilbert, T.R., Nóbrega, J.A., 2003. The use of silica-immobilized brown alga (*Pilayella littoralis*) for metal preconcentration and determination by inductively coupled plasma optical emission spectrometry. Talanta 60, 1131–1140.

Colclough, C., 1982. The impact of primary schooling on economic development: a review of the evidence. World Dev. 10, 167–185.

Das, N.V., Vimala, R., Karthika, P., 2008. Biosorption of heavy metals – an overview. Indian J. Biotechnol. 7, 159–169.

Embrapa Informação Tecnológica, 2006. Plano Nacional de Agroenergia 2006–2011, second ed. Ministério da Agricultura, Pecuária e Abastecimento, Secretaria de Produção e Agroenergia, Brasília. Available in: http://www.agricultura.gov.br/arq_editor/file/Ministerio/planos%20e%20programas/PLANO%20NACIONAL%20DE%20AGROENERGIA.pdf.

Enriquez-Cabot, J., 1998. Genomics and the world's economy. Sci. Mag. 281, 925–926.

Ferreira-Leitão, V., Gottschalk, L.M.F., Ferrara, M.A., Nepomuceno, A.L., Molinari, U.B.C., Bon, E.P.S., 2010. Biomass residues in Brazil: availability and potential uses. Waste Biomass Valor. 1, 65–76.

Glass, D.J., 1998. The 1998 United States Market for Phytoremediation. D. Glass Associates, Needham, MA. pp. 1–139.

Gonzalez, M., Araújo, G.C.L., Pelizaro, C., Menezes, E., Lemos, S., Desousa, G., Nogueira, A., 2008. Coconut coir as biosorbent for Cr (VI) removal from laboratory wastewater. J. Hazard. Mater. 159, 252–256.

Harvard Business Review Analytic Services, Confederação Nacional da Indústria, 2013. Bioeconomia: Uma Agenda Para O Brasil. Confederação Nacional da Indústria, Brasília. pp. 1–40. Available in: http://arquivos.portaldaindustria.com.br/app/conteudo_18/2013/10/10/5091/201310151121222486319o.pdf.

Júnior, C.B., Libânio, J.C., Galinkin, M., Márcio, M., 2009. Agroenergia Da Biomassa Residual: Perspectivas Energéticas, Socioeconômicas e Ambientais, second ed. Technopolitik Editora, Foz do Iguaçu/Brasília.

Koelmel, J., Prasad, M.N.V., Pershell, K., 2015. Bibliometric analysis of phytotechnologies for remediation: global scenario of research and applications. Int. J. Phytoremediation 17, 145–153.

Machado, M.D., Janssens, S., Soares, H.M.V.M., Soares, E.V., 2009. Removal of heavy metals using a brewer's yeast strain of *Saccharomyces cerevisiae*: advantages of using dead biomass. J. Appl. Microbiol. 106, 1792–1804.

Madrid, Y., Cámara, C., 1997. Biological substrates for metals preconcentration and speciation. Trends Anal. Chem. 16, 36–44.

Marques, M., Aguiar, C.R.C., Silva, J.J.L.S., 2011. Desafios técnicos e barreiras sociais, econômicas e regulatórias na fitorremediação de solos contaminados. Revista Brasileira de Ciência do Solo 35, 1–11.

Pesquisa sobre Bioeconomia no Brasil, 2004. 3° Fórum de Bioeconomia: Políticas públicas e Ambiente para Inovação de Negócios no Brasil, disponível em: http://arquivos.portaldaindustria.com.br/app/conteudo_18/2014/10/23/7643/Pesquisa_bioeconomia_2014.pdf.

Pletsch, M., Charlwood, B.V., Araújo, B.S., 1999. Fitorremediação de águas e solos poluídos. Biotecnologia Ciências e Desenvolvimento 11, 26–29.

Prasad, M.N.V., 2011. In: A State-of-the-Art Report on Bioremediation, its Applications to Contaminated Sites in India. Ministry of Environment & Forests, GOI New Delhi, pp. 90.

Psacharopoulos, G., 1994. Returns to investment in education: a global update. World Dev. 22, 1325–1343.

Pugatch Consilium, 2014. Building the bioeconomy examining National Biotechnology Industry development strategies: a briefing paper, pp. 1–74. Available in: http://www.pugatch-consilium.com/reports/Building_The_Bioeconomy_PugatchConsiliumApril%202014DD.pdf.

Raupp, M.A., 2009. Boa educação básica para a melhor educação científica. In: Werthein, J., Cunha, C. (Eds.), Ensino de Ciências e Desenvolvimento: O Que Pensam Os Cientistas". UNESCO, Brasíli, pp. 195–199. Available in: http://unesdoc.unesco.org/images/0018/001859/185928por.pdf.

Sağ, Y., Kutsal, T., 1996. The selective biosorption of chromium (VI) and copper (II) ions from binary metal mixtures by *R. arrhizus*. Process Biochem. 31, 561–572.

http://www.sbpcnet.org.br/site/noticias/materias/detalhe.php?id=3708 (accessed on 02/02/2015).

The European Plant Science Organisation, 2011. The European Bioeconomy in 2030: Delivering Sustainable Growth by addressing the Grand Societal Challenges. Available in: http://www.epsoweb.org/file/560.

Trama, B., Fernandes, J.D.S., Labuto, G., Oliveira, J.C.F., Viana-Niero, C., Pascon, R.C., Vallim, M.A., 2014. The evaluation of bioremediation potential of a yeast collection isolated from composting. Adv. Microbiol. 4, 796–807.

Vegliò, F., Beolchini, F., 1997. Removal of metals by biosorption: a review. Hydrometallurgy 44, 301–316.

Vieira, R.H.S.F., Volesky, B., 2000. Biosorption: a solution to pollution? Int. Microbiol. 3, 317–324.

Volesky, B., 2004. Sorption and Biosorption, first ed. BV Sorbex, Québec.

Volesky, B., Holan, Z.R., 1995. Biosorption of heavy metals. Biotechnol. Progr. 11, 235–250.

Wase, J., Forster, C., 1997. Biosorbents for Metal Ions, first ed. Taylor and Francis, London.

Wiedmann, T.O., Schandl, H., Lenzen, M., Moran, D., Suh, S., West, J., Kanemot, K., 2015. The material footprint of nations. Proc. Natl. Acad. Sci. 112 (20), 6271–6276. http://dx.doi.org/10.1073/pnas.1220362110.

PHYTOREMEDIATION OF SOIL AND GROUNDWATER: ECONOMIC BENEFITS OVER TRADITIONAL METHODOLOGIES

23

Edward Gatliff[1], P. James Linton[2], Douglas J. Riddle[3], Paul R. Thomas[4]

Applied Natural Sciences, Inc., Hamilton, Ohio, USA[1]
Geosyntec, Clearwater, Florida, USA[2]
RELLC, Mountain Center, California, USA[3]
Thomas Consultants, Inc., Cincinnati, Ohio, USA[4]

1 PHYTOREMEDIATION HISTORY

Phytoremediation is the use of plants in environmental restoration. It can refer to applications ranging from treatment wetlands to urban green roof systems. The term *phytoremediation* is used here to describe environmental restoration of soils and groundwater using trees.

The general application of phytoremediation began in the early 1990s and was performed concurrently with active research at that time. Nearly all of the applications were applied to hazardous substances of low risk and thus low potential for impact to the public health and safety. Accordingly, most early applications and research focused on remediation of:

- Agricultural chemicals (Banuelos, 1994; Burken and Schnoor, 1996; Jordahl et al., 1995; Schnoor and Licht, 1991);
- Heavy metals (Baker et al., 1991; Chaney et al., 1997);
- Trinitrotoluene (TNT);
- Petrochemicals (Banks et al., 1994); and
- Volatile organic compounds (VOCs) (Ferro et al., 1996; Rock, 1996).

The history of phytoremediation requires the discussion of roughly concurrent events in the mid- to late 1980s. Two groups were considering the feasibility of remediating agricultural chemical sites in the midwestern United States using plants. These groups included Edward Gatliff and Paul Thomas on sites in Illinois and Jerry Schnoor and Louis Licht in Iowa. One reason that these early field applications were possible was that agricultural chemical sites were regulated differently than other waste sites. In the states of Illinois, Minnesota, and Iowa, for example, agricultural chemical sites fell under the jurisdiction of the respective states' departments of agriculture as opposed to the state regulatory agencies that oversaw typical properties with hazardous waste issues. The departments of agriculture were generally receptive to some of the early applications of plant-based bioremediation. Early plantings confirmed

Copyright © 2016 Elsevier Inc. All rights reserved.

that herbicides and levels of excess nutrients could be reduced by establishing vegetative cover (Gatliff, 1997; Thomas and Buck, 1999).

One very important finding from the early applications was that hybrid poplar trees could be deep planted in soils contaminated with herbicides and salts. In these situations, the root system of the poplar tree would be buried deeper than the near-surface contaminated soil where conditions would prevent the germination of seed and/or impede growth of shallow-rooted plants. Both the *Populus* and *Salix* genuses have rooting characteristics that allow this type of deep planting.

At about the same time as the early work with agricultural chemicals, Scott Cunningham of Dupont identified hyperaccumulator plants appearing spontaneously at sites containing soils contaminated with heavy metals. Analysis of some of these plants showed that levels of heavy metals in the tissues were extremely high, which indicated the possibility that metals could actually be mined from shallow soils using plants.

Other concurrent work was being performed by Department of Defense researchers who found that TNT sites that had been dormant for a number of years showed substantial reductions in soil contaminant concentrations. These reductions were associated with encroaching vegetative growth around the periphery of the sites as the concentration of TNT in soil was reduced. Similar conclusions were drawn with regard to organic chemical sites both by early research and forensic examination of contaminated sites and facilities that had become naturally vegetated. Since 1990, phytoremediation has been progressing with researchers attempting to catch up with field applications that often demonstrated very positive reductions in contaminant concentrations (Erickson et al., 1994; Fletcher et al., 1995; McCutcheon, 1996; Negri et al., 1996). Much of the early field applications of phytoremediation were possible because of the "voluntary" nature of these pilot projects and because there was generally no immediate risk to human health or the environment that would require removal and other intrusive types of remediation.

2 TRADITIONAL VERSUS DESIGNED AND CONSTRUCTED SYSTEMS

Traditional phytoremediation as discussed here involves the use of plants (usually *Populus* and/or *Salix* tree species) planted as live cuttings, unrooted whips, bare-rooted whips, or bare-rooted trees. The potential objective of phytoremediation is the reduction of contaminant mass in soil and/or groundwater. This contaminant reduction is achieved by:

- Rhizodegradation (contaminant degradation/transformation in the root zone through cometabolism with microbes or through enzyme reactions);
- Phytodegradation (uptake of contaminant by plant followed by degradation or transformation in plant tissues);
- Phytovolatilazation (uptake of contaminants followed by translocation to leaves and transpiration);
- Phytosequestration (immobilization of contaminants in the near-root zone); and
- Phytoaccumulation (uptake followed by sequestration of unmodified contaminants in plant tissues).

The remediation objectives are achieved by planting trees in open holes or trenches so that the roots will be in contact with the capillary fringe (the zone of partial saturation in contact with groundwater). If the capillary fringe is too deep, irrigation may be necessary for the trees to survive until roots reach the capillary fringe or become sufficiently established to allow survival from the consumption of water from rainfall events. Irrigation has the potential to be counterproductive to phytoremediation function if the consumption and treatment of groundwater is an objective. If the goal is soil remediation, the plant roots must be in contact with the impacted soils. If the goal is groundwater remediation, the plants must be able to exert a hydraulic influence over the impacted groundwater in order to move the water into the root zone of the plants where it is subject to remediation mechanisms.

Designed and constructed systems were initially used to force the root systems of plants (typically trees) to develop to deep groundwater. In New Jersey in 1990-1991, the authors designed and constructed a prototype system to allow trees to primarily utilize groundwater about 5 m below ground surface in a climate where annual rainfall averages 1140 mm/year; an amount that easily satisfies the water requirements of the trees. To make the system functional and timely, access to rainwater had to be limited, and roots would be required to reach the top of the aquifer by penetrating 5 m of fairly dense sandy clay subsoil in a short time. To overcome these challenges, 90-cm-diameter boreholes were developed to the top of the aquifer, cased with metal corrugated drainage pipe that extended 15 cm above the ground surface and backfilled with topsoil. Relatively large hybrid poplar trees, about 40-60 mm caliper, were installed in these cased holes. The system forced the trees to develop roots vertically and reach the capillary fringe of the aquifer at about 4 m below ground surface within the first year (as determined by the substantial increase in the size of the uppermost leaves, which indicates luxury consumption of water) (Figure 1).

Since the early 1990s, the prototype system has been refined and substantially enhanced to allow targeting of specific horizons of the vadose and saturated zones. In addition, the authors have trademarked the terms TreeMediation and TreeWell to identify their designed and constructed systems. These refinements have substantially increased the efficacy of the system in many ways. The most significant innovation allows plants to utilize water from specific subsurface horizons in which contamination is migrating. Trichloroethylene (TCE) and other organic compounds that are heavier than water are commonly found near the base of permeable aquifers. Plants drawing water from the top of these aquifers will generally have little effect on contaminant concentrations that are in deeper horizons (Figures 2 and 3).

FIGURE 1

Boreholes were developed to the top of the aquifer, cased with metal corrugated drainage pipe.

FIGURE 2

Plants roots penetrating into aquifers (Clean and contaminated).

Other refinements allow the plants to utilize contaminated water that would normally be phytotoxic. As the tree pumps water from the soil column, groundwater passes through bioreactor media in the soil column as it flows upward toward the root system. Depending on the constituent(s) and the residence time, there can be substantial contaminant reduction before the groundwater solution reaches the roots thereby reducing or eliminating phytotoxic effects (Figure 4).

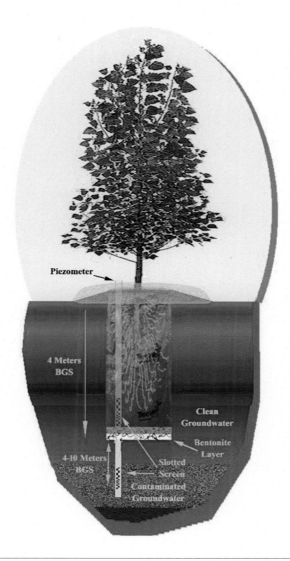

FIGURE 3

Plants drawing water from the top of the aquifers. Please note the bentonite later between clean and contaminated water.

3 OVERCOMING THE LIMITATIONS OF PHYTOREMEDIATION

Designed and constructed systems significantly expand the opportunities available for phytoremediation. While limitations remain, this chapter will focus on the opportunities in order to offset many preconceived biases with regard to the limitations of phytoremediation.

FIGURE 4

The tree act as hydraulic system to pump groundwater. Ground water passes through bioreactor media in the soil column as it flows upward toward the shoot system. Depending on the constituent(s) and the residence time, there can be substantial contaminant reduction before the groundwater solution reaches the roots thereby reducing phytotoxic effects.

3.1 TIGHT FORMATIONS

Phytoremediation has a distinct advantage over pump and treat systems in tight formations. The ability of the root system to explore and utilize capillary water is a significant advantage to soil and groundwater cleanup in tight formations. Pumping systems only extract free water and must rely on pulse pumping to impact contaminants held in the capillary solution. Roots actually pump from the capillary system thereby directly treating contaminants in the capillary solution.

3.2 SINKING CONTAMINANTS

Sinking contaminants would be impossible to treat with traditional phytoremediation systems. However, designed and constructed systems can target these contaminants in many situations. Obviously, tighter formations would enhance the viability of a designed and constructed system over a pump and treat system but opportunities are not limited to tight formations. While phytoremediation systems are limited in water use capacity, they still offer treatment-free water removal.

3.3 HIGH-YIELDING OR FAST-MOVING AQUIFERS

Opportunities do exist for phytoremediation in high-yielding or fast-moving aquifers, especially with the designed and constructed systems. One approach effectively creates a biobarrier by having multiple rows of the bioreactor columns that extend vertically through the aquifer media. These series of bioreactor columns must have comparable or higher porosity than the surrounding media to insure groundwater pass-through. The pumping by the trees further enhances the system by creating a hydraulic gradient toward the column as well as enhancing the flushing of the column. While this approach is feasible, the elevated installation costs may prove to be a limitation.

3.4 PHYTOTOXICITY

Phytotoxicity is a significant issue regarding the potential for treating highly contaminated soil and groundwater. However, some mitigation techniques can be successfully employed depending on the constituent of issue.

3.4.1 Organic contaminant levels

There are effective mitigation techniques available for organic contaminants depending on the constituent of issue. In some cases, selecting the right plant is all that is required. As noted earlier, in situ pretreatment is also possible by selecting the right treatment media.

3.4.2 Salt levels

High salt levels are a problem at many industrial sites. While treatment opportunities are limited or nonexistent, plant selection offers an effective means of overcoming this issue. Halophytes or facultative halophytes can perform well in conditions with fairly high salt levels.

3.4.3 Metals

Not only is the remediation of metals difficult or highly impractical for phytoremediation systems, they are also quite toxic for many plants. As with elevated salt levels, there are plants that can deal with potentially phytotoxic levels of metal constituents, and there are pretreatment systems that can be employed especially with metals in solution but opportunities to use plants for metals remediation remain quite limited.

---

4 CASE HISTORIES

4.1 OCONEE, ILLINOIS: REMEDIATION OF AGRICULTURAL CHEMICALS IN SOIL

An 0.4-ha agricultural chemical dealership near Oconee, Illinois, was closed in 1986 in response to neighbor concerns about chemical releases. Subsequent site characterization activities indicated that soil and groundwater were impacted by spills of liquid chemical fertilizers and herbicides. In 1987, phytoremediation of site soils was proposed as an alternative to excavation and disposal. The site was planted with corn in 1988. Corn was selected because a large component of the soil contamination was herbicide to which corn was known to have resistance. Unfortunately, the salts and other herbicides present in the soil prevented a significant portion of the corn from germinating and the corn that did germinate was adversely impacted. In 1989, an 8-cm-thick layer of sawdust was tilled into the surface of the site to provide a source of organic matter, to improve soil structure, to mitigate the effects of salinity, and to provide a nitrogen sink.

In 1991, 2-m-tall hybrid poplar trees were planted with bare roots located approximately 1 m below the surface where it was known that the salinity and herbicide concentration would be significantly less than at the surface. The trees grew well, and in 1995 an irrigation system was installed that extracted contaminated groundwater from the downgradient end of the site. The intent of the irrigation system was to recirculate groundwater impacted with herbicides and fertilizers in order to reduce the mass of contaminant available for downgradient transport. It was understood that the salinity of the irrigation water would ultimately result in tree mortality and the intent was to allow the salts to reach a level at which the trees no longer survived, then to stop irrigating and allow the salts to naturally flush out of the system. The irrigation system was shut down in 2002, and trees were replanted in 2008. Soil sampling in 2011 showed that the contaminants in site soils had been reduced to the point that the Illinois Environmental Protection Agency requires no further soil remediation. Figure 5 shows the concentration reduction in soil nitrate nitrogen (upper 0.5 m) between 1987 and 2011.

Phytoremediation of soil at Oconee eliminated the need for excavation and disposal resulting in a cost savings of approximately $375,000 (1987 cost basis) for the 0.4-ha site. The cost savings account for phytoremediation costs (including construction and monitoring) of approximately $200,000.

4.2 ABERDEEN PESTICIDE DUMPSITES: REMEDIATION OF LINDANE IN GROUNDWATER

Pilot planting and full-scale phytoremediation systems were installed in Aberdeen, North Carolina, at the Aberdeen Pesticide Dumps Superfund Site (APDS) in 1997 and 1998, respectively. The record of decision (ROD) issued in 1993 originally specified excavation and thermal desorption of contaminated soils, replacement of treated soils back into the excavations from which they had been removed, and a mechanical groundwater extraction and treatment (pump and treat) system to contain and remediate groundwater contaminated with chlorinated volatile organic compounds (CVOCs) and residual pesticides.

The APDS site resulted from the operation of a pesticide reformulation and packaging facility where pesticides from various manufacturers were combined with other compounds to act as carriers and to reduce the percentage of active ingredient to desired levels. VOCs used in formulation as well as off-specification pesticide products were disposed of in trenches on and near the facility, resulting in soil and groundwater contamination. The primary site remedy involved the excavation of soils contaminated with hexachlorocyclohexane (Lindane), which were thermally treated and returned to the excavation (Figure 6).

One of the functions of the pump and treat system was to be the remediation of any residual Lindane in the treated soil or in contaminated soil that might have been missed during excavation. The potentially responsible parties (PRP) group was successful in obtaining a modification to the ROD through explanation of significant difference (ESD). The ESD changed the remedy to include phytoremediation

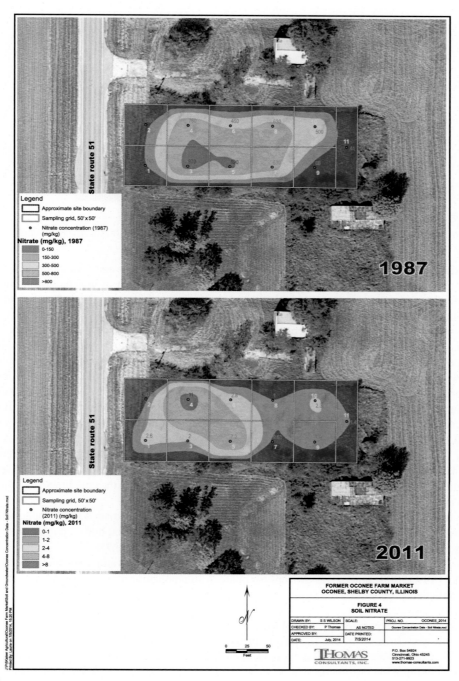

FIGURE 5

Shows reduction in soil nitrate nitrogen concentration between 1987 and 2011.

FIGURE 6

Soils contaminated with hexachlorocyclohexane (Lindane), were excavated and thermally treated.

and natural attenuation, significantly reducing the scope of groundwater extraction and treatment. Phytoremediation has been used principally for hydraulic containment of the shallow aquifer and for "polishing" treatment to remove residual organic contaminants following the removal of most of the contaminated soil in 1998.

The pilot phytoremediation system installed in 1997 was approximately 0.1 ha in size. The pilot plot was located on a slope where the depth to water ranged from approximately 2 m at the bottom of the slope to approximately 5 m at the top of the slope. Bare-rooted hybrid poplar trees ranging from 2 to 5 m were planted in 0.3-m-diameter holes ranging in depth from 1 to 4 m. The primary objective of the pilot system was to establish whether or not the trees, particularly the trees planted in the up-slope locations, would survive without irrigation. In order for the phytoremediation system to function as designed, it would be necessary for the trees to obtain a significant portion of the water they needed for survival from near the surface of the water table. Given the extremely high sand content of site soils (greater than 90%), hybrid poplar trees would not be expected to survive on percolating precipitation. Local plants that survive where depth to groundwater exceeded 2 m is limited to those species that are highly adapted to the sandy soil conditions. The pilot project resulted in the conclusion that the selected species and planting methods would function as required.

The full-scale phytoremediation system installation was performed over a 6-week period in March and April of 1998. Approximately 1.75 ha were planted with bare-rooted hybrid poplar trees to depths ranging from 1.5 to 4 m depending on the depth to groundwater (Figure 7).

Sap flow sensors were used to quantify the volume of groundwater consumed by the system in 1998, 1999, 2000, and 2012. An approximate water consumption rate of 1 L/m^2 of leaf area was established. Measurements from 2013 suggest that peak leaf area was achieved by year three (Figure 8).

Excavations were performed in 2013 to assess the extent of tree root development. In particular the excavations were to confirm that the deeply planted trees (4 m) were actually rooted into the capillary fringe. The excavations showed that the trees planted in 1998 to a depth of 4 m had developed roots at depth and continue to consume groundwater as required for the proper function of the phytoremediation system.

FIGURE 7

The full-scale phytoremediation system was established in March and April of 1998. Approximately 1.75 ha were planted with bare-rooted hybrid poplar trees to depths ranging from 1.5 to 4 m depending on the depth to groundwater for removal of ground water contaminants.

FIGURE 8

Sap flow sensors were used to quantify the volume of groundwater consumed by the trees. Approximate water consumption rate of the planted trees was about 1 L/m 2 of leaf area.

The pilot planting, including reporting, cost approximately $50,000. The full-scale planting cost approximately $350,000, including design, construction, and engineering oversight. Subsequent maintenance, monitoring, and reporting costs associated with the phytoremediation system have totaled approximately $425,000 over 17 years. The use of phytoremediation technology as a substitute for the original pump and treat system is estimated to have saved from $15 to $17 million over the 17-year period since the system was installed.

4.3 SARASOTA: REMEDIATION OF A 1,4-DIOXANE PLUME IN FRACTURED BEDROCK

The Sarasota site is located in west-central Florida near the Gulf Coast and was operated as a manufacturing facility for speed and proximity sensors during the 1970s through approximately 2008. During that period of operation, the facility utilized chlorinated solvents trichloroethene (TCE) and tetrachloroethene (TCA) in the process, and employed an on-site recovery still to recycle the solvent products. No catastrophic releases were recorded; however, accumulated spills and other small releases over time have resulted in groundwater impacts in multiple areas of the site.

1,4-Dioxane is a cyclic ether that was used as a stabilizer in TCA to prevent the degradation of the solvent during storage and use, at concentrations of up to 4% by volume. It is a Lewis base with electrons available for sharing, and is subsequently highly soluble (miscible) in water. The ring structure and position of the oxygen molecules in the ring make 1,4-dioxane highly stable and relatively immune to both abiotic and biotic transformation under normal environmental conditions. These characteristics also prevent 1,4-dioxane from readily sorbing to the soil matrix or other media, and it tends to move in the groundwater at a higher rate than the associated solvents and their breakdown products. This generally results in a mature plume configuration with residual solvent and daughter products close to the source area, and a dilute 1,4-dioxane "halo" that can potentially extend for a considerable distance beyond the residual solvent plume.

These same characteristics also make 1,4-dioxane difficult to recover and treat. The general extent of the dilute plume can require a significant extraction system to capture and contain the plume. Standard air stripping is not effective due to the miscible nature of the compound. Sorption media, such as activated carbon, are ineffective for removal, and the structure of the ring requires considerable energy to break. Treatment systems designed to treat 1,4-dioxane generally require an aggressive (and expensive) component, such as ultraviolet photolysis or chemical oxidation.

The configuration of the groundwater contaminant plume at the Sarasota site generally fits that described earlier, with very little residual degradation products of the chlorinated solvent and an associated extensive, dilute 1,4-dioxane plume downgradient from the "source" area, extending off site. Several smaller residual plumes that may have initially been connected with the main plume are also present on adjacent parcels. In addition, the geochemical changes associated with the biodegradation associated with the solvent component mobilized arsenic from the aquifer matrix.

The "main plume" extends onto an adjoining property, generally beneath an area of what was a distressed wetland overrun with nonnative invasive tree and understory species. A portion of the plume with concentrations of 1,4-dioxane greater than the Natural Attenuation Monitoring (NAM) default concentration, technically considered a source area, remains at the upgradient end of the plume located beneath a low, intermittently inundated area of native oak trees. Lithology within the area of the plume consists of approximately 5-8 ft of silty/sandy soil grading to a more silty layer. A low permeability, fractured limestone is beneath the silt to a depth of up to 12 ft. This is underlain by a tight calcareous clay.

In 2006 an extraction and treatment system was installed at the site to control migration of contaminants further downgradient, and to eventually reduce concentrations to levels that would allow site closure. The system consisted of groundwater extraction wells and an extraction trench, conventional air stripping, photochemical oxidation (ultraviolet light and peroxide), ion exchange, followed by discharge through an infiltration gallery.

The system was designed to operate at approximately 50 gallons per minute (GPM) and was initially effective at both hydraulic containment and mass removal. Low groundwater recovery rates and limits to volume that could be discharged, due to low hydraulic conductivity of the aquifer matrix, resulted in a much lower operation condition (10 GPM) that dramatically reduced the potential efficiency of the system. Mass removal rates had become asymptotic with contaminant concentrations remaining well above cleanup target requirements. Operation and maintenance (O&M) for this system had been in excess of $300,000 year^{-1}, and operation of this system would have been required for many years to reach the remedial requirements for this site.

The political and regulatory climate surrounding this site would not allow site closure or long-term monitoring options without some form of ongoing active remediation. A feasibility study was conducted to evaluate numerous alternatives that had potential for application to the site, and the TreeWell system was selected for further evaluation on the basis of:

1. A high probability of success under the site conditions;
2. The engineered approach is an active remedial alternative with a low projected O&M expense component (essentially landscape maintenance); and
3. It will remain an active system for the life of the trees.

Additional studies were then conducted to confirm the applicability of this technology, and to provide data for the engineering design. High-resolution sampling and lithology evaluation determined that the bulk of the hydraulic flow at the site is through the fractured rock, and that the contaminant impacts are within this zone, most likely due to back-diffusion from the underlying calcareous clay. Agronomic sampling indicated that soil conditions and chemistry would support the application of the TreeWell system. A groundwater flow model was also developed to evaluate the potential for hydraulic capture using variable numbers of TreeWell trees, anticipated evapotranspiration rates at different stages of growth, and different targeted extraction depths (Figure 9).

The resulting design included 154 TreeWell units spaced on 20-ft centers within a 2.5-acre portion of the property containing the distressed wetland. The wetland was initially cleared of the overgrowth of nonnative invasive species, and the TreeWell units were installed to target the depth corresponding to the fractured rock zone. The TreeWell units were then planted with native species adapted to the conditions at the site (slash pine, willow, sycamore, cypress), with inherent resistance to pests and diseases. The small "source area" was also isolated from the downgradient plume using an impermeable barrier wall, and additional TreeWell trees were installed within this area to supplement the existing oak trees.

Installation was completed in March 2013. The initial effects of the installation were seen within the first quarter following that installation and were well established by the end of the first year. Groundwater flow direction, previously to the west-northwest, has been altered in response to a hydraulic low created by the planting area, and now flow is coming into this area from all directions— downgradient flow has been reversed.

The TreeWell system is also removing contaminant mass. The IMW-10, a monitoring well within the midpoint of the main plume, had historically been approximately one order of magnitude above

FIGURE 9

TreeWell system at near Sarasota, Florida showing modification of groundwater flow regime - comparison from March 2013 to 2014. Evapotranspiration rates were dependent upon the stages of plant growth and different extraction depths.

the remedial goal for 1,4-dioxane of 3.2 μg/L. By the end of the first year, concentrations detected in this well had dropped below the remedial goal and have remained at this level. Concentrations of 1,4-dioxane in IMW-24R, located downgradient from the source and a few yards outside of the planting area, historically two orders of magnitude above the remedial target, have been reduced to less than 10 μg/L (Figures 10 and 11).

These trends have continued through the second year of "operation" and have demonstrated that:

1. Hydraulic capture has been achieved, and
2. Mass reduction is underway.

The effects seen in the first two seasons have been consistent with those predicted by the groundwater flow modeling. The initial planting used a species mix that was somewhat experimental to determine which species would do best under the site conditions that would also adapt to growth in the TreeWell system. A small percentage of the trees required replacement following the first growing season, but the planting is now established and should require little maintenance beyond weed control, occasional fertilization, and pruning.

The success of the TreeWell system enabled the Florida Department of Environmental Protection to issue a natural attenuation with monitoring order for the site and allowed shutdown of the remedial system in July 2014. Groundwater modeling has predicted that conditions at the site will allow a Risk-Based Conditional Closure by 2020 (or 7 years from installation).

The installation and operation of the interim groundwater pump and treat system was essentially mandated by the regulatory agency. As might be expected given the circumstances, the economics of the system were not optimal. Both the capital and operations costs were also significantly increased

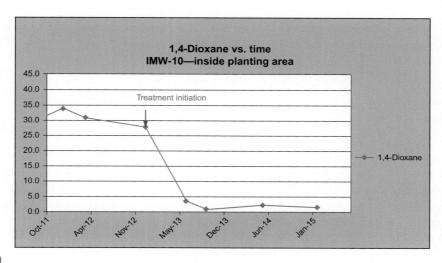

FIGURE 10

The TreeWell system is also removed 1,4-Dioxane (1,4-Dioxane vs time with in IMW-10, a monitoring well).

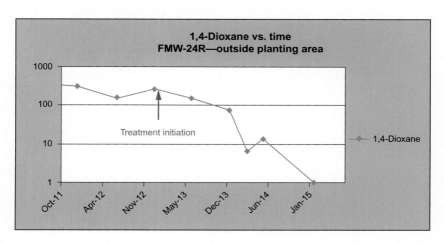

FIGURE 11

The TreeWell system is also removed 1,4-Dioxane (1,4-Dioxane vs time, outside planting area (FMW-24R).

by the requirement that all treated groundwater returned to the site infiltration galleries had to meet groundwater cleanup target levels (GCTLs; i.e., the drinking water standards). As such, naturally occurring compounds had to be treated as well as the constituents of concern. This required the installation of additional treatment media.

In terms of operational data, the interim pump and treat system operated for a period of approximately 8 years from 2006 through 2014. During this period the 1,4-dioxane mass (plume) was reduced by approximately 80% by the extraction (and treatment) of 8,540,547 gallons of groundwater (June 2006 through mid-July 2014). The average groundwater withdrawal rate of the system was 2883 Gallons per day (GDP) for the 2962 days of the operating period. Actual yearly averages are shown in Table 1.

Table 1 Interim pump and treat system operated for a period of approximately 8 years from 2006 through 2014. During this period the 1,4-dioxane mass (plume) was reduced by approximately 80%

Year	2006	2007	2008	2009	2010	2011	2012	2013	2014
Gallons per year	427,330	1,239,770	1,181,000	929,170	1,133,730	1,637,000	1,144,100	689,900	158,547
Gallons per day	2374	3397	3236	2546	3106	4485	3135	1890	672

Based on the extraction volume and the measured influent and treated effluent concentrations, the system removed 2.5 kg of 1,4-dioxane; 1.1 kg of arsenic; and 0.63 kg of CVOCs during operations. It is also estimated that between one and two pore volumes were extracted (between 4.3 and 7.0 million gallons) in the area of the 1,4-dioxane plume. Pore volume estimates were based on effective porosities of either 15% or 25%. It is worth noting that significantly lower extraction rates occurred post-2012.

System costs inclusive of design, construction, operation, and maintenance until system shutdown in July 2014 were $4.24 million. On a gallon-treated basis this equates to $0.50 per gallon. Average O&M costs for the period from 2007 through 2013 (full years of operations) were $314K year^{-1}. Figure 12 provides a summary of capital and O&M costs for the system operating period.

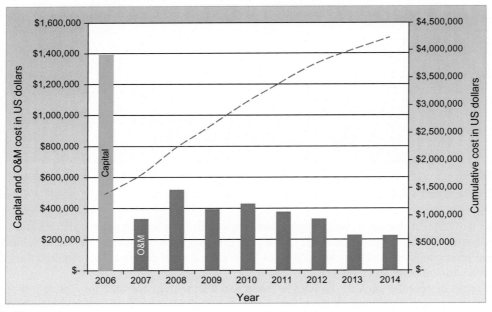

FIGURE 12

Summary of capital and Operation and Maintenance (O&M) costs in US $ for the system operating period.

The system was installed with the primary goal of achieving natural attenuation default concentrations (NADCs) in the groundwater. Once NADCs were achieved, and it could be demonstrated that no rebound occurred, the intent was to turn off the system and (hopefully) begin monitored natural attenuation (MNA). However, once NADCs were achieved, the length of time predicted to be required for MNA to remediate groundwater to the GCTLs became problematic, both in terms of cost and general acceptability to the agency. While by no means universally defined by the regulatory community, the generally accepted time period for MNA to achieve remediation goals is typically on the order of 5 years. In this case, in excess of 20 years was more likely.

A number of options were considered in the evaluation of the path forward. The goal of the effort was to select an option that would reduce the timeframe required for MNA or eliminate MNA entirely. Importantly, the estimated MNA 20 year timeframe served as the principal baseline for comparing the costs of possible remedies on a net present value (NPV) basis.

In the end, the enhancement of extraction infrastructure and continued operation of the interim system were selected to be compared to a designed and constructed phytoremediation system. In the case of the continued operation of the existing system it was assumed that after extraction enhancements were completed, the system would operate for a minimum of 2 years (based on pore volume removal) and MNA would follow. In the case of the designed and constructed phytoremediation, the performance of the system is expected to achieve GCTLs without the need for a period of MNA. In simple terms, the capital cost of enhancing the existing system combined with the anticipated 2-year minimum operating timeframe was comparable to the cost of installation of a designed and constructed phytoremediation system. Therefore, the O&M cost of the designed and constructed phytoremediation system was able to be directly compared to the O&M cost of MNA.

The designed and constructed phytoremediation system is expected to achieve GCTLs in 7-12 years following implementation. The range in timeframe is based on the predicted pore volume extraction rate. Figure 13 provides the comparison of the NPV cost of designed and constructed phytoremediation at 7 and 12 years after planting with the NPV cost of 20 years of MNA (note: inflation assumed at 3%). As can be seen in Figure 13, the anticipated completion of phytoremediation in 2020 results in a NPV cost of $636K. This compares with the estimated NPV cost of MNA of $1432K.

There were a number of clear advantages to the implementation of the designed and constructed phytoremediation system. Besides providing a broader groundwater capture zone than the "enhanced" existing system option, the system, as designed, outperforms the extraction rates achieved by the pump and treat system. Based on the current groundwater elevation contours, the installed phytoremediation system has already outperformed the previous system within the first 2 years.

For illustrative purposes, it is worth evaluating how the designed and constructed phytoremediation system would have performed if installed in 2006 instead of the interim pump and treat system. Figure 14 has been prepared to provide a comparison to the capital and O&M costs presented in Figure 14 for the interim pump and treat system. Figure 14 utilizes actual capital costs for the installation of the designed and constructed phytoremediation system as well as current and predicted O&M costs.

Based on the current groundwater data, we know that the designed and constructed phytoremediation system is conservatively capable of achieving withdrawal of one pore volume (based on original plume size) in 1-2 years once the trees have reached an age of 2 years. If a 2-year startup period is allowed for establishment of the trees, then it can be assumed that the system would be able to achieve the same level of withdrawal (i.e., one to two pore volumes) in an additional 2-4 years as compared to the 8 years that was required of the pump and treat system. Therefore, the phytoremediation system

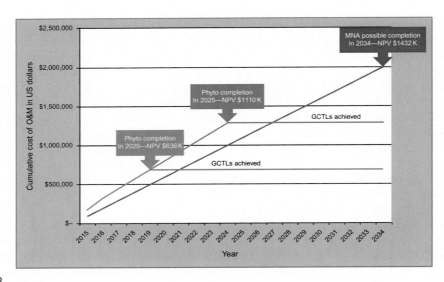

FIGURE 13

Comparison of the net present value (NPV) cost of designed and constructed phytoremediation at 7 and 12 years after planting with the NPV cost of 20 years of monitored natural attenuation (MNA) (note: inflation assumed at 3%).

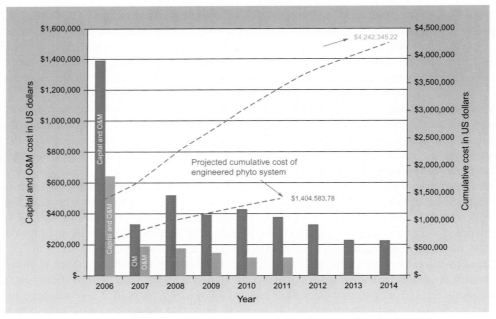

FIGURE 14

Projected cumulative cost of engineered phytosystem for 2006–2014.

would have been able to achieve the same level of cleanup in 4-6 years and at substantially lower annual O&M costs. Figures 12–14 demonstrates that a potential savings on the order of $2.83 million could have been realized if a designed and constructed phytoremediation system was implemented in 2006 instead of the pump and treat system.

REFERENCES

Baker, A.J.M., Reeves, R.D., McGrath, S.P., 1991. In situ decontamination of heavy metal polluted soils using crops of metal-accumulating plants—a feasibility study. In: Hinchee, R.E., Oflenbuttel, R.F. (Eds.), In Situ Bioreclamation: Applications and Investigations for Hydrocarbon and Contaminated Site Remediation. Battelle Memorial Institute, Columbus, OH, pp. 600–605.

Banks, M.K., Schwab, A.P., Govindaraju, R.S., Chen, Z., 1994. Abstract: bioremediation of petroleum contaminated soil using vegetation—a technology transfer project. In: Erickson, L.E., Tillison, D.L., Grant, S.C., McDonald, J.P. (Eds.), Proceedings of the 9th Annual Conference on Hazardous Waste Remediation, June 8-10, 1994, Bozeman, MT. Montana State University, Bozeman, MT, p. 264.

Banuelos, G.S., 1994. Extended abstract: managing high levels of B and Se with trace element accumulator crops. Symposium, In: Tedder, D.W., American Chemical Society. Division of Industrial and Engineering Chemistry (Eds.), In: Emerging Technologies in Hazardous Waste Management VI, 1994 Book of Abstracts, September 19-21, 1994, Atlanta, GA, vol. 2. American Chemical Society, Washington, DC, pp. 1344–1347.

Burken, J.G., Schnoor, J.L., 1996. Phytoremediation: plant uptake of atrazine and role of plant exudates. J. Environ. Eng. 122, 958–963.

Chaney, R.L., Malik, M., Li, Y.M., Brown, S.L., Brewer, E.P., Angle, J.S., Baker, A.J.M., 1997. Phytoremediation of soil metals. Curr. Opin. Biotechnol. 8 (3), 279.

Erickson, L.E., Banks, M.K., Davis, L.C., Schwab, A.P., Muralidharan, N., Reilley, K., Tracy, J.C., 1994. Using vegetation to enhance in situ bioremediation. Environ. Prog. 13, 226–231.

Ferro, A., Kennedy, J., Nelson, S., Jauregui, G., McFarland, B., Doucette, W., Bugbee, B., 1996. Uptake and biodegradation of volatile petroleum hydrocarbons in planted systems. In: International Phytoremediation Conference, May 8-10, 1996, Arlington, VA. International Business Communications, Southborough, MA.

Fletcher, J.S., Donnelly, P.K., Hegde, R.S., 1995. Plant assisted PCB degradation. In: Proceedings/Abstracts of the Fourteenth Annual Symposium, Current Topics in Plant Biochemistry, Physiology, and Molecular Biology— Will Plants Have a Role in Bioremediation?, April 19-22, 1995, Columbia, MO. Interdisciplinary Plant Group, University of Missouri, Columbia, MO, pp. 42–43.

Gatliff, E.G., 1997. Making the industry connection—considerations and justifications for the commercial utilization of phytoremediation. In: IBC's Second Annual Conference on Phytoremediation, June 18-19, 1997, Seattle, WA. International Business Communications, Southborough, MA.

Jordahl, J.L., Licht, L.A., Schnoor, J.L., 1995. Poster abstract: riparian poplar tree buffer impact on agricultural non-point source pollution. In: Erickson, L.E., Tillison, D.L., Grant, S.C., McDonald, J.P. (Eds.), Proceedings of the 10th Annual Conference on Hazardous Waste Research, May 23-24, 1995, Manhattan, KS, p. 238.

McCutcheon, S.C., 1996. Abstract: ecological engineering: new roles for phytotransformation and biochemistry. In: International Phytoremediation Conference, May 8-10, 1996, Arlington, VA. International Business Communications, Southborough, MA.

Negri, M.C., Hinchman, R.R., Gatliff, E.G., 1996. Phytoremediation: using green plants to clean up contaminated soil, groundwater, and wastewater. In: International Phytoremediation Conference, May 8-10, 1996, Arlington, VA. International Business Communications, Southborough, MA.

Rock, S., 1996. Phytoremediation of organic compounds: mechanisms of action and target contaminants. In: Kovalick, W.W., Olexsey, R. (Eds.), Workshop on Phytoremediation of Organic Wastes, December 17-19, 1996, Ft. Worth, TX. An RTDF meeting summary.

Schnoor, J.L., Licht, L.A., 1991. Deep-rooted poplar trees as an innovative treatment technology for pesticide and toxic organics removal from groundwater. In: Program Summary FY 1991, Hazardous Substance Research Centers Program. USEPA 21R-1005.

Thomas, P.R., Buck, J.K., 1999. Agronomic management for phytoremediation. In: Leeson, A., Alleman, B.C. (Eds.), Phytoremediation and Innovative Strategies for Specialized Remedial Applications. Battelle Press, Columbus, OH.

PHYTOMANAGEMENT OF POLYCYCLIC AROMATIC HYDROCARBONS AND HEAVY METALS-CONTAMINATED SITES IN ASSAM, NORTH EASTERN STATE OF INDIA, FOR BOOSTING BIOECONOMY

H. Sarma[1], M.N.V. Prasad[2]

N N Saikia College, Titabar, Assam, India[1]

University of Hyderabad, Hyderabad, Telangana, India[2]

1 INTRODUCTION

The Upper Brahmaputra Valley Agro Climatic Zone of Assam State in India covers an area of 16,013 km^2 encompassing five districts: Golaghat, Jorhat, Sivasagar, Dibrugarh, and Tinsukia (Figure 1). This valley is ideally suited for rice and tea cultivation, and these crops make a significant contribution to India's national economy. It was in the upper Brahmaputra Valley where crude oil was discovered in 1867 at Digboi and the first oil well in Asia was drilled. Since then, several oil wells have been drilled in various parts of upper Assam. Most oil-drilling sites are located at the fringes of human settlements with paddy fields and tea gardens, resulting in hydrocarbon contamination. Hydrocarbon-contaminated soil from oil exploration activities creates a threat to crop production and has a detrimental effect on the bioeconomic system by making land unsuitable for agriculture and other economic purposes. This problem is of particular concern in India, the second most populated country in the world, where agriculture is the primary economic base to feed the people. Traditional methods for soil remediation are often expensive due to mass involvement of labor, energy, and infrastructure (Jeremy et al., 2015). As a result, *in situ* phytotechnology has been developed as a cheaper alternative to conventional methods in areas where remediation would otherwise not be put into practice. In the last few decades many research projects have explored the role of beneficial plants and microbes in bioremediation practices (Frenzel et al., 2010; Mazzeo et al., 2010; Bacosa and Inoue, 2015). However, despite encouraging

Bioremediation and Bioeconomy. http://dx.doi.org/10.1016/B978-0-12-802830-8.00024-1
Copyright © 2016 Elsevier Inc. All rights reserved.

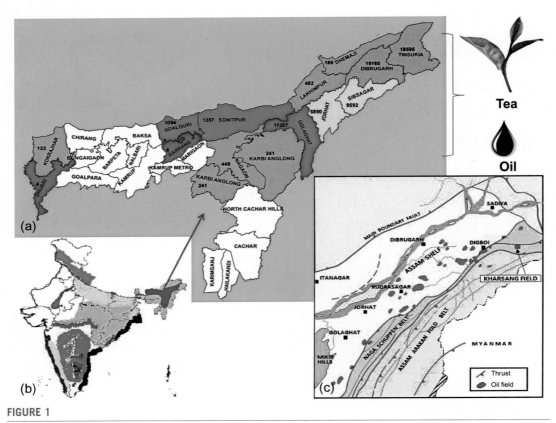

FIGURE 1

(a) Northeastern part India showing Assam State (red colour) which is rich in tea, rice, and crude oil; (b) Indian subcontinent with 12 distinct ecoregions showing the northeastern part, which is a treasurehouse of nature's capital (biotic and abiotic) resources; (c) locations of oil fields in upper Assam. Rice paddies and tea plantations are the most commonly cultivated for bioeconomy in the region.

results in the greenhouse, the practice of *in situ* bioremediation is limited. There are many reasons such as low soil temperature, varying pH levels, lack of synergy between plants-microbe association, type of contaminates and their concentration, soil properties, and so on. Although many technical issues exist with this phytotechnology, many reports have been published currently showing that by using proper high-end technology, degraded land mass can be made fertile again, which will boost the nation's bio-economy. The objective of this study is to provide an inventory of some plant-microbe consortia that can potentially be used for *in situ* bioremediation of crude oil-contaminated soil in the tropical environment of Assam. Emphasis has also been made on developing efficient methods accessible to economically marginalized small-scale tea growers and paddy farmers in the state.

The majority of the rural peoples live on less than US $1 a day in Asia (World Bank, 2001, 2004; FAO, 2005; CIFOR, 2002). These people sustain their livelihoods from the forests through land-extensive cultivation, logging, and exploitation of nonwood products, and sometimes these activities degrade the natural environment (Sarma and Sarma, 2008). It has been estimated that two-thirds of the world's ecosystems are shrinking and are considered degraded as a result of indiscriminate use of renewable and nonrenewable resources, mismanagement, contamination, and failure to look after these resources (Nellemann and Corcoran, 2010). India's recognition as one of the four "mega-diversity" countries of Asia was derived largely from three of its most important biodiversity "hotspots": the Himalayas, Indo Burma, and Western Ghats. Assam is one of the critical priority biodiversity hotspots of India under Himalayan and Indo-Burma hotspots. The major forest types of Assam have received international conservation efforts, based on the levels of endemism, medicinal plant diversity, and human threat. The total wildlife protected areas consist of 3925 km^2 out of the total 28,748 km^2 forest area of the state and comprise 18 wildlife sanctuaries and 5 national parks (Sarma et al., 2011). The northeastern region of India forms a unique biogeographic province comprising major biomes recognized in the world; it harbors about 50% ($\pm$8500 sp) of the floristic wealth of India and 40% of them are endemic (Mao et al., 2009). This region is regarded as the place of origin of progenitors of many cultivated crops.

The Upper Brahmaputra Valley Agro Climatic Zone is a land rich in both bioresources and natural resources, especially petroleum, natural gases and coal, tea, and rice. Crude oil drilling, mining of coal, deforestation, floods, etc., have severely degraded the land ecosystem in this zone and have had a negative impact on bioresources; that is, tea plantation and rice cultivation. It has been recorded that coal mining and crude oil drilling are the major causes in the transformation of fertile, cultivable land into wasteland, as crude oil exploration and mining activity generate a vast quantity of solid waste (Bora et al., 2011) that is deposited on the surface and occupies huge areas of land (Table 1).

In upper Assam, while many hectares of land have been transformed into wasteland due to drilling of crude oil and coal mining, there is still much potential for agriculture as evidenced by the number of tea gardens and rice fields. The available data show the annual average rice production in this zone estimated at 1637 kg ha^{-1} (Ahmed et al., 2014). Recently in rural Assam farmers have also started planting tea in unutilized and underutilized uplands and thus have brought about huge socioeconomic changes in Assam. In the last two decades, the number of small tea growers has swelled to an impressive number of 65,000 (http://aastga.org). A small tea grower means one whose land under tea cultivation does not exceed 10 ha. Almost 9,00000 people are engaged directly and indirectly with tea cultivation. Around 2.5 lakh hectares of land has been covered by small-scale tea plantations. The contribution of small tea growers is about 29% of the total tea produced by Assam, which is ~14% of the total tea production of India (http://aastga.org). Most of the farmers have typical homesteads where they cultivate vegetables, horticulture crops, trees, and bamboo. This zone is also suitable for the cultivation of other plantation crops such as areca nuts, coconuts, bananas, lemons, etc. The soil type in this zone is mostly sandy loam. The total number of farmers and farm families living in this zone is around 581,014 as per the Agriculture Census 2005–2006 (Tables 2 and 3).

Today the need for cultivation in industrial contaminated sites using bioremediation is gaining attention since it is an important task to feed the growing populations in many developing countries of the world like India and China. Such wasteland utilization would not only be economical, but may also result in foreign exchange earnings and control of environmental pollution. But it will be always a matter of dispute whether crops cultivated in contaminated areas are safe for human consumption, since most plants have capacities to uptake metals and polycyclic aromatic hydrocarbons (PAHs) from contaminated soil and therefore there is a risk of these pollutants entering the food chain. As such, proper care should be taken during cultivation in such contaminated sites.

The sites where accidental seepage occurs become desertlike in nature and disturb the rice ecosystem (Figure 2). These activities directly lead to fragmented landscapes, loss of cultivated land, and imbalances in beneficial soil microbes finally resulting in the overall loss of crop production and

Table 1 Land Utilization Statistics (Area in Hectares) of Upper Brahmaputra Valley Agro Climatic Zone

District	Geo-graphical Area	Forest	Area Not Available for Culti-vation	Permanent Pastures and Other Grazing Land	Land Under Miscell-aneous Trees, Groves, etc.	Culti-vable Waste-land	Fallow Land	Net Cropped Area	Gross Cropped Area	Area Sown more than once	Cropping Intensity (%)
Jorhat	851,000	21,904	110,567	4406	9024	6686	12,273	120,240	173,020	52,780	144
Golaghat	350,200	156,905	51,232	8314	8217	5801	4555	119,046	180,097	61,051	151
Sivasagar	266,800	30,465	56,151	7330	20,061	1820	7164	136,822	155,062	18,220	113
Tinsukia	379,000	131,595	114,883	3560	19,786	1586	2876	104,714	147,936	43,222	141
Dibrugarh	338,100	23,341	142,488	6170	16,883	7126	3276	139,498	158,917	19,419	114

Source: Statistical Hand Book, Assam, 2011.

Table 2 Crop Profile of the Upper Brahmaputra Valley Zone

Agro Climatic Zone	District	Major Field Crops Cultivated	Horticulture Crops-Fruits	Plantation Crops
Upper Brahmaputra Valley Zone	Jorhat	Winter rice, Blackgram	Banana	Areca nut, Tea
	Sivasagar	Winter rice, Autumn rice, Summer rice	Banana, Orange, Pineapple	Tea, Areca nut, Coconut
	Golaghat	Terrace rice cultivation, Wet rice cultivation, Maize	Banana, Lemon, Pineapple	Tea, Areca nut, Coconut
	Dibrugarh	Winter paddy, Autumn paddy	Areca nut, Banana, Assam lemon	Tea, Black pepper
	Tinsukia	Paddy, Maize, Blackgram	Banana, Khasi mandarin, Pineapple	Tea, Areca nut, Coconut

Table 3 Breakdown by District of Small Tea Gardens, Area Under Paddy Cultivation, and Farmers/Farming Families

District	Small Tea Gardens	Area Under Paddy ('000 ha)	Farmers/Farming Families
Jorhat	5890	83.10	194,423
Sivasagar	9592	107.18	113,440
Golaghat	11,287	11.036	78,189
Dibrugarh	19,160	77.462	93,605
Tinsukia	18,595	68.434	101,357

Source: Statistical Hand Book, Assam, 2011.

economic loss. The indirect effects can be loss of the water-holding capacity of soil, imbalances in physiochemical properties, discharge of carcinogenic pollutants to ambient air and perennial water sources, loss of biodiversity, and ultimately loss of economic wealth (Bora et al., 2011).

In a tropical country like India, beneficial plant microbe technology to decontaminate PAH-infiltrated soil is feasible because of favorable soil temperatures; thus *in situ* bioremediation can potentially be an appropriate technology for this region. This is especially true for Assam, which has substantial bioresources, but where remediation in crude oil-contaminated areas (of little economic value) cannot be implemented with more expensive methods.

2 OIL SPILLS: CAUSES AND CONCERNS

Crude oil exploration by the petroleum industry and its transport contaminate the environment with PAHs and heavy metals due to oil spillage and leakage from pipelines, refining process, oil field installations, petroleum plants, liquid fuel distribution and storage devices, transportation equipment for petroleum products, airports, and illegal drillings in pipelines (Tiwari et al., 2011; Roy et al., 2013;

FIGURE 2

(a) Crude oil exploration in Borhola oil field (Jorhat), (b) crude oil seepage through Namti drilling site (Sivasagar), (c) rice field contamination near Borhola group (Jorhat), and (d) crude oil spills in tea garden causing oil contamination.

Auta et al., 2014). These spills, leaks, and other releases of petroleum often result in large-scale contamination of soil and groundwater (Gargouri et al., 2013; Figure 2).

Crude oil is a naturally occurring composite organic liquid with natural gas present in underground reserves formed millions of years ago. Chemically it consists of saturated noncyclic hydrocarbons, cyclic hydrocarbons, alkenes, aromatics, sulfur compounds, nitrogen-oxygen compounds, and heavy metals (Cote, 1976), and the refined products of crude oil such as petrol, diesel, etc., are major sources of energy. The refining process of crude oil generates huge quantities of toxic and persistent biodegradable pollutants like PAHs and nonbiodegradable pollutants like heavy metals (chromium, lead, nickel, cadmium, cobalt, etc.). The refining process also discharges other hazardous compounds like phenols (cresols and xylenols), sulfides, ammonia, suspended solids, cyanides, and nitrogen compounds that contaminate the ecosystem (Tiwari et al., 2011; Cote, 1976). Of all these contaminants, the presence of PAHs and heavy metals is most disturbing due to their toxic effects on living organisms as they contain mutagenic and carcinogenic properties (Vagi et al., 2005). These pollutants have considerable environmental impacts; they present substantial hazards to all living beings and destroy the soil health, which may take years or even decades to restore.

Several researches have been done on ways to reduce ecological threats of crude oil contamination and its risks to living organisms and ecological systems. Biodegradation of hydrocarbons by microorganisms has been recognized as one leading, effective, and environment-friendly method by which crude oil is eliminated from contaminated sites (Harayama et al., 2004). This is acceptable when the contamination occurs in noncultivated land but it becomes a serious issue when contaminants spread to cultivated areas. There are several instances where crude oil influx in tea gardens and paddy fields has caused considerable damage. In general, crude oil contamination reduces plant growth in various ways and also affects their biochemical and physiological parameters (Ogbo et al., 2009; Omosun et al., 2008). These damaged areas become unsuitable for cultivation for long periods unless they are remediated through proper methods as was seen in the course of our visit (Figure 3).

It is evident that during the transportation process of crude oil from drilling sites, high pressure generated in the pipeline causes leakage; this seepage into the adjacent soil and water ecosystem is detrimental to the surrounding crops, especially tea and rice. Furthermore, contamination might also take place due to seepage of crude oil from effluent pits and from group gathering stations and oil-collecting stations where crude oil is stored for refining purposes.

FIGURE 3

(a–c). Visible oil spills in rice paddies. Rice (*Oryza sativa*) being a grass, its rhizosphere is involved in degradation of PAH (see also Chapter 11 in this book).

During the course of our visit it was observed that almost invariably the tea growers and rice cultivators have not been keeping any records about their economic loss due to crude oil contamination in the periphery of oil exploration sites. Besides, most of the farmers are also not aware of the importance of soil testing as they have very little knowledge of scientific practices of growing tea. In addition to that, on quite a few occasions it has also been noticed that farmers have resorted to indiscriminate use of pesticides either in sublethal or in overdose (Ajeigbe et al., 2012). This is a major concern for the bioeconomy in this zone, and in order to mitigate the aforesaid shortcoming agricultural extension activities and awareness programs are urgently needed. In fact, the issue of persistent organic pollutant residues is very dynamic in nature, and hence tea and rice growers need to be trained periodically about safe cultivation practices including health hazards of pesticides. Otherwise the tea produced in this region will fail quality control tests and remain unsold in international markets, which will cause significant economic loss. In this situation the combined application of native plants and microbes has tremendous opportunity to decontaminate the cultivated areas. This is all the more so because bacterial strains able to degrade crude oil are known to be ubiquitous in nature since they use these contaminants as a carbon source (Through a process called microbial catabolisim). For this reason, biodegradation of crude oil by bacteria is considered a useful tool in the reclamation of oil spill areas (Masakorala et al., 2013).

3 BENEFICIAL PLANT-MICROBE INTERACTION IN BIODEGRADATION

Organic contaminants like PAHs can be removed from the soil through adsorption, volatilization, photolysis, and chemical degradation, but these are expensive processes (Lu et al., 2014). On the other hand, plant-microbial degradation is considered as one of the prospective low-cost removal procedures (Yuan et al., 2002). Plants have a range of cellular mechanisms that make them capable of resisting high concentrations of organic pollutants without displaying their toxic effects; such plants often accumulate and convert these organics into less toxic metabolites. Furthermore, the release of root exudates and enzymes in rhizosphere in the presence of organic carbon in the soil stimulates faster degradation of organic contaminants. On the other hand, for remediation of metal contaminants, some plants have shown the potential role for phytoextraction (contaminants uptake and accumulation into above-ground biomass through xylem-conducting system), rhizofiltration (metals filtering from water through root systems), phytostabilization (stabilizing the contaminants by erosion control). *Axonopus compressus* (Sw.) P. Beauv (Bordoloi et al., 2012), *Cyperus brevifolius* (Rottb.) Hassk (Basumatary et al., 2012), *Cyperus rotundus* Linn. (Basumatary et al., 2013) have already been identified as model plants showing optimum performance when planted in the hydrocarbon-contaminated soil of upper Assam. Figure 4 shows a photo of a hyperaccumulator (*Cyperus odoratus* L.) along with an illustration depicting the process of phytoremediation.

Researchers have looked into more efficient ways of bioremediation though the results attained to date leave much to be desired. As of now, it has been seen that due to the lesser bioavailability of PAHs only a few of them could be adsorbed by the plants. Many researchers have therefore explored microbe-based technologies and have made progress in PAH degradation studies (Shin et al., 1999; Juhasz and Naidu, 2000). Both beneficial plants and microbes capable of degrading PAHs were isolated and characterized from natural environment in recent times as shown in Table 4.

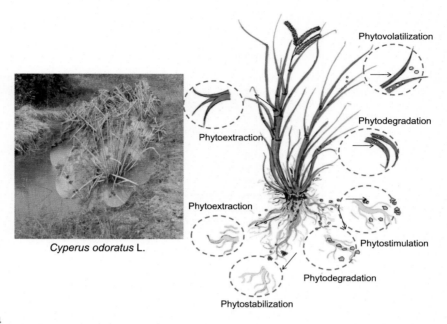

Cyperus odoratus L.

FIGURE 4

Grass (*Oryza sativa*) and sedge (*Cyperus rotundus*, a weed in paddy field) association is beneficial for biodegradation of PAH associated with microbial consortia (Sarma and Prasad, 2015).

Table 4 Beneficial Plant-Microbe Interactions for Degradation of PAHs and Heavy Metals

Plants	Bacterial Strain	Names of PAHs/HM	References
1. *Lolium perenne, Bouteloua gracilis Artemisia frigida, Banksia integrifolia, Lupinus albus, Lupinus luteus, Bidens pilosa, Alternanthera ficoidea, Taraxacum officinale, Aus deflexu, Phaseolus aureus, Triticum durum, Camellia sinensis, Morus rubra*, and *Avena barbata*	*Pseudomonas* sp., *Rhizobium* sp., *Arthrobacter* sp., *Nocardia* sp., *Streptomyces* sp., *Burkholderia* sp.	Phenanthrene and α-pyrene	Chaudhry et al. (2005)
2. *Viola baoshanensis, Sedum alfredii, Rumex crispus, Helianthus annus, Anthyllis vulneraria, Festuca arvernensis, Koeleria vallesiana, Armeria arenaria*, and *Lupinus albus*	*Pseudomonas putida, Pseudomonas fluorescens, Chlorella vulgaris, Methylobacterium oryzae, Berknolderia* sp., *Pseudomonas aeruginosa, Citrobacter* sp., *Zooglea* sp., *Arthrobacter* sp., *Ochrobactrum* sp., *Serratia* sp., *Bacillus subtilis*	Naphthalene, anthracene, fluoranthene, pyrene, benzo(a)pyrene, As, Zn, Cu, Ni, Cr, Cd, Hg, and Pb	Shukla et al. (2010)

(Continued)

Table 4   Beneficial Pant-Microbe Interactions for Degradation of PAHs and Heavy Metals—cont'd

Plants	Bacterial Strain	Names of PAHs/HM	References
3. *Bouteloua gracilis, Cyanodon dactylon, Elymus canadensis, Festuca arundinacea, Festuca rubra,* and *Melilotus officinalis*	*Mycobacterium* sp., *Haemophilus* sp., *Rhodococcus* sp., *Paenibacillus* sp., *Sphingomonas paucimobilis, Agrobacterium* sp., *Burkholderia* sp., *Rhodococcus* sp., and *Mycobacterium* sp.	Benzo(a)pyrene, benzo[k]fluoranthene, anthracene, benzo[b] fluoranthene, benzo(e) pyrene, fluoranthene, naphthalene, phenanthrene, and benzo[ghi] pyrene	Haritash and Kaushik (2009)
4. *Bromus hordeaceous, Festuca arundinacea, Trifolium fragiferum, Trifolium hirtum, Vulpia microstachys, Bromus carinatus, Elymus glaucus, Festuca ruba, Hordeum californicum, Leymus triticoides,* and *Nassella pulchra*	*Pseudomonas oleovorans, Aquaspirillum* sp., *Flavobacterium indologenes, Pseudomonas* sp., and *Burkholderia* sp.	Hexadecane, naphthalene, and phenanthrene	Siciliano et al. (2003)
5. *Cordia subcordata, Thespesia populnea, Prosopis pallida, Scaevola serica,* and *Medicago sativa*	*Pseudomonas fluorescens, P. aeruginosa, Bacillus subtilis, Bacillus* sp., *Alcaligenes* sp., *Acinetobacter lwoffi,* and *Flavobacterium* sp.	Naphthalene	Das and Chandran (2011)
6. *Sinapis alba, Lepidium sativum,* and *Sorghum saccharatum*	*Acinetobacter* sp., *Acinetobacter iwoffii, Actynomices* sp., *Actynomices viscosus, Agrobacterium radiobacter,* and *Alcaligenes faecalis*	Benzo(a)pyrene	Coccia et al. (2009)
7. *Nymphaea pubescens, Typha* sp., *Juncus effusus, Phragmites australis,* and *Schoenoplectus validus*	*Acinetobacter* sp., *Alcaligens* sp., *Listeria* sp., *Staphylococcus* sp., *Acinetobacter* sp., *Alcaligens* sp., and *Listeria* sp.	Cu, Zn, Pb, Cd, and Fe	Kabeer et al. (2014)
8. *Brassica napus*	*Micromonospora* sp., *Bacillus* sp., *Arthrobacter* sp., *Leifsonia* sp., *Staphylococcus* sp.	Na, Mg, K, Fe, Cu, Zn, Cd, and Pb	Croes et al. (2013)

4 POTENTIAL MICROBES BOOSTING BIOECONOMY IN RICE AND TEA

The tea industry is the largest agroindustry in Assam and plays a predominant role in its economy. Assam alone produces more than 50% of the country's tea. Assam tea is considered to be the finest tea, with high demand in the global market. The Upper Brahmaputra Valley produces 15% of the world's tea, which is higher than any other tea-producing countries like Kenya and Sri Lanka. Again, crude oil production in Assam, India, is a century-old process; therefore, the environmental contamination due to crude oil exploration is highly alarming (Roy et al., 2013). Juxtaposition of crude oil exploration sites with rice fields or tea plantations is a predominant feature in upper Assam. Therefore, efforts have been made by various R&D laboratories in recent times to develop a feasible bioremediation technique for ecorestoration of crude oil-contaminated areas of Assam. Recently, the Northeast Institute of Science and Technology, Jorhat, which is a research laboratory of India's premier national R&D organization, the Council of Scientific & Industrial Research (CSIR), has made tremendous contributions toward ecorestoration of degraded land through the application of various bioformulations consisting of native strains of bacteria along with plantation of hyperaccumulating species (Bora et al., 2011). Despite these efforts many areas are still contaminated, and there is a need to develop proper policy making by the government and generate awareness among the farmers. Specific efficient plant species have already been used in trial studies for remediation of land degraded by crude oil in many parts of upper Assam's tea gardens and rice fields, which absorb, accumulate, and detoxify PAHs and heavy metals (Bora et al., 2011; Sarma and Prasad, 2015). It is evident that certain plants have capacities to stabilize hydrocarbon-polluted soils and to stimulate soil microbes/microbial consortia in the rhizosphere by releasing root exudates (Table 5). For boosting bioeconomy and withstanding the toxic effects of hydrocarbons, particularly for rice and tea grown in crude oil exploration sites, it becomes necessary to isolate some potential microbes for pilot-scale bioremediation application.

Table 5 Potential PAH Degrading Bacterial Strains

Location	Bacterial Strain	Names of PAHs	References
1. Shenfu Irrigation Area, Liaoning Province, China	*Mycobacterium* sp., *Pseudomonas* sp., *Sphingomonas* sp., and *Rhodococcus* sp.	Anthracene, fluoranthene, benz(a)anthracene, phenanthrene, and pyrene	Li et al. (2007)
2. Tehran Oil Refinery Site, Persian Gulf coasts	*Pseudomonas pudita, Pseudomonas fluorescence, Serratia liquefaciens*, and *Micrococcus strains*	Phenanthrene, benzo[a] pyrene, benzo[a] anthracene, and chrysene	Mohsen et al. (2009)
3. San Diego Bay, California and Central Pacific Ocean	*Pseudomonas* sp., *Rhodococcus* sp., *Mycobacterium* sp., *Burkholderia cocovenenas, Sphingomonas paucimobilis, Pseudomonas fredrikbergensis*, and *Pseudomonas fluorescens*	BTEX, benzo[a] pyrene, phenanthrene, naphthalene, fluoranthene, and benzo[a]anthracene	Bamforth and Singleton (2005)
4. Jubany station, Argentina	*Pseudomonas aeruginosa*	Phenanthrene	Ruberto et al. (2006)
5. Haldia Refinery site of India	*Bacillus weihenstephansis, Bacillus anthracis, Bacillus mycoides*, and *Bacillus thuringiensis*	Phenanthrene, anthracene, benzo(a)pyrene, and fluoranthene	Maiti and Bhattacharyya (2012)

(Continued)

Table 5 Potential PAH Degrading Bacterial Strains—cont'd

Location	Bacterial Strain	Names of PAHs	References
6. Beijing Coking Plant	*Acinetobacter* sp. and *Rhodococcus ruber*	Naphthalene, acenaphtylene, acenaphthene, fluorene, phenanthrene, anthracene, fluoranthene, pyrene, benz[a]anthracene, chrysene, benzo[b] fluoranthene, benzo[k] fluoranthene, benzo[a] pyrene, dibenzo[a,h] anthracene, benzo[ghi] perylene, and indeno[1,2,3-cd]pyrene	Sun et al. (2012)
7. Gas plant site in Australia	*Alcaligenes* sp., *Paenibacillus* sp., *Escherichia coli*, *Pseudomonas* sp., *Pandorea* sp., *Pseudomonas putida*, *Burkholderia* sp., *Burkholderia cepaciadegrades*, *Mycobacterium* sp., *Sphingomonas maltophilia*	Anthracene, phenanthrene, fluoranthene, pyrene, benzo(a)pyrene, chrysene, benzo[b]fluoranthene, benzo[k]fluoranthene, benzo[a]pyrene, chrysene, acenapthene, acenapthylene, indeno[1, 2, 3-cd]pyrene, dibenz[ah] anthracene, and benzo[ghi] perylene	Palanisami et al. (2012)
8. Southern Illinois, USA	*Mycobacterium* sp. and *Xanthomonas ampelina*	Pyrene, benzo[a]pyrene, anthracene, and benz[a] anthracene	Grosser et al. (1991)
9. Oil and Natural Gas Commission of India (ONGC) oil field, Assam	*Bacillus subtilis*, *Pseudomonas aeruginosa*	Pyrene, asphaltene, phenanthrene (PHE), and anthracene	Das and Mukherjee (2007)

The data available, particularly in the oil exploration sites, tea gardens and rice fields (Lakuwa, Geleky, Amguri, and Borhola), show that as many as 39 native crude oil-degrading bacteria have been isolated by Ray et al. (2014) that have a capacity to degrade PAHs. Furthermore, the researchers confirmed that soil quality was also improved through earthworm mortality bioassay and plant tests on rice (*Oryza sativa*) and mung (*Vigna radiata*). These findings have potential for tea and rice farmers and we can conclude that the combined use of crude oil-degrading bacteria along with nutrient supplements could revive crude oil contaminated-soil effectively on a large scale, which is not only a cheaper but also an environmentally friendly process. Soil contaminated by crude oil is a good habitat for potent hydrocarbon degraders of the genus *Lysinibacillus*, *Brevibacillus*, *Bacillus*, *Paenibacillus*, *Stenotrophomonas*, *Alcaligenes*, *Achromobacter*, and *Pseudomonas* strain. These bacteria singly and in consortia might have contributed to improve the quality of hydrocarbon-contaminated soil, which is supported by studies conducted on this aspect.

The scientists of the Northeast Institute of Science and Technology, Jorhat, and the Institute of Advanced Study in Science and Technology (IASST), Guwahati, an autonomous institute under the Department of Science and Technology, Government of India, have been actively engaged in bioremediation of crude oil-contaminated sites in upper Assam oil fields owned by the Oil and Natural Gas Corporation Limited (ONGC) through a CSIR-ONGC joint venture project. The level of heavy metals in crude oil-contaminated soil in some areas of Sivasagar district, Assam, have attained alarming concentrations, such as arsenic (2.43 ppm), cadmium (4.75 ppm), chromium (7.72), mercury (10.63), and lead (7.98), and these elevated concentrations have been successfully mitigated by application of bioformulations developed by NEIST Jorhat (Bora et al., 2011). It further reported that the number of beneficial microbes in soil (e.g., phosphorous solubilizers, sulfur oxidizers, and cellulose degraders) in crude oil-contaminated soil, which showed a decrease due to contamination, showed an increase during and after remediation. A highly alkaline condition of the soil (pH 10.0–10.5) was reported in their study, which might be due to the low presence of these beneficial microbes. NEIST-Jorhat has successfully reclaimed six sites in ONGC oil fields in upper Assam using technology developed by this institute for ecorestoration of crude oil-contaminated sites (Bora et al., 2011). Several crude oil-drilling sites, oil wells located in and around tea plantation areas and paddy fields, are common features in the landscape of Assam. Understandably, crude oil contamination in such tea gardens and fields is very high due to spillage, pipeline leakages, etc., and these affect the soil's physical and biological properties. This reduces the growth and resistance of the plants to biotic and abiotic factors, making them more vulnerable to pathogen infestation (Udo and Fayemi, 1995; De Jong, 1980; Schutzendubel and Polle, 2002). A few consortia are discussed here (see also Table 6) that can be used in agricultural fields as well as in tea gardens to fight crude oil contamination.

Table 6 Consortia That Can Be Used in Agricultural Fields and Tea Gardens to Fight Crude Oil Contamination

Consortia	Type of Contaminates	References
1. *Cyanobacterium* sp., *Synechococcus elongatus, Methanocaldococcus indiensis, Corynebacterium* sp., *Nocardioides* sp., *Gordonia* sp., *Sinorhizobium* sp., *Rhizobium* sp., *Agrobacterium* sp., *Chelatococcus* sp., *Methylobacterium* sp., *Ochrobacter* sp., *Skermanella* sp., *Corynebacterium* sp., *Pseudomonas* sp., *Sinorhizobium* sp., *Brevibacillus* sp., *Rhizobium* sp., *Agrobacterium* sp., *Streptomyces* sp., *Actinomycetales* sp., *Bacillales* sp., *and Rhizobiales* sp.	Pyrene-metabolizing microbial consortia from the plant rhizoplane	Balcom and Crowley (2010)

(Continued)

Table 6   Consortia That Can Be Used in Agricultural Fields and Tea Gardens to Fight Crude Oil Contamination—cont'd

Consortia	Type of Contaminates	References
2. *Betaproteobacteria, Gammaproteobacteria, Bacteroidetes, Alpha proteobacteria, Stenotrophomonas maltophilia, Acidovorax avenae, Lysinibacillus sphaericus, Stenotrophomonas maltophilia, Achromobacter xylosoxidans, Serratia marcescens, Lysinibacillus, Bacillus subtilis, Caulobacter* sp., *Bacillus pumilus, Bacillus* sp., *Stenotrophomonas maltophilia, Erythromicrobium ramosum, Acidovorax avenae, Labrys* sp., *Burkholderia* sp.	Consortia of PAH-degrading bacteria from crops rhizosphere	Ma et al. (2010)
3. *Sphingobacteria, Mesorhizobium* sp., *Alcaligenes* sp., *Bacillus* sp., *Pedobacter* sp., *Paenibacillus* sp., and *Caulobacter* sp.	Consortia for remediation of PAH-contaminated soil	Mao et al. (2012)
4. *Klebsiella oxytoca, Bacillus subtilis, Streptococcus* sp., *Pseudomonas aeruginosa, Bacillus megaterium, Staphylococcus epidermidis, Enterobacter aerogenes, Escherichia coli, Arthrobacter* sp., *Nocardia* sp., *Corynebacterium* sp., *Aspergillus versicolor, Aspergillus niger, Mucor mucedo, Penicillium chrysogenum and Penicillium* sp., *Acinetobacter iwoffii, Nocardia* sp., *Enterobacter agglomerans, Paenibacillus* sp., *Bacillus thuringeinsis, Alcaligenes* sp., *Agrobacterium* sp., *Chromobacterium* sp., *Enterobacter, Proteus* sp., *Rhizobium* sp., *Brevibacterium* sp., *Corynebacterium* sp., and *Micrococcus* sp.	Consortia for remediation of heavy metals and hydrocarbon-contaminated soil	Jaboro et al. (2012)

Many plant microbe consortia have been inventoried and their bioeconomic significance has been outlined. This chapter has demonstrated the tremendous bioremediation capacities of these consortia with respect to PAHs and heavy metals in laboratory as well as field applications for phytomanagement of moderate to heavily-contaminated sites. This is definitely of tremendous bioeconomic significance although, at the same time, it has many limitations, as has already been discussed earlier. The small-scale tea growers and rice farmers of this region should use this technology to enhance their crop productivity. However, it is to be emphasized that, although many consortia have been developed, only those plant microbes that are native strains will show maximum bioremediation capacities due to their easy adaptability as well as acclimatization in contaminated habitats of this region. Interested farmers who would like to apply these consortia could consult with an R&D laboratory that has come up with this innovative technology. Concurrently, the R&D lab should welcome not only oil exploration companies but should also extend their expertise also to economically marginalized farmers for the greater good of the economy.

5 CONCLUSION

Current research has shown that among the different microbes, use of plant growth-promoting rhizobacteria (PGPR) for bioremediation activity is gaining importance due to their differential abilities to degrade and detoxify contaminants and their positive effects on plant growth promotion (Glick, 2010). Some PGPR strains have the ability to produce Indole-3-acetic acid, siderophore, phosphorus solubilization, Hydrogen cyanide production, phosphorus solubilization, and *in vitro* antifungal activity that enhance tea and rice production and, at the same time, remove contaminants from the soil. Sometimes the concentration of contaminants can be so high that the environment becomes toxic to microbial populations. Therefore, the technologist must use advanced bioremediation techniques to modify the environment to make it more habitable for plant microbes.

ACKNOWLEDGMENTS

The authors are grateful to DBT, Government of India, for their research grant under DBT's Twinning program for northeast India (Grant no. BT/489/NE/TBP/2013). Thanks are also due to Dr. Dolikajyoti Sharma and Ms. Parismita Borgohain for their technical support in the preparation of this manuscript.

REFERENCES

Ahmed, T., Chetia, S.K., Chowdhury, R., Ali, S., 2014. Status Paper on Rice in Assam, pp. 1–49, Rice Knowledge Management Portal http://www.rkmp.co.in.

Ajeigbe, H.A., Adamu, R.S., Singh, B.B., 2012. Yield performance of cowpea as influenced by insecticide types and their combinations in the dry savannas of Nigeria. Afr. J. Agric. Res. 7 (44), 5930–5938.

Auta, H.S., Ijah, U.J.J., Mojuetan, M.A., 2014. Bioaugmentation of crude oil contaminated soil using bacterial consortium. Adv. Sci. Focus 2 (1), 26–33.

Bacosa, P.H., Inoue, C., 2015. Polycyclic aromatic hydrocarbons (PAHs) biodegradation potential and diversity of microbial consortia enriched from tsunami sediments in Miyagi, Japan. J. Hazard. Mater. 283, 689–697.

Balcom, N.I., Crowley, E.D., 2010. Isolation and characterization of pyrene metabolizing microbial consortia from the plant rhizoplane. Int. J. Phytoremediation 12, 599–615.

Bamforth, M.S., Singleton, I., 2005. Bioremediation of polycyclic aromatic hydrocarbons: current knowledge and future directions. J. Chem. Technol. Biotechnol. 80, 723–736.

Basumatary, B., Bordoloi, S., Sarma, H.P., 2012. Crude oil-contaminated soil phytoremediation by using *Cyperus brevifolius* (Rottb.) Hassk. Water Air Soil Pollut. 223 (6), 3373–3383.

Basumatary, B., Saikia, R., Das, C.H., Bordoloi, S., 2013. Field note: phytoremediation of petroleum sludge contaminated field using sedge species, *Cyperus rotundus* (Linn.) and *Cyperus brevifolius* (Rottb.) Hassk. Int. J. Phytoremediation 15 (9), 877–888.

Bora, T.C., Dekabaruah, H.P., Saikia, N., Sarma, A., Bezbaruah, L.R., Saikia, R., Yadav, A., Bordoloi, P., 2011. Bioprospecting microbial diversity from north east gene pool. Sci. Cult. 77 (11-12), 446–450.

Bordoloi, S., Basumatary, B., Saikia, R., Das, C.H., 2012. *Axonopus compressus* (Sw.) P. Beauv. A native grass species for phytoremediation of hydrocarbon-contaminated soil in Assam, India. J. Chem. Technol. Biotechnol. 87 (9), 1335–1341.

Chaudhry, Q., Zandstra, B.M., Satish Gupta, S., Joner, J.E., 2005. Utilising the synergy between plants and rhizosphere microorganisms to enhance breakdown of organic pollutants in the environment. Environ. Sci. Pollut. Res. Int. 12 (1), 34–48.

CIFOR, 2002. Woodlands and rural livelihoods in dryland Africa, Indonesia. CIFOR Info Brief 4, 1–4.

Coccia, M.A., Gucci, B.M.P., Lacchetti, I., Beccaloni, E., Paradiso, R., Beccaloni, M., Musmeci, L., 2009. Hydrocarbon contaminated soil treated by bioremediation technology: microbiological and toxicological preliminary findings. Environ. Biotechnol. 5 (2), 61–72.

Cote, R.P., 1976. The Effects of Petroleum Refinery Liquid Wastes on Aquatic Life, with Special Emphasis on the Canadian Environment. National Research Council of Canada, NRC Associate Committee on Scientific Criteria for Environmental Quality, Ottawa, ON. 77 pp.

Croes, S., Weyens, N., Janssen, J., Vercampt, H., Colpaert, V.J., Carleer, R., Vangronsveld, J., 2013. Bacterial communities associated with *Brassica napus* L. grown on trace element-contaminated and non-contaminated fields: a genotypic and phenotypic comparison. Microb. Biotechnol. 6 (4), 371–384.

Das, N., Chandran, P., 2011. Microbial degradation of petroleum hydrocarbon contaminants: an overview. Biotechnol. Res. Int. 2011. http://dx.doi.org/10.4061/2011/941810.

Das, K., Mukherjee, A.K., 2007. Crude petroleum-oil biodegradation efficiency of *Bacillus subtilis* and *Pseudomonas aeruginosa* strains isolated from a petroleum-oil contaminated soil from North-East India. Bioresour. Technol. 98 (7), 1339–1345.

De Jong, E., 1980. Effect of a crude oil spill on cereals. Environ. Pollut. 22, 187–307.

FAO, 2005. The State of Food Insecurity in the World: Eradicating World Hunger—Key to Achieving the Millennium Development Goals. Food and Agriculture Organization of the United Nations, Rome, Italy.

Frenzel, M., Scarlett, A., Rowland, S.J., Galloway, T.S., Burton, S.K., Lappin-Scott, H.M., Booth, A.M., 2010. Complications with remediation strategies involving the biodegradation and detoxification of recalcitrant contaminant aromatic hydrocarbons. Sci. Total. Environ. 408, 4093–4101.

Gargouri, B., Karray, F., Mhiri, N., Aloui, F., Sayadi, S., 2013. Bioremediation of petroleum hydrocarbons-contaminated soil by bacterial consortium isolated from an industrial wastewater treatment plant. J. Chem. Technol. Biotechnol. 8, 1–10.

Glick, B.R., 2010. Using soil bacteria to facilitate phytoremediation. Biotechnol. Adv. 28, 367–374.

Grosser, R.J., Warshawsky, D., Vestal, J.R., 1991. Indigenous and enhanced mineralization of pyrene, benzo[a]pyrene, and carbazole in soils. Appl. Environ. Microbiol. 57 (12), 3462–3469.

Harayama, S., Kasai, Y., Hara, A., 2004. Microbial communities in oil-contaminated sea water. Curr. Opin. Biotechnol. 15, 205–214.

Haritash, A.K., Kaushik, C.P., 2009. Biodegradation aspects of polycyclic aromatic hydrocarbons (PAHs): a review. J. Hazard. Mater. 169, 1–15.

Jaboro, A.G., Akortha, E.E., Obayagbona, O.N., 2012. Susceptibility to heavy metals and hydrocarbonclastic attributes of soil microbiota. Int. J. Agric. Biosci. 2 (5), 206–212.

Jeremy, K., Prasad, M.N.V., Pershell, K., 2015. Bibliometric analysis of phytotechnologies for remediation: global scenario of research and applications. Int. J. Phytoremediation 17 (2), 145–153.

Juhasz, A.L., Naidu, R., 2000. Bioremediation of high molecular weight polycyclic aromatic hydrocarbons: a review of the microbial degradation of benzo[a]pyrene. Int. Biodeterior. Biodegrad. 45, 57–58.

Kabeer, R., Varghese, R., Kannan, V.M., Thomas, J.R., Poulose, S.V., 2014. Rhizosphere bacterial diversity and heavy metal accumulation in *Nymphaea pubescens* in aid of phytoremediation potential. J. BioSci. Biotech. 3 (1), 89–95.

Li, X., Li, P., Lin, X., Zhang, C., Lia, Q., Gong, Z., 2007. Biodegradation of aged polycyclic aromatic hydrocarbons (PAHs) by microbial consortia in soil and slurry phases. J. Hazard. Mater. 150 (1), 21–26.

Lu, J., Guo, C., Zhang, M., Guining, L., Zhi, D., 2014. Biodegradation of single pyrene and mixtures of pyrene by a fusant bacterial strain F14. Int. Biodeterior. Biodegrad. 87, 75–80.

Ma, B., Chen, H., He, Y., Xu, J., 2010. Isolations and consortia of PAH-degrading bacteria from the rhizosphere of four crops in PAH-contaminated field. In: 19th World Congress of Soil Science, Soil Solutions for a Changing World, 1–6 August 2010.

Maiti, A., Bhattacharyya, N., 2012. Biochemical characteristics of a polycyclic aromatic hydrocarbon degrading bacterium isolated from an oil refinery site of West Bengal, India. Adv. Life Sci. Appl. 1 (3), 48–53.

Mao, A.A., Hynniewta, T.M., Sanjappa, M., 2009. Plant wealth of Northeast India with reference to ethnobotany. Indian J. Tradit. Know. 8 (1), 96–103.

Mao, J., Luo, Y., Teng, Y., Li, Z., 2012. Bioremediation of polycyclic aromatic hydrocarbon-contaminated soil by a bacterial consortium and associated microbial community changes. Int. Biodeterior. Biodegrad. 70, 141–147.

Masakorala, K., Yao, J., Cai, M., Chandankere, R., Yuan, H., Chen, H., 2013. Isolation and characterization of a novel phenanthrene (PHE) degrading strain *Psuedomonas* sp. USTB-RU from petroleum contaminated soil. J. Hazard. Mater. 263 (2), 493–500.

Mazzeo, D.E.C., Levy, C.E., De Angelis, D.D.F., Marin-Morales, M.A., 2010. BTEX biodegradation by bacteria from effluents of petroleum refinery. Sci. Total. Environ. 408, 4334–4340.

Mohsen, A., Simin, N., Chimezie, A., 2009. Biodegradation of polycyclic aromatic hydrocarbons (PAHs) in petroleum contaminated soils. Iran. J. Chem. Chem. Eng. 28 (3), 53–59.

Nellemann, C., Corcoran, E., 2010. Dead Planet, Living Planet: Biodiversity and Restoration for Sustainable Development. UNEP/GRID, Arendal, Norway.

Ogbo, E.M., Zibigha, M., Odogu, G., 2009. The effect of crude oil on growth of weed (*Paspalum scrobiculatum* L) phytoremediation potential of the plant. Afr. J. Environ. Sci. Technol. 9, 229–233.

Omosun, G., Markson, A.A., Mbanasor, O., 2008. Growth and anatomy of *Amaranthus hybridus* as affected by different crude oil concentrations. Am. Eur. J. Sci. Res. 1, 70–74.

Palanisami, T., Mallavarapu, M., Ravi, N., 2012. Bioremediation of high molecular weight polyaromatic hydrocarbons co-contaminated with metals in liquid and soil slurries by metal tolerant PAHs degrading bacterial consortium. Biodegradation 23 (6), 823–835.

Roy, A.S., Yenn, R., Singh, A.K., Boruah, H.P.D., Saikia, N., Deka, M., 2013. Bioremediation of crude oil contaminated tea plantation soil using two Pseudomonas aeruginosa strains AS 03 and NA 108. Afr. J. Biotechnol. 12 (19), 2600–2610.

Roy, A.S., Baruah, R., Borah, M., Singh, A.K., Boruah, H.P.D., Saikia, N., Deka, M., Dutta, N., Bora, T.C., 2014. Bioremediation potential of native hydrocarbon degrading bacterial strains in crude oil contaminated soil under microcosm study. International Biodeterioration & Biodegradation 94, 79–89.

Ruberto, M.A.L., Lucas, A.M.R., Vazquez, C.S., Curtosi, A., Mestre, C.M., Pelletier, E., Cormack, M.P.W., 2006. Phenanthrene biodegradation in soils using an Antarctic bacterial consortium. Biorem. J. 10 (4), 191–201.

Sarma, H., Prasad, M.N.V., 2015. Plant-microbe association-assisted removal of heavy metals and degradation of polycyclic aromati hydrocarbons. In: Mukherjee, S. (Ed.), Petroleum Geosciences, Indian Contexts. Springer International Publishing, Switzerland, pp. 219–236. http://dx.doi.org/10.1007/978-3-319-03119-4_10.

Sarma, H., Sarma, C.M., 2008. Alien traditionally used plant species of Manas Biosphere Reserve, Indo-Burma hotspot. Z. Arznei Gewurzpfla. 13 (3), 117–120.

Sarma, H., Tripathi, K.A., Borah, S., Kumar, D., 2011. Updated estimates of wild edible and threatened plants of Assam: a meta-analysis. Int. J. Bot. 6, 414–423.

Schutzendubel, A., Polle, A., 2002. Plant responses to abiotic stresses: heavy metal-induced oxidative stress and protection by mycorrhization. J. Expt. Bot. 53 (372), 1351–1365.

Shin, S.K., Oh, Y.S., Kim, S.J., 1999. Biodegradation of phenanthrene by *Sphingomonas* sp. Strain KH3–2. J. Microbiol. 37, 185–192.

Shukla, P.K., Singh, K.N., Sharma, S., 2010. Bioremediation: developments current practices and perspectives. Genet. Eng. Biotechnol. J. 3, 1–20.

Siciliano, D.S., Germida, J.J., Banks, K., Greer, W.C., 2003. Changes in microbial community composition and function during a polyaromatic hydrocarbon phytoremediation field trial. Appl. Environ. Microbiol. 69, 483–489.

Sun, D.G., Xu, Y., Jin, H.J., Zhong, P.Z., Liu, Y., Luo, M., Liu, P.Z., 2012. Pilot scale ex-situ bioremediation of heavily PAHs-contaminated soil by indigenous microorganisms and bioaugmentation by a PAHs-degrading and bioemulsifier-producing strain. J. Hazard. Mater. 233–234, 72–78.

Tiwari, J.N., Chaturvedi, P., Ansari, N.G., Patel, D.K., Jain, S.K., Murthy, R.C., 2011. Assessment of polycyclic aromatic hydrocarbons (PAH) and heavy metals in the vicinity of an oil refinery in India. Soil Sediment Contam. 20 (3), 315–328.

Udo, E.J., Fayemi, A.A.A., 1995. The effect of oil pollution on soil germination, growth and nutrient uptake of corn. J. Environ. Qual. 4, 537–540.

Vagi, M.C., Kostopoulou, M.N., Petsas, A.S., Lalousi, M.E., Rasouli, C., Lekkas, T.D., 2005. Toxicity of organophoshorous pesticides to the marine alga *Tetraselmis suecica*. Global NEST J. 7 (2), 222–227.

World Bank, 2001. World Development Report 2000/2001: Attacking Poverty. Oxford University Press, Oxford. pp. 1–200.

World Bank, 2004. Sustaining Forests: A Development Strategy. World Bank, Washington, DC.

Yuan, S.Y., Shiung, L.C., Chang, B.V., 2002. Biodegradation of polycyclic aromatic hydrocarbons by inoculated microorganisms in soil. Bull. Environ. Contam. Toxicol. 69, 66–73.

NEW BIOLOGY

ECOCATALYSIS: A NEW APPROACH TOWARD BIOECONOMY

C. Grison¹, V. Escande¹, T.K. Olszewski²

*FRE 3673 – Bioinspired Chemistry and Ecological Innovations – CNRS, University of Montpellier 2,
Stratoz – Cap Alpha, Avenue de l'Europe, 34830 Clapiers, France¹
Organic Chemistry, Wroclaw University of Technology, Wroclaw²*

1 INTRODUCTION

Intensive mining and metallurgical industrial activities are causing strong soil contamination by metal species. This is a very serious problem because soil performs essential functions and largely determines food production and water quality. In addition, trace metals (TM) are among the most harmful compounds and are not biodegradable. Beyond the environmental consequences, the health risks are real: damage to the nervous, renal, lung, and bone tissue are clearly established. A recent example concerns blood lead levels; the first stage of poisoning have been identified in children residents near mining sites. Specific impacts of metal pollution are not only environmental and medical; they directly affect the economic and tourism development of the areas concerned.

Phytoremediation is one of the few interesting solutions for sustainable rehabilitation of degraded or contaminated soils by TM. The most studied are the phytotechnologies phytostabilization and phytoextraction (Mench et al., 2009, 2010; Vangronsveld et al., 2009; Bert et al., 2009).

Past experiences have shown that with phytostabilization it is possible to immobilize contaminants and to contribute to the growth of vegetation in hostile areas. However, this technique favors the spontaneous appearance of plants that sometimes become capable of accumulating the TM. Thus, the evolution over time of revegetated plots is the delicate problem of risk management in duration.

Phytoextraction is a partial remediation of soil and an environmental technology based on accumulation of TM in shoots of hyperaccumulator plants (e.g., *A. murale*, *N. caerulescens*, or *P. gabriellae*) (Figure 1). Recent studies on evaluation of the adaptive performance of these plants showed the presence of hyperaccumulating legume species reinforcing the interest of phytoextraction in ecological restoration programs (Vidal et al., 2009; Grison et al., 2014a, 2015).

The development of phytoextraction is limited by lack of appreciation of contaminated biomass. Without credible opportunity, the aerial parts of hyperaccumulating plants are considered as contaminated waste. Furthermore, the extraction phenomenon of TM by the root system increases the fraction of the soluble elements. The rise of phytoextraction is entirely related to the use of generated biomass. Much remains to be done in the development of phytotechnologies recovery.

The Laboratory of Bio-inspired Chemistry and Environmental Innovations (FRE CNRS Chim-Eco-UM-Stratoz 3673) recently proposed a new valuation phytoextraction, called ecocatalysis. The generated plant waste is recycled through an innovative concept of ecological recycling. Taking

Bioremediation and Bioeconomy. http://dx.doi.org/10.1016/B978-0-12-802830-8.00025-3
Copyright © 2016 Elsevier Inc. All rights reserved.

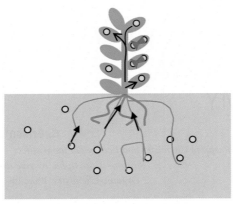

○ ETM (Zn^{2+} or Ni^{2+} or Mn^{2+}, ...)

FIGURE 1

Principle of phytoextraction.

FIGURE 2

Toward a new green channel: from phytoextraction to the eco-friendly preparation of biomolecules via ecocatalysis.

advantage of the remarkable adaptive capacity of certain plants to hyperaccumulate metals, the eco-catalysis is based on the novel use of metal, issued from vegetable species, as reagents and catalysts for fine organic chemical reactions. It allows the preparation of biomolecules in an eco-friendly and bio-inspired approach (Figure 2).

Ecocatalysis created a paradigm shift; biomass from phytoextraction became seen no longer as merely contaminated waste but as a natural source of metals, which has a high added value. This bio-mass is a natural reservoir of precious transition metals useful in organic synthesis. In other words, waste has become useful as a chemical object, innovative and motivating.

Validated by 18 patents, 27 scientific publications, and 8 Innovation Awards, this unusual combina-tion inseparable from the environment, ecology, and chemistry has given rise to new research at the interface of Green Chemistry and Ecological Engineering; it is based on solid achievements and devel-oped in advanced research programs (Escande et al., 2014a,b,c,d,e,f, 2015; Losfeld et al., 2012a,b,c; Thillier et al., 2013; Grison et al., 2013; Garel et al., in press).

This comprehensive approach to scientific ecology has now resulted in the development of a new circular economy and green industry that combines various public and private partners with complementary areas of applications (restoration ecology, mining and chemical industries). Two chemical industry groups, from Europe and Asia, have shown interest in this new process through collaborative agreements. Stratoz, a young innovative company specializing in green chemistry, sets up the necessary tools to the industrial development of this new green industry.

This chapter illustrates how a breakthrough innovation in environmental chemistry, ecocatalysis, managed to stimulate and advance the field of phytoextraction of sites degraded by mining activities, enabling better understanding of the social, environmental, and economic sustainability.

2 A NEW APPROACH TO MULTISITE AND MULTISCALE REHABILITATION BY PHYTOEXTRACTION

The proposed eco-innovation is an integrated and interdisciplinary approach to phytoremediation, which is based on fundamental knowledge in plant and microbial ecology, ecological and environmental chemistry.

The presented results were obtained in public research laboratories in collaboration between research centers (Chemistry Laboratory of Bio-inspired and Ecological Innovations, FRE CNRS-UM-Stratoz 3673), semi-public (New Caledonian Agronomic Institute), and private companies (STRATOZ, Société Le Nickel) who have chosen to combine their own skills in phytoremediation, to develop a new program of chemical recycling of contaminated waste adaptable to the variability and diversity of soil and climatic conditions of the French and New Caledonian degraded mining sites.

The efforts were split into two study sites where the origin of metal waste and expectations of the populations concerning environmental and economic issues are very different, but where the natural adaptation phenomena of certain plants and associated microorganisms are common:

- The Languedoc-Roussillon region, particularly the Gard department, where the touristic town of Saint-Laurent-le-Minier is the site of a former representative pilot mining area where centuries of exploitation have contaminated wide surfaces and adjacent rivers by zinc, cadmium, and lead.
- New Caledonia, a hotspot of biodiversity, home to 3350 plant species, 74% of them endemic, is threatened by the development of nickel mining. Furthermore, depletion of conventional garnierite deposits requires the exploitation of new deposits, potentially increasing the degradation probabilities.

2.1 REMEDIAL PHYTOEXTRACTION ON A FORMER HIGHLY CONTAMINATED MINE SITE: LES AVINIÈRES

2.1.1 What are the characteristics of the pilot site Les Avinières?

The site known as Les Avinières is in the town of Saint-Laurent-le-Minier, Gard. The region is the subject of strong contrasts: It has a low population density (<50 inhabitants per km²) and lies between a national park (Cévennes National Park) and a special area network Natura 2000 conservation. It also has a major history in terms of mining activities: About 1 million tonnes of ore were extracted between 1885 and 1991, making this the largest lead-zinc operation that France has ever known (Leguen et al., 1991). The mine Les Avinières is much older and was active from 1857 to the early 1950s (Figure 3).

FIGURE 3

Mine "Les Avinières" (South of France). (a) Mine drift. (b) Mine tailings and waste rocks. (c) Old settling basin.

The former mine activities in Les Avinières are responsible for significant changes in the environment. The extraction of ore changed the landscape with the formation of a large heap of accumulated waste and TM contents. The purification steps, washing minerals that generated large quantities of high-grade waste TM, contributed to the formation of a zone of very high soil contamination at the site of storage. This area includes all the old settling ponds and is characterized by an almost total lack of vegetation. This part of the site has become phytotoxic (Figure 4). The few plants that survived have adapted to the metal pollution.

These plants are tolerant and possibly are accumulating or hyperaccumulating TM. All the surrounding surfaces have intermediate levels of contamination mainly due to the transportation of minerals, for combustion and displacement of already contaminated material (dust formation, runoff, ash deposition from furnaces …). Wind erosion and water erosion are very significant on this site because of the lack of vegetation; this extreme situation is a major risk of spread of contamination in the environment, especially the soil.

FIGURE 4

Phytotoxic soil on the mining site.

What are the objectives of the program?

The phytoextraction drivers are improving the environmental quality of the site by providing the means to master and manage environmental risks in the long term. The first concrete goals were to limit wind and water erosion of sediments contaminated with the installation of a plantation of local hyperaccumulating plants. These studies were conducted in research programs (ANR ECOTECH "Opportunity (E) 4 CDII" Phytochem, ERDF "GénieEcoChim," Franco-Chinese program Xu Guangqi and PICS CNRS, PIR INGECO CNRS-IRSTEA, PhD scholarship ADEME, Défi CNRS ENVIROMCS).

What are the boundaries of the studied site?

The ponds were used as storage ore. These wastes, a mixture of water and sediment-laden metals, were discharged into settling ponds. The technique then was to let the sediment settle to the bottom before discharging the water into the river. The settling ponds were found to contain very high levels of TM. The contamination is very significant, and the risk of contamination of the entire ecosystem is major. It is therefore in these basins that we chose to perform a large part of our experiments. The three basins occupy a total area of about 7677 m² with a length of 163 m along the river. The first houses are located upstream from the other side of the river at a distance of 130 m (Figure 5).

They were supplemented by important experiments on areas where contamination is lower but not negligible.

Why phytoextraction?

A study of local plant species, considered phytostabilizing, revealed variable but important TM rates. Thus, the phytostabilization of revegetated parts poses a delicate problem of risk of management in duration. Phytoextraction is a partial remediation of soil and an environmental technology by accumulation of TM in the shoots of hyperaccumulating plants. It is this technique that has been studied and developed on the site Les Avinières.

Three settling basins : high contaminationlevel

Flat polluted area: middle contamination level

FIGURE 5

The three settling basins of the old mine of Avinières.

FIGURE 6

(a) *Anthyllis vulneraria* and (b) *Noccaea caerulescens*.

The work on phytoextraction exclusively uses native plants and scrupulously respects the local biodiversity. The hyperaccumulators studied and grown on sites were *Anthyllis vulneraria* and *Noccaea caerulescens* (Figure 6). The first (yellow) is the one of the few legumes known as also being a hyperaccumulator of TM, the second (in white) is listed as the best zinc hyperaccumulator.

Does phytoextraction involve the transfer of TM via pollinating insects?

The phytotoxicity of the soil at Les Avinières limits the passage of animal species on the site. Only a few bees were observed during flowering. It seemed worth checking before the start of the study, if pollinators were able to transfer the TM at pollination, then to bee products, and study the TM transfer bioindicators. A detailed study of these problems has been prepared and published (Losfeld et al., 2014a; Saunier et al., 2013).

The main conclusions are:

Lichens and mosses known to be air pollution bioindicators for TM were sampled: Their concentrations of TM and their lead isotopic signature (one of the main contaminants) permitted assessment of the environmental impacts. In addition, elementary analysis of the bee products of royal jelly, beeswax, and honey from various places near old mines was conducted.

The mining sector represented an obvious problem of contamination by TM from the storage areas of mining waste, but without affecting honey, royal jelly, beeswax, or even bees. The honey was not contaminated by the TM present in the vicinity of mining waste, and the consumption of honey produced in those areas does not appear to be of any particular risk (Table 1).

When it comes to lichens and mosses, the samples taken from different species of *Cladonia rangiformis* or *Parmelia acetabulum* (lichens) and *Scleropodium purum* or *Dicranum scoparium* (mosses) confirmed their potential for atmospheric deposition measurements. Analysis of isotopic signatures of Pb showed that past mining activities were a major source of Pb with other types of anthropogenic impact (Figure 7). Aeolian transport of contaminated dust is the main mechanism for mobilizing TM.

The results obtained prompted us to implement large-scale biospecific cultivation techniques, from harvesting seeds to harvesting the leaves. Our research aimed at studying the phytoextraction on site on a large scale contributing to the environmental improvement of the site through the developed process. Information sought included:

- new knowledge about the bacterial composition of the soil;
- the study of the effect of microbial inoculations (bacteria and symbiotic fungi), on stage seedlings or young shoots, on the growth of plants and their ability to accumulate zinc, growing in greenhouses, and after transfer to contaminated soils and sites;
- chemical valuation of all of these results in organic synthesis.

What is the bacterial composition of the ground at Les Avinières?

A study of bacterial microorganisms in the soil of Les Avinières was performed and led unexpectedly to the isolation and study of a new species of *Rhizobium*.

2.1.2 Discovery of a new species of bacteria in symbiosis with Anthyllis vulneraria

Anthyllis vulneraria is the only legume able to grow on contaminated land sites of Saint-Laurent-le-Minier. The legumes are of great interest for agriculture. They are used in parallel with food crops to enrich the soil with natural fertilizers (e.g., ammonium). Bacteria found in the roots or rhizosphere of legumes in fact produce these natural fertilizers. Introduction of legumes to poor soil enriches the soil with natural fertilizers and so promotes vegetation cover.

Therefore we studied the bacteria roots and rhizosphere of *A. vulneraria* on site at Les Avinières (Figure 8). In 2009, Vidal et al. (Vidal et al., 2009) found a new species of bacteria in the roots of *A. vulneraria*, called *Mesorhizobium metallidurans*. We could not find these bacteria either in the ground or in the roots of the legume. However, we found another new species of bacteria, *Rhizobium metallidurans*, in the roots of *A. vulneraria* and in soil settling ponds 2 and 3, and we continued to study this bacterium. We have first fully characterized it genotypically by sequencing of the DNA (Figure 9). The phylogenetic tree shows the *R. metallidurans* and thus proves that it is a new species of *Rhizobium*.

Table 1 Elemental Analyses of Honey (H), Royal Jelly-1 (RJ), Beeswax (W) (Fresh Weight), and Bees (BH, B) (Dry Mass) (mg kg⁻¹)

Place	Distance from the Mine[a] (m)	Date	Mg	P	Ca	Mn	Zn	As	Cd	Sb	Tl	Pb
Les Avinières	250	April 2011	80.5	40.7	187	5.81	0.806	0.001	0.009	0.002	0.006	0.003
St-Bresson	1500	May 2011	122	78.4	204	9.17	0.556	0.001	0.004	0.001	0.001	0.035
Les Avinières	250	May 2011	125	115	219	6.77	1.4	0.003	0.022	0.003	0.037	0.101
Les Avinières	250	June 2011	44.8	65.6	76.7	3.22	0.905	0.001	0.003	0.001	0.006	0.011
Les Avinières	250	July 2011	163	66.6	277	10.5	0.613	0.001	0.001	0.001	0.012	0.009
St-Bresson	1500	June 2011	101	80.1	200	7.86	0.429	0.008	0.006	0.002	0.013	0.005
St-Bresson	1500	July 2011	66	82.3	168	10.6	0.429	0.002	0.003	0.001	0.019	0.025
Aulas	4000[b]	2011	32.8	59.4	36.4	1.38	<bd	<bd	0.001	<bd	0.012	0.006
Majencoule	7500[c]	2011	145	62.8	206	12.8	<bd	0.001	0.001	0.002	0.003	0.014
Les Avinières	250	May 2011	33.8	61.7	123	0.304	0.906	0.005	0.007	0.002	<bd	0.168
Les Avinières	250	May 2011	167	78.8	305	16.1	1.520	0.012	0.006	0.001	0.013	<bd
Les Avinières	250	May 2011	2047	13,050	1857	93.1	167	0.056	2.9	0.026	0.133	1.44
Les Avinières	250	May 2011	2286	14,290	1783	80.7	172	0.059	2.5	0.025	0.149	0.832

<bd, value below the detection limit.
a Distance to the nearest mine, which is the Avinières, for samples from Saint-Laurent-le-Minier, and Font-Bouillens for those of Saint-Bresson.
b Distance to the mine Bez-et-Esparron.
c Distance to mine Jumeaux in Sumène.

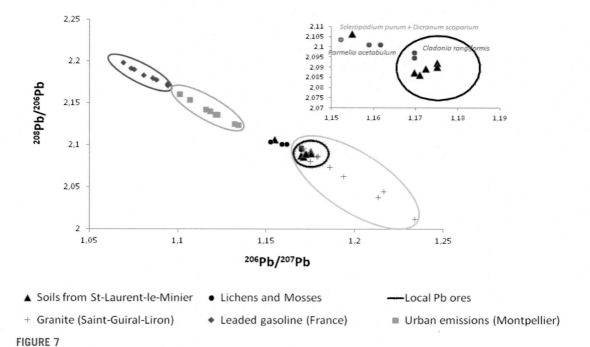

FIGURE 7

Graphical representation of Pb isotopic data.

FIGURE 8

Rhizobium metallidurans sp. nov., a symbiotic heavy metal-resistant bacterium isolated from the *Anthyllis vulneraria* Zn-hyperaccumulator.

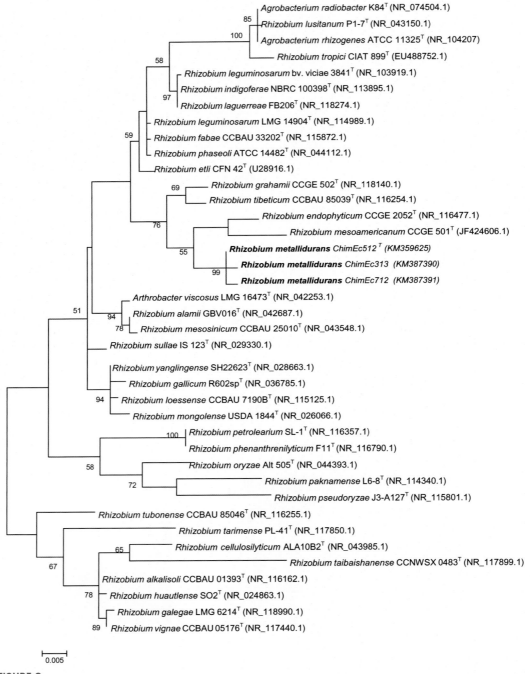

FIGURE 9

Phylogenetic tree (neighbor-joining method) showing the relationship between our strain *Rhizobium metallidurans* ChimEc512T and the most similar reference strains.

What are the adaptation strategies of the newly discovered bacterial species?

The discovery of a new species of bacteria that lives in these extreme conditions poses many questions. It is particularly interesting to understand how the bacteria managed to survive against such a metal stress. A study of the physiological adaptation strategies of this bacteria was carried out by determining the optimum growth conditions, tests for sugar assimilation and amino acids assimilation, antibiotic and heavy metals resistance tests, and determination of the lipid profile.

Figure 10 shows an interesting feature of the bacterial strain. It is capable of withstanding very high concentrations of TM. It is therefore one of the few TM-tolerant bacteria.

We then studied the metabolism of the bacteria in order to stimulate its growth, which enriches the soil with natural fertilizer. The challenge is to specifically stimulate growth of these bacteria in the total bacterial population.

We conducted different growth tests on many minimal media and on a given sugar. We were able to prove that the bacteria do not perform glycolysis as a sugars degradation pathway; instead it uses the Entner-Doudoroff (ED) pathway (Figure 11). This pathway of degradation of sugars is very rare.

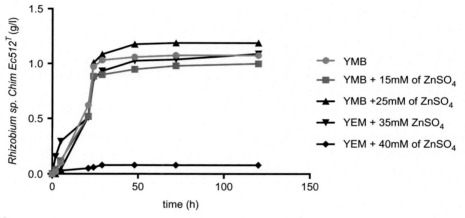

FIGURE 10

Growth kinetics of *Rhizobium metallidurans* on medium enriched with Zn.

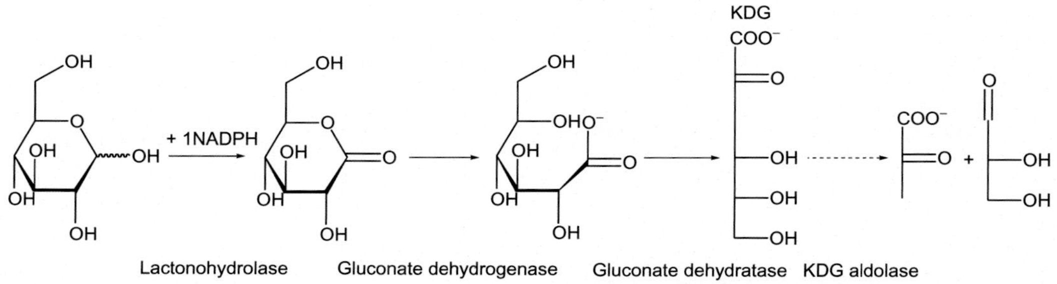

FIGURE 11

Entner-Doudoroff pathway.

Glucose is degraded in this metabolic pathway by a particular intermediary, the 2-keto-3-deoxygluconate (KDG). KDG is present only in this pathway and nowhere else in the cell. It is therefore interesting to be able to feed the bacteria with KDG, which is only present in this metabolic pathway, uncommon in bacteria living in the soil. Thus KDG should specifically stimulate the growth of *R. metallidurans*, allowing an enrichment of soil in natural fertilizer. We still need to know how to synthesize KDG. It is of course not commercially available and its synthesis is virtually unknown.

How to design an optimized synthetic strategy leading to KDG?

From our previous work on the chemical synthesis of ulosoniques higher acid (3-deoxy-D-manno-oct-2-ulosonic acid (KDO), 3-deoxy-D-arabino-2-heptu-losonic acid (DAH)) (Coutrot et al., 1999) we have developed a rapid strategy based on the use of alkyl dihalogenoacetates to lengthen the carbon chain of the aldose with an α-ketoester unit. This is an [n + 2] methodology resulting in an enantiopure ulosonic acid (Grison et al., 2014b).

As the KDG has a D-erythro configuration, the starting saccharide should be a protected D-erythrose. However, due to their instability, the unprotected precursor of type D-erythrose is not commercially available. A compound with the D-threo configuration has been selected as carbohydrate precursor. The synthesis of KDG described here uses in the key step the epimerization of the β position of the D-threose. Epimerization is induced by Swern oxidation of the protected D-threitol followed by diastereoselective type-Darzens condensation (Coutrot et al., 1988).

The potassium enolate-type anion, derived from isopropyl dichloroacetate, is a synthetic equivalent of synthon ("−C(O)COO−"). It was added on the appropriately protected D-threose. The latter was prepared from commercially available 2,3-*O*-isopropylidene-D-threitol (**1**) (Figure 12). The next steps of the synthesis are based on a gentle opening of the epoxide ring with halogen exchange, followed by *in situ* reduction (Figure 13).

A final acid hydrolysis with formic acid (50% solution in water) was performed to give compound **8**. The isopropyl ester was retained to allow the protected KDG **8** to cross the cell membrane of Gram-negative bacteria (Liang et al., 1998; Po-Huang et al., 1997). Additionally, the product **8** was

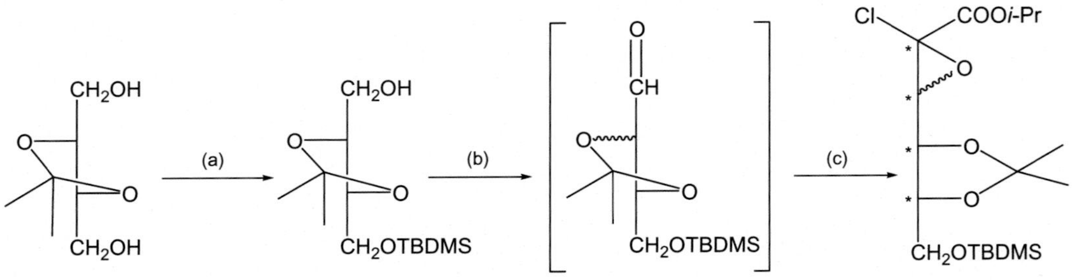

FIGURE 12

Synthesis of compounds 1-5. Conditions and reagents: (a) 1 equiv NaH, THF, 45 min, 5 °C, next 1 equiv TBDMSCl, THF, N_2, 16 h, 20 °C, yield 98%; (b) 1.3 equiv ClC(O)C(O)Cl, 2.6 equiv DMSO, DCM, N_2, 20 min, −78 °C; next 5 equiv TEA, 20 °C, 1 h; (c) 2 equiv K, *i*-PrOH, Et_2O, N_2, 0 °C, 2 h, yield 71%.

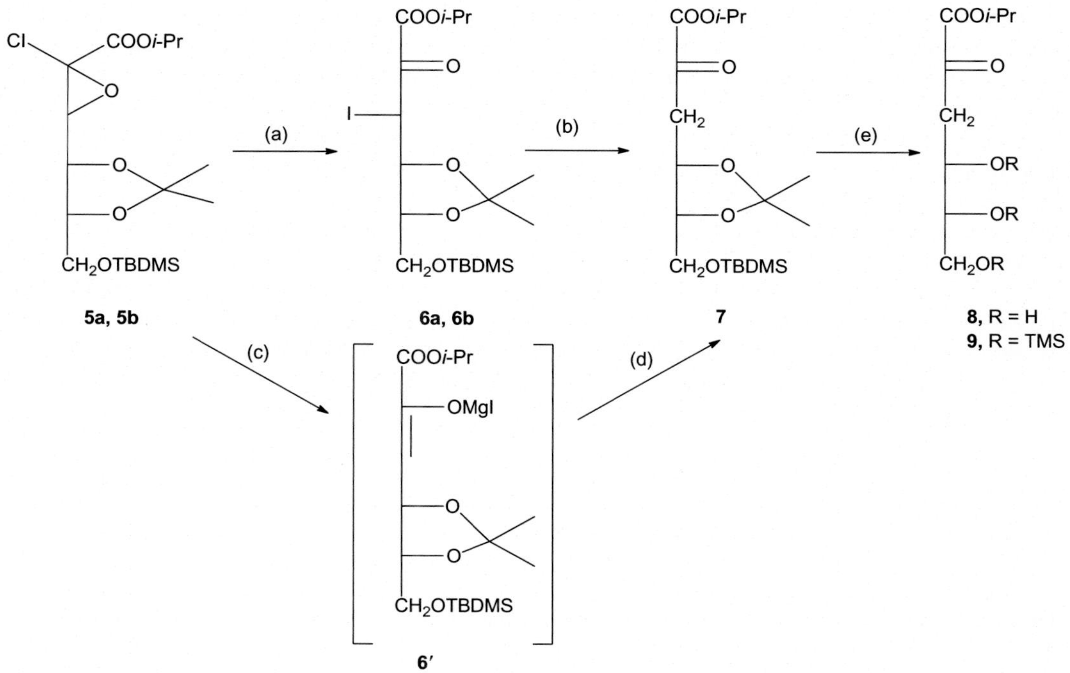

FIGURE 13

Synthesis of compounds 6-9. Conditions and reagents: (a, b; one-pot sequence) 1.3 equiv Mg, Et$_2$O, 1.3 equiv I2, 3 h, 35 °C in the dark, next aq. Na$_2$S$_2$O$_5$ 8%, 30 min, 20 °C, yield 64% (c, d; one-pot) 2.1 equiv Mg, Et$_2$O, 1 equiv I2, 1 h and 35 °C in the dark, next 15 h and 20 °C, yield 78%; (e) formic acid aq. 50%, 20 °C, 24 h, yield 87%.

persilylated to form derivative **9** to allow a better characterization on the molecule by nuclear magnetic resonance, IR (infrared), and gas chromatography mass spectra techniques.

Is KDG capable of inducing a competitive bacterial growth?

The KDG is the key metabolic intermediate of ED pathway. It is specific to this metabolic pathway and cannot be metabolized in a glycolytic pathway. It can be used as a marker of the presence of the ED pathway. In addition, KDG can be used to specifically stimulate the growth of bacteria using the ED pathway and prevent the competition for access to glucose.

In Saint-Laurent-le-Minier, the carbon resources are low due to the high pollution and low abundance of vegetation. An extensive revegetation program started with a hyperaccumulator plant *A. vulneraria* and its guest *Rhizobium metalliduran* (Grison et al., 2015). *R. metallidurans* lives in symbiosis with legume and promotes its growth by producing ammonium from the fixation of atmospheric N$_2$. Selective stimulation of the development of *R. metallidurans* compared to overall microbial population in soil of Saint-Laurent-le-Minier should also stimulate the growth of *A. vulneraria*.

Microbial diversity of polluted soil of Saint-Laurent-le-Minier was studied. A culture-independent approach was chosen to create a 16S DNA library of 303 clones from the rhizosphere of *A. vulneraria*. According to similarity searches on BLASTN, 93% of clones are new species. We chose the nearest cultivated strains (>95% similarity) to compare their growth with *R. metallidurans* on minimal medium with KDG. The evolution of the bacterial population could not be measured in solution because most strains produce extracellular polysaccharides that form aggregates. Comparing the growth was followed on solid medium with glucose as the sole carbon source, for a positive growth control, and with KDG as sole carbon source to demonstrate the existence of the ED pathway (Table 2).

The 75% of the nearest strains have not been able to grow on minimal medium enriched in KDG. The KDG allowed the growth of only 25% of the strains nearest to strains of *R. metallidurans* avoiding competition to assimilate the substrate. In addition, the strains whose growth is promoted by the KDG are all capable of producing ammonium by fixing atmospheric N_2, which enriches the contaminated soil and promotes the growth of hyperaccumulating plants.

Determination of bacterial genes for resistance to heavy metals

The environmental DNA of soil samples has been extracted and cloned in *Escherichia coli* to create a metagenomic library of approximately 273,000 clones. Screening on a medium artificially enriched

Table 2 Comparison of Bacteria Growth in Minimal Medium Containing Glucose and KDG

Sugar-Enriched Medium	Relative Number of Clones	Glucose	KDG
Rhizobium metallidurans[a]	2	+	+
Staphilococcus aureus[b]	0	+	−
Arenimonas metalli	1	+	−
Azotobacter beijerinckii	1	+	+
Bradyrhizobium canariense	3	+	f
Brevibacillus brevis	4	+	−
Burkholderia sp.	2	+	+
Gluconacetobacter diazotrophicus	1	+	−
Limnobacter litoralis	3	−	−
Nitrospira sp.	7	+	−
Nitrosospira multiformis	6	+	−
Rubrobacteridae	1	ND	ND
Segetibacter koreensis	3	+	−
Sphingomonas elodea	2	+	+
Thermobaculum terrenum	1	+	−
Xanthomonas campestris	2	+	−

+, growth; −, no growth; f, low growth; ND, not determined.
[a] *Rhizobium metallidurans has been identified in the soil in the presence of nodules of Anthyllis vulneraria.*
[b] *Staphylococcus aureus is not present in the soil; it represents a negative control.*

in Zn (8 mM $ZnSO_4$) was performed. The 36 clones of *E. coli* were found to be resistant to Zn and therefore potentially containing genes for resistance to heavy metals. The genes were sequenced and remain to be analyzed.

The KDG and the optimum conditions for phytoextraction

A study of microorganisms associated with metallophytes was performed to optimize the phytoextraction. This has been done in collaboration between the laboratory ChimEco (FRE CNRS-UM-Stratoz 3673) and the Valorhiz society. The aim was to observe the effect of microbial inoculations (bacteria and symbiotic fungi), stage seedlings or young shoots, on the growth of plants and their ability to accumulate zinc, once transferred onto substrates enriched in contaminated soil and next directly on site.

Inoculation had no negative effects on growth or accumulation of two metallophytes species. The most significant effect on the growth and accumulation is that of inoculating *A. vulneraria* with *R. metallidurans* bacteria identified in our laboratory.

Does *A. vulneraria* modify the field of phytoextraction of zinc?

The most significant result was that of *A. vulneraria* and its unsuspected abilities for phytoextraction. This plant combines remarkable ability to zinc hyperaccumulation and high biomass. Special study of this zinc hyperaccumulator was performed carefully. The results modify the classic zinc phytoextraction data, often neglected by the absence of metallophytes with developed leaf system.

Phytoextraction technologies are based on hyperaccumulators of Ni from ultramafic soils where nickel is the target metal. Such experiments with Zn hyperaccumulators have not been developed thus far. These species are rare, and the sites for study are often difficult to access because the soils rich in Zn are often subject to mining operations, leading to a degradation of their natural habitats. The mining site Les Avinieres (03°66′50″E; 43°93′13″N) illustrates this problem: Soil pollution is very high (Zn: 156,000 ppm, Pb: 36,354 ppm, Cd: 700 ppm, Tl: 115.1 ppm) but the TM are bioavailable (Grison et al., 2010). Soil fertility is low, the soil texture is not favorable to crops, water retention capacity is reduced. The known Zn hyperaccumulator on this site is *N. caerulescens*. This is a small plant that has low biomass. The other hyperaccumulator of Zn is *A. vulneraria*, identified by Frerot et al. (2006) on the pilot site as without the drawbacks from *N. caerulescens*.

A. vulneraria is definitely an opportunity for phytoextraction of Zn, thanks to its strong ability to hyperaccumulate, agronomic skills, abundant biomass, and its extensive root system capable of increasing the nitrogen-containing organic matter and soil fertility. *A. vulneraria* indeed promotes the development of a unique root zone that can promote the restoration and phytoextraction activities (Grison et al., 2014a). All these aspects have been studied by us for the first time.

A. vulneraria, legume TM-hyperaccumulating plant

About 30 subspecies of *A. vulneraria* were identified. Most of them grow on limestone soils and rocky environments, while others are able to grow on metalliferous soils. Two ecotypes derive from dolomitic limestone. A new distribution of populations of metallicolous and nonmetallicolous were recently discovered in the Cevennes, in the context of this work. Only metallicolous population has been found in this area. It belongs to the *carpatica* subspecies (Figure 14) and grows at the mine site Les Avinières (03°66′50″E; 43°93′13″N). The population consists of more than a thousand individuals, but it is not present on other mining sites in the region.

FIGURE 14

Inflorescence of *Anthyllis vulneraria* subsp. *carpatica*.

Like *A. vulneraria*, many other legumes are also known to tolerate TM and grow on soil rich in metals (Jaffré and L'Huillier, 2010). According to Pajuelo et al. (2007), legumes are not hyperaccumulators. However, some exceptions can be found; legumes can hyperaccumulate Pb (Sahi et al., 2002) and Se (Alford et al., 2012). Therefore the potential for accumulation of *A. vulneraria* subsp. *carpatica* was studied for 5 years and 500 plants (Figure 15). It has been compared to the best reference in the field of Zn hyperaccumulators, namely *N. caerulescens*, naturally present on the site Avinières.

The capacity for accumulation of Zn, Cd, and Pb by *A. vulneraria* and *N. caerulescens* were studied by two independent methods: atomic absorption spectrometry (AAS) and polarography. As shown in Figure 15, Zn concentration is greater than 10,000; both metallophytes can be considered as Zn hyperaccumulators. The concentration of Zn in the aerial parts of *A. vulneraria* is even greater than that of *N. caerulescens*. If one refers to the definition of phytostabilization, the use of *A. vulneraria* for phytostabilization of metalliferous soil is inappropriate. We propose the use of *A. vulneraria* in phytoextraction of Zn.

Moreover, *A. vulneraria* appears to be more selective for metal phytoextraction than *N. caerulescens*, especially vis-à-vis the accumulation of Cd. In view of these results, cultures of *A. vulneraria* were performed under controlled conditions to investigate the profile of Zn accumulation. The concentration of Zn in the leaves of *A. vulneraria* was measured by AAS at different periods after transplantation (Figure 16).

The maximum concentration of Zn, 17.428 ppm, was measured 240 days after transplantation at the level of hyperaccumulation foliar parts of *A. vulneraria* (Figure 17).

An opportunity for phytoextraction of zinc

The estimation of biomass production per hectare may be performed by calculating how many plants are necessary to cover 1 ha with vegetation. Agronomic aspects of phytoextraction were the object of

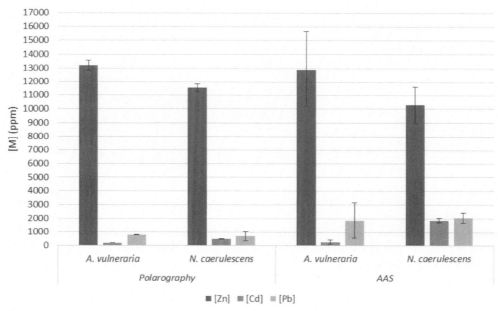

FIGURE 15

Comparison of concentrations (ppm) of Zn, Cd, and Pb in leaf parts of *A. vulneraria* and *N. caerulescens*.

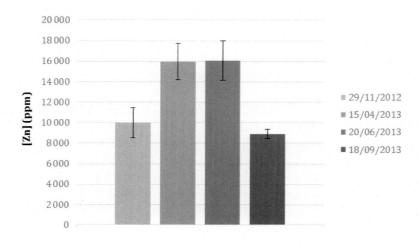

FIGURE 16

Zn concentration in leaves of *A. vulneraria* measured by AAS after 60, 180, 240, and 300 days of transplantation.

FIGURE 17

Anthyllis vulneraria crops on highly contaminated soil.

modeling and calculations detailed by Robinson et al. (2003). In its simplest form, the amount of zinc extracted per hectare per year through phytoextraction process depends primarily on the proportional relationship between the zinc concentration of a crop after phytoextraction and harvestable biomass, therefore:

$$YZn = FZn \times Ybio$$

FZn: average fraction of zinc in the biomass of *A. vulneraria*
Ybio: biomass yield of hyperaccumulator (kg/(ha × year))
YZn: zinc total gain (kg/(ha × year))

If we accept 70% of land cover, the number of plants of *A. vulneraria* per hectare is estimated at 266,160 on the site Les Avinières. The theoretical yield per hectare is 1791 kg ha^{-1} of dry biomass. The average fraction of zinc in the biomass of *A. vulneraria* is estimated at 1.5% to calculate the total gain zinc: foliar parts of *A. vulneraria* are able to extract 27 kg ha^{-1} of Zn in one crop. This result compares with the zinc hyperaccumulator of reference, *N. caerulescens*, the theoretic yield by hectare is 1632 kg ha^{-1} and the average fraction of zinc in *N. caerulescens* is 1%, thus the leaf extract parts of *N. caerulescens* allow to obtain 16.8 kg ha^{-1} of Zn in one crop.

2.2 PHYTOEXTRACTION ON CALEDONIAN MINING SITES: FROM DEGRADATION TO PRODUCTIVE REHABILITATION

2.2.1 An economy toward the exploitation of nickel
In New Caledonia, the key resource of the island is nickel associated with cobalt. Caledonian nickel deposits represent 20-25% of the world's nickel resources, and low-grade laterite could be added to this potential. Nickel mining, which now represents 10% of the gross domestic product (GDP) of the island, could reach 30-40% of GDP in the coming years; it is clearly an economic issue of great importance to this territory.

2.2.2 What are the environmental consequences?
Today all Caledonian mines are open pit; before starting the operation vegetation cover, the topsoil (fertile upper horizon) and other upper horizons must be stripped. If the techniques are continuously

improved, direct ecological impacts are obvious and significant: landscape impact, loss of biodiversity, carbon clearance, etc. New Caledonia is home to over 3350 plant species, 74% of which are endemic to the island. These features make New Caledonia a land of biodiversity that must be preserved. The indirect effects of large-scale mining are also observed: the disruption of water flows associated with soil erosion (Figure 18), which pollutes the lagoon (according to UNESCO) and threatens the coral reef surrounding the island.

The ever-decreasing ore grade, of the order of 3% nickel today, also has the effect of increasing the surface of mining site areas. To date, there have been at least 32,000 ha impacted by past mining operations in New Caledonia. With the growing impact of mining activities, ecological restoration and rehabilitation of degraded areas becomes a major issue of concern to researchers, industry, and government. Despite this awareness, the annual recoverability of the territory is estimated at $30\,\mathrm{ha\,year^{-1}}$. Therefore, the gap between degradation (32,000 ha) and ability to restore ($30\,\mathrm{ha\,year^{-1}}$) is huge.

In this context, our laboratory (Prof. C. Grison, Laboratory of Bio-inspired Chemistry and Environmental Innovations, FRE CNRS 3673 UM Stratoz, Montpellier) has developed a new approach to ecological rehabilitation in order to promote industrial efforts in the field. The research program is based on a sustainable recovery, which takes into account ecological aspects, but for the first time also possible economic opportunities based on the development and use of innovative catalysts derived from plants used as metallophytes in restoration process. In 2010, joint discussions between Eramet (P. Jeanpetit, S. Lepennec), Société Le Nickel (SLN, Alla P., F. Bart, Nicolas C.), the laboratory of Bio-inspired Chemistry and Ecological Innovations (Prof. C. Grison) and the Agronomic Institute Caledonian (IAC, L. L'Huillier, B. Fogliani) led to the joint development of a revegetation program on one of the mine sites of SLN by using hyperaccumulating plants, of Ni (II) and Mn (II). This study is part of a larger project funded by the French National Research Agency (ANR), which combines the CNRS and the IAC. The use of hyperaccumulating plants, of Ni (II) and Mn (II) particularly, abundant in New Caledonia allowed consideration of their valuation in the green chemistry program of CNRS.

FIGURE 18

Erosion caused by mining operations.

2.2.3 What are the objectives of this study?

At the beginning, the main objective was to demonstrate the possibility of developing an integrated system, culture of hyperaccumulators of metals of interest, to produce ecocatalysts for the synthesis of molecules with high value added. The particular aspect studied here concerns the possibility of producing biomass rich in TM (Ni, Mn) by growing hyperaccumulating plants on old mine sites in New Caledonia. Two species known for their good behavior on degraded sites (middle open, soil and water constraints) and germination were selected by the CNRS and IAC partners: *Geissois pruinosa*, nickel hyperaccumulator (Jaffre et al., 1979) and *Grevillea exul* (*exul* varieties and rubiginosa), which accumulates manganese (Jaffre, 1979). The production of seedlings from seeds was selected immediately in order to maintain maximum genetic diversity in the trials set up.

- *Geissois pruinosa*

Geissois pruinosa is a tree that can reach from 3 to 6 m high and is found especially in the southern mountains, only on ultramafic substrates at altitudes of 30-700 m, on various soils except lateritic soils. The species is most often found in groups or as pre-forest but adapts fairly well to various levels of soil moisture, making it possible to consider its use in revegetation of mining sites (Figure 19). The average leaf mineral composition is an average nickel content of 6117 mg kg^{-1} (dry

FIGURE 19

Geissois pruinosa.

matter). Concentrations up to 600 mg kg^{-1} of nickel were observed in *Geissois pruinosa*, making it the most remarkable hyperaccumulator of nickel out of all *Geissois* (threshold hyperaccumulation 10,000 mg kg^{-1} for Ni).

- *Grevillea exul*

Grevillea exul is a shrubby species; the variety *exul* is very common across the south of New Caledonia, but can be found also in the northwest (Figure 20). The average leaf mineral composition is mostly manganese 2590 mg kg^{-1} (dry matter), which situates *Grevillea exul* among the richest manganese species but without being a hyperaccumulator. Individual measurements of manganese content pushed up to 3900 mg kg^{-1} in the *exul* variety and even 6-1200 mg kg^{-1} for variety *rubiginosa* (Jaffre, 1979). The two varieties of *Grevillea exul* showed good results in pre-revegetation trials and are naturally abundant in the deep south of New Caledonia.

2.2.4 Two tests, two objectives

Test 1. Conditions of transplantation on mine sites

The first test was launched in April 2012 and aimed at determining the responses of plant species to different fertilizer treatments when implanted on contaminated mine sites.

Test 2. Effect of nitrogen-fixing symbioses

A second test was set up in December 2012; its goal was to study the impact of additional species with nitrogen-fixing symbioses on the implementation of *Geissois pruinosa*. The advantage of focusing on *Geissois pruinosa* for a test of this type is that this species is less well suited to open areas

FIGURE 20

Grevillea exul.

than *Grevillea exul* and could therefore benefit more from the contribution of additional species. The nitrogen-fixing symbioses are frequently encountered in New Caledonia. Three species in particular have been selected here: *Gymnostoma deplancheanum* (Figure 21) from Casuarinaceae and *Seryanthes calycina* (Figure 22) and *Storckiella pancheri* from legumes.

These species generally have an interest in ecological restoration programs by symbiotic association, they enrich the soil with nitrogen and by rapid growth, and they can provide shade for the development of other species, particularly *Geissois pruinosa*. However, to date neither their

FIGURE 21

Gymnostoma deplancheanum.

FIGURE 22

Seryanthes calycina.

effects nor their associations with TM-hyperaccumulating species have been studied in detail. Those were the objectives of the Test 2, which aimed to determine contribution of these additional species.

2.2.5 Where were tests 1 and 2 carried out?

The first tests were conducted on the Camp des Sapins in Thio (southern province) on SS3 verse (Figure 23) and amphibole (Figure 24).

FIGURE 23

Verse SS3.

FIGURE 24

Amphibole.

Test 1—treatments: fertilizers and doses:
Two soil types were considered for this test: topsoil (30-40 cm) placed on SS3 and amphibole. On each of the two soils, three fertilization treatments were set up:

- Treatment 1: sewage sludge
- Treatment 2: Fertilizer Yates Nutricote 365 days
- Treatment 3: Fertilizer 17-17-17 + Orgasol

In each case, these fertilizers were placed in holes of characteristic dimensions of $30 \times 30 \times 30$ cm and $0.09 \, m^2$ surface per hole. Fertilizer amounts added were adjusted so that the amount of nitrogen introduced was similar in each treatment. The nitrogen inputs are crucial to the growth of seedlings established on degraded soil.

2.2.6 What are the constraints imposed by the nature of the caledonian soil?

Soil conditions on mining lands are quite specific with generally observed deficiencies compared to other types of soil:

- The total content of phosphorus, potassium, and calcium are low with an excess of magnesium that can be toxic;
- This phosphorus can be complexed on the iron oxides present on the surface, making it unavailable to plants;
- The organic material is present in small amounts and deteriorates very slowly; deficiencies in C and N are the result;
- Finally, the presence of elements such as Ni, Cr, or Co are generally toxic for plants, making these poor soils particularly unsuitable for agricultural use.

Detailed planting protocols of hyperaccumulating plants used as part of the partnership between SLN, ChimEco laboratory (CNRS FRE-UM-Stratoz 3673), and the IAC on mine Camp des Sapins in Thio are described in detail in our recent article (Losfeld et al., 2014b).

Today, monitoring for growth and survival of the plants shows encouraging results. The most productive treatments may be selected with respect to the biomass produced and mineral compositions. Finally, all the results obtained depend on the age of the collected leaves. The advantage of this aspect is discussed in more detail in our article (Losfeld et al., 2014c).

The experimental fields established at Camp des Sapins can also be seen as potential seed fields. The gradual introduction of new endemic species hyperaccumulating Ni and Mn in order to extend the biodiversity of species involved in revegetation is envisaged. In this context, it will be important to study the possibility of developing hyperaccumulators of nickel such as *Phyllanthus favieri* or *serpentinus* and other manganese hyperaccumulators such as *Garcinia amplexicaulis*, *Denhamia fournieri*, *Beaupreopsis paniculata*, or *Beauprea gracilis* and *montana*, provided that revegetation sites to which these species are adapted are found.

Remediation of degraded and/or contaminated mining sites is a long-term process. Several aspects need to be taken into consideration such as the status of sites, wise and sustainable planning of operations, growth of plants on degraded soil with respect for local biodiversity, monitoring transplants, the accumulation rate. In such a context, it is clear that economic valuation of remediation is essential to support such efforts over time (Figure 25).

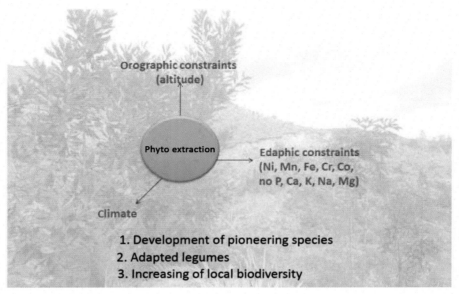

FIGURE 25

Toward a real program of ecological restoration.

3 FROM PHYTOEXTRACTION TO GREEN CHEMISTRY AND VICE VERSA THROUGH ECOCATALYSIS

The valorization of phytoextraction by ecocatalysis is the subject of very few experiments described previously in the literature. The green chemistry process we propose is based on new trends present in the chemical industry based particularly on the application of metal catalysis, and this makes our approach new.

Catalysis in organic chemistry is a demonstrative example of sustainable chemistry in the sense that it can become the engine of a new chemical industry (Corma and Garcia, 2003). Catalysis consumes little first metallic material, thus it is suitable for the proposed approach because it does not require large yields of phytoextraction; 1-5% by MX_n of metal species are generally sufficient to catalyze an organic reaction. Additionally, if the catalyst is supported on a solid, the metal species become recyclable and can be reused.

In addition, metallophyte species concentrate a polymetallic composition. Phytoextraction focuses on a type of TM that is mixed with other metal cations. The study of polymetallic catalysis from metallophytes showed that the presence of several metal cations present does not threaten the catalyst activity. On the contrary, it leads to a gain of stability and catalytic activity; these results are consistent with the tests described in the literature where mixtures of conventional metal species led to the same observations (Corma and Garcia, 2003). In our case, the mixture is natural, and the purification of the proposed catalytic systems is not necessary for the majority of cases. Therefore, the condition of the use of extracts of the TM from metallophytes as ecocatalysts respects the constraints of economic competitiveness.

Finally, the proposed strategy fulfills 2 out of the 7 strategic goals set by the European Commission Innovation 2030 and may constitute a pillar of future development.

– Metal recycling: Projects to make metal recycling viable and effective

– Plant chemistry: Projects using plant chemistry to develop new materials

All aspects are explained in the following section

3.1 ECOCATALYSIS: THE STARTING POINT OF A NEW GREEN ECONOMY?

3.1.1 A project that is part of a sustainable and new chemistry

The concept of green chemistry expressed through the International Network GSCN, a global vision and broader than sustainable chemistry; European chemicals regulation REACH, the report of MEDDEM published on March 23, 2010, placing green chemistry as one of the green industries of the future; and the development of a large scenario of ecologically sustainable chemistry by the Japanese Ministry of Industry (METI) are all witnesses of a rapidly changing discipline. In 2011, declared the International Year of Chemistry, the scientific community demonstrated that it has the tools and the will to contribute effectively to the major problems focused on the environment, energy, resource depletion, and quality of life.

It proposes in particular to effectively contribute to the reduction of waste by developing innovative green technologies for sustainable reuse, reducing the quantities, replacement, and diversification of raw materials in the discipline. One of the pillars of the concept of green chemistry is the use of catalytic systems to replace stoichiometric reagents. A nonenzymatic catalyst is frequently a noble or primary transition metal, a rare earth metal. In February 2010, the Inter-Cluster Foresight and Anticipation of Economic Changes group published a disturbing report on the depletion of many mineral resources; rare earths metals were on that list.

Access to strategic and primary minerals has become crucial from not only the point of view of resource depletion, but also because they are held by a small number of countries, mostly politically unstable. Europe faces a major challenge based on the development of competence and innovation.

In this context ecocatalysis offers the advanced development of a new circular green sector based on recycling of mineral resources (Figure 26). The main objective is to develop an international innovative process of chemical enhancement of phytoextraction technology and biomass contaminated by TM. Taking advantage of the remarkable adaptive capacity of certain plants to hyperaccumulate cations such as Zn^{2+}, Ni^{2+}, Mn^{2+}, Cu^{2+}, Co^{2+}, and Pd^{2+}, the project design is based on the direct use of metal, originated from plants, as reagents or catalysts for use in organic chemical reactions.

Polymetallic extracts that we have obtained offer a unique opportunity to exploit the cooperative catalysis in which synergy and selectivity seem promising. Unimaginable just a few years ago, this new

FIGURE 26

Principles of ecocatalysis.

concept is beginning to emerge as a true revolution in the field of catalysis and related phytotechnologies. It will provide concrete answers to economic and technical changes announced. It already brings new momentum to phytoextraction and green chemistry.

3.1.2 Cooperativity is useful and productive, even in chemistry

The spectacular development of organometallic catalysis during the last 50 years has enabled the preparation of biomolecules whose molecular architecture is becoming more complex. The developed syntheses have often been limited by environmental risks and cost of catalyst. The search for new catalytic processes respecting the principles of green chemistry has become important. Thus, research into the effectiveness of catalysis and catalyst recycling has become of major concern.

The effectiveness of multimetal catalysts is beginning to be addressed by mechanistic studies of supported metals and metal combinations (Haruta et al., 1987; Figure 27). Beyond simple transformations, polymetallic catalysis is particularly suited for multicomponent reactions that are explained by sequential and complex mechanisms.

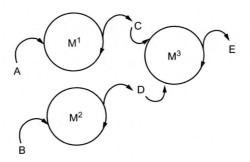

FIGURE 27

Polymetallic ecocatalyst and cooperative catalysis. Catalyst I converts A to B which reacts with D using catalyst III and gives the desired product P. The synthesis of intermediate D could be considered in the same pot from C using catalyst II. This diagram illustrates the possibilities of multicentered catalyst and therefore multimetallic.

The mechanistic study of such complexes is of paramount importance and is extensively studied in the laboratory ChimEco. A crucial feature of ecocatalysts is their polymetallic composition resulting from the combination of transition elements at high concentrations (e.g., Zn^{2+}, Ni^{2+}, Mn^{2+}, Cu^{2+}, and Pd^{2+}) with conventional elements generally necessary for the development of the plants (e.g., Na^+, K^+, Ca^{2+}, Mg^{2+}, and Fe^{3+}). The simultaneous presence of a combination of well-defined active sites results from this variety of metal species. Therefore, the original reaction sequences allow leading to unique selectivity. Indeed, a conventional catalyst could be limited to only influence some of the steps of the reaction process, limiting the opportunities in organic synthesis. Here, the richness of different interactions between species leads to interactions metal/unusual ligand in solution.

Control of metal-ligand interactions can accurately determine the different components of the catalyst and thus result in a very precise understanding of its reactivity. Control of the crucial phase of preparation protocol of ecological catalysts—(1) the assembly of components, (2) the heat treatment, (3) addition of ligands (4) deposit on support—are studied to understand their effects on multiple compositions, structures, textures, and functions induced. The choice of experimental conditions can precisely control the size, shape, composition, structure, chemistry, and finally the surface reactivity.

Today, this green chemistry program is no longer a valuation of remediation technologies. Polymetallic systems obtained from the biomass produced by phytoextraction are original: unusual degrees of oxidation, (Grison et al., 2013) new chemical species, performance (activity, chemo- and stereoselectivity superior to conventional catalysts in a number of reactions). This original work was validated by a series of patents and scientific papers relevant to the aforementioned (Escande et al., 2014a,b,c,d,e,f, 2015; Losfeld et al., 2012a,b,c; Thillier et al., 2013; Grison et al., 2013; Garel et al., in press). The synthetic possibilities of ecocatalysts are far beyond what was originally planned. Many mechanisms of organic synthesis have been revisited by the principle of ecocatalysis (Figure 28). Over 3500 biomolecules could be prepared with success.

Ecocatalysts can be used as heterogeneous catalysts in synthetic transformations allowing access to molecules with high added value for the fine and industrial chemicals (flavors and cosmetics with the "natural" label, drugs and oligomers of biological interest, aromatic heterocycles with highly functionalized chiral structures, key intermediates for various industrial chemical processes and organic materials).

3.1.3 What are concrete examples of the performance of ecocatalysis?

A new approach to pharmaceutical chemistry: from *Psychotria douarrei* to Monastrol

The superiority of the ecology for organic synthesis appears to stem from the ability to accurately select the metal-substrate interaction thanks to the presence of different metal sites with complementary properties. In this context, multicomponent reactions such as the Biginelli reaction are excellent models to study. The chosen example is the synthesis of Monastrol, an interesting heterocycle with antimitotic properties. It is in fact capable of blocking the migration of chromosomes to the cell poles during mitosis (Figure 29).

The preparation of Monastrol requires successive activation of hard and soft base sites to build the final heterocycle.

An evaluation of the overall reactivity of the various ecocatalysts prepared was performed by the average hardness, according to the hard soft acid base (HSAB) theory of Pearson, by the global mineral composition and by Lewis acidity study, performed by IR-ATR. These studies show that the Eco-Ni® ecocatalysts are particularly well suited for the synthesis of Monastrol and much superior to the classical $NiCl_2$ use in the synthesis.

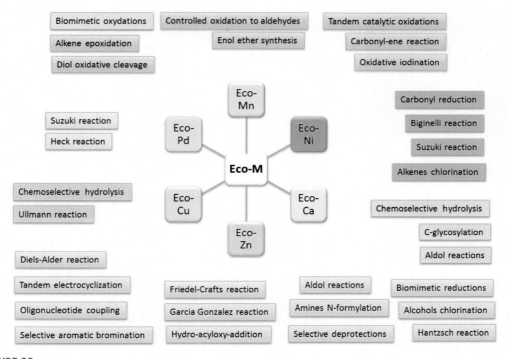

FIGURE 28

Organic synthesis revisited by ecocatalysis.

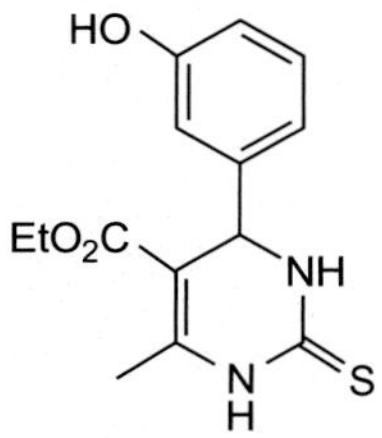

FIGURE 29

Monastrol.

This situation was exploited in our laboratory, particularly in the case of the multicomponent domino Biginelli reaction (Figure 30), where the first step involves catalysis by a hard Lewis acid, forming an alkylidene thiourea, which reacts with the ethyl acetoacetate under the catalysis by a soft Lewis acid. Finally, the last intermediate undergoes cyclization under hard Lewis acid catalysis to arrive at dihydropyrimidones like heterocycles or thioxo-tetrahydropyrimidine. In the case of thiourea, the obtained product is Monastrol molecule.

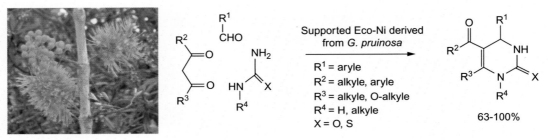

FIGURE 30

Biginelli reaction catalyzed by novel supported ecocatalysts.

Comparison of the activity of the Eco-Ni® to that of classical $NiCl_2$ shows the interest of polymetallic catalysis by various hardness acids, since Monastrol is obtained in 72% yield with Eco-Ni® and only 11% yield with $NiCl_2$ (Figure 31). Furthermore, the starting aldehydes are frequently prone to self-condensation when using $ZnCl_2$ type of salts, which is only slightly observed with the Eco-M®. A possible reason for this result is softer activity mixed transition metal salt/alkali metal or alkaline earth metal, which act as transition metal chloride tanks, gradually released into the reaction medium.

3.1.4 Oxidizing chemistry: alternative solutions according to REACH

Oxidizing chemistry is one of the sectors of chemistry that requires a rapid and concrete innovation to meet the increased demands of European regulations on the restriction and authorization of chemicals (REACH). As part of the requirements of REACH, the chemical industry must develop new methodologies in oxidation reactions; many of them make use of hazardous or toxic reagents or catalysts.

Epoxidation reactions are particularly affected by these requirements, since both reagents commonly used in fine chemical industry, m-chloroperbenzoic acid (mCPBA) and *tert*-butyl hydroperoxide (*t*BuOOH), combine many risks of manipulation and generate pollutant wastes and reactions employing these reagents are not atom economic.

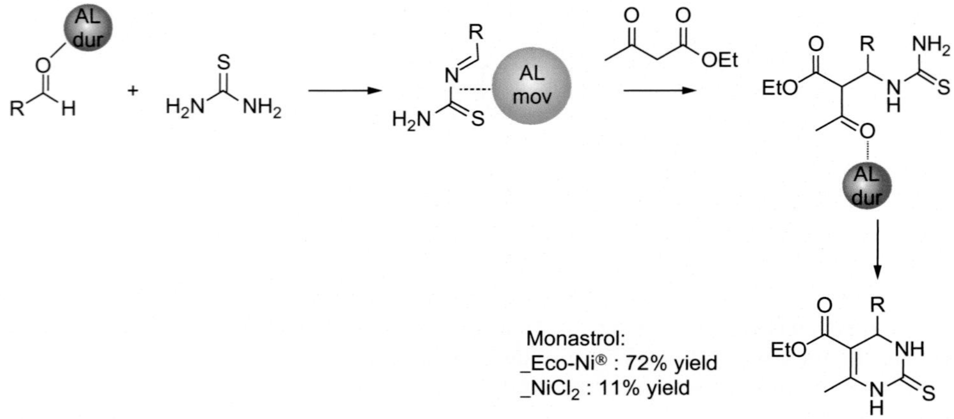

Monastrol:
_Eco-Ni® : 72% yield
_$NiCl_2$: 11% yield

FIGURE 31

Polymetallic catalysis: an advantage for the multicomponent reaction.

In this context, research is underway to develop epoxidation systems using hydrogen peroxide, where water would be the only by-product and which has the advantage of being neither flammable nor toxic. Various bio-inspired Mn-based catalysts have proven effective in these reactions, but they require complex ligands, the preparation of which is difficult and which are therefore particularly expensive. Moreover, the ligands and the bio-inspired Mn-based catalysts reported thus far are usually not recyclable.

A simpler system was proposed by a US team in the year 2000 (Lane and Burgess, 2001), using only simple salts of Mn, involving the formation of Mn (II) peroxycarbonate as epoxidation species (Figure 32). The search for new methods, however, is still required. The researcher has the duty to seek out new solutions and transfer them to the industrial environment, if developable on a large scale.

We sought to determine how the Eco-Mn® was able to catalyze this reaction and possibly make a particular selectivity. Furthermore, it was desirable to obtain a heterogeneous catalytic system, recyclable at the end of reaction.

Inspired by the biological oxidants and their geometric organization in the cytochromes P450, we have prepared a bio-based analog natural heme by electrophilic substitution through ecocatalyst Eco-Mn®. If the basic porphyrin structure of heme is retained, the cation of naturally present iron (II) is substituted by the Eco Mn® enriched in manganese in oxidation state III. Many oxidation reactions were carried out with success. The example of the one-pot synthesis of vanillin from isoeugenol illustrates this possibility: It is based on a one-pot reaction sequence epoxidation/epoxide opening/oxidative cleavage. The preparation of vanillin is remarkably efficient and the catalyst used recyclable (Figure 33). This is an example of green and biosourced chemistry able to reconcile efficiency/speed/naturalness and ecoresponsibility. Green chemistry can be an engine of economic and environmental progress!

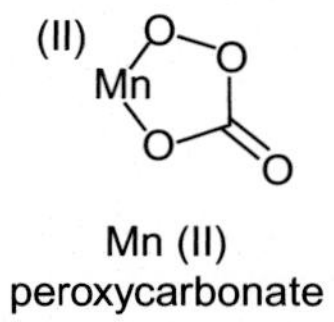

Mn (II)
peroxycarbonate

FIGURE 32

Mn (II) peroxycarbonate.

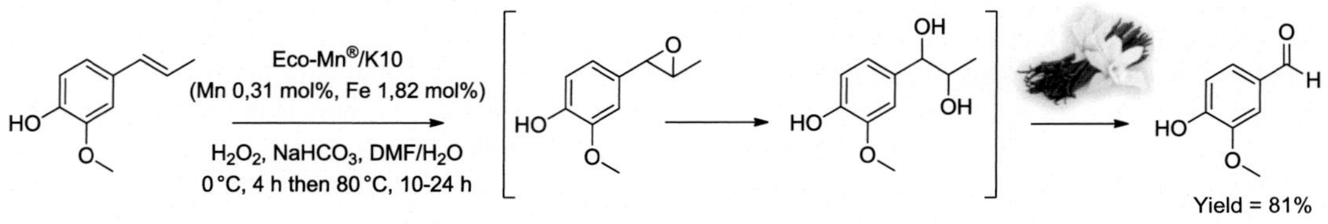

FIGURE 33

Direct synthesis of vanillin catalyzed by Eco-Mn®.

3.1.5 A need to return to nature: biocosmetics or natural cosmetics?

The recent trends in the flavor and fragrances industry are based on an "all natural" approach in which each component (acid carrier ligand, catalyst, reagent, substrate) will be or is a natural substance. These actions allow developing innovative access to molecules that have the "natural" label. This is fundamental for cosmetic applications, especially to European and Asian markets. From a more fundamental point of view, the concept of synthesis is based on the differences in reactivity of complex intermediates formed *in situ* from natural ligands and metal precursors, where the precise control of the physicochemical properties of the surface and reactants used must be ensured, all respecting strict standards of naturalness. These regulatory constraints are a motivation to innovate, to create a different chemistry, bio-inspired and respectable. Today rehabilitation by phytoextraction and rhizofiltration allows us to obtain exceptional biodiversity, source of a new chemodiversity, to develop an innovative ecocatalytic chemistry, more efficient than conventional systems and able to meet the ambitions of the field of biocosmetics (Escande et al., 2014b).

4 CONCLUSION: ECOCATALYSIS, A NEW VISION OF GREEN CHEMISTRY

The concept of ecocatalysis allows us to break free from all the limitations inherent to existing methods via a new generation of environmentally stable and functional materials with more reactive interfaces whose properties can be controlled by the intensity and nature of the interaction metal/substrate, and combined with biodiversity-related phytoextraction efforts.

Our ecocatalysts have a high synthetic potential, very superior performance in a number of transformations when compared to conventional catalysts, and they can be utilized in the synthesis of complex biomolecules of industrial importance. Those ecocatalysts cannot be considered as a complete replacement for catalysts issued from metallurgy, but rather as new tools that incorporate a triple vision of chemistry/ecology/environment.

What are the overall conclusions? What future? In just 6 years, ecocatalysis has been built around a new concept of global ecology that combines reflection and action. A rapid assessment could be the following:

- Eco-design of a new system able to create new products and processes that take advantage of specific features of phytoremediation (phytoextraction and rhizofiltration) and resources generated (waste, plant and bacterial biomass) (Figure 34);
- An original fundamental scaffold based on many organic synthesis results inspired by the founding principles of green chemistry;
- The creation of a new industry capable of rational and useful management of phytoremediation while actively contributing to the achievement of concrete environmental goals.

These results are the prerequisites for the orientation of environmental technology innovations to the socioeconomic interests and ecology of industrial pollution. Stratoz and industrial partners interested in this technology are working together with the Laboratory of Chemistry and Bio-inspired Eco Innovations on the development of this new green industry. One of the partners is in the process of testing the ecocatalysts derived from the biomass in its own production installations.

Beyond its scientific and economic interest, the development of such an approach aims to boost phytoextraction on many polluted or degraded sites by giving it an industrial dimension and social motivation. It is always a very difficult task to be implemented because it combines scientific issues

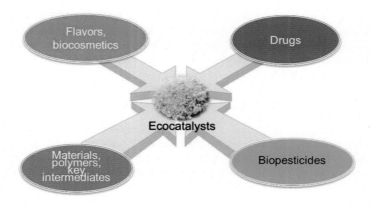

FIGURE 34

Applications derived from ecocatalysis.

(the process is innovative) and difficult human problems (rehousing habitants, redemption of mining lands by the municipality, specific political context). Obtaining approvals by the public authorities and local governments is necessary for the success of the program.

Concrete and advanced engagement with local, regional, and national governments are taken in this direction. The aim is threefold:

- To think about a sustainable rehabilitation of land, compatible with the environment, local people, and relevant authorities;
- To be part of an approach that is consistent with regulation;
- To assess the progressive economic development project involving administration, concerned residents, and industrial success.

It is hoped that the legal and regulatory environments encourage the growth of this new industry, so Stratoz and associated industrial partners can be vectors of sustainable development.

REFERENCES

Alford, E.R., Pilon-Smits, E.A.H., Fakra, S.C., Paschke, M.W., 2012. Selenium hyperaccumulation by Astragalus (Fabaceae) does not inhibit root nodule symbiosis. Am. J. Bot. 99, 1930–1941.

Bert, V., Tack, K., Vialletelle, F., Magnie, M., Berquet, A., Cochet, N., Schoefs, O., 2009. Prospects in biomass valorization from phytoextraction of Cd, Pb and Zn with hyperaccumulators. Recents Progrès en Génie des Procédés 98, 297.

Corma, A., Garcia, H., 2003. Lewis acids: from conventional homogeneous to green homogeneous and heterogeneous catalysis. Chem. Rev. 103, 4307–4365.

Coutrot, P., Grison, C., Tabyaoui, M., Czernecki, S., Valery, J.-M., 1988. Novel application of alkyl dihalogenoacetates; chain extension with an alpha-ketoester unit of carbohydrates. J. Chem. Soc. Chem. Commun, 1515–1516.

Coutrot, P., Dumarçay, S., Finance, C., Tabyaoui, M., Tabyaoui, B., Grison, C., 1999. Investigation of new potent KDO-8-phosphate synthetase inhibitors. Bioorg. Med. Chem. Lett. 9, 949–952.

Escande, V., Garoux, L., Grison, C., Thillier, Y., Debart, F., Vasseur, J.J., Boulanger, C., Grison, C., 2014a. Ecological catalysis and phytoextraction: symbiosis for future. Appl. Catal. B Environ. 146, 279–288.

Escande, V., Olszewski, T., Grison, C., 2014b. From biodiversity to catalytic diversity: how to control the reaction mechanism by the nature of metallophytes. Environ. Sci. Pollut. Res. 22 (8), 5653–5666. http://dx.doi.org/10.1007/s11356-014-3483-6.

Escande, V., Olszewski, T.K., Grison, C., 2014c. Preparation of ecological catalysts derived from Zn hyperaccumulating plants and their catalytic activity in Diels–Alder reaction. C. R. Chim. 17, 731–737.

Escande, V., Olszewski, T.K., Petit, E., Grison, C., 2014d. Biosourced polymetallic catalysts: an efficient means to synthesize underexploited platform molecules from carbohydrates. ChemSusChem 7, 1915–1923.

Escande, V., Renard, B.-L., Grison, C., 2014e. Lewis acid catalysis and Green oxidations: sequential tandem oxidation processes induced by Mn-hyperaccumulating plants. Environ. Sci. Pollut. Res. 22 (8), 5633–5652. http://dx.doi.org/10.1007/s11356-014-3631-z.

Escande, V., Velati, A., Grison, C., 2014f. Ecocatalysis for 2H-chromenes synthesis: an integrated approach for phytomanagement of polluted ecosystems. Environ. Sci. Pollut. Res. 22 (8), 5677–5685. http://dx.doi.org/10.1007/s11356-014-3433-3.

Escande, V., Velati, A., Garel, C., Renard, B.-L., Petit, E., Grison, C., 2015. Phytoextracted mining wastes for ecocatalysis: Eco-Mn, an efficient and eco-friendly plant-based catalyst for reductive amination of ketones. Green Chem. 17, 2188–2199. http://dx.doi.org/10.1039/c4gc02193b.

Frerot, H., Lefebvre, C., Gruber, W., Collin, C., Dos Santos, A., Escarre, J., 2006. Specific interactions between local metallicolous plants improve the phytostabilization of mine soils. Plant Soil 282, 53–65.

Garel, C., Renard, B.-L., Escande, V., Galtayries, A., Hesemann, P., Grison, C. C-C bond formation strategy through ecocatalysis: insights from structural studies and synthetic potential. Appl. Catal., A, in press, 10.1016/j.apcata.2015.01.021.

Grison, C., Escarré, J., Berthommé, M.-L., Couhet-Guichot, J., Grison, C.-M., Hosy, F., 2010. Thlaspi caerulescens, un indicateur de la pollution d'un sol? Actual. Chim. 340, 27–32.

Grison, C., Escande, V., Petit, E., Garoux, L., Boulanger, C., Grison, C., 2013. *Psychotria douarrei* and *Geissois pruinosa*, novel resources for the plant-based catalytic chemistry. RSC Adv. 3, 22340–22345.

Grison, C.M., Mazel, M., Sellini, A., Escande, V., Biton, J., Grison, C., 2014a. The leguminous species *Anthyllis vulneraria* as a Zn-hyperaccumulator and eco-Zn catalyst resources. Environ. Sci. Pollut. Res. 22 (8), 5667–5676. http://dx.doi.org/10.1007/s11356-014-3605-1.

Grison, C.M., Renard, B.L., Grison, C., 2014b. A simple synthesis of 2-keto-3-deoxy-D-erythro-hexonic acid isopropyl ester, a key sugar for the bacterial population living under metallic stress. Bioorg. Chem. 52, 50–55.

Grison, C.M., Jackson, S., Merlot, S., Dobson, A., Grison, C., 2015. *Rhizobium metallidurans* sp. nov., a symbiotic heavy metal resistant bacterium isolated from the *Anthyllis vulneraria* Zn-hyperaccumulator. Int. J. Syst. Evol. Microbiol 65, 1525–1530.

Haruta, M., Kobayashi, T., Sano, H., Yamada, N., 1987. Novel gold catalysts for the oxidation of carbon-monoxide at a temperature far below 0-degrees-C. Chem. Lett. 16, 405–408.

Jaffre, T., 1979. Manganese accumulation by New Caledonian Proteaceae. C.R. Acad. Sci. D Nat. 289, 425–428.

Jaffré, T., L'Huillier, L., 2010. La végétation des roches ultramafiques ou terrains miniers. In: Jaffré, T., L'Huillier, L., Wulff, A. (Eds.), Mines et Environnement en Nouvelle-Calédonie: les milieux sur substrats ultramafiques et leur restauration. IAC, Nouméa, Nouvelle-Calédonie.

Jaffre, T., Brooks, R.R., Trow, J.M., 1979. Hyper-accumulation of nickel by Geissois species. Plant Soil 51, 157–162.

Lane, B.S., Burgess, K., 2001. A cheap, catalytic, scalable, and environmentally benign method for alkene epoxidations. J. Am. Chem. Soc. 123, 2933–2934.

Leguen, M., Orgeval, J.J., Lancelot, J., 1991. Lead isotope behavior in a polyphased Pb-Zn ore deposit – Les Malines (Cevennes, France). Mineral. Deposita 26, 180–188.

Liang, P.-H., Lewis, J., Anderson, K.S., Kohen, A., D'Souza, F.W., Benenson, Y., Baasov, T., 1998. Catalytic mechanism of Kdo8P synthase: transient kinetic studies and evaluation of a putative reaction intermediate. Biochemistry 37, 16390–16399.

Losfeld, G., de la Blache, P.V., Escande, V., Grison, C., 2012a. Zinc hyperaccumulating plants as renewable resources for the chlorination process of alcohols. Green Chem. Lett. Rev. 5, 451–456.

Losfeld, G., Escande, V., de La Blache, P.V., L'Huillier, L., Grison, C., 2012b. Design and performance of supported Lewis acid catalysts derived from metal contaminated biomass for Friedel-Crafts alkylation and acylation. Catal. Today 189, 111–116.

Losfeld, G., Escande, V., Jaffre, T., L'Huillier, L., Grison, C., 2012c. The chemical exploitation of nickel phytoextraction: an environmental, ecologic and economic opportunity for New Caledonia. Chemosphere 89, 907–910.

Losfeld, G., Saunier, J.-B., Grison, C., 2014a. Minor and trace-elements in apiary products from a historical mining district (Les Malines, France). Food Chem. 146, 455–459.

Losfeld, G., Mathieu, R., L'Huillier, L., Fogliani, B., Jaffré, T., Grison, C., 2014b. Phytoextraction from mine spoils: insights from New Caledonia. Environ. Sci. Pollut. Res. 22 (8), 5608–5619. http://dx.doi.org/10.1007/s11356-014-3866-8.

Losfeld, G., L'Huillier, L., Fogliani, B., Coy, S., Grison, C., Jaffré, T., 2014c. Leaf-age and soil-plant relationships: key factors for reporting trace-elements hyperaccumulation by plants and design applications. Environ. Sci. Pollut. Res. 22 (8), 5620–5632. http://dx.doi.org/10.1007/s11356-014-3445-z.

Mench, M., Schwitzguebel, J.P., Schroeder, P., Bert, V., Gawronski, S., Gupta, S., 2009. Assessment of successful experiments and limitations of phytotechnologies: contaminant uptake, detoxification and sequestration, and consequences for food safety. Environ. Sci. Pollut. Res. 16, 876–900.

Mench, M., Lepp, N., Bert, V., Schwitzguebel, J.P., Gawronski, S.W., Schroder, P., Vangronsveld, J., 2010. Successes and limitations of phytotechnologies at field scale: outcomes, assessment and outlook from COST Action 859. J. Soils Sediments 10, 1039–1070.

Pajuelo, E., Carrasco, J.A., Romero, L.C., Chamber, M.A., Gotor, C., 2007. Evaluation of the metal phytoextraction potential of crop legumes. Regulation of the expression of O-acetylserine (thiol) lyase under metal stress. Plant Biol. 9, 672–681.

Po-Huang, L., Kohen, A., Baasov, T., Anderson, K.S., 1997. Catalytic mechanism of KDO8P synthase. Pre-steady-state kinetic analysis using rapid chemical quench flow methods. Bioorg. Med. Chem. Lett. 7, 2463–2468.

Robinson, B., Fernandez, J.E., Madejon, P., Maranon, T., Murillo, J.M., Green, S., Clothier, B., 2003. Phytoextraction: an assessment of biogeochemical and economic viability. Plant Soil 249, 117–125.

Sahi, S.V., Bryant, N.L., Sharma, N.C., Singh, S.R., 2002. Characterization of a lead hyperaccumulator shrub, *Sesbania drummondii*. Environ. Sci. Technol. 36, 4676–4680.

Saunier, J.-B., Losfeld, G., Freydier, R., Grison, C., 2013. Trace elements biomonitoring in a historical mining district (Les Malines, France). Chemosphere 93, 2016–2023.

Thillier, Y., Losfeld, G., Escande, V., Dupouy, C., Vasseur, J.J., Debart, F., Grison, C., 2013. Metallophyte wastes and polymetallic catalysis: a promising combination in green chemistry. The illustrative synthesis of 5'-capped RNA. RSC Adv. 3, 5204–5212.

Vangronsveld, J., Herzig, R., Weyens, N., Boulet, J., Adriaensen, K., Ruttens, A., Thewys, T., Vassilev, A., Meers, E., Nehnevajova, E., van der Lelie, D., Mench, M., 2009. Phytoremediation of contaminated soils and groundwater: lessons from the field. Environ. Sci. Pollut. Res. 16, 765–794.

Vidal, C., Chantreuil, C., Berge, O., Maure, L., Escarre, J., Bena, G., Brunel, B., Cleyet-Marel, J.C., 2009. *Mesorhizobium metallidurans* sp nov., a metal-resistant symbiont of *Anthyllis vulneraria* growing on metallicolous soil in Languedoc, France. Int. J. Syst. Evol. Microbiol. 59, 850–855.

SYNTHETIC BIOLOGY: AN EMERGING FIELD FOR DEVELOPING ECONOMIES

26

J. Koelmel, A. Sebastian, M.N.V. Prasad

University of Hyderabad, Hyderabad, Telangana, India

1 INTRODUCTION

The COP 11, "The Eleventh Meeting of the Conference of the Parties (COP) to the Convention on Biological Diversity," that met in Hyderabad, India, 8-19 October 2012, identified synthetic biology as an emerging area that has attracted the attention of the United Nations and governments. Various governments have agreed to continue monitoring the technology and report back to future meetings of the Convention on Biological Diversity (CBD).[1]

Synthetic biology is an interdisciplinary field that brings together biology and metabolic engineering (Buhk, 2014; Cole, 2014b; Sole, 2015; Yuan and Grotewold, 2015; Edmundson et al., 2014). It is also considered a sold wine in new bottles (Cole, 2014a). It represents advances in biotechnology and goes beyond transferring genes between species to constructing entirely new, self-replicating microorganisms that have the potential (partially proven and partially theoretical) to convert any biomass or carbon feedstock into any product that can be produced by fossil carbons, plus many more (Figure 1). In other words, from the perspective of synthetic biology (synbio), the resource base for the development of marketable "renewable" materials (that are not petroleum based) is not only the world's commercialized 23.8% of annual terrestrial biomass, but also the other 76.2% of annual terrestrial biomass that has, thus far, remained outside the market (Anonymous, 2011, 2014).[2]

2 EMERGING TRENDS IN BIOTECHNOLOGY INDUSTRY AND APPLIED RESEARCH

Globally biotechnology is considered as a multifarious applied research area, which has grown significantly since 2007 (Figure 2). Different fields of biotechnology are attributed with specific colors.

Red biotechnology: Research related to medicine and medical processes. The designing of organisms to manufacture pharmaceutical products like antibiotics and vaccines, the engineering of genetic cures through genomic manipulation, and its use in forensics through DNA profiling.

---

[1] For more details reference is made to "Decision UNEP/CBD/COP/DEC/XI/11 paragraphs 3 and 4; 5 December 2012, "decision adopted by the conference of the parties to the convention on biological diversity at its eleventh meeting, XI/11. new and emerging issues relating to the conservation and sustainable use of biodiversity:" http://www.cbd.int/cop11.
[2] For more background see http://www.etcgroup.org/issues/synthetic-biology.

Bioremediation and Bioeconomy. http://dx.doi.org/10.1016/B978-0-12-802830-8.00026-5
Copyright © 2016 Elsevier Inc. All rights reserved.

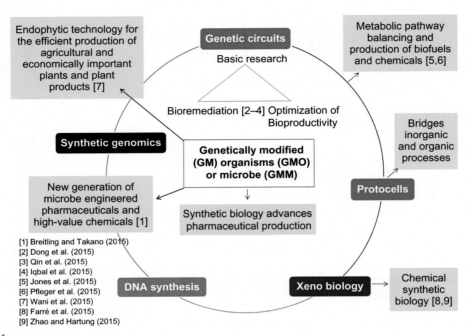

Endophytic technology for the efficient production of agricultural and economically important plants and plant products [7]

Genetic circuits

Basic research

Metabolic pathway balancing and production of biofuels and chemicals [5,6]

Bioremediation [2–4] Optimization of Bioproductivity

Synthetic genomics

Genetically modified (GM) organisms (GMO) or microbe (GMM)

Bridges inorganic and organic processes

Protocells

New generation of microbe engineered pharmaceuticals and high-value chemicals [1]

Synthetic biology advances pharmaceutical production

[1] Breitling and Takano (2015)
[2] Dong et al. (2015)
[3] Qin et al. (2015)
[4] Iqbal et al. (2015)
[5] Jones et al. (2015)
[6] Pfleger et al. (2015)
[7] Wani et al. (2015)
[8] Farré et al. (2015)
[9] Zhao and Hartung (2015)

DNA synthesis

Xeno biology

Chemical synthetic biology [8,9]

FIGURE 1

Major subfields of synthetic biology that are responsible for a wide range of cellular regulation by designing and constructing genetic circuits using metabolic engineering.

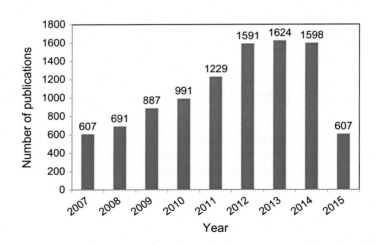

FIGURE 2

Scenario of publications in synthetic biology (search word, synthetic biology; database, Scopus).

White biotechnology: Research related to industrial processes. It involves the use of enzymes and organisms for the processing and production of chemicals, materials, and energy including biofuels.

Green biotechnology: Research related to agriculture processes. It involves the use of environmentally friendly solutions as an alternative to traditional agriculture, horticulture, and animal breeding processes.

Blue biotechnology: Research related to marine and aquatic processes. It involves research on the application of molecular biological methods to marine and freshwater organisms.

Black biotechnology: Research related to bioterrorism. It Involves research on all aspects of bioterrorism related to military, police, surveillance, and counterterrorism.

Gray biotechnology: Research related to the environment.

2.1 METHODS: USING PATENTS AS AN INDICATOR OF BIOTECHNOLOGY GROWTH ACROSS SECTORS AND NATIONS

The first step to navigation is knowing in which direction the ship is already headed. It is important, therefore, to know where resources are being focused in biotechnology sectors over time and geography. In order to determine trends in biotechnology in terms of focus of resources and industrial success, temporal trends within subfields of biotechnology and trends across nations were investigated by querying patents and research articles. Results show that the major sectors of biotechnology are medical (Red Biotechnology) and energy (White Biotechnology) related, with focus in medical biotechnology leveling off compared to normal states of intellectual property growth, and research in bioenergy continuing to grow. There are many minor yet important sectors of biotechnology that are growing rapidly, with much greater focus in developing nations. These include research in enzymes, bioremediation, and bionanotechnology.

Methodology adopted for investigation are as follows: Biotechnologies were subgrouped into seven categories: bioremediation, bioenergy, biological circuit, bioinformatics and omics, medical biotechnology, enzymes, and bionanotechnology and agricultural biotechnology. Certain categories contain simple search terms; for example, bioinformatics and omics consisted of the search terms bioinformatics, genomics, metabolomics, and proteomics. For certain categories such as enzyme biotechnology, search terms were chosen carefully to exclude patents that did not involve the synthesis of enzymes or related technologies (e.g., patents mentioning enzymes for biological relevance but not focused on enzyme technologies). Therefore the search terms were pharmaceutical enzymes, detergent enzymes, and textile enzymes, as well as a list of companies involved in enzyme biotechnologies with patents also containing the keyword enzyme. These companies involved in enzyme research included in our queries were Advanced Enzymes, Novozymes, Rossari Biotech, Zytex, Lumis Enzymes, Kerry Biosciences, Richcore Lifesciences, Anthem Cellutions, Danisco, Enzyme Development Corp., Dyadic International, Excel Industries and Concord Biotech, CHR-Hansen, and Quest International.

Patent data was used as a surrogate of applied research and corporate performance and growth in the biotechnology industry globally and across individual nations. Previous research has shown that patents are a good indicator of corporate performance (Lin et al., 2006), being the most significant indicator of performance in biotechnology sectors in Taiwan (Huang and Huarng, 2015). Changes in number of patents filed is significantly impacted by technological conditions, market conditions, and legal system (e.g., common law producing more documents versus civil law) (Van Zeebroeck et al., 2009).

Therefore direct interpretation of patent growth across time and geography may be more indicative of legal or market changes affecting patent filing numbers than industrial and applied research growth in biotechnology sectors. Therefore the number of patents determined for a specific query was normalized to the total patents for that time period to remove factors influencing patent filing numbers in general. Patent data was obtained from Free Patents Online (FPO) IP Research and Community (FPO, 2014). US patents, European Patent Office (EPO) patents, abstracts of Japan, World Intellectual Property Organization (WIPO), and German Patents were all queried in FPO. FPO's built-in word stemming was used. Using the Expert Search function to query the FPO database, in-depth searches containing keywords covering the topics within each category were established.

For example, for medical biotechnology the search function Search Function 1 was used:

Search Function 1: Medical Biotechnology
(Biopharmaceutical OR "biological medical product" OR "Medical Biotechnology" OR (Medicine AND Biologic) OR (Medicine AND Biotechnology) OR Bio-pharma OR Biosimilars OR "stem cell research")

Data was further broken down and summed up over 4-year time spans from 1990 to 2014. Queries were done for the nations of Brazil, Russia, India, China, and South Africa (BRICS), as well as globally. Searches containing one or more of BRICS nations' names were determined as being part of the BRICS dataset. Therefore an important distinction is that we did not directly determine patents developed in BRICS nations, but only patents referencing BRICS nations.

In order to determine total patents referencing biotechnology, a search was done containing the keyword biotechnology, as well as keywords contained in all other subcategories of biotechnology. Using the resulting data of total biotechnology patents the difference between the total and those found within the categories established earlier was calculated to estimate biotechnology patents not encapsulated by the seven categories chosen.

2.2 RESULTS: BIOTECHNOLOGY GROWTH ACROSS SECTORS AND NATIONS

In BRICS nations, patents involving biotechnologies as a percentage of total intellectual property rights are about one order of magnitude greater than global biotechnology patents. For example, between 1990 and 2014 medical biotechnology and related patents consisted of between about 6% and 13% of total patents in BRICS nations (Figure 3b), while globally during this time span patents in medical biotechnology consisted of between about 0.3% and 1.6% of total patents (Figure 3a). This greater focus on patents in the developing world is seen in application as well. For example, of 5.5 million farmers using transgenic crops in 2001, 75% were growing cotton with insect resistance mainly in China (Toenniessen et al., 2003). In general the high concentration of intellectual property rights in biotechnological sectors is decreasing in BRICS nations. In BRICS nations the percent of total patents in the fields of medical biotechnology, bioinformatics, and omics has decreased from 2000 to 2014 (see Figure 3b). This could be due to growth in other technological and intellectual sectors requiring experts and advancements in infrastructure provided through the countries' development. While there is a decrease in focus on biotechnology, total patents are increasing as well as applications in developing nations. Funding is also increasing in major developing countries; in 2003 about 100 million US dollars was being invested in plant biotechnology in China (Toenniessen et al., 2003; Chong and Xu, 2014), and in 2014 about 400 million US dollars was being invested in plant sciences (Chong and Xu, 2014). This reduction in focus on intellectual property rights in BRICS nations is less pronounced globally.

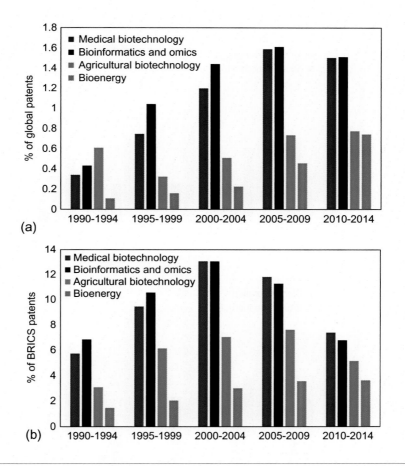

FIGURE 3

Patents from (a) all nations or (b) BRICS nations from 1990 to 2014 falling under major biotechnology sectors normalized to the total number of patents in FPO for the respective 4-year time period.

The global focus on patents in medical biotechnology and the related research field of bioinformatics, metabolomics, genomics, and proteomics have increased from 1990 to about 2005 after which the percent of total patents in the fields of medical biotechnology, bioinformatics, and omics has been stable (see Figure 3a). Agricultural biotechnology follows similar trends to medical biotechnology both globally and for BRICS nations (see Figure 3a and b).

On the contrary to other major biotechnological sectors, the percent of total patents in bioenergy has continually increased from 1990 to 2014 globally (Figure 3a), and begun to stabilize in BRICS nations over the past 8 years (Figure 3b). The United States is a major instigator in driving the production of biomass for energy, with 3% of energy consumption in the US provided by biomass as of 2005. Major current sources of biofuels are agriculturally derived fuel, especially ethanol for subsidized corn production in the US. The Biomass R&D Technical Advisory Committee established by Congress in the US has envisioned a 30% replacement of fossil fuels by biofuels in 2030 (Perlack et al., 2005).

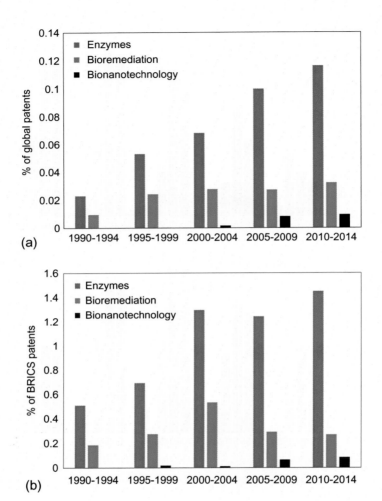

FIGURE 4

Patents from (a) all nations or (b) BRICS nations from 1990 to 2014 falling under minor biotechnology sectors normalized to the total number of patents in FPO for the respective 4-year time period.

In the sector of pharmaceutical enzymes, detergent enzymes, textile enzymes, and other enzyme technologies there has been a steady increase in the percent of global patents (Figure 4a), while patents related to BRICS nations drastically increased between 2000 and 2004 and have remained relatively constant thereafter (Figure 4b). In-depth reviews of enzyme industry in developing nations can be found with special focus on China (Li et al., 2012) and India (Binod et al., 2013). BRICS nations also saw a significant increase in patents in bioremediation (including phytoremediation) in 2000-2004 (Figure 4b). Bionanotechnology is an emerging field with patents only beginning to become significant in number between 2005 and 2009 (see Figure 4a and b).

In the past 4 years (and similarly over the past 24 years), the biomedical industry has dominated biotechnology patents as indicated by bioinformatics and omics, and medical biotechnology consisting of roughly 62% of global biotechnology patents. Growth in the medical biotechnology field has important consequences, especially in developing countries with lower life expectancies. By one estimate pharmaceutical innovation has accounted for 73% of the 1.74 years' increase in life expectancy from 2000 to 2009, globally (Lichtenberg, 2014). Traditional assumptions of high costs of innovative biotechnologies for improving health have been questioned by Daar et al. (2002), who determined the top 10 biotechnologies for improving health in developing nations. The next two dominant biotechnology sectors are agricultural biotechnology (16%) and bioenergy (15%) (Figure 5a). In BRICS nations general trends remain the same, but enzyme technologies (6%) are a much more significant contributor to biotechnologies compared to global enzyme technologies as a percent of total patents (2%). In addition, agriculture also shares a much larger section of the biotechnology intellectual property in BRICS nations (20% versus 16% globally) (Figure 5b). The percent of global patents related to biotechnology contributed by BRICS nations has increased significantly from 1992 to 2013, with the most dramatic increase from 1999 to 2008 (Figure 6). Recently, from 2008 to 2013, the percent of patents contributed by BRICS nations has remained relatively constant (see Figure 6).

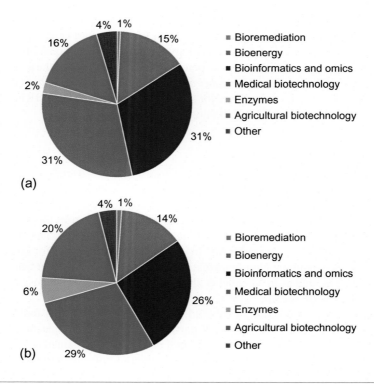

FIGURE 5

Percent by category of patents from different biotechnology sectors for (a) all nations or (b) BRICS nations in the years 2010 to 2014.

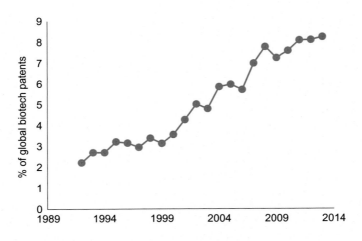

FIGURE 6

BRICS patents related to biotechnology as a percent of the global patents related to biotechnology from 1990 to 2014.

In summary, intellectual property rights for biotechnology are dominated by the medical industry (Red Biotechnology) followed by the agricultural (Green Biotechnology) and energy sectors (a part of White Biotechnology). Most sectors of biotechnology are not experiencing an increase in focus in the past decade, and even decreasing in patents as a percent of all patents filed in the past decade for BRICS nations. Exceptions include rapid growth in emerging sectors such as bioremediation, enzyme research, and bionanotechnology and growth in the major sector of bioenergy. To date BRICS nations continue to have a greater focus on advancing biotechnology than elsewhere globally.

2.3 TRENDS IN PHYTOREMEDIATION

Methodology adopted for investigation of this subject are as follows. To analyze trends for publications in phytoremediation, searches were performed using ScienceDirect. Articles containing the term *phytoremediation* in the keywords, abstract, and/or title were considered as articles on phytoremediation, and data was collected from 1999 to 2014. For comparison across countries the articles needed to contain the term phytoremediation, and the name of the country in affiliations. Data by country and year were normalized to the total articles published by that country in that year for comparison purposes. A trend of total articles containing phytoremediation over time was nearly identical for a search again of Web of Knowledge, as well as in Koelmel et al. (2015) using Sciverse Scopus, and therefore results were deemed as nondatabase specific.

A closer look at phytoremediation, a subfield of bioremediation (which looks at both plant and bacteria), was taken. Bioremediation is a small portion of biotechnology patents (see Figure 5), and globally patents in bioremediation as a percent of total patents have remained relatively constant since 1995 (data not shown). On the contrary publications in the field of phytoremediation as a percent of all publications have increased steadily until 2008, after which the percent of global publications contributed from the field of phytoremediation has remained relatively constant (Figure 9, black line). The total number of publications in phytoremediation has increased over time, as expected for scientific

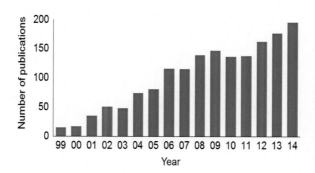

FIGURE 7

Total publications with the term *phytoremediation* in keywords, title, or abstract from 1999 to 2014 in ScienceDirect.

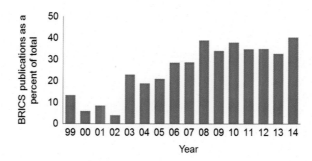

FIGURE 8

Percent of total publications with the term *phytoremediation* contributed from BRICS nations from 1999 to 2014 in ScienceDirect.

research in general (Figure 7). Since 2008 BRICS nations have been a major contributor to publications on phytoremediation, contributing between 30% and 40% of the literature on phytoremediation (Figure 8). Out of the BRICS nations China is the largest contributor of total publications, with 52 publications in 2014, followed by India, with 20. As a percent of total publications, India has the largest focus on phytoremediation (0.09% in 2014) followed by China (0.07% in 2014) (Figure 9).

Publications on phytoremediation in the most prolific BRICS nations India, Brazil, and China are shown in Figure 9a, and can be compared with developed nations: the United States, Japan, and Germany (see Figure 9b). A country's focus on research in phytoremediation is represented as a percent of total publications; in China and India the countries focus on phytoremediation at a significantly higher percentage than the world average (see Figure 9a). In the past three years the United States, Japan, and Germany have phytoremediation research as a percent of total that is significantly below the world average and dropping rapidly, although the percent of publications changes drastically from year to year (see Figure 9b). Further corroboration using Sciverse Scopus and a geographic map of major institutions for phytoremediation development in the countries mentioned have recently been published (Koelmel et al., 2015).

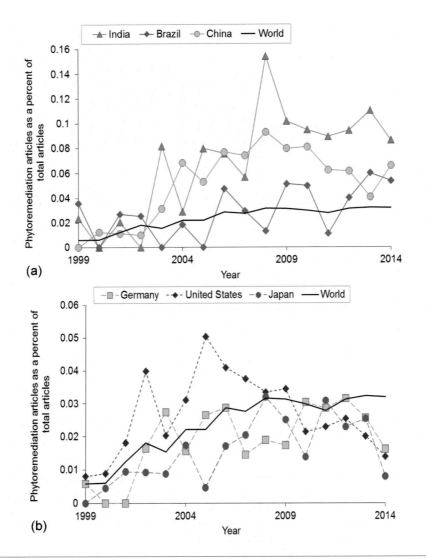

FIGURE 9

Publications containing phytoremediation as a percent of total publications by country from 1999 to 2014 for (a) three BRICS nations and (b) three developed nations compared to the world average.

3 SYNTHETIC BIOLOGY FOR HANDLING BIOWASTE

To optimize the economic return on nonfood crop cultivation, ideally all parts of the plant should be utilized. Biowaste should therefore be treated as an opportunity rather than a problem (Angelini et al., 2014). Instead of costly disposal, new markets could be found for previously discarded biomaterial by improving its quality through conventional processing or biotechnology (Bryksin et al., 2014).

For example, the meal obtained following the extraction of oil from novel crops can be processed into animal feed. Multifunctionality is likely to be particularly important for the large volume of biomass crops that will be processed for energy and biofuels. Cereal straw, for instance, can be used for bioenergy once the grain has been harvested. Multiple uses for crops in all bioproduct projects should be sought at an early stage of development.

The contemporary world is overloaded with waste. Sound policies and affordable technologies are required for handling waste, especially food waste, which is problematic both in developing and developed nations. Some parts of the word are facing frightening food shortages while food is being wasted elsewhere. We are currently living in a world of insecurities, and therefore often we refer to food security, nutritional security, water security, environmental security, energy security, etc. The contemporary world is witnessing a clash between food and fuel. There is increasing demand for fuel while food cost and food waste are gradually increasing. Therefore, there is a need to address waste management and curb pollution due to excessive usage of energy and industrialization. The new innovation wave of biovalorization of biowaste for biomaterials and biofuels is rapidly expanding (Dreschke et al., 2015; Kim et al., 2015; Kumar et al., 2014; Dade-Robertson et al., 2015; Güzel et al., 2014; Pei et al., 2011; Luisi et al., 2014).

Bioeconomy is an intrinsic parameter to the development, production, and use of biological products and processes through biovalorization. Due to land cost and nonavailability of convenient land, landfills are no longer viable in many parts of the world. Therefore, the importance of multiple "Rs" such as Reclamation, Remediation, Reuse, Recycle, Renovation, Resilience, Refuse, Replenish, Rainwater harvest, and Reverence for nature are being increasingly considered in various parts of the worlds. These perspectives are covered in this chapter. Waste can be defined as "right or useful material in the wrong place." Again, biowaste should not be treated as a problem but rather as an opportunity (Fava et al., 2015; Kim et al., 2015; Lin et al., 2013; Patel et al., 2015; Van Wyk, 2011; Vargas et al., 2014).

Biowaste is nature's capital. Disposal and utilization of "biowaste" for production of diverse biofuels (solid, liquid, and gaseous) have gained considerable momentum globally following the biorefinery approach (Angelini et al., 2014; Fava et al., 2015; Patel et al., 2015; Probst et al., 2015). The primary driver in biowaste management is regulatory compliance rather than manufacturing profit. It is an attractive technology in the context of bioeconomy.

Globally domestic, industrial, and food processing wastes are increasing. All this waste can now be used as an invaluable resource for production of a wide variety of materials including energy. Every ton of biowaste is equivalent to 4.5 tons of CO_2 emission contributing to environmental pollution. One approximation shows that in the United States about 45 billion kg of fresh fruits, vegetable, milk, and grain products are lost as waste either in storage, transport, or usage. The US Environmental Protection Agency (EPA) estimates that about US $1 billion are required to dispose of the food waste. In UK about 20 million tons of food is wasted per annum (Anonymous, 2011; Ashish et al., 2014; Lin et al., 2013).

Any waste over a period of time will be stabilized by natural attenuation. The normal practice is to dump waste at designated sites like landfills. A landfill is an area of land that has been specifically engineered to allow for the deposition of waste onto and into it. Landfills may include internal waste disposal sites (where a producer of waste carries out their own waste disposal at the place of production) as well as sites used by many producers. Many landfills are also used for other waste management purposes, such as the temporary storage, consolidation and transfer, or processing of waste material (sorting, treatment, or recycling).

In less developed countries open dumping and other practices lead to major environmental and health concerns. For example, one negative health effect of landfills is toxic ammonia emissions. People who inhale 50,000 parts per billion (ppb) of ammonia for less than 1 day experience minor and temporary irritation. People who inhale 100,000 ppb of ammonia in the air for more than 6 weeks experience eye, nose, and throat irritation. Odor from certain compounds and wastes in landfills also discourages the practice of landfills. Human exposure to landfill odor can cause headaches and nausea. Modern landfills contain liners and have gas collection systems in place to minimize leachate leaching into groundwater and release of odorous, flammable, and greenhouse gas emissions.

While collection of methane gas from landfills is common and often required in developed nations, utilization of this gas is still not always practiced even in developed countries such as the US (instead flares are used converting methane to the less potent greenhouse gas carbon dioxide), and gas collection systems are not 100% efficient. Municipal solid waste landfill sites are a large source of human-related methane emissions and, in some countries such as the US, can account for up to 25% of these emissions. At the same time, methane emissions from landfills represent a lost opportunity to capture and use a significant energy resource. Four main goals of management of biowaste are (a) protecting the environment, (b) promoting partnerships between actors involved in natural resource extraction, industrial production, consumption, and waste management, (c) strengthening the marketplace, and (d) sustaining reuse.

Today with a growing economy, extensive industrialization and extraction of natural resources have resulted in environmental contamination. Large amounts of toxic waste have been dispersed in thousands of contaminated sites spread all over the globe. These industrial barrens represent extreme habitats with adverse and intrinsic effects (Kozlov and Zvereva, 2007). Pollutants often belong to two main classes, inorganic and organic. The challenge is to develop innovative and cost-effective solutions to decontaminate polluted environments. Biowaste can often be processed anaerobically (for energy in the form of methane and compost), aerobically (for compost), or otherwise processed into energy, animal feed, or into other valuable resources. On the other hand, hazardous waste such as persistent organics and heavy metals are much more difficult to deal with and currently high-cost chemical and mechanical treatment is most often used.

Bioremediation is emerging as an invaluable tool for environmental cleanup. Biodiversity is raw material for environmental decontamination (Dreschke et al., 2015; Franzoso et al., 2015; Gianfreda and Rao, 2004). Several articles have been published on various aspects of using biological resources for environmental cleanup starting with a very few in the early 1990s. Often policy and decision makers in academia and civic governance think that bioremediation is a temporary solution of transferring the pollutants and contaminant from one place to another. But, as discussed in detail in the next section, plants can be used to degrade compounds into less toxic molecules (e.g., rhizodegradation), bind up molecules into less bioavailable forms (e.g., phytostabilization and phytochelation), and concentrate toxic molecules for reducing hazardous waste substrate or further refining and recycling of hazardous waste (e.g., phytomining, rhizofiltration, and phytoextraction).

Driven by land scarcity and a knowledge explosion in green chemistry and systems biology, new innovations are being developed to recover contaminated land and reduce future landfills. It is clear that research work that has been conducted during the past decade in the field of biowaste management and bioremediation has resulted in a plethora of scientific literature that leads to a knowledge explosion. Hence, there is need for systems biology approaches for integrating biology and bioremediation with the vast scientific output from fields dealing in toxic pollutants and waste management.

4 ADVANCING BIOREMEDIATION THROUGH SYSTEMS BIOLOGY AND SYNTHETIC BIOLOGY

Biodiversity is the raw material for bioremediation and is an invaluable toolbox for environment and human health protection (Maine et al., 2004; Maiti et al., 2004; Marrs, 1996). Biodiversity can be utilized in bioremediation; for example, by managing natural attenuation of contaminated lands. Bioremediation includes a variety of technologies using plants and microbes to remediate or contain contaminants in soil, groundwater, surface water, or sediments including air. These technologies have become attractive alternatives to conventional cleanup technologies due to relatively low capital costs and the inherent aesthetics.

Environmental decontamination through bioremediation is an integral part of bioeconomy and sustainable development. In recent years there has been growing interest in using biodiversity as raw material for environmental decontamination. On the other hand the volume and diversity of contaminated substrates (water, soil, and air) is increasing due to anthropogenic and technogenic sources. Toxic metals are the most prevalent and are widely used for a variety of needs starting with building materials to information technology. Therefore, the bioavailable fraction of metals is increasing in waste, which can be seen in the much larger portion of metals in developed countries versus developing countries' waste stream.

Systems biology approaches are also implicated in bioremediation and are gaining considerable importance in fostering bioremediation. It is strongly believed that there are three dimensions for the effectiveness of vital bioremediation process; that is, chemical landscape (nutrients-to-be, electron donors/acceptors and stressors), abiotic landscape, and catabolic landscape of which only the catabolic landscape is "genuinely" biological. The chemical landscape has a dynamic interplay with the biological interventions on the abiotic background of the site at stake. This includes humidity, conductivity, temperature, matrix conditions, redox status, etc. (Figure 10).

In order to clean up the widespread environmental pollution by inorganic and organic chemicals, novel methods of decontamination are required. Metabolic pathway engineering is one such emerging subject (Sole, 2015, Yuan and Grotewold, 2015; Farré et al., 2015). Most plants, when encountered

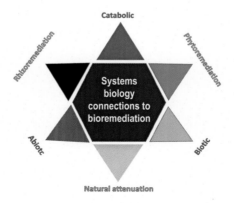

FIGURE 10

Systems biology and synthetic biology connections to bioremediation.

with heavy metals, respond by producing metal-binding peptides such as phytochelatins (PCs) and metallothioneins (MTs). MTs capable of binding to arsenic have been utilized along with an arsenic transporter for the selective removal of arsenic. Research has demonstrated the specificity of the metal-loregulatory protein [MerR] that can be exploited for specific metal removal via cell surface display. Cell surface display also eliminates the metal uptake limitations. Specific examples are used to explain the systems biology approaches in bioremediation.

4.1 HANDLING A POLLUTANT

The removal of a toxic pollutant through bioremediation depends on factors such as bioavailability, biodegradability, biotransformation, and bioaccumulation (van Driesser and Christopher, 2004; Maiti et al., 2004). Hence it is clear that existence of multiple factors determines the efficiency of bioremediation (Prasad and Prasad, 2012) (Figure 11). The very first step in handling a pollutant is the analysis of bioavailability to plants or microbes being used for the remediation program. Bioavailability depends on the specific nature of the pollutant such as solubility, particle size, polarity, toxicity, electrochemical nature of the medium, etc. On the other hand, bioavailability also depends on chemical stimulus (Singer et al., 2003). The chemical stimulus usually arises from the organism used for bioremediation due to the presence of the pollutant. These chemical signals play critical roles in the bioaccumulation of the pollutant through sensing of the lethal dose of pollutant and biologically controlling accumulating based on this sensing. Hence bioavailability and chemical responses act in an interlinked process.

The ability of the organism to remove pollutants during bioremediation could be through biodegradation or bioaccumulation. The processes of biodegradation may be extracellular or intracellular (Anderson et al., 1993). A schematic of these intra- and extracellular mechanisms are shown in Figure 12. During extracellular processes, the organism produces secretions that detoxify the pollutant. This kind of action during bioremediation is common in detoxification of polycyclic aromatic hydrocarbons, textile dyes, pesticides, persistent organic pollutants, environmental persistent pharmaceutical pollutants, and environmental xenobiotics (Shukla et al., 2011). The extracellular secretions occur through a series of cellular processes that result in metabolite channeling for the synthesis of ingredients

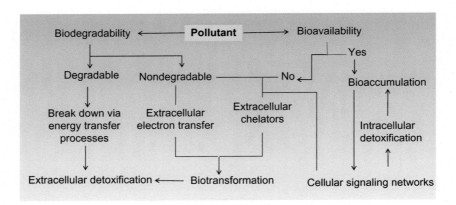

FIGURE 11

Network scheme for handling a pollutant.

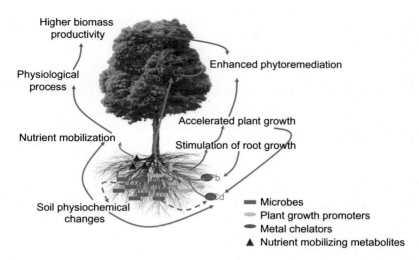

FIGURE 12

Positive interactions in MAP.

in the secretion that detoxify the pollutant. Apart from this, metabolite channeling as well as secretion of the ingredients into the environment are outcomes of complex cellular signaling processes mediated by G-proteins, the phosphoinositol pathway, the adenylate cyclase pathway, etc. (Moller and Chua, 1999). The ingredients in the secretions work either by electron transfer through oxidation reduction or chelation (Gianfreda and Rao, 2004). The ingredients in the secretion may be enzymes, antioxidants, oxidizing agents, carboxylic acids, etc. Peroxidases, laccases, oxygenases, and hydrolytic enzymes are the major enzymes involved in extracellular degradation of the pollutants. Action of Cytochrome P450 through epoxidation of C=C double bonds and hydroxylation of aromatic compounds plays an important role in the detoxification of several organic pollutants, especially chloroorganics. Secretion of carboxylic acids also helps to detoxify the toxic trace elements (TEs) in the soils through chelating action (Pohlmeier, 2004).

Bioremediation of biopermeable pollutants are carried out through intracellular processes. This method is especially applicable to nonbiodegradable compounds and toxic TEs. In this method, the pollutant is entered into the cell and detoxified with the help of products of various metabolic pathways (Sole, 2015; Yuan and Grotewold, 2015). Glutathione S transferase activity is among the chief enzymatic processes that take part in intracellular detoxification of the pollutant (Marrs, 1996). This enzyme catalyzes the conjugation of the reduced form of glutathione to xenobiotic substrates for the purpose of detoxification. On the other hand, toxic TEs are detoxified with the help of PC synthase activity, which produce peptides called PCs that bind with toxic TEs and transport them to the plant vacuole for storage (Clemens, 2001). Apart from PC, carboxylic acid in the plant also chelates toxic TEs and aids in the storage of acid-metal complex in the vacuole. MTs are another class of metal-detoxifying agent in the biological system that binds with toxic TEs and hence avoids toxicity due to free cellular ions (Fernandez et al., 2012). It must be noted that plant-based bioremediation approaches mainly help to remove the nondegradable pollutants while the microbe-based approaches help to immobilize the nondegradable pollutants. Thus handling a pollutant through bioremediation is a multistep process where each step involves assembly of a complex network of actions that evokes the need of system biology for better understanding of the processes involved in bioremediation (Venema, 2001; Veselý et al., 2011; Vymazal et al., 1998).

4.2 INTERACTION NETWORKS IN MICROBE-ASSISTED PHYTOREMEDIATION

Microbe-assisted phytoremediation (MAP) can enhance removal and degradation of toxic pollutants (Juwarkar, 2012). Increase in the number of microbes in the rhizosphere and stimulation of metabolic activity of microbes regulate transformation or plant uptake of pollutants (Anderson et al., 1993). The extracellular matrix of microbes rich in anionic functional groups have high metal-binding capacity; hence the presence of bacteria also helps to immobilize metals in the soil. MTs also help to reduce toxicity of metals accumulated inside the microbe. Microbial activity causes transformation of valence states of metals that can reduce metal toxicity on plants. Siderophore-producing bacteria in the rhizosphere of plants were found to enhance toxic TEs accumulation in plants (Rajkumar et al., 2010; Gurska et al., 2009). Siderophores take part in solubilization of iron. But these high-affinity iron-chelating compounds also chelate elements such as Al, Cd, Cu, Ga, In, Pb, and Zn, as well as with radionuclides including U and Np. This makes microbes that secrete siderophore a potential candidate in the MAP approach.

Positive interactions of microbes with plants are preferred in MAP. Root exudates rich in carbon sources, nutrients, enzymes, etc., make the rhizosphere an ideal habitat for microbial growth (Anderson et al., 1993). A considerable amount of photosynthate is released into soil in the form of sugars, organic acids, and organic compounds such as allelopathic chemicals, phytohormones, phytoalexins, phyto-siderophores, and phytoanticipins. Some of the secondary metabolites of the plant act as inhibitors of bacterial growth, and hence MAP interactions are specific to those plants that do not hinder microbial growth. On the other hand, secondary metabolites such as limonene, cymene, carvone, and pinene are found to help in the degradation of polychlorinated biphenyl (Singer et al., 2003). Hence it is clear those microbes that stimulate synthesis of these compounds in plants help to increase the efficiency of the phytoremediation process.

It must be noted that plant microbe interactions happen with the help of metabolite signals from the plant, which attract microbes toward plant roots. Several genes are reported to play a critical role in MAP. Genes such as alkB (alkane monooxygenase), ndoB (naphthalene dioxygenase), and nid (naphthalene inducible dioxygenase) are reported to play potential roles in the degradation of hydrocarbons with the MAP approach (Oliveira et al., 2015). Auxin synthesis-related genes in bacterial strains were found to enhance root growth of *Lolium perenne*, *Festuca arundinacea*, and *Secale cereal,* which ultimately enhanced degradation of petroleum products (Gurska et al., 2009). Gene responses in plants with microbial inoculation often occur as a cascade. For example, auxin production from microbes and related enhancement of plant growth involve a series of molecular events in plants starting from activation of ubiquitin ligase that degrade AUX/IAA proteins followed by formation of active ARF homodimer (Kovtun et al., 1998). Homodimer binding at the palindromic AuxRE promoter of the auxin early gene ultimately results in auxin-mediated growth or plant development that is also accompanied with series of metabolic and developmental events. All these points are given in order to understand that in the various interactions involved in MAP, the systems biology approach is an inevitable tool. This approach helps to design MAP approaches based on the chemical and physical nature of the pollutant, and hence helps to formulate enhanced pollutant degradation through synthetic biology pathways.

In the past 2 decades transgenic plants for phytoremediation have been produced, but their commercial application is still at infancy. Three main approaches have been developed: (1) transformation with genes from other organisms (mammals, bacteria, etc.); (2) transformation with genes from other plant species; and (3) overexpression of genes from the same plant species. Many encouraging results have been achieved at the laboratory and greenhouse level pending field trials due to regulatory checks (Maestri and Marmiroli, 2011) (see Figure 12).

5 PERSPECTIVES

The industrial revolution and green revolution are the major achievements of civilization in the past century. Simultaneously land use and degradation by pollution is increasing. Natural resources are becoming contaminated. The unprecedented climate change is making profit from natural capital unpredictable. Under such circumstances, synthetic biology offers investors an attractive opportunity: to be able to turn biowaste into a diverse range of profitable products and reclaim valuable degraded land with low-cost remediation solutions.

In the past 2 decades a large body of scientific information has accumulated in the areas of green chemicals, renewable bioproducts, forestry, biomass energy, and novel crops. It includes information on biotechnology with an emphasis on plant genetic engineering for nonfood uses, novel biological crop protection strategies, fermentation, and microbial conversion as well as the production and use of enzymes. This is in addition to chemical, thermal, and mechanical aspects of the processing and use of biological materials in industry for the production of:

Bulk chemicals
Liquid and gaseous biofuels
Cosmetics, drugs, and vaccines
Bioplastics, polymers, and packaging
Nonwood fibers
Wood products
Specialty chemicals
Inputs for integrated crop protection
Organic fertilizers, etc.

6 RISK MANAGEMENT OF SYNTHETIC BIOLOGY APPLICATIONS

The rapid pace of synthetic biology research and product development; the potential environmental release of novel genes, chemicals, and organisms from numerous applications; and the diffuse and diverse nature of the research community are prompting renewed attention on how to design robust ecological risk research programs to investigate such hazards (Kuiken et al., 2014). As with most emerging technologies, opportunities come with potential risks; the potential ecological risks of synthetic biology are particularly salient as academic research and commercial applications often use organisms that may either be intentionally or accidentally released into the environment. In addition, large-scale industrial use of biomass, for example in energy, can pose significant impacts on food prices and challenges for sustainability (Gabrielle et al., 2014), as can be seen in a report from the US Department of Energy and US Department of Agriculture: "Biomass as Feedstock for a Bioenergy and Bioproducts Industry: The Technical Feasibility of a Billion-Ton Annual Supply" (Perlack et al., 2005). Synthetic biology is important to increase the efficiency of product development from biological organisms, and resource conservation and sustainable use are important to retain the natural base for continued growth in biotechnology.

ACKNOWLEDGMENTS

JK thankfully acknowledges the award of USIEF and US Nehru Fulbright Fellowship. Thanks are also due to the University of Hyderabad, Hyderabad, for hosting the tenure of 8 months. A great deal of information was received from the European Union on the internet through the Europa server https://ec.europa.eu/research/agriculture/pdf/towards_know-based_bioeconomy.pdf and Office for Official Publications of the European Communities, 2004, which is thankfully acknowledged.

REFERENCES

Anderson, T.A., Guthrie, E.A., Walton, B.T., 1993. Bioremediation in the rhizosphere. Environ. Sci. Technol. 27, 2630–2636.

Angelini, S., Cerruti, P., Immirzi, B., Santagata, G., Scarinzi, G., Malinconico, M., 2014. From biowaste to bioresource: effect of a lignocellulosic filler on the properties of poly (3-hydroxybutyrate). Int. J. Biol. Macromol. 71, 163–173.

Anonymous, 2011. Global Food Losses and Food Waste—FAO.

Anonymous, 2014. Towards a European knowledge-based bioeconomy, York University 2004. Workshop conclusions on the use of plant biotechnology for the production of industrial biobased products. European Commission, Directorate-General for Research, Information and Communication Unit, B-1049 Brussels. https://ec.europa.eu/research/agriculture/pdf/towards_know-based_bioeconomy.pdf.

Ashish, T., Ashwani, K., Aparna, S.V., Rashmi, H.M., Sunita, G., Virender, K.B., 2014. Synthetic biology: applications in food sector. Crit. Rev. Food Sci. Nutr. http://dx.doi.org/10.1080/10408398.2013.782534.

Binod, P., Palkhiwala, P., Gaikaiwari, R., Nampoothiri, K.M., Duggal, A., Dey, K., Pandey, A., 2013. Industrial enzymes—present status & future perspectives for India. J. Sci. Ind. Res. 72, 271–286.

Breitling, R., Takano, E., 2015. Synthetic biology advances for pharmaceutical production. Curr. Opin. Biotechnol. 35, 46–51.

Bryksin, A.V., Brown, A.C., Baksh, M.M., Finn, M.G., Barker, T.H., 2014. Learning from nature—novel synthetic biology approaches for biomaterial design. Acta Biomater. 10, 1761–1769.

Buhk, H.-J., 2014. Synthetic biology and its regulation in the European Union. New Biotechnol. 31 (6), 528–531.

Chong, K., Xu, Z., 2014. Investment in plant research and development bears fruit in China. Plant Cell Rep. 33, 541–550.

Clemens, S., 2001. Molecular mechanisms of plant metal homeostasis and tolerance. Planta 212, 475–486.

Cole, J.A., 2014a. Synthetic biology: old wine in new bottles with an emerging language that ranges from the sublime to the ridiculous. FEMS Microbiol. Lett. 251, 113–115.

Cole, J., 2014b. Editorial introduction for the Synthetic Biology thematic issue. New Biotechnol. 31 (6), 525–527.

Daar, A.S., Thorsteinsdóttir, H., Martin, D.K., Smith, A.C., Nast, S., Singer, P.A., 2002. Top ten biotechnologies for improving health in developing countries. Nat. Genet. 32, 229–232.

Dade-Robertson, M., Figueroa, C.R., Zhang, M., 2015. Material ecologies for synthetic biology: biomineralization and the state space of design. Comput. Aided Des. 60, 28–39.

Dong, W.H., Zhang, P., Lin, X.Y., Zhang, Y., Tabouré, A., 2015. Natural attenuation of 1,2,4-trichlorobenzene in shallow aquifer at the Luhuagang's landfill site, Kaifeng, China. Sci. Total Environ. 505, 216–222.

Dreschke, G., Probst, M., Walter, A., Pümpel, T., Walde, J., Insam, H., 2015. Lactic acid and methane: improved exploitation of biowaste potential. Bioresour. Technol. 176, 47–55.

Edmundson, M.C., Capeness, M., Horsfall, L., 2014. Exploring the potential of metallic nanoparticles within synthetic biology. New Biotechnol. 31 (6), 572–577.

Farré, G., Twyman, R.M., Christou, P., Capell, T., Zhu, C., 2015. Knowledge-driven approaches for engineering complex metabolic pathways in plants. Curr. Opin. Biotechnol. 32, 54–60.

Fava, F., Totaro, G., Diels, L., Reis, M., Duarte, J., Carioca, O., Poggi-Varaldo, H.M., Ferreira, B.S., 2015. Biowaste biorefinery in Europe: opportunities and research & development needs. New Biotechnol. 32 (1), 100–108.

Fernandez, L.R., Vandenbussche, G., Roosens, N., Govaerts, C., Goormaghtigh, E., Verbruggen, N., 2012. Metal binding properties and structure of a type III metallothionein from the metal hyperaccumulator plant *Noccaea caerulescens*. Biochim. Biophys. Acta, Proteins Proteomics 1824 (9), 1016–1023.

FPO Community, 2014. FPO IP Research and Communities. http://www.freepatentsonline.com/search.html. Accessed 11/16/2014.

Franzoso, F., Tabasso, S., Antonioli, D., Montoneri, E., Persico, P., Laus, M., Mendichi, R., Negre, M., 2015. Films made from poly(vinyl alcohol-co-ethylene) and soluble biopolymers isolated from municipal biowaste. J. Appl. Polym. Sci. 132 (4). art. no. 41359.

Gabrielle, B., Bamière, L., Caldes, N., De Cara, S., Decocq, G., Ferchaud, F., Loyce, C., Pelzer, E., Perez, Y., Wohlfahrt, J., et al., 2014. Paving the way for sustainable bioenergy in Europe: technological options and research avenues for large-scale biomass feedstock supply. Renew. Sust. Energ. Rev. 33, 11–25.

Gianfreda, L., Rao, M.A., 2004. Potential of extra cellular enzymes in remediation of polluted soils: a review. Enzym. Microb. Technol. 35, 339–354.

Gurska, J., Wang, W., Gerhardt, K.E., Khalid, A.M., Isherwood, D.M., Huang, X.-D., Glick, B.R., Greenberg, B.M., 2009. Three year field test of a plant growth promoting rhizobacteria enhanced phytoremediation system at a land farm for treatment of hydrocarbon waste. Environ. Sci. Technol. 43, 4472–4479.

Güzel, F., Sayĝili, H., Sayĝili, G.A., Koyuncu, F., 2014. Elimination of anionic dye by using nanoporous carbon prepared from an industrial biowaste. J. Mol. Liq. 194, 130–140.

Huang, C.W., Huarng, K.H., 2015. Evaluating the performance of biotechnology companies by causal recipes. J. Bus. Res. 68, 851–856.

Iqbal, M., Ahmad, A., Ansari, M.K.A., Qureshi, M.I., Aref, I.M., Khan, P.R., Hegazy, S.S., El-Atta, H., Husen, A., Hakeem, K.R., 2015. Improving the phytoextraction capacity of plants to scavenge metal(loid)-contaminated sites. Environ. Rev. 23, 44–65.

Jones, J.A., Toparlak, T.D., Koffas, M.A.G., 2015. Metabolic pathway balancing and its role in the production of biofuels and chemicals. Curr. Opin. Biotechnol. 33, 52–59.

Juwarkar, A.A., 2012. Microbe-assisted phytoremediation for restoration of biodiversity of degraded lands: a sustainable solution. Proc. Natl. Acad. Sci. India Sect. B Biol. Sci. 82 (2), 313–318.

Kim, N., Park, M., Park, D., 2015. A new efficient forest biowaste as biosorbent for removal of cationic heavy metals. Bioresour. Technol. 175, 629–632.

Koelmel, J., Prasad, M.N.V., Pershell, K., 2015. Bibliometric analysis of phytotechnologies for remediation: global scenario of research and applications. Int. J. Phytoremediation 17, 145–153.

Kovtun, Y., Chiu, W.L., Zeng, W., Sheen, J., 1998. Suppression of auxin signal transduction by a MAPK cascade in higher plants. Nature, 395, 716–720.

Kozlov, M.V., Zvereva, E.L., 2007. Industrial barrens: extreme habitats created by non-ferrous metallurgy. Rev. Environ. Sci. Biotechnol. 6, 231–259.

Kuiken, T., Dana, G., Oye, K., Rejeski, D., 2014. Shaping ecological risk research for synthetic biology. J. Environ. Stud. Sci. 4, 191–199.

Kumar, P., Singh, M., Mehariya, S., Patel, S.K.S., Lee, J., Kalia, V.C., 2014. Ecobiotechnological approach for exploiting the abilities of bacillus to produce co-polymer of polyhydroxyalkanoate. Indian J. Microbiol. 1–7.

Li, S., Yang, X., Yang, S., Zhu, M., Wang, X., 2012. Technology prospecting on enzymes: application, marketing and engineering. Comput. Struct. Biotechnol. J. 2, 1–11.

Lichtenberg, F.R., 2014. Pharmaceutical innovation and longevity growth in 30 developing and high-income countries, 2000–2009. Health Policy Technol. 3, 36–58.

Lin, B.W., Lee, Y., Hung, S.C., 2006. R&D intensity and commercialization orientation effects on financial performance. J. Bus. Res. 59, 679–685.

Lin, C.S.K., Pfaltzgraff, L.A., Herrero-Davila, L., Mubofu, E.B., Abderrahim, S., Clark, J.H., Koutinas, A.K., Kopsahelis, N., Stamatelatou, K., Dickson, F., Thankappan, S., Mohamed, Z., Brocklesby, R., Luque, R., 2013. Food waste as a valuable resource for the production of chemicals, materials and fuels. Current situation and global perspective. Energy Environ. Sci. 6, 426–464.

Luisi, P.L., Chiarabelli, C., Stano, P., 2014. Editorial overview: synthetic biology. Curr. Opin. Chem. Biol. 22, v–vii.

Maestri, E., Marmiroli, N., 2011. Transgenic plants for phytoremediation. Int. J. Phytoremediation 13, 264–279.

Maine, M.A., Suné, N., Lagger, S.C., 2004. Chromium bioaccumulation: comparison of the capacity of two floating aquatic macrophytes. Water Res. 38, 1494–1501.

Maiti, R.K., Pinero, J.L.H., Oreja, J.A.G., Santiago, D.L., 2004. Plant based bioremediation and mechanisms of heavy metal tolerance of plants: a review. Proc. Indian Natl. Sci. Acad. B Biol. Sci. 70, 1–12.

Marrs, K., 1996. The functions and regulation of glutathione S-transferases in plants. Annu. Rev. Plant Physiol. Plant Mol. Biol. 47, 127–158.

Moller, S.G., Chua, N.-H., 1999. Interactions and intersections of plant signaling pathways. J. Mol. Biol. 293, 219–234.

Oliveira, V., Gomes, N.C.M., Almeida, A., Silva, A.M.S., Silva, H., Cunha, A., 2015. Microbe-assisted phytoremediation of hydrocarbons in estuarine environments. Microb. Ecol. 69, 1–12.

Patel, S.K.S., Kumar, P., Singh, M., Lee, J.-K., Kalia, V.C., 2015. Integrative approach to produce hydrogen and polyhydroxybutyrate from biowaste using defined bacterial cultures. Bioresour. Technol. 176, 136–141.

Pei, L., Schmidt, M., Wei, W., 2011. Synthetic biology: an emerging research field in China. Biotechnol. Adv. 29, 804–814.

Perlack, R.D., Wright, L.L., Turhollow, A.F., Graham, R.L., Stokes, B.J., Erbach, D.C., 2005. Biomass as feedstock for a bioenergy and bioproducts industry: the technical feasibility of a billion-ton annual supply.

Pfleger, B.F., Gossing, M., Nielsen, J., 2015. Metabolic engineering strategies for microbial synthesis of oleochemicals. Metab. Eng. 29, 1–11.

Pohlmeier, A., 2004. Metal speciation, chelation and complexing ligands in plants. In: Prasad, M.N.V. (Ed.), Heavy Metal Stress in Plants: From Biomolecules to Ecosystems. Springer, New Delhi.

Prasad, M.N.V., Prasad, R., 2012. Nature's cure for cleanup of contaminated environment—a review of bioremediation strategies. Rev. Environ. Health 28, 181–189.

Probst, M., Walde, J., Pümpel, T., Wagner, A.O., Insam, H., 2015. A closed loop for municipal organic solid waste by lactic acid fermentation. Bioresour. Technol. 175, 142–151.

Qin, K., Struckhoff, G.C., Agrawal, A., Shelley, M.L., Dong, H., 2015. Natural attenuation potential of tricholoroethene in wetland plant roots: role of native ammonium-oxidizing microorganisms. Chemosphere 119, 971–977.

Rajkumar, M., Ae, N., Prasad, M.N.V., Freitas, H., 2010. Potential of siderophore-producing bacteria for improving heavy metal phytoextraction. Trends Biotechnol. 28, 142–149.

Shukla, K.P., Sharma, S., Singh, N.K., Singh, V., Tiwari, K., Singh, S., 2011. Nature and role of root exudates: efficacy in bioremediation. Afr. J. Biotechnol. 10, 9717–9724.

Singer, A.C., Crowley, D.E., Thompson, I.P., 2003. Secondary plant metabolites in phytoremediation and biotransformation. Trends Biotechnol. 21, 123–130.